Springer Optimization and Its Applications

Volume 148

Aims and Scope
Optimization has been expanding in all directions at an astonishing rate during the last few decades. New algorithmic and theoretical techniques have been developed, the diffusion into other disciplines has proceeded at a rapid pace, and our knowledge of all aspects of the field has grown even more profound. At the same time, one of the most striking trends in optimization is the constantly increasing emphasis on the interdisciplinary nature of the field. Optimization has been a basic tool in all areas of applied mathematics, engineering, medicine, economics and other sciences.

The series *Springer Optimization and Its Applications* publishes undergraduate and graduate textbooks, monographs and state-of-the-art expository works that focus on algorithms for solving optimization problems and also study applications involving such problems. Some of the topics covered include nonlinear optimization (convex and nonconvex), network flow problems, stochastic optimization, optimal control, discrete optimization, multi-objective programming, description of software packages, approximation techniques and heuristic approaches.

More information about this series at http://www.springer.com/series/7393

Alexander J. Zaslavski

Turnpike Conditions in Infinite Dimensional Optimal Control

Alexander J. Zaslavski
Department of Mathematics
The Technion – Israel Institute of
Technology
Haifa, Israel

ISSN 1931-6828 ISSN 1931-6836 (electronic)
Springer Optimization and Its Applications
ISBN 978-3-030-20180-7 ISBN 978-3-030-20178-4 (eBook)
https://doi.org/10.1007/978-3-030-20178-4

Mathematics Subject Classification: 49J20, 49J99, 49K20, 49K27

This Springer imprint is published by the registered company Springer Nature Switzerland AG.
The registered company address is: Gewerbestrasse 11, 6330 Cham, Switzerland

Preface

The book is devoted to the study of the turnpike phenomenon arising in optimal control theory. The term was first coined by P. Samuelson in 1948 when he showed that an efficient expanding economy would spend most of the time in the vicinity of a balanced equilibrium path (also called a von Neumann path). To have the turnpike property means, roughly speaking, that the approximate solutions of the problems are determined mainly by the objective function and are essentially independent of the choice of interval and endpoint conditions, except in regions close to the endpoints. The turnpike property discovered by P. Samuelson is well known in the economic literature, where it was studied for various models of economic growth. Usually for these models a turnpike is a singleton.

Now it is well known that the turnpike property is a general phenomenon which holds for large classes of finite-dimensional variational and optimal control problems. In our research, using the Baire category (generic) approach, it was shown that the turnpike property holds for a generic (typical) variational problem [104] and for a generic optimal control problem [117]. According to the generic approach, we say that a property holds for a generic (typical) element of a complete metric space (or the property holds generically) if the set of all elements of the metric space possessing this property contains a G_δ everywhere dense subset of the metric space which is a countable intersection of open everywhere dense sets. This means that the property holds for most elements of the metric space.

In this book we are interested in individual (nongeneric) turnpike results and in sufficient and necessary conditions for the turnpike phenomenon which are of great interest because of their numerous applications in engineering and the economic theory. Sufficient and necessary conditions for the turnpike phenomenon were obtained in our previous research [102, 103, 105, 118] for finite-dimensional variational problems and for discrete-time optimal control problems in compact metric space. In the present book we study sufficient and necessary conditions for the turnpike phenomenon, using the approach developed in [102, 103, 105, 118], for discrete-time optimal control problems in metric spaces, which are not necessarily compact (see Chapters 2–4), and for continuous-time infinite dimensional optimal control problems (see Chapters 5–7). All the results obtained in the book are new.

The monograph contains seven chapters. Chapter 1 is an introduction. In Chapter 2 we study discrete-time autonomous problems. Chapter 3 is devoted to the study of discrete-time nonautonomous problems on subintervals of half-axis. Discrete-time nonautonomous problems on subintervals of axis are analyzed in Chapter 4. Continuous-time autonomous problems are studied in Chapter 5. In Chapter 6 we consider continuous-time nonautonomous problems on subintervals of half-axis. Chapter 7 is devoted to the study of continuous-time nonautonomous problems on subintervals of axis.

Rishon LeZion, Israel
October 19, 2018

Alexander J. Zaslavski

Contents

Chapter 1
Introduction

The study of infinite-dimensional optimal control has been a rapidly growing area of research [1–4, 13–18, 34, 46, 52, 56, 58, 59, 64, 76, 77, 81, 88–91, 100]. In this chapter we present preliminaries which we need in order to study turnpike properties of infinite-dimensional optimal control problems. We discuss Banach space valued functions, unbounded operators, C_0 semigroups, evolution equations, admissible control operators, and a turnpike property for variational problems.

1.1 Banach Space Valued Functions

Let $(X, \|\cdot\|)$ be a Banach space and $a < b$ be real numbers. For any set $E \subset R^1$ define

$$\chi_E(t) = 1 \text{ for all } t \in E \text{ and } \chi_E(t) = 0 \text{ for all } t \in R^1 \setminus E.$$

If a set $E \subset R^1$ is Lebesgue measurable, then its Lebesgue measure is denoted by $|E|$ or by $\text{mes}(E)$.

A function $f : [a, b] \to X$ is called a simple function if there exists a finite collection of Lebesgue measurable sets $E_i \subset [a, b]$, $i \in I$, mutually disjoint, and $x_i \in X$, $i \in I$ such that

$$f(t) = \sum_{i \in I} \chi_{E_i}(t) x_i, \ t \in [a, b].$$

A function $f : [a, b] \to X$ is strongly measurable if there exists a sequence of simple functions $\phi_k : [a, b] \to X$, $k = 1, 2, \ldots$ such that

A. J. Zaslavski, *Turnpike Conditions in Infinite Dimensional Optimal Control*, Springer Optimization and Its Applications 148,
https://doi.org/10.1007/978-3-030-20178-4_1

$$\lim_{k\to\infty} \|\phi_k(t) - f(t)\| = 0, \ t \in [a, b] \text{ almost everywhere (a. e.)}. \tag{1.1}$$

For every simple function $f(\cdot) = \sum_{i\in I} \chi_{E_i}(\cdot)x_i$, where the set I is finite, define its Bochner integral by

$$\int_a^b f(t)dt = \sum_{i\in I} |E_i|x_i.$$

Let $f : [a, b] \to X$ be a strongly measurable function. We say that f is Bochner integrable if there exists a sequence of simple functions $\phi_k : [a, b] \to X$, $k = 1, 2, \ldots$ such that (1.1) holds and the sequence $\{\int_a^b \phi_k(t)dt\}_{k=1}^{\infty}$ strongly converges in X. In this case we define the Bochner integral of the function f by

$$\int_a^b f(t)dt = \lim_{k\to\infty} \int_a^b \phi_k(t)dt.$$

It is known that the integral defined above is independent of the choice of the sequence $\{\phi_k\}_{k=1}^{\infty}$ [64]. Similar to the Lebesgue integral, for any measurable set $E \subset [a, b]$, the Bochner integral of f over E is defined by

$$\int_E f(t)dt = \int_a^b \chi_E(t)f(t)dt.$$

The following result is true (see Proposition 3.4, Chapter 2 of [64]).

Proposition 1.1. *Let $f : [a, b] \to X$ be a strongly measurable function. Then f is Bochner integrable if and only if the function $\|f(\cdot)\|$ is Lebesgue integrable. Moreover, in this case*

$$\|\int_a^b f(t)dt\| \le \int_a^b \|f(t)\|dt.$$

The Bochner integral possesses almost the same properties as the Lebesgue integral. If $f : [a, b] \to X$ is strongly measurable and $\|f(\cdot)\| \in L^p(a, b)$, for some $p \in [1, \infty)$, then we say that $f(\cdot)$ is L^p Bochner integrable. For every $p \ge 1$, the set of all L^p Bochner integrable functions is denoted by $L^p(a, b; X)$ and for every $f \in L^p(a, b; X)$,

$$\|f\|_{L^p(a,b;X)} = (\int_a^b \|f(t)\|^p dt)^{1/p}.$$

Clearly, the set of all Bochner integrable functions on $[a, b]$ is $L^1(a, b; X)$.

Let $a < b$ be real numbers. A function $x : [a, b] \to X$ is absolutely continuous (a. c.) on $[a, b]$ if for each $\epsilon > 0$ there exists $\delta > 0$ such that for each pair of sequences $\{t_n\}_{n=1}^q$, $\{s_n\}_{n=1}^q \subset [a, b]$ satisfying

$$t_n < s_n,\ n = 1, \ldots, q,\ \sum_{n=1}^{q} (s_n - t_n) \le \delta,$$

$$(t_n, s_n) \cap (t_m, s_m) = \emptyset \text{ for all } m, n \in \{1, \ldots, q\} \text{ such that } m \ne n$$

we have

$$\sum_{n=1}^{q} \|x(t_n) - x(s_n)\| \le \epsilon.$$

The following result is true (see Theorem 1.124 of [18]).

Proposition 1.2. *Let X be a reflexive Banach space. Then every a. c. function $x : [a, b] \to X$ is a. e. differentiable on $[a, b]$ and*

$$x(t) = x(a) + \int_a^t (dx/dt)(s)ds,\ t \in [a, b]$$

where $dx/dt \in L^1(a, b; X)$ is the strong derivative of x.

1.2 Unbounded Operators

Let $(X, \|\cdot\|)$ and $(Y, \|\cdot\|)$ be Banach spaces. Denote by $\mathcal{L}(X, Y)$ the set of all linear continuous operators from X to Y. For every $A \in \mathcal{L}(X, Y)$, define

$$\|A\| = \sup\{\|Ax\| : x \in X,\ \|x\| \le 1\}.$$

Set $\mathcal{L}(X) = \mathcal{L}(X, X)$, $X^* = \mathcal{L}(X, R^1)$, and $Y^* = \mathcal{L}(Y, R^1)$. For all $x \in X$, set $Ix = x$. In the space X, we always consider the norm convergence.

Let $\mathcal{D}(A)$ be a linear subspace of X (not necessarily closed), and let $A : \mathcal{D}(A) \to Y$ be a linear operator. We say that A is densely defined if $\mathcal{D}(A)$ is dense in X. We say that A is closed if the graph

$$G(A) = \{(x, y) \in X \times Y : x \in \mathcal{D}(A),\ y = Ax\}$$

of A is closed in $X \times Y$. We say that A is closable if there exists a closed operator $\bar{A} : \mathcal{D}(\bar{A}) \to Y$ such that

$$\mathcal{D}(A) \subset \mathcal{D}(\bar{A}) \subset X,\ \bar{A}x = Ax,\ x \in \mathcal{D}(A).$$

Let the symbol $\langle\cdot,\cdot\rangle$ be referred to as the duality pairing between X^* and X.

Similar to a bounded operator, for any linear operator (not necessarily bounded) $A : \mathcal{D}(A) \subset X \to X$ we still define the resolvent

$$\rho(A) = \{\lambda \in C^1 : (\lambda I - A)^{-1} \in \mathcal{L}(X)\},$$

the spectrum $\sigma(A) = C^1 \setminus \rho(A)$, the point spectrum (or the set of eigenvalues) $\sigma_p(A)$ of A and set $\mathrm{Ran}(A) = \{Ax : x \in \mathcal{D}(A)\}$.

Let $A : \mathcal{D}(A) \subset X \to X$ be densely defined. Clearly, the map

$$x \to \langle Ax, y\rangle, \ x \in \mathcal{D}(A), \ y \in X^*$$

is well-defined. Suppose that $y \in X^*$ and

$$|\langle Ax, y\rangle| \le c_y \|x\| \text{ for all } x \in \mathcal{D}(A).$$

Then the functional $f_y(x) = \langle Ax, y\rangle, \ x \in \mathcal{D}(A)$ can be extended linearly and continuously to the whole X which is the closure of $\mathcal{D}(A)$. Such an extension (still denoted by itself) f_y is in X^* and unique. Hence we obtain

$$\langle Ax, y\rangle = f_y(x) = \langle x, f_y\rangle \text{ for all } x \in \mathcal{D}(A).$$

Define

$$\mathcal{D}(A^*) = \{y \in X^* : \exists c_y \ge 0 \text{ such that } |\langle Ax, y\rangle| \le c_y \|x\| \ \forall x \in \mathcal{D}(A)\},$$

$$A^* y = f_y, \ y \in \mathcal{D}(A^*).$$

Clearly, $A^* : \mathcal{D}(A^*) \subset X^* \to X^*$ is a linear operator satisfying

$$\langle Ax, y\rangle = \langle x, A^* y\rangle \text{ for all } x \in \mathcal{D}(A) \text{ and all } y \in \mathcal{D}(A^*).$$

The mapping A^* is called the adjoint operator of A.

1.3 C_0 Semigroup

Let $(X, \|\cdot\|)$ be a Banach space and let $\{T(t) : t \in [0, \infty)\} \subset \mathcal{L}(X)$. We call $T(\cdot)$ a C_0 semigroup (or a strongly continuous semigroup of operators) on X if the following properties hold:

$$T(0) = I, \tag{1.2}$$

$$T(t+s) = T(s)T(t) \text{ for all } s, t \ge 0, \tag{1.3}$$

$$\lim_{s \to 0} \|T(s)x - x\| = 0, \ x \in X. \tag{1.4}$$

In the case when $T(t)$ is defined for all $t \in R^1$ and (1.3) holds for all $t, s \in R^1$, we call $T(\cdot)$ as C_0 group or a strongly continuous group of operators. Equations (1.3) and (1.4) are usually referred to as the semigroup property and the strong continuity, respectively.

The following result is true (see Proposition 4.7, Chapter 2 of [64]).

Proposition 1.3. *Let $T(\cdot)$ be a C_0 semigroup on X. Then there exist constants $M \geq 1$ and $\omega \in R^1$ such that*

$$\|T(t)\| \leq Me^{\omega t}, \ t \in [0, \infty).$$

Let $T(\cdot)$ be a C_0 semigroup on X, and let

$$\mathcal{D}(A) = \{x \in X : \lim_{t \to 0} t^{-1}(T(t) - I)x \text{ exists}\},$$

$$Ax = \lim_{t \to 0} t^{-1}(T(t) - I)x, \ x \in \mathcal{D}(A).$$

The operator $A : \mathcal{D}(A) \to X$ is called the generator of the semigroup $T(\cdot)$. We also say that A generates the C_0 semigroup $T(\cdot)$. In general, the operator A is not bounded.

The following two results hold.

Proposition 1.4 (Proposition 4.10, Chapter 2 of [64]). *Let $T(\cdot)$ be a C_0 semigroup on X. Then the generator A of $T(\cdot)$ is densely defined closed operator and $\bigcap_{n=1}^{\infty} \mathcal{D}(A^n)$ is dense in X. Furthermore, is $S(\cdot)$ is another C_0 semigroup on X with the same generator A as $T(\cdot)$, then $S(\cdot) = T(\cdot)$.*

Theorem 1.5 ([50, 93] and Theorem 4.11, Chapter 2 of [64]). *Let $A : \mathcal{D}(A) \subset X \to X$ be a linear operator. The following properties are equivalent.*

(i) A generates a C_0 semigroup $T(\cdot)$ on X such that $\|T(t)\| \leq Me^{\omega t}$ for all $t \geq 0$ with some $M \geq 1$ and $\omega \in R^1$.

(ii) A is a densely defined and closed operator such that for the above $M \geq 1$ and $\omega \in R^1$, $\{\lambda \in C^1 : \ Re\,\lambda > \omega\} \subset \rho(A)$ and

$$\|(\lambda I - A)^{-n}\| \leq M(Re\,\lambda - \omega)^{-n}$$

for all integers $n \geq 0$ and all $\lambda \in C^1$ with $Re\,\lambda > \omega$.

(iii) A is a densely defined and closed operator such that for the above $M \geq 1$ and $\omega \in R^1$, there exists a sequence of positive numbers $\lambda_k \to \infty$ as $k \to \infty$ such that for all integers $k \geq 1$, $\lambda_k \in \rho(A)$ and

$$\|(\lambda_k I - A)^{-n}\| \leq M(\lambda_k - \omega)^{-n} \text{ for all integers } n \geq 0 \text{ and all integers } k \geq 1.$$

Because the generator A determines the C_0 semigroup $T(\cdot)$ uniquely and in the case $A \in \mathcal{L}(X)$, the C_0 semigroup has an explicit expression $e^{At} = \sum_{n=0}^{\infty}(n!)^{-1}t^n A^n$ (Proposition 4.9, Chapter 2 of [64]), we denote by e^{At} the C_0 semigroup generated by A.

1.4 Evolution Equations

Let $(X, \|\cdot\|)$ be a Banach space, $A : \mathcal{D}(A) \subset X \to X$ generate a C_0 semigroup e^{At} on X, $T > 0$, $f : [0, T] \to X$ be a Bochner integrable function and $y_0 \in X$. We consider the following evolution equation

$$y'(t) = Ay(t) + f(t),\ t \in [0, T], \qquad (P_0)$$

$$y(0) = y_0.$$

A continuous function $y : [0, T] \to X$ is called a (mild) solution of (P_0) if

$$y(t) = e^{At}y_0 + \int_0^t e^{A(t-s)} f(s)ds,\ t \in [0, T].$$

A continuous function $y : [0, T] \to X$ is called a weak solution of (P_0) if for any $x^* \in \mathcal{D}(A^*)$, $\langle y(\cdot), x^*\rangle$ is an absolutely continuous function on $[0, T]$ and that for all $t \in [0, T]$,

$$\langle y(t), x^*\rangle = \langle y_0, x^*\rangle + \int_0^t [\langle y(s), A^*x^*\rangle + \langle f(s), x^*\rangle]ds.$$

We have the following result.

Proposition 1.6 ([15], Proposition 5.2, Chapter 2 of [64]). *A continuous function $y : [0, T] \to X$ is a solution of (P_0) if and only if it is a weak solution of (P_0).*

The following useful result is also valid (see Proposition 4.14, Chapter 2 of [64]).

Proposition 1.7. *For any $x \in \mathcal{D}(A)$, $e^{At}x \in \mathcal{D}(A)$ for all $t \geq 0$ and*

$$(d/dt)(e^{At}x) = Ae^{At}x = e^{At}Ax \text{ for all } t \geq 0,$$

$$e^{At}x - e^{As}x = \int_s^t e^{Ar}Axdr \text{ for all } t > s \geq 0.$$

1.5 C_0 Groups

Let $(H, \langle \cdot, \cdot \rangle)$ be a Hilbert space equipped with an inner product $\langle \cdot, \cdot \rangle$ which induces the norm $\| \cdot \|$. A one-parametric family $S(t)$, $t \in R^1$ of continuous linear operators from H onto H is a strongly continuous group of continuous linear operators on H or C_0 group on H if

$$S(0)x = x \text{ for all } x \in H,$$

$$S(t_1 + t_2) = S(t_1)S(t_2) \text{ for all } t_1, t_2 \in R^1,$$

$$\lim_{t \to 0} S(t)x = x, \ x \in H.$$

Let $S(t)$, $t \in R^1$ be a C_0 group on H. Then the generator of S is the linear operator $A : \mathcal{D}(A) \subset H \to H$ defined by

$$\mathcal{D}(A) = \{x \in H : \lim_{t \to 0} t^{-1}(S(t) - I)x \text{ exists}\},$$

$$Ax = \lim_{t \to 0} t^{-1}(S(t) - I)x, \ x \in \mathcal{D}(A).$$

The following result is true.

Theorem 1.8 ([34]). *Let $S(\cdot)$ be a C_0 group on H. Then the following assertions hold.*

(i) $\mathcal{D}(A)$ is dense in H and A is a closed linear operator.
(ii) for every $x^0 \in \mathcal{D}(A)$, there exists a unique function $x \in C^1(R^1; H) \cap C^0(R^1; \mathcal{D}(A))$ such that $x(0) = x^0$, $x(t) \in \mathcal{D}(A)$ for all $t \in R^1$ and $(dx/dt)(t) = Ax(t)$ for all $t \in R^1$; moreover, this solution satisfies $x(t) = S(t)x^0$ for all $t \in R^1$.
(iii) $S(t)^*$, $t \in R^1$ is a C_0 group on H, and its generator is the adjoint operator A^* of A.

It is clear that if $S(\cdot)$ is a C_0 group with the generator $A : \mathcal{D}(A) \to H$, then $S(-t)$, $t \in R^1$ is also a C_0 group with the generator $-A : \mathcal{D}(A) \to H$.

Let $A : \mathcal{D}(A) \subset H \to H$. It is known that A and $-A$ are both generators of C_0 semigroups if and only if A is a generator of a C_0 group [91].

Note that a C_0 group is determined by its generator A uniquely and is denoted by e^{At}, $t \in R^1$. Clearly, for all $t \in R^1$, $e^{-At} = e^{A(-t)}$.

1.6 Notation

In this section we collect the notation which will be used in the book.

Let (X, ρ_X) be a metric space equipped with the metric ρ_X which induces the topology in X. For each $x \in X$ and each $r > 0$ set

$$B_X(x, r) = \{y \in X : \rho(x, y) \le r\}.$$

Let (X_i, ρ_{X_i}), $i = 1, 2$ be metric spaces. The set $X_1 \times X_2$ is equipped with the metric

$$\rho_{X_1 \times X_2}((x_1, x_2), (y_1, y_2)) = \rho_{X_1}(x_1, y_1) + \rho_{X_2}(x_2, y_2),$$

$$(x_1, x_2), \ (y_1, y_2) \in X_1 \times X_2.$$

We denote by $\mathrm{mes}(\Omega)$ the Lebesgue measure of a Lebesgue measurable set $\Omega \subset R^1$ and define

$$\chi_\Omega(x) = 1 \text{ for all } x \in \Omega, \ \chi_\Omega(x) = 0 \text{ for all } x \in R^1 \setminus \Omega.$$

For each function $f : X \to [-\infty, \infty]$, where X is nonempty, we set

$$\inf(f) = \inf\{f(x) : x \in X\}.$$

For each $s \in R^1$ set $s_+ = \max\{s, 0\}$, $s_- = \min\{s, 0\}$, $\lfloor s \rfloor = \max\{z : z \text{ is an integer and } z \le s\}$.

If $(X, \|\cdot\|)$ is a normed space, then X is equipped with the metric $\rho_X(x, y) = \|x - y\|$, $x, y \in X$.

Let (X, ρ_X) be a metric space and T be a Lebesgue measurable subset of R^1. A function $f : T \to X$ is called Lebesgue measurable if for any Borel set $D \subset X$ the set $f^{-1}(D)$ is a Lebesgue measurable set.

In the sequel we denote by $\mathrm{Card}(D)$ the cardinality of a set D, suppose that the sum over an empty set is zero and that the infimum over an empty set is ∞.

Let $(X, \langle\cdot,\cdot\rangle_X)$ be a Hilbert space equipped with an inner product $\langle\cdot,\cdot\rangle_X$ which induces the norm $\|\cdot\|_X$. If the space X is understood, we use the notation $\langle\cdot,\cdot\rangle = \langle\cdot,\cdot\rangle_X$ and $\|\cdot\| = \|\cdot\|_X$. Let X be a Banach space equipped with the norm $\|\cdot\|_X$ and X^* is its dual space with the norm $\|\cdot\|_{X^*}$. If the space X is understood, we use the notation $\|\cdot\| = \|\cdot\|_X$. For $x \in X$ and $l \in X^*$, we set $l(x) = \langle l, x\rangle_{X^*,X}$. The symbol $\langle\cdot,\cdot\rangle_{X^*,X}$ is referred to as the duality pairing between X^* and X. When the pair X, X^* is understood, we use the notation $\langle\cdot,\cdot\rangle = \langle\cdot,\cdot\rangle_{X^*,X}$.

1.7 Admissible Control Operators

Let J be an open interval and U be a Hilbert space. Denote by $H^1(J;U)$ the Sobolev space of all locally absolutely continuous functions $z : J \to U$ for which $dz/dt \in L^2(J;U)$. The space $H^2(J;U)$ is the set of all locally absolutely continuous functions $z : J \to U$ for which $dz/dt \in H^1(J;U)$. The space $H_0^1(J;U)$ is the set of all functions in $H^1(J;U)$ which have limits equal to zero at the endpoints of J. Define

$$H_0^2(J;U) = \{h \in H^2(J;U) \cap H_0^1(J;U) : \ dh/dx \in H_0^1(J;U)\}.$$

In this section we collect several useful results of [91].

Proposition 1.9 (Proposition 2.1.2 of [91]). *Let e^{At}, $t \geq 0$ be a C_0 semigroup on a Hilbert space E. Then the mapping $(t, z) \to e^{At}z$, $t \in [0, \infty)$, $z \in E$ is continuous on $[0, \infty) \times E$ in the product topology.*

Let X, Z be Hilbert spaces and $\mathcal{D}(A)$ be a linear subspace of X. A linear operator $A : \mathcal{D}(A) \to Z$ is called closed if its graph $G(A) = \{(f, Af) : f \in \mathcal{D}(A)\}$ is closed in $X \times Z$. If A is closed, then $\mathcal{D}(A)$ is a Hilbert space with the graph norm $\|\cdot\|_{gr}$:

$$\|z\|_{gr}^2 = \|z\|_X^2 + \|Az\|_Z^2, \ z \in \mathcal{D}(A).$$

Proposition 1.10 (Proposition 2.10.1 of [91]). *Let X be a Hilbert space and $A : \mathcal{D}(A) \to X$ be a densely defined linear operator such that $\rho(A) \neq \emptyset$. Then for every $\beta \in \rho(A) \cap R^1$, the space $\mathcal{D}(A)$ with the norm $\|z\|_1 = \|(\beta I - A)(z)\|$, $z \in \mathcal{D}(A)$ is a Hilbert space denoted by X_1. The norms generated as above for different $\beta \in \rho(A) \cap R^1$ are equivalent to the graph norm, and the embedding $X_1 \subset X$ is continuous. If $L \in \mathcal{L}(X)$ satisfies $L\mathcal{D}(A) \subset \mathcal{D}(A)$, then $L \in \mathcal{L}(X_1)$.*

Note that the relation $\rho(A) \neq \emptyset$ implies that the operator A is closed. Let A be as in Proposition 1.10. It is clear that A^* has the same property. We define a Hilbert space $X_1^d = \mathcal{D}(A^*)$ equipped with the norm $\|z\|_1^d = \|(\beta I - A^*)z\|$, $z \in \mathcal{D}(A^*)$, where $\beta \in \rho(A) \cap R^1$, or equivalently, $\beta \in \rho(A^*) \cap R^1$.

Proposition 1.11 (Proposition 2.10.2 of [91]). *Let A be as in Proposition 1.10 and let $\beta \in \rho(A) \cap R^1$. We denote by X_{-1} the completion of X with respect to the norm*

$$\|z\|_{-1} = \|(\beta I - A)^{-1}z\|, \ z \in X. \tag{1.5}$$

The norms generated as above for different $\beta \in \rho(A) \cap R^1$ are equivalent (in particular, X_{-1} is independent of the choice of β). Moreover, X_{-1} is the dual of X_1^d with respect to the pivot space X. If $L \in \mathcal{L}(X)$ satisfies $L^*\mathcal{D}(A^*) \subset \mathcal{D}^*(A)$, then L has a unique extension to an operator $\tilde{L} \in \mathcal{L}(X_{-1})$.

Proposition 1.12 (Proposition 2.10.3 of [91]). *Let $A : \mathcal{D}(A) \to X$ be a densely defined linear operator such that $\rho(A) \neq \emptyset$, $\beta \in \rho(A) \cap R^1$, X_1 be as in Proposition 1.10 and X_{-1} be as in Proposition 1.11. Then $A \in \mathcal{L}(X_1, X)$ and A has a unique extension $\tilde{A} \in \mathcal{L}(X, X_{-1})$. Moreover, the operators $(\beta I - A)^{-1} \in \mathcal{L}(X, X_1)$ and $(\beta I - \tilde{A})^{-1} \in \mathcal{L}(X_{-1}, X)$ are unitary.*

Proposition 1.13 (Proposition 2.10.4 of [91]). *We use the notation from Proposition 1.12 and assume that A generates a C_0 semigroup $S(\cdot)$ on X. Then for every $t \geq 0$, $S(t)$ has a unique extension $\tilde{S}(t) \in \mathcal{L}(X_{-1})$, $\tilde{S}(t)$, $t \geq 0$ is a C_0 semigroup on X_{-1}, and $\tilde{A}$ is its generator.*

Let U, X be Hilbert spaces, $S(\cdot)$ be a C_0 semigroup on X and $A : \mathcal{D}(A) \to X$ be its generator. We use the notation above and denote by A and $S(\cdot)$ also the extension of the generator to X and the extension of the semigroup to X_{-1}, respectively.

Consider the differential equation

$$z'(t) = Az(t) + f(t) \tag{1.6}$$

where $f \in L^1_{loc}([0, \infty); X_{-1})$. A solution of (1.6) in X_{-1} is a function $z \in L^1_{loc}([0, \infty); X) \cap C([0, \infty); X_{-1})$ which satisfies the following equations in X_{-1}:

$$z(t) - z(0) = \int_0^t [Az(s) + f(s)]ds \text{ for all } t \in [0, \infty). \tag{1.7}$$

It is also called a "strong" solution of (1.6) in X_{-1}. Equation (1.7) implies that z is an a. c. function with values in X_{-1} and (1.6) holds for almost every $t \geq 0$, with the derivative computed with respect to the norm of X_{-1}.

We can also define the concept of a "weak" solution of (1.6) in X_{-1} by regarding instead of (1.7) that for every $\phi \in X_1^d$ and every $t \geq 0$,

$$\langle z(t) - z(0), \phi \rangle_{X_{-1}, X_1^d} = \int_0^t [\langle z(s), A^*\phi \rangle_X + \langle f(s), \phi \rangle_{X_{-1}, X_1^d}]ds.$$

These two concepts are equivalent.

Proposition 1.14 (Proposition 4.1.4 of [91]). *Suppose that z is a solution of (1.6) in X_{-1} and let $z_0 = z(0)$. Then z is given by*

$$z(t) = S(t)z_0 + \int_0^t S(t - \sigma)f(\sigma)d\sigma, \ t \geq 0. \tag{1.8}$$

In particular, for every $z_0 \in X$, there exists at most one solution in X_{-1} of (1.6) which satisfies initial condition $z(0) = z_0$.

Note that z satisfying (1.8) is called the mild solution of (1.6) with the initial state $z_0 \in X$.

It is not difficult to see that the following result is valid.

Proposition 1.15. *Let $f \in L^1_{loc}([0,\infty); X_{-1})$, $T > 0$ and $z \in C^0([0,T]; X_{-1})$. The z satisfies (1.8) for all $t \in [0,T]$ in X_{-1} if and only if for all $\xi \in X_1^d = \mathcal{D}(A^*)$ and all $t \in [0,T]$,*

$$\langle z(t), \xi\rangle_{X_{-1},X_1^d} = \langle S(t)z_0, \xi\rangle_{X_{-1},X_1^d} + \int_0^t [\langle S(t-\sigma)f(\sigma), \xi\rangle_{X_{-1},X_1^d}]d\sigma$$

$$= \langle S(t)z_0, \xi\rangle_{X_{-1},X_1^d} + \int_0^t [\langle f(\sigma), S^*(t-\sigma)\xi\rangle_{X_{-1},X_1^d}]d\sigma.$$

Proposition 1.16 (Proposition 12.1.2 of [91]). *Assume that Z_1, Z_2 and Z_3 are Hilbert spaces, $F \in \mathcal{L}(Z_1, Z_3)$, $G \in \mathcal{L}(Z_2, Z_3)$ and*

$$Ran(F) = \{F(z) : z \in Z_1\} \subset Ran(G) = \{G(z) : z \in Z_2\}.$$

Then there exists an operator $L \in \mathcal{L}(Z_1, Z_2)$ such that $F = GL$.

Corollary 1.17. *Assume that Z_1, Z_2 are Hilbert spaces and $G \in \mathcal{L}(Z_1, Z_2)$ satisfies $Ran(G) = Z_2$. Then there exists an operator $L \in \mathcal{L}(Z_2, Z_1)$ such that $GL = I$ - the identity operator in Z_2.*

Let $B \in \mathcal{L}(U, X_{-1})$ and $\tau \geq 0$. Define $\Phi_\tau \in \mathcal{L}(L^2(0,\infty; U), X_{-1})$ by

$$\Phi_\tau u = \int_0^\tau S(\tau-\sigma)Bu(\sigma)d\sigma. \tag{1.9}$$

The operator $B \in \mathcal{L}(U, X_{-1})$ is called an admissible control operator for $S(\cdot)$ if for some $\tau > 0$,

$$\text{Ran}(\Phi_\tau) = \{\Phi_\tau u : u \in L^2(0,\infty; U)\} \subset X.$$

Proposition 1.18 (Proposition 4.2.2 and (4.2.5) of [91]). *Assume that $B \in \mathcal{L}(U, X_{-1})$ is an admissible control operator for $S(\cdot)$. Then for every $t \geq 0$,*

$$\Phi_t \in \mathcal{L}(L^2(0,\infty; U), X)$$

and for all $T > t > 0$, $\|\Phi_t\| \leq \|\Phi_T\|$.

Proposition 1.19 (Proposition 4.2.5 of [91]). *Assume that $B \in \mathcal{L}(U, X_{-1})$ is an admissible control operator for $S(\cdot)$. Then for every $z_0 \in X$ and every $u \in L^2_{loc}(0,\infty; U)$, the initial value problem*

$$z'(t) = Az(t) + Bu(t), \; z(0) = z_0$$

has a unique solution in X_{-1}. *This solution is given by*

$$z(t) = S(t)z_0 + \Phi_t u = S(t)z_0 + \int_0^t S(t-\sigma)Bu(\sigma)d\sigma$$

and it satisfies

$$z \in C([0,\infty); X) \cap H^1_{loc}((0,\infty); X_{-1}).$$

Proposition 1.20 (Proposition 4.2.6 of [91]). *Assume that* $B \in \mathcal{L}(U, X_{-1})$ *is an admissible control operator for* $S(\cdot)$. *Then for every* $z_0 \in X$ *and every* $u \in L^2_{loc}(0,\infty; U)$, *there exists a unique function* $z \in C([0,\infty); X)$ *such that for every* $t \geq 0$ *and every* $\psi \in \mathcal{D}(A^*)$,

$$\langle z(t) - z_0, \psi\rangle_X = \int_0^t [\langle z(\sigma), A^*\psi\rangle_X + \langle u(\sigma), B^*\psi\rangle_U]d\sigma.$$

Propositions 1.18 and 1.19 and Theorem 1.5 imply the following result.

Proposition 1.21. *Assume that* $B \in \mathcal{L}(U, X_{-1})$ *is an admissible control operator for* $S(\cdot)$ *and* $T > 0$. *Then there exists a constant* $c_T > 0$ *such that for every* $z_0 \in X$ *and every* $u \in L^2(0,T;U)$ *the unique solution z of the initial value problem*

$$z'(t) = Az(t) + Bu(t), \ t \in (0,T) \text{ a. e.}, z(0) = z_0$$

satisfies $z \in C([0,T]; X)$ *and that for all* $t \in [0,T]$,

$$\|z(t)\| \leq c_T(\|z_0\| + \|u\|).$$

Proposition 1.18 and Corollary 1.17 imply the following result.

Proposition 1.22. *Assume that* $B \in \mathcal{L}(U, X_{-1})$ *is an admissible control operator for* $S(\cdot)$, $T > 0$ *and* $Ran(\Phi_T) = X$. *Then there exists* $L \in \mathcal{L}(X, L^2(0,T;U))$ *such that* $\Phi_T Lx = x$ *for all* $x \in X$.

Theorem 1.23. *Assume that* $B \in \mathcal{L}(U, X_{-1})$ *is an admissible control operator for a* C_0 *semigroup* $S(\cdot)$, $T > 0$, $Ran(\Phi_T) = X$, $x_f \in X$, $u_f \in U$, $x^*(t) = x_f$, $u^*(t) = u_f$, $t \in [0,\infty)$, *and* $x^*(\cdot)$ *is a unique solution in* X_{-1} *of the initial value problem*

$$z'(t) = Az(t) + Bu^*(t), \ z(0) = x_f. \tag{1.10}$$

Then there exists a constant $c > 0$ *such that for each* $z^0, z^1 \in X$, *there exist* $u \in L^2(0,T;U)$ *and* $z \in C^0([0,T]; X)$ *which is a solution of the problem*

$$z'(t) = Az(t) + Bu(t), \ t \in [0,T] \text{ a. e.}, \ z(0) = z^0$$

in X_{-1} *and satisfy* $z(T) = z^1$,

$$\|u(t) - u_f\| \leq c(\|z^1 - x_f\| + \|z^0 - x_f\|),\ t \in [0, T],$$

$$\|z(t) - x_f\| \leq c(\|z^1 - x_f\| + \|z^0 - x_f\|),\ t \in [0, T].$$

Proof. Let $L \in \mathcal{L}(X, L^2(0, T; U))$ be as guaranteed by Proposition 1.22. Therefore

$$\Phi_T L x = x \text{ for all } x \in X. \tag{1.11}$$

Let $c_T > 0$ be as guaranteed by Proposition 1.21. Set

$$c = c_T + c_T\|L\|. \tag{1.12}$$

Let $z^0, z^1 \in X$. By Proposition 1.19, there exists $\tilde{z} \in C^0([0, T]; X)$ which is a solution of the initial value problem

$$\tilde{z}' = A\tilde{z} + Bu_f,\ t \in (0, T), \tag{1.13}$$

$$\tilde{z}(0) = z^0 \tag{1.14}$$

in X_{-1}. Proposition 1.19 implies that there exists a unique function $\widehat{z} \in C^0([0, T]; X)$ such that

$$\widehat{z}' = A\widehat{z} + B(L(z^1 - \tilde{z}(T)),\ t \in (0, T), \tag{1.15}$$

$$\widehat{z}(0) = 0 \tag{1.16}$$

in X_{-1}. In view of (1.11) and Proposition 1.19,

$$\widehat{z}(T) = S(T)\widehat{z}(0) + \Phi_T(L(z^1 - \tilde{z}(T))) = \Phi_T(L(z^1 - \tilde{z}(T))) = z^1 - \tilde{z}(T). \tag{1.17}$$

Set

$$z = \tilde{z} + \widehat{z}. \tag{1.18}$$

By (1.14) and (1.16)–(1.18),

$$z(0) = \tilde{z}(0) + \widehat{z}(0) = z^0, \tag{1.19}$$

$$z(T) = \tilde{z}(T) + \widehat{z}(T) = z^1. \tag{1.20}$$

It follows from (1.13), (1.15), (1.17), and (1.18) that

$$z' = \tilde{z}' + \widehat{z}' = A(\tilde{z} + \widehat{z}) + B(L(z^1 - \tilde{z}(T)) + u_f)$$

$$= Az + B(L(z^1 - \tilde{z}(T)) + u_f), \ t \in (0, T) \tag{1.21}$$

in X_{-1}. In view of (1.10), (1.13) and (1.14),

$$(\tilde{z} - x^*)' = A(\tilde{z} - x^*), \ t \in (0, T) \tag{1.22}$$

in X_{-1} and

$$(\tilde{z} - x^*)(0) = z^0 - x_f. \tag{1.23}$$

It follows from (1.14), (1.22), (1.23), the choice of c_T and Proposition 1.21 that for each $\tau \in [0, T]$,

$$\|\tilde{z}(\tau) - x_f\| \le c_T \|\tilde{z}(0) - x_f\| = c_T \|z^0 - x_f\|. \tag{1.24}$$

In view of (1.10), (1.19), and (1.21),

$$(z - x^*)' = A(z - x^*) + B((L(z^1 - \tilde{z}(T)))), \ t \in (0, T) \tag{1.25}$$

in X_{-1},

$$(z - x^*)(0) = z^0 - x_f. \tag{1.26}$$

By (1.24),

$$\|L(z^1 - \tilde{z}(T))\| \le \|L\| \|z^1 - \tilde{z}(T)\| \le \|L\|(\|z^1 - x^f\| + \|x_f - \tilde{z}(T)\|)$$

$$\le \|L\|(\|z^1 - x_f\| + c_T \|z^0 - x_f\|). \tag{1.27}$$

Proposition 1.21, the choice of c_T, (1.12) and (1.24)–(1.27) imply that for all $t \in [0, T]$,

$$\|z(t) - x_f\| \le c_T(\|z^0 - x_f\| + \|L(z^1 - \tilde{z}(T))\|)$$

$$\le c_T(\|z^0 - x_f\| + \|L\|(\|z^1 - x_f\| + c_T \|z^0 - x_f\|))$$

$$= \|z^0 - x_f\|(c_T + c_T \|L\|) + c_T \|L\| \|z^1 - x_f\| \le c(\|z^0 - x_f\| + \|z^1 - x_f\|).$$

Theorem 1.23 is proved.

1.8 Examples

Let U, X be Hilbert spaces, $S(\cdot)$ be a C_0 semigroup on X, $A : \mathcal{D}(A) \to X$ be its generator and $B \in \mathcal{L}(U, X_{-1})$ be an admissible control operator for $S(\cdot)$. We say that the pair (A, B) is exactly controllable in a time $\tau > 0$ if $\mathrm{Ran}(\Phi_\tau) = X$ [34, 91]. It is known that the exactly controllability in time $\tau > 0$ is equivalent to the following property:

for each pair $z_0, z_1 \in X$, there exists $u \in L^2(0, \tau; U)$ such that the solution z of the initial value problem

$$z' = Az + Bu, \ z(0) = z_0 \tag{1.28}$$

satisfies $z(\tau) = z_1$.

In this section we consider examples of the pairs (A, B) which are exactly controllable in some time $\tau > 0$. All of these examples were discussed in [91] and [34].

It should be mentioned that in Chapters 5 and 6 our turnpike results will be established for large and general classes of infinite-dimensional optimal control problems. One of this classes contains problems which are defined by an integrand and a pair of operators (A, B) introduced above, which is exactly controllable in some time $\tau > 0$. Therefore, each example of pairs (A, B) considered below gives us a subclass of infinite-dimensional optimal control problems, for which the results of Chapters 5 and 6 hold.

Example 1.24 (Example 11.2.2 of [91]). We consider the problem of controlling the vibrations of an elastic membrane by a force field acting on a part of this membrane. More precisely, let n be a natural number and let $\Omega \subset R^n$ be a bounded open set with $\partial\Omega$ of class C^2 or Ω be a rectangular domain. The physical problem described above can be modeled by the equations

$$\frac{\partial^2 w}{\partial t^2} - \Delta w = u \text{ in } \Omega \times (0, \infty),$$

$$w = 0 \text{ on } \partial\Omega \times (0, \infty),$$

$$w(x, 0) = f(x), \ \frac{\partial w}{\partial t}(x, 0) = g(x) \text{ for } x \in \Omega,$$

where f is the initial displacement and g is the initial velocity. Let $\mathcal{O}$ be a nonempty open subset of Ω and $u \in L^2(0, \infty; L^2(\mathcal{O}))$ be the input function. For any such u, we assume that $u(x, t) = 0$ for all $x \in \Omega \setminus \mathcal{O}$. The equation above can be written in the form (1.28) using the following spaces and operators: $X = H_0^1(\Omega) \times L^2(\Omega)$, $\mathcal{D}(A) = (H^2(\Omega) \cap H_0^1(\Omega)) \times H_0^1(\Omega)$, $U = L^2(\mathcal{O}) \subset L^2(\Omega)$,

$$A \begin{pmatrix} f \\ g \end{pmatrix} = \begin{pmatrix} g \\ \Delta f \end{pmatrix} \text{ for all } \begin{pmatrix} f \\ g \end{pmatrix} \in \mathcal{D}(A), \ Bu = \begin{pmatrix} 0 \\ u \end{pmatrix} \text{ for all } u \in U.$$

Here $B \in \mathcal{L}(U, X)$. It was shown in Example 11.2.2 of [91] that B is an admissible control operator for the C_0 semigroup e^{At} and that the pair (A, B) is exactly controllable in time $\tau > 0$ if Γ and $\mathcal{O}$ satisfy the assumptions of Theorem 7.4.1 of [91] where Γ is a relatively open subset of $\partial\Omega$. In particular, (A, B) is exactly controllable in time $\tau > 0$ if there exist $x_0 \in R^n$ and $\epsilon > 0$ such that

$$N_\epsilon(\{x \in \partial\Omega : (x - x_0) \cdot \nu(x) > 0\}) \subset \text{clos}\mathcal{O}, \ \tau > 2r(x_0),$$

where $r(x_0) = \sup\{\|x - x_0\| : x \in \Omega\}$, $|\cdot|$ is the Euclidean norm in R^n, ν is the unit outward normal vector field in $\partial\Omega$ and $N_\epsilon(D) = \{x \in \Omega : d(x, D) < \epsilon\}$ for any $D \subset \Omega$ with $d(x, D) = \inf\{|x - y| : y \in D\}$.

Example 1.25 (Example 11.2.4 of [91]). Let n be a natural number, let $\Omega \subset R^n$ be a bounded open set with $\partial\Omega$ of class C^2 or let Ω be a rectangular domain, and let $\mathcal{O}$ be a nonempty open subset of Ω. We consider the problem of controlling the vibrations of an elastic plate occupying the domain Ω by a force field acting on $\mathcal{O}$. More precisely, we consider the following initial and boundary initial value problem

$$\frac{\partial^2 w}{\partial t^2} + \Delta^2 w = u \text{ in } \Omega \times (0, \infty),$$

$$w = \Delta w = 0 \text{ on } \partial\Omega \times (0, \infty),$$

$$w(x, 0) = 0, \ \frac{\partial w}{\partial t}(x, 0) = 0 \text{ for } x \in \Omega,$$

where $u \in L^2(0, \infty; L^2(\mathcal{O}))$ is the input function. As usual we assume that $u(x, t) = 0$ for all $x \in \Omega \setminus \mathcal{O}$. The equations above determines a system with the state space $X = (H^2(\Omega) \cap H_0^1(\Omega)) \times L^2(\Omega)$ and the input space $U = L^2(\Omega)$ which is exactly controllable in any time $\tau > 0$, as it was shown in Example 11.2.4 of [91], if the pair $(\Omega, \mathcal{O})$ satisfies one of the assumptions (A1) or (A2) in Example 11.2.3 of [91]. More precisely, let $H = L^2(\Omega)$ and $\mathcal{D}(A_0) = H_1$ be the Sobolev space $H^2(\Omega) \cap H_0^1(\Omega)$, $A_0 : \mathcal{D}(A_0) \to H$ be defined by $A_0\phi = -\Delta\phi$, $\phi \in \mathcal{D}(A_0)$, and let $H_2 = \mathcal{D}(A_0^2)$ be endowed with the graph norm. Let χ be the Hilbert space $H_1 \times H$, consider the dense subset of χ defined by $\mathcal{D}(A) = H_2 \times H_1$ and let the linear operator $\mathcal{A} : \mathcal{D}(\mathcal{A}) \to \chi$ be defined by

$$\mathcal{A} = \begin{pmatrix} 0 & I \\ -A_0^2 & 0 \end{pmatrix}.$$

Then the equations above can be written in the form

$$z' = \mathcal{A}z + Bu, \ z(0) = 0$$

where $B \in \mathcal{L}(U, \chi)$ is defined by $Bu = \begin{pmatrix} 0 \\ u \end{pmatrix}$ for all $u \in U$. It was shown in Example 11.2.4 of [91] that the pair (A, B) is exactly controllable in any time $\tau > 0$ if the pair $(\Omega, \mathcal{O})$ satisfies one of the assumptions (A1) or (A2) in Example 11.2.3 of [91].

Example 1.26 (Example 11.2.6 of [91]). We consider the problem of controlling the vibrations of a string occupying the interval $[0, \pi]$ by means of a force $u(t)$ acting at its left end. The equations describing this problem are formulated as a well-posed boundary control system in subsection 10.2.2 of [91]:

$$\frac{\partial^2 w}{\partial t^2}(x,t) = \frac{\partial^2 w}{\partial x^2}(x,t), \; 0 < x < \pi, \; t \geq 0,$$

$$w(\pi, t) = 0, \; \frac{\partial w}{\partial x}(0,t) = u(t), \; t \geq 0,$$

$$w(x, 0) = f(x), \; \frac{\partial w}{\partial t}(x, 0) = g(x), 0 < x < \pi.$$

For this problem $X = H^1_R(0, \pi) \times L^2(0, \pi)$, $A : \mathcal{D}(A) \to X$ is defined by $\mathcal{D}(A) = \{f \in H^2(0, \pi) \cap H^1_R(0, \pi) : \frac{\partial f}{\partial x}(0) = 0\} \times H^1_R(0, \pi)$,

$$A \begin{pmatrix} f \\ g \end{pmatrix} = \begin{pmatrix} g \\ \frac{d^2 f}{dx^2} \end{pmatrix} \text{ for all } \begin{pmatrix} f \\ g \end{pmatrix} \in \mathcal{D}(A).$$

The control operator B satisfies $B^* \begin{pmatrix} f \\ g \end{pmatrix} = -g(0)$ for all $\begin{pmatrix} f \\ g \end{pmatrix} \in \mathcal{D}(A)$. It was shown in Example 10.2.6 of [91] that the pair (A, B) is exactly controllable in any time $\tau \geq 2\pi$.

Example 1.27 (Example 11.2.7 of [91]). We consider the boundary control of the nonhomogeneous elastic string. The model is described by the equation

$$\frac{\partial^2 w}{\partial t^2}(x,t) = \frac{\partial}{\partial x}(a(x)\frac{\partial w}{\partial x}(x,t)) - b(x)w(x,t), \; 0 < x < \pi, \; t > 0,$$

$$w(0,t) = u(t), \; w(\pi, t) = 0, \; w(\cdot, 0) = f, \; \frac{\partial w}{\partial t}(\cdot, 0) = g.$$

Here $a \in C^2[0, \pi]$, $b \in L^\infty(0, \pi)$, $a(x) \geq m > 0$, $b(x) \geq 0$ for all $x \in [0, \pi]$. These equations correspond to a well-posed boundary control system with state space $X = L^2(0, \pi) \times H^{-1}(0, \pi)$. The generator A is defined by

$$A \begin{pmatrix} f \\ g \end{pmatrix} = \begin{pmatrix} g \\ -A_0 f \end{pmatrix} \text{ for all } \begin{pmatrix} f \\ g \end{pmatrix} \in \mathcal{D}(A) = H_0^1(0, \pi) \times L^2(0, \pi),$$

where $A_0 \in \mathcal{L}(H_0^1(0, \pi), H^{-1}(0, \pi))$ is defined by

$$A_0 f = \frac{d}{dx}(a \frac{df}{dx}) + bf \text{ for all } f \in H_0^1(0, \pi).$$

The control operator B of this system is determined by

$$B^* \begin{pmatrix} \phi \\ \psi \end{pmatrix} = a(0) \frac{d}{dx}(A_0^{-1} \psi)|_{x=0} \text{ for all } \begin{pmatrix} \phi \\ \psi \end{pmatrix} \in \mathcal{D}(A^*) = \mathcal{D}(A).$$

It was shown in Example 11.2.7 of [91] that the pair (A, B) is exactly controllable in any time $\tau \geq 2 \int_0^\pi (a(x))^{-1/2} dx$.

Example 1.28 (Example 11.2.8 of [91]). We consider the problem of controlling the vibrations of a beam occupying the interval $[0, \pi]$ by means of a torque $u(t)$ acting at its left end. The model is described by the initial and boundary value problem

$$\frac{\partial^2 w}{\partial t^2}(x, t) = -\frac{\partial^4 w}{\partial x^4}(x, t), \; 0 < x < \pi, \; t > 0,$$

$$w(0, t) = 0, \; w(\pi, t) = 0, \; \frac{\partial^2 w}{\partial x^2}(0, t) = u(t), \; \frac{\partial^2 w}{\partial x^2}(\pi, t) = 0,$$

$$w(\cdot, 0) = f, \; \frac{\partial w}{\partial t}(\cdot, 0) = g.$$

These equations correspond to a control system where $H = L^2(0, \pi)$, $H_1 = H^2(0, \pi) \cap H_0^1(0, \pi)$, $A_0 : H_1 \to H$ is defined by $A_0 f = -\frac{d^2 f}{dx^2}$ for all $f \in H_1$, $H_{1/2} = H_0^1(0, \pi)$, $H_{-1/2} = H_0^{-1}(0, \pi)$. The unique extensions of A_0 to unitary operators from $H_{1/2}$ onto $H_{-1/2}$ and from H onto H_{-1} are still denoted by A_0. The space $H_{3/2} = A_0^{-1} H_{1/2}$ is

$$H_{3/2} = \{g \in H^3(0, \pi) \cap H_0^1(0, \pi) : \frac{d^2 g}{dx^2}(0) = \frac{d^2 g}{dx^2}(\pi) = 0\}.$$

We set $X = H_{1/2} \times H_{-1/2}$, $\mathcal{D}(A) = H_{3/2} \times H_{1/2}$,

$$A \begin{pmatrix} f \\ g \end{pmatrix} = \begin{pmatrix} g \\ -A_0^2 f \end{pmatrix} \text{ for all } \begin{pmatrix} f \\ g \end{pmatrix} \in \mathcal{D}(A).$$

The control operator B of this system is determined by

$$B^* \begin{pmatrix} f \\ g \end{pmatrix} = -\frac{d}{dx}(A_0^{-1}g)|_{x=0} \text{ for all } \begin{pmatrix} f \\ g \end{pmatrix} \in \mathcal{D}(A^*) = \mathcal{D}(A).$$

It was shown in Example 11.2.8 of [91] that the pair (A, B) is exactly controllable in any time $\tau > 0$.

Example 1.29 (Example 11.2.9 of [91]). We consider the problem of controlling the vibrations of a beam occupying the interval $[0, 1]$ by means of an angular velocity $u(t)$ applied at its left end. The equations describing this problem have been formulated as a well-posed boundary control system in Section 10.5 of [91] as follows:

$$\frac{\partial^2 w}{\partial t^2}(x,t) = -\frac{\partial^4 w}{\partial x^4}(x,t),\ 0 < x < 1,\ t > 0,$$

$$w(0,t) = 0,\ w(1,t) = 0,\ \frac{\partial w}{\partial x}(0,t) = u(t),\ \frac{\partial w}{\partial x}(1,t) = 0,$$

$$w(\cdot,0) = f,\ \frac{\partial f}{\partial t}(\cdot,0) = g.$$

We denote $X = V \times L^2(0,1)$, where $V = \{h \in H^2(0,1) : h(0) = h(1) = \frac{dh}{dx}(1) = 0\}$. The norm on X is defined by

$$\|z\|^2 = \|z_1\|_V^2 + \|z_2\|_{L^2}^2, \text{ where } \|z_1\|_V^2 = \int_0^1 |\frac{d^2 z_1}{dx^2}|^2 dx.$$

Let $Z \subset X$ be defined by $Z = (V \cap H^4(0,1)) \times V$, $L : Z \to X$, $G : Z \to R^1$ be defined by

$$L\begin{pmatrix} 0 & I \\ -\frac{d^4}{dx^4} & 0 \end{pmatrix},\ G\begin{pmatrix} z_1 \\ z_2 \end{pmatrix} = \frac{dz_2}{dx}(0),$$

$$\text{Ker}G = \{z \in Z : G(z) = 0\} = (V \cap H^4(0,1)) \times H_0^2(0,1),\ A = L|\text{Ker}G$$

and a control operator B be defined by

$$B^* \begin{pmatrix} \psi_1 \\ \psi_2 \end{pmatrix} = -\frac{d^2\psi_1}{dx^2}(0), \text{ for all } \begin{pmatrix} \psi_1 \\ \psi_2 \end{pmatrix} \in \mathcal{D}(A^*) = \mathcal{D}(A).$$

It was shown in Example 11.2.9 of [91] that the pair (A, B) is exactly controllable in any time $\tau > 0$.

The next two examples are discussed in [34].

Example 1.30. The transport equation. Let $L > 0$. We consider the linear control system

$$y_t + y_x = 0, \ t \in (0, T), \ x \in (0, L),$$

$$y(t, 0) = u(t), \ t \in (0, T),$$

where $u(t) \in R^1$, $y(t, \cdot) : (0, L) \to R^1$, with $X = L^2(0, L)$, $\mathcal{D}(A) = \{f \in H^1(0, L) : \ f(0) = 0\}$, $Af = -f_x$, $f \in \mathcal{D}(A)$, $U = R^1$ and $B : R^1 \to \mathcal{D}(A^*)'$ defined by $(Bu)z = uz(0)$ for all $u \in R^1$ and all $z \in \mathcal{D}(A^*)$. It was shown in [34] that the pair (A, B) is controllable.

Example 1.31. The Korteweg-de Vries equation.

Let $L > 0$. We consider the linear control system

$$y_t + y_x + y_{xxx} = 0, \ t \in (0, T), \ x \in (0, L),$$

$$y(t, 0) = y(t, L) = 0, \ y_x(t, L) = u(t), \ t \in (0, T),$$

where $u(t) \in R^1$, $y(t, \cdot) : (0, L) \to R^1$, with $X = L^2(0, L)$, $\mathcal{D}(A) = \{f \in H^3(0, L) : \ f(0) = f(L) = f_x(L) = 0\}$, $Af = -f_x - f_{xxx}$, $f \in \mathcal{D}(A)$, $U = R^1$ and $B : R^1 \to \mathcal{D}(A^*)'$ is defined by $(Bu)z = uz_x(L)$ for all $u \in R^1$ and all $z \in \mathcal{D}(A^*)$. It was shown in [34] that the pair (A, B) is controllable.

1.9 Turnpike Property for Variational Problems

The study of the existence and the structure of solutions of optimal control problems and dynamic games defined on infinite intervals and on sufficiently large intervals has been a rapidly growing area of research [7, 8, 21, 29, 36, 38–40, 51, 54, 55, 67, 78, 104, 106, 108, 111, 116, 120, 125, 126, 129, 131, 134] which has various applications in engineering [5, 29, 61, 104], in models of economic growth [6, 10, 26–29, 37, 42, 53, 60, 66, 72, 79, 84–86, 92, 104, 114, 121, 130, 131], in infinite discrete models of solid-state physics related to dislocations in one-dimensional crystals [11, 12, 83, 94], in model predictive control [35, 45], and in the theory of thermodynamical equilibrium for materials [33, 62, 69–71]. Discrete-time problems optimal control problems were considered in [9, 13, 14, 24, 41, 48, 95, 96, 101, 103, 107, 110, 112, 114, 118, 119, 122, 127, 128, 130], finite-dimensional continuous-time problems were analyzed in [20, 22, 23, 25, 30, 60, 63, 65, 68, 80, 99, 102, 105, 117, 132, 133], infinite-dimensional optimal control was studied in [28, 29, 46, 74, 76, 77, 81, 88, 89, 98, 100, 109], while solutions of dynamic games were discussed in [19, 43, 44, 47, 49, 57, 82, 113, 115, 123, 124]. Sufficient and necessary conditions for the turnpike phenomenon were obtained in our previous research [102, 103, 105, 118] for finite-dimensional variational problems and for discrete-time optimal control problems in compact metric space.

In this section, which is based on [102], we discuss the structure of approximate solutions of variational problems with continuous integrands $f : [0, \infty) \times R^n \times R^n \to R^1$ which belong to a complete metric space of functions. We do not impose any convexity assumption. The main result of this section obtained in [102] deals with the turnpike property of variational problems.

We consider the variational problems

$$\int_{T_1}^{T_2} f(t, z(t), z'(t))dt \to \min, \ z(T_1) = x, \ z(T_2) = y, \tag{P}$$

$$z : [T_1, T_2] \to R^n \text{ is an absolutely continuous function,}$$

where $T_1 \geq 0$, $T_2 > T_1$, $x, y \in R^n$ and $f : [0, \infty) \times R^n \times R^n \to R^1$ belongs to a space of integrands described below.

It is well known that the solutions of the problems (P) exist for integrands f which satisfy two fundamental hypotheses concerning the behavior of the integrand as a function of the last argument (derivative): one that the integrand should grow superlinearly at infinity and the other that it should be convex [87]. Moreover, certain convexity assumptions are also necessary for properties of lower semicontinuity of integral functionals which are crucial in most of the existence proofs, although there are some interesting theorems without convexity [31, 73, 75]. For integrands f which do not satisfy the convexity assumption, the existence of solutions of the problems (P) is not guaranteed, and in this situation we consider *δ-approximate solutions*.

Let $T_1 \geq 0$, $T_2 > T_1$, $x, y \in R^n$, $f : [0, \infty) \times R^n \times R^n \to R^1$ be an integrand, and let δ be a positive number. We say that an absolutely continuous (a.c.) function $u : [T_1, T_2] \to R^n$ satisfying $u(T_1) = x$, $u(T_2) = y$ is a δ-approximate solution of the problem (P) if

$$\int_{T_1}^{T_2} f(t, u(t), u'(t))dt \leq \int_{T_1}^{T_2} f(t, z(t), z'(t))dt + \delta$$

for each a.c. function $z : [T_1, T_2] \to R^n$ satisfying $z(T_1) = x$, $z(T_2) = y$.

The main result of [102] deals with the turnpike property of the variational problems (P). As usual, to have this property means, roughly speaking, that the approximate solutions of the problems (P) are determined mainly by the integrand and are essentially independent of the choice of interval and endpoint conditions, except in regions close to the endpoints.

In the classical turnpike theory, it was assumed that a cost function (integrand) is convex. The convexity of the cost function played a crucial role there. In [102] we get rid of convexity of integrands and establish necessary and sufficient conditions for the turnpike property for a space of nonconvex integrands $\mathcal{M}$ described below.

Let us now define the space of integrands. Denote by $|\cdot|$ the Euclidean norm in R^n. Let a be a positive constant, and let $\psi : [0, \infty) \to [0, \infty)$ be an increasing function such that $\psi(t) \to +\infty$ as $t \to \infty$. Denote by $\mathcal{M}$ the set of all continuous functions $f : [0, \infty) \times R^n \times R^n \to R^1$ which satisfy the following assumptions:

A(i) the function f is bounded on $[0, \infty) \times E$ for any bounded set $E \subset R^n \times R^n$;
A(ii) $f(t, x, u) \geq \max\{\psi(|x|), \psi(|u|)|u|\} - a$ for each $(t, x, u) \in [0, \infty) \times R^n \times R^n$;
A(iii) for each $M, \epsilon > 0$, there exist $\Gamma, \delta > 0$ such that

$$|f(t, x_1, u) - f(t, x_2, u)| \leq \epsilon \max\{f(t, x_1, u), f(t, x_2, u)\}$$

for each $t \in [0, \infty)$ and each $u, x_1, x_2 \in R^n$ which satisfy

$$|x_i| \leq M, \ i = 1, 2, \ |u| \geq \Gamma, \quad |x_1 - x_2| \leq \delta;$$

A(iv) for each $M, \epsilon > 0$ there exists $\delta > 0$ such that $|f(t, x_1, u_1) - f(t, x_2, u_2)| \leq \epsilon$ for each $t \in [0, \infty)$ and each $u_1, u_2, x_1, x_2 \in R^n$ which satisfy

$$|x_i|, |u_i| \leq M, \ i = 1, 2, \quad \max\{|x_1 - x_2|, |u_1 - u_2|\} \leq \delta.$$

It is easy to show that an integrand $f = f(t, x, u) \in C^1([0, \infty) \times R^n \times R^n)$ belongs to $\mathcal{M}$ if f satisfies assumption A(ii), and if $\sup\{|f(t, 0, 0)| : t \in [0, \infty)\} < \infty$ and also there exists an increasing function $\psi_0 : [0, \infty) \to [0, \infty)$ such that

$$\sup\{|\partial f/\partial x(t, x, u)|, |\partial f/\partial u(t, x, u)|\} \leq \psi_0(|x|)(1 + \psi(|u|)|u|)$$

for each $t \in [0, \infty)$ and each $x, u \in R^n$.

For the set $\mathcal{M}$ we consider the uniformity which is determined by the following base:

$$E(N, \epsilon, \lambda) = \{(f, g) \in \mathcal{M} \times \mathcal{M} : |f(t, x, u) - g(t, x, u)| \leq \epsilon$$

$$\text{for each } t \in [0, \infty) \text{ and each } x, u \in R^n \text{ satisfying } |x|, |u| \leq N$$

$$\text{and } (|f(t, x, u)| + 1)(|g(t, x, u)| + 1)^{-1} \in [\lambda^{-1}, \lambda]$$

$$\text{for each } t \in [0, \infty) \text{ and each } x, u \in R^n \text{ satisfying } |x| \leq N\},$$

where $N > 0, \epsilon > 0, \lambda > 1$.

It is not difficult to show that the space $\mathcal{M}$ with this uniformity is metrizable (by a metric ρ_w). It is known (see [102]) that the metric space $(\mathcal{M}, \rho_w)$ is complete. The metric ρ_w induces in $\mathcal{M}$ a topology.

We consider functionals of the form

$$I^f(T_1, T_2, x) = \int_{T_1}^{T_2} f(t, x(t), x'(t))dt$$

where $f \in \mathcal{M}, 0 \leq T_1 < T_2 < \infty$ and $x : [T_1, T_2] \to R^n$ is an a.c. function.

For $f \in \mathcal{M}$, $y, z \in R^n$ and numbers T_1, T_2 satisfying $0 \le T_1 < T_2$, we set

$$U^f(T_1, T_2, y, z) = \inf\{I^f(T_1, T_2, x) : x : [T_1, T_2] \to R^n$$

$$\text{is an a.c. function satisfying } x(T_1) = y,\ x(T_2) = z\}.$$

It is easy to see that $-\infty < U^f(T_1, T_2, y, z) < \infty$ for each $f \in \mathcal{M}$, each $y, z \in R^n$ and all numbers T_1, T_2 satisfying $0 \le T_1 < T_2$.

Let $f \in \mathcal{M}$. A locally absolutely continuous (a.c.) function $x : [0, \infty) \to R^n$ is called an (f)-*good function* [104] if for any a.c function $y : [0, \infty) \to R^n$ there is a number M_y such that

$$I^f(0, T, y) \ge M_y + I^f(0, T, x) \text{ for each } T \in (0, \infty).$$

The following result was proved in [102].

Proposition 1.32. *Let $f \in \mathcal{M}$ and let $x : [0, \infty) \to R^n$ be a bounded a.c. function. Then the function x is (f)-good if and only if there is $M > 0$ such that*

$$I^f(0, T, x) \le U^f(0, T, x(0), x(T)) + M \text{ for any } T > 0.$$

Let us now give the precise definition of the turnpike property.

Assume that $f \in \mathcal{M}$. We say that f has the turnpike property, or briefly TP, if there exists a bounded continuous function $X_f : [0, \infty) \to R^n$ which satisfies the following condition:

For each $K, \epsilon > 0$, there exist constants $\delta, L > 0$ such that for each $x, y \in R^n$ satisfying $|x|, |y| \le K$, each $T_1 \ge 0$, $T_2 \ge T_1 + 2L$, and each a.c. function $v :$ $[T_1, T_2] \to R^n$ which satisfies

$$v(T_1) = x,\ v(T_2) = y,\ I^f(T_1, T_2, v) \le U^f(T_1, T_2, x, y) + \delta$$

the inequality $|v(t) - X_f(t)| \le \epsilon$ holds for all $t \in [T_1 + L, T_2 - L]$.

The function X_f is called the turnpike of f.

Assume that $f \in \mathcal{M}$ and $X : [0, \infty) \to R^n$ is a bounded continuous function. How to verify if the integrand f has TP and X is its turnpike? In [102] we introduced two properties (P1) and (P2) and show that f has TP if and only if f possesses the properties (P1) and (P2). The property (P2) means that all (f)-good functions have the same asymptotic behavior while the property (P1) means that if an a.c. function $v : [0, T] \to R^n$ is an approximate solution and T is large enough, then there is $\tau \in [0, T]$ such that $v(\tau)$ is close to $X(\tau)$.

The next theorem is the main result [102].

Theorem 1.33. *Let $f \in \mathcal{M}$ and $X_f : [0, \infty) \to R^n$ be a bounded absolutely continuous function. Then f has the turnpike property with X_f being the turnpike if and only if the following two properties hold:*

(P1) *For each* $K, \epsilon > 0$*, there exist* $\gamma, l > 0$ *such that for each* $T \geq 0$ *and each a.c. function* $w : [T, T + l] \to R^n$ *which satisfies*

$$|w(T)|, |w(T+l)| \leq K, \ I^f(T, T+l, w) \leq U^f(T, T+l, w(T), w(T+l))+\gamma$$

there is $\tau \in [T, T + l]$ *for which* $|X_f(\tau) - v(\tau)| \leq \epsilon$.

(P2) *For each* (f)*-good function* $v : [0, \infty) \to R^n$,

$$|v(t) - X_f(t)| \to 0 \text{ as } t \to \infty.$$

In [102] we proved the following theorem which is an extension of Theorem 1.33.

Theorem 1.34. *Let* $f \in \mathcal{M}$, $X_f : [0, \infty) \to R^n$ *be an* (f)*-good function. Assume that the properties (P1), (P2) hold. Then for each* $K, \epsilon > 0$*, there exist* $\delta, L > 0$ *and a neighborhood* $\mathcal{U}$ *of* f *in* $\mathcal{M}$ *such that for each* $g \in \mathcal{U}$*, each* $T_1 \geq 0$, $T_2 \geq T_1 + 2L$ *and each a.c. function* $v : [T_1, T_2] \to R^n$ *which satisfies*

$$|v(T_1)|, |v(T_2)| \leq K, \ I^g(T_1, T_2, v) \leq U^g(T_1, T_2, v(T_1), v(T_2)) + \delta$$

the inequality $|v(t) - X_f(t)| \leq \epsilon$ *holds for all* $t \in [T_1 + L, T_2 - L]$.

In the present book, we study sufficient and necessary conditions for the turnpike phenomenon, using the approach developed in [102, 103, 105, 118]; for discrete-time optimal control problems in metric spaces, which are not necessarily compact (see Chapters 2–4); and for continuous-time infinite-dimensional optimal control problems (see Chapters 5–7). Our main results have Theorem 1.33 as their prototype.

Since the discovery of the turnpike phenomenon by Paul Samuelson in 1948, different versions of the turnpike property were considered in the literature. In this book as well as in [102, 103, 105, 118], we study the turnpike property introduced and used in our previous research [99, 104, 117, 118, 129]. This turnpike property differs from other versions and has important features. Our turnpike property is a property of approximate solutions, and the turnpike is a nonstationary trajectory. As it was shown in [99, 117], our turnpike property holds for most problems belonging to large classes of variational and optimal control problems.

Chapter 2
Discrete-Time Autonomous Problems

In this chapter we establish sufficient and necessary conditions for the turnpike phenomenon for discrete-time optimal control problems in metric spaces, which are not necessarily compact. For these optimal control problems, the turnpike is a singleton. We also study the structure of approximate solutions on large intervals in the regions close to the endpoints and the existence of solutions of the corresponding infinite horizon optimal control problems.

2.1 Preliminaries and Main Results

Let (E, ρ_E) and (F, ρ_F) be metric spaces. We suppose that $\mathcal{A}$ is a nonempty subset of E, $\mathcal{U} : \mathcal{A} \to 2^F$ is a point to set mapping with a graph $\mathcal{M} = \{(x, u) : x \in \mathcal{A},\ u \in \mathcal{U}(x)\}$, $G : \mathcal{M} \to E$ and $f : \mathcal{M} \to R^1$.

Let $0 \leq T_1 < T_2$ be integers. We denote by $X(T_1, T_2, \mathcal{M}, G)$ the set of all pairs of sequences $(\{x_t\}_{t=T_1}^{T_2}, \{u_t\}_{t=T_1}^{T_2-1})$ such that for each integer $t \in \{T_1, \dots, T_2 - 1\}$, $x_t \in \mathcal{A}$, $u_t \in \mathcal{U}(x_t)$, $x_{t+1} = G(x_t, u_t)$ and which are called trajectory-control pairs. For simplicity we use the notation $X(T_1, T_2) = X(T_1, T_2, \mathcal{M}, G)$ if the pair $(\mathcal{M}, G)$ is understood.

Let $T_1 \geq 0$ be an integer. Denote by $X(T_1, \infty, \mathcal{M}, G)$ (or $X(T_1, \infty)$ if the pair $(\mathcal{M}, G)$ is understood) the set of all pairs of sequences $\{x_t\}_{t=T_1}^{\infty} \subset \mathcal{A}$, $\{u_t\}_{t=T_1}^{\infty} \subset F$ such that for each integer $T_2 > T_1$, $(\{x_t\}_{t=T_1}^{T_2}, \{u_t\}_{t=T_1}^{T_2-1}) \in X(T_1, T_2, \mathcal{M}, G)$. The elements of $X(T_1, \infty, \mathcal{M}, G)$ are called trajectory-control pairs.

Let $0 \leq T_1 < T_2$ be integers. A sequence $\{x_t\}_{t=T_1}^{T_2} \subset \mathcal{A}$ ($\{x_t\}_{t=T_1}^{\infty} \subset \mathcal{A}$, respectively) is called a trajectory if there exists a sequence $\{u_t\}_{t=T_1}^{T_2-1} \subset F$ ($\{u_t\}_{t=T_1}^{\infty} \subset F$, respectively) referred to as a control such that

A. J. Zaslavski, *Turnpike Conditions in Infinite Dimensional Optimal Control*, Springer Optimization and Its Applications 148,
https://doi.org/10.1007/978-3-030-20178-4_2

$$(\{x_t\}_{t=T_1}^{T_2}, \{u_t\}_{t=T_1}^{T_2-1}) \in X(T_1, T_2)$$

$((\{x_t\}_{t=T_1}^{\infty}, \{u_t\}_{t=T_1}^{\infty}) \in X(T_1, \infty)$, respectively).

Let $\theta_0 \in E$, $\theta_1 \in F$, $a_0 > 0$ and let $\psi : [0, \infty) \to [0, \infty)$ be an increasing function such that $\psi(t) \to \infty$ as $t \to \infty$. We suppose that the function f satisfies

$$f(x, u) \geq \psi(\rho_E(x, \theta_0)) - a_0 \text{ for each } (x, u) \in \mathcal{M}. \tag{2.1}$$

For each pair of integers $T_2 > T_1 \geq 0$ and each pair of points $y, z \in \mathcal{A}$, we consider the following problems:

$$\sum_{t=T_1}^{T_2-1} f(x_t, u_t) \to \min, \ (\{x_t\}_{t=T_1}^{T_2}, \{u_t\}_{t=T_1}^{T_2-1}) \in X(T_1, T_2), \ x_{T_1} = y, \ x_{T_2} = z, \tag{P_1}$$

$$\sum_{t=T_1}^{T_2-1} f(x_t, u_t) \to \min, \ (\{x_t\}_{t=T_1}^{T_2}, \{u_t\}_{t=T_1}^{T_2-1}) \in X(T_1, T_2), \ x_{T_1} = y, \tag{P_2}$$

$$\sum_{t=T_1}^{T_2-1} f(x_t, u_t) \to \min, \ (\{x_t\}_{t=T_1}^{T_2}, \{u_t\}_{t=T_1}^{T_2-1}) \in X(T_1, T_2). \tag{P_3}$$

For each pair of integers $T_2 > T_1 \geq 0$ and each pair of points $y, z \in \mathcal{A}$, we define

$$U^f(T_1, T_2, y, z) = \inf\{\sum_{t=T_1}^{T_2-1} f(x_t, u_t) : \ (\{x_t\}_{t=T_1}^{T_2}, \{u_t\}_{t=T_1}^{T_2-1}) \in X(T_1, T_2), \\ x_{T_1} = y, x_{T_2} = z\}, \tag{2.2}$$

$$\sigma^f(T_1, T_2, y) = \inf\{U^f(T_1, T_2, y, h) : h \in \mathcal{A}\}, \tag{2.3}$$

$$\widehat{\sigma}^f(T_1, T_2, y) = \inf\{U^f(T_1, T_2, h, y) : h \in \mathcal{A}\}, \tag{2.4}$$

$$\sigma^f(T_1, T_2) = \inf\{U^f(T_1, T_2, h, y) : h, y \in \mathcal{A}\}. \tag{2.5}$$

We assume that

$$x_f \in \mathcal{A}, \ u_f \in \mathcal{U}(x_f) \text{ and } x_f = G(x_f, u_f). \tag{2.6}$$

This implies that $(\{\bar{x}_t\}_{t=0}^{\infty}, \{\bar{u}_t\}_{t=0}^{\infty}) \in X(0, \infty)$, where $\bar{x}_t = x_f$, $\bar{u}_t = u_f$ for all integers $t \geq 0$. We suppose that the following assumptions hold.

(A1) For each $S_1 > 0$, there exist $S_2 > 0$ and an integer $c > 0$ such that

$$(T_2 - T_1) f(x_f, u_f) \le \sum_{t=T_1}^{T_2-1} f(x_t, u_t) + S_2$$

for each pair of integers $T_1 \ge 0$, $T_2 \ge T_1 + c$ and each $(\{x_t\}_{t=T_1}^{T_2}, \{u_t\}_{t=T_1}^{T_2-1}) \in X(T_1, T_2)$ satisfying $\rho_E(\theta_0, x_j) \le S_1$, $j = T_1, T_2 - 1$.

(A2) There exists an integer $b_f > 0$, and for each $\epsilon > 0$ there exists $\delta > 0$ such that for each $z_i \in \mathcal{A}$, $i = 1, 2$ satisfying $\rho_E(z_i, x_f) \le \delta$, $i = 1, 2$, there exist an integer $\tau \in (0, b_f]$ and $(\{x_t\}_{t=0}^{\tau}, \{u_t\}_{t=0}^{\tau-1}) \in X(0, \tau)$ which satisfies $x_0 = z_1$, $x_\tau = z_2$ and

$$\sum_{t=0}^{\tau-1} f(x_t, u_t) \le \tau f(x_f, u_f) + \epsilon.$$

Section 2.6 contains examples of optimal control problems satisfying assumptions (A1) and (A2). Many examples can also be found in [106–108, 118, 124, 125, 134].

The following result is proved in Section 2.8.

Theorem 2.1.

1. *There exists $S > 0$ such that for each pair of integers $T_2 > T_1 \ge 0$ and each $(\{x_t\}_{t=T_1}^{T_2}, \{u_t\}_{t=T_1}^{T_2-1}) \in X(T_1, T_2)$,*

$$\sum_{t=T_1}^{T_2-1} f(x_t, u_t) + S \ge (T_2 - T_1) f(x_f, u_f).$$

2. *For each $(\{x_t\}_{t=0}^{\infty}, \{u_t\}_{t=0}^{\infty}) \in X(0, \infty)$ either*

$$\sum_{t=0}^{T-1} f(x_t, u_t) - T f(x_f, u_f) \to \infty \text{ as } T \to \infty$$

or

$$\sup\{|\sum_{t=0}^{T-1} f(x_t, u_t) - T f(x_f, u_f)| : T \in \{1, 2, \dots\}\} < \infty. \tag{2.7}$$

Moreover, if (2.7) holds, then the sequence $\{x_t\}_{t=0}^{\infty}$ is bounded.

We say that $(\{x_t\}_{t=0}^{\infty}, \{u_t\}_{t=0}^{\infty}) \in X(0, \infty)$ is $(f, \mathcal{M}, G)$-good ((f)-good if the pair $(\mathcal{M}, G)$ is understood; $(f, \mathcal{M})$-good if G is understood) if (2.7) holds [29, 104, 116, 120, 129, 134]. The next boundedness result is proved in Section 2.8. It has a prototype in [104].

Theorem 2.2. *Let $M_0 > 0$. Then there exists $M_1 > 0$ such that for each pair of integers $T_2 > T_1 \geq 0$ and each $(\{x_t\}_{t=T_1}^{T_2}, \{u_t\}_{t=T_1}^{T_2-1}) \in X(T_1, T_2)$ satisfying*

$$\sum_{t=T_1}^{T_2-1} f(x_t, u_t) \leq (T_2 - T_1) f(x_f, u_f) + M_0$$

the inequality $\rho_E(\theta_0, x_t) \leq M_1$ holds for all $t = T_1, \dots, T_2 - 1$.

Let $L > 0$ be an integer. Denote by $\mathcal{A}_L$ the set of all $z \in \mathcal{A}$ for which there exist an integer $\tau \in (0, L]$ and $(\{x_t\}_{t=0}^{\tau}, \{u_t\}_{t=0}^{\tau-1}) \in X(0, \tau)$ such that

$$x_0 = z, \; x_\tau = x_f, \; \sum_{t=0}^{\tau-1} f(x_t, u_t) \leq L.$$

Denote by $\widehat{\mathcal{A}}_L$ the set of all $z \in \mathcal{A}$ for which there exist an integer $\tau \in (0, L]$ and $(\{x_t\}_{t=0}^{\tau}, \{u_t\}_{t=0}^{\tau-1}) \in X(0, \tau)$ such that

$$x_0 = x_f, \; x_\tau = z, \; \sum_{t=0}^{\tau-1} f(x_t, u_t) \leq L.$$

The following Theorems 2.3–2.5 are also boundedness results. They are proved in Section 2.8.

Theorem 2.3. *Let $L > 0$ be an integer and $M_0 > 0$. Then there exists $M_1 > 0$ such that for each integer $T_1 \geq 0$, each integer $T_2 \geq T_1 + 2L$ and each $(\{x_t\}_{t=T_1}^{T_2}, \{u_t\}_{t=T_1}^{T_2-1}) \in X(T_1, T_2)$ satisfying*

$$x_{T_1} \in \mathcal{A}_L, \; x_{T_2} \in \widehat{\mathcal{A}}_L,$$

$$\sum_{t=T_1}^{T_2-1} f(x_t, u_t) \leq U^f(T_1, T_2, x_{T_1}, x_{T_2}) + M_0$$

the inequality $\rho_E(\theta_0, x_t) \leq M_1$ holds for all $t = T_1, \dots, T_2 - 1$.

Theorem 2.4. *Let $L > 0$ be an integer and $M_0 > 0$. Then there exists $M_1 > 0$ such that for each integer $T_1 \geq 0$, each integer $T_2 \geq T_1 + L$, and each $(\{x_t\}_{t=T_1}^{T_2}, \{u_t\}_{t=T_1}^{T_2-1}) \in X(T_1, T_2)$ satisfying*

$$x_{T_1} \in \mathcal{A}_L, \sum_{t=T_1}^{T_2-1} f(x_t, u_t) \le \sigma^f(T_1, T_2, x_{T_1}) + M_0$$

the inequality $\rho_E(\theta_0, x_t) \le M_1$ *holds for all* $t = T_1, \dots, T_2 - 1$.

Theorem 2.5. *Let* $L > 0$ *be an integer and* $M_0 > 0$. *Then there exists* $M_1 > 0$ *such that for each integer* $T_1 \ge 0$, *each integer* $T_2 \ge T_1 + L$, *and each* $(\{x_t\}_{t=T_1}^{T_2}, \{u_t\}_{t=T_1}^{T_2-1}) \in X(T_1, T_2)$ *satisfying*

$$x_{T_2} \in \widehat{\mathcal{A}}_L, \sum_{t=T_1}^{T_2-1} f(x_t, u_t) \le \widehat{\sigma}^f(T_1, T_2, x_{T_2}) + M_0$$

the inequality $\rho_E(\theta_0, x_t) \le M_1$ *holds for all* $t = T_1, \dots, T_2 - 1$.

We say that the triplet $(f, \mathcal{M}, G)$ (or f if the pair $(\mathcal{M}, G)$ is understood; $(f, \mathcal{M})$ if G is understood) possesses the turnpike property (or TP for short) if for each $\epsilon > 0$ and each $M > 0$, there exist $\delta > 0$ and an integer $L > 0$ such that for each integer $T_1 \ge 0$, each integer $T_2 \ge T_1 + 2L$, and each $(\{x_t\}_{t=T_1}^{T_2}, \{u_t\}_{t=T_1}^{T_2-1}) \in X(T_1, T_2)$ which satisfies

$$\sum_{t=T_1}^{T_2-1} f(x_t, u_t) \le \min\{\sigma^f(T_1, T_2) + M, U^f(T_1, T_2, x_{T_1}, x_{T_2}) + \delta\},$$

we have

$$\rho_E(x_t, x_f) \le \epsilon, \; t = T_1 + L, \dots, T_2 - L.$$

Moreover, if $\rho_E(x_{T_1}, x_f) \le \delta$, then $\rho_E(x_t, x_f) \le \epsilon$ for all $t = T_1, \dots, T_2 - L$, and if $\rho_E(x_{T_2}, x_f) \le \delta$, then $\rho_E(x_t, x_f) \le \epsilon$ for all $t = T_1 + L, \dots, T_2$.

Theorem 2.1 implies the following result.

Proposition 2.6. *Assume that* f *has TP and that* $\epsilon, M > 0$. *Then there exist* $\delta > 0$ *and an integer* $L > 0$ *such that for each integer* $T_1 \ge 0$, *each integer* $T_2 \ge T_1 + 2L$, *and each* $(\{x_t\}_{t=T_1}^{T_2}, \{u_t\}_{t=T_1}^{T_2-1}) \in X(T_1, T_2)$ *which satisfies*

$$\sum_{t=T_1}^{T_2-1} f(x_t, u_t) \le \min\{(T_2 - T_1) f(x_f, u_f) + M, U^f(T_1, T_2, x_{T_1}, x_{T_2}) + \delta\},$$

there exist integers $\tau_1 \in [T_1, T_1 + L]$, $\tau_2 \in [T_2 - L, T_2]$ *such that*

$$\rho_E(x_t, x_f) \le \epsilon, \; t = \tau_1, \dots, \tau_2.$$

Moreover, if $\rho_E(x_{T_1}, x_f) \le \delta$, *then* $\tau_1 = T_1$ *and if* $\rho_E(x_{T_2}, x_f) \le \delta$, *then* $\tau_2 = T_2$.

The following turnpike results are proved in Section 2.8.

Theorem 2.7. *Assume that f has TP, that $L > 0$ is an integer, and that $\epsilon > 0$. Then there exist $\delta > 0$ and an integer $L_0 > L$ such that for each integer $T_1 \geq 0$, each integer $T_2 \geq T_1 + 2L_0$, and each $(\{x_t\}_{t=T_1}^{T_2}, \{u_t\}_{t=T_1}^{T_2-1}) \in X(T_1, T_2)$ which satisfies*

$$x_{T_1} \in \mathcal{A}_L, \ x_{T_2} \in \widehat{\mathcal{A}}_L,$$

$$\sum_{t=T_1}^{T_2-1} f(x_t, u_t) \leq U^f(T_1, T_2, x_{T_1}, x_{T_2}) + \delta,$$

there exist integers $\tau_1 \in [T_1, T_1 + L_0]$, $\tau_2 \in [T_2 - L_0, T_2]$ such that

$$\rho_E(x_t, x_f) \leq \epsilon, \ t = \tau_1, \ldots, \tau_2.$$

Moreover, if $\rho_E(x_{T_1}, x_f) \leq \delta$, then $\tau_1 = T_1$, and if $\rho_E(x_{T_2}, x_f) \leq \delta$, then $\tau_2 = T_2$.

Theorem 2.8. *Assume that f has TP, that $L > 0$ is an integer, and that $M, \epsilon > 0$. Then there exist $\delta > 0$ and an integer $L_0 > L$ such that for each integer $T_1 \geq 0$, each integer $T_2 \geq T_1 + 2L_0$, and each $(\{x_t\}_{t=T_1}^{T_2}, \{u_t\}_{t=T_1}^{T_2-1}) \in X(T_1, T_2)$ which satisfies*

$$x_{T_1} \in \mathcal{A}_L,$$

$$\sum_{t=T_1}^{T_2-1} f(x_t, u_t) \leq \min\{\sigma^f(T_1, T_2, x_{T_1}) + M, \ U^f(T_1, T_2, x_{T_1}, x_{T_2}) + \delta\},$$

there exist integers $\tau_1 \in [T_1, T_1 + L_0]$, $\tau_2 \in [T_2 - L_0, T_2]$ such that

$$\rho_E(x_t, x_f) \leq \epsilon, \ t = \tau_1, \ldots, \tau_2.$$

Moreover, if $\rho_E(x_{T_1}, x_f) \leq \delta$, then $\tau_1 = T_1$, and if $\rho_E(x_{T_2}, x_f) \leq \delta$, then $\tau_2 = T_2$.

The next theorem is our main result in this chapter. It is proved in Section 2.9.

Theorem 2.9. *f has TP if and only if the following properties hold:*

(P1) for each (f)-good pair of sequences $(\{x_t\}_{t=0}^{\infty}, \{u_t\}_{t=0}^{\infty}) \in X(0, \infty)$,

$$\lim_{t \to \infty} \rho(x_t, x_f) = 0;$$

(P2) for each $\epsilon > 0$ and each $M > 0$, there exist $\delta > 0$ and an integer $L > 0$ such that for each $(\{x_t\}_{t=0}^{L}, \{u_t\}_{t=0}^{L-1}) \in X(0, L)$ which satisfies

$$\sum_{t=0}^{L-1} f(x_t, u_t) \leq \min\{U^f(0, L, x_0, x_L) + \delta, \ Lf(x_f, u_f) + M\},$$

there exists an integer $s \in [0, L]$ such that $\rho_E(x_s, x_f) \leq \epsilon$.

We say that the triplet $(f, \mathcal{M}, G)$ (or f if the pair $(\mathcal{M}, G)$ is understood; $(f, \mathcal{M})$ if G is understood) possesses the weak turnpike property (or WTP for short) if for each $\epsilon > 0$ and each $M > 0$, there exist natural numbers Q, l such that for each integer $T_1 \geq 0$, each integer $T_2 l \geq T_1 + lQ$, and each $(\{x_t\}_{t=T_1}^{T_2}, \{u_t\}_{t=T_1}^{T_2-1}) \in X(T_1, T_2)$ which satisfies

$$\sum_{t=T_1}^{T_2-1} f(x_t, u_t) \leq (T_2 - T_1) f(x_f, u_f) + M,$$

there exist finite sequences of natural numbers $\{a_i\}_{i=1}^q$, $\{b_i\}_{i=1}^q \subset \{T_1, \dots, T_2\}$ such that an integer $q \leq Q$,

$$0 \leq b_i - a_i \leq l, \; i = 1, \dots, q,$$

$$b_i \leq a_{i+1} \text{ for all integers } i \text{ satisfying } 1 \leq i < q,$$

$$\rho_E(x_t, x_f) \leq \epsilon \text{ for all integers } t \in [T_1, T_2] \setminus \cup_{i=1}^q [a_i, b_i].$$

WTP was studies in [95, 96] for discrete-time unconstrained problems, in [97] for variational problems and in [114] for discrete-time constrained problems.

The next result is proved in Section 2.11.

Theorem 2.10. *f has WTP if and only if f has TP.*

2.2 Lower Semicontinuity Property

We say that the triplet $(f, \mathcal{M}, G)$ (or f if the pair $(\mathcal{M}, G)$ is understood; $(f, \mathcal{M})$ if G is understood) possesses the lower semicontinuity property (LSC property for short) if for each pair of integers $T_2 > T_1 \geq 0$ and each sequence

$$(\{x_t^{(j)}\}_{t=T_1}^{T_2}, \{u_t^{(j)}\}_{t=T_1}^{T_2-1}) \in X(T_1, T_2), \; j = 1, 2, \dots$$

which satisfies

$$\sup\{\sum_{t=T_1}^{T_2-1} f(x_t^{(j)}, u_t^{(j)}) : \; j = 1, 2, \dots\} < \infty,$$

there exist a strictly increasing sequence of natural numbers $\{j_k\}_{k=1}^\infty$ and $(\{x_t\}_{t=T_1}^{T_2}, \{u_t\}_{t=T_1}^{T_2-1}) \in X(T_1, T_2)$ such that for any $t \in \{T_1, \dots, T_2 - 1\}$,

$$x_t^{(j_k)} \to x_t \text{ as } k \to \infty,$$

$$\sum_{t=T_1}^{T_2-1} f(x_t, u_t) \le \liminf_{j\to\infty} \sum_{t=T_1}^{T_2-1} f(x_t^{(j)}, u_t^{(j)}).$$

LSC property plays an important role in the calculus of variations and optimal control theory [32].

Theorem 2.11. *Assume that f possesses LSC property. Then f has TP if and only if f has (P1).*

Theorem 2.11 follows from Theorem 2.9 and the next proposition which is proved in Section 2.12.

Proposition 2.12. *Assume that f has LSC property and (P1). Then f has (P2).*

A pair $(\{x_t\}_{t=0}^{\infty}, \{u_t\}_{t=0}^{\infty}) \in X(0,\infty)$ is called $(f, \mathcal{M}, G)$-overtaking optimal (or (f)-overtaking optimal if the pair $(\mathcal{M}, G)$ is understood; $(f, \mathcal{M})$-overtaking optimal if G is understood) [29,104,120] if for every $(\{y_t\}_{t=0}^{\infty}, \{v_t\}_{t=0}^{\infty}) \in X(0,\infty)$ satisfying $x_0 = y_0$,

$$\limsup_{T\to\infty} [\sum_{t=0}^{T-1} f(x_t, u_t) - \sum_{t=0}^{T-1} f(y_t, v_t)] \le 0.$$

A pair $(\{x_t\}_{t=0}^{\infty}, \{u_t\}_{t=0}^{\infty}) \in X(0,\infty)$ is called $(f, \mathcal{M}, G)$-weakly optimal (or (f)-weakly optimal if the pair $(\mathcal{M}, G)$ is understood; $(f, \mathcal{M})$-weakly optimal if G is understood) [29, 104, 120] if for every $(\{y_t\}_{t=0}^{\infty}, \{v_t\}_{t=0}^{\infty}) \in X(0,\infty)$ satisfying $x_0 = y_0$,

$$\liminf_{T\to\infty} [\sum_{t=0}^{T-1} f(x_t, u_t) - \sum_{t=0}^{T-1} f(y_t, v_t)] \le 0.$$

A pair $(\{x_t\}_{t=0}^{\infty}, \{u_t\}_{t=0}^{\infty}) \in X(0,\infty)$ is called $(f, \mathcal{M}, G)$-minimal (or (f)-minimal if the pair $(\mathcal{M}, G)$ is understood; $(f, \mathcal{M})$-minimal if G is understood) [12, 94] if for every integer $T \ge 1$,

$$\sum_{t=0}^{T-1} f(x_t, u_t) = U^f(0, T, x_0, x_T).$$

In infinite horizon optimal control, the main goal is to show the existence of solutions using the optimality criterions above. The next result is proved in Section 2.14.

Theorem 2.13. *Assume that f has (P1) and LSC property and that*

$$(\{x_t\}_{t=0}^{\infty}, \{u_t\}_{t=0}^{\infty}) \in X(0,\infty)$$

is (f)-good. Then there exists an (f)-overtaking optimal pair

$$(\{x_t^*\}_{t=0}^{\infty}, \{u_t^*\}_{t=0}^{\infty}) \in X(0, \infty)$$

such that $x_0^* = x_0$.

The following theorem is also proved in Section 2.14.

Theorem 2.14. *Assume that f has (P1) and LSC property,*

$$(\{\tilde{x}_t\}_{t=0}^{\infty}, \{\tilde{u}_t\}_{t=0}^{\infty}) \in X(0, \infty)$$

is (f)-good and that $(\{x_t^*\}_{t=0}^{\infty}, \{u_t^*\}_{t=0}^{\infty}) \in X(0, \infty)$ satisfies $x_0^* = \tilde{x}_0$. Then the following conditions are equivalent:

(i) $(\{x_t^*\}_{t=0}^{\infty}, \{u_t^*\}_{t=0}^{\infty})$ is (f)-overtaking optimal;
(ii) $(\{x_t^*\}_{t=0}^{\infty}, \{u_t^*\}_{t=0}^{\infty})$ is (f)-weakly optimal;
(iii) $(\{x_t^*\}_{t=0}^{\infty}, \{u_t^*\}_{t=0}^{\infty})$ is (f)-minimal and (f)-good;
(iv) $(\{x_t^*\}_{t=0}^{\infty}, \{u_t^*\}_{t=0}^{\infty})$ is (f)-minimal and satisfies $\lim_{t\to\infty} \rho_E(x_t, x_f) = 0$;
(v) $(\{x_t^*\}_{t=0}^{\infty}, \{u_t^*\}_{t=0}^{\infty})$ is (f)-minimal and satisfies $\liminf_{t\to\infty} \rho_E(x_t, x_f) = 0$.

The following three results are proved in Section 2.15.

Theorem 2.15. *Assume that f has (P1) and LSC property and that $\epsilon > 0$. Then there exists $\delta > 0$ such that the following assertions hold.*

(i) For every $z \in \mathcal{A}$ satisfying $\rho_E(z, x_f) \le \delta$, there exists an (f)-overtaking optimal pair $(\{x_t\}_{t=0}^{\infty}, \{u_t\}_{t=0}^{\infty}) \in X(0, \infty)$ which satisfies $x_0 = z$.
(ii) If an (f)-overtaking optimal pair $(\{x_t\}_{t=0}^{\infty}, \{u_t\}_{t=0}^{\infty}) \in X(0, \infty)$ satisfies $\rho_E(x_0, x_f) \le \delta$, then $\rho_E(x_t, x_f) \le \epsilon$ for all integers $t \ge 0$.

Theorem 2.16. *Assume that f has (P1) and LSC property, that $\epsilon > 0$, and that $L > 0$ is an integer. Then there exists an integer $\tau_0 > 0$ such that for every (f)-overtaking optimal pair $(\{x_t\}_{t=0}^{\infty}, \{u_t\}_{t=0}^{\infty}) \in X(0, \infty)$ which satisfies $x_0 \in \mathcal{A}_L$, the inequality $\rho_E(x_t, x_f) \le \epsilon$ holds for all integers $t \ge \tau_0$.*

Theorem 2.17. *Assume that f has (P1) and LSC property,*

$$(\{x_t\}_{t=0}^{\infty}, \{u_t\}_{t=0}^{\infty}) \in X(0, \infty)$$

is an (f)-overtaking optimal and (f)-good pair and that integers $t_2 > t_1 \ge 0$ satisfy $x_{t_1} = x_{t_2}$. Then $x_t = x_f$ for all integers $t \ge t_1$.

In the sequel we use the following result which easily follows from (P1) and (A2).

Proposition 2.18. *Let f have (P1) and $z \in \mathcal{A}$. There exists an (f)-good pair*

$$(\{x_t\}_{t=0}^{\infty}, \{u_t\}_{t=0}^{\infty}) \in X(0, \infty)$$

satisfying $x_0 = z$ if and only if $z \in \cup\{\mathcal{A}_L : L = 1, 2, \dots\}$.

LSC property implies the following result.

Proposition 2.19. *Let f have LSC property, $T_2 > T_1 \geq 0$ and $L > 0$ be integers.*

1. *There exists $(\{x_t\}_{t=T_1}^{T_2}, \{u_t\}_{t=T_1}^{T_2-1}) \in X(T_1, T_2)$ such that*

$$\sum_{t=T_1}^{T_2-1} f(x_t, u_t) \leq \sum_{t=T_1}^{T_2-1} f(y_t, v_t) \text{ for all } (\{y_t\}_{t=T_1}^{T_2}, \{v_t\}_{t=T_1}^{T_2-1}) \in X(T_1, T_2).$$

2. *If $T_2 - T_1 \geq 2L$, $z_1 \in \mathcal{A}_L$, $z_2 \in \widehat{\mathcal{A}}_L$, then there exists*

$$(\{x_t\}_{t=T_1}^{T_2}, \{u_t\}_{t=T_1}^{T_2-1}) \in X(T_1, T_2)$$

such that $x_{T_i} = z_i$, $i = 1, 2$ and

$$\sum_{t=T_1}^{T_2-1} f(x_t, u_t) = U^f(T_1, T_2, z_1, z_2).$$

3. *If $T_2 - T_1 \geq L$, $z \in \mathcal{A}_L$, then there exists $(\{x_t\}_{t=T_1}^{T_2}, \{u_t\}_{t=T_1}^{T_2-1}) \in X(T_1, T_2)$ such that $x_{T_1} = z$ and*

$$\sum_{t=T_1}^{T_2-1} f(x_t, u_t) = \sigma^f(T_1, T_2, z).$$

4. *If $T_2 - T_1 \geq L$, $z \in \widehat{\mathcal{A}}_L$, then there exists $(\{x_t\}_{t=T_1}^{T_2}, \{u_t\}_{t=T_1}^{T_2-1}) \in X(T_1, T_2)$ such that $x_{T_2} = z$ and*

$$\sum_{t=T_1}^{T_2-1} f(x_t, u_t) = \widehat{\sigma}^f(T_1, T_2, z).$$

Theorem 2.20. *f has (P1) and (P2) if and only if the following property holds:*

(i) for each $\epsilon > 0$ and each $M > 0$, there exists an integer $L > 0$ such that for each pair of integers $T_1 \geq 0$, $T_2 \geq T_1 + L$ and each $(\{x_t\}_{t=T_1}^{T_2}, \{u_t\}_{t=T_1}^{T_2-1}) \in X(T_1, T_2)$ which satisfies

$$\sum_{t=T_1}^{T_2-1} f(x_t, u_t) \le (T_2 - T_1) f(x_f, u_f) + M,$$

the inequality

$$Card(\{t \in \{T_1, \dots, T_2\} : \rho_E(x_t, x_f) > \epsilon\}) \le L$$

holds.

Proof. In view of (2.10) and Theorem 2.9, (P1) and (P2) imply (i). Clearly, (i) implies (P2). In order to show that (i) implies (P1), it is sufficient to note that if $(\{x_t\}_{t=0}^{\infty}, \{u_t\}_{t=0}^{\infty}) \in X(0, \infty)$ is (f)-good, then for some $M > 0$,

$$\sum_{t=0}^{T-1} f(x_t, u_t) \le T f(x_f, u_f) + M \text{ for all integers } T \ge 1$$

and

$$\text{Card}(\{t \in \{0, 1, \dots\} : \rho_E(x_t, x_f) > \epsilon\}) < \infty.$$

2.3 Perturbed Problems

In this section we use the following assumption.

(A3) There exists an integer $b_f > 0$, and for each $\epsilon > 0$ there exists $\delta > 0$ such that for each $z_i \in \mathcal{A}$, $i = 1, 2$ satisfying $\rho_E(z_i, x_f) \le \delta$, $i = 1, 2$, there exist an integer $\tau \in (0, b_f]$ and $(\{x_t\}_{t=0}^{\tau}, \{u_t\}_{t=0}^{\tau-1}) \in X(0, \tau)$ which satisfies $x_0 = z_1$, $x_\tau = z_2$ and

$$\sum_{t=0}^{\tau-1} f(x_t, u_t) \le \tau f(x_f, u_f) + \epsilon, \ \rho(x_t, x_f) \le \epsilon, \ t = 0, \dots, \tau.$$

Clearly, (A3) implies (A2). Assume that $\phi : E \times E \to [0, 1]$ is a continuous function satisfying $\phi(x, x) = 0$ for all $x \in E$ and such that the following property holds:

(a) for each $\epsilon > 0$, there exists $\delta > 0$ such that if $x, y \in E$ and $\rho_E(x, y) \le \delta$, then $\phi(x, y) \le \epsilon$, and if $x, y \in E$ satisfies $\phi(x, y) \le \delta$, then $\rho_E(x, y) \le \epsilon$.

For each $r \in (0, 1)$ set

$$f_r(x, u) = f(x, u) + r\phi(x, x_f), \ (x, u) \in \mathcal{M}. \tag{2.8}$$

Clearly, for any $r \in (0, 1]$, if (A3) holds, then (A1) and (A3) hold for f_r with $(x_{f_r}, u_{f_r}) = (x_f, u_f)$.

Theorem 2.21. *Let (A3) hold and $r \in (0, 1]$. Then f_r has (P1) and (P2).*

Theorem 2.21 is proved in Section 2.16.

2.4 A Subclass of Problems

For each $x \in \mathcal{A}$ define

$$\widehat{\mathcal{U}}(x) = \{y \in \mathcal{A} : \text{ there exists } u \in \mathcal{U}(x) \text{ such that } y = G(x, u)\},$$

$$\widehat{\mathcal{M}} = \{(x, y) \in \mathcal{A} \times \mathcal{A} : y \in \widehat{\mathcal{U}}(x)\}, \ \widehat{G}(x, y) = y, \ (x, y) \in \widehat{\mathcal{M}}$$

and for every $(x, y) \in \widehat{\mathcal{M}}$ define

$$\widehat{f}(x, y) = \inf\{f(x, u) : u \in \mathcal{U}(x), \ y = G(x, u)\}.$$

Clearly, $(\widehat{f}, \widehat{\mathcal{M}}, \widehat{G})$ satisfies (A1) and (A2), and $(\widehat{f}, \widehat{\mathcal{M}}, \widehat{G})$ has TP if and only if $(f, \mathcal{M}, G)$ has TP.

Suppose that $(F, \rho_F) = (E, \rho_E)$ and

$$G(x, u) = u, \ (x, u) \in \mathcal{M}. \tag{2.9}$$

We continue to use the assumptions and definitions introduced in Section 2.1 but with some modifications which are given below. We assume that

$$x_f = u_f, \tag{2.10}$$

for all integers $T_2 > T_1 \geq 0$, $X(T_1, T_2)$ is the set of all sequences $\{x_t\}_{t=T_1}^{T_2} \subset \mathcal{A}$ satisfying $(x_t, x_{t+1}) \in \mathcal{M}$ for all $t = T_1, \dots, T_2 - 1$ and $X(T_1, \infty)$ is the set of all sequences $\{x_t\}_{t=T_1}^{\infty} \subset \mathcal{A}$ satisfying $(x_t, x_{t+1}) \in \mathcal{M}$ for all integers $t \geq T_1$.

For each pair of integers $T_2 > T_1 \geq 0$ and each pair of points $y, z \in \mathcal{A}$, we define

$$U^f(T_1, T_2, y, z) = \inf\{\sum_{t=T_1}^{T_2-1} f(x_t, u_t) : \{x_t\}_{t=T_1}^{T_2} \in X(T_1, T_2),$$

$$x_{T_1} = y, \ x_{T_2} = z\},$$

$$\sigma^f(T_1, T_2, y) = \inf\{U^f(T_1, T_2, y, h) : h \in \mathcal{A}\},$$

$$\widehat{\sigma}^f(T_1, T_2, y) = \inf\{U^f(T_1, T_2, h, y) : h \in \mathcal{A}\},$$

$$\sigma^f(T_1, T_2) = \inf\{U^f(T_1, T_2, h, y) : h, y \in \mathcal{A}\}.$$

Define

$$\bar{\mathcal{M}} = \{(y, x) \in \mathcal{A} \times \mathcal{A} : (x, y) \in \mathcal{M}\} \tag{2.11}$$

and for all $(y, x) \in \bar{\mathcal{M}}$ set

$$\bar{f}(y, x) = f(x, y). \tag{2.12}$$

Let $T_2 > T_1 \geq 0$ be integers and $\{x_t\}_{t=T_1}^{T_2} \in X(T_1, T_2)$. Set

$$\bar{x}_t = x_{T_2-t+T_1},\ t = T_1, \ldots, T_2. \tag{2.13}$$

By (2.12) and (2.13), for all $t = T_1, \ldots, T_2 - 1$,

$$(\bar{x}_t, \bar{x}_{t+1}) = (x_{T_2-t+T_1}, x_{T_2-t+T_1-1}) \in \bar{\mathcal{M}}, \tag{2.14}$$

$$\sum_{t=T_1}^{T_2-1} \bar{f}(\bar{x}_t, \bar{x}_{t+1}) = \sum_{t=T_1}^{T_2-1} \bar{f}(x_{T_2-t+T_1}, x_{T_2-t+T_1-1})$$

$$= \sum_{t=T_1}^{T_2-1} f(x_{T_2-t+T_1-1}, x_{T_2-t+T_1}) = \sum_{t=T_1}^{T_2-1} f(x_t, x_{t+1}). \tag{2.15}$$

Let $T_2 > T_1 \geq 0$ be integers. Denote by $\bar{X}(T_1, T_2)$ the set of all sequences $\{x_t\}_{t=T_1}^{T_2} \subset \mathcal{A}$ satisfying $(x_t, x_{t+1}) \in \bar{\mathcal{M}}$ for all $t = T_1, \ldots, T_2 - 1$ and by $\bar{X}(T_1, \infty)$ the set of all sequences $\{x_t\}_{t=T_1}^{\infty} \subset \mathcal{A}$ satisfying $(x_t, x_{t+1}) \in \bar{\mathcal{M}}$ for all integers $t \geq T_1$. Suppose that for all $(x, y) \in \mathcal{M}$,

$$f(x, y) \geq \max\{\psi(\rho_E(x, \theta_0)),\ \psi(\rho_E(y, \theta_0))\} - a_0. \tag{2.16}$$

For each pair of integers $T_2 > T_1 \geq 0$ and each pair of points $y, z \in \mathcal{A}$, we define

$$U_-^f(T_1, T_2, y, z) = \inf\{\sum_{t=T_1}^{T_2-1} \bar{f}(x_t, x_{t+1}) :\ \{x_t\}_{t=T_1}^{T_2} \in \bar{X}(T_1, T_2),$$

$$x_{T_1} = y,\ x_{T_2} = z\},$$

$$\sigma_-^f(T_1, T_2, y) = \inf\{U_-^f(T_1, T_2, y, h) : h \in \mathcal{A}\},$$

$$\widehat{\sigma}_-^f(T_1, T_2, y) = \inf\{U_-^f(T_1, T_2, h, y) : h \in \mathcal{A}\},$$

$$\sigma_{-}^{f}(T_1, T_2) = \inf\{U_{-}^{f}(T_1, T_2, h, y) : h, y \in \mathcal{A}\}.$$

Relations (2.12), (2.13), and (2.15) imply the following result.

Proposition 2.22. *Let $S_2 > S_1 \geq 0$ be integers, $M \geq 0$ and $\{x_t^{(i)}\}_{t=S_1}^{S_2} \in X(S_1, S_2)$, $i = 1, 2$. Then*

$$\sum_{t=S_1}^{S_2-1} f(x_t^{(1)}, x_{t+1}^{(1)}) \geq \sum_{t=S_1}^{S_2-1} f(x_t^{(2)}, x_{t+1}^{(2)}) - M$$

$$\text{if and only if } \sum_{t=S_1}^{S_2-1} \bar{f}(\bar{x}_t^{(1)}, \bar{x}_{t+1}^{(1)}) \geq \sum_{t=S_1}^{S_2-1} \bar{f}(\bar{x}_t^{(2)}, \bar{x}_{t+1}^{(2)}) - M.$$

Proposition 2.22 implies the following result.

Proposition 2.23. *Let $S_2 > S_1 \geq 0$ be integers, $M \geq 0$ and $\{x_t\}_{t=S_1}^{S_2} \in X(S_1, S_2)$. Then the following assertions hold:*

$$\sum_{t=S_1}^{S_2-1} f(x_t, x_{t+1}) \leq \sigma^f(S_1, S_2) + M \text{ if and only if}$$

$$\sum_{t=S_1}^{S_2-1} \bar{f}(\bar{x}_t, \bar{x}_{t+1}) \leq \sigma_{-}^{f}(S_1, S_2) + M;$$

$$\sum_{t=S_1}^{S_2-1} f(x_t, x_{t+1}) \leq U^f(S_1, S_2, x_{S_1}, x_{S_2}) + M$$

$$\text{if and only if } \sum_{t=S_1}^{S_2-1} \bar{f}(\bar{x}_t, \bar{x}_{t+1}) \leq U_{-}^{f}(S_1, S_2, \bar{x}_{S_1}, \bar{x}_{S_2}) + M;$$

$$\sum_{t=S_1}^{S_2-1} f(x_t, x_{t+1}) \leq \sigma^f(S_1, S_2, x_{S_1}) + M \text{ if and only if}$$

$$\sum_{t=S_1}^{S_2-1} \bar{f}(\bar{x}_t, \bar{x}_{t+1}) \leq \widehat{\sigma}_{-}^{f}(S_1, S_2, \bar{x}_{S_2}) + M;$$

$$\sum_{t=S_1}^{S_2-1} f(x_t, x_{t+1}) \le \widehat{\sigma}^f(S_1, S_2, x_{S_2}) + M \textit{ if and only if}$$

$$\sum_{t=S_1}^{S_2-1} \bar{f}(\bar{x}_t, \bar{x}_{t+1}) \le \sigma_-^f(S_1, S_2, \bar{x}_{S_1}) + M.$$

Clearly, $(x_f, x_f) \in \bar{\mathcal{M}}$. In view of Assertion 1 of Theorem 2.1 and Proposition 2.22, (A1) and (A2) hold for $(\bar{f}, \bar{\mathcal{M}})$, and if (A3) holds for $(f, \mathcal{M})$, then (A3) holds for $(\bar{f}, \bar{\mathcal{M}})$ too.

We suppose that $(f, \mathcal{M})$ has (P1) and (P2). Then by Theorem 2.9, $(f, \mathcal{M})$ has TP. Proposition 2.23 implies that $(\bar{f}, \bar{\mathcal{M}})$ has TP too. Clearly, (P1) and (P2) hold for $(\bar{f}, \bar{\mathcal{M}})$. Clearly, if $(f, \mathcal{M})$ possesses LSC property, then it holds also for $(\bar{f}, \bar{\mathcal{M}})$.

By Theorem 2.1, there exists $S_* > 0$ such that for each pair of integers $T_2 > T_1 \ge 0$ and each $\{x_t\}_{t=T_1}^{T_2} \in X(T_1, T_2)$,

$$\sum_{t=T_1}^{T_2-1} f(x_t, x_{t+1}) \ge (T_2 - T_1) f(x_f, u_f) - S_*. \tag{2.17}$$

For all $z \in \mathcal{A} \setminus \cup\{\mathcal{A}_L : L = 1, 2, \dots\}$ set

$$\pi^f(z) = \infty. \tag{2.18}$$

Let

$$z \in \cup\{\mathcal{A}_L : L = 1, 2, \dots\}. \tag{2.19}$$

Define

$$\pi^f(z) = \inf\{\liminf_{T\to\infty}[\sum_{t=0}^{T-1} f(x_t, x_{t+1}) - Tf(x_f, x_f)] :$$

$$\{x_t\}_{t=0}^{\infty} \in X(0, \infty),\ x_0 = z\}. \tag{2.20}$$

By (2.17) and (2.19),

$$-S_* \le \pi^f(z) < \infty. \tag{2.21}$$

There exists an integer $L > 0$ such that

$$z \in \mathcal{A}_L. \tag{2.22}$$

In view of (2.22), there exist $\{\tilde{x}_t\}_{t=0}^{\infty} \in X(0, \infty)$ and an integer $\tau \in \{1, \dots, L\}$ such that

$$\tilde{x}_0 = z, \ \tilde{x}_t = x_f \text{ for all integers } t \geq \tau, \ \sum_{t=0}^{\tau-1} f(\tilde{x}_t, \tilde{x}_{t+1}) \leq L.$$

Then for all integers $T > \tau$,

$$\sum_{t=0}^{T-1} f(\tilde{x}_t, \tilde{x}_{t+1}) - Tf(x_f, u_f) \leq L + (T - \tau) f(x_f, x_f) - Tf(x_f, x_f)$$

$$\leq L(1 + |f(x_f, x_f)|),$$

$$\pi^f(z) \leq L(1 + |f(x_f, x_f)|) \text{ for all } z \in \mathcal{A}_L. \tag{2.23}$$

Definition (2.20) implies the following result.

Proposition 2.24.

1. *Let $S > T \geq 0$ be integers and $\{x_t\}_{t=T}^{S} \in X(T, S)$ satisfy $\pi^f(x_T), \ \pi^f(x_S) < \infty$. Then*

$$\pi^f(x_T) \leq \sum_{t=T}^{S-1} f(x_t, x_{t+1}) - (S - T) f(x_f, u_f) + \pi^f(x_S). \tag{2.24}$$

2. *Let $\{x_t\}_{t=0}^{\infty} \in X(0, \infty)$ be (f)-good. Then for each pair of integers $S > T \geq 0$, (2.24) holds.*

The following three propositions are proved in Section 2.17.

Proposition 2.25. $\pi^f(x_f) = 0$.

Proposition 2.26. *The function $\pi^f : \mathcal{A} \to R^1 \cup \{\infty\}$ is finite in a neighborhood of x_f and continuous at x_f.*

Proposition 2.27. *For each $M > 0$, the set $\{z \in \mathcal{A} : \pi^f(z) \leq M\}$ is bounded,*

Corollary 2.28.

$$\pi^f(z) \to \infty \text{ as } z \in \mathcal{A} \text{ and } \rho_E(z, \theta_0) \to \infty. \tag{2.25}$$

Let $T_2 > T_1 \geq 0$ be integers and $\{x_t\}_{t=T_1}^{T_2} \in X(T_1, T_2)$ satisfy $\pi^f(x_{T_1}) < \infty$. Define

$$\Gamma^f(\{x_t\}_{t=T_1}^{T_2}) = \sum_{t=T_1}^{T_2-1} f(x_t, x_{t+1}) - (T_2 - T_1) f(x_f, x_f) - \pi^f(x_{T_1}) + \pi^f(x_{T_2}). \tag{2.26}$$

Proposition 2.24 implies that

$$0 \leq \Gamma^f(\{x_t\}_{t=T_1}^{T_2}) \leq \infty. \tag{2.27}$$

The next result is proved in Section 2.17.

Proposition 2.29. *Let $\{x_t\}_{t=0}^{\infty} \in X(0, \infty)$ and $x_0 \in \cup\{\mathcal{A}_L : L = 1, 2, \dots\}$. Then $\{x_t\}_{t=0}^{\infty}$ is (f)-good if and only if*

$$\sup\{\Gamma^f(\{x_t\}_{t=0}^{T}) : T = 1, 2, \dots\} = \lim_{T \to \infty} \Gamma^f(\{x_t\}_{t=0}^{T}) < \infty.$$

If $\{x_t\}_{t=0}^{\infty}$ is (f)-good, then

$$\lim_{T \to \infty} [\sum_{t=0}^{T-1} f(x_t, x_{t+1}) - Tf(x_f, x_f)] = \pi^f(x_0) + \lim_{T \to \infty} \Gamma^f(\{x_t\}_{t=0}^{T}).$$

Corollary 2.30. *Let $z \in \mathcal{A}$ satisfy $\pi^f(z) < \infty$ and $\epsilon > 0$. Then there exists an (f)-good $\{x_t\}_{t=0}^{\infty} \in X(0, \infty)$ such that $x_0 = z$ and*

$$\lim_{T \to \infty} \Gamma^f(\{x_t\}_{t=0}^{T}) \leq \epsilon.$$

Set

$$\inf(\pi^f) = \inf\{\pi^f(z) : z \in \mathcal{A}\}.$$

Proposition 2.25 and (2.21) imply that $\inf(\pi^f)$ is finite. Set

$$\mathcal{A}_f = \{z \in \mathcal{A} : \pi^f(z) \leq \inf(\pi^f) + 1\}. \tag{2.28}$$

The next result is proved in Section 2.17.

Proposition 2.31. *There exists an integer $M_0 > 0$ such that $\mathcal{A}_f \subset \mathcal{A}_{M_0}$.*

Corollary 2.30 implies the following result.

Proposition 2.32. *Let $z \in \cup\{\mathcal{A}_L : L = 1, 2, \dots\}$ and $\epsilon > 0$. Then there exists $\{x_t\}_{t=0}^{\infty} \in X(0, \infty)$ such that $x_0 = z$ and*

$$\Gamma^f(\{x_t\}_{t=0}^{T}) \leq \epsilon \textit{ for all integers } T > 0.$$

The next result if proved in Section 2.17.

Proposition 2.33. *Let $M_0 > 0$ be an integer. Then the set $\mathcal{A}_{M_0}$ is bounded.*

2.5 Structure of Solutions in the Regions Close to the Endpoints

We continue to study the problem introduced in Section 2.4 and assume that f possesses (P1) and LSC property. Then f has (P2) and TP. Let

$$z \in \cup\{\mathcal{A}_L : L = 1, 2, \ldots\}. \tag{2.29}$$

Denote by $\Lambda(f, z)$ the set of all (f)-overtaking optimal sequences $\{x_t\}_{t=0}^{\infty} \in X(0, \infty)$ such that $x_0 = z$. In view of Theorem 2.13 and (2.29), $\Lambda(f, z) \neq \emptyset$. Theorem 2.1 implies that any element of $\Lambda(f, z)$ is (f)-good. Equation (2.20) implies the following result.

Proposition 2.34. *For every $z \in \cup\{\mathcal{A}_L : L = 1, 2, \ldots\}$ and every $\{x_t\}_{t=0}^{\infty} \in \Lambda(f, z)$,*

$$\pi^f(z) = \liminf_{T\to\infty}[\sum_{t=0}^{T-1} f(x_t, x_{t+1}) - Tf(x_f, x_f)].$$

The next result follows from (P1) and Propositions 2.25, 2.26, and 2.34.

Proposition 2.35. *Let $\{x_t\}_{t=0}^{\infty} \in X(0, \infty)$ be (f)-good. Then for each pair of integers $S > T \geq 0$,*

$$\pi^f(x_T) = \sum_{t=T}^{S-1} f(x_t, x_{t+1}) - (S - T)f(x_f, x_f) + \pi^f(x_S)$$

if and only if $\{x_t\}_{t=0}^{\infty}$ is (f)-overtaking optimal.

Proposition 2.36. *Let $\{x_t\}_{t=0}^{\infty} \in X(0, \infty)$ be (f)-overtaking optimal and (f)-good. Then*

$$\pi^f(x_0) = \lim_{T\to\infty}[\sum_{t=0}^{T-1} f(x_t, x_{t+1}) - Tf(x_f, x_f)].$$

Proof. Propositions 2.25, 2.26, and 2.35 and (P1) imply that for all integers $T > 0$,

$$\sum_{t=0}^{T-1} f(x_t, x_{t+1}) - Tf(x_f, x_f) = \pi^f(x_0) - \pi^f(x_T) \to \pi^f(x_0) \text{ as } T \to \infty.$$

Proposition 2.36 is proved.

The next result is proved in Section 2.18.

Proposition 2.37. *$\pi^f : \mathcal{A} \to R^1 \cup \{\infty\}$ is a lower semicontinuous function on $\mathcal{A}$.*

It is clear that all the results obtained for the pair $(f, \mathcal{M})$ also hold for the pair $(\bar{f}, \bar{\mathcal{M}})$

For all $z \in \mathcal{A} \setminus \cup\{\widehat{\mathcal{A}}_L : L = 1, 2, \dots\}$ set

$$\pi_-^f(z) = \infty \tag{2.30}$$

and for all $z \in \cup\{\widehat{\mathcal{A}}_L : L = 1, 2, \dots\}$ define

$$\pi_-^f(z) = \inf\{\liminf_{T\to\infty}[\sum_{t=0}^{T-1} \bar{f}(x_t, x_{t+1}) - Tf(x_f, x_f)] :$$

$$\{x_t\}_{t=0}^{\infty} \in \bar{X}(0, \infty),\ x_0 = z\}. \tag{2.31}$$

Let $T_2 > T_1 \geq 0$ be integers and $\{x_t\}_{t=T_1}^{T_2} \in \bar{X}(T_1, T_2)$. Define

$$\Gamma_-^f(\{x_t\}_{t=T_1}^{T_2}) = \sum_{t=T_1}^{T_2-1} \bar{f}(x_t, x_{t+1}) - (T_2 - T_1)f(x_f, x_f) - \pi_-^f(x_{T_1}) + \pi_-^f(x_{T_2}). \tag{2.32}$$

In view of (2.27),

$$0 \leq \Gamma_-^f(\{x_t\}_{t=T_1}^{T_2}) \leq \infty. \tag{2.33}$$

The following two results are proved in Section 2.20. They describe the structure of approximate solutions in the regions close to the endpoints.

Theorem 2.38. *Let L_0, M be natural numbers and $\epsilon \in (0, 1)$. Then there exist $\delta > 0$ and an integer $L_1 > L_0$ such that for each integer $T \geq L_1$ and each $\{x_t\}_{t=0}^{T} \in X(0, T)$ which satisfies*

$$x_0 \in \mathcal{A}_M, \sum_{t=0}^{T-1} f(x_t, x_{t+1}) \leq \sigma^f(0, T, x_0) + \delta,$$

the inequalities

$$\pi_-^f(x_T) \leq \inf(\pi_-^f) + \epsilon,\ \Gamma_-^f(\{\bar{x}_t\}_{t=0}^{L_0}) \leq \epsilon$$

hold, where $\bar{x}_t = x_{T-t}$, $t = 0, \dots, L_0$.

Theorem 2.39. *Let $L_0 > 0$ be an integer and $\epsilon \in (0, 1)$. Then there exist $\delta > 0$ and an integer $L_1 > L_0$ such that for each integer $T \geq L_1$ and each $\{x_t\}_{t=0}^{T} \in X(0, T)$ which satisfies*

$$\sum_{t=0}^{T-1} f(x_t, x_{t+1}) \le \sigma^f(0, T) + \delta,$$

the inequalities

$$\pi^f(x_0) \le \inf(\pi^f) + \epsilon, \ \Gamma^f(\{x_t\}_{t=0}^{L_0}) \le \epsilon,$$

$$\pi_-^f(x_T) \le \inf(\pi_-^f) + \epsilon, \ \Gamma_-^f(\{\bar{x}_t\}_{t=0}^{L_0}) \le \epsilon$$

hold, where $\bar{x}_t = x_{T-t}$, $t = 0, \dots, T$.

The following two results are proved in Section 2.21. They also describe the structure of approximate solutions in the regions close to the endpoints.

Theorem 2.40. *Let* $(f, \mathcal{M})$ *have LSC property,* $L_0, M > 0$ *be integers and* $\epsilon \in (0, 1)$. *Then there exist* $\delta > 0$ *and an integer* $L_1 > L_0$ *such that for each integer* $T \ge L_1$ *and each* $\{x_t\}_{t=0}^T \in X(0, T)$ *which satisfies*

$$x_0 \in \mathcal{A}_M, \ \sum_{t=0}^{T-1} f(x_t, x_{t+1}) \le \sigma^f(0, T, x_0) + \delta,$$

there exists an $(\bar{f}, \bar{\mathcal{M}})$*-overtaking optimal* $\{x_t^*\}_{t=0}^\infty \in \bar{X}(0, \infty)$ *such that*

$$\pi_-^f(x_0^*) = \inf(\pi_-^f), \ \rho_E(x_t^*, x_{T-t}) \le \epsilon, \ t = 0, \dots, L_0.$$

Theorem 2.41. *Let* $(f, \mathcal{M})$ *have LSC property,* $L_0 > 0$ *be an integer and* $\epsilon \in (0, 1)$. *Then there exist* $\delta > 0$ *and an integer* $L_1 > L_0$ *such that for each integer* $T \ge L_1$ *and each* $\{x_t\}_{t=0}^T \in X(0, T)$ *which satisfies*

$$\sum_{t=0}^{T-1} f(x_t, x_{t+1}) \le \sigma^f(0, T) + \delta,$$

there exist an $(f, \mathcal{M})$*-overtaking optimal* $\{x_t^{*,1}\}_{t=0}^\infty \in X(0, \infty)$ *and an* $(\bar{f}, \bar{\mathcal{M}})$*-overtaking optimal* $\{x_t^{*,2}\}_{t=0}^\infty \in \bar{X}(0, \infty)$ *such that*

$$\pi^f(x_0^{*,1}) = \inf(\pi^f), \ \pi_-^f(x_0^{*,2}) = \inf(\pi_-^f),$$

$$\rho_E(x_t, x_t^{*,1}) \le \epsilon, \ \rho_E(x_{T-t}, x_t^{*,2}) \le \epsilon, \ t = 0, \dots, L_0.$$

2.6 Examples

Example 2.42. We consider an example which is a particular case of the problem introduced in Section 2.1 and continue to assume that $x_f \in \mathcal{A}$, $u_f \in \mathcal{U}(x_f)$, $x_f = G(x_f, u_f)$. We suppose that G is continuous at (x_f, u_f) and that the following assumption holds.

(B1) For each $\epsilon > 0$, there exists $\delta > 0$ such that for each $z_i \in \mathcal{A}$, $i = 1, 2$ satisfying $\rho_E(z_i, x_f) \leq \delta$, $i = 1, 2$, there exists $u \in \mathcal{U}(z_1)$ such that $\rho_F(u, u_f) \leq \epsilon$ and $z_2 = G(z_1, u)$.

Let $a_1 > 0$, $\psi_1 : [0, \infty) \to [0, \infty)$ be an increasing function such that $\psi_1(t) \to \infty$ as $t \to \infty$, $\mu \in R^1$, a function $\pi : E \to R^1$ be continuous at x_f, a function $L : \mathcal{M} \to [0, \infty)$ be upper semicontinuous at (x_f, u_f) and for all $(x, u) \in \mathcal{M}$,

$$L(x, u) = 0 \text{ if and only if } x = x_f, u = u_f,$$

$$f(x, u) = \mu + L(x, u) + \pi(x) - \pi(G(x, u)), \tag{2.34}$$

$$\pi(x) \geq -a_1 \text{ for all } x \in E, \tag{2.35}$$

$$L(x, u) - \pi(G(x, u)) \geq -a_1 + \psi_1(\rho_E(\theta_0, x)), \ (x, u) \in \mathcal{M}. \tag{2.36}$$

Clearly,

$$\mu = f(x_f, u_f). \tag{2.37}$$

Let $T_2 > T_1 \geq 0$ be integers and $(\{x_t\}_{t=T_1}^{T_2}, \{u_t\}_{t=T_1}^{T_2-1}) \in X(T_1, T_2)$. By (2.34),

$$\sum_{t=T_1}^{T_2-1} f(x_t, x_{t+1}) = (T_2 - T_1)\mu(f) + \sum_{t=T_1}^{T_2-1} L(x_t, u_t) + \pi(x_{T_1}) - \pi(G(x_{T_2-1}, u_{T_2-1})). \tag{2.38}$$

It is easy to see that (2.1) holds for some $a_0 > 0$ and an increasing $\psi : [0, \infty) \to [0, \infty)$ with $\lim_{t\to\infty} \psi(t) = \infty$. By (2.35)–(2.38), (A1) holds. It follows from (B1) that (A2) and (A3) hold.

Example 2.43. We consider an example which is a particular case of the problem introduced in Section 2.1 and continue to assume that $x_f \in \mathcal{A}$, $u_f \in \mathcal{U}(x_f)$, $x_f = G(x_f, u_f)$. We suppose that G is continuous at (x_f, u_f) and that the following assumption holds.

(B1) There exists an integer $d_f > 0$, and for each $\epsilon > 0$ there exists $\delta > 0$ such that for each $z_i \in \mathcal{A}$, $i = 1, 2$ satisfying $\rho_E(z_i, x_f) \leq \delta$, $i = 1, 2$, there exists $(\{x_t\}_{t=0}^{d_f}, \{u_t\}_{t=0}^{d_f-1}) \in X(0, d_f)$ which satisfies $x_0 = z_1$, $x_{d_f} = z_2$ and $\rho(u_t, u_f) \leq \epsilon$, $t = 0, \ldots, d_f - 1$.

Let $a_1 > 0$, $\psi_1 : [0, \infty) \to [0, \infty)$ be an increasing function such that $\psi_1(t) \to \infty$ as $t \to \infty$, $\mu \in R^1$, a function $\pi : E \to R^1$ be continuous at x_f, a function $L : \mathcal{M} \to [0, \infty)$ be upper semicontinuous at (x_f, u_f) and that for all $(x, u) \in \mathcal{M}$,

$$L(x, u) = 0 \text{ if and only if } x = x_f, u = u_f,$$

$$f(x, u) = \mu + L(x, u) + \pi(x) - \pi(G(x, u)), \tag{2.39}$$

$$\pi(x) \geq -a_1 \text{ for all } x \in E, \tag{2.40}$$

$$L(x, u) - \pi(G(x, u)) \geq -a_1 + \psi_1(\rho_E(\theta_0, x)),\ (x, u) \in \mathcal{M}. \tag{2.41}$$

Clearly,

$$\mu = f(x_f, u_f). \tag{2.42}$$

Let $T_2 > T_1 \geq 0$ be integers and $(\{x_t\}_{t=T_1}^{T_2}, \{u_t\}_{t=T_1}^{T_2-1}) \in X(T_1, T_2)$. By (2.39),

$$\sum_{t=T_1}^{T_2-1} f(x_t, u_t) = (T_2 - T_1)\mu + \sum_{t=T_1}^{T_2-1} L(x_t, u_t) + \pi(x_{T_1}) - \pi(G(x_{T_2-1}, u_{T_2-1})). \tag{2.43}$$

It follows from (2.39)–(2.41) that (2.1) holds with an appropriate choice of a_0, ψ. In view of (2.40), (2.42), and (2.43), (A1) holds. By (2.41), (B1), and the continuity of G at (x_f, u_f), (A2) and (A3) hold.

Assume now that π is bounded on bounded subsets of E and that the following property holds:

(B2) for each $M, \epsilon > 0$ there exists $\delta > 0$ such that for each $(x, u) \in \mathcal{M}$ which satisfies

$$\rho_E(\theta_0, x) \leq M \text{ and } L(x, u) \leq \delta$$

we have $\rho_E(x, x_f) \leq \epsilon$.

We claim that f has TP. In view of Theorems 2.9 and 2.20, it is sufficient to show that property (i) of Theorem 2.20 holds.

Let $\epsilon, M > 0$ and $M_1 > 0$ be as guaranteed by Theorem 2.2 with $M_0 = M$. Since π is bounded on bounded sets, there exists $M_2 > 0$ such that

$$|\pi(z)| \leq M_2 \text{ for all } z \in B_E(\theta_0, M_1).$$

Let $\delta \in (0, 1)$ be as guaranteed by (B2) with $M = M_1$ and

$$L = \lfloor 3 + \delta^{-1}(2M_2 + M + a_0 + |f(x_f, u_f)|) \rfloor.$$

Assume that $T_1 \geq 0$, $T_2 \geq T_1 + L$ are integers and that $(\{x_t\}_{t=T_1}^{T_2}, \{u_t\}_{t=T_1}^{T_2-1}) \in X(T_1, T_2)$ satisfies

$$\sum_{t=T_1}^{T_2-1} f(x_t, u_t) \leq (T_2 - T_1) f(x_f, u_f) + M.$$

Combined with Theorem 2.2 and the choice of M_1, this implies that

$$\rho_E(\theta_0, x_t) \leq M_1, \ t = T_1, \dots, T_2 - 1.$$

It is not difficult to see that

$$\sum_{t=T_1}^{T_2-2} f(x_t, u_t) \leq \sum_{t=T_1}^{T_2-1} f(x_t, u_t) - f(x_{T_2-1}, u_{T_2-1}) \leq (T_2 - T_1) f(x_f, u_f) + M + a_0.$$

Together with (2.39), (2.42), the choice of M_2 and δ, and (B2), this implies that

$$M + a_0 + |f(x_f, u_f)| \geq \sum_{t=T_1}^{T_2-2} (f(x_t, u_t) - f(x_f, u_f))$$

$$= \sum_{t=T_1}^{T_2-2} L(x_t, u_t) + \pi(x_{T_1}) - \pi(x_{T_2-1}) \geq \sum_{t=T_1}^{T_2-2} L(x_t, u_t) - 2M_2$$

and

$$2M_2 + M + a_0 + |f(x_f, u_f)| \geq \sum_{t=T_1}^{T_2-2} L(x_t, u_t)$$

$$\geq \delta \mathrm{Card}(\{t \in \{T_1, \dots, T_2 - 2\} : \ L(x_t, u_t) > \delta\})$$

$$\geq \delta \mathrm{Card}(\{t \in \{T_1, \dots, T_2 - 2\} : \ \rho_E(x_t, x_f) > \epsilon\}).$$

This implies that

$$\mathrm{Card}(\{t \in \{T_1, \dots, T_2 - 1\} : \ \rho_E(x_t, x_f) > \epsilon\})$$

$$\leq 1 + \delta^{-1}(2M_2 + M + a_0 + |f(x_f, u_f)|) < L$$

and that property (i) of Theorem 2.20 holds. Therefore f has TP.

Example 2.44. We consider a particular case of the problem introduced in Section 2.1. Let $\mathcal{A}$ be a closed compact set, $\mathcal{M}$ be a closed compact set, G be continuous, and f be lower semicontinuous. Then LSC property holds.

Example 2.45. We consider a particular case of the problem introduced in Section 2.1. Assume that for each $M > 0$, the set $\{(x,u) \in \mathcal{M} : \rho_E(x,\theta_0) \le M\}$ is closed and compact, G is continuous, and f is lower semicontinuous. It is not difficult to see that LSC property holds.

Example 2.46. We consider an example which is a particular case of the problem introduced in Section 2.4 and continue to assume that $(E,\rho_E) = (F,\rho_F)$, $G(x,u) = u$, $(x,u) \in \mathcal{M}$, $u_f = x_f$, and $(x_f,u_f) \in \mathcal{M}$ and assume that the following assumption holds.

(B1) There exists $\epsilon_0 > 0$ such that for each $z_i \in \mathcal{A}$, $i = 1,2$ satisfying $\rho_E(z_i,x_f) \le \epsilon_0$, $i = 1,2$ we have $(z_1,z_2) \in \mathcal{M}$.

Let $a_1 > 0$, $\psi_1 : [0,\infty) \to [0,\infty)$ be an increasing function such that $\psi_1(t) \to \infty$ as $t \to \infty$, $\mu \in R^1$, a function $\pi : E \to R^1$ be continuous at x_f, a function $L : \mathcal{M} \to [0,\infty)$ be upper semicontinuous at (x_f,x_f) and that for all $(x,u) \in \mathcal{M}$,

$$L(x,u) = 0 \text{ if and only if } x = x_f,\ u = x_f,$$

$$f(x,u) = \mu + L(x,u) + \pi(x) - \pi(u), \tag{2.44}$$

$$\pi(x) \ge -a_1 \text{ for all } x \in E, \tag{2.45}$$

$$L(x,u) - \pi(u) \ge -a_1 + \psi_1(\rho_E(\theta_0,x)),\ (x,u) \in \mathcal{M}. \tag{2.46}$$

Let $T_2 > T_1 \ge 0$ be integers and $\{x_t\}_{t=T_1}^{T_2} \in X(T_1,T_2)$. By (2.44),

$$\sum_{t=T_1}^{T_2-1} f(x_t,x_{t+1}) = (T_2 - T_1)\mu + \sum_{t=T_1}^{T_2-1} L(x_t,u_t) + \pi(x_{T_1}) - \pi(x_{T_2}). \tag{2.47}$$

It follows from (2.44)–(2.46) that (2.1) holds for an appropriate choice of $a_0 > 0$ and ψ. By (2.45)–(2.47), (A1) holds. It follows from (B1) and upper semicontinuity of L at (x_f,x_f) that (A2) and (A3) hold.

Example 2.47. We consider an example which is a particular case of the problem introduced in Section 2.4 and continue to assume that $(E,\rho_E) = (F,\rho_F)$, $G(x,u) = u$, $(x,u) \in \mathcal{M}$, $u_f = x_f$, $(x_f,x_f) \in \mathcal{M}$ and assume that the following assumption holds.

(B1) There exists an integer $d_f > 0$, and for each $\epsilon > 0$ there exists $\delta > 0$ such that for each $z_i \in \mathcal{A}$, $i = 1,2$ satisfying $\rho_E(z_i,x_f) \le \delta$, $i = 1,2$, there exists $\{x_t\}_{t=0}^{d_f} \in X(0,d_f)$ which satisfies $x_0 = z_1$, $x_{d_f} = z_2$ and $\rho(x_t,x_f) \le \epsilon$, $t = 0,\dots,d_f$.

Let $a_1 > 0$, $\psi_1 : [0, \infty) \to [0, \infty)$ be an increasing function such that $\psi_1(t) \to \infty$ as $t \to \infty$, $\mu \in R^1$, a function $\pi : E \to R^1$ be continuous at x_f, a function $L : \mathcal{M} \to [0, \infty)$ be upper semicontinuous at (x_f, x_f) and for all $(x, u) \in \mathcal{M}$,

$$L(x, u) = 0 \text{ if and only if } x = x_f, u = x_f,$$

$$f(x, u) = \mu + L(x, u) + \pi(x) - \pi(u), \tag{2.48}$$

$$\pi(x) \geq -a_1 \text{ for all } x \in E,$$

$$L(x, u) - \pi(u) \geq -a_1 + \psi_1(\rho_E(\theta_0, x)), \ (x, u) \in \mathcal{M}. \tag{2.49}$$

Let $T_2 > T_1 \geq 0$ be integers and $\{x_t\}_{t=T_1}^{T_2} \in X(T_1, T_2)$. By (2.48), equation (2.47) holds. It follows from (2.48) and (2.49) that (2.1) holds with an appropriate choice of a_0, ψ. In view of (2.47) and (2.49), (A1) holds. By (B1), (2.48), the upper semicontinuity of L at (x_f, x_f) and continuity of π at x_f, (A2) and (A3) hold.

2.7 Auxiliary Results for Theorem 2.1

Lemma 2.48. *There exist integers $S, c_0 > 0$ such that for each pair of integers $T_1 \geq 0$, $T_2 \geq T_1 + c_0$ and each $(\{x_t\}_{t=T_1}^{T_2}, \{u_t\}_{t=T_1}^{T_2-1}) \in X(T_1, T_2)$,*

$$(T_2 - T_1) f(x_f, u_f) \leq \sum_{t=T_1}^{T_2-1} f(x_t, u_t) + S. \tag{2.50}$$

Proof. There exists an integer $S_1 > 0$ such that

$$\psi(S_1) > |f(x_f, u_f)| + a_0 + 1. \tag{2.51}$$

By (A1), there exist integers $S_2 > 0$, $c_0 > 0$ such that

$$(T_2 - T_1) f(x_f, u_f) \leq \sum_{t=T_1}^{T_2-1} f(x_t, u_t) + S_2$$

for each pair of integers $T_1 \geq 0$, $T_2 \geq T_1 + c_0$ and each $(\{x_t\}_{t=T_1}^{T_2}, \{u_t\}_{t=T_1}^{T_2-1}) \in X(T_1, T_2)$ satisfying $\rho_E(\theta_0, x_j) \leq S_1$, $j = T_1, T_2 - 1$.

Fix an integer

$$S > S_2 + (c_0 + 2)(|f(x_f, u_f)| + a_0 + 2). \tag{2.52}$$

Assume that $T_1 \geq 0$, $T_2 \geq T_1 + c_0$ are integers and that $(\{x_t\}_{t=T_1}^{T_2}, \{u_t\}_{t=T_1}^{T_2-1}) \in X(T_1, T_2)$. We show that (2.50) is true. Assume that

$$\rho_E(\theta_0, x_t) \geq S_1, \; t = T_1, \ldots, T_2 - 1. \tag{2.53}$$

By (2.1), (2.51), and (2.53), for all $t = T_1, \ldots, T_2 - 1$,

$$f(x_t, u_t) \geq -a_0 + \psi(\rho_E(\theta_0, x_t)) \geq -a_0 + \psi(S_1) > |f(x_f, u_f)| + 1,$$

$$\sum_{t=T_1}^{T_2-1} f(x_t, u_t) \geq (T_2 - T_1)(|f(x_f, u_f)| + 1) \geq (T_2 - T_1) f(x_f, u_f)$$

and (2.50) holds. Assume that

$$\min\{\rho_E(\theta_0, x_t) : \; t = T_1, \ldots, T_2 - 1\} < S_1. \tag{2.54}$$

Set

$$\tau_1 = \min\{t \in \{T_1, \ldots, T_2 - 1\} : \; \rho_E(\theta_0, x_t) \leq S_1\}, \tag{2.55}$$

$$\tau_2 = \max\{t \in \{T_1, \ldots, T_2 - 1\} : \; \rho_E(\theta_0, x_t) \leq S_1\}. \tag{2.56}$$

By (2.54)–(2.56), τ_1, τ_2 are well-defined, $\tau_1 \leq \tau_2$ and

$$\rho_E(\theta_0, x_{\tau_i}) \leq S_1, \; i = 1, 2. \tag{2.57}$$

There are two cases:

$$\tau_2 - \tau_1 \geq c_0; \tag{2.58}$$

$$\tau_2 - \tau_1 < c_0. \tag{2.59}$$

Assume that (2.58) holds. It follows from (2.57), (2.58), and the choice of S_2 and c_0 that

$$(\tau_2 - \tau_1 + 1) f(x_f, u_f) \leq \sum_{t=\tau_1}^{\tau_2-1} f(x_t, u_t) + S_2. \tag{2.60}$$

By (2.1), (2.51), (2.55), and (2.56), for each $t \in (\{T_1, \ldots, \tau_1\} \setminus \{\tau_1\}) \cup (\{\tau_2 + 1, \ldots, T_2\} \setminus \{T_2\})$,

$$f(x_t, u_t) \geq -a_0 + \psi(\rho_E(\theta_0, x_t)) \geq -a_0 + \psi(S_1) > |f(x_f, u_f)| + 1. \tag{2.61}$$

In view of (2.52), (2.60), and (2.61),

$$(T_2 - T_1)f(x_f, u_f) \le (\tau_1 - T_1)f(x_f, u_f)$$

$$+(\tau_2 - \tau_1 + 1)f(x_f, u_f) + (T_2 - \tau_2 - 1)f(x_f, u_f)$$

$$\le \sum\{f(x_t, u_t) : t \in \{T_1, \dots, \tau_1\} \setminus \{\tau_1\}\}$$

$$+\sum_{t=\tau_1}^{\tau_2} f(x_t, u_t) + S_2 + \sum\{f(x_t, u_t) : t \in \{\tau_2 + 1, \dots, T_2\} \setminus \{T_2\}\}$$

$$= \sum_{t=T_1}^{T_2-1} f(x_t, u_t) + S_2 \le \sum_{t=T_1}^{T_2-1} f(x_t, u_t) + S.$$

Assume that (2.59) holds. Note that (2.61) is true. By (2.1), (2.52), (2.59), and (2.61),

$$\sum_{t=T_1}^{T_2-1} f(x_t, u_t) = \sum\{f(x_t, u_t) : t \in \{T_1, \dots, \tau_1\} \setminus \{\tau_1\}\}$$

$$+\sum_{t=\tau_1}^{\tau_2} f(x_t, u_t) + \sum\{f(x_t, u_t) : t \in \{\tau_2 + 1, \dots, T_2\} \setminus \{T_2\}\}$$

$$\ge (|f(x_f, u_f)| + 1)\mathrm{Card}(\{t \in \{T_1, \dots, \tau_1\} \setminus \{\tau_1\}\}) - a_0(\tau_2 - \tau_1 + 1)$$

$$+(|f(x_f, u_f)| + 1)\mathrm{Card}(\{t \in \{\tau_2 + 1, \dots, T_2\} \setminus \{T_2\}\})$$

$$\ge -a_0(c_0 + 1) + (|f(x_f, u_f)| + 1)(T_2 - T_1 - c_0 - 2)$$

$$\ge |f(x_f, u_f)|(T_2 - T_1) - (c_0 + 2)(a_0 + |f(x_f, u_f)| + 2) \ge (T_2 - T_1)f(x_f, u_f) - S.$$

Lemma 2.48 is proved.

The next auxiliary result easily follows from (2.1).

Lemma 2.49. *Let $M_0 > 0$ and τ_0 be natural number. Then there exists $M_1 > 0$ such that for each integer $T_1 \ge 0$, each $T_2 \in \{T_1, \dots, T_1 + \tau_0\}$, and each $(\{x_t\}_{t=T_1}^{T_2}, \{u_t\}_{t=T_1}^{T_2-1}) \in X(T_1, T_2)$ which satisfies $\sum_{t=T_1}^{T_2-1} f(x_t, u_t) \le M_0$, the inequality $\rho_E(x_t, \theta_0) \le M_1$ holds for all $t = T_1, \dots, T_2 - 1$,*

2.8 Proofs of Theorems 2.1–2.4

Proof of Theorem 2.1. Assertion 1 follows from Lemma 2.48 and (2.1). Let us prove Assertion 2. Let $S > 0$ be as guaranteed by Assertion 1. Assume that (2.7) does not hold and that $Q > 0$. Since (2.7) does not hold, Assertion 1 implies that there exists an integer $T_Q > 0$ such that

$$\sum_{t=0}^{T_Q-1} f(x_t, u_t) - T_Q f(x_f, u_f) > Q. \tag{2.62}$$

By (2.62) and the choice of S, for each integer $T > T_Q$,

$$\sum_{t=0}^{T-1} f(x_t, u_t) = \sum_{t=0}^{T_Q-1} f(x_t, u_t) + \sum_{t=T_Q}^{T-1} f(x_t, u_t)$$

$$\geq T_Q f(x_f, u_f) + Q + (T - T_Q) f(x_f, u_f) - S = T f(x_f, u_f) + Q - S.$$

Since Q is any positive number, we obtain that $\sum_{t=0}^{T-1} f(x_t, u_t) - T f(x_f, u_f) \to \infty$ as $T \to \infty$.

Assume that (2.7) holds. Then there exists $S_1 > 0$ such that for each integer $T > 0$,

$$|\sum_{t=0}^{T-1} f(x_t, u_t) - T f(x_f, u_f)| < S_1. \tag{2.63}$$

By (2.63), for each pair of integers $T_2 > T_1 > 0$,

$$|\sum_{t=T_1}^{T_2-1} f(x_t, u_t) - (T_2 - T_1) f(x_f, u_f)| < 2S_1. \tag{2.64}$$

In view of (2.64), for each integer $T \geq 0$,

$$\sum_{t=T}^{T+9} f(x_t, u_t) \leq 10|f(x_f, u_f)| + 2S_1. \tag{2.65}$$

Lemma 2.49 and (2.65) imply that the sequence $\{x_t\}_{t=0}^{\infty}$ is bounded. Thus Assertion 2 holds. Theorem 2.1 is proved.

Proof of Theorem 2.2. By Theorem 2.1, there exists $S_0 > 0$ such that the following property holds:

(i) for each pair of integers $T_2 > T_1 \geq 0$ and each $(\{x_t\}_{t=T_1}^{T_2}, \{u_t\}_{t=T_1}^{T_2-1}) \in X(T_1, T_2)$,

$$\sum_{t=T_1}^{T_2-1} f(x_t, u_t) + S_0 \geq (T_2 - T_1) f(x_f, u_f).$$

There exists $M_1 > 0$ such that

$$\psi(M_1) > a_0 + 2S_0 + |f(x_f, u_f)| + 1 + M_0. \tag{2.66}$$

Let $T_2 > T_1 \geq 0$ be integers and $(\{x_t\}_{t=T_1}^{T_2}, \{u_t\}_{t=T_1}^{T_2-1}) \in X(T_1, T_2)$ satisfy

$$\sum_{t=T_1}^{T_2-1} f(x_t, u_t) \leq (T_2 - T_1) f(x_f, u_f) + M_0. \tag{2.67}$$

We show that

$$\rho_E(\theta_0, x_t) \leq M_1, \ t = T_1, \ldots, T_2 - 1.$$

Assume the contrary. Then there exists $j \in \{T_1, \ldots, T_2 - 1\}$ such that

$$\rho_E(\theta_0, x_j) > M_1. \tag{2.68}$$

By (2.1) and (2.68),

$$f(x_j, u_j) \geq \psi(\rho_E(\theta_0, x_j)) - a_0 \geq \psi(M_1) - a_0. \tag{2.69}$$

Property (i), (2.66), and (2.69) imply that

$$\begin{aligned}
\sum_{t=T_1}^{T_2-1} f(x_t, u_t) &= \sum \{f(x_t, u_t) : \ t \in \{T_1, \ldots, j\} \setminus \{j\}\} + f(x_j, u_j) \\
&\quad + \{f(x_t, u_t) : \ t \in \{j, \ldots, T_2 - 1\} \setminus \{j\}\} \\
&\geq f(x_f, u_f)(\mathrm{Card}(\{T_1, \ldots, j\} \setminus \{j\}) - S_0 + \psi(M_1) - a_0 \\
&\quad + f(x_f, u_f)(\mathrm{Card}(\{j, \ldots, T_2 - 1\} \setminus \{j\}) - S_0 \\
&= f(x_f, u_f)((j - T_1) + (T_2 - 1 - j)) - a_0 - 2S_0 + \psi(M_1) \\
&= f(x_f, u_f)(T_2 - T_1) + \psi(M_1) - a_0 - 2S_0 - |f(x_f, u_f)| \\
&> M_0 + 1 + f(x_f, u_f)(T_2 - T_1).
\end{aligned}$$

This contradicts (2.67). The contradiction we have reached proves Theorem 2.2.

Theorem 2.3 follows from Theorem 2.2 and the following proposition.

Proposition 2.50. *Let $L > 0$ be an integer. Then for each pair of integers $T_1 \geq 0$, $T_2 \geq T_1 + 2L$ and each pair of points $z_1 \in \mathcal{A}_L$, $z_2 \in \widehat{\mathcal{A}}_L$,*

$$U^f(T_1, T_2, z_1, z_2) \leq (T_2 - T_1)f(x_f, u_f) + (2L + 2)(2 + a_0).$$

Proof. Let $T_1 \geq 0$, $T_2 \geq T_1 + 2L$ be integers, $z_1 \in \mathcal{A}_L$, $z_2 \in \widehat{\mathcal{A}}_L$. By the definition of $\mathcal{A}_L$, $\widehat{\mathcal{A}}_L$, there exist integers $\tau_1 \in (0, L]$, $\tau_2 \in (0, L]$ and $(\{x_t\}_{t=T_1}^{T_2}, \{u_t\}_{t=T_1}^{T_2-1}) \in X(T_1, T_2)$ such that

$$x_{T_1} = z_1,\ x_t = x_f,\ t = T_1 + \tau_1, \dots, T_2 - \tau_2,$$

$$u_t = u_f,\ t \in \{T_1 + \tau_1, \dots, T_2 - \tau_2\} \setminus \{T_2 - \tau_2\},\ x_{T_2} = z_2,$$

$$\sum_{t=T_1}^{T_1+\tau_1-1} f(x_t, u_t) \leq L,\ \sum_{t=T_2-\tau_2}^{T_2-1} f(x_t, u_t) \leq L.$$

In view of the relations above and (2.1),

$$U^f(T_1, T_2, z_1, z_2) \leq \sum_{t=T_1}^{T_2-1} f(x_t, u_t)$$

$$\leq 2L + \sum \{f(x_t, u_t) :\ t \in \{T_1 + \tau_1, \dots, T_2 - \tau_2\} \setminus \{T_2 - \tau_2\}\}$$

$$\leq 2L + (T_2 - T_1)f(x_f, u_f) + (2L + 2)a_0.$$

Proposition 2.50 is proved.

Theorem 2.4 follows from Theorem 2.2 and the following proposition.

Proposition 2.51. *Let $L > 0$ be an integer, $T_1 \geq 0$, $T_2 \geq T_1 + L$ be integers, $z_1 \in \mathcal{A}_L$. Then*

$$\sigma^f(T_1, T_2, z_1) \leq (T_2 - T_1)f(x_f, u_f) + (L + 1)(1 + a_0).$$

Proof. By the definition of $\mathcal{A}_L$, there exist an integer $\tau_1 \in (0, L]$ and $(\{x_t\}_{t=T_1}^{T_2}, \{u_t\}_{t=T_1}^{T_2-1}) \in X(T_1, T_2)$ such that

$$x_{T_1} = z_1,\ x_t = x_f,\ t = T_1 + \tau_1, \dots, T_2,\ u_t = u_f,\ t \in \{T_1 + \tau_1, \dots, T_2\} \setminus \{T_2\},$$

$$\sum_{t=T_1}^{T_1+\tau_1-1} f(x_t, u_t) \leq L.$$

In view of the relations above and (2.1),

$$\sigma^f(T_1, T_2, z_1) \le \sum_{t=T_1}^{T_2-1} f(x_t, u_t)$$

$$\le L + (T_2 - T_1) f(x_f, u_f) + (L+1)a_0.$$

Proposition 2.51 is proved.

Theorem 2.5 follows from Theorem 2.2 and the following proposition.

Proposition 2.52. *Let $L > 0$ be an integer, $T_1 \ge 0$, $T_2 \ge T_1 + L$ be integers, $z_1 \in \widehat{\mathcal{A}}_L$. Then*

$$\widehat{\sigma}^f(T_1, T_2, z_1) \le (T_2 - T_1) f(x_f, u_f) + (L+1)(1 + a_0).$$

Proof. By the definition of $\widehat{\mathcal{A}}_L$, there exist an integer $\tau \in (0, L]$ and

$$(\{x_t\}_{t=T_1}^{T_2}, \{u_t\}_{t=T_1}^{T_2-1}) \in X(T_1, T_2)$$

such that

$$x_t = x_f,\ t = T_1, \dots, T_2 - \tau,\ u_t = u_f,\ t = T_1, \dots, T_2 - \tau - 1,$$

$$x_{T_2} = z_1,\ \sum_{t=T_2-\tau}^{T_2-1} f(x_t, u_t) \le L.$$

In view of the relations above and (2.1),

$$\widehat{\sigma}^f(T_1, T_2, z_1) \le \sum_{t=T_1}^{T_2-1} f(x_t, u_t)$$

$$\le L + (T_2 - T_1) f(x_f, u_f) + (L+1)a_0.$$

Proposition 2.52 is proved.

Theorem 2.7 follows from Propositions 2.6 and 2.50, while Theorem 2.8 follows from Propositions 2.6 and 2.51.

2.9 Proof of Theorem 2.9

In view of Theorem 2.1, TP implies (P2). We show that TP implies (P1). Assume that TP holds, $(\{x_t\}_{t=0}^{\infty}, \{u_t\}_{t=0}^{\infty}) \in X(0, \infty)$ is (f)-good and $\epsilon > 0$. There exists $S > 0$ such that for each integer $T > 0$,

$$|\sum_{t=0}^{T-1} f(x_t, u_t) - Tf(x_f, u_f)| < S.$$

This implies that for each pair of integers $T_2 > T_1 \geq 0$,

$$|\sum_{t=T_1}^{T_2-1} f(x_t, u_t) - (T_2 - T_1)f(x_f, u_f)| < 2S. \tag{2.70}$$

Let $\delta > 0$. We show that there exists an integer $T_\delta > 0$ such that for each integer $T > T_\delta$,

$$\sum_{t=T_\delta}^{T-1} f(x_t, u_t) \leq U^f(T_\delta, T, x_{T_\delta}, x_T) + \delta.$$

Assume the contrary. Then for each integer $T \geq 0$, there exists an integer $S > T$ such that

$$\sum_{t=T}^{S-1} f(x_t, u_t) > U^f(T, S, x_T, x_S) + \delta.$$

This implies that there exists a strictly increasing sequence of integers $\{T_i\}_{i=0}^{\infty}$ such that

$$T_0 = 0 \tag{2.71}$$

and for every integer $i \geq 0$,

$$\sum_{t=T_i}^{T_{i+1}-1} f(x_t, u_t) > U^f(T_i, T_{i+1}, x_{T_i}, x_{T_{i+1}}) + \delta. \tag{2.72}$$

By (2.71) and (2.72), there exists $(\{y_t\}_{t=0}^{\infty}, \{v_t\}_{t=0}^{\infty}) \in X(0, \infty)$ such that for every integer $i \geq 0$,

$$y_{T_i} = x_{T_i}, \tag{2.73}$$

$$\sum_{t=T_i}^{T_{i+1}-1} f(x_t, u_t) > \sum_{t=T_i}^{T_{i+1}-1} f(y_t, v_t) + \delta. \tag{2.74}$$

In view of (2.70), (2.73), and (2.74), for each integer $k \geq 1$,

$$\sum_{t=0}^{T_k-1} f(y_t, v_t) - T_k f(x_f, u_f) = \sum_{t=0}^{T_k-1} f(y_t, v_t) - \sum_{t=0}^{T_k-1} f(x_t, u_t)$$

$$+ \sum_{t=0}^{T_k-1} f(x_t, u_t) - T_k f(x_f, u_f) \leq -k\delta + 2S \to -\infty \text{ as } k \to \infty.$$

This contradicts Theorem 2.1. The contradiction we have reached proves that the following property holds:

(i) for each $\delta > 0$ there exists an integer $T_\delta > 0$ such that for each integer $T > T_\delta$,

$$\sum_{t=T_\delta}^{T-1} f(x_t, u_t) \leq U^f(T_\delta, T, x_{T_\delta}, x_T) + \delta.$$

Proposition 2.6 implies that there exist $\delta > 0$ and an integer $L > 0$ such that the following property holds:

(ii) for each integer $S_1 \geq 0$, each integer $S_2 \geq S_1 + 2L$, and each $(\{z_t\}_{t=S_1}^{S_2}, \{\xi_t\}_{t=S_1}^{S_2-1}) \in X(S_1, S_2)$ which satisfies

$$\sum_{t=S_1}^{S_2-1} f(z_t, \xi_t) \leq \min\{(S_2 - S_1) f(x_f, u_f) + 2S, U^f(S_1, S_2, z_{S_1}, z_{S_2}) + \delta\},$$

we have

$$\rho_E(z_t, x_f) \leq \epsilon, \; t = S_1 + L, \ldots, S_2 - L.$$

Let an integer T_δ be as guaranteed by (i). Properties (i) and (ii), (2.70), and the choice of T_δ imply that for each integer $T \geq T_\delta + 2L$,

$$\rho_E(x_t, x_f) \leq \epsilon \text{ for all integers } t \in [T_\delta + L, T - L].$$

Therefore

$$\rho_E(x_t, x_f) \leq \epsilon \text{ for all integers } t \geq T_\delta + L$$

and (P1) holds. Thus TP implies (P1).

Lemma 2.53. *Assume that (P1) holds and that $\epsilon > 0$. Then there exists $\delta > 0$ such that for each integer $T_1 \geq 0$, each integer $T_2 \geq T_1 + 2b_f$, and each $(\{x_t\}_{t=T_1}^{T_2}, \{u_t\}_{t=T_1}^{T_2-1}) \in X(T_1, T_2)$ which satisfies*

$$\rho_E(x_{T_i}, x_f) \leq \delta, \; i = 1, 2,$$

$$\sum_{t=T_1}^{T_2-1} f(x_t, u_t) \le U^f(T_1, T_2, x_{T_1}, x_{T_2}) + \delta,$$

the inequality $\rho_E(x_t, x_f) \le \epsilon$ *holds for all* $t = T_1, \dots, T_2$.

Proof. By (A2), for each integer $k \ge 1$, there exists $\delta_k \in (0, 2^{-k})$ such that the following property holds:

(iii) for each $z \in \mathcal{A}$ satisfying $\rho_E(z, x_f) \le \delta_k$, there exist integers $\tau_1, \tau_2 \in (0, b_f]$ and

$$(\{x_t^{(1)}\}_{t=0}^{\tau_1}, \{u_t^{(1)}\}_{t=0}^{\tau_1-1}) \in X(0, \tau_1),\ (\{x_t^{(2)}\}_{t=0}^{\tau_2}, \{u_t^{(2)}\}_{t=0}^{\tau_2-1}) \in X(0, \tau_2)$$

which satisfy

$$x_0^{(1)} = z,\ x_{\tau_1}^{(1)} = x_f,\ x_0^{(2)} = x_f,\ x_{\tau_2}^{(2)} = z,$$

$$\sum_{t=0}^{\tau_i-1} f(x_t^{(i)}, u_t^{(i)}) \le \tau_i f(x_f, u_f) + 2^{-k},\ i = 1, 2.$$

Assume that the lemma does not hold. Then for each integer $k \ge 1$, there exist an integer $T_k \ge 2b_f$ and

$$(\{x_t^{(k)}\}_{t=0}^{T_k}, \{u_t^{(k)}\}_{t=0}^{T_k-1}) \in X(0, T_k)$$

such that

$$\rho_E(x_0^{(k)}, x_f) \le \delta_k,\ \rho_E(x_{T_k}^{(k)}, x_f) \le \delta_k, \tag{2.75}$$

$$\sum_{t=0}^{T_k-1} f(x_t^{(k)}, u_t^{(k)}) \le U^f(0, T_k, x_0^{(k)}, x_{T_k}^{(k)}) + \delta_k, \tag{2.76}$$

$$\sup\{\rho_E(x_t^{(k)}, x_f) : t \in \{0, \dots, T_k\}\} > \epsilon. \tag{2.77}$$

Let $k \ge 1$ be an integer. Property (iii) and (2.75) imply that there exist integers $\tau_{k,1}, \tau_{k,2} \in (0, b_f]$ and

$$(\{\tilde{x}_t^{(k)}\}_{t=0}^{T_k+\tau_{k,1}+\tau_{k,2}}, \{\tilde{u}_t^{(k)}\}_{t=0}^{T_k+\tau_{k,1}+\tau_{k,2}-1}) \in X(0, T_k + \tau_{k,1} + \tau_{k,2})$$

such that

$$\tilde{x}_0^{(k)} = x_f,\ \tilde{x}_{\tau_{k,1}}^{(k)} = x_0^{(k)}, \tag{2.78}$$

$$\sum_{t=0}^{\tau_{k,1}-1} f(\tilde{x}_t^{(k)}, \tilde{u}_t^{(k)}) \le \tau_{k,1} f(x_f, u_f) + 2^{-k}, \tag{2.79}$$

$$\tilde{x}_{\tau_{k,1}+t}^{(k)} = x_t^{(k)},\ t = 0, \dots, T_k,\ \tilde{u}_{\tau_{k,1}+t}^{(k)} = u_t^{(k)},\ t = 0, \dots, T_k - 1, \tag{2.80}$$

$$\tilde{x}_{T_k+\tau_{k,1}+\tau_{k,2}}^{(k)} = x_f, \tag{2.81}$$

$$\sum_{t=T_k+\tau_{k,1}}^{T_k+\tau_{k,1}+\tau_{k,2}-1} f(\tilde{x}_t^{(k)}, \tilde{u}_t^{(k)}) \le \tau_{k,2} f(x_f, u_f) + 2^{-k}. \tag{2.82}$$

By (2.76), (2.79), (2.80), and (2.82),

$$\sum_{t=0}^{T_k+\tau_{k,1}+\tau_{k,2}-1} f(\tilde{x}_t^{(k)}, \tilde{u}_t^{(k)})$$

$$\le (\tau_{k,1} + \tau_{k,2}) f(x_f, u_f) + 2^{-k+1} + \sum_{t=0}^{T_k-1} f(x_t^{(k)}, u_t^{(k)})$$

$$\le (\tau_{k,1} + \tau_{k,2}) f(x_f, u_f) + U^f(0, T_k, x_0^{(k)}, x_{T_k}^{(k)}) + \delta_k + 2^{-k+1}. \tag{2.83}$$

Property (iii) and (2.75) imply that there exist integers $\tau_{k,3}, \tau_{k,4} \in (0, b_f]$ and $(\{\widehat{x}_t^{(k)}\}_{t=0}^{T_k}, \{\widehat{u}_t^{(k)}\}_{t=0}^{T_k-1}) \in X(0, T_k)$ such that

$$\widehat{x}_0^{(k)} = x_0^{(k)},\ \widehat{x}_{T_k}^{(k)} = x_{T_k}^{(k)}, \tag{2.84}$$

$$\widehat{x}_t^{(k)} = x_f,\ t = \tau_{k,3}, \dots, T_k - \tau_{k,4}, \tag{2.85}$$

$$\widehat{u}_t^{(k)} = u_f,\ t \in \{\tau_{k,3}, \dots, T - \tau_{k,4}\} \setminus \{T - \tau_{k,4}\}, \tag{2.86}$$

$$\sum_{t=0}^{\tau_{k,3}-1} f(\widehat{x}_t^{(k)}, \widehat{u}_t^{(k)}) \le \tau_{k,3} f(x_f, u_f) + 2^{-k}, \tag{2.87}$$

$$\sum_{t=T_k-\tau_{k,4}}^{T_k-1} f(\widehat{x}_t^{(k)}, \widehat{u}_t^{(k)}) \le \tau_{k,4} f(x_f, u_f) + 2^{-k}. \tag{2.88}$$

In view of (2.84)–(2.88),

$$U^f(0, T_k, x_0^{(k)}, x_{T_k}^{(k)}) \le \sum_{t=0}^{T_k-1} f(\widehat{x}_t^{(k)}, \widehat{u}_t^{(k)}) \le T_k f(x_f, u_f) + 2^{-k+1}. \tag{2.89}$$

By (2.83) and (2.89),

$$\sum_{t=0}^{T_k+\tau_{k,1}+\tau_{k,2}-1} f(\tilde{x}_t^{(k)}, \tilde{u}_t^{(k)}) \le (T_k + \tau_{k,1} + \tau_{k,2}) f(x_f, u_f) + 3 \cdot 2^{-k+1}. \tag{2.90}$$

By (2.78) and (2.81), there exists $(\{x_t\}_{t=0}^{\infty}, \{u_t\}_{t=0}^{\infty}) \in X(0, \infty)$ such that

$$x_t = \tilde{x}_t^{(1)}, \; t = 0, \dots, T_1 + \tau_{1,1} + \tau_{1,2}, \tag{2.91}$$

$$u_t = \tilde{u}_t^{(1)}, \; t = 0, \dots, T_1 + \tau_{1,1} + \tau_{1,2} - 1 \tag{2.92}$$

and for each integer $k \ge 1$,

$$x_{\sum_{i=1}^k (T_i + \tau_{i,1} + \tau_{i,2}) + t} = \tilde{x}_t^{(k+1)}, \; t = 0, \dots, T_{k+1} + \tau_{k+1,1} + \tau_{k+1,2}, \tag{2.93}$$

$$u_{\sum_{i=1}^k (T_i + \tau_{i,1} + \tau_{i,2}) + t} = \tilde{u}_t^{(k+1)}, \; t = 0, \dots, T_{k+1} + \tau_{k+1,1} + \tau_{k+1,2} - 1. \tag{2.94}$$

In view of (2.91)–(2.94), for each integer $k \ge 1$,

$$\begin{aligned} &\sum \{f(x_t, u_t) : \; t = 0, \dots, \sum_{i=1}^k (T_i + \tau_{i,1} + \tau_{i,2}) - 1\} \\ &= \sum_{i=1}^k \sum_{t=0}^{T_i+\tau_{i,1}+\tau_{i,2}-1} f(\tilde{x}_t^{(i)}, \tilde{u}_t^{(i)}) \\ &\le \sum_{i=1}^k (T_i + \tau_{i,1} + \tau_{i,2}) f(x_f, u_f) + 3 \sum_{i=1}^k 2^{-i+1} \\ &\le \sum_{i=1}^k (T_i + \tau_{i,1} + \tau_{i,2}) f(x_f, u_f) + 6. \end{aligned} \tag{2.95}$$

Theorem 2.1 and (2.95) imply that $(\{x_t\}_{t=0}^{\infty}, \{u_t\}_{t=0}^{\infty}) \in X(0, \infty)$ is (f)-good. Together with (P1) this implies that $\lim_{t\to\infty} \rho_E(x_t, x_f) = 0$. On the other hand in view of (2.77) and (2.93), $\limsup_{t\to\infty} \rho_E(x_t, x_f) > \epsilon$. The contradiction we have reached completes the proof of Lemma 2.53.

Completion of the Proof of Theorem 2.9. Assume that properties (P1) and (P2) hold. Let $\epsilon, M > 0$. By Lemma 2.53, there exists $\delta_0 > 0$ such that the following property holds:

(iv) for each integer $T_1 \geq 0$, each integer $T_2 \geq T_1 + 2b_f$, and each $(\{x_t\}_{t=T_1}^{T_2}, \{u_t\}_{t=T_1}^{T_2-1}) \in X(T_1, T_2)$ which satisfies

$$\rho_E(x_{T_i}, x_f) \leq \delta_0,\ i = 1, 2,$$

$$\sum_{t=T_1}^{T_2-1} f(x_t, u_t) \leq U^f(T_1, T_2, x_{T_1}, x_{T_2}) + \delta_0,$$

the inequality $\rho_E(x_t, x_f) \leq \epsilon$ holds for all $t = T_1, \ldots, T_2$.

By Theorem 2.1, there exists $S_0 > 0$ such that the following property holds:

(v) for each pair of integers $T_2 > T_1 \geq 0$ and each $(\{x_t\}_{t=T_1}^{T_2}, \{u_t\}_{t=T_1}^{T_2-1}) \in X(T_1, T_2)$, we have

$$\sum_{t=T_1}^{T_2-1} f(x_t, u_t) + S_0 \geq (T_2 - T_1) f(x_f, u_f).$$

In view of (P2), there exist $\delta \in (0, \delta_0)$ and an integer $L_0 > 0$ such that the following property holds:

(vi) for each $(\{x_t\}_{t=0}^{L_0}, \{u_t\}_{t=0}^{L_0-1}) \in X(0, L_0)$ which satisfies

$$\sum_{t=0}^{L_0-1} f(x_t, u_t) \leq \min\{U^f(0, L_0, x_0, x_{L_0}) + \delta,\ L_0 f(x_f, u_f) + 2S_0 + M\},$$

there exists an integer $t_0 \in [0, L_0]$ such that $\rho_E(x_{t_0}, x_f) \leq \delta_0$.

Set

$$L = L_0 + b_f. \tag{2.96}$$

Assume that $T_1 \geq 0$ and $T_2 \geq T_1 + 2L$ are integers and that

$$(\{x_t\}_{t=T_1}^{T_2}, \{u_t\}_{t=T_1}^{T_2-1}) \in X(T_1, T_2)$$

satisfies

$$\sum_{t=T_1}^{T_2-1} f(x_t, u_t) \leq \min\{\sigma^f(T_1, T_2) + M, U^f(T_1, T_2, x_{T_1}, x_{T_2}) + \delta\}. \tag{2.97}$$

Theorem 2.1, properties (iv) and (v), and (2.97) imply that for each pair of integers $Q_1, Q_2 \in [T_1, T_2]$ satisfying $Q_1 < Q_2$,

$$\sum_{t=Q_1}^{Q_2-1} f(x_t, u_t) = \sum_{t=T_1}^{T_2-1} f(x_t, u_t) - \sum\{f(x_t, u_t) : t \in \{T_1, \dots, Q_1\} \setminus \{Q_1\}\}$$

$$- \sum\{f(x_t, u_t) : t \in \{Q_2, \dots, T_2\} \setminus \{T_2\}\}$$

$$\leq (T_2 - T_1) f(x_f, u_f) + M - (Q_1 - T_1) f(x_f, u_f)$$

$$+ S_0 - (T_2 - Q_2) f(x_f, u_f) + S_0$$

$$= (Q_2 - Q_1) f(x_f, u_f) + M + 2S_0.$$

In particular,

$$\sum_{t=0}^{L_0-1} f(x_t, u_t) \leq L_0 f(x_f, u_f) + 2S_0 + M, \tag{2.98}$$

$$\sum_{t=T_2-L_0}^{T_2-1} f(x_t, u_t) \leq L_0 f(x_f, u_f) + 2S_0 + M. \tag{2.99}$$

By (2.97)–(2.99) and property (vi), there exist integers $\tau_1 \in [T_1, T_1 + L_0]$, $\tau_2 \in [T_2 - L_0, T_2]$ such that

$$\rho_E(x_{\tau_i}, x_f) \leq \delta_0, \ i = 1, 2. \tag{2.100}$$

If $\rho_E(x_{T_1}, x_f) \leq \delta$, then we may assume that $\tau_1 = T_1$, and if $\rho_E(x_{T_2}, x_f) \leq \delta$, then we may assume that $\tau_2 = T_2$. In view of (2.97),

$$\sum_{t=\tau_1}^{\tau_2-1} f(x_t, u_t) \leq U^f(\tau_1, \tau_2, x_{\tau_1}, x_{\tau_2}) + \delta. \tag{2.101}$$

It follows from (2.96) that

$$\tau_2 - \tau_1 \geq T_2 - T_1 - 2L_0 \geq 2L - 2L_0 \geq 2b_f. \tag{2.102}$$

Property (iv) and (2.100)–(2.102) imply that

$$\rho_E(x_t, x_f) \leq \epsilon, \ t = \tau_1, \dots, \tau_2.$$

Theorem 2.9 is proved.

2.10 An Auxiliary Result for Theorem 2.10

Lemma 2.54. *Assume that (P2) holds and that* $M, \epsilon > 0$. *Then there exists a natural number* L *such that for each* $(\{x_t\}_{t=0}^{L}, \{u_t\}_{t=0}^{L-1}) \in X(0, L)$ *which satisfies*

$$\sum_{t=0}^{L-1} f(x_t, u_t) \le Lf(x_f, u_f) + M,$$

the following inequality holds:

$$\min\{\rho_E(x_t, x_f) : t = 0, \dots, L\} \le \epsilon.$$

Proof. By Theorem 2.1, there exists $S_0 > 0$ such that the following property holds:

(i) for each pair of integers $T_2 > T_1 \ge 0$ and each $(\{x_t\}_{t=T_1}^{T_2}, \{u_t\}_{t=T_1}^{T_2-1}) \in X(T_1, T_2)$,

$$\sum_{t=T_1}^{T_2-1} f(x_t, u_t) + S_0 \ge (T_2 - T_1) f(x_f, u_f).$$

By (P2), there exist $\delta_0 \in (0, \epsilon)$ and an integer $L_0 > 0$ such that the following property holds:

(ii) for each $(\{x_t\}_{t=0}^{L_0}, \{u_t\}_{t=0}^{L_0-1}) \in X(0, L_0)$ which satisfies

$$\sum_{t=0}^{L_0-1} f(x_t, u_t) \le \min\{U^f(0, L_0, x_0, x_{L_0}) + \delta_0,$$

$$L_0 f(x_f, u_f) + 2S_0 + 2M + |f(x_f, u_f)|\},$$

we have

$$\min\{\rho_E(x_t, x_f) : t = 0, \dots, L_0\} \le \epsilon.$$

Choose an integer

$$q_0 > (M + S_0)\delta^{-1} \tag{2.103}$$

and set

$$L = q_0 L_0. \tag{2.104}$$

Let $(\{x_t\}_{t=0}^{L}, \{u_t\}_{t=0}^{L-1}) \in X(0, L)$ satisfy

$$\sum_{t=0}^{L-1} f(x_t, u_t) \le Lf(x_f, u_f) + M. \tag{2.105}$$

We show that

$$\min\{\rho_E(x_t, x_f) : t = 0, \dots, L\} \le \epsilon.$$

Assume the contrary. Then

$$\rho_E(x_t, x_f) > \epsilon, \ t = 0, \dots, L. \tag{2.106}$$

Let an integer $i \in \{0, \dots, q_0 - 1\}$. Property (i) and (2.105) imply that

$$\begin{aligned}\sum_{t=iL_0}^{(i+1)L_0-1} f(x_t, u_t) &= \sum_{t=0}^{L-1} f(x_t, u_t) - \sum\{f(x_t, u_t) : t \in \{0, \dots, iL_0\} \setminus \{iL_0\}\} \\ &\quad - \sum\{f(x_t, u_t) : t \in \{(i+1)L_0, \dots, L\} \setminus \{L\}\} \\ &\le Lf(x_f, u_f) + M - iL_0 f(x_f, u_f) + S_0 - (L - (i+1)L_0) f(x_f, u_f) + S_0 \\ &= L_0 f(x_f, u_f) + M + 2S_0. \end{aligned} \tag{2.107}$$

By (2.106), (2.107), and property (ii),

$$\sum_{t=iL_0}^{(i+1)L_0-1} f(x_t, u_t) > U^f(iL_0, (i+1)L_0, x_{iL_0}, x_{(i+1)L_0}) + \delta_0. \tag{2.108}$$

It follows from (2.104) and (2.108) that there exists $(\{y_t\}_{t=0}^{L}, \{v_t\}_{t=0}^{L-1}) \in X(0, L)$ such that $y_{iL_0} = x_{iL_0}$, $i = 0, \dots, q_0$ and for all $i = 0, \dots, q_0 - 1$,

$$\sum_{t=iL_0}^{(i+1)L_0-1} f(x_t, u_t) > \sum_{t=iL_0}^{(i+1)L_0-1} f(y_t, v_t) + \delta_0. \tag{2.109}$$

In view of (2.105), (2.109), and property (i),

$$Lf(x_f, u_f) + M \ge \sum_{t=0}^{L-1} f(x_t, u_t)$$

$$\geq \sum_{t=0}^{L-1} f(y_t, v_t) + q_0\delta \geq Lf(x_f, u_f) - S_0 + q_0\delta,$$

$$q_0 \leq (M + S_0)\delta^{-1}.$$

This contradicts (2.103). The contradiction we have reached proves Lemma 2.54.

2.11 Proof of Theorem 2.10

Assume that f has TP. In view of Theorem 2.9, (P1) and (P2) hold. We show that WTP holds. Let $\epsilon, M > 0$. By Theorem 2.1, there exists $S_0 > 0$ such that the following property holds:

(i) for each pair of integers $\tau_2 > \tau_1 \geq 0$ and each $(\{y_t\}_{t=\tau_1}^{\tau_2}, \{v_t\}_{t=\tau_1}^{\tau_2-1}) \in X(\tau_1, \tau_2)$,

$$\sum_{t=\tau_1}^{\tau_2-1} f(y_t, v_t) + S_0 \geq (\tau_2 - \tau_1) f(x_f, u_f).$$

TP implies that there exist $\delta \in (0, \epsilon)$ and an integer $L_0 > 0$ such that the following property holds:

(ii) for each integer $\tau_1 \geq 0$, each integer $\tau_2 \geq \tau_1 + 2L_0$, and each $(\{y_t\}_{t=\tau_1}^{\tau_2}, \{v_t\}_{t=\tau_1}^{\tau_2-1}) \in X(\tau_1, \tau_2)$ which satisfies

$$\sum_{t=\tau_1}^{\tau_2-1} f(y_t, v_t) \leq \min\{\sigma^f(\tau_1, \tau_2) + M + 3S_0, U^f(\tau_1, \tau_2, y_{\tau_1}, y_{\tau_2}) + \delta\},$$

we have

$$\rho_E(y_t, x_f) \leq \epsilon, \; t = \tau_1 + L_0, \ldots, \tau_2 - L_0.$$

Set

$$l = 2L_0 + 4. \tag{2.110}$$

Choose a natural number

$$Q \geq 6 + 6(M + S_0)\delta^{-1}. \tag{2.111}$$

Assume that $T_1 \geq 0$, $T_2 \geq T_1 + lQ$ are integers and $(\{x_t\}_{t=T_1}^{T_2}, \{u_t\}_{t=T_1}^{T_2-1}) \in X(T_1, T_2)$ satisfies

$$\sum_{t=T_1}^{T_2-1} f(x_t, u_t) \leq (T_2 - T_1) f(x_f, u_f) + M. \tag{2.112}$$

Property (i) and (2.112) imply that for each pair of integers $\tau_1, \tau_2 \in [T_1, T_2]$ satisfying $\tau_1 < \tau_2$,

$$\sum_{t=\tau_1}^{\tau_2-1} f(x_t, u_t) = \sum_{t=T_1}^{T_2-1} f(x_t, u_t) - \sum \{f(x_t, u_t) : t \in \{T_1, \ldots, \tau_1\} \setminus \{\tau_1\}\}$$

$$- \sum \{f(x_t, u_t) : t \in \{\tau_2, \ldots, T_2\} \setminus \{T_2\}\}$$

$$\leq M + (T_2 - T_1) f(x_f, u_f) - (\tau_1 - T_1) f(x_f, u_f) + S_0 - (T_2 - \tau_2) f(x_f, u_f) + S_0$$

$$\leq M + 2S_0 + (\tau_2 - \tau_1) f(x_f, u_f). \tag{2.113}$$

Set

$$t_0 = T_1. \tag{2.114}$$

If

$$\sum_{t=T_1}^{T_2-1} f(x_t, u_t) \leq U^f(T_1, T_2, x_{T_1}, x_{T_2}) + \delta,$$

then we set $t_1 = T_2$. Assume that

$$\sum_{t=T_1}^{T_2-1} f(x_t, u_t) > U^f(T_1, T_2, x_{T_1}, x_{T_2}) + \delta. \tag{2.115}$$

Set

$$t_1 = \min\{t \in \{T_1 + 1, \ldots, T_2\} : \sum_{s=T_1}^{t-1} f(x_s, u_s) > U^f(T_1, t, x_{T_1}, x_t) + \delta\}.$$

Assume that $k \geq 1$ is an integer and we defined a sequence of integers $\{t_i\}_{i=0}^k \subset [T_1, T_2]$ such that

$$t_0 < t_1 \cdots < t_k,$$

for each integer $i = 1, \dots, k$ if $t_i < T_2$, then

$$\sum_{s=t_{i-1}}^{t_i-1} f(x_s, u_s) - U^f(t_{i-1}, t_i, x_{t_{i-1}}, x_{t_i}) > \delta, \tag{2.116}$$

and if $t_i - 1 > t_{i-1}$, then

$$\sum_{s=t_{i-1}}^{t_i-2} f(x_s, u_s) - U^f(t_{i-1}, t_i - 1, x_{t_{i-1}}, x_{t_i-1}) \leq \delta. \tag{2.117}$$

(Note that in view of (2.115) and the choice of t_1, our assumption holds for $k = 1$.)

By (2.116), there exists $(\{y_t\}_{t=T_1}^{T_2}, \{v_t\}_{t=T_1}^{T_2-1}) \in X(T_1, T_2)$ such that

$$y_{t_i} = x_{t_i},\ i = 1, \dots, k, \tag{2.118}$$

$$\sum_{s=t_{i-1}}^{t_i-1} f(x_s, u_s) - \sum_{s=t_{i-1}}^{t_i-1} f(y_s, v_s) > \delta,\ i \in \{1, \dots, k\} \setminus \{k\}, \tag{2.119}$$

$$y_t = x_t,\ t \in \{t_k, \dots, T_2\},\ v_t = u_t,\ t \in \{t_k, \dots, T_2\} \setminus \{T_2\}. \tag{2.120}$$

Property (i), (2.112), and (2.118)–(2.120) imply that

$$(T_2 - T_1) f(x_f, u_f) + M \geq \sum_{t=T_1}^{T_2-1} f(x_t, u_t) \geq \sum_{t=T_1}^{T_2-1} f(y_t, v_t) + \delta(k-1)$$

$$\geq (T_2 - T_1) f(x_f, u_f) - S_0 + \delta(k-1),$$

$$k \leq 1 + \delta^{-1}(M + S_0). \tag{2.121}$$

If $t_k = T_2$, then the construction of the sequence is completed. Assume that $t_k < T_2$. If

$$\sum_{s=t_k}^{T_2-1} f(x_s, u_s) \leq U^f(t_k, T_2, x_{t_k}, x_{T_2}) + \delta,$$

then we set $t_{k+1} = T_2$, and the construction of the sequence is completed. Assume that

$$\sum_{s=t_k}^{T_2-1} f(x_s, u_s) > U^f(t_k, T_2, x_{t_k}, x_{T_2}) + \delta.$$

Set

$$t_{k+1} = \min\{t \in \{t_k+1, \dots, T_2\} : \sum_{s=t_k}^{t-1} f(x_s, u_s) > U^f(t_k, t, x_{t_k}, x_t) + \delta\}.$$

It is not difficult to see that the assumption made for k also holds for $k+1$. By induction we constructed a finite sequence $\{t_i\}_{i=0}^{q} \subset [T_1, T_2]$ such that $T_1 = t_0 < t_1 \cdots < t_q = T_2$, for each integer $i = 1, \dots, q$ if $t_i < T_2$, then (2.116) holds and if $t_i - 1 > t_{i-1}$, then (2.117) holds. In view of (2.121),

$$q \le 1 + \delta^{-1}(M + S_0). \tag{2.122}$$

Assume that

$$i \in \{0, \dots, q-1\},\ t_{i+1} - t_i \ge 2L_0 + 1. \tag{2.123}$$

Then (2.117) holds. Property (i) and (2.113) imply that

$$\begin{aligned}\sum_{s=t_i}^{t_{i+1}-1} f(x_s, u_s) &\le M + 2S_0 + (t_{i+1} - t_i - 1) f(x_f, u_f)\\ &\le M + 3S_0 + \sigma^f(t_i, t_{i+1} - 1). \end{aligned}\tag{2.124}$$

It follows from (2.117), (2.123), (2.124), and property (ii) that

$$\rho_E(x_t, x_f) \le \epsilon,\ t = t_i + L_0, \dots, t_{i+1} - 1 - L_0. \tag{2.125}$$

In view of (2.125),

$$\begin{aligned}&\{t \in \{T_1, \dots, T_2\} :\ \rho_E(x_t, x_f) > \epsilon\}\\ &\subset \cup\{\{t_i, \dots, t_i + L_0\} \cup \{t_{i+1} - 1 - L_0, \dots, t_{i+1}\} :\\ &\quad i \in \{0, \dots, q-1\} \text{ and } t_{i+1} - t_i \ge 2L_0 + 1\}\\ &\cup\{\{t_i, \dots, t_{i+1}\} :\ i \in \{0, \dots, q-1\} \text{ and } t_{i+1} - t_i < 2L_0 + 1\}.\end{aligned}$$

In view of (2.111) and (2.122), the number of intervals in the right-hand side of the relation above does not exceed $3q \le 3(1 + \delta^{-1}(M + S_0)) < Q$, and their maximal length does not exceed $2L_0 + 2 < l$. Thus WTP holds.

Assume that WTP holds. We show that TP holds. In view of Theorem 2.9, it is sufficient to show that (P1) and (P2) hold. It is clear that (P2) holds. Let us show that (P1) holds. Let $(\{x_t\}_{t=0}^{\infty}, \{u_t\}_{t=0}^{\infty}) \in X(0, \infty)$ be (f)-good. There exists $M_0 > 0$ such that for each integer $T \ge 1$,

$$|\sum_{t=0}^{T-1} f(x_t, u_t) - Tf(x_f, u_f)| \le M_0. \tag{2.126}$$

By (2.126), for each pair of integers $T_2 > T_1 \ge 0$,

$$\sum_{t=T_1}^{T_2-1} f(x_t, u_t) - (T_2 - T_1)f(x_f, u_f) = \sum_{t=0}^{T_2-1} f(x_t, u_t) - T_2 f(x_f, u_f)$$

$$-\sum\{f(x_t, u_t) : t \in \{0, \dots, T_1\} \setminus \{T_1\}\} - T_1 f(x_f, u_f) \le 2M_0. \tag{2.127}$$

Let $\epsilon > 0$ and natural numbers l, Q be as guaranteed by WTP with ϵ and $M = 2M_0$. Set

$$\Omega = \{t \in \{0, 1, \dots\} : \rho_E(x_t, x_f) > \epsilon\}.$$

We show that Ω is bounded. Assume that contrary. Then there exists a sequence of natural numbers $\{t_i\}_{i=1}^{\infty}$ such that for all integers $i \ge 1$,

$$t_{i+1} - t_i \ge 2l + 2, \ \rho_E(x_{t_i}, x_f) > \epsilon. \tag{2.128}$$

Let $k \ge 1 + Q$ be an integer. It follows from (2.128) that

$$t_k \ge t_{Q+1} \ge Q(2l + 2). \tag{2.129}$$

By (2.127), (2.129), WTP, and the choice of Q, l, there exist an integer $q \le Q$ and finite sequences of integers $\{a_i\}_{i=1}^{q}$, $\{b_i\}_{i=1}^{q} \subset [0, t_k]$ such that

$$0 \le b_i - a_i \le l, \ i = 1, \dots, q, \tag{2.130}$$

$$b_i \le a_{i+1} \text{ for all integers } i \text{ satisfying } 1 \le i < q,$$

$$\rho_E(x_t, x_f) \le \epsilon \text{ for all integers } t \in \{0, \dots, t_k\} \setminus \cup_{i=1}^{q}[a_i, b_i]. \tag{2.131}$$

In view of (2.128) and (2.130),

$$\{t_i\}_{i=1}^{k} \subset \cup_{i=1}^{q}[a_i, b_i]. \tag{2.132}$$

By (2.128), (2.130), and (2.132), $k \le q \le Q$, a contradiction. The contradiction we have reached proves that Ω is bounded. Since ϵ is an arbitrary positive number, we conclude that (P1) holds. This completes the proof of Theorem 2.10.

2.12 Proof of Proposition 2.12

Assume that (P2) does not hold. Then there exist ϵ, $M > 0$ such that for each natural number k, there exists $(\{x_t^{(k)}\}_{t=0}^k, \{u_t^{(k)}\}_{t=0}^{k-1}) \in X(0,k)$ which satisfies

$$\sum_{t=0}^{k-1} f(x_t^{(k)}, u_t^{(k)}) \leq kf(x_f, u_f) + M, \tag{2.133}$$

$$\rho_E(x_t^{(k)}, x_f) > \epsilon, \ t = 0, \ldots, k. \tag{2.134}$$

Let $S_0 > 0$ be as guaranteed by Assertion 1 of Theorem 2.1. By the choice of S_0 and (2.133), for each integer $k \geq 1$ and each pair of integers $\tau_1, \tau_2 \in [0,k]$ satisfying $\tau_1 < \tau_2$,

$$\sum_{t=\tau_1}^{\tau_2-1} f(x_t^{(k)}, u_t^{(k)}) - (\tau_2 - \tau_1) f(x_f, u_f) = \sum_{t=0}^{k-1} f(x_t^{(k)}, u_t^{(k)}) - kf(x_f, u_f)$$

$$-\sum \{f(x_t^{(k)}, u_t^{(k)}) : t \in \{0, \ldots, \tau_1\} \setminus \{\tau_1\}\} + \tau_1 f(x_f, u_f)$$

$$-\sum \{f(x_t^{(k)}, u_t^{(k)}) : t \in \{\tau_2, \ldots, k\} \setminus \{k\}\} + (k - \tau_2) f(x_f, u_f) \leq M + 2S_0. \tag{2.135}$$

By (2.1), (2.135), and LSC property, extracting subsequences, using the diagonalization process, and re-indexing, we may assume that for each integer $i \geq 1$, there exists $\lim_{k\to\infty} f(x_{i-1}^{(k)}, u_{i-1}^{(k)})$ and there exists $(\{y_{i-1}^{(i)}, y_i^{(i)}\}, \{v_{i-1}^{(i)}\}) \in X(i-1, i)$ such that

$$x_{i-1}^{(k)} \to y_{i-1}^{(i)}, \ x_i^{(k)} \to y_i^{(i)} \text{ as } k \to \infty, \tag{2.136}$$

$$f(y_{i-1}^{(i)}, v_{i-1}^{(i)}) \leq \lim_{k\to\infty} f(x_{i-1}^{(k)}, u_{i-1}^{(k)}). \tag{2.137}$$

In view of (2.136), there exists $(\{y_t\}_{t=0}^\infty, \{v_t\}_{t=0}^\infty) \in X(0,\infty)$ such that for each integer $i \geq 1$,

$$(y_{i-1}, v_{i-1}) = (y_{i-1}^{(i)}, v_{i-1}^{(i)}). \tag{2.138}$$

It follows from (2.137), (2.138), and (2.140) that for every integer $m \geq 1$

$$\sum_{i=0}^{m-1} f(y_i, v_i) = \sum_{i=1}^{m} f(y_{i-1}^{(i)}, v_{i-1}^{(i)}) = \sum_{i=1}^{m} \lim_{k\to\infty} f(x_{i-1}^{(k)}, u_{i-1}^{(k)})$$

$$= \lim_{k\to\infty} \sum_{t=1}^{m} f(x_{i-1}^{(k)}, u_{i-1}^{(k)}) \le M + 2S_0 + mf(x_f, u_f).$$

Together with Theorem 2.1, this implies that $(\{y_t\}_{t=0}^{\infty}, \{v_t\}_{t=0}^{\infty})$ is (f)-good. Property (P1) implies that there exists an integer $\tau_0 > 0$ such that

$$\rho_E(y_t, x_f) \le \epsilon/4 \text{ for all integers } t \ge \tau_0 - 1. \tag{2.139}$$

It follows from (2.136), (2.138), and (2.139) that

$$\rho_E(y_t^{(\tau_0)}, x_f) \le \epsilon/4,\ t = \tau_0 - 1, \tau_0. \tag{2.140}$$

By (2.136), there exists an integer $k_0 \ge 1 + \tau_0$ such that for all integers $k \ge k_0$,

$$\rho_E(y_t^{(\tau_0)}, x_t^{(k)}) \le \epsilon/4,\ t = \tau_0 - 1, \tau_0. \tag{2.141}$$

In view of (2.140) and (2.141), for all integers $k \ge k_0$ and all $t \in \{\tau_0 - 1, \tau_0\}$,

$$\rho_E(x_t^{(k)}, x_f) \le \rho_E(x_t^{(k)}, y_t^{(\tau_0)}) + \rho_E(y_t^{(\tau_0)}, x_f) \le \epsilon/2.$$

This contradicts (2.134). The contradiction we have reached proves that (P2) holds and completes the proof of Proposition 2.12.

2.13 Auxiliary Results for Theorems 2.13 and 2.14

Lemma 2.55. *Let $\epsilon > 0$. Then there exists $\delta > 0$ such that for each $(\{x_t\}_{t=0}^{b_f}, \{u_t\}_{t=0}^{b_f-1}) \in X(0, b_f)$ which satisfies $\rho_E(x_0, x_f) \le \delta$, $\rho_E(x_{b_f}, x_f) \le \delta$, the following inequality holds:*

$$\sum_{t=0}^{b_f-1} f(x_t, u_t) \ge b_f f(x_f, u_f) - \epsilon.$$

Proof. By (A2), there exists $\delta > 0$ such that the following property holds:

(i) for each $z \in \mathcal{A}$ satisfying $\rho_E(z, x_f) \le \delta$, there exist integers $\tau_1 \in (0, b_f]$, $\tau_2 \in (0, b_f]$ and $(\{\tilde{x}_t^{(1)}\}_{t=0}^{\tau_1}, \{\tilde{u}_t^{(1)}\}_{t=0}^{\tau_1-1}) \in X(0, \tau_1)$, $(\{\tilde{x}_t^{(2)}\}_{t=0}^{\tau_2}, \{\tilde{u}_t^{(2)}\}_{t=0}^{\tau_2-1}) \in X(0, \tau_2)$ such that

$$\tilde{x}_0^{(1)} = z\ , \tilde{x}_{\tau_1}^{(1)} = x_f,\ \sum_{t=0}^{\tau_1-1} f(\tilde{x}_t^{(1)}, \tilde{u}_t^{(1)}) \le \tau_1 f(x_f, u_f) + \epsilon/4,$$

$$\tilde{x}_0^{(2)} = x_f, \tilde{x}_{\tau_2}^{(2)} = z, \sum_{t=0}^{\tau_2-1} f(\tilde{x}_t^{(2)}, \tilde{u}_t^{(2)}) \leq \tau_2 f(x_f, u_f) + \epsilon/4.$$

Assume that $(\{x_t\}_{t=0}^{b_f}, \{u_t\}_{t=0}^{b_f-1}) \in X(0, b_f)$ and

$$\rho_E(x_0, x_f) \leq \delta, \ \rho_E(x_{b_f}, x_f) \leq \delta. \tag{2.142}$$

By property (i), there exists $(\{y_t\}_{t=0}^{\infty}, \{v_t\}_{t=0}^{\infty}) \in X(0, \infty)$ such that

$$y_{t+3b_f} = y_t, \ v_{t+3b_f} = v_t \text{ for all integers } t \geq 0,$$

$$y_0 = x_f, \ y_{b_f} = x_0, \ \sum_{t=0}^{b_f-1} f(y_t, v_t) \leq b_f f(x_f, u_f) + \epsilon/4,$$

$$y_t = x_{t-b_f}, \ t \in \{b_f, \dots, 2b_f\}, \ v_t = u_{t-b_f}, \ t \in \{b_f, \dots, 2b_f - 1\},$$

$$y_{3b_f} = x_f, \ \sum_{t=2b_f}^{3b_f-1} f(y_t, v_t) \leq b_f f(x_f, u_f) + \epsilon/4.$$

In view of the relations above and Theorem 2.1,

$$3b_f f(x_f, u_f) \leq \sum_{t=0}^{3b_f-1} f(y_t, v_t)$$

$$= \sum_{t=0}^{b_f-1} f(y_t, v_t) + \sum_{t=0}^{b_f-1} f(x_t, u_t) + \sum_{t=2b_f}^{3b_f-1} f(y_t, v_t),$$

$$\sum_{t=0}^{b_f-1} f(x_t, u_t) \geq 3b_f f(x_f, u_f) - 2b_f f(x_f, u_f) - \epsilon/2 \geq b_f f(x_f, u_f) - \epsilon.$$

Lemma 2.55 is proved.

Proposition 2.56. *Assume that f has LSC property and (P1) and that $z \in \cup\{\mathcal{A}_L : L = 1, 2, \dots\}$. Then there exists an (f)-good and (f)-minimal pair $(\{x_t^*\}_{t=0}^{\infty}, \{u_t^*\}_{t=0}^{\infty}) \in X(0, \infty)$ such that $x_0^* = z$.*

Proof. Theorem 2.1 implies that f possesses (P2) and TP. There exists an integer $L_0 > 0$ such that

$$z \in \mathcal{A}_{L_0}. \tag{2.143}$$

It follows from Theorem 2.1 that there exists $S_0 > 0$ such that for each pair of integers $T_2 > T_1 \geq 0$ and each $(\{x_t\}_{t=T_1}^{T_2}, \{u_t\}_{t=T_1}^{T_2-1}) \in X(T_1, T_2)$,

$$\sum_{t=T_1}^{T_2-1} f(x_t, u_t) + S_0 \geq (T_2 - T_1) f(x_f, u_f). \tag{2.144}$$

LSC property and (2.143) imply that for each integer $k \geq L_0$ there exists $(\{x_t^{(k)}\}_{t=0}^{k}, \{u_t^{(k)}\}_{t=0}^{k-1}) \in X(0, k)$ satisfying

$$x_0^{(k)} = z, \tag{2.145}$$

$$\sum_{t=0}^{k-1} f(x_t^{(k)}, u_t^{(k)}) = \sigma^f(0, k, z). \tag{2.146}$$

In view of (2.143), for each integer $k \geq L_0$,

$$\sigma^f(0, k, z) \leq L_0 + f(x_f, u_f)(k - L_0). \tag{2.147}$$

By (2.144), (2.146), and (2.147), for each integer $k \geq L_0$ and each pair of integers $T_1, T_2 \in [0, k]$ satisfying $T_1 < T_2$,

$$\sum_{t=T_1}^{T_2-1} f(x_t^{(k)}, u_t^{(k)}) = \sum_{t=0}^{k-1} f(x_t^{(k)}, u_t^{(k)})$$

$$-\sum\{f(x_t^{(k)}, u_t^{(k)}) : t \in \{0, \dots, T_1\} \setminus \{T_1\}\}$$

$$-\sum\{f(x_t^{(k)}, u_t^{(k)}) : t \in \{T_2, \dots, k\} \setminus \{k\}\}$$

$$\leq L + f(x_f, u_f)(k - L_0) - T_1 f(x_f, u_f) + S_0 - (k - T_2) f(x_f, u_f) + S_0$$

$$\leq (T_2 - T_1) f(x_f, u_f) + L_0(1 - f(x_f, u_f)) + 2S_0. \tag{2.148}$$

By (2.1), (2.148), and LSC property, extracting subsequences, using the diagonalization process, and re-indexing, we obtain that there exists a strictly increasing sequence of natural numbers $\{k_p\}_{p=1}^{\infty}$ such that $k_1 \geq L_0$ and for each integer $i \geq 0$, there exists $\lim_{p\to\infty} f(x_i^{(k_p)}, u_i^{(k_p)})$ and there exists $(\{y_i^{(i)}, y_{i+1}^{(i)}\}, \{v_i^{(i)}\}) \in X(i, i+1)$ such that

$$x_i^{(k_p)} \to y_i^{(i)},\ x_{i+1}^{(k_p)} \to y_{i+1}^{(i)} \text{ as } p \to \infty, \tag{2.149}$$

$$f(y_i^{(i)}, v_i^{(i)}) \le \lim_{p\to\infty} f(x_i^{(k_p)}, u_i^{(k_p)}). \tag{2.150}$$

In view of (2.149), there exists $(\{x_t^*\}_{t=0}^\infty, \{u_t^*\}_{t=0}^\infty) \in X(0,\infty)$ such that for each integer $i \ge 1$,

$$(x_i^*, u_i^*) = (y_i^{(i)}, v_i^{(i)}). \tag{2.151}$$

It follows from (2.148), (2.150), and (2.151) that for every integer $q \ge 1$

$$\sum_{i=0}^{q-1} f(x_t^*, u_t^*) = \lim_{p\to\infty} \sum_{t=0}^{q-1} f(x_t^{(k_p)}, u_t^{(k_p)}) \le qf(x_f,u_f) + L_0(1-f(x_f,u_f)) + 2S_0. \tag{2.152}$$

Theorem 2.1 and (2.152) imply that $(\{x_t^*\}_{t=0}^\infty, \{u_t^*\}_{t=0}^\infty)$ is (f)-good. In view of (2.145), (2.149), and (2.151),

$$x_0^* = z. \tag{2.153}$$

In order to complete the proof of the proposition, it is sufficient to show that $(\{x_t^*\}_{t=0}^\infty, \{u_t^*\}_{t=0}^\infty)$ is (f)-minimal. Assume the contrary. Then there exist $\Delta > 0$, an integer $\tau_0 \ge 1$, and $(\{y_t\}_{t=0}^{\tau_0}, \{v_t\}_{t=0}^{\tau_0-1}) \in X(0,\tau_0)$ such that

$$y_0 = x_0^*,\ y_{\tau_0} = x_{\tau_0}^*, \tag{2.154}$$

$$\sum_{t=0}^{\tau_0-1} f(x_t^*, u_t^*) > \sum_{t=0}^{\tau_0-1} f(y_t, v_t) + 2\Delta. \tag{2.155}$$

By (A2) and Lemma 2.55, there exists $\delta > 0$ such that the following properties hold:

(ii) for each $\xi \in \mathcal{A}$ satisfying $\rho_E(\xi, x_f) \le \delta$, there exist integers $\tau_1, \tau_2 \in (0, b_f]$ and

$$(\{\tilde{x}_t^{(1)}\}_{t=0}^{\tau_1}, \{\tilde{u}_t^{(1)}\}_{t=0}^{\tau_1-1}) \in X(0,\tau_1),\ (\{\tilde{x}_t^{(2)}\}_{t=0}^{\tau_2}, \{\tilde{u}_t^{(2)}\}_{t=0}^{\tau_2-1}) \in X(0,\tau_2)$$

such that

$$\tilde{x}_0^{(1)} = \xi\ ,\tilde{x}_{\tau_1}^{(1)} = x_f,\ \sum_{t=0}^{\tau_1-1} f(\tilde{x}_t^{(1)}, \tilde{u}_t^{(1)}) \le \tau_1 f(x_f,u_f) + \Delta/8,$$

$$\tilde{x}_0^{(2)} = x_f\ ,\tilde{x}_{\tau_2}^{(2)} = \xi,\ \sum_{t=0}^{\tau_2-1} f(\tilde{x}_t^{(2)}, \tilde{u}_t^{(2)}) \le \tau_2 f(x_f,u_f) + \Delta/8;$$

(iii) for each $(\{x_t\}_{t=0}^{b_f}, \{u_t\}_{t=0}^{b_f-1}) \in X(0, b_f)$ which satisfies $\rho_E(x_0, x_f) \leq \delta$, $\rho_E(x_{b_f}, x_f) \leq \delta$, we have

$$\sum_{t=0}^{b_f-1} f(x_t, u_t) \geq b_f f(x_f, u_f) - \Delta/8.$$

Theorems 2.1 and 2.11, TP, (P1), LSC property, (2.146), and (2.147) imply that there exists an integer $L_1 > L_0$ such that for each integer $k \geq L_0 + 2L_1$,

$$\rho_E(x_t^{(k)}, x_f) \leq \delta,\ t \in \{L_1, \ldots, k - L_1\}. \tag{2.156}$$

In view of (2.149) and (2.151),

$$\rho_E(x_t^*, x_f) \leq \delta \text{ for all integers } t \geq L_1. \tag{2.157}$$

By (2.150) and (2.151), there exists a natural number p_0 such that

$$k_{p_0} > L_0 + 2L_1 + 2\tau_0 + 2 + 2b_f, \tag{2.158}$$

$$\sum_{t=0}^{\tau_0+L_1-1} f(x_t^*, u_t^*) \leq \sum_{t=0}^{\tau_0+L_1-1} f(x_t^{(k_{p_0})}, u_t^{(k_{p_0})}) + \Delta/2. \tag{2.159}$$

Property (ii), (2.154), and (2.157) imply that there is $(\{x_t\}_{t=0}^{k_{p_0}}, \{u_t\}_{t=0}^{k_{p_0}-1}) \in X(0, k_{p_0})$ such that

$$x_t = y_t,\ t = 0, \ldots, \tau_0,\ u_t = v_t,\ t = 0, \ldots, \tau_0 - 1, \tag{2.160}$$

$$x_t = x_t^*,\ t = \tau_0, \ldots, \tau_0 + L_1,\ u_t = u_t^*,\ t = \tau_0, \ldots, \tau_0 + L_1 - 1, \tag{2.161}$$

$$x_{\tau_0+L_1+b_f} = x_f,\quad \sum_{t=\tau_0+L_1}^{\tau_0+L_1+b_f-1} f(x_t, u_t) \leq b_f f(x_f, u_f) + \Delta/8, \tag{2.162}$$

$$x_t = x_t^{(k_{p_0})},\ t = \tau_0 + L_1 + 2b_f, \ldots, k_{p_0},$$

$$u_t = u_t^{(k_{p_0})},\ t = \tau_0 + L_1 + 2b_f, \ldots, k_{p_0} - 1, \tag{2.163}$$

$$\sum_{t=\tau_0+L_1+b_f}^{\tau_0+L_1+2b_f-1} f(x_t, u_t) \leq b_f f(x_f, u_f) + \Delta/8. \tag{2.164}$$

By (2.145), (2.146), (2.153), (2.154), and (2.160),

$$\sum_{t=0}^{k_{p_0}-1} f(x_t, u_t) \geq \sum_{t=0}^{k_{p_0}-1} f(x_t^{(k_{p_0})}, u_t^{(k_{p_0})}). \tag{2.165}$$

It follows from (2.155), (2.156), (2.158)–(2.160), (2.163), (2.165), and property (iii) that

$$\begin{aligned}
0 &\leq \sum_{t=0}^{k_{p_0}-1} f(x_t, u_t) - \sum_{t=0}^{k_{p_0}-1} f(x_t^{(k_{p_0})}, u_t^{(k_{p_0})}) \\
&= \sum_{t=0}^{\tau_0+L_1+2b_f-1} f(x_t, u_t) - \sum_{t=0}^{\tau_0+L_1+2b_f-1} f(x_t^{(k_{p_0})}, u_t^{(k_{p_0})}) \\
&\leq \sum_{t=0}^{\tau_0-1} f(y_t, v_t) + 2b_f f(x_f, u_f) + \Delta/4 + \sum_{t=\tau_0}^{\tau_0+L_1-1} f(x_t^*, u_t^*) \\
&\quad - \sum_{t=0}^{\tau_0+L_1-1} f(x_t^{(k_{p_0})}, u_t^{(k_{p_0})}) - \sum_{t=\tau_0+L_1}^{\tau_0+L_1+b_f-1} f(x_t^{(k_{p_0})}, u_t^{(k_{p_0})}) \\
&\quad - \sum_{t=\tau_0+L_1+b_f}^{\tau_0+L_1+2b_f-1} f(x_t^{(k_{p_0})}, u_t^{(k_{p_0})}) \\
&\leq \sum_{t=0}^{\tau_0-1} f(y_t, v_t) + \sum_{t=\tau_0}^{\tau_0+L_1-1} f(x_t^*, u_t^*) + 2b_f f(x_f, u_f) + \Delta/4 \\
&\quad - \sum_{t=0}^{\tau_0+L_1-1} f(x_t^{(k_{p_0})}, u_t^{(k_{p_0})}) - 2b_f f(x_f, u_f) + \Delta/4 \\
&\leq \sum_{t=0}^{\tau_0-1} f(x_t^*, u_t^*) - 2\Delta + \sum_{t=\tau_0}^{\tau_0+L_1-1} f(x_t^*, u_t^*) \\
&\quad - \sum_{t=0}^{\tau_0+L_1-1} f(x_t^{(k_{p_0})}, u_t^{(k_{p_0})}) + \Delta/2 \leq -\Delta,
\end{aligned}$$

a contradiction. The contradiction we have reached completes the proof of Proposition 2.56.

2.14 Proofs of Theorems 2.13 and 2.14

Proof of Theorem 2.14. Clearly, (i) implies (ii). In view of Theorem 2.1, (ii) implies (iii). By (P1), (iii) implies (iv). Evidently (iv) implies (v). We show that (v) implies (iii). Assume that $(\{x_t^*\}_{t=0}^{\infty}, \{u_t^*\}_{t=0}^{\infty}) \in X(0, \infty)$ is (f)-minimal and satisfies

$$\liminf_{t\to\infty} \rho_E(x_f, x_t^*) = 0. \tag{2.166}$$

(P1) implies that

$$\lim_{t\to\infty} \rho_E(\tilde{x}_t, x_f) = 0. \tag{2.167}$$

In view of Theorem 2.1, there exists $S_0 > 0$ such that for all natural numbers T,

$$|\sum_{t=0}^{T-1} f(\tilde{x}_t, \tilde{u}_t) - Tf(x_f, u_f)| \leq S_0. \tag{2.168}$$

By (A2), there exists $\delta > 0$ such that the following property holds:

for each $z \in \mathcal{A}$ satisfying $\rho_E(z, x_f) \leq \delta$, there exist integers $\tau_1, \tau_2 \in (0, b_f]$ and $(\{x_t^{(i)}\}_{t=0}^{\tau_i}, \{u_t^{(i)}\}_{t=0}^{\tau_i-1}) \in X(0, \tau_i)$, $i = 1, 2$ which satisfy

$$x_0^{(1)} = z,\ x_{\tau_1}^{(1)} = x_f,\ x_0^{(2)} = x_f,\ x_{\tau_2}^{(2)} = z,$$

$$\sum_{t=0}^{\tau_i-1} f(x_t^{(i)}, u_t^{(i)}) \leq \tau_i f(x_f, u_f) + 1,\ i = 1, 2.$$

In view of (2.166) and (2.167), there exists an increasing sequence of natural numbers $\{t_k\}_{k=1}^{\infty}$ such that

$$\rho_E(\tilde{x}_t, x_f) \leq \delta \text{ for all integers } t \geq t_0, \tag{2.169}$$

$$\lim_{k\to\infty} t_k = \infty,$$

$$\rho_E(x_{t_k+2b_f}^*, x_f) \leq \delta,\ k = 1, 2, \dots. \tag{2.170}$$

Let $k \geq 1$ be an integer. By (A2), the choice of δ, (2.169), and (2.170), there exists $(\{y_t\}_{t=0}^{t_k+2b_f}, \{v_t\}_{t=0}^{t_k+2b_f-1}) \in X(0, t_k + 2b_f)$ such that

$$y_t = \tilde{x}_t,\ t = 0, \dots, t_k,\ v_t = \tilde{u}_t,\ t = 0, \dots, t_k - 1 \tag{2.171}$$

$$y_{t_k+b_f} = x_f, \tag{2.172}$$

$$\sum_{t=t_k}^{t_k+b_f-1} f(y_t, v_t) \le b_f f(x_f, u_f) + 1, \tag{2.173}$$

$$y_{t_k+2b_f} = x^*_{t_k+2b_f}, \tag{2.174}$$

$$\sum_{t=t_k+b_f}^{t_k+2b_f-1} f(y_t, v_t) \le b_f f(x_f, u_f) + 1. \tag{2.175}$$

Property (v), (2.168), and (2.171)–(2.175) imply that

$$\sum_{t=0}^{t_k+2b_f-1} f(x_t^*, u_t^*) \le \sum_{t=0}^{t_k+2b_f-1} f(y_t, v_t) \le \sum_{t=0}^{t_k-1} f(\tilde{x}_t, \tilde{u}_t) + 2b_f f(x_f, u_f) + 2$$

$$\le t_k f(x_f, u_f) + S_0 + 2b_f f(x_f, u_f) + 2 = (t_k + 2b_f) f(x_f, u_f) + 2 + S_0.$$

Together with Theorem 2.1, this implies that $(\{x_t^*\}_{t=0}^\infty, \{u_t^*\}_{t=0}^\infty)$ is (f)-good and (iii) holds.

We show that (iii) implies (i). Assume that $(\{x_t^*\}_{t=0}^\infty, \{u_t^*\}_{t=0}^\infty) \in X(0, \infty)$ is (f)-minimal and (f)-good. (P1) implies that

$$\lim_{t\to\infty} \rho_E(x_t^*, x_f) = 0. \tag{2.176}$$

In view of Theorem 2.1, there exists $S_0 > 0$ such that

$$|\sum_{t=0}^{T-1} f(x_t^*, u_t^*) - T f(x_f, u_f)| \le S_0, \ T \in \{1, 2, \dots\}.$$

Let $(\{x_t\}_{t=0}^\infty, \{u_t\}_{t=0}^\infty) \in X(0, \infty)$ satisfy

$$x_0 = x_0^*. \tag{2.177}$$

We show that

$$\limsup_{T\to\infty} [\sum_{t=0}^{T-1} f(x_t^*, u_t^*) - \sum_{t=0}^{T-1} f(x_t, u_t)] \le 0. \tag{2.178}$$

In view of Theorem 2.1 and (iii), we may assume that $(\{x_t\}_{t=0}^\infty, \{u_t\}_{t=0}^\infty)$ is (f)-good. (P1) implies that

$$\lim_{t\to\infty} \rho_E(x_t, x_f) = 0. \tag{2.179}$$

Let $\epsilon > 0$. By (A2) and Lemma 2.55, there exists $\delta \in (0, \epsilon)$ such that the following property holds:

(a) for each $z \in \mathcal{A}$ satisfying $\rho_E(z, x_f) \le \delta$, there exist integers $\tau_1, \tau_2 \in (0, b_f]$ and $(\{x_t^{(i)}\}_{t=0}^{\tau_i}, \{u_t^{(i)}\}_{t=0}^{\tau_i-1}) \in X(0, \tau_i)$, $i = 1, 2$ satisfying

$$x_0^{(1)} = z\ , x_{\tau_1}^{(1)} = x_f,\ x_0^{(2)} = x_f\ , x_{\tau_2}^{(2)} = z,$$

$$\sum_{t=0}^{\tau_i-1} f(x_t^{(i)}, u_t^{(i)}) \le \tau_i f(x_f, u_f) + \epsilon/8,\ i = 1, 2;$$

(b) for each $(\{y_t\}_{t=0}^{b_f}, \{v_t\}_{t=0}^{b_f-1}) \in X(0, b_f)$ which satisfies $\rho_E(y_0, x_f) \le \delta$, $\rho_E(y_{b_f}, x_f) \le \delta$, we have

$$\sum_{t=0}^{b_f-1} f(y_t, v_t) \ge b_f f(x_f, u_f) - \epsilon/8.$$

It follows from (2.176) and (2.179) that there exists an integer $\tau_0 > 0$ such that

$$\rho_E(x_t, x_f) \le \delta,\ \rho_E(x_t^*, x_f) \le \delta \text{ for all integers } t \ge \tau_0. \tag{2.180}$$

Let $T \ge \tau_0$ be an integer. Property (a) and (2.180) imply that there exists $(\{y_t\}_{t=0}^{T+2b_f}, \{v_t\}_{t=0}^{T+2b_f-1}) \in X(0, T + 2b_f)$ which satisfies

$$y_t = x_t,\ t = 0, \dots, T,\ v_t = u_t,\ t = 0, \dots, T-1,\ y_{T+b_f} = x_f, \tag{2.181}$$

$$\sum_{t=T}^{T+b_f-1} f(y_t, v_t) \le b_f f(x_f, u_f) + \epsilon/8, \tag{2.182}$$

$$y_{T+2b_f} = x^*_{T+2b_f},\ \sum_{t=T+b_f}^{T+2b_f-1} f(y_t, v_t) \le b_f f(x_f, u_f) + \epsilon/8. \tag{2.183}$$

By property (b) and (2.180),

$$\sum_{t=T+b_f}^{T+2b_f-1} f(x_t^*, u_t^*),\ \sum_{t=T+b_f}^{T+2b_f-1} f(x_t, u_t) \ge b_f f(x_f, u_f) - \epsilon/8. \tag{2.184}$$

It follows from property (iii), (2.177), and (2.181)–(2.184) that

$$\sum_{t=0}^{T-1} f(x_t^*, u_t^*) + 2b_f f(x_f, u_f) - \epsilon/4 \le \sum_{t=0}^{T+2b_f-1} f(x_t^*, u_t^*)$$

$$\le \sum_{t=0}^{T+2b_f-1} f(y_t, v_t) \le \sum_{t=0}^{T-1} f(x_t, u_t) + 2b_f f(x_f, u_f) + \epsilon/4$$

and

$$\sum_{t=0}^{T-1} f(x_t^*, u_t^*) \le \sum_{t=0}^{T-1} f(x_t, u_t) + \epsilon/2 \text{ for all integers } T \ge \tau_0.$$

Since ϵ is any positive number, we conclude that (2.178) holds and that the pair $(\{x_t^*\}_{t=0}^{\infty}, \{u_t^*\}_{t=0}^{\infty})$ is (f)-overtaking optimal. Thus (iii) implies (i) and Theorem 2.14 is proved.

Theorem 2.14 and Proposition 2.56 imply Theorem 2.13.

2.15 Proofs of Theorems 2.15–2.17

Proof of Theorem 2.15. Theorem 2.11 implies TP. Lemma 2.53, TP, (P1), and (A2) imply that there exists $\delta \in (0, \epsilon)$ such that the following properties hold:

(a) for each $z \in \mathcal{A}$ satisfying $\rho_E(z, x_f) \le \delta$, there exist an integer $\tau_0 \in (0, b_f]$ and $(\{x_t^{(0)}\}_{t=0}^{\tau_0}, \{u_t^{(0)}\}_{t=0}^{\tau_0-1}) \in X(0, \tau_0)$ such that

$$x_0^{(0)} = z\ , x_{\tau_0}^{(0)} = x_f, \ \sum_{t=0}^{\tau_0-1} f(x_t^{(0)}, u_t^{(0)}) \le \tau_0 f(x_f, u_f) + 1;$$

(b) for each integer $T_1 \ge 0$, each integer $T_2 \ge T_1 + 2b_f$, and each $(\{y_t\}_{t=T_1}^{T_2}, \{v_t\}_{t=T_1}^{T_2-1}) \in X(T_1, T_2)$ which satisfies

$$\rho_E(y_{T_i}, x_f) \le \delta, \ i = 1, 2,$$

$$\sum_{t=T_1}^{T_2-1} f(y_t, v_t) \le U^f(T_1, T_2, y_{T_1}, y_{T_2}) + \delta,$$

we have

$$\rho_E(y_t, x_f) \le \epsilon, \ t = T_1, \dots, T_2.$$

Assertion (i) follows from (a) and Theorem 2.13. We prove Assertion (ii). Assume that $(\{x_t\}_{t=0}^{\infty}, \{u_t\}_{t=0}^{\infty}) \in X(0, \infty)$ is (f)-overtaking optimal and that

$$\rho_E(x_0, x_f) \le \delta. \tag{2.185}$$

Theorem 2.14 and (P1) imply that

$$\lim_{t\to\infty} \rho_E(x_t, x_f) = 0.$$

There exists an integer $t_0 > 2b_f$ such that for each integer $t \ge t_0$,

$$\rho_E(x_t, x_f) \le \delta. \tag{2.186}$$

Let $T \ge t_0$ be an integer. It follows from property (b), (2.185), and (2.186) that

$$\rho_E(x_t, x_f) \le \epsilon,\ t = 0, \dots, T.$$

Therefore Assertion (ii) holds and Theorem 2.15 is proved.

Proof of Theorem 2.16. In view of Theorem 2.11, TP holds. Together with Proposition 2.6, this implies that there exists an integer $\tau_0 > 0$ such that the following property holds:

(c) for each integer $T \ge 2\tau_0$ and each $(\{x_t\}_{t=0}^{T}, \{u_t\}_{t=0}^{T-1}) \in X(0, T)$ which satisfies

$$\sum_{t=0}^{T-1} f(x_t, u_t) = U^f(0, T, x_0, x_T),$$

$$\sum_{t=0}^{T-1} f(x_t, u_t) \le Tf(x_f, x_f) + L(1 + |f(x_f, u_f)|) + 2,$$

we have

$$\rho_E(x_t, x_f) \le \epsilon,\ t = \tau_0, \dots, T - \tau_0.$$

Let $(\{x_t\}_{t=0}^{\infty}, \{u_t\}_{t=0}^{\infty}) \in X(0, \infty)$ be (f)-overtaking optimal and

$$x_0 \in \mathcal{A}_L. \tag{2.187}$$

Theorem 2.13 and (2.187) imply that

$$\lim_{t\to\infty} \rho_E(x_t, x_f) = 0. \tag{2.188}$$

In view of (2.188), there exists an integer $s_0 \in (0, L]$ and $(\{y_t\}_{t=0}^{\infty}, \{v_t\}_{t=0}^{\infty}) \in X(0, \infty)$ such that

$$y_0 = x_0,\ y_{s_0} = x_f,\ \sum_{t=0}^{s_0-1} f(y_t, v_t) \le L, \tag{2.189}$$

$$y_t = x_f,\ u_t = u_f \text{ for all integers } t \ge s_0. \tag{2.190}$$

By (2.189) and (2.190), for all sufficiently large integers $T > L + 2\tau_0$,

$$\sum_{t=0}^{T-1} f(x_t, u_t) \le \sum_{t=0}^{T-1} f(y_t, v_t) + 1 \le 1 + L + \sum_{t=s_0}^{T-1} f(y_t, v_t)$$
$$\le 1 + L + (T - s_0) f(x_f, u_f) \le T f(x_f, u_f) + 1 + L(1 + |f(x_f, u_f)|)$$

and in view of property (c),

$$\rho_E(x_t, x_f) \le \epsilon,\ t = \tau_0, \ldots, T - \tau_0.$$

Since the relation above holds for all sufficiently large natural numbers T, we conclude that

$$\rho_E(x_t, x_f) \le \epsilon \text{ for all integers } t \ge \tau_0.$$

This completes the proof of Theorem 2.16.

Proof of Theorem 2.17. We may assume without loss of generality that $t_1 = 0$. Clearly, there exists $(\{y_t\}_{t=0}^{\infty}, \{v_t\}_{t=0}^{\infty}) \in X(0, \infty)$ such that

$$y_t = x_t,\ t = 0, \ldots, t_2,\ v_t = u_t,\ t = 0, \ldots, t_2 - 1,$$

$$y_{t+t_2} = y_t,\ v_{t+t_2} = v_t \text{ for all integers } t \ge 0. \tag{2.191}$$

Theorem 2.1 and (2.191) imply that

$$\sum_{t=0}^{t_2-1} f(y_t, v_t) \ge t_2 f(x_f, u_f). \tag{2.192}$$

We show that

$$\sum_{t=0}^{t_2-1} f(y_t, v_t) = t_2 f(x_f, u_f). \tag{2.193}$$

Assume the contrary. Then in view of (2.191) and (2.192),

$$\sum_{t=0}^{t_2-1} f(x_t, u_t) = \sum_{t=0}^{t_2-1} f(y_t, v_t) > t_2 f(x_f, u_f). \tag{2.194}$$

Set

$$\Delta = \sum_{t=0}^{t_2-1} f(x_t, u_t) - t_2 f(x_f, u_f). \tag{2.195}$$

By (A2) and Lemma 2.55, there exists $\delta \in (0, 1)$ such that the following properties hold:

(d) for each $z \in \mathcal{A}$ satisfying $\rho_E(z, x_f) \le \delta$, there exist integers $\tau_1, \tau_2 \in (0, b_f]$ and $(\{x_t^{(i)}\}_{t=0}^{\tau_i}, \{u_t^{(i)}\}_{t=0}^{\tau_i-1}) \in X(0, \tau_i)$, $i = 1, 2$ satisfying

$$x_0^{(1)} = z\,,\, x_{\tau_1}^{(1)} = x_f,\ x_0^{(2)} = x_f\,,\, x_{\tau_2}^{(2)} = z,$$

$$\sum_{t=0}^{\tau_i-1} f(x_t^{(i)}, u_t^{(i)}) \le \tau_i f(x_f, u_f) + \Delta/8,\ i = 1, 2;$$

(e) for each $(\{z_t\}_{t=0}^{b_f}, \{\xi_t\}_{t=0}^{b_f-1}) \in X(0, b_f)$ which satisfies $\rho_E(z_0, x_f) \le \delta$, $\rho_E(z_{b_f}, x_f) \le \delta$, we have

$$\sum_{t=0}^{b_f-1} f(z_t, \xi_t) \ge b_f f(x_f, u_f) - \Delta/8.$$

It follows from Theorem 2.14 and (P1) that there exists an integer $T_0 > 0$ such that

$$\rho_E(x_t, x_f) \le \delta \text{ for all integers } t \ge T_0. \tag{2.196}$$

Choose an integer

$$T_1 > 2T_0 + 2t_2 + 2b_f + 4. \tag{2.197}$$

By (d) and (2.196), there exists $(\{\widehat{x}_t\}_{t=0}^{\infty}, \{\widehat{u}_t\}_{t=0}^{\infty}) \in X(0, \infty)$ such that

$$\widehat{x}_t = x_{t+t_2},\ t = 0, \dots, T_0 + t_2 + 2, \tag{2.198}$$

$$\widehat{u}_t = u_{t+t_2},\ t = 0, \dots, T_0 + t_2 + 1, \tag{2.199}$$

$$\sum_{t=T_0+t_2+2}^{T_0+t_2+b_f+1} f(\widehat{x}_t, \widehat{u}_t) \le b_f f(x_f, u_f) + \Delta/8, \tag{2.200}$$

$$\widehat{x}_{T_0+t_2+b_f+2} = x_f, \tag{2.201}$$

$$\widehat{x}_t = x_f,\ t = T_0 + t_2 + 2 + b_f, \dots, T_0 + 2t_2 + 2 + b_f, \tag{2.202}$$

$$\widehat{u}_t = u_f,\ t = T_0 + t_2 + 2 + b_f, \dots, T_0 + 2t_2 + 1 + b_f, \tag{2.203}$$

$$\widehat{x}_{T_0+2t_2+2b_f+2} = x_{T_0+2t_2+2b_f+2}, \tag{2.204}$$

$$\sum_{t=T_0+2t_2+2+b_f}^{T_0+2t_2+2b_f+1} f(\widehat{x}_t, \widehat{u}_t) \le b_f f(x_f, u_f) + \Delta/8, \tag{2.205}$$

$$\widehat{x}_t = x_t,\ \widehat{u}_t = u_t \text{ for all integers } t \ge T_0 + 2t_2 + 2 + 2b_f. \tag{2.206}$$

In view of (2.198), (2.206), and the relation $x_0 = x_{t_2}$,

$$\sum_{t=0}^{T_0+2t_2+2b_f+1} f(x_t, u_t) \le \sum_{t=0}^{T_0+2t_2+2b_f+1} f(\widehat{x}_t, \widehat{u}_t). \tag{2.207}$$

It follows from (2.198)–(2.200), (2.202), (2.203), and (2.205) that

$$\begin{aligned}\sum_{t=0}^{T_0+2t_2+2b_f+1} f(\widehat{x}_t, \widehat{u}_t) &\le \sum_{t=t_2}^{T_0+2t_2+1} f(x_t, u_t) + b_f f(x_f, u_f) + \Delta/8 \\ &\quad + t_2 f(x_f, u_f) + b_f f(x_f, u_f) + \Delta/8 \\ &= \sum_{t=t_2}^{T_0+2t_2+1} f(x_t, u_t) + f(x_f, u_f)(t_2 + 2b_f) + \Delta/4. \end{aligned} \tag{2.208}$$

Property (e), (2.195), and (2.196) imply that

$$\begin{aligned}\sum_{t=0}^{T_0+2t_2+2b_f+1} f(x_t, u_t) &= \sum_{t=0}^{t_2-1} f(x_t, u_t) + \sum_{t=t_2}^{T_0+2t_2+1} f(x_t, u_t) \\ &\quad + \sum_{t=T_0+2t_2+2}^{T_0+2t_2+b_f+1} f(x_t, u_t) + \sum_{t=T_0+2t_2+2+b_f}^{T_0+2t_2+2b_f+1} f(x_t, u_t)\end{aligned}$$

$$\geq t_2 f(x_f, u_f) + \Delta + \sum_{t=t_2}^{T_0+2t_2+1} f(x_t, u_t) + 2b_f f(x_f, u_f) - \Delta/4. \qquad (2.209)$$

By (2.207)–(2.209),

$$0 \leq \sum_{t=0}^{T_0+2t_2+2b_f+1} f(\widehat{x}_t, \widehat{u}_t) - \sum_{t=0}^{T_0+2t_2+1+2b_f} \leq -\Delta + \Delta/8,$$

a contradiction. The contradiction we have reached proves that

$$\sum_{t=0}^{t_2-1} f(y_t, v_t) = t_2 f(x_f, u_f). \qquad (2.210)$$

Theorem 2.1, (2.191), and (2.210) imply that $(\{y_t\}_{t=0}^{\infty}, \{v_t\}_{t=0}^{\infty})$ is (f)-good. By (P1), $\lim_{t\to\infty} \rho_E(y_t, x_f) = 0$. Together with (2.191) this implies that $y_t = x_f$ for all integers $t \geq 0$ and that $x_t = x_f$ for all $t \in \{0, \ldots, t_2\}$. Combined with Theorem 2.15, this implies that $x_t = x_f$ for all integers $t \geq 0$. Theorem 2.17 is proved.

2.16 Proof of Theorem 2.21

We show that (P1) holds. Let $(\{x_t\}_{t=0}^{\infty}, \{u_t\}_{t=0}^{\infty}) \in X(0, \infty)$ be (f_r)-good. There exists $S_0 > 0$ such that for all integers $T \geq 1$,

$$S_0 > |\sum_{t=0}^{T-1} f_r(x_t, u_t) - T f_f(x_f, u_f)|$$

$$= |\sum_{t=0}^{T-1} f(x_t, u_t) - T f(x_f, u_f) + r \sum_{t=0}^{T-1} \phi(x_t, x_f)|. \qquad (2.211)$$

Theorem 2.1 and (2.211) imply that

$$\Delta := \lim_{T\to\infty} \sum_{t=0}^{T-1} \phi(x_t, x_f) < \infty. \qquad (2.212)$$

Property (a) and (2.212) imply (P1). We show that (P2) holds. Let $M, \epsilon > 0$. By property (a) there exists $\gamma \in (0, 1)$ such that

$$\text{if } \xi \in \mathcal{A},\ \phi(\xi, x_f) \leq \gamma,\ \text{then } \rho_E(\xi, x_f) \leq \epsilon. \qquad (2.213)$$

By Theorem 2.1, there exists $S_0 > 0$ such that for each pair of integers $T_2 > T_1 \geq 0$ and each $(\{x_t\}_{t=T_1}^{T_2}, \{u_t\}_{t=T_1}^{T_2-1}) \in X(T_1, T_2)$, we have

$$\sum_{t=T_1}^{T_2-1} f(x_t, u_t) + S_0 \geq (T_2 - T_1) f(x_f, u_f). \tag{2.214}$$

Choose a natural number

$$L > (r\gamma)^{-1}(M + S_0). \tag{2.215}$$

Let $(\{x_t\}_{t=0}^{L}, \{u_t\}_{t=0}^{L-1}) \in X(0, L)$ satisfy

$$\sum_{t=0}^{L-1} f_r(x_t, u_t) \leq L f(x_f, u_f) + M. \tag{2.216}$$

We show that there exists an integer $s \in \{0, \ldots, L\}$ such that $\rho_E(x_s, x_f) \leq \epsilon$. Assume the contrary. By (2.213), for all $t = 0, \ldots, L$,

$$\rho_E(x_s, x_f) > \epsilon, \ \phi(x_s, x_f) \geq \gamma. \tag{2.217}$$

It follows from (2.8), (2.214), (2.216), and (2.217) that

$$M + L f(x_f, u_f) \geq \sum_{t=0}^{L-1} f_r(x_t, u_t) = \sum_{t=0}^{L-1} f(x_t, u_t) + r \sum_{t=0}^{L-1} \phi(x_t, x_f)$$

$$\geq L f(x_f, x_f) - S_0 + r L \gamma,$$

$$L \leq (\gamma r)^{-1}(M + S_0).$$

This contradicts (2.215). The contradiction we have reached proves that there exists an integer $s \in \{0, \ldots, L\}$ such that $\rho_E(x_s, x_f) \leq \epsilon$. Therefore (P2) holds for f_r and Theorem 2.21 is proved.

2.17 Proofs of Propositions 2.25–2.27, 2.29, and 2.31

Proof of Proposition 2.25. In view of (2.20), $\pi^f(x_f) \leq 0$. Assume that $\pi^f(x_f) < 0$. By (2.20), there exists $\{x_t^{(0)}\}_{t=0}^{\infty} \in X(0, \infty)$ such that

$$x_0^{(0)} = x_f, \tag{2.218}$$

$$\liminf_{T\to\infty}[\sum_{t=0}^{T-1} f(x_t^{(0)}, x_{t+1}^{(0)}) - Tf(x_f, x_f)] < 2^{-1}\pi^f(x_f). \tag{2.219}$$

Theorem 2.1 and (2.219) imply that $\{x_t^{(0)}\}_{t=0}^{\infty}$ is (f)-good. In view of (P1),

$$\lim_{t\to\infty} \rho_E(x_t^{(0)}, x_f) = 0. \tag{2.220}$$

It follows from (2.219) that there exists a strictly increasing sequence of natural number $\{T_k\}_{k=1}^{\infty}$ such that

$$\sum_{t=0}^{T_k-1} f(x_t^{(0)}, x_{t+1}^{(0)}) - T_k f(x_f, x_f) < 2^{-1}\pi^f(x_f),\ k = 1, 2, \ldots. \tag{2.221}$$

By (A2), there exists $\delta \in (0, 1)$ such that the following property holds:

(i) for each $z \in \mathcal{A}$ satisfying $\rho_E(z, x_f) \le \delta$, there exist integers $\tau_1, \tau_2 \in (0, b_f]$ and $\{y_t^{(i)}\}_{t=0}^{\tau_i} \in X(0, \tau_i)$, $i = 1, 2$ which satisfy

$$y_0^{(1)} = z,\ y_{\tau_1}^{(1)} = x_f,\ y_0^{(2)} = x_f,\ y_{\tau_2}^{(2)} = z,$$

$$\sum_{t=0}^{\tau_i-1} f(y_t^{(i)}, y_{t+1}^{(i)}) \le \tau_i f(x_f, x_f) - \pi^f(x_f)/8,\ i = 1, 2.$$

In view of (2.220), there exists an integer $\tau_* > 0$ such that

$$\rho_E(x_t^{(0)}, x_f) \le \delta \text{ for all integers } t \ge \tau_*. \tag{2.222}$$

Let k be a natural number such that $T_k > \tau_*$. Property (i), (2.218), and (2.222) imply that there exists $\{y_t\}_{t=0}^{\infty} \in X(0, \infty)$ such that

$$y_{t+T_k+b_f} = y_t \text{ for all integers } t \ge 0, \tag{2.223}$$

$$y_t = x_t^{(0)},\ t = 0, \ldots, T_k,\ y_{T_k+b_f} = x_f, \tag{2.224}$$

$$\sum_{t=T_k}^{T_k+b_f-1} f(y_t, y_{t+1}) \le b_f f(x_f, u_f) - \pi^f(x_f)/8. \tag{2.225}$$

By (2.221), (2.223), and (2.225),

$$\sum_{t=0}^{T_k+b_f-1} f(y_t, y_{t+1}) = \sum_{t=0}^{T_k-1} f(y_t, y_{t+1}) + \sum_{t=T_k}^{T_k+b_f-1} f(y_t, y_{t+1})$$

$$< T_k f(x_f, x_f) + 2^{-1}\pi^f(x_f) + b_f f(x_f, x_f) - \pi^f(x_f)/8$$

$$\leq (T_k + b_f) f(x_f, x_f) + \pi^f(x_f)/4,$$

$$\liminf_{T\to\infty}[\sum_{t=0}^{T-1} f(y_t, y_{t+1}) - T f(x_f, x_f)] = -\infty,$$

a contradiction which proves that $\pi^f(x_f) = 0$ and Proposition 2.25 itself.

Proof of Proposition 2.26. Let $\epsilon > 0$. By (A2), there exists $\delta > 0$ such that the following property holds:

(ii) for each $z \in \mathcal{A}$ satisfying $\rho_E(z, x_f) \leq \delta$, there exist integers $\tau_1, \tau_2 \in [1, b_f]$ and $\{x_t^{(i)}\}_{t=0}^{\tau_i} \in X(0, \tau_i)$, $i = 1, 2$ which satisfy

$$x_0^{(1)} = z,\ x_{\tau_1}^{(1)} = x_f,\ x_0^{(2)} = x_f,\ x_{\tau_2}^{(2)} = z,$$

$$\sum_{t=0}^{\tau_i-1} f(x_t^{(i)}, x_{t+1}^{(i)}) \leq \tau_i f(x_f, u_f) + \epsilon/8,\ i = 1, 2.$$

Let $z \in \mathcal{A}$ satisfy

$$\rho_E(z, x_f) \leq \delta \tag{2.226}$$

and let integers $\tau_1, \tau_2 \in (0, b_f]$ and $\{x_t^{(i)}\}_{t=0}^{\tau_i} \in X(0, \tau_i)$, $i = 1, 2$ be as guaranteed by property (ii). Set

$$\tilde{x}_t^{(1)} = x_t^{(1)},\ t = 0, \dots, \tau_1 - 1,\ \tilde{x}_t^{(1)} = x_f \text{ for all integers } t \geq \tau_1. \tag{2.227}$$

By (2.20) and (2.227),

$$\pi^f(z) \leq \liminf_{T\to\infty}[\sum_{t=0}^{T-1} f(\tilde{x}_t^{(1)}, \tilde{x}_{t+1}^{(1)}) - T f(x_f, x_f)]$$

$$= \sum_{t=0}^{\tau_1-1} f(x_t^{(1)}, x_{t+1}^{(1)}) - \tau_1 f(x_f, x_f) \leq \epsilon/8. \tag{2.228}$$

Propositions 2.24 and 2.25, property (ii), and (2.228) imply that

$$0 = \pi^f(x_f) \le \sum_{t=0}^{\tau_2-1} f(x_t^{(2)}, x_{t+1}^{(2)}) - \tau_2 f(x_f, x_f) + \pi^f(z) \le \epsilon/8 + \pi^f(z),$$

$$\pi^f(z) \ge -\epsilon/8.$$

Thus $|\pi^f(z)| \le \epsilon/8$. Proposition 2.26 is proved.

Proof of Proposition 2.27. Let $M > 0$ and suppose that the set $\{z \in \mathcal{A} : \pi^f(z) \le M\}$ is nonempty. Propositions 2.25 and 2.26 imply that there exists $\delta > 0$ such that for each $z \in \mathcal{A}$ satisfying $\rho_E(z, x_f) \le \delta$, $\pi^f(z)$ is finite and

$$|\pi^f(z)| \le 1. \tag{2.229}$$

By Theorem 2.20, there exists an integer $L_0 > 0$ such that the following property holds:

(iii) for each integer $T \ge L_0$ and each $\{x_t\}_{t=0}^T \in X(0, T)$ which satisfies

$$\sum_{t=0}^{T-1} f(x_t, x_{t+1}) \le Tf(x_f, u_f) + M + 4$$

and each $S \in \{0, \dots, T - L_0\}$, we have

$$\min\{\rho_E(x_t, x_f) : t = S, \dots, S + L_0\} \le \delta.$$

Lemma 2.49 implies that there exists $M_1 > 0$ such that the following property holds:

(iv) for each $S \in \{1, \dots, L_0 + 1\}$ and each $\{x_t\}_{t=0}^S \in X(0, S)$ which satisfies

$$\sum_{t=0}^{S-1} f(x_t, x_{t+1}) \le (1 + L_0)|f(x_f, u_f)| + M + 4,$$

we have

$$\rho_E(\theta_0, x_t) \le M_1, \ t = 0, \dots, S - 1.$$

Assume that $z \in \mathcal{A}$ satisfies

$$\pi^f(z) \le M. \tag{2.230}$$

In view of (2.20) and (2.230), there exists $\{x_t\}_{t=0}^{\infty} \in X(0, \infty)$ such that

$$x_0 = z, \tag{2.231}$$

$$\liminf_{T\to\infty}[\sum_{t=0}^{T-1} f(x_t, x_{t+1}) - Tf(x_f, x_f)] < \pi^f(z) + 1 \le M + 1. \tag{2.232}$$

By property (iii) and (2.232), there exists

$$t_0 \in \{1, \dots, L_0 + 1\} \tag{2.233}$$

such that

$$\rho_E(x_{t_0}, x_f) \le \delta. \tag{2.234}$$

It follows from (2.229) and (2.234) that

$$\pi^f(x_{t_0}) \le 1. \tag{2.235}$$

Proposition 2.24, (2.230)–(2.232), and (2.235) imply that

$$\pi^f(x_0) \le \sum_{t=0}^{t_0-1} f(x_t, x_{t+1}) - t_0 f(x_f, x_f) + \pi^f(x_{t_0})$$

$$\le \liminf_{T\to\infty}[\sum_{t=0}^{T-1} f(x_t, x_{t+1}) - Tf(x_f, x_f)] < M + 1.$$

Together with (2.235) this implies that

$$\sum_{t=0}^{t_0-1} f(x_t, x_{t+1}) \le t_0 f(x_f, x_f) + M + 2. \tag{2.236}$$

By (2.233) and (2.236), $\rho_E(z, \theta_0) = \rho_E(x_0, \theta_0) \le M_1$. Proposition 2.27 is proved.

Proof of Proposition 2.29. Assume that $\{x_t\}_{t=0}^{\infty} \in X(0, \infty)$ is (f)-good. Propositions 2.25 and 2.26, (P1), and (2.20) imply that

$$\lim_{t\to\infty} \rho_E(x_t, x_f) = 0, \ \lim_{t\to\infty} \pi^f(x_t) = 0 \tag{2.237}$$

and $\pi^f(x_t) < \infty$ for all integers $t \geq 0$. It follows from (2.26) and (2.27) that

$$\sup\{\Gamma^f(\{x_t\}_{t=0}^T) : T = 1, 2, \dots\} = \lim_{T \to \infty} \Gamma^f(\{x_t\}_{t=0}^T)$$

$$= \lim_{T \to \infty} (\sum_{t=0}^{T-1} f(x_t, x_{t+1}) - Tf(x_f, x_f) - \pi^f(x_0)) < \infty.$$

Assume that

$$\Delta := \lim_{T \to \infty} \Gamma^f(\{x_t\}_{t=0}^T) < \infty. \tag{2.238}$$

It follows from (2.21), (2.26), (2.27), and (2.238) that for all integers $T > 0$, $\pi^f(x_T) < \infty$, and

$$\sum_{t=0}^{T-1} f(x_t, x_{t+1}) - Tf(x_f, x_f) = \Gamma^f(\{x_t\}_{t=0}^T)$$

$$+\pi^f(x_0) - \pi^f(x_T) \leq \Delta + \pi^f(x_0) + S_*.$$

In view of Theorem 2.1, $\{x_t\}_{t=0}^\infty$ is (f)-good.

Assume that $\{x_t\}_{t=0}^\infty$ is (f)-good. By (2.26), (2.27), and (2.237), for all integers $T \geq 1$,

$$\sum_{t=0}^{T-1} f(x_t, x_{t+1}) - Tf(x_f, x_f) = \Gamma^f(\{x_t\}_{t=0}^T) + \pi^f(x_0) - \pi^f(x_T)$$

$$\to \lim_{T \to \infty} \Gamma^f(\{x_t\}_{t=0}^T) + \pi^f(x_0).$$

Proposition 2.29 is proved.

Proof of Proposition 2.31. Proposition 2.27 and (2.28) imply that there exists $M_1 > 0$ such that

$$\mathcal{A}_f \subset \{z \in \mathcal{A} : \rho_E(z, \theta_0) \leq M_1\}. \tag{2.239}$$

By Theorem 2.2, there exists $M_2 > 0$ such that the following property holds:

(v) for each integer $T > 0$ and each $\{x_t\}_{t=0}^T \in X(0, T)$ satisfying

$$\sum_{t=0}^{T-1} f(x_t, x_{t+1}) \leq Tf(x_f, x_f) + |\inf(\pi^f)| + 3,$$

the inequality $\rho_E(x_f, x_t) \leq M_2$ holds for all $t = 0, \dots, T-1$.

Theorem 2.1 implies that there exists $c_1 > 0$ such that for each integer $T > 0$ and each $\{x_t\}_{t=0}^T \in X(0, T)$,

$$\sum_{t=0}^{T-1} f(x_t, x_{t+1}) \geq Tf(x_f, u_f) - c_1. \tag{2.240}$$

In view of (A2), there exists $\delta \in (0, 1)$ such that the following property holds:

(vi) for each $z \in \mathcal{A}$ satisfying $\rho_E(z, x_f) \leq \delta$, there exist an integer $\tau \in [1, b_f]$ and $\{x_t\}_{t=0}^{\tau} \in X(0, \tau)$ which satisfies

$$x_0 = z, \ x_\tau = x_f, \ \sum_{t=0}^{\tau-1} f(x_t, x_{t+1}) \leq \tau f(x_f, u_f) + 1.$$

By Theorem 2.20, there exists an integer $L_0 \geq 1$ such that the following property holds:

(vii) for each integer $T_1 \geq 0$, each integer $T_2 \geq T_1 + L_0$, and each $\{x_t\}_{t=T_1}^{T_2} \in X(T_1, T_2)$ which satisfies

$$\sum_{t=T_1}^{T_2-1} f(x_t, x_{t+1}) - (T_2 - T_1) f(x_f, u_f) \leq |\inf(\pi^f)| + 3 + c_1$$

there exists $t_0 \in \{0, \ldots, L_0\}$ such that $\rho_E(x_{t_0}, x_f) \leq \delta$.

Set

$$M_0 = (L_0 + b_f)(1 + |f(x_f, x_f)|) + |\inf(\pi^f)| + 4 + c_1. \tag{2.241}$$

Let

$$z \in \mathcal{A}_f. \tag{2.242}$$

In view of (2.28) and (2.242),

$$\pi^f(z) \leq \inf(\pi^f) + 1. \tag{2.243}$$

Proposition 2.29, (2.20), and (2.243) imply that there exists $\{x_t\}_{t=0}^{\infty} \in X(0, \infty)$ such that

$$x_0 = z, \tag{2.244}$$

$$\liminf_{T\to\infty} [\sum_{t=0}^{T-1} f(x_t, x_{t+1}) - Tf(x_f, x_f)]$$

$$= \lim_{T\to\infty} \Gamma^f(\{x_t\}_{t=0}^T) + \pi^f(z) \le \pi^f(z) + 1 \le \inf(\pi^f) + 2. \tag{2.245}$$

By (2.245), there exists a natural number T_0 such that for all integers $T \ge T_0$,

$$\sum_{t=0}^{T-1} f(x_t, x_{t+1}) - Tf(x_f, x_f) \le \inf(\pi^f) + 3. \tag{2.246}$$

Property (v) and (2.246) imply that

$$\rho_E(x_t, x_f) \le M_2 \text{ for all integers } t \ge 0. \tag{2.247}$$

It follows from (2.240) and (2.246) that for every integer $T \in (0, T_0)$,

$$\sum_{t=0}^{T-1} f(x_t, x_{t+1}) - Tf(x_f, x_f) = \sum_{t=0}^{T_0-1} f(x_t, x_{t+1}) - T_0 f(x_f, x_f)$$

$$-(\sum\{f(x_t, x_{t+1}) : t \in \{T, \dots, T_0\} \setminus \{T_0\}\} - (T - T_0) f(x_f, x_f))$$

$$\le \inf(\pi^f) + 3 + c_1.$$

Together with (2.246) this implies that for all integers $T > 0$,

$$\sum_{t=0}^{T-1} f(x_t, x_{t+1}) - Tf(x_f, x_f) \le \inf(\pi^f) + 3 + c_1. \tag{2.248}$$

Property (vii) and (2.248) imply that there exists an integer

$$t_0 \in [0, L_0] \tag{2.249}$$

such that

$$\rho_E(x_{t_0}, x_f) \le \delta. \tag{2.250}$$

It follows from property (vi) and (2.250) that there exists $\{y_t\}_{t=0}^{t_0+b_f} \in X(0, t_0 + b_f)$ such that

$$y_t = x_t,\ t = 0, \dots, t_0,\ y_{t_0+b_f} = x_f, \tag{2.251}$$

$$\sum_{t=t_0}^{t_0+b_f-1} f(y_t, y_{t+1}) \le b_f f(x_f, u_f) + 1. \tag{2.252}$$

In view of (2.241) and (2.249),

$$t_0 + b_f \le L_0 + b_f \le M_0. \tag{2.253}$$

By (2.241), (2.248), (2.249), (2.251), and (2.252),

$$\sum_{t=0}^{t_0+b_f-1} f(y_t, y_{t+1}) \le t_0 f(x_f, x_f) + \inf(\pi^f) + 3 + c_1 + b_f f(x_f, u_f) + 1$$

$$\le |f(x_f, x_f)|(L_0 + b_f) + |\inf(\pi^f)| + 4 + c_1 \le M_0.$$

Together with (2.244), (2.251), and (2.253), this implies that $z \in \mathcal{A}_{M_0}$. Proposition 2.31 is proved.

Proof of Proposition 2.33. Let $z \in \mathcal{A}_{M_0}$. There exists $\{x_t\}_{t=0}^{\infty} \in X(0, \infty)$ such that

$$x_0 = z,\ x_t = x_f \text{ for all integers } t \ge M_0,$$

$$\sum_{t=0}^{M_0-1} f(x_t, x_{t+1}) \le M_0 + M_0|f(x_f, u_f)|.$$

By the relations above,

$$\pi^f(z) \le \liminf_{T\to\infty}[\sum_{t=0}^{T-1} f(x_t, x_{t+1}) - Tf(x_f, x_f)]$$

$$\le M_0(1 + |f(x_f, x_f)|) + M_0|f(x_f, x_f)|.$$

Thus $\mathcal{A}_{M_0} \subset \{z \in \mathcal{A} :\ \pi^f(z) \le 2M_0(1 + |f(x_f, u_f)|)\}$ which is bounded by Proposition 2.27. This completes the proof of Proposition 2.33.

2.18 Proof of Proposition 2.37

Let $\{x^{(k)}\}_{k=1}^{\infty} \subset \mathcal{A}$, $x \in \mathcal{A}$ and

$$x = \lim_{k\to\infty} x^{(k)}. \tag{2.254}$$

We show that $\pi^f(x) \le \liminf_{k\to\infty} \pi^f(x^{(k)})$. Extracting subsequences and reindexing, we may assume without loss of generality that there exists

$$\lim_{k\to\infty} \pi^f(x^{(k)}) < \infty \tag{2.255}$$

and that $\pi^f(x^{(k)}) < \infty$ for all integers $k \geq 1$. For each integer $k \geq 1$, there exists an $(f, \mathcal{M})$-overtaking optimal $\{y_t^{(k)}\}_{t=0}^{\infty} \in X(0, \infty)$ such that

$$y_0^{(k)} = x^{(k)}. \tag{2.256}$$

Proposition 2.36 implies that for every integer $k \geq 1$,

$$\pi^f(x^{(k)}) = \lim_{T\to\infty} [\sum_{t=0}^{T-1} f(y_t^{(k)}, y_{t+1}^{(k)}) - Tf(x_f, x_f)]. \tag{2.257}$$

It follows from Proposition 2.35, (2.21), and (2.255) that for every integer $T \geq 1$, the sequence $\{\sum_{t=0}^{T-1} f(y_t^{(k)}, y_{t+1}^{(k)})\}_{k=1}^{\infty}$ is bounded. By LSC property, extracting subsequences, using the diagonalization process, and re-indexing, we may assume that there exists $\{y_t\}_{t=0}^{\infty} \in X(0, \infty)$ and a subsequence $\{y_t^{(i_k)}\}_{t=0}^{\infty}$, $k = 1, 2, \ldots$ such that for every integer $t \geq 0$,

$$y_t^{(i_k)} \to y_t \text{ as } k \to \infty, \tag{2.258}$$

for every integer $T \geq 1$, there exists

$$\lim_{k\to\infty} \sum_{t=0}^{T-1} f(y_t^{(i_k)}, y_{t+1}^{(i_k)}), \tag{2.259}$$

$$\sum_{t=0}^{T-1} f(y_t, y_{t+1}) \leq \lim_{k\to\infty} \sum_{t=0}^{T-1} f(y_t^{(i_k)}, y_{t+1}^{(i_k)}). \tag{2.260}$$

Let $\epsilon > 0$. By Propositions 2.25 and 2.26, there exists $\delta > 0$ such that for each $\xi \in \mathcal{A}$ satisfying $\rho_E(\xi, x_f) \leq \delta$, we have

$$|\pi^f(\xi)| \leq \epsilon/2. \tag{2.261}$$

Since the sequences $\{y_t^{(k)}\}_{t=0}^{\infty}$, $k = 1, 2, \ldots$ are $(f, \mathcal{M})$-overtaking optimal, it follows from (2.255), (2.257), and Proposition 2.6 that there exists an integer $L_0 > 0$ such that for each integer $k \geq 1$ and each integer $t \geq L_0$,

$$\rho_E(y_t^{(k)}, x_f) \leq \delta. \tag{2.262}$$

Proposition 2.35, (2.261), and (2.262) imply that for each integer $k \geq 1$ and each integer $T \geq L_0$,

$$|\pi^f(y_T^{(k)})| \leq \epsilon/2, \tag{2.263}$$

$$\sum_{t=0}^{T-1} f(y_t^{(k)}, y_{t+1}^{(k)}) = Tf(x_f, x_f) + \pi^f(y_0^{(k)}) - \pi^f(y_T^{(k)})$$

$$\leq Tf(x_f, x_f) + \pi^f(y_0^{(k)}) + \epsilon/2. \tag{2.264}$$

In view of (2.255), (2.256), (2.260), and (2.264), for each integer $T \geq L_0$,

$$\sum_{t=0}^{T-1} f(y_t, y_{t+1}) \leq Tf(x_f, x_f) + \lim_{k\to\infty} \pi^f(y_0^{(k)}) + \epsilon/2,$$

$$\sum_{t=0}^{T-1} f(y_t, y_{t+1}) - Tf(x_f, x_f) \leq \lim_{k\to\infty} \pi^f(x^{(k)}) + \epsilon/2. \tag{2.265}$$

By (2.254), (2.256), and (2.258), $y_0 = x$. Together with (2.265) this implies that $\pi^f(x) \leq \lim_{k\to\infty} \pi^f(x^{(k)}) + \epsilon/2$. Since ϵ is any positive number, this completes the proof of Proposition 2.37.

2.19 Auxiliary Results for Theorems 2.38 and 2.39

Since the results obtained for the pair $(f, \mathcal{M})$ also hold for the pair $(\bar{f}, \bar{\mathcal{M}})$, in view of Theorem 2.8, we have the following result.

Proposition 2.57. *Assume that $L, M > 0$ are integers and that $\epsilon > 0$. Then there exist $\delta > 0$ and an integer $L_0 > L$ such that for each integer $T_1 \geq 0$, each integer $T_2 \geq T_1 + 2L_0$, and each $\{x_t\}_{t=T_1}^{T_2} \in \bar{X}(T_1, T_2)$ which satisfies*

$$x_{T_1} \in \widehat{\mathcal{A}}_L,$$

$$\sum_{t=T_1}^{T_2-1} \bar{f}(x_t, x_{t+1}) \leq \min\{\sigma_-^f(T_1, T_2, x_{T_1}) + M,\ U_-^f(T_1, T_2, x_{T_1}, x_{T_2}) + \delta\},$$

there exist integers $\tau_1 \in [T_1, T_1 + L_0]$, $\tau_2 \in [T_2 - L_0, T_2]$ such that

$$\rho_E(x_t, x_f) \leq \epsilon,\ t = \tau_1, \ldots, \tau_2.$$

Moreover, if $\rho_E(x_{T_1}, x_f) \leq \delta$, then $\tau_1 = T_1$, and if $\rho_E(x_{T_2}, x_f) \leq \delta$, then $\tau_2 = T_2$.

Proposition 2.57 implies the following result.

Proposition 2.58. *Assume that $L, M > 0$ are integers and that $\epsilon > 0$. Then there exist $\delta > 0$ and an integer $L_0 > 0$ such that for each integer $T_1 \geq 0$, each integer $T_2 \geq T_1 + 2L_0$, and each $\{x_t\}_{t=T_1}^{T_2} \in X(T_1, T_2)$ which satisfies*

$$x_{T_2} \in \widehat{\mathcal{A}}_L,$$

$$\sum_{t=T_1}^{T_2-1} f(x_t, x_{t+1}) \le \min\{\widehat{\sigma}^f(T_1, T_2, x_{T_2}) + M,\ U^f(T_1, T_2, x_{T_1}, x_{T_2}) + \delta\},$$

there exist integers $\tau_1 \in [T_1, T_1 + L_0]$, $\tau_2 \in [T_2 - L_0, T_2]$ *such that*

$$\rho_E(x_t, x_f) \le \epsilon,\ t = \tau_1, \dots, \tau_2.$$

Moreover, if $\rho_E(x_{T_1}, x_f) \le \delta$, *then* $\tau_1 = T_1$, *and if* $\rho_E(x_{T_2}, x_f) \le \delta$, *then* $\tau_2 = T_2$.

2.20 Proofs of Theorems 2.38 and 2.39

We prove the following result.

Theorem 2.59. *Let* L_0, M *be natural numbers and* $\epsilon \in (0, 1)$. *Then there exist* $\delta > 0$ *and an integer* $L_1 > L_0$ *such that for each integer* $T \ge L_1$ *and each* $\{x_t\}_{t=0}^T \in X(0, T)$ *which satisfies*

$$x_T \in \widehat{\mathcal{A}}_M,\ \sum_{t=0}^{T-1} f(x_t, x_{t+1}) \le \widehat{\sigma}^f(0, T, x_T) + \delta,$$

the inequalities

$$\pi^f(x_0) \le \inf(\pi^f) + \epsilon,\ \Gamma^f(\{x_t\}_{t=0}^{L_0}) \le \epsilon$$

hold.

Proof. By Propositions 2.25 and 2.26, Lemma 2.55, and (A2), there exists $\delta_1 \in (0, \epsilon/4)$ such that:

(i) for each $z \in \mathcal{A}$ satisfying $\rho_E(z, x_f) \le 2\delta_1$, we have $|\pi^f(z)| \le \epsilon/16$;
(ii) for each $\{x_t\}_{t=0}^{b_f} \in X(0, b_f)$ which satisfies

$$\rho_E(x_0, x_f) \le 2\delta_1\ ,\ \rho_E(x_{b_f}, x_f) \le 2\delta_1,$$

we have

$$\sum_{t=0}^{b_f-1} f(x_t, x_{t+1}) \ge b_f f(x_f, x_f) - \epsilon/16;$$

(iii) for each $z_i \in \mathcal{A}$, $i = 1, 2$ satisfying $\rho_E(z_i, x_f) \le 2\delta_1$, $i = 1, 2$, there exist an integer $\tau \in (0, b_f]$ and $\{x_t\}_{t=0}^{\tau} \in X(0, \tau)$ which satisfies $x_0 = z_1$, $x_\tau = z_2$ and

$$\sum_{t=0}^{\tau-1} f(x_t, x_{t+1}) \le \tau f(x_f, x_f) + \epsilon/16.$$

By Proposition 2.58, there exist $\delta_2 \in (0, \delta_1/8)$ and an integer $l_0 > 0$ such that the following property holds:

(iv) for each integer $T_1 \ge 0$, each integer $T_2 \ge T_1 + 2l_0$, and each $\{x_t\}_{t=T_1}^{T_2} \in X(T_1, T_2)$ which satisfies

$$x_{T_2} \in \widehat{\mathcal{A}}_M,$$

$$\sum_{t=T_1}^{T_2-1} f(x_t, x_{t+1}) \le \widehat{\sigma}^f(T_1, T_2, x_{T_2}) + \delta_2,$$

we have

$$\rho_E(x_t, x_f) \le \delta_1,\ t = T_1 + l_0, \dots, T_2 - l_0.$$

Corollary 2.30 and (2.21) imply that there exists $z_* \in \mathcal{A}$ such that

$$\pi^f(z_*) \le \inf(\pi^f) + \delta_2/8, \tag{2.266}$$

and $\{x_t^*\}_{t=0}^{\infty} \in X(0, \infty)$ for which

$$x_0^* = z_*,\ \lim_{T\to\infty} \Gamma^f(\{x_t^*\}_{t=0}^{T}) \le \delta_2/8. \tag{2.267}$$

By Proposition 2.29, (P1), (2.266), and (2.267), there exists an integer $l_1 > 0$ such that

$$\rho_E(x_t^*, x_f) \le \delta_2/8 \text{ for all integers } t \ge l_1. \tag{2.268}$$

Choose

$$\delta \in (0, \delta_2/4) \tag{2.269}$$

and an integer

$$L_1 > 2L_0 + 2l_0 + 2l_1 + 2b_f + 8. \tag{2.270}$$

Assume that $T \geq L_1$ is an integer and that $\{x_t\}_{t=0}^{T} \in X(0, T)$ satisfies

$$x_T \in \widehat{\mathcal{A}}_M, \tag{2.271}$$

$$\sum_{t=0}^{T-1} f(x_t, x_{t+1}) \leq \widehat{\sigma}^f(0, T, x_T) + \delta. \tag{2.272}$$

By (2.269)–(2.272),

$$\rho_E(x_t, x_f) \leq \delta_1, \ t = l_0, \ldots, T - l_0. \tag{2.273}$$

In view of (2.270) and (2.273),

$$[l_0 + l_1 + L_0, l_0 + l_1 + L_0 + 2b_f + 8] \subset [l_0, T - l_0 - l_1 - L_0], \tag{2.274}$$

$$\rho_E(x_t, x_f) \leq \delta_1, \ t = l_0 + l_1 + L_0, \ldots, l_0 + l_1 + L_0 + 2b_f + 8. \tag{2.275}$$

Property (iii), (2.268), and (2.275) imply that there exist integers $\tau_1, \tau_2 \in (0, b_f]$ and $\{x_t^{(1)}\}_{t=0}^{T} \in X(0, T)$ such that

$$x_t^{(1)} = x_t^*, \ t = 0, \ldots, l_0 + l_1 + L_0 + 4, \ x^{(1)}_{l_0+l_1+L_0+4+\tau_1} = x_f,$$

$$\sum_{t=l_0+l_1+L_0+4}^{l_0+l_1+L_0+3+\tau_1} f(x_t^{(1)}, x_{t+1}^{(1)}) \leq \tau_1 f(x_f, u_f) + \epsilon/16,$$

$$x^{(1)}_{l_0+l_1+L_0+4+2b_f-\tau_2} = x_f,$$

$$x^{(1)}_{l_0+l_1+L_0+4+2b_f} = x_{l_0+l_1+L_0+4+2b_f},$$

$$\sum_{t=l_0+l_1+L_0+4+2b_f-\tau_2}^{l_0+l_1+L_0+3+2b_f} f(x_t^{(1)}, x_{t+1}^{(1)}) \leq \tau_2 f(x_f, u_f) + \epsilon/16,$$

$$x_t^{(1)} = x_f, \ t = l_0 + l_1 + L_0 + 4 + \tau_1, \ldots, l_0 + l_1 + L_0 + 4 + 2b_f - \tau_2,$$

$$x_t^{(1)} = x_t, \ t = l_0 + l_1 + L_0 + 4 + 2b_f, \ldots, T. \tag{2.276}$$

It follows from (2.272) and (2.276) that

$$\delta \geq \sum_{t=0}^{T-1} f(x_t, x_{t+1}) - \sum_{t=0}^{T-1} f(x_t^{(1)}, x_{t+1}^{(1)})$$

$$= \sum_{t=0}^{l_0+l_1+L_0+3+2b_f} f(x_t, x_{t+1}) - \sum_{t=0}^{l_0+l_1+L_0+3+2b_f} f(x_t^{(1)}, x_{t+1}^{(1)})$$

$$= \sum_{t=0}^{l_0+l_1+L_0+3} f(x_t, x_{t+1}) + \sum_{t=l_0+l_1+L_0+4}^{l_0+l_1+L_0+3+b_f} f(x_t, x_{t+1})$$

$$+ \sum_{t=l_0+l_1+L_0+4+b_f}^{l_0+l_1+L_0+3+2b_f} f(x_t, x_{t+1})$$

$$- \sum_{t=0}^{l_0+l_1+L_0+3} f(x_t^*, x_{t+1}^*) - \tau_1 f(x_f, x_f) - \epsilon/16$$

$$- \tau_2 f(x_f, x_f) - \epsilon/16 - f(x_f, x_f)(2b_f - \tau_1 - \tau_2). \tag{2.277}$$

Property (iii) and (2.275) imply that

$$\sum_{t=l_0+l_1+L_0+4}^{l_0+l_1+L_0+3+b_f} f(x_t, x_{t+1}) \geq b_f f(x_f, x_f) - \epsilon/16, \tag{2.278}$$

$$\sum_{t=l_0+l_1+L_0+4+b_f}^{l_0+l_1+L_0+3+2b_f} f(x_t, x_{t+1}) \geq b_f f(x_f, x_f) - \epsilon/16. \tag{2.279}$$

In view of (2.277)–(2.279),

$$\delta \geq \sum_{t=0}^{l_0+l_1+L_0+3} f(x_t, x_{t+1}) + 2b_f f(x_f, x_f) - \epsilon/8$$

$$- \sum_{t=0}^{l_0+l_1+L_0+3} f(x_t^*, x_{t+1}^*) - 2b_f f(x_f, x_f) - \epsilon/8. \tag{2.280}$$

By property (ii), (2.26), (2.266), (2.267), (2.275), and (2.280),

$$\sum_{t=0}^{l_0+l_1+L_0+3} f(x_t, x_{t+1}) \le \epsilon/4 + \epsilon/16 + \sum_{t=0}^{l_0+l_1+L_0+3} f(x_t^*, x_{t+1}^*)$$

$$= 5\epsilon/16 + \Gamma^f(\{x_t^*\}_{t=0}^{l_0+l_1+L_0+4}) + (l_0 + l_1 + L_0 + 4) f(x_f, x_f)$$

$$+\pi^f(x_0^*) - \pi^f(x^*_{l_0+l_1+L_0+4})$$

$$\le \Gamma^f(\{x_t^*\}_{t=0}^{l_0+l_1+L_0+4}) + (l_0 + l_1 + L_0 + 4) f(x_f, x_f) + \pi^f(x_0^*) + 6\epsilon/16$$

$$\le \delta_2/8 + (l_0 + l_1 + L_0 + 4) f(x_f, x_f) + \pi^f(x_0^*) + 3\epsilon/8$$

$$\le \inf(\pi^f) + \delta_2/4 + (l_0 + l_1 + L_0 + 4) f(x_f, x_f) + 3\epsilon/8. \tag{2.281}$$

It follows from (2.26), (2.275), (2.281), and property (i) that

$$\inf(\pi^f) + (l_0 + l_1 + L_0 + 4) f(x_f, x_f) \ge -3\epsilon/8 - \delta_2/4 + \sum_{t=0}^{l_0+l_1+L_0+3} f(x_t, x_{t+1})$$

$$\ge -\epsilon/2 + \Gamma^f(\{x_t\}_{t=0}^{l_0+l_1+L_0+4}) + (l_0 + l_1 + L_0 + 4) f(x_f, x_f)$$

$$+\pi^f(x_0) - \pi^f(x_{l_0+l_1+L_0+4})$$

$$\ge -\epsilon/2 + \Gamma^f(\{x_t\}_{t=0}^{l_0+l_1+L_0+4}) + (l_0 + l_1 + L_0 + 4) f(x_f, x_f) + \pi^f(x_0) - \epsilon/16,$$

$$\pi^f(x_0) + \Gamma^f(\{x_t\}_{t=0}^{l_0+l_1+L_0+4}) \le 9\epsilon/16 + \inf(\pi^f).$$

The inequality above implies that

$$\pi^f(x_0) \le \inf(\pi^f) + \epsilon, \quad \Gamma^f(\{x_t\}_{t=0}^{l_0+l_1+L_0+4}) \le \epsilon.$$

Theorem 2.59 is proved.

Theorem 2.38 follows from Theorem 2.59, applied for $(\bar{f}, \bar{\mathcal{M}})$, and Proposition 2.23. Theorem 2.40 follows from Proposition 2.26 and Theorem 2.38 and Theorem 2.59.

2.21 Proofs of Theorem 2.40 and 2.41

We prove the following result.

Theorem 2.60. *Let $(f, \mathcal{M})$ have LSC property, $L_0, M > 0$ be integers and $\epsilon \in (0, 1)$. Then there exist $\delta > 0$ and an integer $L_1 > L_0$ such that for each integer $T \geq L_1$ and each $\{x_t\}_{t=0}^{T} \in X(0, T)$ which satisfies*

$$x_T \in \widehat{\mathcal{A}}_M, \ \sum_{t=0}^{T-1} f(x_t, x_{t+1}) \leq \widehat{\sigma}^f(0, T, x_T) + \delta,$$

there exists an $(f, \mathcal{M})$-overtaking optimal $\{x_t^*\}_{t=0}^{\infty} \in X(0, \infty)$ such that

$$\pi^f(x_0^*) = \inf(\pi^f), \ \rho_E(x_t^*, x_t) \leq \epsilon, \ t = 0, \ldots, L_0.$$

Note that Theorem 2.40 follows from Theorem 2.60, applied for $(\bar{f}, \bar{\mathcal{M}})$, and Proposition 2.23. Theorem 2.41 follows from Theorems 2.40 and 2.60 and Proposition 2.6.

Proposition 2.61. *Let $(f, \mathcal{M})$ have LSC property, $L_0 > 0$ be an integer, and $\epsilon \in (0, 1)$. Then there exist $\delta \in (0, \epsilon)$ such that for each $\{x_t\}_{t=0}^{L_0} \in X(0, L_0)$ which satisfies*

$$\pi^f(x_0) \leq \inf(\pi^f) + \delta, \ \Gamma^f(\{x_t\}_{t=0}^{L_0}) \leq \delta,$$

there exists an $(f, \mathcal{M})$-overtaking optimal $\{x_t^*\}_{t=0}^{\infty} \in X(0, \infty)$ such that

$$\pi^f(x_0^*) = \inf(\pi^f), \ \rho_E(x_t^*, x_t) \leq \epsilon, \ t = 0, \ldots, L_0.$$

Proof. Assume that the proposition does not hold. Then there exist a sequence $\{\delta_k\}_{k=1}^{\infty} \subset (0, 1]$ and a sequence $\{x_t^{(k)}\}_{t=0}^{L_0} \in X(0, L_0)$, $k = 1, 2, \ldots$ such that

$$\lim_{k \to \infty} \delta_k = 0 \tag{2.282}$$

and that for all integers $k \geq 1$,

$$\pi^f(x_0^{(k)}) \leq \inf(\pi^f) + \delta_k, \tag{2.283}$$

$$\Gamma^f(\{x_t^{(k)}\}_{t=0}^{L_0}) \leq \delta_k \tag{2.284}$$

and the following property holds:

(i) for each $(f, \mathcal{M})$-overtaking optimal $\{y_t\}_{t=0}^{\infty} \in X(0, \infty)$ satisfying $\pi^f(y_0) = \inf(\pi^f)$, we have $\max\{\rho_E(x_t^{(k)}, y_t) : t = 0, \ldots, L_0\} > \epsilon$

In view of (2.26), (2.282)–(2.284), and the boundedness from below of the function π^f (see (2.21)), the sequence $\{\sum_{t=0}^{L_0-1} f(x_t^{(k)}, x_{t+1}^{(k)})\}_{k=1}^{\infty}$ is bounded. By LSC property, extracting a subsequence and re-indexing if necessary, we may assume without loss of generality that there exists $\{x_t\}_{t=0}^{L_0} \in X(0, L_0)$ such that

$$x_t^{(k)} \to x_t \text{ as } k \to \infty \text{ for all } t = 0, \ldots, L_0, \tag{2.285}$$

$$\sum_{t=0}^{L_0-1} f(x_t, x_{t+1}) \le \liminf_{k\to\infty} \sum_{t=0}^{L_0-1} f(x_t^{(k)}, x_{t+1}^{(k)}). \tag{2.286}$$

The lower semicontinuity of π^f (see Proposition 2.37) and (2.285) imply that

$$\pi^f(x_0) \le \liminf_{k\to\infty} \pi^f(x_0^{(k)}) = \inf(\pi^f), \tag{2.287}$$

$$\pi^f(x_0) = \inf(\pi^f). \tag{2.288}$$

By the lower semicontinuity of π^f (see Proposition 2.37) and (2.285),

$$\pi^f(x_{L_0}) \le \liminf_{k\to\infty} \pi^f(x_{L_0}^{(k)}). \tag{2.289}$$

It follows from (2.26), (2.27), (2.282)–(2.284), and (2.286)–(2.288) that

$$\sum_{t=0}^{L_0-1} f(x_t, x_{t+1}) - L_0 f(x_f, x_f) - \pi^f(x_0) + \pi^f(x_{L_0})$$

$$\le \liminf_{k\to\infty}\Big(\sum_{t=0}^{L_0-1} f(x_t^{(k)}, x_{t+1}^{(k)}) - L_0 f(x_f, x_f)\Big) - \lim_{k\to\infty} \pi^f(x_0^{(k)}) + \liminf_{k\to\infty} \pi^f(x_{L_0}^{(k)})$$

$$\le \liminf_{k\to\infty}\Big(\sum_{t=0}^{L_0-1} f(x_t^{(k)}, x_{t+1}^{(k)}) - L_0 f(x_f, x_f) - \pi^f(x_0^{(k)}) + \pi^f(x_{L_0}^{(k)})\Big) \le 0. \tag{2.290}$$

In view of (2.26), (2.27), and (2.290),

$$\sum_{t=0}^{L_0-1} f(x_t, x_{t+1}) - L_0 f(x_f, x_f) - \pi^f(x_0) + \pi^f(x_{L_0}) = 0. \tag{2.291}$$

Theorem 2.13 and the relation $\pi^f(x_{T_0}) < \infty$ imply that there exists an $(f, \mathcal{M})$-overtaking optimal $\{\tilde{x}_t\}_{t=0}^{\infty} \in X(0, \infty)$ such that

$$\tilde{x}_0 = x_{L_0}. \tag{2.292}$$

For all integers $t > L_0$ set

$$x_t = \tilde{x}_{t-L_0}. \tag{2.293}$$

It is clear that $\{x_t\}_{t=0}^{\infty} \in X(0, \infty)$ is (f)-good. By Proposition 2.35, (2.26), (2.27), (2.291), and (2.292), for all integers $S > 0$,

$$\sum_{t=0}^{S-1} f(x_t, x_{t+1}) - Sf(x_f, x_f) - \pi^f(x_0) + \pi^f(x_S) = 0. \tag{2.294}$$

In view of (2.288) and (2.294), $\{x_t\}_{t=0}^{\infty} \in X(0, \infty)$ is $(f, \mathcal{M})$-overtaking optimal and (2.288) holds. By (2.285), for all sufficiently large natural numbers k, $\rho_E(x_t^{(k)}, x_t) \leq \epsilon/2$ for all $t = 0, \ldots, L_0$. This contradicts property (i). The contradiction we have reached proves Proposition 2.61.

Theorem 2.60 follows from Proposition 2.61 and Theorem 2.59.

2.22 The First Bolza Problem

We consider the control system introduced in Section 2.4 and use the notation, definitions, and assumptions used there and in Section 2.5. Let $a_1 > 0$. Denote by $\mathfrak{A}(\mathcal{A})$ the set of all lower semicontinuous functions $h : \mathcal{A} \to R^1$ which are bounded on bounded subsets of $\mathcal{A}$ and such that

$$h(z) \geq -a_1 \text{ for all } z \in \mathcal{A}. \tag{2.295}$$

The set $\mathfrak{A}(\mathcal{A})$ is equipped with the uniformity which is determined by the base

$$\mathcal{E}(N, \epsilon) = \{(h_1, h_2) : \mathfrak{A}(\mathcal{A}) \times \mathfrak{A}(\mathcal{A}) : \ |h_1(z) - h_2(z) \leq \epsilon$$

$$\text{for all } z \in \mathcal{A} \text{ satisfying } \rho_E(z, \theta_0) \leq N\}, \tag{2.296}$$

where $N, \epsilon > 0$. It is not difficult to see that the uniform space $\mathfrak{A}(\mathcal{A})$ is metrizable and complete.

For each pair of integers $T_2 > T_1 \geq 0$, each $y \in \mathcal{A}$, and each $h \in \mathfrak{A}(\mathcal{A})$, we define

$$\sigma^{f,h}(T_1, T_2, y) = \inf\{\sum_{t=T_1}^{T_2-1} f(x_t, x_{t+1}) + h(x_{T_2}) :$$

$$\{x_t\}_{t=T_1}^{T_2} \in X(T_1, T_2),\ x_{T_1} = y\}, \tag{2.297}$$

$$\widehat{\sigma}^{f,h}(T_1, T_2, y) = \inf\{\sum_{t=T_1}^{T_2-1} f(x_t, x_{t+1}) + h(x_{T_1}) :$$
$$\{x_t\}_{t=T_1}^{T_2} \in X(T_1, T_2),\ x_{T_2} = y\}, \tag{2.298}$$

$$\sigma_{-}^{f,h}(T_1, T_2, y) = \inf\{\sum_{t=T_1}^{T_2-1} \bar{f}(x_t, x_{t+1}) + h(x_{T_2}) :$$
$$\{x_t\}_{t=T_1}^{T_2} \in \bar{X}(T_1, T_2),\ x_{T_1} = y\}, \tag{2.299}$$

$$\widehat{\sigma}_{-}^{f,h}(T_1, T_2, y) = \inf\{\sum_{t=T_1}^{T_2-1} \bar{f}(x_t, x_{t+1}) + h(x_{T_1}) :$$
$$\{x_t\}_{t=T_1}^{T_2} \in \bar{X}(T_1, T_2),\ x_{T_2} = y\}. \tag{2.300}$$

The next result follows from (2.12), (2.13), and (2.15).

Proposition 2.62. *Let $S_2 > S_1 \geq 0$ be integers, $M \geq 0$, $h \in \mathfrak{A}(\mathcal{A})$ and $\{x_t\}_{t=S_1}^{S_2} \in X(S_1, S_2)$. Then the following assertions hold:*

$$\sum_{t=S_1}^{S_2-1} f(x_t, x_{t+1}) + h(x_{S_2}) \leq \sigma^{f,h}(S_1, S_2, x_{S_1}) + M \text{ if and only if}$$

$$\sum_{t=S_1}^{S_2-1} \bar{f}(\bar{x}_t, \bar{x}_{t+1}) + h(\bar{x}_{S_1}) \leq \widehat{\sigma}_{-}^{f,h}(S_1, S_2, \bar{x}_{S_2}) + M;$$

$$\sum_{t=S_1}^{S_2-1} f(x_t, x_{t+1}) + h(x_{S_1}) \leq \widehat{\sigma}^{f,h}(S_1, S_2, x_{S_2}) + M \text{ if and only if}$$

$$\sum_{t=S_1}^{S_2-1} \bar{f}(\bar{x}_t, \bar{x}_{t+1}) + h(\bar{x}_{S_2}) \leq \sigma_{-}^{f,h}(S_1, S_2, \bar{x}_{S_1}) + M.$$

The next result is proved in Section 2.23.

Theorem 2.63. *Assume that $(f, \mathcal{M})$ has TP, that $M, L > 0$ are integers, and that $\epsilon > 0$. Then there exist $\delta > 0$ and an integer $L_0 > L$ such that for each integer $T_1 \geq 0$, each integer $T_2 \geq T_1 + 2L_0$, each $h \in \mathfrak{A}(\mathcal{A})$ satisfying $|h(x_f)| \leq M$, and each $\{x_t\}_{t=T_1}^{T_2} \in X(T_1, T_2)$ which satisfies*

$$x_{T_1} \in \mathcal{A}_L,$$

$$\sum_{t=T_1}^{T_2-1} f(x_t, u_t) + h(x_{T_2}) \le \sigma^{f,h}(T_1, T_2, x_{T_1}) + \delta,$$

there exist integers $\tau_1 \in [T_1, T_1 + L_0]$, $\tau_2 \in [T_2 - L_0, T_2]$ *such that*

$$\rho_E(x_t, x_f) \le \epsilon,\ t = \tau_1, \ldots, \tau_2.$$

Moreover, if $\rho_E(x_{T_1}, x_f) \le \delta$ *then* $\tau_1 = T_1$ *and if* $\rho_E(x_{T_2}, x_f) \le \delta$ *then* $\tau_2 = T_2$.

Note that Theorem 2.63 is valid for the triplet $(\bar{f}, h, \bar{\mathcal{M}})$ too, and combined with Proposition 2.62, this implies the following result.

Theorem 2.64. *Assume that* $(f, \mathcal{M})$ *has TP, that* $M, L > 0$ *are integers, and that* $\epsilon > 0$. *Then there exist* $\delta > 0$ *and an integer* $L_0 > L$ *such that for each integer* $T_1 \ge 0$, *each integer* $T_2 \ge T_1 + 2L_0$, *each* $h \in \mathfrak{A}(\mathcal{A})$ *satisfying* $|h(x_f)| \le M$, *and each* $\{x_t\}_{t=T_1}^{T_2} \in X(T_1, T_2)$ *which satisfies*

$$x_{T_2} \in \widehat{\mathcal{A}}_L,$$

$$\sum_{t=T_1}^{T_2-1} f(x_t, x_{t+1}) + h(x_{T_1}) \le \widehat{\sigma}^{f,h}(T_1, T_2, x_{T_2}) + \delta,$$

there exist integers $\tau_1 \in [T_1, T_1 + L_0]$, $\tau_2 \in [T_2 - L_0, T_2]$ *such that*

$$\rho_E(x_t, x_f) \le \epsilon,\ t = \tau_1, \ldots, \tau_2.$$

Moreover, if $\rho_E(x_{T_1}, x_f) \le \delta$, *then* $\tau_1 = T_1$, *and if* $\rho_E(x_{T_2}, x_f) \le \delta$, *then* $\tau_2 = T_2$.

Let $g \in \mathfrak{A}(\mathcal{A})$ be given. By (2.21), (2.25), and (2.295), $\pi^f + g : \mathcal{A} \to R^1 \cup \{\infty\}$ is bounded from below and satisfies

$$(\pi^f + g)(z) \to \infty \text{ as } z \in \mathcal{A},\ \rho_E(z, \theta_0) \to \infty. \tag{2.301}$$

We prove the following results.

Theorem 2.65. *Let* $(f, \mathcal{M})$ *have TP,* L_0, M *be natural numbers and* $\epsilon \in (0, 1)$. *Then there exist* $\delta > 0$, *an integer* $L_1 > L_0$, *and a neighborhood* $\mathfrak{U}$ *of* g *in* $\mathfrak{A}(\mathcal{A})$ *such that for each integer* $T \ge L_1$, *each* $h \in \mathfrak{U}$, *and each* $\{x_t\}_{t=0}^{T} \in X(0, T)$ *which satisfies*

$$x_0 \in \mathcal{A}_M,\ \sum_{t=0}^{T-1} f(x_t, x_{t+1}) + h(x_T) \le \sigma^{f,h}(0, T, x_0) + \delta,$$

the inequalities

$$(g+\pi_-^f)(x_T) \le \inf(g+\pi_-^f)+\epsilon,\ \Gamma_-^f(\{\bar{x}_t\}_{t=0}^{L_0}) \le \epsilon$$

hold where $\bar{x}_t = x_{T-t}$, $t = 0, \dots, T$.

Theorem 2.66. *Let* $(f, \mathcal{M})$ *have TP,* L_0, M *be natural numbers and* $\epsilon \in (0,1)$. *Then there exist* $\delta > 0$, *an integer* $L_1 > L_0$, *and a neighborhood* $\mathfrak{U}$ *of* g *in* $\mathfrak{A}(\mathcal{A})$ *such that for each integer* $T \ge L_1$, *each* $h \in \mathfrak{U}$, *and each* $\{x_t\}_{t=0}^T \in X(0,T)$ *which satisfies*

$$x_T \in \widehat{\mathcal{A}}_M,\ \sum_{t=0}^{T-1} f(x_t, x_{t+1}) + h(x_0) \le \widehat{\sigma}^{f,h}(0, T, x_T) + \delta,$$

the inequalities

$$(g+\pi^f)(x_0) \le \inf(g+\pi^f)+\epsilon,\ \Gamma^f(\{x_t\}_{t=0}^{L_0}) \le \epsilon$$

hold.

Theorem 2.66 is proved in Section 2.24. It is also valid for $(\bar{f}, \bar{\mathcal{M}})$. Together with Proposition 2.62, this implies Theorem 2.65.

Theorem 2.67. *Let* $(f, \mathcal{M})$ *have TP and LSC property,* $L_0, M > 0$ *be integers and* $\epsilon \in (0,1)$. *Then there exist* $\delta > 0$, *an integer* $L_1 > L_0$, *and a neighborhood* $\mathfrak{U}$ *of* g *in* $\mathfrak{A}(\mathcal{A})$ *such that for each integer* $T \ge L_1$, *each* $h \in \mathfrak{U}$, *and each* $\{x_t\}_{t=0}^T \in X(0,T)$ *which satisfies*

$$x_0 \in \mathcal{A}_M,\ \sum_{t=0}^{T-1} f(x_t, x_{t+1}) + h(x_T) \le \sigma^{f,h}(0, T, x_0) + \delta,$$

there exists an $(\bar{f}, \bar{\mathcal{M}})$*-overtaking optimal* $\{x_t^*\}_{t=0}^\infty \in X(0,\infty)$ *such that*

$$(\pi_-^f + g)(x_0^*) = \inf(\pi_-^f + g),\ \rho_E(x_t^*, x_{T-t}) \le \epsilon,\ t = 0, \dots, L_0.$$

Theorem 2.68. *Let* $(f, \mathcal{M})$ *have TP and LSC property,* $M, L_0 > 0$ *be integers and* $\epsilon \in (0,1)$. *Then there exist* $\delta > 0$, *an integer* $L_1 > L_0$, *and a neighborhood* $\mathfrak{U}$ *of* g *in* $\mathfrak{A}(\mathcal{A})$ *such that for each integer* $T \ge L_1$, *each* $h \in \mathfrak{U}$, *and each* $\{x_t\}_{t=0}^T \in X(0,T)$ *which satisfies*

$$x_T \in \widehat{\mathcal{A}}_M,\ \sum_{t=0}^{T-1} f(x_t, x_{t+1}) + h(x_0) \le \widehat{\sigma}^{f,h}(0, T, x_T) + \delta,$$

there exist an $(f, \mathcal{M})$*-overtaking optimal* $\{x_t^*\}_{t=0}^{\infty} \in X(0, \infty)$ *such that*

$$(\pi^f + g)(x_0^*) = \inf(\pi^f + g), \ \rho_E(x_t, x_t^*) \leq \epsilon, \ t = 0, \dots, L_0.$$

We prove only Theorem 2.68. It is also valid for $(\bar{f}, \bar{\mathcal{M}})$. Together with Proposition 2.62, this implies Theorem 2.67. Theorem 2.68 follows from Theorem 2.66 and the following result which is proved in Section 2.25.

Proposition 2.69. *Let* $(f, \mathcal{M})$ *have TP and LSC property,* $L_0 > 0$ *be an integer, and* $\epsilon \in (0, 1)$*. Then there exist* $\delta \in (0, \epsilon)$ *such that for each* $\{x_t\}_{t=0}^{L_0} \in X(0, L_0)$ *which satisfies*

$$(\pi^f + g)(x_0) \leq \inf(\pi^f + g) + \delta, \ \Gamma^f(\{x_t\}_{t=0}^{L_0}) \leq \delta,$$

there exists an $(f, \mathcal{M})$*-overtaking optimal* $\{x_t^*\}_{t=0}^{\infty} \in X(0, \infty)$ *such that*

$$(\pi^f + g)(x_0^*) = \inf(\pi^f + g), \ \rho_E(x_t^*, x_t) \leq \epsilon, \ t = 0, \dots, L_0.$$

2.23 Proof of Theorem 2.63

By Proposition 2.6, there exist $\delta \in (0, 1)$ and an integer $L_0 > L$ such that the following property holds:

(i) for each integer $T_1 \geq 0$, each integer $T_2 \geq T_1 + 2L_0$, and each $\{x_t\}_{t=T_1}^{T_2} \in X(T_1, T_2)$ which satisfies

$$\sum_{t=T_1}^{T_2-1} f(x_t, u_t) \leq \min\{(T_2-T_1)f(x_f, x_f)+a_1+M+a_0+1+2(1+|f(x_f, x_f)|),$$

$$U^f(T_1, T_2, x_{T_1}, x_{T_2}) + \delta\},$$

there exist integers $\tau_1 \in [T_1, T_1 + L_0]$, $\tau_2 \in [T_2 - L_0, T_2]$ such that

$$\rho_E(x_t, x_f) \leq \epsilon, \ t = \tau_1, \dots, \tau_2.$$

Moreover, if $\rho_E(x_{T_1}, x_f) \leq \delta$, then $\tau_1 = T_1$, and if $\rho_E(x_{T_2}, x_f) \leq \delta$, then $\tau_2 = T_2$.

Assume that $T_1 \geq 0$, $T_2 \geq T_1 + 2L_0$ are integers, $h \in \mathfrak{A}(\mathcal{A})$ satisfies

$$|h(x_f)| \leq M, \tag{2.302}$$

and $\{x_t\}_{t=T_1}^{T_2} \in X(T_1, T_2)$ satisfies

$$x_{T_1} \in \mathcal{A}_L, \tag{2.303}$$

$$\sum_{t=T_1}^{T_2-1} f(x_t, x_{t+1}) + h(x_{T_2}) \le \sigma^{f,h}(T_1, T_2, x_{T_1}) + \delta. \tag{2.304}$$

By (2.304),

$$\sum_{t=T_1}^{T_2-1} f(x_t, x_{t+1}) \le U^f(T_1, T_2, x_{T_1}, x_{T_2}) + \delta. \tag{2.305}$$

In view of (2.303), there exists $\{y_t\}_{t=T_1}^{T_2} \in X(T_1, T_2)$ such that

$$y_{T_1} = x_{T_1}, \ y_t = x_f \text{ for all integers } t \in [L, T_2], \tag{2.306}$$

$$\sum_{t=T_1}^{T_1+L-1} f(y_t, y_{t+1}) \le L + L|f(x_f, x_f)|. \tag{2.307}$$

It follows from (2.295), (2.304), and (2.306) that

$$-a_1 + \sum_{t=T_1}^{T_2-1} f(x_t, x_{t+1}) \le \sum_{t=T_1}^{T_2-1} f(x_t, x_{t+1}) + h(x_{T_2}) \le \sum_{t=T_1}^{T_2-1} f(y_t, y_{t+1}) + h(x_f) + 1$$

$$\le h(x_f) + 1 + L(1 + |f(x_f, x_f)|) + (T_2 - T_1 - L) f(x_f, x_f). \tag{2.308}$$

By (2.302) and (2.308),

$$\sum_{t=T_1}^{T_2-1} f(x_t, x_{t+1}) \le a_1 + 1 + M + 2L(1 + |f(x_f, x_f)|) + (T_2 - T_1) f(x_f, x_f). \tag{2.309}$$

In view of (2.305) and (2.309), there exist integers $\tau_1 \in [T_1, T_1 + L_0]$, $\tau_2 \in [T_2 - L_0, T_2]$ such that

$$\rho_E(x_t, x_f) \le \epsilon, \ t = \tau_1, \dots, \tau_2.$$

Moreover, if $\rho_E(x_{T_1}, x_f) \le \delta$, then $\tau_1 = T_1$, and if $\rho_E(x_{T_2}, x_f) \le \delta$, then $\tau_2 = T_2$. Theorem 2.63 is proved.

2.24 Proof of Theorem 2.66

By Propositions 2.25 and 2.26, Lemma 2.55, and (A2), there exists $\delta_1 \in (0, \epsilon/4)$ such that:

(i) for each $z \in \mathcal{A}$ satisfying $\rho_E(z, x_f) \le 2\delta_1$, we have $|\pi^f(z)| \le \epsilon/16$;

(ii) for each $\{x_t\}_{t=0}^{b_f} \in X(0, b_f)$ which satisfies $\rho_E(x_0, x_f) \le 2\delta_1$, $\rho_E(x_{b_f}, x_f) \le 2\delta_1$, we have

$$\sum_{t=0}^{b_f-1} f(x_t, x_{t+1}) \ge b_f f(x_f, u_f) - \epsilon/16;$$

(iii) for each $z_i \in \mathcal{A}$, $i = 1, 2$ satisfying $\rho_E(z_i, x_f) \le 2\delta_1$, $i = 1, 2$, there exist an integer $\tau \in (0, b_f]$ and $\{x_t\}_{t=0}^{\tau} \in X(0, \tau)$ which satisfies $x_0 = z_1$, $x_\tau = z_2$ and

$$\sum_{t=0}^{\tau-1} f(x_t, x_{t+1}) \le \tau f(x_f, x_f) + \epsilon/16.$$

By (2.25), (2.295), and (2.296), there exists a neighborhood $\mathfrak{U}_1$ of g in $\mathfrak{A}(\mathcal{A})$ such that for each $h \in \mathfrak{U}_1$,

$$|\inf(\pi^f + g) - \inf(\pi^f + h)| \le \delta_1/16. \tag{2.310}$$

Theorem 2.64 implies that there exist $\delta_2 \in (0, \delta_1/8)$, an integer $l_0 > M$, and a neighborhood $\mathfrak{U}_2$ of g in $\mathfrak{A}(\mathcal{A})$ such that the following property holds:

(iv) for each integer $T_1 \ge 0$, each integer $T_2 \ge T_1 + 2l_0$, each $h \in \mathfrak{U}_2$, and each $\{x_t\}_{t=T_1}^{T_2} \in X(T_1, T_2)$ which satisfies

$$x_{T_2} \in \widehat{\mathcal{A}}_M,$$

$$\sum_{t=T_1}^{T_2-1} f(x_t, x_{t+1}) + h(x_{T_1}) \le \widehat{\sigma}^{fh}(T_1, T_2, x_{T_2}) + \delta_2,$$

we have

$$\rho_E(x_t, x_f) \le \delta_1, \ t = T_1 + l_0, \dots, T_2 - l_0.$$

There exists $z_* \in \mathcal{A}$ such that

$$(\pi^f + g)(z_*) \le \inf(\pi^f + g) + \delta_2/8. \tag{2.311}$$

Corollary 2.30 implies that there exists $\{x_t^*\}_{t=0}^{\infty} \in X(0,\infty)$ for which

$$x_0^* = z_*, \lim_{T\to\infty} \Gamma^f(\{x_t^*\}_{t=0}^T) \le \delta_2/8. \tag{2.312}$$

In view of (P1) and (2.311),

$$\lim_{t\to\infty} \rho_E(x_t^*, x_f) = 0. \tag{2.313}$$

In view of Proposition 2.33, the sets $\mathcal{A}_M$ and $\widehat{\mathcal{A}}_M$ are bounded. Equation (2.313) implies that there exists an integer $l_1 > 0$ such that

$$\rho_E(x_t^*, x_f) \le \delta_2/8 \text{ for all integers } t \ge l_1. \tag{2.314}$$

By Theorem 3.2, there exists $\tilde{M} > 0$ such that for each $\{x_t\}_{t=0}^{l_0+l_1+L_0+4} \in X(0, l_0 + l_1 + L_0 + 4)$ satisfying

$$\sum_{t=0}^{l_0+l_1+L_0+3} f(x_t, x_{t+1}) \le (l_0 + l_1 + L_0 + 4) f(x_f, x_f) + a_1 + 2 + \inf(\pi^f + g),$$

we have

$$\rho_E(x_t, \theta_0) \le \tilde{M},\ t = 0, \dots, l_0 + l_1 + L_0 + 3. \tag{2.315}$$

In view of (2.296), there exists a neighborhood $\mathfrak{U}$ of g in $\mathfrak{A}(\mathcal{A})$ such that

$$\mathfrak{U} \subset \mathfrak{U}_1 \cap \mathfrak{U}_2, \tag{2.316}$$

$$|h(x_0^*) - g(x_0^*)| \le \delta_1/16 \text{ for all } h \in \mathfrak{U}, \tag{2.317}$$

$$|h(z) - g(z)| \le \delta_1/16 \text{ for all } h \in \mathfrak{U} \text{ and all } z \in \mathcal{A}_M \cup (\mathcal{A} \cap B_E(\theta_0, \tilde{M})). \tag{2.318}$$

Choose

$$\delta \in (0, \delta_2/4) \tag{2.319}$$

and an integer

$$L_1 > 2L_0 + 2l_0 + 2l_1 + 2b_f + 8. \tag{2.320}$$

Assume that $T \ge L_1$ is an integer,

$$h \in \mathfrak{U}, \tag{2.321}$$

$\{x_t\}_{t=0}^T \in X(0, T)$ satisfies

$$x_T \in \widehat{\mathcal{A}}_M, \tag{2.322}$$

$$\sum_{t=0}^{T-1} f(x_t, x_{t+1}) + h(x_0) \le \widehat{\sigma}^{f,h}(0, T, x_T) + \delta. \tag{2.323}$$

Property (iv), (2.316), and (2.319)–(2.323) imply that

$$\rho_E(x_t, x_f) \le \delta_1, \ t = l_0, \dots, T - l_0. \tag{2.324}$$

In view of (2.320),

$$[l_0 + l_1 + L_0, l_0 + l_1 + L_0 + 2b_f + 8] \subset [l_0, T - l_0 - l_1 - L_0]. \tag{2.325}$$

It follows from (2.324) and (2.325) that

$$\rho_E(x_t, x_f) \le \delta_1, \ t = l_0 + l_1 + L_0, \dots, l_0 + l_1 + L_0 + 2b_f + 8. \tag{2.326}$$

Property (iii), (2.314), (2.320), and (2.326) imply that there exist integers $\tau_1, \tau_2 \in (0, b_f]$ and $\{x_t^{(1)}\}_{t=0}^T \in X(0, T)$ such that

$$x_t^{(1)} = x_t^*, \ t = 0, \dots, l_0 + l_1 + L_0 + 4, \tag{2.327}$$

$$x_{l_0+l_1+L_0+4+\tau_1}^{(1)} = x_f, \tag{2.328}$$

$$\sum_{t=l_0+l_1+L_0+4}^{l_0+l_1+L_0+3+\tau_1} f(x_t^{(1)}, x_{t+1}^{(1)}) \le \tau_1 f(x_f, u_f) + \epsilon/16, \tag{2.329}$$

$$x_{l_0+l_1+L_0+4+2b_f}^{(1)} = x_{l_0+l_1+L_0+4+2b_f}, \tag{2.330}$$

$$x_{l_0+l_1+L_0+4+2b_f-\tau_2}^{(1)} = x_f, \tag{2.331}$$

$$\sum_{t=l_0+l_1+L_0+4+2b_f-\tau_2}^{l_0+l_1+L_0+3+2b_f} f(x_t^{(1)}, x_{t+1}^{(1)}) \le \tau_2 f(x_f, u_f) + \epsilon/16, \tag{2.332}$$

$$x_t^{(1)} = x_f, \ t = l_0 + l_1 + L_0 + 4 + \tau_1, \dots, l_0 + l_1 + L_0 + 4 + 2b_f - \tau_2, \tag{2.333}$$

$$x_t^{(1)} = x_t, \ t = l_0 + l_1 + L_0 + 4 + 2b_f, \dots, T. \tag{2.334}$$

By (2.329) and (2.334),

$$\sum_{t=0}^{T-1} f(x_t, x_{t+1}) + h(x_0) \le \sum_{t=0}^{T-1} f(x_t^{(1)}, x_{t+1}^{(1)}) + h(x_0^{(1)}) + \delta. \tag{2.335}$$

It follows from (2.333)–(2.335) that

$$\delta \ge \sum_{t=0}^{T-1} f(x_t, x_{t+1}) + h(x_0) - \sum_{t=0}^{T-1} f(x_t^{(1)}, x_{t+1}^{(1)}) - h(x_0^{(1)})$$

$$= \sum_{t=0}^{l_0+l_1+L_0+3+2b_f} f(x_t, x_{t+1}) + h(x_0) - \sum_{t=0}^{l_0+l_1+L_0+3+2b_f} f(x_t^{(1)}, x_{t+1}^{(1)}) - h(x_0^{(1)})$$

$$= \sum_{t=0}^{l_0+l_1+L_0+3} f(x_t, x_{t+1}) + \sum_{t=l_0+l_1+L_0+4}^{l_0+l_1+L_0+3+2b_f} f(x_t, x_{t+1}) + h(x_0)$$

$$- \sum_{t=0}^{l_0+l_1+L_0+3} f(x_t^{(1)}, x_{t+1}^{(1)}) - \sum_{t=l_0+l_1+L_0+4}^{l_0+l_1+L_0+3+\tau_1} f(x_t^{(1)}, x_{t+1}^{(1)})$$

$$- f(x_f, x_f)(2b_f - \tau_1 - \tau_2) - \sum_{t=l_0+l_1+L_0+4+2b_f-\tau_2}^{l_0+l_1+L_0+3+2b_f} f(x_t^{(1)}, x_{t+1}^{(1)}) - h(x_0^{(1)}). \tag{2.336}$$

Property (ii) and (2.326) imply that

$$\sum_{t=l_0+l_1+L_0+4}^{l_0+l_1+L_0+3+2b_f} f(x_t, x_{t+1}) \ge 2b_f f(x_f, x_f) - \epsilon/8. \tag{2.337}$$

By (2.26), (2.311), (2.312), (2.317), (2.319), (2.326), (2.327), (2.329), (2.332), (2.333), (2.336), (2.337), and property (i),

$$\sum_{t=0}^{l_0+l_1+L_0+3} f(x_t, x_{t+1}) + h(x_0) \le \delta + \epsilon/4 + \sum_{t=0}^{l_0+l_1+L_0+3} f(x_t^{(1)}, x_{t+1}^{(1)}) + h(x_0^*)$$

$$= \delta + \epsilon/4 + \sum_{t=0}^{l_0+l_1+L_0+3} f(x_t^*, x_{t+1}^*) + h(x_0^*)$$

$$= \delta + \epsilon/4 + \Gamma^f(\{x_t^*\}_{t=0}^{l_0+l_1+L_0+3}) + (l_0 + l_1 + L_0 + 4) f(x_f, x_f)$$

$$+\pi^f(x_0^*) - \pi^f(x^*_{l_0+l_1+L_0+4}) + h(x_0^*)$$

$$\leq \Gamma^f(\{x_t^*\}_{t=0}^{l_0+l_1+L_0+3}) + (l_0+l_1+L_0+4)f(x_f, x_f) + (\pi^f+h)(x_0^*) + \delta + \epsilon/4 + \epsilon/16$$

$$\leq (l_0 + l_1 + L_0 + 4)f(x_f, x_f) + \delta_2/8 + \delta + \epsilon/4 + \epsilon/16 + (\pi^f + h)(x_0^*)$$

$$\leq (l_0 + l_1 + L_0 + 4)f(x_f, x_f) + \epsilon/64 + \epsilon/64 + 5\epsilon/16 + (\pi^f + g)(x_0^*) + \delta_1/16$$

$$\leq (l_0 + l_1 + L_0 + 4)f(x_f, x_f) + \inf(\pi^f + g) + \delta_2/8 + \delta_1/16 + \epsilon/32 + 5\epsilon/16$$

$$\leq (l_0 + l_1 + L_0 + 4)f(x_f, x_f) + \inf(\pi^f + g) + 3\epsilon/8. \tag{2.338}$$

In view of (2.295) and (2.338),

$$\sum_{t=0}^{l_0+l_1+L_0+3} f(x_t, x_{t+1}) \leq (l_0 + l_1 + L_0 + 4)f(x_f, x_f) + \inf(\pi^f + g) + a_1 + 2. \tag{2.339}$$

It follows from (2.315) and (2.339) that

$$\rho_E(x_t, \theta_0) \leq \tilde{M},\ t = 0, \dots, l_0 + l_1 + L_0 + 3. \tag{2.340}$$

It follows from (2.318), (2.321), and (2.340) that

$$|h(x_0) - g(x_0)| \leq \delta_1/16 \leq \epsilon/16. \tag{2.341}$$

By (2.338) and (2.341),

$$\sum_{t=0}^{l_0+l_1+L_0+3} f(x_t, x_{t+1}) + g(x_0)$$

$$\leq (l_0 + l_1 + L_0 + 4)f(x_f, x_f) + \inf(\pi^f + g) + 3\epsilon/8 + \epsilon/16. \tag{2.342}$$

In view of (2.26) and (2.340),

$$\Gamma^f(\{x_t\}_{t=0}^{l_0+l_1+L_0+4}) + \pi^f(x_0) - \pi^f(x_{l_0+l_1+L_0+4}) + g(x_0)$$

$$\leq \inf(\pi^f + g) + 3\epsilon/8 + \epsilon/16. \tag{2.343}$$

Property (i) and (2.326) imply that

$$|\pi^f(x_{l_0+l_1+L_0+4})| \leq \epsilon/16. \tag{2.344}$$

It follows from (2.343) and (2.344) that

$$\Gamma^f(\{x_t\}_{t=0}^{l_0+l_1+L_0+4}) + (\pi^f + g)(x_0) \le \inf(\pi^f + g) + \epsilon/2. \tag{2.345}$$

By (2.27) and (2.345),

$$(\pi^f + g)(x_0) \le \inf(\pi^f + g) + \epsilon, \; \Gamma^f(\{x_t\}_{t=0}^{l_0+l_1+L_0+4}) \le \epsilon.$$

Theorem 2.66 is proved.

2.25 Proof of Proposition 2.69

Assume that the proposition does not hold. Then there exist a sequence $\{\delta_k\}_{k=1}^{\infty} \subset (0, 1]$ and a sequence $\{x_t^{(k)}\}_{t=0}^{L_0} \in X(0, L_0)$, $k = 1, 2, \dots$ such that

$$\lim_{k\to\infty} \delta_k = 0 \tag{2.346}$$

and that for all integers $k \ge 1$,

$$(\pi^f + g)(x_0^{(k)}) \le \inf(\pi^f + g) + \delta_k, \tag{2.347}$$

$$\Gamma^f(\{x_t^{(k)}\}_{t=0}^{L_0}) \le \delta_k \tag{2.348}$$

and that the following property holds:

(i) for each $(f, \mathcal{M})$-overtaking optimal $\{y_t\}_{t=0}^{\infty} \in X(0, \infty)$ satisfying $(\pi^f + g)(y_0) = \inf(\pi^f + g)$, we have $\max\{\rho_E(x_t^{(k)}, y_t) : t = 0, \dots, L_0\} > \epsilon$.

By (2.21), (2.295), and (2.347), the sequence

$$\{\pi^f(x_0^{(k)})\}_{k=1}^{\infty} \text{ is bounded.} \tag{2.349}$$

Proposition 2.27 and (2.349) imply that the sequence $\{x_0^{(k)}\}_{k=1}^{\infty}$ is bounded. In view of (2.20), (2.26), (2.348), and (2.349), the sequence

$$\{\sum_{t=0}^{L_0-1} f(x_t^{(k)}, x_{t+1}^{(k)})\}_{k=1}^{\infty}$$

is bounded. By LSC property, extracting a subsequence and re-indexing if necessary, we may assume without loss of generality that there exists $\{x_t\}_{t=0}^{L_0} \in X(0, L_0)$ such that

$$x_t^{(k)} \to x_t \text{ as } k \to \infty \text{ for all } t = 0, \dots, L_0, \tag{2.350}$$

$$\sum_{t=0}^{L_0-1} f(x_t, x_{t+1}) \le \liminf_{k\to\infty} \sum_{t=0}^{L_0-1} f(x_t^{(k)}, x_{t+1}^{(k)}). \tag{2.351}$$

By Proposition 2.37, (2.347), (2.350), and the lower semicontinuity of $\pi^f, \pi^f + g$,

$$\pi^f(x_0) \le \liminf_{k\to\infty} \pi^f(x_0^{(k)}),\ g(x_0) \le \liminf_{k\to\infty} g(x_0^{(k)}), \tag{2.352}$$

$$(\pi^f + g)(x_0) \le \liminf_{k\to\infty} (\pi^f + g)(x_0^{(k)}) \le \inf(\pi^f + g), \tag{2.353}$$

$$(\pi^f + g)(x_0) = \inf(\pi^f + g) = \lim_{k\to\infty} (\pi^f + g)(x_0^{(k)}). \tag{2.354}$$

In view of (2.352) and (2.354),

$$g(x_0) = \lim_{k\to\infty} g(x_0^{(k)}),\ \pi^f(x_0) = \lim_{k\to\infty} \pi^f(x_0^{(k)}). \tag{2.355}$$

By (2.350) and the lower semicontinuity of $\pi^f + g$,

$$(\pi^f + g)(x_{L_0}) \le \liminf_{k\to\infty} (\pi^f + g)(x_{L_0}^{(k)}). \tag{2.356}$$

It follows from (2.26), (2.27), (2.346), (2.348), (2.351), and the lower semicontinuity of π^f that

$$\sum_{t=0}^{L_0-1} f(x_t, x_{t+1}) - L_0 f(x_f, x_f) - \pi^f(x_0) + \pi^f(x_{L_0})$$

$$\le \liminf_{k\to\infty} \Big(\sum_{t=0}^{L_0-1} f(x_t^{(k)}, x_{t+1}^{(k)}) - L_0 f(x_f, x_f)\Big) - \lim_{k\to\infty} \pi^f(x_0^{(k)}) + \liminf_{k\to\infty} \pi^f(x_{L_0}^{(k)})$$

$$\le \liminf_{k\to\infty} \Big(\sum_{t=0}^{L_0-1} f(x_t^{(k)}, x_{t+1}^{(k)}) - L_0 f(x_f, x_f) - \pi^f(x_0^{(k)}) + \pi^f(x_{L_0}^{(k)})\Big) \le 0,$$

$$\sum_{t=0}^{L_0-1} f(x_t, x_{t+1}) - L_0 f(x_f, x_f) - \pi^f(x_0) + \pi^f(x_{L_0}) = 0. \tag{2.357}$$

Theorem 2.13 and (2.357) imply that there exists an $(f, \mathcal{M})$-overtaking optimal $\{\tilde{x}_t\}_{t=0}^{\infty} \in X(0, \infty)$ such that

$$\tilde{x}_0 = x_{L_0}. \tag{2.358}$$

For all integers $t > L_0$ set

$$x_t = \tilde{x}_{t-L_0}. \tag{2.359}$$

It is clear that $\{x_t\}_{t=0}^{\infty} \in X(0, \infty)$ is (f)-good. By Proposition 2.25, (2.26), (2.27), (2.359), and (2.365), for all integers $S > 0$,

$$\sum_{t=0}^{S-1} f(x_t, x_{t+1}) - Sf(x_f, x_f) - \pi^f(x_0) + \pi^f(x_S) = 0$$

and $\{x_t\}_{t=0}^{\infty} \in X(0, \infty)$ is $(f, \mathcal{M})$-overtaking optimal. It follows from (2.346), (2.347) and (2.356) that $(\pi^f + g)(x_0) = \inf(\pi^f + g)$. By (2.350), for all sufficiently large natural numbers k, $\rho_E(x_t^{(k)}, x_t) \le \epsilon/2$ for all $t = 0, \dots, L_0$. This contradicts property (i). The contradiction we have reached proves Proposition 2.69.

2.26 The Second Bolza Problem

We consider the control system introduced and studied in Section 2.4 and in Section 2.5 and use the notation, definitions, and assumptions used there. Let $a_1 > 0$. Denote by $\mathfrak{A}$ the set of all lower semicontinuous functions $h : \mathcal{A} \times \mathcal{A} \to R^1$ which are bounded on bounded subsets of $\mathcal{A} \times \mathcal{A}$ and such that

$$h(z_1, z_2) \ge -a_1 \text{ for all } z_1, z_2 \in \mathcal{A}. \tag{2.360}$$

The set $\mathfrak{A}$ is equipped with the uniformity which is determined by the base

$$\mathcal{E}(N, \epsilon) = \{(h_1, h_2) : \mathfrak{A} \times \mathfrak{A} : \ |h_1(z_1, z_2) - h_2(z_1, z_2)| \le \epsilon$$
$$\text{for all } z_1, z_2 \in \mathcal{A} \text{ satisfying } \rho_E(z_i, \theta_0) \le N,\ i = 1, 2\}, \tag{2.361}$$

where $N, \epsilon > 0$. It is not difficult to see that the uniform space $\mathfrak{A}$ is metrizable and complete.

For each pair of integers $T_2 > T_1 \ge 0$ and each $h \in \mathfrak{A}$, we define

$$\sigma^{f,h}(T_1, T_2) = \inf\{\sum_{t=T_1}^{T_2-1} f(x_t, x_{t+1}) + h(x_{T_1}, x_{T_2}) : \{x_t\}_{t=T_1}^{T_2} \in X(T_1, T_2)\}. \tag{2.362}$$

The next result is proved in Section 2.27.

Theorem 2.70. *Assume that $(f, \mathcal{M})$ has TP, $M, \epsilon > 0$. Then there exist $\delta > 0$ and an integer $L > 0$ such that for each integer $T_1 \ge 0$, each integer $T_2 \ge T_1 + 2L$, each $h \in \mathfrak{A}$ satisfying $|h(x_f, x_f)| \le M$, and each $\{x_t\}_{t=T_1}^{T_2} \in X(T_1, T_2)$ which satisfies*

$$\sum_{t=T_1}^{T_2-1} f(x_t, x_{t+1}) + h(x_{T_1}, x_{T_2}) \le \sigma^{f,h}(T_1, T_2) + \delta,$$

there exist integers $\tau_1 \in [T_1, T_1 + L]$, $\tau_2 \in [T_2 - L, T_2]$ *such that*

$$\rho_E(x_t, x_f) \le \epsilon, \ t = \tau_1, \dots, \tau_2.$$

Moreover, if $\rho_E(x_{T_1}, x_f) \le \delta$, *then* $\tau_1 = T_1$, *and if* $\rho_E(x_{T_2}, x_f) \le \delta$, *then* $\tau_2 = T_2$.

Let $h \in \mathfrak{A}$. Define

$$\psi_h(z_1, z_2) = \pi^f(z_1) + \pi_-^f(z_2) + h(z_1, z_2), \ z_1, z_2 \in \mathcal{A}. \tag{2.363}$$

By (2.21), (2.25), (2.360), and Proposition 2.37, $\psi_h : \mathcal{A} \times \mathcal{A} \to R^1 \cup \{\infty\}$ is bounded from below and satisfies

$$\psi_h(z_1, z_2) \to \infty \text{ as } z_1, z_2 \in \mathcal{A}, \rho_E(z_1, \theta_0) + \rho_E(z_2, \theta_0) \to \infty. \tag{2.364}$$

The next theorem is proved in Section 2.28.

Theorem 2.71. *Let* $(f, \mathcal{M})$ *have TP,* $g \in \mathfrak{A}$, L_0 *be a natural number and* $\epsilon \in (0, 1)$. *Then there exist* $\delta > 0$, *an integer* $L_1 > L_0$, *and a neighborhood* $\mathfrak{U}$ *of* g *in* $\mathfrak{A}$ *such that for each integer* $T \ge L_1$, *each* $h \in \mathfrak{U}$, *and each* $\{x_t\}_{t=0}^T \in X(0, T)$ *which satisfies*

$$\sum_{t=0}^{T-1} f(x_t, x_{t+1}) + h(x_0, x_T) \le \sigma^{f,h}(0, T) + \delta,$$

the inequalities

$$\psi_g(x_0, x_T) \le \inf(\psi_g) + \epsilon, \ \Gamma^f(\{x_t\}_{t=0}^{L_0}) \le \epsilon, \ \Gamma_-^f(\{\bar{x}_t\}_{t=0}^{L_0}) \le \epsilon$$

hold where $\bar{x}_t = x_{T-t}$, $t = 0, \dots, T$.

Theorem 2.72. *Let* $(f, \mathcal{M})$ *have TP and LSC property,* $g \in \mathfrak{A}$, $L_0 > 0$ *be an integer and* $\epsilon > 0$. *Then there exist* $\delta > 0$, *an integer* $L_1 > L_0$, *and a neighborhood* $\mathfrak{U}$ *of* g *in* $\mathfrak{A}$ *such that for each integer* $T \ge L_1$, *each* $h \in \mathfrak{U}$, *and each* $\{x_t\}_{t=0}^T \in X(0, T)$ *which satisfies*

$$\sum_{t=0}^{T-1} f(x_t, x_{t+1}) + h(x_0, x_T) \le \sigma^{f,h}(0, T) + \delta,$$

there exist an $(f, \mathcal{M})$*-overtaking optimal* $\{x_t^{*,1}\}_{t=0}^\infty \in X(0, \infty)$ *and an* $(\bar{f}, \bar{\mathcal{M}})$*-overtaking optimal* $\{x_t^{*,2}\}_{t=0}^\infty \in \bar{X}(0, \infty)$ *such that*

$$\psi_g(x_0^{*,1}, x_0^{*,2}) = \inf(\psi_g),\ \rho_E(x_t^{*,1}, x_t) \le \epsilon,\ \rho_E(x_t^{*,2}, x_{T-t}) \le \epsilon,\ t = 0, \dots, L_0.$$

Theorem 2.72 follows from Theorem 2.71 and the following result which is proved in Section 2.29.

Proposition 2.73. *Let $(f, \mathcal{M})$ have TP and LSC property, $g \in \mathfrak{A}$, $L_0 > 0$ be an integer and $\epsilon \in (0, 1)$. Then there exist $\delta > 0$ such that for each $\{x_t^{(1)}\}_{t=0}^{L_0} \in X(0, L_0)$ and each $\{x_t^{(2)}\}_{t=0}^{L_0} \in \bar{X}(0, L_0)$ which satisfy*

$$\psi_g(x_0^{(1)}, x_0^{(2)}) \le \inf(\psi_g) + \delta,\ \Gamma^f(\{x_t^{(1)}\}_{t=0}^{L_0}) \le \delta,\ \Gamma_-^f(\{x_t^{(2)}\}_{t=0}^{L_0}) \le \delta,$$

there exist an $(f, \mathcal{M})$-overtaking optimal $\{y_t^{(1)}\}_{t=0}^{\infty} \in X(0, \infty)$ and an $(\bar{f}, \bar{\mathcal{M}})$-overtaking optimal $\{y_t^{(2)}\}_{t=0}^{\infty} \in \bar{X}(0, \infty)$ such that

$$\psi_g(y_0^{(1)}, y_0^{(2)}) = \inf(\psi_g),\ \rho_E(x_t^{(i)}, y_t^{(i)}) \le \epsilon,\ i = 1, 2,\ t = 0, \dots, L_0.$$

2.27 Proof of Theorem 2.70

By Proposition 2.6, there exist $\delta \in (0, 1)$ and an integer $L > 0$ such that the following property holds:

(i) for each integer $T_1 \ge 0$, each integer $T_2 \ge T_1 + 2L$, and each $\{x_t\}_{t=T_1}^{T_2} \in X(T_1, T_2)$ which satisfies

$$\sum_{t=T_1}^{T_2-1} f(x_t, u_t) \le \min\{(T_2 - T_1) f(x_f, u_f) + a_1 + M + a_0 + 1,$$

$$U^f(T_1, T_2, x_{T_1}, x_{T_2}) + \delta\},$$

there exist integers $\tau_1 \in [T_1, T_1 + L]$, $\tau_2 \in [T_2 - L, T_2]$ such that

$$\rho_E(x_t, x_f) \le \epsilon,\ t = \tau_1, \dots, \tau_2.$$

Moreover, if $\rho_E(x_{T_1}, x_f) \le \delta$, then $\tau_1 = T_1$, and if $\rho_E(x_{T_2}, x_f) \le \delta$, then $\tau_2 = T_2$.

Assume that $T_1 \ge 0$, $T_2 \ge T_1 + 2L$ are integers, $h \in \mathfrak{A}$ satisfies

$$|h(x_f, x_f)| \le M, \tag{2.365}$$

and $\{x_t\}_{t=T_1}^{T_2} \in X(T_1, T_2)$ satisfies

$$\sum_{t=T_1}^{T_2-1} f(x_t, x_{t+1}) + h(x_{T_1}, x_{T_2}) \le \sigma^{f,h}(T_1, T_2) + \delta. \tag{2.366}$$

By (2.366),

$$\sum_{t=T_1}^{T_2-1} f(x_t, x_{t+1}) \le U^f(T_1, T_2, x_{T_1}, x_{T_2}) + \delta. \tag{2.367}$$

It follows from (2.360), (2.365), and (2.366) that

$$-a_1 + \sum_{t=T_1}^{T_2-1} f(x_t, x_{t+1}) \le \sum_{t=T_1}^{T_2-1} f(x_t, x_{t+1}) + h(x_{T_1}, x_{T_2})$$

$$\le h(x_f, x_f) + 1 + (T_2 - T_1) f(x_f, x_f),$$

$$\sum_{t=T_1}^{T_2-1} f(x_t, x_{t+1}) \le (T_2 - T_1) f(x_f, x_f) + M + a_1 + 1. \tag{2.368}$$

In view of (2.367), (2.368), and property (i), there exist integers $\tau_1 \in [T_1, T_1 + L]$, $\tau_2 \in [T_2 - L, T_2]$ such that

$$\rho_E(x_t, x_f) \le \epsilon, \ t = \tau_1, \dots, \tau_2.$$

Moreover, if $\rho_E(x_{T_1}, x_f) \le \delta$, then $\tau_1 = T_1$, and if $\rho_E(x_{T_2}, x_f) \le \delta$, then $\tau_2 = T_2$. Theorem 2.70 is proved.

2.28 Proof of Theorem 2.71

By Propositions 2.25 and 2.26, Lemma 2.55, and (A2), there exists $\delta_1 \in (0, \epsilon/16)$ such that:

(i) for each $z \in \mathcal{A}$ satisfying $\rho_E(z, x_f) \le 2\delta_1$, we have $|\pi^f(z)| \le \epsilon/16$, $|\pi_-^f(z)| \le \epsilon/16$;
(ii) for each $\{x_t\}_{t=0}^{b_f} \in X(0, b_f)$ which satisfies $\rho_E(x_0, x_f) \le 2\delta_1$, $\rho_E(x_{b_f}, x_f) \le 2\delta_1$, we have

$$\sum_{t=0}^{b_f-1} f(x_t, x_{t+1}) \geq b_f f(x_f, u_f) - \epsilon/16;$$

(iii) for each $z_i \in \mathcal{A}$, $i = 1, 2$ satisfying $\rho_E(z_i, x_f) \leq 2\delta_1$, $i = 1, 2$, there exist an integer $\tau \in (0, b_f]$ and $\{x_t\}_{t=0}^{\tau} \in X(0, \tau)$ which satisfies $x_0 = z_1$, $x_\tau = z_2$ and

$$\sum_{t=0}^{\tau-1} f(x_t, x_{t+1}) \leq \tau f(x_f, u_f) + \epsilon/16.$$

By (2.21), (2.360), (2.361), (2.363), and (2.364), there exists a neighborhood $\mathfrak{U}_1$ of g in $\mathfrak{A}$ such that for each $h \in \mathfrak{U}_1$,

$$|\inf(\psi_h) - \inf(\psi_g)| \leq \delta_1/16. \tag{2.369}$$

Theorem 2.70 implies that there exist $\delta_2 \in (0, \delta_1/8)$, an integer $l_0 > 0$, and a neighborhood $\mathfrak{U}_2$ of g in $\mathfrak{A}$ such that the following property holds:

(iv) for each integer $T_1 \geq 0$, each integer $T_2 \geq T_1 + 2l_0$, each $h \in \mathfrak{U}_2$, and each $\{x_t\}_{t=T_1}^{T_2} \in X(T_1, T_2)$ which satisfies

$$\sum_{t=T_1}^{T_2-1} f(x_t, x_{t+1}) + h(x_{T_1}, x_{T_2}) \leq \sigma^{f,h}(T_1, T_2) + \delta_2,$$

we have

$$\rho_E(x_t, x_f) \leq \delta_1, \ t = T_1 + l_0, \dots, T_2 - l_0.$$

There exist $z_{*,1}, z_{*,2} \in \mathcal{A}$ such that

$$\psi_g(z_{*,1}, z_{*,2}) \leq \inf(\psi_g) + \delta_2/8. \tag{2.370}$$

Corollary 2.30, (2.363), and (2.370) imply that there exist $\{x_t^{*,1}\}_{t=0}^{\infty} \in X(0, \infty)$, $\{x_t^{*,2}\}_{t=0}^{\infty} \in \bar{X}(0, \infty)$ for which

$$x_0^{*,i} = z_{*,i}, \ i = 1, 2, \tag{2.371}$$

$$\lim_{T \to \infty} \Gamma^f(\{x_t^{*,1}\}_{t=0}^{T}) \leq \delta_2/8, \tag{2.372}$$

$$\lim_{T \to \infty} \Gamma_-^f(\{x_t^{*,2}\}_{t=0}^{T}) \leq \delta_2/8. \tag{2.373}$$

In view of (P1), Proposition 2.29, (2.372), and (2.373),

$$\lim_{t\to\infty} \rho_E(x_t^{*,i}, x_f) = 0, \ i = 1, 2. \tag{2.374}$$

Equation (2.374) implies that there exists an integer $l_1 > 0$ such that

$$\rho_E(x_t^{*,i}, x_f) \le \delta_2/8, \ i = 1, 2 \text{ for all integers } t \ge l_1. \tag{2.375}$$

In view of (2.25), (2.361), and (2.363), there is a neighborhood $\mathfrak{U}$ of g in $\mathfrak{A}$ such that

$$\mathfrak{U} \subset \mathfrak{U}_1 \cap \mathfrak{U}_2, \tag{2.376}$$

$$|h(x_0^{*,1}, x_0^{*,2}) - g(x_0^{*,1} x_0^{*,2})| \le \delta_1/16 \text{ for all } h \in \mathfrak{U}, \tag{2.377}$$

and for all $h \in \mathfrak{U}$ and all $z_1, z_2 \in \mathfrak{A}$ satisfying

$$\psi_h(z_1, z_2) \le \inf(\psi_g) + 4,$$

we have

$$|\psi_g(z_1, z_2) - \psi_h(z_1, z_2)| \le \epsilon/64. \tag{2.378}$$

Choose

$$\delta \in (0, \delta_2/8) \tag{2.379}$$

and an integer

$$L_1 > 4L_0 + 4l_0 + 4l_1 + 4b_f + 16. \tag{2.380}$$

Assume that $T \ge L_1$ is an integer,

$$h \in \mathfrak{U}, \tag{2.381}$$

$\{x_t\}_{t=0}^T \in X(0, T)$ satisfies

$$\sum_{t=0}^{T-1} f(x_t, x_{t+1}) + h(x_0, x_T) \le \sigma^{f,h}(0, T) + \delta. \tag{2.382}$$

Property (iv), (2.376), and (2.379)–(2.382) imply that

$$\rho_E(x_t, x_f) \le \delta_1, \ t = l_0, \dots, T - l_0. \tag{2.383}$$

Set

$$\bar{x}_t = x_{T-t},\ t = 0, \ldots, T. \tag{2.384}$$

Property (iii), (2.375), (2.380), (2.381), and (2.383) imply that there exist integers $\tau_1, \tau_2, \tau_3, \tau_4 \in (0, b_f]$ and $\{x_t^{(1)}\}_{t=0}^T \in X(0, T)$ such that

$$x_t^{(1)} = x_t^{*,1},\ t = 0, \ldots, l_0 + l_1 + L_0 + 4, \tag{2.385}$$

$$x^{(1)}_{l_0+l_1+L_0+4+\tau_1} = x_f, \quad \sum_{t=l_0+l_1+L_0+4}^{l_0+l_1+L_0+3+\tau_1} f(x_t^{(1)}, x_{t+1}^{(1)}) \le \tau_1 f(x_f, x_f) + \epsilon/16, \tag{2.386}$$

$$x^{(1)}_{l_0+l_1+L_0+4+2b_f} = x_{l_0+l_1+L_0+4+2b_f},\ x^{(1)}_{l_0+l_1+L_0+4+2b_f-\tau_2} = x_f, \tag{2.387}$$

$$\sum_{t=l_0+l_1+L_0+4+2b_f-\tau_2}^{l_0+l_1+L_0+3+2b_f} f(x_t^{(1)}, x_{t+1}^{(1)}) \le \tau_2 f(x_f, x_f) + \epsilon/16, \tag{2.388}$$

$$x_t^{(1)} = x_f,\ t = l_0 + l_1 + L_0 + 4 + \tau_1, \ldots, l_0 + l_1 + L_0 + 4 + 2b_f - \tau_2, \tag{2.389}$$

$$x_{T-t}^{(1)} = x_t^{*,2},\ t = 0, \ldots, l_0 + l_1 + L_0 + 4, \tag{2.390}$$

$$x^{(1)}_{T-l_0-l_1-L_0-4-\tau_3} = x_f, \tag{2.391}$$

$$\sum_{t=T-l_0-l_1-L_0-4-\tau_3}^{T-l_0-l_1-L_0-5} f(x_t^{(1)}, x_{t+1}^{(1)}) \le \tau_3 f(x_f, x_f) + \epsilon/16, \tag{2.392}$$

$$x^{(1)}_{T-l_0-l_1-L_0-4-2b_f} = x_{T-l_0-l_1-L_0-4-2b_f}, \tag{2.393}$$

$$x^{(1)}_{T-l_0-l_1-L_0-4-2b_f+\tau_4} = x_f, \tag{2.394}$$

$$\sum_{t=T-l_0-l_1-L_0-4-2b_f}^{T-l_0-l_1-L_0-5-2b_f+\tau_4} f(x_t^{(1)}, x_{t+1}^{(1)}) \le \tau_4 f(x_f, x_f) + \epsilon/16, \tag{2.395}$$

$$x_t^{(1)} = x_f,\ t = T - l_0 - l_1 - L_0 - 4 - 2b_f + \tau_4, \ldots, T - l_0 - l_1 - L_0 - 4 - \tau_3, \tag{2.396}$$

$$x_t^{(1)} = x_t,\ t = l_0 + l_1 + L_0 + 4 + 2b_f, \ldots, T - l_0 - l_1 - L_0 - 4 - 2b_f. \tag{2.397}$$

By (2.382) and (2.397),

$$\sum_{t=0}^{T-1} f(x_t, x_{t+1}) + h(x_0, x_T) \le \sum_{t=0}^{T-1} f(x_t^{(1)}, x_{t+1}^{(1)}) + h(x_0^{(1)}, x_T^{(1)}) + \delta. \tag{2.398}$$

It follows from (2.397) and (2.398) that

$$\begin{aligned}
\delta &\ge \sum_{t=0}^{T-1} f(x_t, x_{t+1}) + h(x_0, x_T) - \sum_{t=0}^{T-1} f(x_t^{(1)}, x_{t+1}^{(1)}) - h(x_0^{(1)}, x_T^{(1)}) \\
&= \sum_{t=0}^{l_0+l_1+L_0+3} f(x_t, x_{t+1}) + \sum_{t=l_0+l_1+L_0+4}^{l_0+l_1+L_0+3+b_f} f(x_t, x_{t+1}) \\
&\quad + \sum_{t=l_0+l_1+L_0+4+b_f}^{l_0+l_1+L_0+3+2b_f} f(x_t, x_{t+1}) \\
&\quad + \sum_{t=T-l_0-l_1-L_0-4-2b_f}^{T-l_0-l_1-L_0-5-b_f} f(x_t, x_{t+1}) + \sum_{t=T-l_0-l_1-L_0-4-b_f}^{T-l_0-l_1-L_0-5} f(x_t, x_{t+1}) \\
&\quad + \sum_{t=T-l_0-l_1-L_0-4}^{T-l} f(x_t, x_{t+1}) - h(x_0, x_T) \\
&\quad - \sum_{t=0}^{l_0+l_1+L_0+3} f(x_t^{(1)}, x_{t+1}^{(1)}) - \sum_{t=l_0+l_1+L_0+4}^{l_0+l_1+L_0+3+\tau_1} f(x_t^{(1)}, x_{t+1}^{(1)}) \\
&\quad - \sum_{t=l_0+l_1+L_0+4+\tau_1}^{l_0+l_1+L_0+2b_f-\tau_2+3} f(x_t^{(1)}, x_{t+1}^{(1)}) - \sum_{t=l_0+l_1+L_0+4+2b_f-\tau_2}^{l_0+l_1+L_0+3+2b_f} f(x_t^{(1)}, x_{t+1}^{(1)}) \\
&\quad - \sum_{t=T-l_0-l_1-L_0-4-2b_f}^{T-l_0-l_1-L_0-5-2b_f+\tau_4} f(x_t^{(1)}, x_{t+1}^{(1)}) - \sum_{t=T-l_0-l_1-L_0-4-2b_f+\tau_4}^{T-l_0-l_1-L_0-5-\tau_3} f(x_t^{(1)}, x_{t+1}^{(1)}) \\
&\quad - \sum_{t=T-l_0-l_1-L_0-4-\tau_3}^{T-l_0-l_1-L_0-5} f(x_t^{(1)}, x_{t+1}^{(1)}) - \sum_{t=T-l_0-l_1-L_0-4}^{T-l} f(x_t^{(1)}, x_{t+1}^{(1)}) - h(x_0^{(1)}, x_T^{(1)}).
\end{aligned} \tag{2.399}$$

Property (ii), (2.15), (2.380), (2.381), (2.383)–(2.386), (2.397), and (2.399) imply that

$$\delta \geq \sum_{t=0}^{l_0+l_1+L_0+3} f(x_t, x_{t+1}) + 2(b_f f(x_f, x_f) - \epsilon/16) + 2(b_f f(x_f, x_f) - \epsilon/16)$$

$$+ \sum_{t=0}^{l_0+l_1+L_0+3} \bar{f}(\bar{x}_t, \bar{x}_{t+1}) + h(x_0, \bar{x}_0)$$

$$- \sum_{t=0}^{l_0+l_1+L_0+3} f(x_t^{*,1}, x_{t+1}^{*,1}) - \tau_1 f(x_f, x_f) - \epsilon/16$$

$$-(2b_f - \tau_1 - \tau_2) f(x_f, x_f) - \tau_2 f(x_f, x_f) - \epsilon/16$$

$$-\tau_4 f(x_f, x_f) - \epsilon/16 - (2b_f - \tau_3 - \tau_4) f(x_f, x_f) - \tau_3 f(x_f, x_f) - \epsilon/16$$

$$- \sum_{t=0}^{l_0+l_1+L_0+3} \bar{f}(x_t^{*,2}, x_{t+1}^{*,2}) - h(x_0^{*,1}, x_0^{*,2})$$

$$\geq \sum_{t=0}^{l_0+l_1+L_0+3} f(x_t, x_{t+1}) + \sum_{t=0}^{l_0+l_1+L_0+3} \bar{f}(\bar{x}_t, \bar{x}_{t+1}) + h(x_0, \bar{x}_0)$$

$$- \sum_{t=0}^{l_0+l_1+L_0+3} f(x_t^{*,1}, x_{t+1}^{*,1}) - \sum_{t=0}^{l_0+l_1+L_0+3} \bar{f}(x_t^{*,2}, x_{t+1}^{*,2}) - h(x_0^{*,1}, x_0^{*,2}) - \epsilon/2. \tag{2.400}$$

By property (i), (2.363), (2.372), (2.373), (2.375), (2.377), (2.380), (2.381), (2.383), (2.288), and (2.400),

$$\delta + \epsilon/2 \geq \Gamma^f(\{x_t\}_{t=0}^{l_0+l_1+L_0+4})$$

$$+(l_0 + l_1 + L_0 + 4) f(x_f, x_f) + \pi^f(x_0) - \pi^f(x_{l_0+l_1+L_0+4})$$

$$+\Gamma_-^f(\{\bar{x}_t\}_{t=0}^{l_0+l_1+L_0+4}) + (l_0 + l_1 + L_0 + 4) f(x_f, x_f)$$

$$+\pi_-^f(\bar{x}_0) - \pi_-^f(\bar{x}_{l_0+l_1+L_0+4}) + h(x_0, \bar{x}_0)$$

$$-\Gamma^f(\{x_t^{*,1}\}_{t=0}^{l_0+l_1+L_0+4}) - (l_0 + l_1 + L_0 + 4) f(x_f, x_f)$$

$$-\pi^f(x_0^{*,1}) + \pi^f(x_{l_0+l_1+L_0+4}^{*,1})$$

$$-\Gamma_{-}^{f}(\{x_t^{*,2}\}_{t=0}^{l_0+l_1+L_0+4}) - (l_0+l_1+L_0+4)f(x_f,x_f)$$

$$-\pi_{-}^{f}(x_0^{*,2}) + \pi_{-}^{f}(x_{l_0+l_1+L_0+4}^{*,2}) - h(x_0^{*,1}, x_0^{*,2})$$

$$\geq \Gamma^{f}(\{x_t\}_{t=0}^{l_0+l_1+L_0+4}) + \Gamma_{-}^{f}(\{\bar{x}_t\}_{t=0}^{l_0+l_1+L_0+4})$$

$$+\pi^{f}(x_0) + \pi_{-}^{f}(\bar{x}_0) + h(x_0,\bar{x}_0) - \epsilon/8$$

$$-\Gamma^{f}(\{x_t^{*,1}\}_{t=0}^{l_0+l_1+L_0+4}) - \Gamma_{-}^{f}(\{x_t^{*,2}\}_{t=0}^{l_0+l_1+L_0+4})$$

$$-\pi^{f}(x_0^{*,1}) - \pi_{-}^{f}(x_0^{*,2}) - h(x_0^{*,1}, x_0^{*,2}) - \epsilon/8$$

$$\geq \Gamma^{f}(\{x_t\}_{t=0}^{l_0+l_1+L_0+4}) + \Gamma_{-}^{f}(\{\bar{x}_t\}_{t=0}^{l_0+l_1+L_0+4}) + \psi_h(x_0,\bar{x}_0)$$

$$-\epsilon/4 - \delta_1/4 - \psi_h(x_0^{*,1}, x_0^{*,2})$$

$$\geq \Gamma^{f}(\{x_t\}_{t=0}^{l_0+l_1+L_0+4}) + \Gamma_{-}^{f}(\{\bar{x}_t\}_{t=0}^{l_0+l_1+L_0+4}) + \psi_h(x_0,\bar{x}_0)$$

$$-\epsilon/4 - \delta_1/4 - \delta_1/16 - \psi_g(x_0^{*,1}, x_0^{*,2}). \tag{2.401}$$

It follows from (2.370), (2.371), and (2.401) that

$$\Gamma^{f}(\{x_t\}_{t=0}^{l_0+l_1+L_0+4}) + \Gamma_{-}^{f}(\{\bar{x}_t\}_{t=0}^{l_0+l_1+L_0+4}) + \psi_h(x_0,\bar{x}_0)$$

$$\leq \delta + \epsilon/2 + \delta_1/4 + \delta_1/16 + \psi_g(x_0^{*,1}, x_0^{*,2}) + \epsilon/4$$

$$\leq 3\epsilon/4 + \epsilon/8 + \inf(\psi_g). \tag{2.402}$$

By (2.369), (2.381), and (2.402),

$$\Gamma^{f}(\{x_t\}_{t=0}^{l_0+l_1+L_0+4}) + \Gamma_{-}^{f}(\{\bar{x}_t\}_{t=0}^{l_0+l_1+L_0+4}) + \psi_h(x_0,\bar{x}_0)$$

$$\leq 3\epsilon/4 + \epsilon/8 + \inf(\psi_h) + \delta_1/16 \leq \inf(\phi_h) + 3\epsilon/4 + \epsilon/8 + \epsilon/64. \tag{2.403}$$

In view of (2.26), (2.27), (2.369), and (2.403),

$$\Gamma^{f}(\{x_t\}_{t=0}^{l_0+l_1+L_0+4}),\ \Gamma_{-}^{f}(\{\bar{x}_t\}_{t=0}^{l_0+l_1+L_0+4}) \leq \epsilon,$$

$$\psi_h(x_0,\bar{x}_0) \leq \inf(\psi_h) + (57/64)\epsilon \leq \inf(\psi_g) + (58/64)\epsilon. \tag{2.404}$$

By (2.378), (2.381), and (2.404),

$$\psi_g(x_0, \bar{x}_0) \le \psi_h(x_0, \bar{x}_0) + \epsilon/64 \le \inf(\psi_g) + \epsilon.$$

Theorem 2.71 is proved.

2.29 Proof of Proposition 2.73

Assume that the proposition does not hold. Then there exist a sequence $\{\delta_k\}_{k=1}^{\infty} \subset (0, 1]$ and sequences $\{x_t^{(k,1)}\}_{t=0}^{L_0} \in X(0, L_0)$, $k = 1, 2, \dots$ and $\{x_t^{(k,2)}\}_{t=0}^{L_0} \in \bar{X}(0, L_0)$, $k = 1, 2, \dots$ such that

$$\lim_{k\to\infty} \delta_k = 0 \tag{2.405}$$

and that for all integers $k \ge 1$,

$$\Gamma^f(\{x_t^{(k,1)}\}_{t=0}^{L_0}) \le \delta_k,\ \Gamma_-^f(\{x_t^{(k,2)}\}_{t=0}^{L_0}) \le \delta_k, \tag{2.406}$$

$$\psi_g(x_0^{(k,1)}, x_0^{(k,2)}) \le \inf(\psi_g) + \delta_k, \tag{2.407}$$

and that the following property holds:

(i) for each $(f, \mathcal{M})$-overtaking optimal $\{y_t^{(1)}\}_{t=0}^{\infty} \in X(0, \infty)$ and each $(\bar{f}, \bar{\mathcal{M}})$-overtaking optimal $\{y_t^{(2)}\}_{t=0}^{\infty} \in \bar{X}(0, \infty)$ satisfying $\psi_g(y_0^{(1)}, y_0^{(2)}) = \inf(\psi_g)$, we have

$$\max\{\rho_E(x_t^{(k,1)}, y_t^{(1)}) + \rho_E(x_t^{(k,2)}, y_t^{(2)}) : t = 0, \dots, L_0\} > \epsilon.$$

By (2.21), (2.360), and (2.407), the sequences

$$\{\pi^f(x_0^{(k,1)})\}_{k=1}^{\infty},\ \{\pi_-^f(x_0^{(k,2)})\}_{k=1}^{\infty} \text{ are bounded.} \tag{2.408}$$

It follows from (2.25) and (2.408) that

$$\{\pi^f(x_{L_0}^{(k,1)})\}_{k=1}^{\infty},\ \{\pi_-^f(x_{L_0}^{(k,2)})\}_{k=1}^{\infty} \text{ are bounded.} \tag{2.409}$$

In view of (2.20), (2.26), (2.405), (2.406), and (2.409), the sequences

$$\{\sum_{t=0}^{L_0-1} f(x_t^{(k,1)}, x_{t+1}^{(k,1)})\}_{k=1}^{\infty},\ \{\sum_{t=0}^{L_0-1} \bar{f}(x_t^{(k,2)}, x_{t+1}^{(k,2)})\}_{k=1}^{\infty} \text{ are bounded.} \tag{2.410}$$

By LSC property and (2.410), extracting a subsequence and re-indexing if necessary, we may assume without loss of generality that there exist $\{y_t^{(1)}\}_{t=0}^{L_0} \in X(0, L_0)$, $\{y_t^{(2)}\}_{t=0}^{L_0} \in \bar{X}(0, L_0)$ such that

$$x_t^{(k,i)} \to y_t^{(i)} \text{ as } k \to \infty \text{ in } E \text{ for all } t = 0, \dots, L_0,\ i = 1, 2, \tag{2.411}$$

$$\sum_{t=0}^{L_0-1} f(y_t^{(1)}, y_{t+1}^{(1)}) \le \liminf_{k\to\infty} \sum_{t=0}^{L_0-1} f(x_t^{(k,1)}, x_{t+1}^{(k,1)}), \tag{2.412}$$

$$\sum_{t=0}^{L_0-1} \bar{f}(y_t^{(2)}, y_{t+1}^{(2)}) \le \liminf_{k\to\infty} \sum_{t=0}^{L_0-1} \bar{f}(x_t^{(k,2)}, x_{t+1}^{(k,2)}). \tag{2.413}$$

In view of (2.363), (2.405), (2.407), and (2.411),

$$\psi_g(y_0^{(1)}, y_0^{(2)}) \le \liminf_{k\to\infty} \psi_g(x_0^{(k,1)}, x_0^{(k,2)}) = \inf(\psi_g), \tag{2.414}$$

$$\psi_g(y_0^{(1)}, y_0^{(2)}) = \inf(\psi_g). \tag{2.415}$$

Proposition 2.37 and (2.411) imply that

$$\pi^f(y_0^{(1)}) \le \liminf_{k\to\infty} \pi^f(x_0^{(k,1)}),\ \pi_-^f(y_0^{(2)}) \le \liminf_{k\to\infty} \pi_-^f(x_0^{(k,2)}), \tag{2.416}$$

$$g(y_0^{(1)}, y_0^{(2)}) \le \liminf_{k\to\infty} g(x_0^{(k,1)}, x_0^{(k,2)}). \tag{2.417}$$

It follows from (2.363), (2.407), and (2.415)–(2.417) that

$$\pi^f(y_0^{(1)}) = \lim_{k\to\infty} \pi^f(x_0^{(k,1)}),\ \pi_-^f(y_0^{(2)}) = \lim_{k\to\infty} \pi_-^f(x_0^{(k,2)}), \tag{2.418}$$

$$g(y_0^{(1)}, y_0^{(2)}) = \lim_{k\to\infty} g(x_0^{(k,1)}, x_0^{(k,2)}). \tag{2.419}$$

Proposition 2.37 and (2.411) imply that

$$\pi^f(y_{L_0}^{(1)}) \le \liminf_{k\to\infty} \pi^f(x_{L_0}^{(k,1)}),\ \pi_-^f(y_{L_0}^{(2)}) \le \liminf_{k\to\infty} \pi_-^f(x_{L_0}^{(k,2)}), \tag{2.420}$$

It follows from (2.26), (2.406), (2.412), (2.418), and (2.420) that

$$\sum_{t=0}^{L_0-1} f(y_t^{(1)}, y_{t+1}^{(1)}) - L_0 f(x_f, x_f) - \pi^f(y_0^{(1)}) + \pi^f(y_{L_0}^{(1)})$$

$$\leq \liminf_{k\to\infty} \sum_{t=0}^{L_0-1} f(x_t^{(k,1)}, x_{t+1}^{(k,1)}) - L_0 f(x_f, x_f)$$

$$- \lim_{k\to\infty} \pi^f(x_0^{(k,1)}) + \liminf_{k\to\infty} \pi^f(x_{L_0}^{(k,1)})$$

$$\leq \liminf_{k\to\infty}\Big(\sum_{t=0}^{L_0-1} f(x_t^{(k,1)}, x_{t+1}^{(k,1)}) - L_0 f(x_f, x_f)$$
$$-\pi^f(x_0^{(k,1)}) + \pi^f(x_{L_0}^{(k,1)})\Big) \leq 0. \tag{2.421}$$

It follows from (2.26), (2.405), (2.406), (2.413), (2.418), and (2.420) that

$$\sum_{t=0}^{L_0-1} \bar{f}(y_t^{(2)}, y_{t+1}^{(2)}) - L_0 f(x_f, x_f) - \pi_-^f(y_0^{(2)}) + \pi_-^f(y_{L_0}^{(2)})$$

$$\leq \liminf_{k\to\infty} \sum_{t=0}^{L_0-1} \bar{f}(x_t^{(k,2)}, x_{t+1}^{(k,2)}) - L_0 f(x_f, x_f)$$

$$- \lim_{k\to\infty} \pi_-^f(x_0^{(k,2)}) + \liminf_{k\to\infty} \pi_-^f(x_{L_0}^{(k,2)})$$

$$\leq \liminf_{k\to\infty}\Big(\sum_{t=0}^{L_0-1} \bar{f}(x_t^{(k,2)}, x_{t+1}^{(k,2)}) - L_0 f(x_f, x_f)$$
$$-\pi_-^f(x_0^{(k,2)}) + \pi_-^f(x_{L_0}^{(k,2)})\Big) \leq 0. \tag{2.422}$$

By (2.26), (2.27), (2.421), and (2.422),

$$\sum_{t=0}^{L_0-1} f(y_t^{(1)}, y_{t+1}^{(1)}) - L_0 f(x_f, x_f) - \pi^f(y_0^{(1)}) + \pi^f(y_{L_0}^{(1)}) = 0, \tag{2.423}$$

$$\sum_{t=0}^{L_0-1} \bar{f}(y_t^{(2)}, y_{t+1}^{(2)}) - L_0 f(x_f, x_f) - \pi_-^f(y_0^{(2)}) + \pi_-^f(y_{L_0}^{(2)}) = 0. \tag{2.424}$$

In view of (2.415), (2.423), and (2.424),

$$\pi^f(y_{L_0}^{(1)}),\ \pi_-^f(y_{L_0}^{(2)}) \text{ are finite.} \tag{2.425}$$

Theorem 2.13 and (2.425) imply that there exist an $(f, \mathcal{M})$-overtaking optimal $\{\tilde{y}_t^{(1)}\}_{t=0}^{\infty} \in X(0, \infty)$ and $(\bar{f}, \bar{\mathcal{M}})$-overtaking optimal $\{\tilde{y}_t^{(2)}\}_{t=0}^{\infty} \in \bar{X}(0, \infty)$ such that

$$\tilde{y}_0^{(i)} = y_{L_0}^{(i)}, \ i = 1, 2. \tag{2.426}$$

For all integers $t > L_0$ and $i = 1, 2$ set

$$y_t^{(i)} = \tilde{y}_{t-L_0}^{(i)}.$$

Proposition 2.25, (2.26), (2.27), and (2.423)–(2.426) imply that for all integers $T_2 > T_1 \geq 0$,

$$\sum_{t=T_1}^{T_2-1} f(y_t^{(1)}, y_{t+1}^{(1)}) - (T_2 - T_1) f(x_f, x_f) - \pi^f(y_{T_1}^{(1)}) + \pi^f(y_{T_2}^{(1)}) = 0, \tag{2.427}$$

$$\sum_{t=T_1}^{T_2-1} \bar{f}(y_t^{(2)}, y_{t+1}^{(2)}) - (T_2 - T_1) f(x_f, x_f) - \pi_-^f(y_{T_1}^{(2)}) + \pi_-^f(y_{T_2}^{(2)}) = 0. \tag{2.428}$$

Proposition 2.35, (2.425), (2.427), and (2.428) imply that $\{y_t^{(1)}\}_{t=0}^{\infty} \in X(0, \infty)$ is $(f, \mathcal{M})$-overtaking optimal and $\{y_t^{(2)}\}_{t=0}^{\infty} \in \bar{X}(0, \infty)$ is $(\bar{f}, \bar{\mathcal{M}})$-overtaking optimal. It follows from (2.411) that for all sufficiently large natural numbers k and $i = 1, 2$, $\rho_E(x_t^{(k,i)}, y_t^{(i)}) \leq \epsilon/4$ for all $t = 0, \ldots, L_0$. Combined with (2.415) this contradicts property (i). The contradiction we have reached proves Proposition 2.73.

Chapter 3
Discrete-Time Nonautonomous Problems on the Half-Axis

In this chapter we establish sufficient and necessary conditions for the turnpike phenomenon for discrete-time nonautonomous problems on subintervals of half-axis in metric spaces, which are not necessarily compact. For these optimal control problems, the turnpike is not a singleton. We also study the existence of solutions of the corresponding infinite horizon optimal control problems.

3.1 Preliminaries and Main Results

Let (E, ρ_E) and (F, ρ_F) be metric spaces. We suppose that $\mathcal{A}$ is a nonempty subset of $\{0, 1, \dots, \} \times E, \mathcal{U} : \mathcal{A} \to 2^F$ is a point to set mapping with a graph

$$\mathcal{M} = \{(t, x, u) : (t, x) \in \mathcal{A}, \ u \in \mathcal{U}(t, x)\}, \tag{3.1}$$

$G : \mathcal{M} \to E$ and $f : \mathcal{M} \to R^1$.

Let $0 \leq T_1 < T_2$ be integers. We denote by $X(T_1, T_2)$ the set of all pairs of sequences $(\{x_t\}_{t=T_1}^{T_2}, \{u_t\}_{t=T_1}^{T_2-1})$ such that for each integer $t \in \{T_1, \dots, T_2\}$,

$$(t, x_t) \in \mathcal{A}, \tag{3.2}$$

for each integer $t \in \{T_1, \dots, T_2 - 1\}$,

$$u_t \in \mathcal{U}(t, x_t), \tag{3.3}$$

$$x_{t+1} = G(t, x_t, u_t) \tag{3.4}$$

and which are called trajectory-control pairs.

A. J. Zaslavski, *Turnpike Conditions in Infinite Dimensional Optimal Control*, Springer Optimization and Its Applications 148,
https://doi.org/10.1007/978-3-030-20178-4_3

Let $T_1 \geq 0$ be an integer. Denote by $X(T_1, \infty)$ the set of all pairs of sequences $\{x_t\}_{t=T_1}^{\infty} \subset E$, $\{u_t\}_{t=T_1}^{\infty} \subset F$ such that for each integer $T_2 > T_1$, $(\{x_t\}_{t=T_1}^{T_2}, \{u_t\}_{t=T_1}^{T_2-1}) \in X(T_1, T_2)$. Elements of $X(T_1, \infty)$ are called trajectory-control pairs. Let $0 \leq T_1 < T_2$ be integers. A sequence $\{x_t\}_{t=T_1}^{T_2} \subset E$ ($\{x_t\}_{t=T_1}^{\infty} \subset E$, respectively) is called a trajectory if there exists a sequence $\{u_t\}_{t=T_1}^{T_2-1} \subset F$ ($\{u_t\}_{t=T_1}^{\infty} \subset F$, respectively) referred to as a control such that $(\{x_t\}_{t=T_1}^{T_2}, \{u_t\}_{t=T_1}^{T_2-1}) \in X(T_1, T_2)$ ($(\{x_t\}_{t=T_1}^{\infty}, \{u_t\}_{t=T_1}^{\infty}) \in X(T_1, \infty)$, respectively).

Let $\theta_0 \in E$, $\theta_1 \in F$, $a_0 > 0$ and let $\psi : [0, \infty) \to [0, \infty)$ be an increasing function such that

$$\psi(t) \to \infty \text{ as } t \to \infty. \tag{3.5}$$

We suppose that the function f satisfies

$$f(t, x, u) \geq \psi(\rho_E(x, \theta_0)) - a_0 \text{ for each } (t, x, u) \in \mathcal{M}. \tag{3.6}$$

For each pair of integers $T_2 > T_1 \geq 0$ and each pair of points $y, z \in E$ satisfying $(T_1, y), (T_2, z) \in \mathcal{A}$, we consider the following problems:

$$\sum_{t=T_1}^{T_2-1} f(t, x_t, u_t) \to \min, \ (\{x_t\}_{t=T_1}^{T_2}, \{u_t\}_{t=T_1}^{T_2-1}) \in X(T_1, T_2), \ x_{T_1} = y, \ x_{T_2} = z, \tag{P_1}$$

$$\sum_{t=T_1}^{T_2-1} f(t, x_t, u_t) \to \min, \ (\{x_t\}_{t=T_1}^{T_2}, \{u_t\}_{t=T_1}^{T_2-1}) \in X(T_1, T_2), \ x_{T_1} = y, \tag{P_2}$$

$$\sum_{t=T_1}^{T_2-1} f(t, x_t, u_t) \to \min, \ (\{x_t\}_{t=T_1}^{T_2}, \{u_t\}_{t=T_1}^{T_2-1}) \in X(T_1, T_2). \tag{P_3}$$

For each pair of integers $T_2 > T_1 \geq 0$ and each pair of points $y, z \in E$ satisfying $(T_1, y), (T_2, z) \in \mathcal{A}$, we define

$$U^f(T_1, T_2, y, z) = \inf\{\sum_{t=T_1}^{T_2-1} f(t, x_t, u_t) : \ (\{x_t\}_{t=T_1}^{T_2}, \{u_t\}_{t=T_1}^{T_2-1}) \in X(T_1, T_2),$$

$$x_{T_1} = y, \ x_{T_2} = z\}, \tag{3.7}$$

$$\sigma^f(T_1, T_2, y) = \inf\{U^f(T_1, T_2, y, h) : \ h \in E, \ (T_2, h) \in \mathcal{A}\}, \tag{3.8}$$

$$\widehat{\sigma}^f(T_1, T_2, z) = \inf\{U^f(T_1, T_2, h, z) : h \in E, \ (T_1, h) \in \mathcal{A}\}, \tag{3.9}$$

$$\sigma^f(T_1, T_2) = \inf\{U^f(T_1, T_2, h, \xi) : h, \xi \in E, \ (T_1, h), (T_2, \xi) \in \mathcal{A}\}. \tag{3.10}$$

We suppose that $b_f > 0$ is an integer, $(\{x_t^f\}_{t=0}^{\infty}, \{u_t^f\}_{t=0}^{\infty}) \in X(0, \infty)$,

$$\{x_t^f : t = 0, 1, \ldots\} \text{ is bounded }, \tag{3.11}$$

$$\Delta_f := \sup\{|f(t, x_t^f, u_t^f)| : t = 0, 1, \ldots\} < \infty \tag{3.12}$$

and the the following assumptions hold.

(A1) For each $S_1 > 0$, there exist $S_2 > 0$ and an integer $c > 0$ such that

$$\sum_{t=T_1}^{T_2-1} f(t, x_t^f, u_t^f) \le \sum_{t=T_1}^{T_2-1} f(t, x_t, u_t) + S_2$$

for each pair of integers $T_1 \ge 0$, $T_2 \ge T_1 + c$ and each $(\{x_t\}_{t=T_1}^{T_2}, \{u_t\}_{t=T_1}^{T_2-1}) \in X(T_1, T_2)$ satisfying $\rho_E(\theta_0, x_j) \le S_1$, $j = T_1, T_2$.

(A2) For each $\epsilon > 0$, there exists $\delta > 0$ such that for each $(T_i, z_i) \in \mathcal{A}$, $i = 1, 2$ satisfying $\rho_E(z_i, x_{T_i}^f) \le \delta$, $i = 1, 2$ and $T_2 \ge b_f$, there exist integers $\tau_1, \tau_2 \in (0, b_f]$,

$$(\{x_t^{(1)}\}_{t=T_1}^{T_1+\tau_1}, \{u_t^{(1)}\}_{t=T_1}^{T_1+\tau_1-1}) \in X(T_1, T_1 + \tau_1),$$

$$(\{x_t^{(2)}\}_{t=T_2-\tau_2}^{T_2}, \{u_t^{(2)}\}_{t=T_2-\tau_2}^{T_2-1}) \in X(T_2 - \tau_2, T_2)$$

such that

$$x_{T_1}^{(1)} = z_1, \; x_{T_1+\tau_1}^{(1)} = x_{T_1+\tau_1}^f,$$

$$\sum_{t=T_1}^{T_1+\tau_1-1} f(t, x_t^{(1)}, u_t^{(1)}) \le \sum_{t=T_1}^{T_1+\tau_1-1} f(t, x_t^f, u_t^f) + \epsilon,$$

$$x_{T_2}^{(2)} = z_2, \; x_{T_2-\tau_2}^{(2)} = x_{T_2-\tau_2}^f,$$

$$\sum_{t=T_2-\tau_2}^{T_2-1} f(t, x_t^{(2)}, u_t^{(2)}) \le \sum_{t=T_2-\tau_2}^{T_2-1} f(t, x_t^f, u_t^f) + \epsilon.$$

Section 3.6 contains examples of optimal control problems satisfying assumptions (A1) and (A2). Many examples can also be found in [106–108, 118, 124, 125, 134].

3.2 The Boundedness Results

The following result is proved in Section 3.8.

Theorem 3.1.

1. *There exists $S > 0$ such that for each pair of integers $T_2 > T_1 \geq 0$ and each $(\{x_t\}_{t=T_1}^{T_2}, \{u_t\}_{t=T_1}^{T_2-1}) \in X(T_1, T_2)$,*

$$\sum_{t=T_1}^{T_2-1} f(t, x_t, u_t) + S \geq \sum_{t=T_1}^{T_2-1} f(t, x_t^f, u_t^f).$$

2. *For each $(\{x_t\}_{t=0}^{\infty}, \{u_t\}_{t=0}^{\infty}) \in X(0, \infty)$ either*

$$\sum_{t=0}^{T-1} f(t, x_t, u_t) - \sum_{t=0}^{T-1} f(t, x_t^f, u_t^f) \to \infty \text{ as } T \to \infty$$

or

$$\sup\{|\sum_{t=0}^{T-1} f(t, x_t, u_t) - \sum_{t=0}^{T-1} f(t, x_t^f, u_t^f)| : T \in \{1, 2, \dots\}\} < \infty. \tag{3.13}$$

Moreover, if (3.13) *holds, then*

$$\sup\{\rho_E(x_t, \theta_0) : t = 0, 1, \dots\} < \infty.$$

We say that $(\{x_t\}_{t=0}^{\infty}, \{u_t\}_{t=0}^{\infty}) \in X(0, \infty)$ is (f)-goof if (3.13) holds [29, 104, 116, 120, 129, 134]. The next boundedness result is proved in Section 3.8. It has a prototype in [104].

Theorem 3.2. *Let $c > 0$ be an integer and $M_0 > 0$. Then there exists $M_1 > 0$ such that for each pair of integers $T_1 \geq 0$, $T_2 \geq T_1 + c$ and each $(\{x_t\}_{t=T_1}^{T_2}, \{u_t\}_{t=T_1}^{T_2-1}) \in X(T_1, T_2)$ satisfying*

$$\sum_{t=T_1}^{T_2-1} f(t, x_t, u_t) \leq \sum_{t=T_1}^{T_2-1} f(t, x_t^f, u_t^f) + M_0$$

the inequality $\rho_E(\theta_0, x_t) \leq M_1$ holds for all $t = T_1, \dots, T_2 - 1$.

Let $L > 0$ be an integer. Denote by $\mathcal{A}_L$ the set of all $(S, z) \in \mathcal{A}$ for which there exist an integer $\tau \in (0, L]$ and $(\{x_t\}_{t=S}^{S+\tau}, \{u_t\}_{t=S}^{S+\tau-1}) \in X(S, S + \tau)$ such that

$$x_S = z, \; x_{S+\tau} = x_{S+\tau}^f, \; \sum_{t=S}^{S+\tau-1} f(t, x_t, u_t) \leq L.$$

Denote by $\widehat{\mathcal{A}}_L$ the set of all $(S, z) \in \mathcal{A}$ such that $S \geq L$ and there exist an integer $\tau \in (0, L]$ and $(\{x_t\}_{t=S-\tau}^{S}, \{u_t\}_{t=S-\tau}^{S-1}) \in X(S-\tau, S)$ satisfying

$$x_{S-\tau} = x_{S-\tau}^f,\ x_S = z,\ \sum_{t=S-\tau}^{S-1} f(t, x_t, u_t) \leq L.$$

The following Theorems 3.3–3.5 are also boundedness results. They are proved in Section 3.8.

Theorem 3.3. *Let $L > 0$ be an integer and $M_0 > 0$. Then there exists $M_1 > 0$ such that for each integer $T_1 \geq 0$, each integer $T_2 \geq T_1 + 2L$ and each $(\{x_t\}_{t=T_1}^{T_2}, \{u_t\}_{t=T_1}^{T_2-1}) \in X(T_1, T_2)$ satisfying*

$$(T_1, x_{T_1}) \in \mathcal{A}_L,\ (T_2, x_{T_2}) \in \widehat{\mathcal{A}}_L,$$

$$\sum_{t=T_1}^{T_2-1} f(t, x_t, u_t) \leq U^f(T_1, T_2, x_{T_1}, x_{T_2}) + M_0$$

the inequality $\rho_E(\theta_0, x_t) \leq M_1$ holds for all $t = T_1, \ldots, T_2 - 1$.

Theorem 3.4. *Let $L > 0$ be an integer and $M_0 > 0$. Then there exists $M_1 > 0$ such that for each integer $T_1 \geq 0$, each integer $T_2 \geq T_1 + L$, and each $(\{x_t\}_{t=T_1}^{T_2}, \{u_t\}_{t=T_1}^{T_2-1}) \in X(T_1, T_2)$ satisfying*

$$(T_1, x_{T_1}) \in \mathcal{A}_L,\ \sum_{t=T_1}^{T_2-1} f(t, x_t, u_t) \leq \sigma^f(T_1, T_2, x_{T_1}) + M_0$$

the inequality $\rho_E(\theta_0, x_t) \leq M_1$ holds for all $t = T_1, \ldots, T_2 - 1$.

Theorem 3.5. *Let $L > 0$ be an integer and $M_0 > 0$. Then there exists $M_1 > 0$ such that for each integer $T_1 \geq 0$, each integer $T_2 \geq T_1 + L$ and each $(\{x_t\}_{t=T_1}^{T_2}, \{u_t\}_{t=T_1}^{T_2-1}) \in X(T_1, T_2)$ satisfying*

$$(T_2, x_{T_2}) \in \widehat{\mathcal{A}}_L,\ \sum_{t=T_1}^{T_2-1} f(t, x_t, u_t) \leq \widehat{\sigma}^f(T_1, T_2, x_{T_2}) + M_0$$

the inequality $\rho_E(\theta_0, x_t) \leq M_1$ holds for all $t = T_1, \ldots, T_2 - 1$.

3.3 Turnpike Properties

We say that f possesses the turnpike property (or TP for short) if for each $\epsilon > 0$ and each $M > 0$, there exist $\delta > 0$ and an integer $L > 0$ such that for each integer $T_1 \geq 0$, each integer $T_2 \geq T_1 + 2L$, and each $(\{x_t\}_{t=T_1}^{T_2}, \{u_t\}_{t=T_1}^{T_2-1}) \in X(T_1, T_2)$ which satisfies

$$\sum_{t=T_1}^{T_2-1} f(t, x_t, u_t) \leq \min\{\sigma^f(T_1, T_2) + M, U^f(T_1, T_2, x_{T_1}, x_{T_2}) + \delta\},$$

there exist integers $\tau_1 \in [T_1, T_1 + L]$, $\tau_2 \in [T_2 - L, T_2]$ such that

$$\rho_E(x_t, x_t^f) \leq \epsilon, \ t = \tau_1 \dots, \tau_2.$$

Moreover, if $\rho_E(x_{T_2}, x_{T_2}^f) \leq \delta$, then $\tau_2 = T_2$ and if $\rho_E(x_{T_1}, x_{T_1}^f) \leq \delta$, $T_1 \geq L$, then $\tau_1 = T_1$.

We say that f possesses the strong turnpike property (or STP for short) if for each $\epsilon > 0$ and each $M > 0$, there exist $\delta > 0$ and an integer $L > 0$ such that for each integer $T_1 \geq 0$, each integer $T_2 \geq T_1 + 2L$, and each $(\{x_t\}_{t=T_1}^{T_2}, \{u_t\}_{t=T_1}^{T_2-1}) \in X(T_1, T_2)$ which satisfies

$$\sum_{t=T_1}^{T_2-1} f(t, x_t, u_t) \leq \min\{\sigma^f(T_1, T_2) + M, U^f(T_1, T_2, x_{T_1}, x_{T_2}) + \delta\},$$

there exist integers $\tau_1 \in [T_1, T_1 + L]$, $\tau_2 \in [T_2 - L, T_2]$ such that

$$\rho_E(x_t, x_t^f) \leq \epsilon, \ t = \tau_1 \dots, \tau_2.$$

Moreover, if $\rho_E(x_{T_2}, x_{T_2}^f) \leq \delta$, then $\tau_2 = T_2$ and if $\rho_E(x_{T_1}, x_{T_1}^f) \leq \delta$, then $\tau_1 = T_1$.

Theorem 3.1 implies the following two results.

Theorem 3.6. *Assume that f has TP and that $\epsilon, M > 0$. Then there exist $\delta > 0$ and an integer $L > 0$ such that for each integer $T_1 \geq 0$, each integer $T_2 \geq T_1 + 2L$, and each $(\{x_t\}_{t=T_1}^{T_2}, \{u_t\}_{t=T_1}^{T_2-1}) \in X(T_1, T_2)$ which satisfies*

$$\sum_{t=T_1}^{T_2-1} f(t, x_t, u_t) \leq \min\{\sum_{t=T_1}^{T_2-1} f(t, x_t^f, u_t^f) + M, U^f(T_1, T_2, x_{T_1}, x_{T_2}) + \delta\},$$

there exist integers $\tau_1 \in [T_1, T_1 + L]$, $\tau_2 \in [T_2 - L, T_2]$ *such that*

$$\rho_E(x_t, x_t^f) \le \epsilon, \ t = \tau_1, \dots, \tau_2.$$

Moreover, if $\rho_E(x_{T_2}, x_{T_2}^f) \le \delta$, *then* $\tau_2 = T_2$ *and if* $\rho_E(x_{T_1}, x_{T_1}^f) \le \delta$, $T_1 \ge L$, *then* $\tau_1 = T_1$.

Theorem 3.7. *Assume that* f *has STP and that* $\epsilon, M > 0$. *Then there exist* $\delta > 0$ *and an integer* $L > 0$ *such that for each integer* $T_1 \ge 0$, *each integer* $T_2 \ge T_1 + 2L$, *and each* $(\{x_t\}_{t=T_1}^{T_2}, \{u_t\}_{t=T_1}^{T_2-1}) \in X(T_1, T_2)$ *which satisfies*

$$\sum_{t=T_1}^{T_2-1} f(t, x_t, u_t) \le \min\{\sum_{t=T_1}^{T_2-1} f(t, x_t^f, u_t^f) + M, U^f(T_1, T_2, x_{T_1}, x_{T_2}) + \delta\},$$

there exist integers $\tau_1 \in [T_1, T_1 + L]$, $\tau_2 \in [T_2 - L, T_2]$ *such that*

$$\rho_E(x_t, x_t^f) \le \epsilon, \ t = \tau_1, \dots, \tau_2.$$

Moreover, if $\rho_E(x_{T_2}, x_{T_2}^f) \le \delta$, *then* $\tau_2 = T_2$ *and if* $\rho_E(x_{T_1}, x_{T_1}^f) \le \delta$, *then* $\tau_1 = T_1$.

Proposition 3.8. *Let* $L > 0$ *be an integer,* $T_1 \ge 0$, $T_2 \ge T_1 + 2L$ *be integers,* $(T_1, z_1) \in \mathcal{A}_L$, $(T_2, z_2) \in \widehat{\mathcal{A}}_L$. *Then*

$$U^f(T_1, T_2, z_1, z_2) \le \sum_{t=T_1}^{T_2-1} f(t, x_t^f, u_t^f) + 2L(1 + a_0).$$

Proof. By the definition of $\mathcal{A}_L, \widehat{\mathcal{A}}_L$, there exist integers $\tau_1 \in (0, L]$, $\tau_2 \in (0, L]$ and $(\{y_t\}_{t=T_1}^{T_2}, \{v_t\}_{t=T_1}^{T_2-1}) \in X(T_1, T_2)$ such that

$$y_{T_1} = z_1, \ y_{T_2} = z_2, \ y_t = x_t^f, \ t = T_1 + \tau_1, \dots, T_2 - \tau_2,$$

$$v_t = u_t^f, \ t \in \{T_1 + \tau_1, \dots, T_2 - \tau_2\} \setminus \{T_2 - \tau_2\},$$

$$\sum_{t=T_1}^{T_1+\tau_1-1} f(t, y_t, v_t) \le L, \quad \sum_{t=T_2-\tau_2}^{T_2-1} f(t, y_t, v_t) \le L.$$

In view of the relations above,

$$U^f(T_1, T_2, z_1, z_2) \le \sum_{t=T_1}^{T_2-1} f(t, y_t, v_t)$$

$$\leq 2L + \sum \{f(t, x_t^f, u_t^f) : t \in \{T_1 + \tau_1, \ldots, T_2 - \tau_2\} \setminus \{T_2 - \tau_2\}\}$$

$$\leq 2L + 2a_0 L + \sum_{t=T_1}^{T_2-1} f(t, x_t^f, u_t^f).$$

Proposition 3.8 is proved.

Proposition 3.8 and Theorems 3.6 and 3.7 imply the following two results.

Theorem 3.9. *Assume that f has TP, $L_0 > 0$ is an integer, and $\epsilon > 0$. Then there exist $\delta > 0$ and an integer $L > L_0$ such that for each integer $T_1 \geq 0$, each integer $T_2 \geq T_1 + 2L$, and each $(\{x_t\}_{t=T_1}^{T_2}, \{u_t\}_{t=T_1}^{T_2-1}) \in X(T_1, T_2)$ which satisfies*

$$(T_1, x_{T_1}) \in \mathcal{A}_{L_0}, \; (T_2, x_{T_2}) \in \widehat{\mathcal{A}}_{L_0},$$

$$\sum_{t=T_1}^{T_2-1} f(t, x_t, u_t) \leq U^f(T_1, T_2, x_{T_1}, x_{T_2}) + \delta,$$

there exist integers $\tau_1 \in [T_1, T_1 + L]$, $\tau_2 \in [T_2 - L, T_2]$ such that

$$\rho_E(x_t, x_t^f) \leq \epsilon, \; t = \tau_1, \ldots, \tau_2.$$

Moreover, if $\rho_E(x_{T_2}, x_{T_2}^f) \leq \delta$, then $\tau_2 = T_2$, and if $\rho_E(x_{T_1}, x_{T_1}^f) \leq \delta$, $T_1 \geq L$, then $\tau_1 = T_1$.

Theorem 3.10. *Assume that f has STP, $L_0 > 0$ is an integer, and $\epsilon > 0$. Then there exist $\delta > 0$ and an integer $L > L_0$ such that for each integer $T_1 \geq 0$, each integer $T_2 \geq T_1 + 2L$, and each $(\{x_t\}_{t=T_1}^{T_2}, \{u_t\}_{t=T_1}^{T_2-1}) \in X(T_1, T_2)$ which satisfies*

$$(T_1, x_{T_1}) \in \mathcal{A}_{L_0}, \; (T_2, x_{T_2}) \in \widehat{\mathcal{A}}_{L_0},$$

$$\sum_{t=T_1}^{T_2-1} f(t, x_t, u_t) \leq U^f(T_1, T_2, x_{T_1}, x_{T_2}) + \delta,$$

there exist integers $\tau_1 \in [T_1, T_1 + L]$, $\tau_2 \in [T_2 - L, T_2]$ such that

$$\rho_E(x_t, x_t^f) \leq \epsilon, \; t = \tau_1, \ldots, \tau_2.$$

Moreover, if $\rho_E(x_{T_2}, x_{T_2}^f) \leq \delta$, then $\tau_2 = T_2$, and if $\rho_E(x_{T_1}, x_{T_1}^f) \leq \delta$, then $\tau_1 = T_1$.

Proposition 3.11. *Let $L > 0$ be an integer, $(T_1, z) \in \mathcal{A}_L$, $T_2 \geq T_1 + L$ be an integer. Then*

$$\sigma^f(T_1, T_2, z) \leq \sum_{t=T_1}^{T_2-1} f(t, x_t^f, u_t^f) + L(1 + a_0).$$

Proof. By the definition of $\mathcal{A}_L$, there exist an integer $\tau \in (0, L]$ and

$$(\{y_t\}_{t=T_1}^{T_2}, \{v_t\}_{t=T_1}^{T_2-1}) \in X(T_1, T_2)$$

such that

$$y_{T_1} = z,\ y_t = x_t^f,\ t = T_1 + \tau, \ldots, T_2,\ v_t = u_t^f,\ t \in \{T_1 + \tau, \ldots, T_2\} \setminus \{T_2\},$$

$$\sum_{t=T_1}^{T_1+\tau-1} f(t, y_t, v_t) \leq L.$$

In view of the relations above,

$$\sigma^f(T_1, T_2, z) \leq \sum_{t=T_1}^{T_2-1} f(t, y_t, v_t)$$

$$\leq L + \sum\{f(t, x_t^f, u_t^f) : t \in \{T_1 + \tau, \ldots, T_2\} \setminus \{T_2\}\}$$

$$\leq L + a_0 L + \sum_{t=T_1}^{T_2-1} f(t, x_t^f, u_t^f).$$

Proposition 3.11 is proved.

Proposition 3.11 and Theorems 3.6, 3.7 imply the following two results.

Theorem 3.12. *Assume that f has TP, $L_0 > 0$ is an integer, and $\epsilon > 0$. Then there exist $\delta > 0$ and an integer $L > L_0$ such that for each integer $T_1 \geq 0$, each integer $T_2 \geq T_1 + 2L$, and each $(\{x_t\}_{t=T_1}^{T_2}, \{u_t\}_{t=T_1}^{T_2-1}) \in X(T_1, T_2)$ which satisfies*

$$(T_1, x_{T_1}) \in \mathcal{A}_{L_0},\ \sum_{t=T_1}^{T_2-1} f(t, x_t, u_t) \leq \sigma^f(T_1, T_2, x_{T_1}) + \delta,$$

there exist integers $\tau_1 \in [T_1, T_1 + L]$, $\tau_2 \in [T_2 - L, T_2]$ such that

$$\rho_E(x_t, x_t^f) \leq \epsilon,\ t = \tau_1, \ldots, \tau_2.$$

Moreover, if $\rho_E(x_{T_2}, x_{T_2}^f) \leq \delta$, then $\tau_2 = T_2$, and if $\rho_E(x_{T_1}, x_{T_1}^f) \leq \delta$, $T_1 \geq L$, then $\tau_1 = T_1$.

Theorem 3.13. *Assume that f has STP, $L_0 > 0$ is an integer, and $\epsilon > 0$. Then there exist $\delta > 0$ and an integer $L > L_0$ such that for each integer $T_1 \geq 0$, each integer $T_2 \geq T_1 + 2L$, and each $(\{x_t\}_{t=T_1}^{T_2}, \{u_t\}_{t=T_1}^{T_2-1}) \in X(T_1, T_2)$ which satisfies*

$$(T_1, x_{T_1}) \in \mathcal{A}_L, \sum_{t=T_1}^{T_2-1} f(t, x_t, u_t) \leq \sigma^f(T_1, T_2, x_{T_1}) + \delta,$$

there exist integers $\tau_1 \in [T_1, T_1 + L]$, $\tau_2 \in [T_2 - L, T_2]$ such that

$$\rho_E(x_t, x_t^f) \leq \epsilon, \ t = \tau_1, \ldots, \tau_2.$$

Moreover, if $\rho_E(x_{T_2}, x_{T_2}^f) \leq \delta$, then $\tau_2 = T_2$, and if $\rho_E(x_{T_1}, x_{T_1}^f) \leq \delta$, then $\tau_1 = T_1$.

Proposition 3.14. *Let $L > 0$ be an integer, $T_1 \geq 0$ be an integer, $(T_2, z) \in \widehat{\mathcal{A}}_L$ satisfy $T_2 \geq T_1 + L$. Then*

$$\widehat{\sigma}^f(T_1, T_2, z) \leq \sum_{t=T_1}^{T_2-1} f(t, x_t^f, u_t^f) + L(1 + a_0).$$

Proof. By the definition of $\widehat{\mathcal{A}}_L$, there exist an integer $\tau \in (0, L]$ and

$$(\{y_t\}_{t=T_1}^{T_2}, \{v_t\}_{t=T_1}^{T_2-1}) \in X(T_1, T_2)$$

such that

$$y_{T_2} = z, \ y_t = x_t^f, \ t = T_1, \ldots, T_2 - \tau, \ v_t = u_t^f, \ t \in \{T_1, \ldots, T_2 - \tau\} \setminus \{T_2 - \tau\},$$

$$\sum_{t=T_2-\tau}^{T_2-1} f(t, y_t, v_t) \leq L.$$

In view of the relations above,

$$\widehat{\sigma}^f(T_1, T_2, z) \leq \sum_{t=T_1}^{T_2-1} f(t, y_t, v_t)$$

$$\leq L + \sum \{f(t, x_t^f, u_t^f) : \ t \in \{T_1, \ldots, T_2 - \tau\} \setminus \{T_2 - \tau\}\}$$

$$\leq L + a_0 L + \sum_{t=T_1}^{T_2-1} f(t, x_t^f, u_t^f).$$

Proposition 3.14 is proved.

Proposition 3.14 and Theorems 3.6 and 3.7 imply the following two results.

Theorem 3.15. *Assume that f has TP, $L_0 > 0$ is an integer, and $\epsilon > 0$. Then there exist $\delta > 0$ and an integer $L > L_0$ such that for each integer $T_1 \geq 0$, each integer $T_2 \geq T_1 + 2L$, and each $(\{x_t\}_{t=T_1}^{T_2}, \{u_t\}_{t=T_1}^{T_2-1}) \in X(T_1, T_2)$ which satisfies*

$$(T_2, x_{T_2}) \in \widehat{\mathcal{A}}_{L_0}, \ \sum_{t=T_1}^{T_2-1} f(t, x_t, u_t) \leq \widehat{\sigma}^f(T_1, T_2, x_{T_2}) + \delta,$$

there exist integers $\tau_1 \in [T_1, T_1 + L]$, $\tau_2 \in [T_2 - L, T_2]$ such that

$$\rho_E(x_t, x_t^f) \leq \epsilon, \ t = \tau_1, \ldots, \tau_2.$$

Moreover, if $\rho_E(x_{T_2}, x_{T_2}^f) \leq \delta$, then $\tau_2 = T_2$, and if $\rho_E(x_{T_1}, x_{T_1}^f) \leq \delta$, $T_1 \geq L$, then $\tau_1 = T_1$.

Theorem 3.16. *Assume that f has STP, $L_0 > 0$ is an integer, and $\epsilon > 0$. Then there exist $\delta > 0$ and an integer $L > L_0$ such that for each integer $T_1 \geq 0$, each integer $T_2 \geq T_1 + 2L$, and each $(\{x_t\}_{t=T_1}^{T_2}, \{u_t\}_{t=T_1}^{T_2-1}) \in X(T_1, T_2)$ which satisfies*

$$(T_2, x_{T_2}) \in \widehat{\mathcal{A}}_{L_0}, \ \sum_{t=T_1}^{T_2-1} f(t, x_t, u_t) \leq \widehat{\sigma}^f(T_1, T_2, x_{T_2}) + \delta,$$

there exist integers $\tau_1 \in [T_1, T_1 + L]$, $\tau_2 \in [T_2 - L, T_2]$ such that

$$\rho_E(x_t, x_t^f) \leq \epsilon, \ t = \tau_1, \ldots, \tau_2.$$

Moreover, if $\rho_E(x_{T_2}, x_{T_2}^f) \leq \delta$ then $\tau_2 = T_2$ and if $\rho_E(x_{T_1}, x_{T_1}^f) \leq \delta$, then $\tau_1 = T_1$.

The next theorem is our main result in this chapter. It is proved in Section 3.10.

Theorem 3.17. *f has TP if and only if the following properties hold:*

(P1) for each (f)-good pair of sequences $(\{x_t\}_{t=0}^{\infty}, \{u_t\}_{t=0}^{\infty}) \in X(0, \infty)$,

$$\lim_{t \to \infty} \rho_E(x_t, x_t^f) = 0;$$

(P2) for each $\epsilon > 0$ and each $M > 0$, there exist $\delta > 0$ and an integer $L > 0$ such that for each integer $T \geq 0$ and each $(\{x_t\}_{t=T}^{T+L}, \{u_t\}_{t=T}^{T+L-1}) \in X(T, T + L)$ which satisfies

$$\sum_{t=T}^{T+L-1} f(t, x_t, u_t) \le \min\{U^f(T, T+L, x_T, x_{T+L})$$

$$+\delta, \sum_{t=T}^{T+L-1} f(t, x_t^f, u_t^f) + M\},$$

there exists an integer $s \in [T, T+L]$ such that $\rho_E(x_s, x_s^f) \le \epsilon$.

We say that f possesses the weak turnpike property (or WTP for short) if for each $\epsilon > 0$ and each $M > 0$, there exist natural numbers Q, l such that for each integer $T_1 \ge 0$, each integer $T_2 \ge T_1 + lQ$, and each $(\{x_t\}_{t=T_1}^{T_2}, \{u_t\}_{t=T_1}^{T_2-1}) \in X(T_1, T_2)$ which satisfies

$$\sum_{t=T_1}^{T_2-1} f(t, x_t, u_t) \le \sum_{t=T_1}^{T_2-1} f(t, x_t^f, u_t^f) + M,$$

there exist finite sequences of integers $\{a_i\}_{i=1}^q$, $\{b_i\}_{i=1}^q \subset \{T_1, \dots, T_2\}$ such that an integer $q \le Q$,

$$0 \le b_i - a_i \le l, \ i = 1, \dots, q,$$

$$b_i \le a_{i+1} \text{ for all integers } i \text{ satisfying } 1 \le i < q,$$

$$\rho_E(x_t, x_t^f) \le \epsilon \text{ for all integers } t \in [T_1, T_2] \setminus \cup_{i=1}^q [a_i, b_i].$$

WTP was studies in [95, 96] for discrete-time unconstrained problems, in [97] for variational problems, and in [114] for discrete-time constrained problems.

The next result is proved in Section 3.12.

Theorem 3.18. *f has WTP if and only if f has (P1) and (P2).*

3.4 Lower Semicontinuity Property and Infinite Horizon Problems

We say that f possesses lower semicontinuity property (or LSC property for short) if for each pair of integers $T_2 > T_1 \ge 0$ and each sequence $(\{x_t^{(j)}\}_{t=T_1}^{T_2}, \{u_t^{(j)}\}_{t=T_1}^{T_2-1}) \in X(T_1, T_2)$, $j = 1, 2, \dots$ which satisfies

$$\sup\{\sum_{t=T_1}^{T_2-1} f(t, x_t^{(j)}, u_t^{(j)}) : \ j = 1, 2, \dots\} < \infty,$$

there exist a subsequence $(\{x_t^{(j_k)}\}_{t=T_1}^{T_2}, \{u_t^{(j_k)}\}_{t=T_1}^{T_2-1})$, $k = 1, 2, \dots$ and

$$(\{x_t\}_{t=T_1}^{T_2}, \{u_t\}_{t=T_1}^{T_2-1}) \in X(T_1, T_2)$$

such that for any $t \in \{T_1, \dots, T_2\}$,

$$x_t^{(j_k)} \to x_t \text{ as } k \to \infty,$$

$$\sum_{t=T_1}^{T_2-1} f(t, x_t, u_t) \le \liminf_{j\to\infty} \sum_{t=T_1}^{T_2-1} f(t, x_t^{(j)}, u_t^{(j)}).$$

LSC property plays an important role in the calculus of variations and optimal control theory [32].

Let $S \ge 0$ be an integer. A pair $(\{x_t\}_{t=S}^{\infty}, \{u_t\}_{t=S}^{\infty}) \in X(S, \infty)$ is called (f)-overtaking optimal [29, 104, 120] if for every $(\{y_t\}_{t=S}^{\infty}, \{v_t\}_{t=S}^{\infty}) \in X(S, \infty)$ satisfying $x_S = y_S$,

$$\limsup_{T\to\infty} [\sum_{t=S}^{T-1} f(t, x_t, u_t) - \sum_{t=S}^{T-1} f(t, y_t, v_t)] \le 0.$$

A pair $(\{x_t\}_{t=S}^{\infty}, \{u_t\}_{t=S}^{\infty}) \in X(S, \infty)$ is called (f)-weakly optimal [29, 104, 120] if for every $(\{y_t\}_{t=S}^{\infty}, \{v_t\}_{t=S}^{\infty}) \in X(S, \infty)$ satisfying $x_S = y_S$,

$$\liminf_{T\to\infty} [\sum_{t=S}^{T-1} f(t, x_t, u_t) - \sum_{t=S}^{T-1} f(t, y_t, v_t)] \le 0.$$

A pair $(\{x_t\}_{t=S}^{\infty}, \{u_t\}_{t=S}^{\infty}) \in X(S, \infty)$ is called (f)-minimal [12, 94] if for every integer $T > S$,

$$\sum_{t=S}^{T-1} f(t, x_t, u_t) = U^f(S, T, x_S, x_T).$$

In infinite horizon optimal control, the main goal is to show the existence of solutions using the optimality criterions above. The next result is proved in Section 3.14.

Theorem 3.19. *Assume that f has (P1), (P2), and LSC property, $S \ge 0$ is an integer, and $(\{x_t\}_{t=S}^{\infty}, \{u_t\}_{t=S}^{\infty}) \in X(S, \infty)$ is (f)-good. Then there exists an (f)-overtaking optimal pair $(\{x_t^*\}_{t=S}^{\infty}, \{u_t^*\}_{t=S}^{\infty}) \in X(S, \infty)$ such that $x_S^* = x_S$.*

The following theorem is also proved in Section 3.14.

Theorem 3.20. *Assume that f has (P1), (P2), and LSC property, $S \geq 0$ is an integer, $(\{\tilde{x}_t\}_{t=S}^{\infty}, \{\tilde{u}_t\}_{t=S}^{\infty}) \in X(S, \infty)$ is (f)-good, and*

$$(\{x_t^*\}_{t=S}^{\infty}, \{u_t^*\}_{t=S}^{\infty}) \in X(S, \infty)$$

satisfies $x_S^* = \tilde{x}_S$. Then the following conditions are equivalent:

(i) $(\{x_t^*\}_{t=S}^{\infty}, \{u_t^*\}_{t=S}^{\infty})$ is (f)-overtaking optimal;
(ii) $(\{x_t^*\}_{t=S}^{\infty}, \{u_t^*\}_{t=S}^{\infty})$ is (f)-weakly optimal;
(iii) $(\{x_t^*\}_{t=S}^{\infty}, \{u_t^*\}_{t=S}^{\infty})$ is (f)-minimal and (f)-good;
(iv) $(\{x_t^*\}_{t=S}^{\infty}, \{u_t^*\}_{t=S}^{\infty})$ is (f)-minimal and satisfies $\lim_{t\to\infty} \rho_E(x_t, x_t^f) = 0$;
(v) $(\{x_t^*\}_{t=S}^{\infty}, \{u_t^*\}_{t=S}^{\infty})$ is (f)-minimal and satisfies

$$\liminf_{t\to\infty} \rho_E(x_t, x_t^f) = 0.$$

The next result easily follows from Theorems 3.9 and 3.17.

Theorem 3.21. *Assume that f has (P1), (P2), and LSC property, $L > 0$ be an integer, and $\epsilon > 0$. Then there exists an integer $\tau_0 > 0$ such that for each integer $T_0 \geq 0$ and each (f)-overtaking optimal pair $(\{x_t\}_{t=T_0}^{\infty}, \{u_t\}_{t=T_0}^{\infty}) \in X(T_0, \infty)$ satisfying $(T_0, x_{T_0}) \in \mathcal{A}_L$,*

$$\rho_E(x_t, x_t^f) \leq \epsilon \text{ for all integers } t \geq T_0 + \tau_0.$$

In the sequel we use the following result which easily follows from (P1) and (A2).

Proposition 3.22. *Let f have (P1) and $(T_0, z_0) \in \mathcal{A}$. There exists an (f)-good pair $(\{x_t\}_{t=T_0}^{\infty}, \{u_t\}_{t=T_0}^{\infty}) \in X(T_0, \infty)$ satisfying $x_{T_0} = z_0$ if and only if $(T_0, z_0) \in \cup\{\mathcal{A}_L : L = 1, 2, \dots\}$.*

LSC property implies the following result.

Proposition 3.23. *Let f have LSC property, $T_2 > T_2 \geq 0$, and $L > 0$ be integers.*

1. *If $X(T_1, T_2) \neq \emptyset$, then there exists $(\{x_t\}_{t=T_1}^{T_2}, \{u_t\}_{t=T_1}^{T_2-1}) \in X(T_1, T_2)$ such that*

$$\sum_{t=T_1}^{T_2-1} f(t, x_t, u_t) \leq \sum_{t=T_1}^{T_2-1} f(t, y_t, v_t) \text{ for all } (\{y_t\}_{t=T_1}^{T_2}, \{v_t\}_{t=T_1}^{T_2-1}) \in X(T_1, T_2).$$

2. *If $T_2 - T_1 \geq 2L$, $(T_1, z_1) \in \mathcal{A}_L$, $(T_2, z_2) \in \widehat{\mathcal{A}}_L$, then there exists $(\{x_t\}_{t=T_1}^{T_2}, \{u_t\}_{t=T_1}^{T_2-1}) \in X(T_1, T_2)$ such that $x_{T_i} = z_i$, $i = 1, 2$ and*

$$\sum_{t=T_1}^{T_2-1} f(t, x_t, u_t) = U^f(T_1, T_2, z_1, z_2).$$

3. *If* $T_2 - T_1 \geq L$, $(T_1, z) \in \mathcal{A}_L$, *then there exists* $(\{x_t\}_{t=T_1}^{T_2}, \{u_t\}_{t=T_1}^{T_2-1}) \in X(T_1, T_2)$ *such that* $x_{T_1} = z$ *and*

$$\sum_{t=T_1}^{T_2-1} f(t, x_t, u_t) = \sigma^f(T_1, T_2, z).$$

4. *If* $T_2 - T_1 \geq L$, $(T_2, z) \in \widehat{\mathcal{A}}_L$, *then there exists* $(\{x_t\}_{t=T_1}^{T_2}, \{u_t\}_{t=T_1}^{T_2-1}) \in X(T_1, T_2)$ *such that* $x_{T_2} = z$ *and*

$$\sum_{t=T_1}^{T_2-1} f(t, x_t, u_t) = \widehat{\sigma}^f(T_1, T_2, z).$$

The following result is proved in Section 3.16. It provides necessary and sufficient conditions for STP.

Theorem 3.24. *Let f have LSC property. If f has STP, then (P1), (P2), and the following property hold:*

(P3) *there exists $\{\tilde{u}_t^f\}_{t=0}^{\infty} \subset F$ such that $(\{x_t^f\}_{t=0}^{\infty}, \{\tilde{u}_t^f\}_{t=0}^{\infty}) \in X(0, \infty)$ is (f)-overtaking optimal and for each (f)-overtaking optimal*

$$(\{y_t\}_{t=0}^{\infty}, \{v_t\}_{t=0}^{\infty}) \in X(0, \infty)$$

satisfying $y_0 = x_0^f$, the equality $y_t = x_t^f$ holds for all integers $t \geq 0$.

If (P1), (P2), and (P3) hold and $\tilde{u}_t^f = u_t^f$ for all integers $t \geq 0$, then f has STP.

The next result follows from (P1) and STP.

Theorem 3.25. *Assume that f has STP and $\epsilon > 0$. Then there exists $\delta > 0$ such that for every integer $T_1 \geq 0$ and every (f)-overtaking optimal pair $(\{x_t\}_{t=T_1}^{\infty}, \{u_t\}_{t=T_1}^{\infty}) \in X(T_1, \infty)$ satisfying $\rho_E(x_{T_1}, x_{T_1}^f) \leq \delta$, the inequality $\rho_E(x_t, x_t^f) \leq \epsilon$ holds for all integers $t \geq T_1$.*

3.5 Perturbed Problems

In this section we suppose that the following assumption holds:

(A3) For each $\epsilon > 0$, there exists $\delta > 0$ such that for each $(T_i, z_i) \in \mathcal{A}$, $i = 1, 2$ satisfying $\rho_E(z_i, x_{T_i}^f) \leq \delta$, $i = 1, 2$ and $T_2 \geq b_f$, there exist integers $\tau_1, \tau_2 \in (0, b_f]$,

$$(\{x_t^{(1)}\}_{t=T_1}^{T_1+\tau_1}, \{u_t^{(1)}\}_{t=T_1}^{T_1+\tau_1-1}) \in X(T_1, T_1 + \tau_1),$$

$$(\{x_t^{(2)}\}_{t=T_2-\tau_2}^{T_2}, \{u_t^{(2)}\}_{t=T_2-\tau_2}^{T_2-1}) \in X(T_2-\tau_2, T_2)$$

such that

$$x_{T_1}^{(1)} = z_1,\ x_{T_1+\tau_1}^{(1)} = x_{T_1+\tau_1}^f,$$

$$\sum_{t=T_1}^{T_1+\tau_1-1} f(t, x_t^{(1)}, u_t^{(1)}) \le \sum_{t=T_1}^{T_1+\tau_1-1} f(t, x_t^f, u_t^f) + \epsilon,$$

$$\rho(x_t^{(1)}, x_t^f) \le \epsilon,\ t = T_1, \dots, T_1+\tau_1,$$

$$x_{T_2}^{(2)} = z_2,\ x_{T_2-\tau_2}^{(2)} = x_{T_2-\tau_2}^f,$$

$$\sum_{t=T_2-\tau_2}^{T_2-1} f(t, x_t^{(2)}, u_t^{(2)}) \le \sum_{t=T_2-\tau_2}^{T_2-1} f(t, x_t^f, u_t^f) + \epsilon,$$

$$\rho(x_t^{(2)}, x_t^f) \le \epsilon,\ t = T_2-\tau_2, \dots, T_2.$$

Clearly, (A3) implies (A2). Assume that $\phi : E \times E \to [0, 1]$ is a continuous function satisfying $\phi(x, x) = 0$ for all $x \in E$ and such that the following property holds:

(i) for each $\epsilon > 0$, there exists $\delta > 0$ such that if $x, y \in E$ and $\phi(x, y) \le \delta$, then $\rho_E(x, y) \le \epsilon$.

For each $r \in (0, 1)$ set

$$f_r(t, x, u) = f(t, x, u) + r\phi(x, x_t^f),\ (t, x, u) \in \mathcal{M}. \tag{3.14}$$

Clearly, for any $r \in (0, 1)$, (A1), (A3) hold for f_r with $(x_t^{f_r}, u_t^{f_r}) = (x_t^f, u_t^f)$, $t = 0, 1, \dots$. The next result is proved in Section 3.17.

Theorem 3.26. *Let $r \in (0, 1)$. Then f_r has TP, (P1), and (P2). If*

$$(\{x_t^f\}_{t=0}^\infty, \{u_t^f\}_{t=0}^\infty)$$

is (f)-minimal, then f has (P3) and $\{\tilde{u}_t^f\}_{t=0}^\infty = \{u_t^f\}_{t=0}^\infty$.

In the sequel we use the following result for which we assume only that (A2) holds and does not need (A3).

Proposition 3.27. *Let $\gamma > 0$. Then there exists $\delta > 0$ such that if $(T, z_1) \in \mathcal{A}$, $(T + 2b_f, z_2) \in \mathcal{A}$ satisfy*

$$\rho_E(z_1, x_T^f) \le \delta,\ \rho_E(z_2, x_{T+2b_f}^f) \le \delta, \tag{3.15}$$

then there exists

$$(\{x_t\}_{t=T}^{T+2b_f}, \{u_t\}_{t=T}^{T+2b_f-1}) \in X(T, T+2b_f)$$

such that

$$x_T = z_1,\ x_{T+2b_f} = z_2,\ \sum_{t=T}^{T+2b_f-1} f(t, x_t, u_t) \le \sum_{t=T}^{T+2b_f-1} f(t, x_t^f, u_t^f) + \gamma.$$

Proof. Set $\epsilon = \gamma/2$. Let $\delta > 0$ be as guaranteed by (A2). Let $(T, z_1) \in \mathcal{A}$, $(T + 2b_f, z_2) \in \mathcal{A}$ and (3.15) hold. By (A2), there exist integers $\tau_1, \tau_2 \in (0, b_f]$,

$$(\{x_t^{(1)}\}_{t=T}^{T+\tau_1}, \{u_t^{(1)}\}_{t=T}^{T+\tau_1-1}) \in X(T, T+\tau_1),$$

$$(\{x_t^{(2)}\}_{t=T+2b_f-\tau_2}^{T+2b_f}, \{u_t^{(2)}\}_{t=T+2b_f-\tau_2}^{T+2b_f-1}) \in X(T+2b_f-\tau_2, T+2b_f)$$

such that

$$x_T^{(1)} = z_1,\ x_{T+\tau_1}^{(1)} = x_{T+\tau_1}^f,$$

$$x_{T+2b_f}^{(2)} = z_2,\ x_{T+2b_f-\tau_2}^{(2)} = x_{T+2b_f-\tau_2}^f,$$

$$\sum_{t=T}^{T+\tau_1-1} f(t, x_t^{(1)}, u_t^{(1)}) \le \sum_{t=T}^{T+\tau_1-1} f(t, x_t^f, u_t^f) + \gamma/2,$$

$$\sum_{t=T+2b_f-\tau_2}^{T+2b_f-1} f(t, x_t^{(2)}, u_t^{(2)}) \le \sum_{t=T+2b_f-\tau_2}^{T+2b_f-1} f(t, x_t^f, u_t^f) + \gamma/2.$$

Define

$$x_t = x_t^{(1)},\ t = T \dots, T+\tau_1,\ u_t = u_t^{(1)},\ t = T \dots, T+\tau_1-1,$$

$$x_t = x_t^f,\ t = T+\tau_1 \dots, T+2b_f-\tau_2,\ u_t = u_t^f,$$

$$t \in \{T+\tau_1 \dots, T+2b_f-\tau_2\} \setminus \{T+2b_f-\tau_2\},$$

$$x_t = x_t^{(2)},\ t = T+2b_f-\tau_2, \dots, T,\ u_t = u_t^{(2)},\ t = T+2b-\tau_2, \dots, T-1.$$

By the relations above,

$$\sum_{t=T}^{T+2b_f-1} f(t, x_t, u_t) \le \sum_{t=T}^{T+2b_f-1} f(t, x_t^f, u_t^f) + \gamma.$$

Proposition 3.27 is proved.

3.6 Examples

Example 3.28. We consider an example which is a particular case of the problem introduced in Section 3.1.

Let $a_1 > 0$, $\psi_1 : [0, \infty) \to [0, \infty)$ be an increasing function such that $\psi_1(t) \to \infty$ as $t \to \infty$, $\mu : \{0, 1, \dots\} \to R^1$, $\pi : \{0, 1, \dots\} \times E \to R^1$, $L : \mathcal{M} \to [0, \infty)$,

$$\pi(t, x) \ge -a_1 \text{ for all } (t, x) \in \{0, 1, \dots\} \times E,$$

π be bounded on bounded subsets of $\{0, 1, \dots\} \times E$,

$$\mu(t) \ge -a_1 \text{ for all } t \in \{0, 1, \dots\},$$

and for all $(t, x, u) \in \mathcal{M}$,

$$L(t, x, u) - \pi(t, G(t, x, u)) \ge -a_1 + \psi_1(\rho_E(\theta_0, x)), \tag{3.16}$$

$$L(t, x, u) = 0 \text{ if and only if } x = x_t^f,\ u = u_t^f, \tag{3.17}$$

$$f(t, x, u) = \mu(t) + L(t, x, u) + \pi(t, x) - \pi(t + 1, G(t, x, u)). \tag{3.18}$$

By the relation above, for all $(t, x, u) \in \mathcal{M}$.

$$f(t, x, u) \ge -3a_1 + \psi_1(\rho_E(\theta_0, x)) \tag{3.19}$$

and (3.6) holds with $a_0 = 3a_1$, $\psi_0 = \psi_1$.

Let $T_2 > T_1 \ge 0$ be integers and $(\{x_t\}_{t=T_1}^{T_2}, \{u_t\}_{t=T_1}^{T_2-1}) \in X(T_1, T_2)$. It is not difficult to see that

$$\sum_{t=T_1}^{T_2-1} f(t, x_t, u_t) = \sum_{t=T_1}^{T_2-1} \mu(t) + \sum_{t=T_1}^{T_2-1} L(t, x_t, u_t) + \sum_{t=T_1}^{T_2-1} (\pi(t, x_t) - \pi(t + 1, x_{t+1}))$$

$$= \sum_{t=T_1}^{T_2-1} \mu(t) + \sum_{t=T_1}^{T_2-1} L(t, x_t, u_t) + \pi(T_1, x_{T_1}) - \pi(T_2, x_{T_2}), \tag{3.20}$$

$$\sum_{t=T_1}^{T_2-1} f(t, x_t^f, u_t^f) = \sum_{t=T_1}^{T_2-1} \mu(t) + \pi(T_1, x_{T_1}^f) - \pi(T_2, x_{T_2}^f) \tag{3.21}$$

and

$$\sum_{t=T_1}^{T_2-1} f(t, x_t, u_t) - \sum_{t=T_1}^{T_2-1} f(t, x_t^f, u_t^f)$$

$$\geq \pi(T_1, x_{T_1}) - \pi(T_2, x_{T_2}) - \pi(T_1, x_{T_1}^f) + \pi(T_2, x_{T_2}^f). \tag{3.22}$$

Since π is bounded on bounded subsets of $\{0, 1, \dots, \} \times E$, we conclude that (A1) holds.

We suppose that the following assumptions hold:

(B1) for each $\epsilon > 0$, there exists $\delta > 0$ such that for each integer $T \geq 0$, each pair $(T, z_1), (T+1, z_2) \in \mathcal{A}$ satisfying $\rho_E(z_1, x_T^f), \rho_E(z_2, x_{T+1}^f) \leq \delta$, there exists $u \in \mathcal{U}(T, z_1)$ such that

$$\rho_F(u, u_T^f) \leq \epsilon, \ z_2 = G(T, z_1, u);$$

(B2) for each $\epsilon > 0$, there exists $\delta > 0$ such that for each $(T, z, \xi) \in \mathcal{M}$ satisfying $\rho_E(z, x_T^f) \leq \delta$, $\rho_F(\xi, u_T^f) \leq \delta$, we have $f(T, z, \xi) \leq f(T, x_T^f, u_T^f) + \epsilon$.
It is clear that (B1) and (B2) imply (A2) and (A3).

Example 3.29. We consider an example which is a particular case of the problem introduced in Section 3.1. Let $a_1 > 0$, $\psi_1 : [0, \infty) \to [0, \infty)$ be an increasing function such that $\psi_1(t) \to \infty$ as $t \to \infty$, $\mu : \{0, 1, \dots\} \to R^1$, $\pi : \{0, 1, \dots\} \times E \to R^1$, $L : \mathcal{M} \to [0, \infty)$,

$$\pi(t, x) \geq -a_1 \text{ for all } (t, x) \in \{0, 1, \dots\} \times E,$$

π be bounded on bounded subsets of $\{0, 1, \dots\} \times E$,

$$\mu(t) \geq -a_1 \text{ for all } t \in \{0, 1, \dots\},$$

and for all $(t, x, u) \in \mathcal{M}$, (3.16)–(3.18) hold. Then as it was shown in Example 2.28, (A1) and (3.6) hold with $a_0 = 3a_1$, $\psi_0 = \psi_1$.

We suppose that there exists a natural number d_f such that the following assumptions hold:

(B3) for each $\epsilon > 0$, there exists $\delta > 0$ such that for each integer $T \geq 0$, each pair $(T, z_1), (T + d_f, z_2) \in \mathcal{A}$ satisfying $\rho_E(z_1, x_T^f), \rho_E(z_2, x_{T+d_f}^f) \leq \delta$, there exists $(\{x_t\}_{t=T}^{T+d_f}, \{u_t\}_{t=T}^{T+d_f-1}) \in X(T, T + d_f)$ such that

$$x_T = z_1, \ x_{T+d_f} = z_2, \ \rho(u_t, u_t^f) \leq \epsilon, t = T, \ldots, T + d_f - 1;$$

(B4) for each $\epsilon > 0$, there exists $\delta > 0$ such that for each $(T, z) \in \mathcal{A}$ satisfying $\rho_E(z, x_T^f) \leq \delta$ and each $u \in \mathcal{U}(T, z)$ satisfying $\rho_F(u, u_T^f) \leq \delta$, we have

$$\rho_E(x_{T+1}^f, G(T, z, u)) \leq \epsilon, \ f(T, z, u) \leq f(T, x_T^f, u_T^f) + \epsilon.$$

(B3) and (B4) imply (A2) and (A3).

Assume now that for any nonempty bounded set $\Omega \subset E$, the function π is bounded on $\{0, 1, \ldots, \} \times \Omega$, the function μ is bounded, and the following property holds:

(B5) for each $M, \epsilon > 0$, there exists $\delta > 0$ such that for each $(t, x, u) \in \mathcal{M}$ which satisfies

$$\rho_E(\theta_0, x) \leq M \text{ and } L(t, x, u) \leq \delta$$

we have $\rho_E(x, x_t^f) \leq \epsilon$.

We claim that f has TP. In view of Theorem 3.17, it is sufficient to show that f possesses (P1) and (P2). Let us show that f has (P1). Assume that $(\{x_t\}_{t=0}^{\infty}, \{u_t\}_{t=0}^{\infty}) \in X(0, \infty)$ is (f)-good. Theorem 3.1 and the boundedness assumption on π imply that

$$\sup\{\rho_E(\theta_0, x_t) : \ t = 0, 1, \ldots\} < \infty,$$

$$\sup\{|\pi(x_t)| : \ t = 0, 1, \ldots\} < \infty,$$

$$\sup\{|\sum_{t=0}^{T-1} f(t, x_t, u_t) - \sum_{t=0}^{T} f(t, x_t^f, u_t^f)| : \ T \in \{1, 2, \ldots\}\} < \infty,$$

$$\sup\{|\pi(x_t^f)| : \ t = 0, 1, \ldots\} < \infty.$$

Combined with (3.17) and (3.18), these relations imply that

$$\sup\{\sum_{t=0}^{T-1} L(t, x_t, u_t) : \ T \in \{1, 2, \ldots\}\}$$

$$= \sup\{\sum_{t=0}^{T-1} f(t, x_t, u_t) - \sum_{t=0}^{T-1} f(t, x_t^f, u_t^f)$$

$$-\pi(0, x_0) + \pi(T, x_T) + \pi(0, x_0^f) - \pi(T, x_T^f) : T \in \{1, 2, \dots\}\} < \infty.$$

Together with (B5) this implies that $\lim_{t\to\infty} \rho_E(x_t, x_t^f) = 0$ and (P1) holds.

We claim that f has (P2). Let $\epsilon, M > 0$ and $M_1 > 0$ be as guaranteed by Theorem 3.2 with $c = 1$ and $M_0 = M$. Since π satisfies the boundedness assumption, there exists $M_2 > 0$ such that

$$|\pi(t, z)| \le M_2 \text{ for all integers } t \ge 0 \text{ and all } z \in B_E(\theta_0, M_1).$$

Let $\delta \in (0, 1)$ be as guaranteed by (B5) with $M = M_1$. Choose an integer

$$L > 3 + \delta^{-1}(M + 3a_1 + \sup\{|\mu(t)| : t = 0, 1, \dots\}$$
$$+6 \sup\{|\pi(t, z)| : t \in \{0, 1, \dots\}, z \in B_E(\theta_0, M)\}).$$

Assume that $T \ge 0$ is an integer and that $(\{x_t\}_{t=T}^{T+L}, \{u_t\}_{t=T}^{T+L-1}) \in X(T, T + L)$ satisfies

$$\sum_{t=T}^{T+L-1} f(t, x_t, u_t) \le \sum_{t=T}^{T+L-1} f(t, x_t^f, u_t^f) + M.$$

Combined with Theorem 3.2 and the choice of M_1, this implies that

$$\rho_E(\theta_0, x_t) \le M_1, \ t = T, \dots, T + L - 1,$$
$$\rho_E(\theta_0, x_t^f) \le M_1, \ t = 0, 1, \dots.$$

It is not difficult to see that

$$\sum_{t=T}^{T+L-2} f(t, x_t, u_t) \le \sum_{t=T}^{T+L-1} f(t, x_t, u_t) - f(T + L - 1, x_{T+L-1}, u_{T+L-1})$$
$$\le \sum_{t=T}^{T+L-1} f(t, x_t^f, u_t^f) + M + 3a_1$$
$$\le \sum_{t=T}^{T+L-2} f(t, x_t^f, u_t^f) + M + 3a_1 + \sup\{|f(s, x_s^f, u_s^f)| : s = 0, 1, \dots\}$$
$$\le \sum_{t=T}^{T+L-2} f(t, x_t^f, u_t^f) + M + 3a_1 + \sup\{|\mu(t)| : t = 0, 1, \dots\}$$
$$+2 \sup\{|\pi(t, z)| : t \in \{0, 1, \dots\}, z \in B_E(\theta_0, M)\}.$$

Combined with (3.17) and (3.18), this implies that

$$M + 3a_1 + \sup\{|\mu(t)| : t = 0, 1, \dots\}$$

$$+2\sup\{|\pi(t, z)| : t \in \{0, 1, \dots\}, z \in B_E(\theta_0, M)\}$$

$$\ge \sum_{t=T}^{T+L-2} f(t, x_t, u_t) - \sum_{t=T}^{T+L-2} f(t, x_t^f, u_t^f)$$

$$= \sum_{t=T}^{T+L-2} L(t, x_t, u_t) + \pi(T, x_T) - \pi(T + L - 1, x_{T+L-1})$$

$$-\pi(T, x_T^f) - \pi(T + L - 1, x_{T+L-1}^f)$$

$$\ge \sum_{t=T}^{T+L-2} L(t, x_t, u_t) - 4\sup\{|\pi(t, z)| : t \in \{0, 1, \dots\}, z \in B_E(\theta_0, M)\}.$$

Together with the choice of δ and (B5), this implies that

$$M + 3a_1 + \sup\{|\mu(t)| : t = 0, 1, \dots\}$$

$$+6\sup\{|\pi(t, z)| : t \in \{0, 1, \dots\}, z \in B_E(\theta_0, M)\}$$

$$\ge \sum_{t=T}^{T+L-2} L(t, x_t, u_t)$$

$$\ge \delta \mathrm{Card}(\{t \in \{T, \dots, T + L - 2\} : L(t, x_t, u_t) > \delta\})$$

$$\ge \delta \mathrm{Card}(\{t \in \{T, \dots, T + L - 2\} : \rho_E(x_t, x_t^f) > \epsilon\})$$

and

$$\mathrm{Card}(\{t \in \{T, \dots, T + L - 1\} : \rho_E(x_t, x_t^f) > \epsilon\})$$

$$\le 1 + \delta^{-1}(M + 3a_1 + \sup\{|\mu(t)| : t = 0, 1, \dots\}$$

$$+6\sup\{|\pi(t, z)| : t \in \{0, 1, \dots\}, z \in B_E(\theta_0, M)\}) < L - 2.$$

Therefore (P2) holds and f has TP.

Example 3.30. We consider a particular case of the problem introduced in Section 3.1. Assume that for each integer $t \ge 0$, the set $\{x \in E : (t, x) \in \mathcal{A}\}$ is a closed compact set, $\{(x, u) \in E \times F : (t, x, u) \in \mathcal{M}\}$ is a closed compact set, $G(t, \cdot, \cdot) : \{(x, u) \in E \times F : (t, x, u) \in \mathcal{M}\} \to E$ is continuous and that $f(t, \cdot, \cdot) : \{(x, u) \in E \times F : (t, x, u) \in \mathcal{M}\} \to R^1$ is lower semicontinuous. Then LSC property holds.

Example 3.31. We consider a particular case of the problem introduced in Section 3.1. Assume that for each $M > 0$ and each integer $t \geq 0$, the set $\{(x, u) \in E \times F : (t, x, u) \in \mathcal{M}, \rho_E(x, \theta_0) \leq M\}$ is a closed and compact, $G(t, \cdot, \cdot) : \{(x, u) \in E \times F : (t, x, u) \in \mathcal{M}\} \to E$ is continuous and $f(t, \cdot, \cdot) : \{(x, u) \in E \times F : (t, x, u) \in \mathcal{M}\} \to R^1$ is lower semicontinuous. Then LSC property holds.

3.7 Auxiliary Results for Theorems 3.1 and 3.2

Lemma 3.32. *There exist $S > 0$ and an integer $c_0 \geq 1$ such that for each pair of integers $T_1 \geq 0$, $T_2 \geq T_1 + c_0$, and each $(\{x_t\}_{t=T_1}^{T_2}, \{u_t\}_{t=T_1}^{T_2-1}) \in X(T_1, T_2)$,*

$$\sum_{t=T_1}^{T_2-1} f(t, x_t^f, u_t^f) \leq \sum_{t=T_1}^{T_2-1} f(t, x_t, u_t) + S. \tag{3.23}$$

Proof. In view of (3.5), there exists $S_1 > 0$ such that

$$\psi(S_1) > a_0 + 1 + \Delta_f. \tag{3.24}$$

By (A1), there exist $S_2 > 0$ and an integer $c_0 > 0$ such that

$$\sum_{t=T_1}^{T_2-1} f(t, x_t^f, u_t^f) \leq \sum_{t=T_1}^{T_2-1} f(t, x_t, u_t) + S_2$$

for each pair of integers $T_1 \geq 0$, $T_2 \geq T_1 + c_0$, and each $(\{x_t\}_{t=T_1}^{T_2}, \{u_t\}_{t=T_1}^{T_2-1}) \in X(T_1, T_2)$ satisfying $\rho_E(\theta_0, x_j) \leq S_1$, $j = T_1, T_2$.

Fix an integer

$$S \geq S_2 + 2 + (c_0 + 2)(2a_0 + |\Delta_f|). \tag{3.25}$$

Assume that $T_1 \geq 0$, $T_2 \geq T_1 + c_0$ are integers and that

$$(\{x_t\}_{t=T_1}^{T_2}, \{u_t\}_{t=T_1}^{T_2-1}) \in X(T_1, T_2).$$

We show that (3.23) is true. Assume that

$$\rho_E(\theta_0, x_t) \geq S_1, \ t = T_1, \dots, T_2 - 1. \tag{3.26}$$

By (3.6) and (3.26), for all $t = T_1, \dots, T_2 - 1$,

$$f(t, x_t, u_t) \geq -a_0 + \psi(\rho_E(\theta_0, x_t)) \geq -a_0 + \psi(S_1). \tag{3.27}$$

It follows from (3.12), (3.24), and (3.27),

$$\sum_{t=T_1}^{T_2-1} f(t, x_t, u_t) \geq (T_2 - T_1)(\psi(S_1) - a_0)$$

$$\geq (T_2 - T_1)(\Delta_f + 1) \geq \sum_{t=T_1}^{T_2-1} f(t, x_t^f, u_t^f) + 1$$

and (3.23) holds. Assume that

$$\min\{\rho_E(\theta_0, x_t) : \ t = T_1, \dots, T_2 - 1\} < S_1. \tag{3.28}$$

Set

$$\tau_1 = \min\{t \in \{T_1, \dots, T_2 - 1\} : \ \rho_E(\theta_0, x_t) \leq S_1\},$$
$$\tau_2 = \max\{t \in \{T_1, \dots, T_2 - 1\} : \ \rho_E(\theta_0, x_t) \leq S_1\}. \tag{3.29}$$

Clearly, τ_1, τ_2 are well-defined, $\tau_1 \leq \tau_2$ and

$$\rho_E(\theta_0, x_{\tau_i}) \leq S_1, \ i = 1, 2. \tag{3.30}$$

There are two cases:

$$\tau_2 - \tau_1 \geq c_0; \tag{3.31}$$

$$\tau_2 - \tau_1 < c_0. \tag{3.32}$$

Assume that (3.31) holds. It follows from (3.29), (3.31), and the choice of S_2 and c_0 that

$$\sum_{t=\tau_1}^{\tau_2-1} f(t, x_t^f, u_t^f) \leq \sum_{t=\tau_1}^{\tau_2-1} f(t, x_t, u_t) + S_2. \tag{3.33}$$

By (3.6), (3.12), (3.24), and (3.29), for each integer $t \in ([T_1, \tau_1] \cup [\tau_2, T_2]) \setminus \{\tau_1, \tau_2, T_2\}$,

$$\rho_E(\theta_0, x_t) > S_1, \ f(t, x_t, u_t) \geq -a_0 + \psi(\rho_E(\theta_0, x_t)) \geq -a_0 + \psi(S_1). \tag{3.34}$$

It follows from (3.34) that

$$\sum\{f(t,x_t,u_t):\ t\in\{T_1,\dots,\tau_1\}\setminus\{\tau_1\}\}\ge(\tau_1-T_1)(\psi(S_1)-a_0)$$
$$\ge\sum\{f(t,x_t^f,u_t^f):\ t\in\{T_1,\dots,\tau_1\}\setminus\{\tau_1\}\},\tag{3.35}$$

$$\sum\{f(t,x_t,u_t):\ t\in\{\tau_2,\dots,T_2\}\setminus\{\tau_2,T_2\}\}$$
$$\ge\sum\{f(t,x_t^f,u_t^f):\ t\in\{\tau_2,\dots,T_2\}\setminus\{\tau_2,T_2\}\}.\tag{3.36}$$

In view of (3.6), (3.12), (3.25), (3.33), (3.35), and (3.36),

$$\sum_{t=T_1}^{T_2-1} f(t,x_t,u_t)\ge\sum_{t=T_1}^{T_2-1} f(t,x_t^f,u_t^f)-S_2-\Delta_f-a_0\ge\sum_{t=T_1}^{T_2-1} f(t,x_t^f,u_t^f)-S$$

and (3.23) holds.

Assume that (3.32) holds. By (3.6), (3.24), (3.25), (3.29), and (3.32),

$$\sum_{t=T_1}^{T_2-1} f(t,x_t,u_t)=\sum\{f(t,x_t,u_t):\ t\in\{T_1,\dots,\tau_1\}\setminus\{\tau_1\}\}$$
$$+\sum\{f(t,x_t,u_t):\ t\in\{\tau_1,\dots,\tau_2\}\}$$
$$+\sum\{f(t,x_t,u_t):\ t\in\{\tau_2+1,\dots,T_2+1\}\setminus\{T_2,T_2+1\}\}$$
$$\ge(\psi(S_1)-a_0)(\tau_1-T_1)-(c_0+1)a_0+(\psi(S_1)-a_0)(T_2-\tau_2-1)$$
$$\ge\Delta_f(\tau_1-T_1)+\Delta_f(\tau_2-\tau_1)-|\Delta_f|c_0-(c_0+1)a_0+\Delta_f(T_2-\tau_2)-|\Delta_f|$$
$$=\Delta_f(T_2-T_1)-(c_0+1)a_0-(c_0+1)|\Delta_f|\ge\sum_{t=T_1}^{T_2-1} f(t,x_t^f,u_t^f)-S$$

and (3.23) is true. Lemma 3.32 is proved.

The next auxiliary result easily follows from (3.5) and (3.6).

Lemma 3.33. *Let $M_0>0$ and τ_0 be a natural number. Then there exists $M_1>M_0$ such that for each integer $T_1\ge 0$, each integer $T_2\in(T_1,T_1+\tau_0]$, and each $(\{x_t\}_{t=T_1}^{T_2},\{u_t\}_{t=T_1}^{T_2-1})\in X(T_1,T_2)$ which satisfies $\sum_{t=T_1}^{T_2-1} f(t,x_t,u_t)\le M_0$, the inequality $\rho_E(x_t,\theta_0)\le M_1$ holds for all $t=T_1,\dots,T_2-1$,*

3.8 Proofs of Theorems 3.1–3.5

Assertion 1 of Theorem 3.1 follows from Lemmas 3.32, (3.6), and (3.12). Let us prove Assertion 2. Assume that there exists a sequence of natural numbers $\{t_k\}_{k=1}^{\infty}$ such that

$$t_k \to \infty \text{ as } k \to \infty,$$

$$\sum_{t=0}^{t_k-1} f(t, x_t, u_t) - \sum_{t=0}^{t_k-1} f(t, x_t^f, u_t^f) \to \infty \text{ as } k \to \infty. \tag{3.37}$$

Let a number S be as guaranteed by Assertion 1 and let $Q > 0$. In view of (3.37), there exists a natural number k_0 such that for each integer $k \geq k_0$,

$$\sum_{t=0}^{t_k-1} f(t, x_t, u_t) - \sum_{t=0}^{t_k-1} f(t, x_t^f, u_t^f) > Q + S. \tag{3.38}$$

Let $T > t_{k_0}$ be an integer. By (3.38), the choice of S and Assertion 1 of Theorem 3.1,

$$\sum_{t=0}^{T-1} f(t, x_t, u_t) - \sum_{t=0}^{T-1} f(t, x_t^f, u_t^f) = \sum_{t=0}^{t_{k_0}-1} f(t, x_t, u_t) - \sum_{t=0}^{t_{k_0}-1} f(t, x_t^f, u_t^f)$$

$$+ \sum_{t=t_{k_0}}^{T-1} f(t, x_t, u_t) - \sum_{t=t_{k_0}}^{T-1} f(t, x_t^f, u_t^f) \geq Q + S - S = Q.$$

Since Q is any positive number, we obtain that

$$\lim_{T\to\infty} [\sum_{t=0}^{T-1} f(t, x_t, u_t) - \sum_{t=0}^{T-1} f(t, x_t^f, u_t^f)] = \infty.$$

Assume that

$$\sup\{|\sum_{t=0}^{T-1} f(t, x_t, u_t) - \sum_{t=0}^{T-1} f(t, x_t^f, u_t^f)| : t = 1, 2, \ldots\} < \infty.$$

This implies that there exists an integer $S_1 > 0$ such that for each pair of integers $T_2 > T_1 \geq 0$,

$$|\sum_{t=T_1}^{T_2-1} f(t, x_t, u_t) - \sum_{t=T_1}^{T_2-1} f(t, x_t^f, u_t^f)| < S_1.$$

Together with (3.12) the inequality above implies that for all integers $t \geq 0$, we have $f(t, x_t, u_t) \leq \Delta_f + S_1$. Combined with (3.5) and (3.6), this implies that

$$\sup\{\rho_E(x_t, \theta_0) : t = 0, 1, 2, \ldots\} < \infty.$$

Theorem 3.1 is proved.

Proof of Theorem 3.2. We may assume that $c = 1$. By Theorem 3.1, there exists $S_0 > 0$ such that the following property holds:

(i) for each pair of integers $T_2 > T_1 \geq 0$ and each $(\{x_t\}_{t=T_1}^{T_2}, \{u_t\}_{t=T_1}^{T_2-1}) \in X(T_1, T_2)$,

$$\sum_{t=T_1}^{T_2-1} f(t, x_t, u_t) + S_0 \geq \sum_{t=T_1}^{T_2-1} f(t, x_t^f, u_t^f).$$

In view of (3.5) and (3.6), there exists $M_1 > 0$ such that for each $(t, x, u) \in \mathcal{M}$ satisfying $f(t, x, u) \leq 2S_0 + M_0 + 2 + |\Delta_f|$, we have

$$\rho_E(x, \theta_0) \leq M_1. \tag{3.39}$$

Let $T_2 > T_1 \geq 0$ be integers and $(\{x_t\}_{t=T_1}^{T_2}, \{u_t\}_{t=T_1}^{T_2-1}) \in X(T_1, T_2)$ satisfy

$$\sum_{t=T_1}^{T_2-1} f(t, x_t, u_t) \leq \sum_{t=T_1}^{T_2-1} f(t, x_t^f, u_t^f) + M_0. \tag{3.40}$$

We show that

$$\rho_E(\theta_0, x_t) \leq M_1, \ t = T_1, \ldots, T_2 - 1.$$

Assume the contrary. Then there exists $t_0 \in \{T_1, \ldots, T_2 - 1\}$ such that

$$\rho_E(\theta_0, x_{t_0}) > M_1. \tag{3.41}$$

Property (i), (3.12), and (3.40) imply that

$$f(t_0, x_{t_0}, u_{t_0}) = \sum_{t=T_1}^{T_2-1} f(t, x_t, u_t) - \sum\{f(t, x_t, u_t) : t \in \{T_1, \ldots, t_0\} \setminus \{t_0\}\}$$

$$-\sum\{f(t, x_t, u_t) : t \in \{t_0, \ldots, T_2 - 1\} \setminus \{t_0\}\} \leq \sum_{t=T_1}^{T_2-1} f(t, x_t^f, u_t^f) + M_0$$

$$-\sum\{f(t, x_t^f, u_t^f) : t \in \{T_1, \ldots, t_0\} \setminus \{t_0\}\} + S_0$$

$$-\sum\{f(t, x_t^f, u_t^f) : t \in \{t_0, \dots, T_2 - 1\} \setminus \{t_0\}\} + S_0$$

$$\leq \Delta_f + M_0 + 2S_0.$$

In view of the relation above and (3.39), $\rho_E(\theta_0, x_{t_0}) \leq M_1$. This contradicts (3.41). The contradiction we have reached proves Theorem 3.2.

Theorems 3.3–3.5 follow from Theorem 3.2 and Propositions 3.8, 3.11, and 3.14, respectively.

3.9 An Auxiliary Result for Theorem 3.17

Lemma 3.34. *Let* $(\{x_t\}_{t=0}^{\infty}, \{u_t\}_{t=0}^{\infty}) \in X(0, \infty)$ *be* (f)*-good and* $\delta > 0$. *Then there exists an integer* $T_\delta > 0$ *such that for each integer* $T > T_\delta$,

$$\sum_{t=T_\delta}^{T-1} f(t, x_t, u_t) \leq U^f(T_\delta, T, x_{T_\delta}, x_T) + \delta.$$

Proof. Assume the contrary. Then for each integer $T \geq 0$, there exists an integer $S > T$ such that

$$\sum_{t=T}^{S-1} f(t, x_t, u_t) > U^f(T, S, x_T, x_S) + \delta.$$

This implies that there exists a strictly increasing sequence of integers $\{T_i\}_{i=0}^{\infty}$ such that $T_0 = 0$ and for every integer $i \geq 0$,

$$\sum_{t=T_i}^{T_{i+1}-1} f(t, x_t, u_t) > U^f(T_i, T_{i+1}, x_{T_i}, x_{T_{i+1}}) + \delta. \tag{3.42}$$

By (3.42), there exists $(\{y_t\}_{t=0}^{\infty}, \{v_t\}_{t=0}^{\infty}) \in X(0, \infty)$ such that for every integer $i \geq 0$, $y_{T_i} = x_{T_i}$ and

$$\sum_{t=T_i}^{T_{i+1}-1} f(t, x_t, u_t) > \sum_{t=T_i}^{T_{i+1}-1} f(t, y_t, v_t) + \delta. \tag{3.43}$$

Since $(\{x_t\}_{t=0}^{\infty}, \{u_t\}_{t=0}^{\infty})$ is (f)-good there exists $S > 0$ such that for each pair of integers $T_2 > T_1 \geq 0$,

$$|\sum_{t=T_1}^{T_2-1} f(t, x_t, u_t) - \sum_{t=T_1}^{T_2-1} f(t, x_t^f, u_t^f)| < S. \tag{3.44}$$

In view of (3.43) and (3.44), for every integer $k \geq 1$,

$$\sum_{t=0}^{T_k-1} f(t, y_t, v_t) - \sum_{t=T_1}^{T_k-1} f(t, x_t^f, u_t^f) = \sum_{t=0}^{T_k-1} f(t, y_t, v_t) - \sum_{t=T_1}^{T_k-1} f(t, x_t, u_t)$$

$$+ \sum_{t=0}^{T_k-1} f(t, x_t, u_t) - \sum_{t=T_1}^{T_k-1} f(t, x_t^f, u_t^f) \leq -k\delta + S \to -\infty \text{ as } k \to \infty.$$

This contradicts Theorem 3.1. The contradiction we have reached proves Lemma 3.34.

3.10 Proof of Theorem 3.17

First we show that TP implies (P1) and (P2). In view of Theorem 3.1, TP implies (P2). We show that TP implies (P1). Assume that TP holds, $(\{x_t\}_{t=0}^{\infty}, \{u_t\}_{t=0}^{\infty}) \in X(0, \infty)$ is (f)-good and $\epsilon > 0$. There exists $S > 0$ such that for each integer $T > 0$,

$$|\sum_{t=0}^{T-1} f(t, x_t, u_t) - \sum_{t=0}^{T-1} f(t, x_t^f, u_t^f)| < S.$$

This implies that for each pair of integers $T_2 > T_1 \geq 0$,

$$|\sum_{t=T_1}^{T_2-1} f(t, x_t, u_t) - \sum_{t=T_1}^{T_2-1} f(t, x_t^f, u_t^f)| < 2S. \tag{3.45}$$

Theorem 3.6 implies that there exist $\delta > 0$ and an integer $L > 0$ such that the following property holds:

(i) for each integer $S_1 \geq 0$, each integer $S_2 \geq S_1 + 2L$ and each $(\{z_t\}_{t=S_1}^{S_2}, \{\xi_t\}_{t=S_1}^{S_2-1}) \in X(S_1, S_2)$ which satisfies

$$\sum_{t=S_1}^{S_2-1} f(t, z_t, \xi_t) \leq \min\{\sum_{t=S_1}^{S_2-1} f(t, x_t^f, u_t^f) + 2S, U^f(S_1, S_2, z_{S_1}, z_{S_2}) + \delta\}$$

we have

$$\rho_E(z_t, x_t^f) \leq \epsilon, \; t = S_1 + L, \ldots, S_2 - L.$$

Let an integer $T_\delta > 0$ be as guaranteed by Lemma 3.34. Equation (3.45), Lemma 3.34, and the choice of T_δ imply that for each integer $T \geq T_\delta + 2L$,

$$\rho_E(x_t, x_t^f) \leq \epsilon \text{ for all integers } t \geq T_\delta + L.$$

Thus TP implies (P1).

Lemma 3.35. *Assume that (P1) holds and that $\epsilon > 0$. Then there exist $\delta > 0$ and an integer $L > 0$ such that for each integer $T_1 \geq L$, each integer $T_2 \geq T_1 + 2b_f$ and each $(\{x_t\}_{t=T_1}^{T_2}, \{u_t\}_{t=T_1}^{T_2-1}) \in X(T_1, T_2)$ which satisfies*

$$\rho_E(x_{T_i}, x_{T_i}^f) \leq \delta,\ i = 1, 2,$$

$$\sum_{t=T_1}^{T_2-1} f(t, x_t, u_t) \leq U^f(T_1, T_2, x_{T_1}, x_{T_2}) + \delta$$

the inequality $\rho_E(x_t, x_t^f) \leq \epsilon$ holds for all $t = T_1, \dots, T_2$.

Proof. By (A2), for each integer $k \geq 1$, there exists $\delta_k \in (0, 2^{-k})$ such that the following property holds:

(ii) for each $(T_i, z_i) \in \mathcal{A}$, $i = 1, 2$ satisfying $\rho_E(z_i, x_{T_i}^f) \leq \delta_k$, $i = 1, 2$ and $T_2 \geq b_f$ there exist integers $\tau_1, \tau_2 \in (0, b_f]$,

$$(\{x_t^{(1)}\}_{t=T_1}^{T_1+\tau_1}, \{u_t^{(1)}\}_{t=T_1}^{T_1+\tau_1-1}) \in X(T_1, T_1 + \tau_1),$$

$$(\{x_t^{(2)}\}_{t=T_2-\tau_2}^{T_2}, \{u_t^{(2)}\}_{t=T_2-\tau_2}^{T_2-1}) \in X(T_2 - \tau_2, T_2)$$

such that

$$x_{T_1}^{(1)} = z_1,\ x_{T_1+\tau_1}^{(1)} = x_{T_1+\tau_1}^f,\ x_{T_2}^{(2)} = z_2,\ x_{T_2-\tau_2}^{(2)} = x_{T_2-\tau_2}^f,$$

$$\sum_{t=T_1}^{T_1+\tau_1-1} f(t, x_t^{(1)}, u_t^{(1)}) \leq \sum_{t=T_1}^{T_1+\tau_1-1} f(t, x_t^f, u_t^f) + 2^{-k},$$

$$\sum_{t=T_2-\tau_2}^{T_2-1} f(t, x_t^{(2)}, u_t^{(2)}) \leq \sum_{t=T_2-\tau_2}^{T_2-1} f(t, x_t^f, u_t^f) + 2^{-k}.$$

Assume that the lemma does not hold. Then for each integer $k \geq 1$, there exist integers $T_{k,1} \geq k + b_f$, $T_{k,2} \geq T_{k,1} + 2b_f$ and $(\{x_t^{(k)}\}_{t=T_{k,1}}^{T_{k,2}}, \{u_t^{(k)}\}_{t=T_{k,1}}^{T_{k,2}-1}) \in X(T_{k,1}, T_{k,2})$ such that

$$\rho_E(x_{T_i}^{(k)}, x_{T_i}^f) \le \delta_k, \ i = 1, 2, \tag{3.46}$$

$$\sum_{t=T_{k,1}}^{T_{k,2}-1} f(t, x_t^{(k)}, u_t^{(k)}) \le U^f(T_{k,1}, T_{k,2}, x_{T_{k,1}}^{(k)}, x_{T_{k,2}}^{(k)}) + \delta_k, \tag{3.47}$$

$$\max(\{\rho_E(x_t^{(k)}, x_t^f) : \ t \in \{T_{k,1}, \dots, T_{k,2}\}) > \epsilon. \tag{3.48}$$

Extracting a subsequence and re-indexing, we may assume without loss of generality that for each integer $k \ge 1$,

$$T_{k+1,1} \ge T_{k,2} + 4b_f. \tag{3.49}$$

Let $k \ge 1$ be an integer. Property (ii) and (3,46) imply that there exist integers $\tau_{k,1}, \tau_{k,2} \in (0, b_f]$ and $(\{\tilde{x}_t^{(k)}\}_{t=T_{k,1}-\tau_{k,1}}^{T_{k,2}+\tau_{k,2}}, \{\tilde{u}_t^{(k)}\}_{t=T_{k,1}-\tau_{k,1}}^{T_{k,2}+\tau_{k,2}-1}) \in X(T_{k,1} - \tau_{k,1}, T_{k,2} + \tau_{k,2})$ such that

$$\tilde{x}_t^{(k)} = x_t^{(k)}, \ t = T_{k,1}, \dots, T_{k,2}, \ \tilde{u}_t^{(k)} = u_t^{(k)}, \ t = T_{k,1}, \dots, T_{k,2} - 1, \tag{3.50}$$

$$\tilde{x}_{T_{k,1}-\tau_{k,1}}^{(k)} = x_{T_{k,1}-\tau_{k,1}}^f, \ \tilde{x}_{T_{k,2}+\tau_{k,2}}^{(k)} = x_{T_{k,2}+\tau_{k,2}}^f, \tag{3.51}$$

$$\sum_{t=T_{k,1}-\tau_{k,1}}^{T_{k,1}-1} f(t, \tilde{x}_t^{(k)}, \tilde{u}_t^{(k)}) \le \sum_{t=T_{k,1}-\tau_{k,1}}^{T_{k,1}-1} f(t, x_t^f, u_t^f) + 2^{-k}, \tag{3.52}$$

$$\sum_{t=T_{k,2}}^{T_{k,2}+\tau_{k,2}-1} f(t, \tilde{x}_t^{(k)}, \tilde{u}_t^{(k)}) \le \sum_{t=T_{k,2}}^{T_{k,2}+\tau_{k,2}-1} f(t, x_t^f, u_t^f) + 2^{-k}. \tag{3.53}$$

Property (ii) and (3.46) imply that there exist integers $\tau_{k,3}, \tau_{k,4} \in (0, b_f]$ and $(\{\widehat{x}_t^{(k)}\}_{t=T_{k,1}}^{T_{k,2}}, \{\widehat{u}_t^{(k)}\}_{t=T_{k,1}}^{T_{k,2}-1}) \in X(T_{k,1}, T_{k,2})$ such that

$$\widehat{x}_{T_{k,1}}^{(k)} = x_{T_{k,1}}^{(k)}, \ \widehat{x}_{T_{k,2}}^{(k)} = x_{T_{k,2}}^{(k)}, \tag{3.54}$$

$$\widehat{x}_t^{(k)} = x_t^f, \ t = T_{k,1} + \tau_{k,3} \dots, T_{k,2} - \tau_{k,4},$$

$$\widehat{u}_t^{(k)} = u_t^f, \ t \in \{T_{k,1} + \tau_{k,3}, \dots, T_{k,2} - \tau_{k,4}\} \setminus \{T_{k,2} - \tau_{k,4}\}, \tag{3.55}$$

$$\sum_{t=T_{k,1}}^{T_{k,1}+\tau_{k,3}-1} f(t, \widehat{x}_t^{(k)}, \widehat{u}_t^{(k)}) \le \sum_{t=T_{k,1}}^{T_{k,1}+\tau_{k,3}-1} f(t, x_t^f, u_t^f) + 2^{-k}, \tag{3.56}$$

$$\sum_{t=T_{k,2}-\tau_{k,4}}^{T_{k,2}-1} f(t, \widehat{x}_t^{(k)}, \widehat{u}_t^{(k)}) \le \sum_{t=T_{k,2}-\tau_{k,4}}^{T_{k,2}-1} f(t, x_t^f, u_t^f) + 2^{-k}. \tag{3.57}$$

By (3.54)–(3.57),

$$U^f(T_{k,1}, T_{k,2}, x_{T_{k,1}}^{(k)}, x_{T_{k,2}}^{(k)}) \le \sum_{t=T_{k,1}}^{T_{k,2}-1} f(t, \widehat{x}_t^{(k)}, \widehat{u}_t^{(k)})$$

$$\le \sum_{t=T_{k,1}}^{T_{k,2}-1} f(t, x_t^f, u_t^f) + 2^{-k+1}. \tag{3.58}$$

It follows from (3.47) and (3.58) that

$$\sum_{t=T_{k,1}}^{T_{k,2}-1} f(t, x_t^{(k)}, u_t^{(k)}) \le \sum_{t=T_{k,1}}^{T_{k,2}-1} f(t, x_t^f, u_t^f) + 2^{-k+2}. \tag{3.59}$$

By (3.52), (3.53), (3.56), and (3,59),

$$\sum_{t=T_{k,1}-\tau_{k,1}}^{T_{k,2}+\tau_{k,2}-1} f(t, \tilde{x}_t^{(k)}, \tilde{u}_t^{(k)}) \le \sum_{t=T_{k,1}-\tau_{k,1}}^{T_{k,1}-1} f(t, x_t^f, u_t^f) + \sum_{t=T_{k,1}}^{T_{k,2}-1} f(t, x_t^{(k)}, u_t^{(k)})$$

$$+ \sum_{t=T_{k,2}}^{T_{k,2}+\tau_{k,2}-1} f(t, x_t^f, u_t^f) + 2^{-k+1} \le \sum_{t=T_{k,1}-\tau_{k,1}}^{T_{k,2}+\tau_{k,2}-1} f(t, x_t^f, u_t^f) + 2^{-k+3}. \tag{3.60}$$

By (3.49) and (3.51), there exists $(\{x_t\}_{t=0}^{\infty}, \{u_t\}_{t=0}^{\infty}) \in X(0, \infty)$ such that for every integer $k \ge 1$,

$$x_t = \tilde{x}_t^{(k)}, \ t = T_{k,1} - \tau_{k,1}, \dots, T_{k,2} + \tau_{k,2},$$

$$u_t = \tilde{u}_t^{(k)}, \ t = T_{k,1} - \tau_{k,1}, \dots, T_{k,2} + \tau_{k,2} - 1, \tag{3.61}$$

$$x_t = x_t^f, \ t \in \{0, 1, \dots\} \setminus \cup_{k=0}^{\infty} \{T_{k,1} - \tau_{k,1}, \dots, T_{k,2} + \tau_{k,2}\}, \tag{3.62}$$

$$u_t = u_t^f, \ t \in \{0, 1, \dots\} \setminus \cup_{k=0}^{\infty} [T_{k,1} - \tau_{k,1}, \dots, T_{k,2} + \tau_{k,2}). \tag{3.63}$$

It follows from (3.60)–(3.63) that

$$\sum \{f(t, x_t, u_t) : \ t = 0, \dots, T_{k,2} + \tau_{k,2} - 1\}$$

$$-\sum\{f(t, x_t^f, u_t^f) : t = 0, \dots, T_{k,2} + \tau_{k,2} - 1\}$$

$$= \sum_{i=1}^{k} \sum_{t=T_{i,1}-\tau_{i,1}}^{T_{i,2}+\tau_{i,2}-1} f(t, \tilde{x}_t^{(i)}, \tilde{u}_t^{(i)}) - \sum_{i=1}^{k} \sum_{t=T_{i,1}-\tau_{i,1}}^{T_{i,2}+\tau_{i,2}-1} f(t, x_t^f, u_t^f) \leq \sum_{i=1}^{k} 2^{-i+3} < \infty.$$

Theorem 3.1 implies that $(\{x_t\}_{t=0}^{\infty}, \{u_t\}_{t=0}^{\infty})$ is (f)-good. By (P1),

$$\lim_{t\to\infty} \rho_E(x_t, x_t^f) = 0.$$

On the other hand, in view of (3.48)–(3.50), $\limsup_{t\to\infty} \rho_E(x_t, x_t^f) \geq \epsilon$. The contradiction we have reached completes the proof of Lemma 3.35.

Completion of the Proof of Theorem 3.17. Assume that properties (P1) and (P2) hold. Let $\epsilon, M > 0$. By Lemma 3.35, there exist $\delta_0 > 0$ and an integer $L_0 > 0$ such that the following property holds:

(iii) for each integer $T_1 \geq L_0$, each integer $T_2 \geq T_1 + 2b_f$ and each $(\{x_t\}_{t=T_1}^{T_2}, \{u_t\}_{t=T_1}^{T_2-1}) \in X(T_1, T_2)$ which satisfies

$$\rho_E(x_{T_i}, x_{T_i}^f) \leq \delta_0,\ i = 1, 2,$$

$$\sum_{t=T_1}^{T_2-1} f(t, x_t, u_t) \leq U^f(T_1, T_2, x_{T_1}, x_{T_2}) + \delta_0$$

the inequality $\rho_E(x_t, x_t^f) \leq \epsilon$ holds for all $t = T_1, \dots, T_2$.

By Theorem 3.1, there exists $S_0 > 0$ such that the following property holds:

(iv) for each pair of integers $T_2 > T_1 \geq 0$ and each $(\{x_t\}_{t=T_1}^{T_2}, \{u_t\}_{t=T_1}^{T_2-1}) \in X(T_1, T_2)$, we have

$$\sum_{t=T_1}^{T_2-1} f(t, x_t, u_t) + S_0 \geq \sum_{t=T_1}^{T_2-1} f(t, x_t^f, u_t^f).$$

In view of (P2), there exist $\delta \in (0, \delta_0)$ and an integer $L_1 > 0$ such that the following property holds:

(v) for each integer $T \geq 0$ and each $(\{x_t\}_{t=T}^{T+L_1}, \{u_t\}_{t=T}^{T+L_1-1}) \in X(T, T+L_1)$ which satisfies

$$\sum_{t=T}^{T+L_1-1} f(t, x_t, u_t) \leq \min\{U^f(T, T + L_1, x_T, x_{T+L_1}) + \delta,$$

$$\sum_{t=T}^{T+L_1-1} f(t, x_t^f, u_t^f) + 2S_0 + M\},$$

there exists an integer $s \in [0, L_1]$ such that $\rho_E(x_{T+s}, x_{T+s}^f) \le \delta_0$.

Set

$$L = L_0 + L_1 + b_f. \tag{3.64}$$

Assume that $T_1 \ge 0$ and $T_2 \ge T_1 + 2L$ are integers and that

$$(\{x_t\}_{t=T_1}^{T_2}, \{u_t\}_{t=T_1}^{T_2-1}) \in X(T_1, T_2) \tag{3.65}$$

satisfies

$$\sum_{t=T_1}^{T_2-1} f(t, x_t, u_t) \le \min\{\sigma^f(T_1, T_2) + M, U^f(T_1, T_2, x_{T_1}, x_{T_2}) + \delta\}. \tag{3.66}$$

In view of (3.66),

$$\sum_{t=T_1}^{T_2-1} f(t, x_t, u_t) \le \sum_{t=T_1}^{T_2-1} f(t, x_t^f, u_t^f) + M. \tag{3.67}$$

Properties (iv) and (3.67) imply that for each pair of integers $Q_1, Q_2 \in [T_1, T_2]$ satisfying $Q_1 < Q_2$,

$$\sum_{t=Q_1}^{Q_2-1} f(t, x_t, u_t) = \sum_{t=T_1}^{T_2-1} f(t, x_t, u_t) - \sum\{f(t, x_t, u_t) : t \in \{T_1, \dots, Q_1\} \setminus \{Q_1\}\}$$

$$-\sum\{f(t, x_t, u_t) : t \in \{Q_2, \dots, T_2\} \setminus \{T_2\}\}$$

$$\le \sum_{t=T_1}^{T_2-1} f(t, x_t^f, u_t^f) + M - \sum\{f(t, x_t^f, u_t^f) : t \in \{T_1, \dots, Q_1\} \setminus \{Q_1\}\} + S_0$$

$$-\sum\{f(t, x_t^f, u_t^f) : t \in \{Q_2, \dots, T_2\} \setminus \{T_2\}\} + S_0 = \sum_{t=Q_1}^{Q_2-1} f(t, x_t^f, u_t^f) + M + 2S_0. \tag{3.68}$$

By (3.68),

$$\sum\{f(t,x_t,u_t): t=\max\{T_1,L_0\},\dots,\{T_1,L_0\}+L_1-1\}$$

$$\le \sum\{f(t,x_t^f,u_t^f): t=\max\{T_1,L_0\},\dots,\{T_1,L_0\}+L_1-1\}+M+2S_0, \tag{3.69}$$

$$\sum_{t=T_2-L_1}^{T_2-1} f(t,x_t,u_t) \le \sum_{t=T_2-L_1}^{T_2-1} f(t,x_t^f,u_t^f)+M+2S_0. \tag{3.70}$$

It follows from (3.64), (3.66), (3.69), (3.70), and property (v) that there exist integers $\tau_1 \in [\max\{T_1,L_0\},\max\{T_1,L_0\}+L_1]$, $\tau_2 \in [T_2-L_1,T_2]$ such that

$$\rho_E(x_{\tau_i},x_{\tau_i}^f) \le \delta_0,\ i=1,2. \tag{3.71}$$

If $\rho_E(x_{T_2},x_{T_2}^f) \le \delta$, then we may assume that $\tau_2 = T_2$ and if $T_1 \ge L_0$ and $\rho_E(x_{T_1},x_{T_1}^f) \le \delta$, then we may assume that $\tau_1 = T_1$. In view of (3.64),

$$\tau_2-\tau_1 \ge T_2-T_1-2L_1-L_0 \ge L_0+2b_f. \tag{3.72}$$

Property (iii), (3.66), (3.71), and (3.72) imply that

$$\rho_E(x_t,x_t^f) \le \epsilon,\ t=\tau_1,\dots,\tau_2.$$

Theorem 3.17 is proved.

3.11 An Auxiliary Result

Lemma 3.36. *Assume that (P2) holds and that $M,\epsilon > 0$. Then there exists a natural number L such that for each integer $T \ge 0$ and each*

$$(\{x_t\}_{t=T}^{T+L},\{u_t\}_{t=T}^{T+L-1}) \in X(T,T+L)$$

which satisfies

$$\sum_{t=T}^{T+L-1} f(t,x_t,u_t) \le \sum_{t=T}^{T+L-1} f(t,x_t^f,u_t^f)+M$$

the following inequality holds:

$$\min\{\rho_E(x_t,x_t^f): t=T,\dots,T+L\} \le \epsilon.$$

Proof. By Theorem 3.1, there exists $S_0 > 0$ such that the following property holds:

(i) for each pair of integers $T_2 > T_1 \geq 0$ and each $(\{x_t\}_{t=T_1}^{T_2}, \{u_t\}_{t=T_1}^{T_2-1}) \in X(T_1, T_2)$,

$$\sum_{t=T_1}^{T_2-1} f(t, x_t, u_t) + S_0 \geq \sum_{t=T_1}^{T_2-1} f(t, x_t^f, u_t^f).$$

By (P2), there exist $\delta_0 \in (0, \epsilon)$ and an integer $L_0 > 0$ such that the following property holds:

(ii) for each integer $T \geq 0$ and each $(\{x_t\}_{t=T}^{T+L_0}, \{u_t\}_{t=T}^{T+L_0-1}) \in X(T, L_0 + T)$ which satisfies

$$\sum_{t=T}^{T+L_0-1} f(t, x_t, u_t) \leq \min\{U^f(T, T + L_0, x_T, x_{T+L_0}) + \delta_0,$$

$$\sum_{t=T}^{T+L_0-1} f(t, x_t^f, u_t^f) + 2S_0 + M\},$$

we have

$$\min\{\rho_E(x_t, x_t^f) : t = T, \dots, T + L_0\} \leq \epsilon.$$

Choose an integer

$$q_0 > (M + S_0)\delta_0^{-1} \tag{3.73}$$

and set

$$L = q_0 L_0. \tag{3.74}$$

Let $T \geq 0$ be an integer and $(\{x_t\}_{t=T}^{T+L}, \{u_t\}_{t=T}^{T+L-1}) \in X(T, T + L)$ satisfy

$$\sum_{t=T}^{T+L-1} f(t, x_t, u_t) \leq \sum_{t=T}^{T+L-1} f(t, x_t^f, u_t^f) + M. \tag{3.75}$$

We show that

$$\min\{\rho_E(x_t, x_t^f) : t = T, \dots, T + L\} \leq \epsilon.$$

Assume the contrary. Then

$$\rho_E(x_t, x_t^f) > \epsilon, \ t = T, \dots, T + L. \tag{3.76}$$

Let an integer $i \in \{0, \dots, q_0 - 1\}$. Property (i), (3.74), and (3.75) imply that

$$\sum_{t=T+iL_0}^{T+(i+1)L_0-1} f(t, x_t, u_t) = \sum_{t=T}^{T+L-1} f(t, x_t, u_t)$$

$$-\sum\{f(t, x_t, u_t) : t \in \{T, \dots, T+iL_0\} \setminus \{T+iL_0\}\}$$

$$-\sum\{f(t, x_t, u_t) : t \in \{T+(i+1)L_0, \dots, T+L\} \setminus \{T+L\}\}$$

$$\le \sum_{t=T}^{T+L-1} f(t, x_t^f, u_t^f) + M$$

$$-\sum\{f(t, x_t^f, u_t^f) : t \in \{T, \dots, T+iL_0\} \setminus \{T+iL_0\}\} + S_0$$

$$-\sum\{f(t, x_t^f, u_t^f) : t \in \{T+(i+1)L_0, \dots, T+L\} \setminus \{T+L\}\} + S_0$$

$$= \sum_{t=T+iL_0}^{T+(i+1)L_0-1} f(t, x_t^f, u_t^f) + M + 2S_0. \tag{3.77}$$

By (3.76), (3.77), and property (ii),

$$\sum_{t=T+iL_0}^{T+(i+1)L_0-1} f(t, x_t, u_t) > U^f(T+iL_0, T+(i+1)L_0, x_{T+iL_0}, x_{T+(i+1)L_0}) + \delta_0. \tag{3.78}$$

It follows from (3.78) that there exists $(\{y_t\}_{t=T}^{T+L}, \{v_t\}_{t=T}^{T+L-1}) \in X(T, T+L)$ such that

$$y_{T+iL_0} = x_{T+iL_0},\ i = 0, \dots, q_0 \tag{3.79}$$

and for all $i = 0, \dots, q_0 - 1$,

$$\sum_{t=T+iL_0}^{T+(i+1)L_0-1} f(t, x_t, u_t) > \sum_{t=T+iL_0}^{T+(i+1)L_0-1} f(t, y_t, v_t) + \delta_0. \tag{3.80}$$

In view of (3.74), (3.75), (3.80), and property (i),

$$\sum_{t=T}^{T+L-1} f(t, x_t^f, u_t^f) + M \ge \sum_{t=T}^{T+L-1} f(t, x_t, u_t) \ge \sum_{t=T}^{T+L-1} f(t, y_t, v_t) + q_0\delta_0$$

$$\geq \sum_{t=T}^{T+L-1} f(t, x_t^f, u_t^f) - S_0 + q_0\delta_0,$$

$$q_0 \leq (M + S_0)\delta_0^{-1}.$$

This contradicts (3.73). The contradiction we have reached proves Lemma 3.36.

3.12 Proof of Theorem 3.18

Assume that f has (P1) and (P2). In view of Theorem 3.17, f has TP. We show that WTP holds. Let $\epsilon, M > 0$. By Theorem 3.1, there exists $S_0 > 0$ such that the following property holds:

(i) for each pair of integers $\tau_2 > \tau_1 \geq 0$ and each $(\{y_t\}_{t=\tau_1}^{\tau_2}, \{v_t\}_{t=\tau_1}^{\tau_2-1}) \in X(\tau_1, \tau_2)$,

$$\sum_{t=\tau_1}^{\tau_2-1} f(t, y_t, v_t) + S_0 \geq \sum_{t=\tau_1}^{\tau_2-1} f(t, x_t^f, u_t^f).$$

Theorem 3.17 and TP imply that there exist $\delta \in (0, \epsilon)$ and an integer $L_0 > 0$ such that the following property holds:

(ii) for each integer $\tau_1 \geq 0$, each integer $\tau_2 \geq \tau_1 + 2L_0$, and each $(\{y_t\}_{t=\tau_1}^{\tau_2}, \{v_t\}_{t=\tau_1}^{\tau_2-1}) \in X(\tau_1, \tau_2)$ which satisfies

$$\sum_{t=\tau_1}^{\tau_2-1} f(t, y_t, v_t) \leq \min\{\sigma^f(\tau_1, \tau_2) + M + 3S_0, U^f(\tau_1, \tau_2, y_{\tau_1}, y_{\tau_2}) + \delta\},$$

we have

$$\rho_E(y_t, x_t^f) \leq \epsilon, \; t = \tau_1 + L_0, \dots, \tau_2 - L_0.$$

Set

$$l = 2L_0 + 1. \tag{3.81}$$

Choose a natural number

$$Q \geq 2 + 2(M + S_0)\delta^{-1}. \tag{3.82}$$

Assume that $T_1 \geq 0$, $T_2 \geq T_1 + lQ$ are integers and $(\{x_t\}_{t=T_1}^{T_2}, \{u_t\}_{t=T_1}^{T_2-1}) \in X(T_1, T_2)$ satisfies

$$\sum_{t=T_1}^{T_2-1} f(t, x_t, u_t) \leq \sum_{t=T_1}^{T_2-1} f(t, x_t^f, u_t^f) + M. \tag{3.83}$$

Property (i) and (3.83) imply that for each pair of integers $\tau_1, \tau_2 \in [T_1, T_2]$ satisfying $\tau_1 < \tau_2$,

$$\begin{aligned}
\sum_{t=\tau_1}^{\tau_2-1} f(t, x_t, u_t) &= \sum_{t=T_1}^{T_2-1} f(t, x_t, u_t) - \sum\{f(t, x_t, u_t) : t \in \{T_1, \dots, \tau_1\} \setminus \{\tau_1\}\} \\
&\quad - \sum\{f(t, x_t, u_t) : t \in \{\tau_2, \dots, T_2\} \setminus \{T_2\}\} \\
&\leq \sum_{t=T_1}^{T_2-1} f(t, x_t^f, u_t^f) + M - \sum\{f(t, x_t^f, u_t^f) : t \in \{T_1, \dots, \tau_1\} \setminus \{\tau_1\}\} + S_0 \\
&\quad - \sum\{f(t, x_t^f, u_t^f) : t \in \{\tau_2, \dots, T_2\} \setminus \{T_2\}\} + S_0 \\
&= \sum_{t=\tau_1}^{\tau_2-1} f(t, x_t^f, u_t^f) + M + 2S_0.
\end{aligned} \tag{3.84}$$

Set

$$t_0 = T_1. \tag{3.85}$$

If

$$\sum_{t=T_1}^{T_2-1} f(t, x_t, u_t) \leq U^f(T_1, T_2, x_{T_1}, x_{T_2}) + \delta,$$

then we set $t_1 = T_2$. Assume that

$$\sum_{t=T_1}^{T_2-1} f(t, x_t, u_t) > U^f(T_1, T_2, x_{T_1}, x_{T_2}) + \delta.$$

Set

$$t_1 = \min\{t \in \{T_1 + 1, \dots, T_2\} : \sum_{\tau=T_1}^{t-1} f(\tau, x_\tau, u_\tau) - U^f(T_1, t, x_{T_1}, x_t) > \delta\}. \tag{3.86}$$

Clearly, t_1 is well-defined.

Assume that $k \ge 1$ is an integer and we defined a sequence of integers $\{t_i\}_{i=0}^{k} \subset [T_1, T_2]$ such that

$$t_0 < t_1 \cdots < t_k,$$

for each integer $i = 1, \dots, k$ if $t_i - t_{i-1} \ge 2$, then

$$\sum_{t=t_{i-1}}^{t_i-2} f(t, x_t, u_t) - U^f(t_{i-1}, t_i - 1, x_{t_{i-1}}, x_{t_i-1}) \le \delta \tag{3.87}$$

and if $t_i < T_2$, then

$$\sum_{t=t_{i-1}}^{t_i-1} f(t, x_t, u_t) - U^f(t_{i-1}, t_i, x_{t_{i-1}}, x_{t_i}) > \delta. \tag{3.88}$$

(Note that in view of (3.86), our assumption holds for $k = 1$.)

By (3.88), there exists $(\{y_t\}_{t=T_1}^{T_2}, \{v_t\}_{t=T_1}^{T_2-1}) \in X(T_1, T_2)$ such that

$$y_{t_i} = x_{t_i},\ i = 0, \dots, k, \tag{3.89}$$

$$\sum_{t=t_{i-1}}^{t_i-1} f(t, x_t, u_t) - \sum_{t=t_{i-1}}^{t_i-1} f(t, y_t, v_t) > \delta,\ i \in \{1, \dots, k\} \setminus \{k\}, \tag{3.90}$$

$$y_t = x_t,\ t \in \{t_{k-1}, \dots, T_2\},\ v_t = u_t,\ t \in \{t_{k-1}, \dots, T_2\} \setminus \{T_2\}. \tag{3.91}$$

Property (i), (3.83), and (3.89)–(3.91) imply that

$$\sum_{t=T_1}^{T_2-1} f(t, x_t^f, u_t^f) + M \ge \sum_{t=T_1}^{T_2-1} f(t, x_t, u_t) \ge \sum_{t=T_1}^{T_2-1} f(t, y_t, v_t) + \delta(k-1)$$

$$\ge \sum_{t=T_1}^{T_2-1} f(t, x_t^f, u_t^f) - S_0 + \delta(k-1),$$

$$k \le 1 + \delta^{-1}(M + S_0). \tag{3.92}$$

If $t_k = T_2$, then the construction of the sequence is completed. Assume that $t_k < T_2$. If

$$\sum_{t=t_k}^{T_2-1} f(t, x_t, u_t) \le U^f(t_k, T_2, x_{t_k}, x_{T_2}) + \delta,$$

then we set $t_{k+1} = T_2$ and the construction of the sequence is completed. Assume that

$$\sum_{t=t_k}^{T_2-1} f(t, x_t, u_t) > U^f(t_k, T_2, x_{t_k}, x_{T_2}) + \delta.$$

Set

$$t_{k+1} = \min\{t \in \{t_k + 1, \dots, T_2\} : \sum_{\tau=t_k}^{t-1} f(\tau, x_\tau, u_\tau) - U^f(t_k, t, x_{t_k}, x_t) > \delta\}.$$

Clearly, t_{k+1} is well-defined, and the assumption made for k also holds for $k+1$. By induction (see (3.87) and (3.88)), we constructed a finite sequence $\{t_i\}_{i=0}^{q} \subset [T_1, T_2]$ such that

$$q \le 1 + \delta^{-1}(M + S_0), \tag{3.93}$$

$T_1 = t_0 < t_1 \cdots < t_q = T_2$, for each integer $i = 1, \dots, q$ if $t_i - t_{i-1} \ge 2$, then

$$\sum_{t=t_{i-1}}^{t_i-2} f(t, x_t, u_t) - U^f(t_{i-1}, t_i - 1, x_{t_{i-1}}, x_{t_i-1}) \le \delta \tag{3.94}$$

and if $t_i < T_2$, then

$$\sum_{t=t_{i-1}}^{t_i-1} f(t, x_t, u_t) - U^f(t_{i-1}, t_i, x_{t_{i-1}}, x_{t_i}) > \delta. \tag{3.95}$$

Assume that

$$i \in \{0, \dots, q-1\},\ t_{i+1} - t_i \ge 2L_0 + 1. \tag{3.96}$$

By (3.94) and (3.96),

$$\sum_{t=t_i}^{t_{i+1}-2} f(t, x_t, u_t) - U^f(t_i, t_{i+1} - 1, x_{t_i}, x_{t_{i+1}-1}) \le \delta. \tag{3.97}$$

Property (i), (3.84), and (3.96) imply that

$$\sum_{t=t_i}^{t_{i+1}-2} f(t, x_t, u_t) \le \sum_{t=t_i}^{t_{i+1}-2} f(t, x_t^f, u_t^f) + M + 2S_0$$

$$\le M + 3S_0 + \sigma^f(t_i, t_{i+1} - 1). \tag{3.98}$$

It follows from (3.96)–(3.98) and property (ii) that

$$\rho_E(x_t, x_t^f) \le \epsilon,\ t = t_i + L_0, \dots, t_{i+1} - 1 - L_0. \tag{3.99}$$

In view of (3.99),

$$\{t \in \{T_1, \dots, T_2\} :\ \rho_E(x_t, x_t^f) > \epsilon\}$$

$$\subset \cup\{\{t_i, \dots, t_i + L_0 - 1\} \cup \{t_{i+1} - L_0, \dots, t_{i+1}\} :$$

$$i \in \{0, \dots, q-1\} \text{ and } t_{i+1} - t_i \ge 2L_0 + 1\}$$

$$\cup\{\{t_i, \dots, t_{i+1}\} :\ i \in \{0, \dots, q-1\} \text{ and } t_{i+1} - t_i < 2L_0 + 1\}.$$

In view of (3.81), (3.82), and (3.93), WTP holds.

Assume that WTP holds. It is easy to see that (P1) and (P2) hold. This completes the proof of Theorem 3.18.

3.13 Auxiliary Results for Theorems 3.19

Lemma 3.37. *Let $\epsilon > 0$. Then there exist $\delta > 0$ and an integer $L > 0$ such that for each pair of integers $T_2 > T_1 \ge L$ and each $(\{x_t\}_{t=T_1}^{T_2}, \{u_t\}_{t=T_1}^{T_2-1}) \in X(T_1, T_2)$ which satisfies $\rho_E(x_{T_1}, x_{T_1}^f) \le \delta$, $\rho_E(x_{T_2}, x_{T_2}^f) \le \delta$, the following inequality holds:*

$$\sum_{t=T_1}^{T_2-1} f(t, x_t, u_t) \ge \sum_{t=T_1}^{T_2-1} f(t, x_t^f, u_t^f) - \epsilon.$$

Proof. By (A2), there exists $\delta > 0$ such that the following property holds:

(i) for each $(T, z) \in \mathcal{A}$ satisfying $\rho_E(z, x_T^f) \le \delta$, there exist an integer $\tau_1 \in (0, b_f]$, and $(\{\tilde{x}_t^{(1)}\}_{t=T}^{T+\tau_1}, \{\tilde{u}_t^{(1)}\}_{t=T}^{T+\tau_1-1}) \in X(T, T+\tau_1)$, such that

$$\tilde{x}_T^{(1)} = z,\ \tilde{x}_{T+\tau_1}^{(1)} = x_{T+\tau_1}^f,$$

$$\sum_{t=T}^{T+\tau_1-1} f(t, \tilde{x}_t^{(1)}, \tilde{u}_t^{(1)}) \le \sum_{t=T}^{T+\tau_1-1} f(t, x_t^f, u_t^f) + \epsilon/4$$

and if $T \ge b_f$, there exist an integer $\tau_2 \in (0, b_f]$ and

$$(\{\tilde{x}_t^{(2)}\}_{t=T-\tau_2}^{T}, \{\tilde{u}_t^{(2)}\}_{t=T-\tau_2}^{T-1}) \in X(T-\tau_2, T)$$

such that

$$\tilde{x}^{(2)}_{T-\tau_2} = x^f_{T-\tau_2}, \; \tilde{x}^{(2)}_T = z,$$

$$\sum_{t=T-\tau_2}^{T-1} f(t, \tilde{x}^{(2)}_t, \tilde{u}^{(2)}_t) \le \sum_{t=T-\tau_2}^{T-1} f(t, x^f_t, u^f_t) + \epsilon/4.$$

Lemma 3.34 implies that there exists an integer $L > 2b_f$ such that for each pair of integers $T_2 > T_1 \ge L - b_f$,

$$\sum_{t=T_1}^{T_2-1} f(t, x^f_t, u^f_t) < U^f(T_1, T_2, x^f_{T_1}, x^f_{T_2}) + \epsilon/4. \tag{3.100}$$

Assume that $T_2 > T_1 \ge L$ are integers, $(\{x_t\}_{t=T_1}^{T_2}, \{u_t\}_{t=T_1}^{T_2-1}) \in X(T_1, T_2)$ and

$$\rho_E(x_{T_i}, x^f_{T_i}) \le \delta, \; i = 1, 2. \tag{3.101}$$

By property (i) and (3.101), there exist integers $\tau_1, \tau_2 \in (0, b_f]$ and

$$(\{y_t\}_{t=T_1-\tau_1}^{T_2+\tau_2}, \{v_t\}_{t=T_1-\tau_1}^{T_2+\tau_2-1}) \in X(T_1 - \tau_1, T_2 + \tau_2)$$

such that

$$y_{T_1-\tau_1} = x^f_{T_1-\tau_1}, \; y_t = x_t, \; t = T_1, \dots, T_2, \; v_t = u_t, \; t = T_1, \dots, T_2 - 1, \tag{3.102}$$

$$\sum_{t=T_1-\tau_1}^{T_1-1} f(t, y_t, v_t) \le \sum_{t=T_1-\tau_1}^{T_1-1} f(t, x^f_t, u^f_t) + \epsilon/4, \tag{3.103}$$

$$y_{T_2+\tau_2} = x^f_{T_2+\tau_2}, \; \sum_{t=T_2}^{T_2+\tau_2-1} f(t, y_t, v_t) \le \sum_{t=T_2}^{T_2+\tau_2-1} f(t, x^f_t, u^f_t) + \epsilon/4. \tag{3.104}$$

In view of (3.100)–(3.102) and (3.104),

$$\sum_{t=T_1-\tau_1}^{T_2+\tau_2-1} f(t, x^f_t, u^f_t) < \sum_{t=T_1-\tau_1}^{T_2+\tau_2-1} f(t, y_t, v_t) + \epsilon/4. \tag{3.105}$$

It follows from (3.102)–(3.105) that

$$\sum_{t=T_1}^{T_2-1} f(t, x_t, u_t) = \sum_{t=T_1-\tau_1}^{T_2+\tau_2-1} f(t, y_t, v_t) - \sum_{t=T_1-\tau_1}^{T_1-1} f(t, y_t, v_t) - \sum_{t=T_2}^{T_2+\tau_2-1} f(t, y_t, v_t)$$

$$\geq \sum_{t=T_1-\tau_1}^{T_2+\tau_2-1} f(t, x_t^f, u_t^f) - \epsilon/4 - \sum_{t=T_1-\tau_1}^{T_1-1} f(t, x_t^f, u_t^f) - \epsilon/4 - \sum_{t=T_2}^{T_2+\tau_2-1} f(t, x_t^f, u_t^f) - \epsilon/4$$

$$= \sum_{t=T_1}^{T_2-1} f(t, x_t^f, u_t^f) - 3\epsilon/4.$$

Lemma 3.37 is proved.

Proposition 3.38. *Assume that f has LSC property, (P1), and (P2) and that $(T_0, z_0) \in \cup\{\mathcal{A}_L : L = 1, 2, \ldots\}$. Then there exists an (f)-good and (f)-minimal pair $(\{x_t^*\}_{t=T_0}^{\infty}, \{u_t^*\}_{t=T_0}^{\infty}) \in X(T_0, \infty)$ such that $x_{T_0}^* = z$.*

Proof. There exists an integer $L_0 > 0$ such that

$$(T_0, z_0) \in \mathcal{A}_{L_0}. \tag{3.106}$$

It follows from Theorem 3.1 that there exists $S_0 > 0$ such that for each pair of integers $T_2 > T_1 \geq 0$ and each $(\{x_t\}_{t=T_1}^{T_2}, \{u_t\}_{t=T_1}^{T_2-1}) \in X(T_1, T_2)$,

$$\sum_{t=T_1}^{T_2-1} f(t, x_t, u_t) + S_0 \geq \sum_{t=T_1}^{T_2-1} f(t, x_t^f, u_t^f). \tag{3.107}$$

Fix an integer $k_0 \geq L_0$. LSC property and (3.106) imply that for each integer $k \geq k_0$ there exists $(\{x_t^{(k)}\}_{t=T_0}^{T_0+k}, \{u_t^{(k)}\}_{t=T_0}^{T_0+k-1}) \in X(T_0, T_0 + k)$ satisfying

$$x_{T_0}^{(k)} = z_0, \tag{3.108}$$

$$\sum_{t=T_0}^{T_0+k-1} f(t, x_t^{(k)}, u_t^{(k)}) = \sigma^f(T_0, T_0 + k, z_0). \tag{3.109}$$

In view of (3.6) and (3.106), for each integer $k \geq k_0$,

$$\sigma^f(T_0, T_0 + k, z_0) \leq L_0 + \sum_{t=T_0}^{T_0+k-1} f(t, x_t^f, u_t^f) + a_0 L_0. \tag{3.110}$$

By (3.107), (3.109), and (3.110), for each integer $k \geq k_0$ and each pair of integers $T_1, T_2 \in [T_0, T_0 + k]$ satisfying $T_1 < T_2$,

$$\sum_{t=T_1}^{T_2-1} f(t, x_t^{(k)}, u_t^{(k)}) = \sum_{t=T_0}^{T_0+k-1} f(t, x_t^{(k)}, u_t^{(k)})$$

$$-\sum\{f(t, x_t^{(k)}, u_t^{(k)}) : t \in \{T_0, \ldots, T_1\} \setminus \{T_1\}\}$$

$$-\sum\{f(t, x_t^{(k)}, u_t^{(k)}) : t \in \{T_2, \ldots, T_0 + k\} \setminus \{T_0 + k\}\}$$

$$\leq \sum_{t=T_0}^{T_0+k-1} f(t, x_t^f, u_t^f) + L_0(1 + a_0)$$

$$-\sum\{f(t, x_t^f, u_t^f) : t \in \{T_0, \ldots, T_1\} \setminus \{T_1\}\} + S_0$$

$$-\sum\{f(t, x_t^f, u_t^f) : t \in \{T_2, \ldots, T_0 + k\} \setminus \{T_0 + k\}\} + S_0$$

$$= \sum_{t=T_1}^{T_2-1} f(t, x_t^f, u_t^f) + 2S_0 + L_0(1 + a_0). \tag{3.111}$$

By (3.12), (3.111), and LSC property, extracting subsequences, using the diagonalization process, and re-indexing, we obtain that there exists a strictly increasing sequence of natural numbers $\{k_p\}_{p=1}^{\infty}$ such that $k_1 \geq k_0$, and for each integer $t \geq T_0$, there exists $\lim_{p\to\infty} f(t, x_t^{(k_p)}, u_t^{(k_p)})$, and there exists $(\{x_t^*\}_{t=T_0}^{\infty}, \{u_t^*\}_{t=T_0}^{\infty}) \in X(T_0, \infty)$ such that for each integer $t \geq T_0$,

$$x_t^{(k_p)} \to x_t^* \text{ as } p \to \infty, \tag{3.112}$$

$$f(t, x_t^*, u_t^*) \leq \lim_{p\to\infty} f(t, x_t^{(k_p)}, u_t^{(k_p)}). \tag{3.113}$$

In view of (3.111) and (3.113) for every integer $q \geq 1$

$$\sum_{t=T_0}^{T_0+q-1} f(t, x_t^*, u_t^*) \leq \lim_{p\to\infty} \sum_{t=T_0}^{T_0+q-1} f(t, x_t^{(k_p)}, u_t^{(k_p)})$$

$$\leq \sum_{t=T_0}^{T_0+q-1} f(t, x_t^f, u_t^f) + 2S_0 + L_0(1 + a_0). \tag{3.114}$$

Theorem 3.1 and (3.114) imply that $(\{x_t^*\}_{t=T_0}^{\infty}, \{u_t^*\}_{t=T_0}^{\infty})$ is (f)-good. In view of (3.108) and (3.112),

$$x_{T_0}^* = z_0. \tag{3.115}$$

In order to complete the proof of the proposition, it is sufficient to show that $(\{x_t^*\}_{t=T_0}^{\infty}, \{u_t^*\}_{t=T_0}^{\infty})$ is (f)-minimal. Assume the contrary. Then there exist $\Delta > 0$, an integer $\tau_0 \geq 1$ and $(\{y_t\}_{t=T_0}^{T_0+\tau_0}, \{v_t\}_{t=T_0}^{T_0+\tau_0-1}) \in X(T_0, T_0+\tau_0)$ such that

$$y_{T_0} = x_{T_0}^*, \ y_{T_0+\tau_0} = x_{T_0+\tau_0}^*, \tag{3.116}$$

$$\sum_{t=T_0}^{T_0+\tau_0-1} f(t, x_t^*, u_t^*) > \sum_{t=T_0}^{T_0+\tau_0-1} f(t, y_t, v_t) + 2\Delta. \tag{3.117}$$

By (A2) and Lemma 3.37, there exist $\delta > 0$ and an integer $L_1 > 0$ such that the following properties hold:

(ii) for each $(T, \xi) \in \mathcal{A}$ satisfying $\rho_E(\xi, x_T^f) \leq \delta$, there exist an integer $\tau_1 \in (0, b_f]$ and $(\{\tilde{x}_t^{(1)}\}_{t=T}^{T+\tau_1}, \{\tilde{u}_t^{(1)}\}_{t=T}^{T+\tau_1-1}) \in X(T, T+\tau_1)$, such that

$$\tilde{x}_T^{(1)} = \xi, \ \tilde{x}_{T+\tau_1}^{(1)} = x_{T+\tau_1}^f,$$

$$\sum_{t=T}^{T+\tau_1-1} f(t, \tilde{x}_t^{(1)}, \tilde{u}_t^{(1)}) \leq \sum_{t=T}^{T+\tau_1-1} f(t, x_t^f, u_t^f) + \Delta/8$$

and if $T \geq b_f$, then there exist an integer $\tau_2 \in (0, b_f]$ and

$$(\{\tilde{x}_t^{(2)}\}_{t=T-\tau_2}^{T}, \{\tilde{u}_t^{(2)}\}_{t=T-\tau_2}^{T-1}) \in X(T-\tau_2, T)$$

such that

$$\tilde{x}_{T-\tau_2}^{(2)} = x_{T-\tau_2}^f, \tilde{x}_T^{(2)} = \xi,$$

$$\sum_{t=T-\tau_2}^{T-1} f(t, \tilde{x}_t^{(2)}, \tilde{u}_t^{(2)}) \leq \sum_{t=T-\tau_2}^{T-1} f(t, x_t^f, u_t^f) + \Delta/8;$$

(iii) for each pair of integers $T_2 > T_1 \geq L_1$ and each $(\{x_t\}_{t=T_1}^{T_2}, \{u_t\}_{t=T_1}^{T_2-1}) \in X(T_1, T_2)$ which satisfies $\rho_E(x_{T_1}, x_{T_1}^f) \leq \delta$, $\rho_E(x_{T_2}, x_{T_2}^f) \leq \delta$, the following inequality holds:

$$\sum_{t=T_1}^{T_2-1} f(t, x_t, u_t) \geq \sum_{t=T_1}^{T_2-1} f(t, x_t^f, u_t^f) - \Delta/8.$$

Theorems 3.1 and 3.17, (P1), (P2), (3.106), (3.108), and (3.110) imply that there exists an integer $L_2 > L_0 + L_1$ such that for each integer $k \geq k_0 + 2L_2$,

$$\rho_E(x_t^{(k)}, x_t^f) \le \delta,\ t \in \{T_0 + L_2, \ldots, T_0 + k - L_2\}, \tag{3.118}$$

$$\rho_E(x_t^*, x_t^f) \le \delta \text{ for all integers } t \ge T_0 + L_2. \tag{3.119}$$

By (3.113), there exists a natural number p_0 such that

$$k_{p_0} > k_0 + 2L_2 + 2L_1 + 2\tau_0 + 2 + 2b_f + 2T_0, \tag{3.120}$$

$$\sum_{t=T_0}^{T_0+\tau_0+L_2-1} f(t, x_t^*, u_t^*) \le \sum_{t=T_0}^{T_0+\tau_0+L_2-1} f(t, x_t^{(k_{p_0})}, u_t^{(k_{p_0})}) + \Delta/2. \tag{3.121}$$

Property (ii), (3.116), and (3.118)–(3.120) imply that there exists

$$(\{x_t\}_{t=T_0}^{T_0+k_{p_0}}, \{u_t\}_{t=T_0}^{T_0+k_{p_0}-1}) \in X(T_0, T_0 + k_{p_0})$$

such that

$$x_t = y_t,\ t = T_0, \ldots, T_0 + \tau_0,\ u_t = v_t,\ t = T_0, \ldots, T_0 + \tau_0 - 1, \tag{3.122}$$

$$x_t = x_t^*,\ t = T_0+\tau_0, \ldots, T_0+\tau_0+L_2,\ u_t = u_t^*,\ t = T_0+\tau_0, \ldots, T_0+\tau_0+L_2-1, \tag{3.123}$$

$$x_{T_0+\tau_0+L_2+b_f} = x^f_{T_0+\tau_0+L_2+b_f}, \tag{3.124}$$

$$\sum_{t=T_0+\tau_0+L_2}^{T_0+\tau_0+L_2+b_f-1} f(t, x_t, u_t) \le \sum_{t=T_0+\tau_0+L_2}^{T_0+\tau_0+L_2+b_f-1} f(t, x_t^f, u_t^f) + \Delta/8, \tag{3.125}$$

$$x_t = x_t^{(k_{p_0})},\ t = T_0 + \tau_0 + L_2 + 2b_f, \ldots, k_{p_0} + T_0, \tag{3.126}$$

$$u_t = u_t^{(k_{p_0})},\ t = T_0 + \tau_0 + L_2 + 2b_f, \ldots, k_{p_0} + T_0 - 1, \tag{3.127}$$

$$\sum_{t=T_0+\tau_0+L_2+b_f}^{T_0+\tau_0+L_2+2b_f-1} f(t, x_t, u_t) \le \sum_{t=T_0+\tau_0+L_2+b_f}^{T_0+\tau_0+L_2+2b_f-1} f(t, x_t^f, u_t^f) + \Delta/8. \tag{3.128}$$

It follows from (3.108), (3.109), (3.115), (3.116) and (3.122),

$$\sum_{t=T_0}^{T_0+k_{p_0}-1} f(t, x_t, u_t) \ge \sum_{t=T_0}^{T_0+k_{p_0}-1} f(t, x_t^{(k_{p_0})}, u_t^{(k_{p_0})}). \tag{3.129}$$

By (3.117), (3.118), (3.120), (3.122), (3.123), (3.125)–(3.129) and property (iii),

$$0 \le \sum_{t=T_0}^{T_0+k_{p0}-1} f(t, x_t, u_t) - \sum_{t=T_0}^{T_0+k_{p0}-1} f(t, x_t^{(k_{p0})}, u_t^{(k_{p0})})$$

$$= \sum_{t=T_0}^{T_0+\tau_0-1} f(t, y_t, v_t) + \sum_{t=T_0+\tau_0}^{T_0+\tau_0+L_2-1} f(t, x_t^*, u_t^*)$$

$$+ \sum_{t=T_0+\tau_0+L_2}^{T_0+\tau_0+L_2+b_f-1} f(t, x_t^f, u_t^f) + \Delta/8$$

$$+ \sum_{t=T_0+\tau_0+L_2+b_f}^{T_0+\tau_0+L_2+2b_f-1} f(t, x_t^f, u_t^f) + \Delta/8 - \sum_{t=T_0}^{T_0+\tau_0+L_2-1} f(t, x_t^{(k_{p0})}, u_t^{(k_{p0})})$$

$$- \sum_{t=T_0+\tau_0+L_2}^{T_0+\tau_0+L_2+b_f-1} f(t, x_t^{(k_{p0})}, u_t^{(k_{p0})}) - \sum_{t=T_0+\tau_0+L_2+b_f}^{T_0+\tau_0+L_2+2b_f-1} f(t, x_t^{(k_{p0})}, u_t^{(k_{p0})})$$

$$\le \sum_{t=T_0}^{T_0+\tau_0-1} f(t, y_t, v_t) + \sum_{t=T_0+\tau_0}^{T_0+\tau_0+L_2-1} f(t, x_t^*, u_t^*)$$

$$+ \sum_{t=T_0+\tau_0+L_2}^{T_0+\tau_0+L_2+b_f-1} f(t, x_t^f, u_t^f) + \Delta/8 + \sum_{t=T_0+\tau_0+L_2+b_f}^{T_0+\tau_0+L_2+2b_f-1} f(t, x_t^f, u_t^f) + \Delta/8$$

$$- \sum_{t=T_0}^{T_0+\tau_0+L_2-1} f(t, x_t^{(k_{p0})}, u_t^{(k_{p0})}) - \sum_{t=T_0+\tau_0+L_2+1}^{T_0+\tau_0+L_2+b_f-1} f(t, x_t^f, u_t^f) + \Delta/8$$

$$- \sum_{t=T_0+\tau_0+L_2+b_f}^{T_0+\tau_0+L_2+2b_f-1} f(t, x_t^f, u_t^f) + \Delta/8$$

$$= \sum_{t=T_0}^{T_0+\tau_0-1} f(t, y_t, v_t) + \sum_{t=T_0+\tau_0}^{T_0+\tau_0+L_2-1} f(t, x_t^*, u_t^*)$$

$$- \sum_{t=T_0}^{T_0+\tau_0+L_2-1} f(t, x_t^{(k_{p0})}, u_t^{(k_{p0})}) + \Delta/2$$

$$< \sum_{t=T_0}^{T_0+\tau_0-1} f(t, x_t^*, u_t^*) - 2\Delta + \sum_{t=T_0+\tau_0}^{T_0+\tau_0+L_2-1} f(t, x_t^*, u_t^*)$$

$$- \sum_{t=T_0}^{T_0+\tau_0+L_2-1} f(t, x_t^{(k_{p_0})}, u_t^{(k_{p_0})}) + \Delta/2 \le -2\Delta + \Delta/2 + \Delta/2,$$

a contradiction. The contradiction we have reached completes the proof of Proposition 3.38.

3.14 Proofs of Theorems 3.19 and 3.20

Proof of Theorem 3.20. Clearly, (i) implies (ii). In view of Theorem 3.1, (ii) implies (iii). By (P1), (iii) implies (iv). Evidently (iv) implies (v). We show that (v) implies (iii). Assume that $(\{x_t^*\}_{t=S}^{\infty}, \{u_t^*\}_{t=S}^{\infty}) \in X(S, \infty)$ is (f)-minimal and satisfies

$$\liminf_{t\to\infty} \rho_E(x_t^f, x_t^*) = 0. \tag{3.130}$$

(P1) implies that

$$\lim_{t\to\infty} \rho_E(\tilde{x}_t, x_t^f) = 0. \tag{3.131}$$

There exists $S_0 > 0$ such that for all natural numbers $T > S$,

$$|\sum_{t=S}^{T-1} f(t, \tilde{x}_t, \tilde{u}_t) - \sum_{t=S}^{T-1} f(t, x_t^f, u_t^f)| \le S_0. \tag{3.132}$$

By (A2), there exists $\delta > 0$ such that the following property holds:

(a) for each $(T, z) \in \mathcal{A}$ satisfying $\rho_E(z, x_T^f) \le \delta$ there exist an integer $\tau_1 \in (0, b_f]$ and $(\{x_t^{(1)}\}_{t=T}^{T+\tau_1}, \{u_t^{(1)}\}_{t=T}^{T+\tau_1-1}) \in X(T, T+\tau_1)$ satisfying

$$x_T^{(1)} = z,\ x_{T+\tau_1}^{(1)} = x_{T+\tau_1}^f,$$

$$\sum_{t=T}^{T+\tau_1-1} f(t, x_t^{(1)}, u_t^{(1)}) \le \sum_{t=T}^{T+\tau_1-1} f(t, x_t^f, u_t^f) + 1$$

and if $T \ge b_f$, then there exist an integer $\tau_2 \in (0, b_f]$ and

$$(\{x_t^{(2)}\}_{t=T-\tau_2}^{T}, \{u_t^{(2)}\}_{t=T-\tau_2}^{T-1}) \in X(T-\tau_2, T)$$

satisfying

$$x^{(2)}_{T-\tau_2} = x^f_{T-\tau_2},\ x^{(2)}_T = z,$$

$$\sum_{t=T-\tau_2}^{T-1} f(t, x_t^{(2)}, u_t^{(2)}) \le \sum_{t=T-\tau_2}^{T-1} f(t, x_f, u_f) + 1.$$

In view of (3.130) and (3.131), there exists a strictly increasing sequence of natural numbers $\{t_k\}_{k=1}^{\infty}$ such that

$$\rho_E(x^*_{t_k+2b_f}, x^f_{t_k+2b_f}) \le \delta,\ k = 1, 2, \dots. \tag{3.133}$$

$$\rho_E(\tilde{x}_t, x_t^f) < \delta \text{ for all integers } t \ge t_0. \tag{3.134}$$

Let $k \ge 1$ be an integer. By property (a) and (3.134), there exists

$$(\{y_t\}_{t=S}^{t_k+2b_f}, \{v_t\}_{t=S}^{t_k+2b_f-1}) \in X(S, t_k + 2b_f)$$

such that

$$y_t = \tilde{x}_t,\ t = S, \dots, t_k,\ v_t = \tilde{u}_t,\ t = S, \dots, t_k - 1 \tag{3.135}$$

$$y_{t_k+b_f} = x^f_{t+b_f},\ \sum_{t=t_k}^{t_k+b_f-1} f(t, y_t, v_t) \le \sum_{t=t_k}^{t_k+b_f-1} f(t, x_t^f, u_t^f) + 1, \tag{3.136}$$

$$y_{t_k+2b_f} = x^*_{t_k+2b_f},\ \sum_{t=t_k+b_f}^{t_k+2b_f-1} f(t, y_t, v_t) \le \sum_{t=t_k+b_f}^{t_k+2b_f-1} f(t, x_t^f, u_t^f) + 1. \tag{3.137}$$

It follows from (3.132) and (3.135)–(3.137) that

$$\sum_{t=S}^{t_k+2b_f-1} f(t, x_t^*, u_t^*) \le \sum_{t=S}^{t_k+2b_f-1} f(t, y_t, v_t)$$

$$\le \sum_{t=S}^{t_k-1} f(t, \tilde{x}_t, \tilde{u}_t) + \sum_{t=t_k}^{t_k+2b_f-1} f(t, x_t^f, u_t^f) + 2$$

$$\le \sum_{t=S}^{t_k+2b_f-1} f(t, x_t^f, u_t^f) + S_0 + 2.$$

Together with Theorem 3.1, this implies that $(\{x_t^*\}_{t=S}^\infty, \{u_t^*\}_{t=S}^\infty)$ is (f)-good and (iii) holds.

We show that (iii) implies (i). Assume that $(\{x_t^*\}_{t=S}^\infty, \{u_t^*\}_{t=S}^\infty) \in X(S, \infty)$ is (f)-minimal and (f)-good. (P1) implies that

$$\lim_{t\to\infty} \rho_E(x_t^*, x_t^f) = 0. \tag{3.138}$$

There exists $S_0 > 0$ such that

$$|\sum_{t=S}^{T-1} f(t, x_t^*, u_t^*) - \sum_{t=S}^{T-1} f(t, x_t^f, u_t^f)| \le S_0 \text{ for all integers } T > S. \tag{3.139}$$

Let $(\{x_t\}_{t=S}^\infty, \{u_t\}_{t=S}^\infty) \in X(S, \infty)$ satisfy

$$x_S = x_S^*. \tag{3.140}$$

We show that

$$\limsup_{T\to\infty} [\sum_{t=S}^{T-1} f(t, x_t^*, u_t^*) - \sum_{t=S}^{T-1} f(t, x_t, u_t)] \le 0.$$

In view of Theorem 3.1, we may assume that $(\{x_t\}_{t=S}^\infty, \{u_t\}_{t=S}^\infty)$ is (f)-good. (P1) implies that

$$\lim_{t\to\infty} \rho_E(x_t, x_t^f) = 0. \tag{3.141}$$

Let $\epsilon > 0$. By (A2) and Lemma 3.37, there exist $\delta \in (0, \epsilon)$ and an integer $L_1 > 0$ such that the following properties hold:

(b) for each $(T, z) \in \mathcal{A}$ satisfying $\rho_E(z, x_T^f) \le \delta$, there exist an integer $\tau_1 \in (0, b_f]$ and $(\{x_t^{(1)}\}_{t=T}^{T+\tau_1}, \{u_t^{(1)}\}_{t=T}^{T+\tau_1-1}) \in X(T, T + \tau_1)$ satisfying

$$x_T^{(1)} = z\,, x_{T+\tau_1}^{(1)} = x_{T+\tau_1}^f,$$

$$\sum_{t=T}^{T+\tau_1-1} f(t, x_t^{(1)}, u_t^{(1)}) \le \sum_{t=T}^{T+\tau_1-1} f(t, x_t^f, u_t^f) + \epsilon/8$$

and if $T \ge b_f$, then there exist an integer $\tau_2 \in (0, b_f]$ and

$$(\{x_t^{(2)}\}_{t=T-\tau_2}^{T}, \{u_t^{(2)}\}_{t=T-\tau_2}^{T-1}) \in X(T - \tau_2, T)$$

satisfying

$$x^{(2)}_{T-\tau_2} = x^f_{T-\tau_2},\ x^{(2)}_T = z,$$

$$\sum_{t=T-\tau_2}^{T-1} f(t, x^{(2)}_t, u^{(2)}_t) \le \sum_{t=T-\tau_2}^{T-1} f(t, x^f_t, u^f_t) + \epsilon/8;$$

(c) for each pair of integers $T_2 > T_1 \ge L_1$ and each $(\{y_t\}_{t=T_1}^{T_2}, \{v_t\}_{t=T_1}^{T_2-1}) \in X(T_1, T_2)$ which satisfies $\rho_E(y_{T_i}, x^f_{T_i}) \le \delta$, $i = 1, 2$, we have

$$\sum_{t=T_1}^{T_2-1} f(t, y_t, v_t) \ge \sum_{t=T_1}^{T_2-1} f(t, x^f_t, y^f_t) - \epsilon/8.$$

It follows from (3.138) and (3.141) that there exists an integer $\tau_0 > S$ such that

$$\rho_E(x_t, x^f_t) \le \delta,\ \rho_E(x^*_t, x^f_t) \le \delta \text{ for all integers } t \ge \tau_0. \tag{3.142}$$

Let $T \ge \tau_0 + L_1$ be an integer. Property (b) and (3.142) imply that there exists $(\{y_t\}_{t=S}^{T+2b_f}, \{v_t\}_{t=S}^{T+2b_f-1}) \in X(S, T + 2b_f)$ which satisfies

$$y_t = x_t,\ t = S, \dots, T,\ v_t = u_t,\ t = S, \dots, T-1,\ y_{T+b_f} = x^f_{T+b_f}, \tag{3.143}$$

$$\sum_{t=T}^{T+b_f-1} f(t, y_t, v_t) \le \sum_{t=T}^{T+b_f-1} f(t, x^f_t, u^f_t) + \epsilon/8, \tag{3.144}$$

$$y_{T+2b_f} = x^*_{T+2b_f}, \tag{3.145}$$

$$\sum_{t=T+b_f}^{T+2b_f-1} f(t, y_t, v_t) \le \sum_{t=T+b_f}^{T+2b_f-1} f(t, x^f_t, u^f_t) + \epsilon/8. \tag{3.146}$$

By property (c) and (3.142),

$$\sum_{t=T}^{T+2b_f-1} f(t, x^*_t, u^*_t) \ge \sum_{t=T}^{T+2b_f-1} f(t, x^f_t, u^f_t) - \epsilon/8. \tag{3.147}$$

It follows from (3.140), (3.143), and (3.145)–(3.147) that

$$\sum_{t=S}^{T-1} f(t, x^*_t, u^*_t) + \sum_{t=T}^{T+2b_f-1} f(t, x^f_t, u^f_t) - \epsilon/8 \le \sum_{t=S}^{T+2b_f-1} f(t, x^*_t, u^*_t)$$

$$\leq \sum_{t=S}^{T+2b_f-1} f(t, y_t, v_t) = \sum_{t=S}^{T-1} f(t, x_t, u_t) + \sum_{t=T}^{T+2b_f-1} f(t, x_t^f, u_t^f) + \epsilon/4$$

and

$$\sum_{t=S}^{T-1} f(t, x_t^*, u_t^*) \leq \sum_{t=S}^{T-1} f(t, x_t, u_t) + \epsilon$$

for all integers $T \geq \tau_0 + L_1$. Since ϵ is any positive number, we conclude that $(\{x_t^*\}_{t=S}^{\infty}, \{u_t^*\}_{t=S}^{\infty})$ is (f)-overtaking optimal and (i) holds. Theorem 3.20 is proved.
Theorem 3.19 follows from Proposition 3.38 and Theorem 3.20.

3.15 An Auxiliary Result for Theorem 3.24

Lemma 3.39. *Let $\Delta > 0$. Then there exists $\delta > 0$ such that for each $(T_1, z_1), (T_2, z_2) \in \mathcal{A}$ satisfying*

$$T_2 \geq T_1 + 2b_f, \tag{3.148}$$

$$\rho_E(z_i, x_{T_i}^f) \leq \delta, \ i = 1, 2 \tag{3.149}$$

the following inequality holds:

$$U^f(T_1, T_2, z_1, z_2) \leq \sum_{t=T_1}^{T_2-1} f(t, x_t^f, u_t^f) + \Delta.$$

Proof. Let $\delta > 0$ be as guaranteed by (A2) with $\epsilon = \Delta/4$. Let $(T_1, z_1), (T_2, z_2) \in \mathcal{A}$ satisfy (3.148) and (3.149). By (3.148), (3.149), and (A2) with $\epsilon = \Delta/4$, there exists $(\{y_t\}_{t=T_1}^{T_2}, \{v_t\}_{t=T_1}^{T_2-1}) \in X(T_1, T_2)$ such that

$$y_{T_1} = z_1, \ y_{T_2} = z_2, \tag{3.150}$$

$$y_t = x_t^f, \ t = T_1+b_f, \ldots, T_2-b_f, \ v_t = u_t^f, \ t \in \{T_1+b_f, \ldots, T_2-b_f\} \setminus \{T_2-b_f\}, \tag{3.151}$$

$$\sum_{t=T_1}^{T_1+b_f-1} f(t, y_t, v_t) \leq \sum_{t=T_1}^{T_1+b_f-1} f(t, x_t^f, u_t^f) + \Delta/4, \tag{3.152}$$

$$\sum_{t=T_2-b_f}^{T_2-1} f(t, y_t, v_t) \leq \sum_{t=T_2-b_f}^{T_2-1} f(t, x_t^f, u_t^f) + \Delta/4. \tag{3.153}$$

In view of (3.150)–(3.153),

$$U^f(T_1, T_2, z_1, z_2) \le \sum_{t=T_1}^{T_2-1} f(t, y_t, v_t) \le \sum_{t=T_1}^{T_2-1} f(t, x_t^f, u_t^f) + \Delta/2.$$

Lemma 3.39 is proved.

3.16 Proof of Theorem 3.24

Assume that STP holds. Theorem 3.17 implies that (P1) and (P2) hold. Let us show that (P3) holds. By Theorem 3.19, there exists an (f)-overtaking optimal $(\{x_t^*\}_{t=0}^{\infty}, \{u_t^*\}_{t=0}^{\infty}) \in X(0, \infty)$ such that $x_0^* = x_0^f$. Assume that $(\{\tilde{x}_t\}_{t=0}^{\infty}, \{\tilde{u}_t\}_{t=0}^{\infty}) \in X(0, \infty)$ is (f)-overtaking optimal and that $\tilde{x}_0 = x_0^f$. By STP, (P1) and Theorem 3.1, $\tilde{x}_t = x_t^f$ for all integers $t \ge 0$.

Assume that (P1), (P2), and (P3) hold and that $\tilde{u}_t = u_t^f$ for all integers $t \ge 0$. Thus, $(\{x_t^f\}_{t=0}^{\infty}, \{u_t^f\}_{t=0}^{\infty}) \in X(0, \infty)$ is (f)-overtaking optimal.

Lemma 3.40. *Let $\epsilon \in (0, 1)$. Then there exists $\delta > 0$ such that for each pair of integers $T_1 \ge 0$, $T_2 \ge T_1 + 3b_f$ and each $(\{x_t\}_{t=T_1}^{T_2}, \{u_t\}_{t=T_1}^{T_2-1}) \in X(T_1, T_2)$ satisfying*

$$\rho_E(x_{T_i}, x_{T_i}^f) \le \delta, \ i = 1, 2$$

$$\sum_{t=T_1}^{T_2-1} f(t, x_t, u_t) \le U^f(T_1, T_2, x_{T_1}, x_{T_2}) + \delta,$$

the inequality $\rho_E(x_t, x_t^f) \le \epsilon$ holds for all $t = T_1, \dots, T_2$.

Proof. Lemma 3.35 and (A2) imply that there exist $\delta_0 \in (0, \epsilon/4)$ and an integer $L_0 > 0$ such that

(i) (A2) holds with $\epsilon = 1$ and $\delta = \delta_0$;
(ii) for each pair of integers $T_1 \ge L_0$, $T_2 \ge T_1 + 2b_f$ and each

$$(\{x_t\}_{t=T_1}^{T_2}, \{u_t\}_{t=T_1}^{T_2-1}) \in X(T_1, T_2)$$

satisfying

$$\rho_E(x_{T_i}, x_{T_i}^f) \le \delta_0, \ i = 1, 2$$

$$\sum_{t=T_1}^{T_2-1} f(t, x_t, u_t) \le U^f(T_1, T_2, x_{T_1}, x_{T_2}) + \delta_0$$

the inequality $\rho_E(x_t, x_t^f) \le \epsilon$ holds for all $t = T_1, \dots, T_2$.

Theorem 3.1 implies that there exists $S_0 > 0$ such that for each pair of integers $T_2 > T_1 \ge 0$ and each $(\{x_t\}_{t=T_1}^{T_2}, \{u_t\}_{t=T_1}^{T_2-1}) \in X(T_1, T_2)$,

$$\sum_{t=T_1}^{T_2-1} f(t, x_t, u_t) + S_0 \ge \sum_{t=T_1}^{T_2-1} f(t, x_t^f, u_t^f). \tag{3.154}$$

It follows from (P2) and Lemma 3.36 that there exist a natural number L_1 such that the following property holds:

(iii) for each integer $T \ge 0$ and each $(\{x_t\}_{t=T}^{T+L_1}, \{u_t\}_{t=T}^{T+L_1-1}) \in X(T, T + L_1)$ satisfying

$$\sum_{t=T}^{T+L_1-1} f(t, x_t, u_t) \le \sum_{t=T}^{T+L_1-1} f(t, x_t^f, u_t^f) + 2S_0 + 3$$

we have

$$\min\{\rho_E(x_t, x_t^f) : t = T, \dots, T + L_1\} \le \delta_0.$$

Consider a sequence $\{\delta_i\}_{i=1}^{\infty} \subset (0, 1)$ such that

$$\delta_i < 2^{-1}\delta_{i-1}, \ i = 2, 3, \dots. \tag{3.155}$$

Assume that the lemma does not hold. Then for each integer $i \ge 1$, there exist integers $T_{i,1} \ge 0$, $T_{i,2} \ge T_{i,1} + 3b_f$, and $(\{x_t^{(i)}\}_{t=T_{i,1}}^{T_{i,2}}, \{u_t^{(i)}\}_{t=T_{i,1}}^{T_{i,2}-1}) \in X(T_{i,1}, T_{i,2})$ such that

$$\rho_E(x_{T_{i,j}}^{(i)}, x_{T_{i,j}}^f) \le \delta_i, \ j = 1, 2, \tag{3.156}$$

$$\sum_{t=T_{i,1}}^{T_{i,2}-1} f(t, x_t^{(i)}, u_t^{(i)}) \le U^f(T_{i,1}, T_{i,2}, x_{T_{i,1}}^{(i)}, x_{T_{i,2}}^{(i)}) + \delta_i \tag{3.157}$$

and $t_i \in \{T_{i,1}, \dots, T_{i,2}\}$ for which

$$\rho_E(x_{t_i}^{(i)}, x_{t_i}^f) > \epsilon. \tag{3.158}$$

Let i be a natural number. Property (ii) and (3.155)–(3.158) imply that

$$T_{i,1} < L_0. \tag{3.159}$$

We show that

$$t_i \le 2b_f + 1 + L_1 + L_0. \tag{3.160}$$

Property (i), (A2) with $\epsilon = 1$, $\delta = \delta_0$, (3.155), and (3.156) imply that there exists $(\{y_t^{(i)}\}_{t=T_{i,1}}^{T_{i,2}}, \{v_t^{(i)}\}_{t=T_{i,1}}^{T_{i,2}-1}) \in X(T_{i,1}, T_{i,2})$ such that

$$y_{T_{i,1}}^{(i)} = x_{T_{i,1}}^{(i)}, \ y_{T_{i,2}}^{(i)} = x_{T_{i,2}}^{(i)}, \tag{3.161}$$

$$y_t^{(i)} = x_t^f, \ t \in \{T_{i,1} + b_f, \dots, T_{i,2} - b_f\}, \tag{3.162}$$

$$v_t^{(i)} = u_t^f, \ t \in \{T_{i,1} + b_f, \dots, T_{i,2} - b_f - 1\}, \tag{3.163}$$

$$\sum_{t=T_{i,1}}^{T_{i,1}+b_f-1} f(t, y_t^{(i)}, v_t^{(i)}) \le \sum_{t=T_{i,1}}^{T_{i,1}+b_f-1} f(t, x_t^f, u_t^f) + 1, \tag{3.164}$$

$$\sum_{t=T_{i,2}-b_f}^{T_{i,2}-1} f(t, y_t^{(i)}, v_t^{(i)}) \le \sum_{t=T_{i,2}-b_f}^{T_{i,2}-1} f(t, x_t^f, u_t^f) + 1. \tag{3.165}$$

It follows from (3.155), (3.157), and (3.161)–(3.165) that

$$\sum_{t=T_{i,1}}^{T_{i,2}-1} f(t, x_t^{(i)}, u_t^{(i)}) \le \sum_{t=T_{i,1}}^{T_{i,2}-1} f(t, y_t^{(i)}, v_t^{(i)}) + 1$$

$$\le \sum_{t=T_{i,1}}^{T_{i,2}-1} f(t, x_t^f, u_t^f) + 3. \tag{3.166}$$

In view of (3.154), (3.166), for each pair of integers $S_1, S_2 \in [T_{i,1}, T_{i,2}]$ satisfying $S_1 < S_2$,

$$\sum_{t=S_1}^{S_2-1} f(t, x_t^{(i)}, u_t^{(i)}) = \sum_{t=T_{i,1}}^{T_{i,2}-1} f(t, x_t^{(i)}, u_t^{(i)})$$

$$-\sum\{f(t, x_t^{(i)}, u_t^{(i)}) : \ t \in \{T_{i,1}, \dots, S_1\} \setminus \{S_1\}\}$$

$$-\sum\{f(t,x_t^{(i)},u_t^{(i)}):\ t\in\{S_2,\dots,T_{i,2}\}\setminus\{T_{i,2}\}\}$$

$$\leq\sum_{t=T_{i,1}}^{T_{i,2}-1} f(t,x_t^f,u_t^f)+3-\sum\{f(t,x_t^f,u_t^f):\ t\in\{T_{i,1},\dots,S_1\}\setminus\{S_1\}\}+S_0$$

$$-\sum\{f(t,x_t^f,u_t^f):\ t\in\{S_2,\dots,T_{i,2}\}\setminus\{T_{i,2}\}\}+S_0\leq\sum_{t=S_1}^{S_2-1} f(t,x_t^f,u_t^f)+3+2S_0. \tag{3.167}$$

Assume that (3.160) does not hold. Then

$$t_i>2b_f+1+L_1+L_0. \tag{3.168}$$

In view of (3.168),

$$[t_i-2b_f-1-L_1,t_i-1-2b_f]\subset[L_0,\infty). \tag{3.169}$$

Consider the pair $(\{x_t^{(i)}\}_{t=t_i-L_1-2b_f-1}^{t_i-2b_f-1},\{u_t^{(i)}\}_{t=t_i-L_1-2b_f-1}^{t_i-2b_f-2})\in X(t_i-L_1-2b_f-1,t_i-2b_f-1)$. Property (iii) and (3.167) imply that there exists

$$\tilde{t}\in\{t_i-L_1-2b_f-1,\dots,t_i-2b_f-1\} \tag{3.170}$$

such that

$$\rho_E(x_{\tilde{t}},x_{\tilde{t}}^f)\leq\delta_0. \tag{3.171}$$

By property (ii), (3.155)–(3.157), and (3.169)–(3.171),

$$\rho_E(x_t,x_t^f)\leq\epsilon,\ t=\tilde{t},\dots,T_{i,2}$$

and in particular, $\rho_E(x_{t_i},x_{t_i}^f)\leq\epsilon$. This contradicts (3.158). The contradiction we have reached proves (3.160).

In view of (3.12) and (3.167), there exists $M_1>0$ such that

$$f(t,x_t^{(i)},u_t^{(i)})\leq M_1 \text{ for each integer } i\geq 1 \text{ each } t\in\{T_{i,1},\dots,T_{i,2}-1\}. \tag{3.172}$$

It follows from (3.5), (3.6), and (3.172) that there exists $M_2>0$ such that

$$\rho_E(x_t^{(i)},\theta_0)\leq M_2,\ i=1,2,\dots,\ t\in\{T_{i,1},\dots,T_{i,2}-1\}. \tag{3.173}$$

Extracting a subsequence and re-indexing, we may assume without loss of generality that

$$T_{i,1} = T_{1,1} \text{ for all integers } i \geq 1, \tag{3.174}$$

$$t_i = t_1 \text{ for all integers } i \geq 1, \tag{3.175}$$

$$\text{there exists } \lim_{i\to\infty} T_{i,2}. \tag{3.176}$$

There are two cases:

$$\lim_{i\to\infty} T_{i,2} < \infty, \tag{3.177}$$

$$\lim_{i\to\infty} T_{i,2} = \infty. \tag{3.178}$$

Assume that (3.177) holds. Then extracting a subsequence and re-indexing we may assume that

$$T_{i,2} = T_{1,2} \text{ for all integers } i \geq 1. \tag{3.179}$$

In view of (3.174), (3.176), (3.179), and LSC property, extracting a subsequence and re-indexing, we may assume without loss of generality that there exists $(\{\widehat{x}_t\}_{t=T_{1,1}}^{T_{1,2}}, \{\widehat{u}_t\}_{t=T_{1,1}}^{T_{1,2}-1}) \in X(T_{1,1}, T_{1,2})$ such that

$$x_t^{(i)} \to \widehat{x}_t \text{ as } i \to \infty \text{ for all } t \in \{T_{1,1}, \ldots, T_{1,2}\}, \tag{3.180}$$

there exists $\lim_{i\to\infty} \sum_{t=T_{1,1}}^{T_{1,2}-1} f(t, x_t^{(i)}, u_t^{(i)})$ and

$$\sum_{t=T_{1,1}}^{T_{1,2}-1} f(t, \widehat{x}_t, \widehat{u}_t) \leq \lim_{i\to\infty} \sum_{t=T_{1,1}}^{T_{1,2}-1} f(t, x_t^{(i)}, u_t^{(i)}). \tag{3.181}$$

It follows from (3.155), (3.156), (3.158), (3.175), and (3.180) that

$$\widehat{x}_{T_{1,1}} = x^f_{T_{1,1}}, \ \widehat{x}_{T_{1,2}} = x^f_{T_{1,2}}, \ \rho_E(\widehat{x}_{t_1}, x^f_{t_1}) \geq \epsilon. \tag{3.182}$$

Lemmas 3.39, (3.155)–(3.157), and (3.181) imply that

$$\sum_{t=T_{1,1}}^{T_{1,2}-1} f(t, \widehat{x}_t, \widehat{u}_t) \leq \liminf_{i\to\infty} U^f(T_{1,1}, T_{1,2}, x^{(i)}_{T_{1,1}}, x^{(i)}_{T_{1,2}}) \leq \sum_{t=T_{1,1}}^{T_{1,2}-1} f(t, x^f_t, u^f_t). \tag{3.183}$$

It is clear that (3.182) and (3.183) contradict (P3). The contradiction we have reached proves that (3.177) does not hold. Therefore (3.178) holds.

In view of (3.172) and LSC property, extracting a subsequence, using the diagonalization process, and re-indexing, we may assume without loss of generality that for every integer $t \ge T_{1,1}$, there exists $\lim_{i\to\infty} f(t, x_t^{(i)}, u_t^{(i)})$ and there exists $(\{\widehat{x}_t\}_{t=T_{1,1}}^{\infty}, \{\widehat{u}_t\}_{t=T_{1,1}}^{\infty}) \in X(T_{1,1}, \infty)$ such that for every integer $t \ge T_{1,1}$,

$$\widehat{x}_t = \lim_{i\to\infty} x_t^{(i)}, \tag{3.184}$$

$$f(t, \widehat{x}_t, \widehat{u}_t) \le \lim_{i\to\infty} f(t, x_t^{(i)}, u_t^{(i)}). \tag{3.185}$$

By (3.155), (3.156), (3.158), (3.174), (3.175), and (3.184),

$$\widehat{x}_{T,1} = x^f_{T,1},\ \rho_E(x^f_{t_1}, \widehat{x}_{t_1}) \ge \epsilon. \tag{3.186}$$

For all nonnegative integers $t < T_{1,1}$, set

$$\widehat{x}_t = x_t^f.$$

Let $\Delta > 0$. Lemma 3.39 implies that there exists $\delta > 0$ such that the following property holds:

(iv) for each pair of integers $\tau_1 \ge 0$, $\tau_2 \ge \tau_1 + 2b_f$ and each

$$(\{x_t\}_{t=\tau_1}^{\tau_2}, \{u_t\}_{t=\tau_1}^{\tau_2 - 1}) \in X(\tau_1, \tau_2)$$

satisfying

$$\rho_E(x_{\tau_i}, x^f_{\tau_i}) \le \delta,\ i = 1, 2,$$

$$\sum_{t=\tau_1}^{\tau_2-1} f(t, x_t, u_t) \le U^f(\tau_1, \tau_2, x_{\tau_1}, x_{\tau_2}) + \delta$$

we have

$$\sum_{t=\tau_1}^{\tau_2-1} f(t, x_t, u_t) \le \sum_{t=\tau_1}^{\tau_2-1} f(t, x_t^f, u_t^f) + \Delta/2.$$

In view of (3.155), there exists an integer $k_0 \ge 1$ such that

$$\delta_k < \delta \text{ for all integers } k \ge k_0. \tag{3.187}$$

By (3.156), (3.157), (3.174), and (3.187), the following property holds:

(v) for each integer $k \geq k_0$,

$$\sum_{t=T_{1,1}}^{T_{k,2}-1} f(t, x_t^{(k)}, u_t^{(k)}) \leq \sum_{t=T_{1,1}}^{T_{k,2}-1} f(t, x_t^f, u_t^f) + \Delta/2.$$

It follows from (3.167) and (3.185) that for each pair of integers $S_1, S_2 \in [T_{1,1}, \infty)$ satisfying $S_1 < S_2$,

$$\sum_{t=S_1}^{S_2-1} f(t, \widehat{x}_t, \widehat{u}_t) \leq \sum_{t=S_1}^{S_2-1} f(t, x_t^f, u_t^f) + 3 + S_0. \tag{3.188}$$

Theorem 3.1 and (3.188) imply that $(\{\widehat{x}_t\}_{t=0}^{\infty}, \{\widehat{u}_t\}_{t=0}^{\infty})$ is (f)-good. (P1) implies that

$$\lim_{t \to \infty} \rho_E(\widehat{x}_t, x_t^f) = 0. \tag{3.189}$$

In view of (3.189), there exists an integer $\tau_0 > 0$ such that

$$\rho_E(\widehat{x}_t, x_t^f) \leq \delta/4 \text{ for all integers } t \geq \tau_0. \tag{3.190}$$

Let $q \geq 1$ be an integer satisfying

$$T_{q,2} > \tau_0. \tag{3.191}$$

By (3.184), there exists an integer $k_1 \geq k_0$ such that for each integer $k \geq k_1$,

$$T_{k,2} > T_{q,2}, \ \rho_E(\widehat{x}_{T_{q,2}}, x_{T_{q,2}}^{(k)}) \leq \delta/4. \tag{3.192}$$

Assume that an integer $k \geq k_1$. Then (3.192) holds. By (3.156), (3.157), (3.174), (3.187), and (3.190)–(3.192),

$$\sum_{t=T_{1,1}}^{T_{q,2}-1} f(t, x_t^{(k)}, u_t^{(k)}) \leq U^f(T_{1,1}, T_{q,2}, x_{T_{1,1}}^{(k)}, x_{T_{q,2}}^{(k)}) + \delta, \tag{3.193}$$

$$\rho_E(x_{T_{1,1}}^{(k)}, x_{T_{1,1}}^f) \leq \delta_k \leq \delta, \tag{3.194}$$

$$\rho_E(x_{T_{q,2}}^{(k)}, x_{T_{q,2}}^f) \leq \rho_E(x_{T_{q,2}}^{(k)}, \widehat{x}_{T_{q,2}}) + \rho_E(\widehat{x}_{T_{q,2}}, x_{T_{q,2}}^f) \leq \delta/4 + \delta/4. \tag{3.195}$$

Property (iv), (3.174) and (3.193)–(3.195) imply that

$$\sum_{t=T_{1,1}}^{T_{q,2}-1} f(t, x_t^{(k)}, u_t^{(k)}) \le \sum_{t=T_{1,1}}^{T_{q,2}-1} f(t, x_t^f, u_t^f) + \Delta/2. \tag{3.196}$$

In view of (3.185),

$$\sum_{t=T_{1,1}}^{T_{q,2}-1} f(t, \widehat{x}_t, \widehat{u}_t) \le \sum_{t=T_{1,1}}^{T_{q,2}-1} f(t, x_t^f, u_t^f) + \Delta/2.$$

Since the relation above holds for every integer $q \ge 1$ satisfying $T_{q,2} > \tau_0$, we have

$$\liminf_{T\to\infty}\Big(\sum_{t=T_{1,1}}^{T-1} f(t, \widehat{x}_t, \widehat{u}_t) - \sum_{t=T_{1,1}}^{T-1} f(t, x_t^f, u_t^f)\Big) \le \Delta.$$

Since Δ is an arbitrary positive number, the inequality above implies that

$$\liminf_{T\to\infty}\Big(\sum_{t=T_{1,1}}^{T-1} f(t, \widehat{x}_t, \widehat{u}_t) - \sum_{t=T_{1,1}}^{T-1} f(t, x_t^f, u_t^f)\Big) \le 0$$

and $(\{\widehat{x}_t\}_{t=0}^{\infty}, \{\widehat{u}_t\}_{t=0}^{\infty})$ is (f)-weakly optimal. Theorem 3.20 implies that $(\{\widehat{x}_t\}_{t=0}^{\infty}, \{\widehat{u}_t\}_{t=0}^{\infty})$ is (f)-overtaking optimal. In view of (P3), $\widehat{x}_t = x_t^f$ for all integers $t \ge 0$. This contradicts (3.186). The contradiction we have reached completes the proof of Lemma 3.40.

Completion of the Proof of Theorem 3.24. Theorem 3.17 implies TP. Lemma 3.40 and TP imply STP.

3.17 Proof of Theorem 3.26

We show that f_r possesses (P2). Let $\epsilon, M > 0$. Theorem 3.1 implies that there exists $M_0 > 0$ such that for each pair of integers $S_2 > S_1 \ge 0$ and each $(\{x_t\}_{t=S_1}^{S_2}, \{u_t\}_{t=S_1}^{S_2-1}) \in X(S_1, S_2)$, we have

$$\sum_{t=S_1}^{S_2-1} f(t, x_t, u_t) + M_0 \ge \sum_{t=S_1}^{S_2-1} f(t, x_t^f, u_t^f). \tag{3.197}$$

Property (i) (see Section 3.5) implies that there exists $\delta > 0$ such that the following property holds:

(a) for each $x_1, x_2 \in E$ satisfying $\phi(x_1, x_2) \le \delta$, we have $\rho_E(x_1, x_2) \le \epsilon$.

Choose a natural number

$$L > 2(r\delta)^{-1}(M + M_0). \tag{3.198}$$

Assume that a integer $T \ge 0$ and $(\{x_t\}_{t=T}^{T+L}, \{u_t\}_{t=T}^{T+L-1}) \in X(T, T + L)$ satisfy

$$\sum_{t=T}^{T+L-1} f_r(t, x_t, u_t) \le \sum_{t=T}^{T+L-1} f_r(t, x_t^f, u_t^f) + M. \tag{3.199}$$

It follows from (3.14) and (3.197)–(3.199) that

$$\sum_{t=T}^{T+L-1} f(t, x_t, u_t) + r \sum_{t=T}^{T+L-1} \phi(x_t, x_t^f) = \sum_{t=T}^{T+L-1} f_r(t, x_t, u_t)$$

$$\le \sum_{t=T}^{T+L-1} f_r(t, x_t^f, u_t^f) + M = \sum_{t=T}^{T+L-1} f(t, x_t^f, u_t^f) + M$$

$$\le \sum_{t=T}^{T+L-1} f(t, x_t, u_t) + M + M_0,$$

$$r \sum_{t=T}^{T+L-1} \phi(x_t, x_t^f) \le M_0 + M,$$

$$\min\{\phi(x_t, x_t^f) : t \in \{T, \dots, T + L - 1\}\} \le (Lr)^{-1}(M_0 + M) < \delta$$

and there exists an integer $\tau \in [T, T + L - 1]$ such that $\phi(x_\tau, x_\tau^f) < \delta$. Property (a) implies that $\rho_E(x_\tau, x_\tau^f) < \epsilon$. Therefore f_r has (P2).

We show that f_r has (P1). Let $T_0 \ge 0$ be an integer and let

$$(\{x_t\}_{t=T_0}^{\infty}, \{u_t\}_{t=T_0}^{\infty}) \in X(T_0, \infty)$$

be (f_r)-good. We claim that

$$\lim_{t\to\infty} \rho_E(x_t, x_t^f) = 0. \tag{3.200}$$

Assume that (3.200) does not hold. Then there exist $\epsilon > 0$ and a sequence of natural numbers $\{t_k\}_{k=1}^{\infty} \subset [T_0, \infty)$ such that for each integer $k \ge 1$,

$$t_{k+1} \ge t_k + 8, \ \rho_E(x_{t_k}, x_{t_k}^f) > \epsilon. \tag{3.201}$$

There exists $M_0 > 0$ such that for each integer $T > T_0$,

$$M_0 > |\sum_{t=T_0}^{T-1} f_r(t, x_t, u_t) - \sum_{t=T_0}^{T-1} f_r(t, x_t^f, u_t^f)|$$

$$= |\sum_{t=T_0}^{T-1} f(t, x_t, u_t) - \sum_{t=T_0}^{T-1} f(t, x_t^f, u_t^f) + r \sum_{t=T_0}^{T-1} \phi(x_t, x_t^f)|. \quad (3.202)$$

Set

$$\Delta = \sum_{t=T_0}^{\infty} \phi(x_t, x_t^f). \quad (3.203)$$

Theorem 3.1, (3.202), and (3.203) imply that

$$\Delta < \infty. \quad (3.204)$$

In view of (3.204), for every $\delta > 0$ the set $\{t \text{ is an integer } : t \geq T_0,\ \phi(x_t, x_t^f) \geq \delta\}$ is finite. Together with property (a), this implies that the set

$$\{t \text{ is an integer } : t \geq T_0,\ \rho_E(x_t, x_t^f) > \epsilon\}$$

is finite for any $\epsilon > 0$. This implies that $\lim_{t\to\infty} \rho_E(x_t, x_t^f) = 0$. Thus f_r has (P1).

Assume that $(\{x_t^f\}_{t=0}^{\infty}, \{u_t^f\}_{t=0}^{\infty})$ is (f)-minimal. Theorem 3.20 and (3.14) imply that $(\{x_t^f\}_{t=0}^{\infty}, \{u_t^f\}_{t=0}^{\infty})$ is (f_r)-minimal and (f_r)-overtaking optimal.

Let $T_0 \geq 0$ be an integer and $(\{x_t\}_{t=T_0}^{\infty}, \{u_t\}_{t=T_0}^{\infty}) \in X(T_0, \infty)$ be (f_r)-overtaking optimal and

$$x_{T_0} = x_{T_0}^f. \quad (3.205)$$

Property (P1) and (3.205) imply that

$$\lim_{t\to\infty} \rho_E(x_t, x_t^f) = 0. \quad (3.206)$$

We show that $x_t = x_t^f$ for all integers $t \geq T_0$. Assume the contrary. Then there exists an integer $t_0 > T_0$ such that

$$\gamma := \rho(x_{t_0}, x_{t_0}^f) > 0. \quad (3.207)$$

Property (a) implies that there exists $\delta_0 > 0$ such that the following property holds:

$$\text{if } z_1, z_2 \in E,\ \phi(z_1, z_2) \leq \delta_0, \text{ then } \rho_E(z_1, z_2) \leq \gamma/4. \quad (3.208)$$

In view of (3.207) and (3.208),

$$\phi(x_{t_0}, x^f_{t_0}) > \delta_0. \tag{3.209}$$

By (3.14), (3.205) and (3.209),

$$0 = \lim_{T\to\infty}[\sum_{t=T_0}^{T-1} f_r(t, x_t, u_t) - \sum_{t=T_0}^{T-1} f_r(t, x^f_t, u^f_t)]$$

$$= \lim_{T\to\infty}[\sum_{t=T_0}^{T-1} f(t, x_t, u_t) + r\sum_{t=T_0}^{T-1} \phi(x_t, x^f_t) - \sum_{t=T_0}^{T-1} f(t, x^f_t, u^f_t)]$$

$$= \lim_{T\to\infty}[\sum_{t=T_0}^{T-1} f(t, x_t, u_t) - \sum_{t=T_0}^{T-1} f(t, x^f_t, u^f_t)] + r\sum_{t=T_0}^{\infty} \phi(x_t, x^f_t)$$

$$\geq \lim_{T\to\infty}[\sum_{t=T_0}^{T-1} f(t, x_t, u_t) - \sum_{t=T_0}^{T-1} f(t, x^f_t, u^f_t)] + \delta_0. \tag{3.210}$$

By (3.210), there exists an integer $T_1 > T_0$ such that for each integer $T \geq T_1$,

$$\sum_{t=T_0}^{T-1} f(t, x_t, u_t) - \sum_{t=T_0}^{T-1} f(t, x^f_t, u^f_t) \leq -\delta_0/2. \tag{3.211}$$

It follows from (A2) that there exists $\delta_1 > 0$ such that the following property holds:

(b) for each $(S, z) \in \mathcal{A}$ satisfying $\rho_E(z, x^f_S) \leq \delta_1$, there exist an integer $\tau \in (0, b_f]$,

$$(\{y_t\}_{t=S}^{S+\tau}, \{v_t\}_{t=S}^{S+\tau-1}) \in X(S, S+\tau)$$

such that

$$y_S = z,\ y_{S+\tau} = x^f_{S+\tau},$$

$$\sum_{t=S}^{S+\tau-1} f(t, y_t, v_t) \leq \sum_{t=S}^{S+\tau-1} f(t, x^f_t, u^f_t) + 8^{-1}\delta_0.$$

In view of (3.206), there exists an integer $T_2 > T_1$ such that

$$\rho_E(x_t, x_t^f) \le \delta_1 \text{ for all integers } t \ge T_2. \tag{3.212}$$

Property (b) and (3.212) imply that there exists

$$(\{y_t\}_{t=T_0}^{T_2+t_0+b_f}, \{v_t\}_{t=T_0}^{T_2+t_0+b_f-1}) \in X(T_0, T_2 + t_0 + b_f)$$

such that

$$y_t = x_t,\ t = T_0, \dots, T_2 + t_0,\ v_t = u_t,\ t = T_0, \dots, T_2 + t_0 - 1, \tag{3.213}$$

$$y_{T_2+t_0+b_f} = x^f_{T_2+t_0+b_f}, \tag{3.214}$$

$$\sum_{t=T_2+t_0}^{T_2+t_0+b_f-1} f(t, y_t, v_t) \le \sum_{t=T_2+t_0}^{T_2+t_0+b_f-1} f(t, x_t^f, u_t^f) + 8^{-1}\delta_0. \tag{3.215}$$

In view of (3.205) and (3.213),

$$y_{T_0} = x^f_{T_0}. \tag{3.216}$$

It follows from (3.211), (3.213), and (3.215) that

$$\sum_{t=T_0}^{T_2+t_0+b_f-1} f(t, y_t, v_t) \le \sum_{t=T_0}^{T_2+t_0-1} f(t, x_t, u_t) + \sum_{t=T_2+t_0}^{T_2+t_0+b_f-1} f(t, x_t^f, u_t^f) + 8^{-1}\delta_0$$

$$\le \sum_{t=T_0}^{T_2+t_0-1} f(t, x_t^f, u_t^f) - \delta_0/2 + \sum_{t=T_2+L_0}^{T_2+t_0+b_f-1} f(t, x_t^f, u_t^f) + 8^{-1}\delta_0$$

$$= \sum_{t=T_0}^{T_2+t_0+b_f-1} f(t, x_t^f, u_t^f) - 3\delta_0/8.$$

In view of (3.214) and (3.216), $(\{x_t^f\}_{t=0}^{\infty}, \{u_t^f\}_{t=T}^{\infty})$ is not (f)-minimal, a contradiction. The contradiction we have reached proves that $x_t = x_t^f$ for all integers $t \ge T_0$ and (P3) holds for f_r. This completes the proof of Theorem 3.26.

Chapter 4
Discrete-Time Nonautonomous Problems on Axis

In this chapter we establish sufficient and necessary conditions for the turnpike phenomenon for discrete-time nonautonomous problems on subintervals of axis in metric spaces, which are not necessarily compact. For these optimal control problems the turnpike is not a singleton. We also study the existence of solutions of the corresponding infinite horizon optimal control problems.

4.1 Preliminaries and Main Results

Denote by **Z** the set of all integers.

Let (E, ρ_E) and (F, ρ_F) be metric spaces. We suppose that $\mathcal{A}$ is a nonempty subset of $\mathbf{Z} \times E, \mathcal{U} : \mathcal{A} \to 2^F$ is a point to set mapping with a graph

$$\mathcal{M} = \{(t, x, u) : (t, x) \in \mathcal{A}, u \in \mathcal{U}(t, x)\},$$

$G : \mathcal{M} \to E$ and $f : \mathcal{M} \to R^1$.

Let $T_1 < T_2$ be integers. We denote by $X(T_1, T_2)$ the set of all pairs of sequences $(\{x_t\}_{t=T_1}^{T_2}, \{u_t\}_{t=T_1}^{T_2-1})$ such that for each integer $t \in \{T_1, \dots, T_2\}$,

$$(t, x_t) \in \mathcal{A},$$

for each integer $t \in \{T_1, \dots, T_2 - 1\}$,

$$u_t \in \mathcal{U}(t, x_t), \tag{4.1}$$

$$x_{t+1} = G(t, x_t, u_t) \tag{4.2}$$

and which are called trajectory-control pairs.

A. J. Zaslavski, *Turnpike Conditions in Infinite Dimensional Optimal Control*, Springer Optimization and Its Applications 148,
https://doi.org/10.1007/978-3-030-20178-4_4

Let T_1 be an integer. Denote by $X(T_1, \infty)$ the set of all pairs of sequences $\{x_t\}_{t=T_1}^{\infty} \subset E$, $\{u_t\}_{t=T_1}^{\infty} \subset F$, which are called trajectory-control pairs, such that for each integer $T_2 > T_1$, $(\{x_t\}_{t=T_1}^{T_2}, \{u_t\}_{t=T_1}^{T_2-1}) \in X(T_1, T_2)$. We denote by $X(-\infty, \infty)$ the set of all pairs of sequences $\{x_t\}_{t=-\infty}^{\infty} \subset E$, $\{u_t\}_{t=-\infty}^{\infty} \subset F$, which are called trajectory-control pairs, such that for each pair of integers $T_2 > T_1$, $(\{x_t\}_{t=T_1}^{T_2}, \{u_t\}_{t=T_1}^{T_2-1}) \in X(T_1, T_2)$.

Let $T_1 < T_2$ be integers. A sequence $\{x_t\}_{t=T_1}^{T_2} \subset E$ ($\{x_t\}_{t=T_1}^{\infty} \subset E$, $\{x_t\}_{t=-\infty}^{\infty} \subset E$, respectively) is called a trajectory if there exists a sequence $\{u_t\}_{t=T_1}^{T_2-1} \subset F$ ($\{u_t\}_{t=T_1}^{\infty} \subset F$, $\{u_t\}_{t=-\infty}^{\infty} \subset F$, respectively) referred to as a control such that $(\{x_t\}_{t=T_1}^{T_2}, \{u_t\}_{t=T_1}^{T_2-1}) \in X(T_1, T_2)$ ($(\{x_t\}_{t=T_1}^{\infty}, \{u_t\}_{t=T_1}^{\infty}) \in X(T_1, \infty)$, $(\{x_t\}_{t=-\infty}^{\infty}, \{u_t\}_{t=-\infty}^{\infty}) \in X(-\infty, \infty)$, respectively).

Let $\theta_0 \in E$ and $a_0 > 0$ and let $\psi : [0, \infty) \to [0, \infty)$ be an increasing function such that

$$\psi(t) \to \infty \text{ as } t \to \infty. \tag{4.3}$$

We suppose that the function f satisfies

$$f(t, x, u) \geq \psi(\rho_E(x, \theta_0)) - a_0 \text{ for each } (t, x, u) \in \mathcal{M}. \tag{4.4}$$

For each pair of integers $T_2 > T_1$ and each pair of points $y, z \in E$ satisfying $(T_1, y), (T_2, z) \in \mathcal{A}$ we consider the following problems:

$$\sum_{t=T_1}^{T_2-1} f(t, x_t, u_t) \to \min, \ (\{x_t\}_{t=T_1}^{T_2}, \{u_t\}_{t=T_1}^{T_2-1}) \in X(T_1, T_2), \ x_{T_1} = y, \ x_{T_2} = z, \tag{P_1}$$

$$\sum_{t=T_1}^{T_2-1} f(t, x_t, u_t) \to \min, \ (\{x_t\}_{t=T_1}^{T_2}, \{u_t\}_{t=T_1}^{T_2-1}) \in X(T_1, T_2), \ x_{T_1} = y, \tag{P_2}$$

$$\sum_{t=T_1}^{T_2-1} f(t, x_t, u_t) \to \min, \ (\{x_t\}_{T_1}^{T_2}, \{u_t\}_{T_1}^{T_2-1}) \in X(T_1, T_2). \tag{P_3}$$

For each pair of integers $T_2 > T_1$ and each pair of points $y, z \in E$ satisfying $(T_1, y), (T_2, z) \in \mathcal{A}$ we define

$$U^f(T_1, T_2, y, z) = \inf\{\sum_{t=T_1}^{T_2-1} f(t, x_t, u_t) : \ (\{x_t\}_{t=T_1}^{T_2}, \{u_t\}_{t=T_1}^{T_2-1}) \in X(T_1, T_2),$$

$$x_{T_1} = y, \ x_{T_2} = z\}, \tag{4.5}$$

$$\sigma^f(T_1, T_2, y) = \inf\{U^f(T_1, T_2, y, h) : \ h \in E, \ (T_2, h) \in \mathcal{A}\}, \tag{4.6}$$

$$\widehat{\sigma}^f(T_1, T_2, z) = \inf\{U^f(T_1, T_2, h, z) : h \in E,\ (T_1, h) \in \mathcal{A}\}, \tag{4.7}$$

$$\sigma^f(T_1, T_2) = \inf\{U^f(T_1, T_2, h, \xi) : h, \xi \in E,\ (T_1, h), (T_2, \xi) \in \mathcal{A}\}. \tag{4.8}$$

We suppose that $b_f > 0$ is an integer,

$$(\{x_t^f\}_{t=-\infty}^{\infty}, \{u_t^f\}_{t=-\infty}^{\infty}) \in X(-\infty, \infty),$$

$$\{x_t^f : t \in \mathbf{Z}\} \text{ is bounded,} \tag{4.9}$$

$$\Delta_f := \sup\{|f(t, x_t^f, u_t^f)| : t \in \mathbf{Z}\} < \infty \tag{4.10}$$

and that the following assumptions hold.

(A1) For each $S_1 > 0$ there exist $S_2 > 0$ and an integer $c > 0$ such that

$$\sum_{t=T_1}^{T_2-1} f(t, x_t^f, u_t^f) \le \sum_{t=T_1}^{T_2-1} f(t, x_t, u_t) + S_2$$

for each pair of integers T_1, $T_2 \ge T_1 + c$ and each $(\{x_t\}_{t=T_1}^{T_2}, \{u_t\}_{t=T_1}^{T_2-1}) \in X(T_1, T_2)$ satisfying $\rho_E(\theta_0, x_j) \le S_1$, $j = T_1, T_2$.

(A2) For each $\epsilon > 0$ there exists $\delta > 0$ such that for each $(T, z) \in \mathcal{A}$, satisfying $\rho_E(z, x_T^f) \le \delta$ there exist integers $\tau_1, \tau_2 \in (0, b_f]$,

$$(\{x_t^{(1)}\}_{t=T}^{T+\tau_1}, \{u_t^{(1)}\}_{t=T}^{T+\tau_1-1}) \in X(T, T + \tau_1),$$

$$(\{x_t^{(2)}\}_{t=T-\tau_2}^{T}, \{u_t^{(2)}\}_{t=T-\tau_2}^{T-1}) \in X(T - \tau_2, T)$$

such that

$$x_T^{(1)} = z,\ x_{T+\tau_1}^{(1)} = x_{T+\tau_1}^f,$$

$$\sum_{t=T}^{T+\tau_1-1} f(t, x_t^{(1)}, u_t^{(1)}) \le \sum_{t=T}^{T+\tau_1-1} f(t, x_t^f, u_t^f) + \epsilon,$$

$$x_T^{(2)} = z,\ x_{T-\tau_2}^{(2)} = x_{T-\tau_2}^f,$$

$$\sum_{t=T-\tau_2}^{T-1} f(t, x_t^{(2)}, u_t^{(2)}) \le \sum_{t=T-\tau_2}^{T-1} f(t, x_t^f, u_t^f) + \epsilon,$$

Many examples of optimal control problems satisfying assumptions (A1) and (A2) can be found in [106–108, 118, 124, 125, 134].

The following result is proved in Section 4.6.

Theorem 4.1.

1. There exists $S > 0$ such that for each pair of integers $T_2 > T_1$ and each $(\{x_t\}_{t=T_1}^{T_2}, \{u_t\}_{t=T_1}^{T_2-1}) \in X(T_1, T_2)$,

$$\sum_{t=T_1}^{T_2-1} f(t, x_t, u_t) + S \geq \sum_{t=T_1}^{T_2-1} f(t, x_t^f, u_t^f).$$

2. For each $(\{x_t\}_{t=-\infty}^{\infty}, \{u_t\}_{t=-\infty}^{\infty}) \in X(-\infty, \infty)$ either

$$\sum_{t=-T}^{T-1} f(t, x_t, u_t) - \sum_{t=-T}^{T-1} f(t, x_t^f, u_t^f) \to \infty \text{ as } T \to \infty$$

or

$$\sup\{|\sum_{t=-T}^{T-1} f(t, x_t, u_t) - \sum_{t=-T}^{T-1} f(t, x_t^f, u_t^f)| : T \in \{1, 2, \dots\}\} < \infty. \quad (4.11)$$

Moreover, if (4.11) *holds, then*

$$\sup\{\rho_E(x_t, \theta_0) : t \in \mathbf{Z}\} < \infty.$$

We say that $(\{x_t\}_{t=-\infty}^{\infty}, \{u_t\}_{t=-\infty}^{\infty}) \in X(-\infty, \infty)$ is (f)-good [29, 104, 116, 120, 129, 134] if

$$\sup\{|\sum_{t=-T}^{T-1} f(t, x_t, u_t) - \sum_{t=-T}^{T-1} f(t, x_t^f, u_t^f)| : T \in \{1, 2, \dots\}\} < \infty.$$

4.2 Boundedness Results

The next result is proved in Section 4.6.

Theorem 4.2. *Let $M_0 > 0$. Then there exists $M_1 > 0$ such that for each pair of integers $T_2 > T_1$ and each $(\{x_t\}_{t=T_1}^{T_2}, \{u_t\}_{t=T_1}^{T_2-1}) \in X(T_1, T_2)$ satisfying*

$$\sum_{t=T_1}^{T_2-1} f(t, x_t, u_t) \le \sum_{t=T_1}^{T_2-1} f(t, x_t^f, u_t^f) + M_0$$

the inequality $\rho_E(\theta_0, x_t) \le M_1$ *holds for all* $t = T_1, \dots, T_2 - 1$.

Let $L > 0$ be an integer. Denote by $\mathcal{A}_L$ the set of all $(S, z) \in \mathcal{A}$ for which there exist an integer $\tau \in (0, L]$ and $(\{x_t\}_{t=S}^{S+\tau}, \{u_t\}_{t=S}^{S+\tau-1}) \in X(S, S+\tau)$ such that

$$x_S = z, \ x_{S+\tau} = x_{S+\tau}^f, \ \sum_{t=S}^{S+\tau-1} f(t, x_t, u_t) \le L.$$

Denote by $\widehat{\mathcal{A}}_L$ the set of all $(S, z) \in \mathcal{A}$ such that there exist an integer $\tau \in (0, L]$ and $(\{x_t\}_{t=S-\tau}^{S}, \{u_t\}_{t=S-\tau}^{S-1}) \in X(S-\tau, S)$ satisfying

$$x_{S-\tau} = x_{S-\tau}^f, \ x_S = z, \ \sum_{t=S-\tau}^{S-1} f(t, x_t, u_t) \le L.$$

The next three results are proved in Section 4.7.

Theorem 4.3. *Let* $L > 0$ *be an integer and* $M_0 > 0$. *Then there exists* $M_1 > 0$ *such that for each integer* T_1, *each integer* $T_2 \ge T_1 + 2L$ *and each* $(\{x_t\}_{t=T_1}^{T_2}, \{u_t\}_{t=T_1}^{T_2-1}) \in X(T_1, T_2)$ *satisfying*

$$(T_1, x_{T_1}) \in \mathcal{A}_L, \ (T_2, x_{T_2}) \in \widehat{\mathcal{A}}_L,$$

$$\sum_{t=T_1}^{T_2-1} f(t, x_t, u_t) \le U^f(T_1, T_2, x_{T_1}, x_{T_2}) + M_0$$

the inequality $\rho_E(\theta_0, x_t) \le M_1$ *holds for all* $t = T_1, \dots, T_2 - 1$.

Theorem 4.4. *Let* $L > 0$ *be an integer and* $M_0 > 0$. *Then there exists* $M_1 > 0$ *such that for each integer* T_1, *each integer* $T_2 \ge T_1 + L$ *and each* $(\{x_t\}_{t=T_1}^{T_2}, \{u_t\}_{t=T_1}^{T_2-1}) \in X(T_1, T_2)$ *satisfying*

$$(T_1, x_{T_1}) \in \mathcal{A}_L, \ \sum_{t=T_1}^{T_2-1} f(t, x_t, u_t) \le \sigma^f(T_1, T_2, x_{T_1}) + M_0$$

the inequality $\rho_E(\theta_0, x_t) \le M_1$ *holds for all* $t = T_1, \dots, T_2 - 1$.

Theorem 4.5. *Let* $L > 0$ *be an integer and* $M_0 > 0$. *Then there exists* $M_1 > 0$ *such that for each integer* T_1, *each integer* $T_2 \ge T_1 + L$ *and each* $(\{x_t\}_{t=T_1}^{T_2}, \{u_t\}_{t=T_1}^{T_2-1}) \in X(T_1, T_2)$ *satisfying*

$$(T_2, x_{T_2}) \in \widehat{\mathcal{A}}_L, \ \sum_{t=T_1}^{T_2-1} f(t, x_t, u_t) \le \widehat{\sigma}^f(T_1, T_2, x_{T_2}) + M_0$$

the inequality $\rho_E(\theta_0, x_t) \le M_1$ *holds for all* $t = T_1, \dots, T_2 - 1$.

Theorem 4.6. *For each integer* T_0 *and each* $(\{x_t\}_{t=T_0}^{\infty}, \{u_t\}_{t=T_0}^{\infty}) \in X(T_0, \infty)$ *either*

$$\sum_{t=T_0}^{T-1} f(t, x_t, u_t) - \sum_{t=T_0}^{T-1} f(t, x_t^f, u_t^f) \to \infty \text{ as } T \to \infty$$

or

$$\sup\{|\sum_{t=T_0}^{T-1} f(t, x_t, u_t) - \sum_{t=T_0}^{T-1} f(t, x_t^f, u_t^f)| : T > T_0 \text{ is an integer}\} < \infty. \tag{4.12}$$

Moreover, if (4.12) *holds, then* $\{x_t : t \in \mathbf{Z}, \ t \ge T_0\}$ *is bounded.*

Theorem 4.6 easily follows from Assertion 1 of Theorem 4.1 and Theorem 4.2.

Let T_0 be an integer. We say that $(\{x_t\}_{t=T_0}^{\infty}, \{u_t\}_{t=T_0}^{\infty}) \in X(T_0, \infty)$ is (f)-good [29, 104, 116, 120, 129, 134] if (4.12) holds.

Let T_0 be an integer. We say that $(\{x_t\}_{t=T_0}^{\infty}, \{u_t\}_{t=T_0}^{\infty}) \in X(T_0, \infty)$ is (f)-minimal if for each integer $T > T_0$,

$$\sum_{t=T_0}^{T-1} f(t, x_t, u_t) = U^f(T_0, T, x_{T_0}, x_T).$$

We say that $(\{x_t\}_{t=-\infty}^{\infty}, \{u_t\}_{t=-\infty}^{\infty}) \in X(-\infty, \infty)$ is (f)-minimal [12, 94] if for each pair of integers $T_2 > T_1$,

$$\sum_{t=T_1}^{T_2-1} f(t, x_t, u_t) = U^f(T_1, T_2, x_{T_1}, x_{T_2}).$$

4.3 Turnpike Results

We say that f possesses the strong turnpike property (or STP for short) if for each $\epsilon > 0$ and each $M > 0$, there exist $\delta > 0$ and an integer $L > 0$ such that for each integer T_1, each integer $T_2 \ge T_1 + 2L$, and each $(\{x_t\}_{t=T_1}^{T_2}, \{u_t\}_{t=T_1}^{T_2-1}) \in X(T_1, T_2)$ which satisfies

$$\sum_{t=T_1}^{T_2-1} f(t, x_t, u_t) \le \min\{\sigma^f(T_1, T_2) + M, U^f(T_1, T_2, x_{T_1}, x_{T_2}) + \delta\}$$

there exist integers $\tau_1 \in [T_1, T_1 + L]$, $\tau_2 \in [T_2 - L, T_2]$ such that

$$\rho_E(x_t, x_t^f) \le \epsilon, \; t = \tau_1 \dots, \tau_2.$$

Moreover, if $\rho_E(x_{T_2}, x_{T_2}^f) \le \delta$, then $\tau_2 = T_2$, and if $\rho_E(x_{T_1}, x_{T_1}^f) \le \delta$, then $\tau_1 = T_1$.

We say that f possesses lower semicontinuity property (or LSC property for short) if for each pair of integers $T_2 > T_1$ and each sequence $(\{x_t^{(j)}\}_{t=T_1}^{T_2}, \{u_t^{(j)}\}_{t=T_1}^{T_2-1}) \in X(T_1, T_2)$, $j = 1, 2, \dots$ which satisfies

$$\sup\{\sum_{t=T_1}^{T_2-1} f(t, x_t^{(j)}, u_t^{(j)}) : \; j = 1, 2, \dots\} < \infty$$

there exist a subsequence $(\{x_t^{(j_k)}\}_{t=T_1}^{T_2}, \{u_t^{(j_k)}\}_{t=T_1}^{T_2-1})$, $k = 1, 2, \dots$ and

$$(\{x_t\}_{t=T_1}^{T_2}, \{u_t\}_{t=T_1}^{T_2-1}) \in X(T_1, T_2)$$

such that for any $t \in \{T_1, \dots, T_2\}$,

$$x_t^{(j_k)} \to x_t \text{ as } k \to \infty,$$

$$\sum_{t=T_1}^{T_2-1} f(t, x_t, u_t) \le \liminf_{j\to\infty} \sum_{t=T_1}^{T_2-1} f(t, x_t^{(j)}, u_t^{(j)}).$$

LSC property plays an important role in the calculus of variations and optimal control theory [32].

The next theorem is the main result of this chapter. It is proved in Sections 4.9 and 4.10.

Theorem 4.7. *Let f have LSC property. If f has STP, then the following properties hold:*

(P1) *for each (f)-good pair $(\{x_t\}_{t=-\infty}^{\infty}, \{u_t\}_{t=-\infty}^{\infty}) \in X(-\infty, \infty)$,*

$$\lim_{t\to\infty} \rho(x_t, x_t^f) = 0, \quad \lim_{t\to-\infty} \rho(x_t, x_t^f) = 0;$$

(P2) *for each $\epsilon > 0$ and each $M > 0$, there exist $\delta > 0$ and an integer $L > 0$ such that for each integer T and each $(\{x_t\}_{t=T}^{T+L}, \{u_t\}_{t=T}^{T+L-1}) \in X(T, T+L)$ which satisfies*

$$\sum_{t=T}^{T+L-1} f(t, x_t, u_t) \le \min\{U^f(T, T+L, x_T, x_{T+L})$$

$$+\delta, \sum_{t=T}^{T+L-1} f(t, x_t^f, u_t^f) + M\},$$

there exists an integer $s \in [T, T+L]$ *such that* $\rho_E(x_s, x_s^f) \le \epsilon$.

(P3) *there exists* $\{\tilde{u}_t^f\}_{t=-\infty}^{\infty} \subset F$ *such that* $(\{x_t^f\}_{t=-\infty}^{\infty}, \{\tilde{u}_t^f\}_{t=-\infty}^{\infty}) \in X(-\infty, \infty)$ *is* (f)*-good and* (f)*-minimal and for each* (f)*-good and* (f)*-minimal* $(\{x_t\}_{t=-\infty}^{\infty}, \{u_t\}_{t=-\infty}^{\infty}) \in X(-\infty, \infty)$, *the equality* $x_t = x_t^f$ *holds for all integers* t.

If (P1), (P2), and (P3) hold and $\tilde{u}_t^f = u_t^f$ *for all integers* $t \ge 0$, *then* f *has STP.*

The next result is proved in Section 4.11.

Theorem 4.8. f *has properties (P1) and (P2) if and only if the following property holds:*

(P4) *for each* $\epsilon > 0$ *and each* $M > 0$ *there exist* $\delta > 0$ *and an integer* $L > 0$ *such that for each integer* T_1, *each integer* $T_2 > T_1$ *and each* $(\{x_t\}_{t=T_1}^{T_2}, \{u_t\}_{t=T_1}^{T_2-1}) \in X(T_1, T_2)$ *which satisfies*

$$\sum_{t=T_1}^{T_2-1} f(t, x_t, u_t) \le \min\{\sigma^f(T_1, T_2) + M, U^f(T_1, T_2, x_{T_1}, x_{T_2}) + \delta\}$$

if $T_2 \ge (T_1)_+ + 2L$, *then* $\rho_E(x_t, x_t^f) \le \epsilon$, $t = (T_1)_+ + L, \ldots, T_2 - L$, *and if* $T_1 \le (T_2)_- - 2L$, *then* $\rho_E(x_t, x_t^f) \le \epsilon$, $t = T_1 + L, \ldots, (T_2)_- - L$.

Theorem 4.9. *Assume that* f *has STP, that* $L_0 > 0$ *is an integer, and that* $\epsilon > 0$. *Then there exist* $\delta > 0$ *and an integer* $L > L_0$ *such that for each integer* T_1, *each integer* $T_2 \ge T_1 + 2L$ *and each* $(\{x_t\}_{t=T_1}^{T_2}, \{u_t\}_{t=T_1}^{T_2-1}) \in X(T_1, T_2)$ *which satisfies*

$$(T_1, x_{T_1}) \in \mathcal{A}_{L_0}, \ (T_2, x_{T_2}) \in \widehat{\mathcal{A}}_{L_0},$$

$$\sum_{t=T_1}^{T_2-1} f(t, x_t, u_t) \le U^f(T_1, T_2, x_{T_1}, x_{T_2}) + \delta$$

there exist integers $\tau_1 \in [T_1, T_1 + L]$, $\tau_2 \in [T_2 - L, T_2]$ *such that*

$$\rho_E(x_t, x_f) \le \epsilon, \ t = \tau_1, \ldots, \tau_2.$$

Moreover, if $\rho_E(x_{T_2}, x_{T_2}^f) \le \delta$ *then* $\tau_2 = T_2$ *and if* $\rho_E(x_{T_1}, x_{T_1}^f) \le \delta$, *then* $\tau_1 = T_1$.

Theorem 4.9 follows from STP and the following result which is proved in Section 4.7.

Proposition 4.10. *Let $L > 0$ be an integer. There exists $M_0 > 0$ such that for each $(T_1, z_1) \in \mathcal{A}_L$, $(T_2, z_2) \in \widehat{\mathcal{A}}_L$ satisfying $T_2 \geq T_1 + 2L$,*

$$U^f(T_1, T_2, z_1, z_2) \leq \sigma^f(T_1, T_2) + M_0.$$

Theorem 4.11. *Assume that f has STP, $L_0 > 0$ is an integer and that $\epsilon > 0$. Then there exist $\delta > 0$ and an integer $L > L_0$ such that for each integer T_1, each integer $T_2 \geq T_1 + 2L$ and each $(\{x_t\}_{t=T_1}^{T_2}, \{u_t\}_{t=T_1}^{T_2-1}) \in X(T_1, T_2)$ which satisfies*

$$(T_1, x_{T_1}) \in \mathcal{A}_{L_0}, \ \sum_{t=T_1}^{T_2-1} f(t, x_t, u_t) \leq \sigma^f(T_1, T_2, x_{T_1}) + \delta$$

there exist integers $\tau_1 \in [T_1, T_1 + L]$, $\tau_2 \in [T_2 - L, T_2]$ such that

$$\rho_E(x_t, x_t^f) \leq \epsilon, \ t = \tau_1, \ldots, \tau_2.$$

Moreover, if $\rho_E(x_{T_2}, x_{T_2}^f) \leq \delta$ then $\tau_2 = T_2$ and if $\rho_E(x_{T_1}, x_{T_1}^f) \leq \delta$, then $\tau_1 = T_1$.

Theorem 4.11 follows from STP and the next result which is proved in Section 4.7.

Proposition 4.12. *Let $L_0 > 0$ be an integer. Then there exists $M > 0$ such that for each $(T_1, z_1) \in \mathcal{A}_{L_0}$, each integer $T_2 \geq T_1 + L_0$,*

$$\sigma^f(T_1, T_2, z_1) \leq \sigma^f(T_1, T_2) + M.$$

We say that f possesses the weak turnpike property (or WTP for short) if for each $\epsilon > 0$ and each $M > 0$, there exist natural numbers Q, l such that for each integer T_1, each integer $T_2 \geq T_1 + lQ$, and each $(\{x_t\}_{t=T_1}^{T_2}, \{u_t\}_{t=T_1}^{T_2-1}) \in X(T_1, T_2)$ which satisfies

$$\sum_{t=T_1}^{T_2-1} f(t, x_t, u_t) \leq \sum_{t=T_1}^{T_2-1} f(t, x_t^f, u_t^f) + M$$

there exist finite sequences of integers $\{a_i\}_{i=1}^q$, $\{b_i\}_{i=1}^q \subset \{T_1, \ldots, T_2\}$ such that an integer $q \leq Q$,

$$0 \leq b_i - a_i \leq l, \ i = 1, \ldots, q,$$

$$b_i \leq a_{i+1} \text{ for all integers } i \text{ satisfying } 1 \leq i < q,$$

$$\rho_E(x_t, x_t^f) \leq \epsilon \text{ for all integers } t \in [T_1, T_2] \setminus \cup_{i=1}^q [a_i, b_i].$$

The next result is proved in Section 4.12.

Theorem 4.13. *f has WTP if and only if f has (P1), (P2).*

Let T_0 be an integer. A pair $(\{x_t\}_{t=T_0}^{\infty}, \{u_t\}_{t=T_0}^{\infty}) \in X(T_0, \infty)$ is called (f)-overtaking optimal if $(\{x_t\}_{t=T_0}^{\infty}, \{u_t\}_{t=T_0}^{\infty})$ is (f)-good and for every

$$(\{y_t\}_{t=T_0}^{\infty}, \{v_t\}_{t=T_0}^{\infty}) \in X(T_0, \infty)$$

satisfying $y_{T_0} = x_{T_0}$,

$$\limsup_{T\to\infty}[\sum_{t=T_0}^{T-1} f(t, x_t, u_t) - \sum_{t=T_0}^{T-1} f(t, y_t, v_t)] \le 0.$$

A pair $(\{x_t\}_{t=T_0}^{\infty}, \{u_t\}_{t=T_0}^{\infty}) \in X(T_0, \infty)$ is called (f)-weakly optimal if $(\{x_t\}_{t=T_0}^{\infty}, \{u_t\}_{t=T_0}^{\infty})$ is (f)-good and for every $(\{y_t\}_{t=T_0}^{\infty}, \{v_t\}_{t=T_0}^{\infty}) \in X(T_0, \infty)$ satisfying $y_{T_0} = x_{T_0}$,

$$\liminf_{T\to\infty}[\sum_{t=T_0}^{T-1} f(t, x_t, u_t) - \sum_{t=T_0}^{T-1} f(t, y_t, v_t)] \le 0.$$

The next result is proved by the same scheme as Theorem 3.19. Its proof is omitted.

Theorem 4.14. *Assume that f has (P1), (P2), and LSC property, that S is an integer, and that $(\{x_t\}_{t=S}^{\infty}, \{u_t\}_{t=S}^{\infty}) \in X(S, \infty)$ is (f)-good. Then there exists an (f)-overtaking optimal pair $(\{x_t^*\}_{t=S}^{\infty}, \{u_t^*\}_{t=S}^{\infty}) \in X(S, \infty)$ such that $x_S^* = x_S$.*

The following theorem is proved by the same scheme as Theorem 3.20. Its proof is omitted.

Theorem 4.15. *Assume that f has (P1), (P2), and LSC property, that S is an integer, and that $(\{\tilde{x}_t\}_{t=S}^{\infty}, \{\tilde{u}_t\}_{t=S}^{\infty}) \in X(S, \infty)$ is (f)-good and that $(\{x_t^*\}_{t=S}^{\infty}, \{u_t^*\}_{t=S}^{\infty}) \in X(S, \infty)$ satisfies $x_S^* = \tilde{x}_S$. Then the following conditions are equivalent:*

(i) $(\{x_t^*\}_{t=S}^{\infty}, \{u_t^*\}_{t=S}^{\infty})$ *is (f)-overtaking optimal;*
(ii) $(\{x_t^*\}_{t=S}^{\infty}, \{u_t^*\}_{t=S}^{\infty})$ *is (f)-weakly optimal;*
(iii) $(\{x_t^*\}_{t=S}^{\infty}, \{u_t^*\}_{t=S}^{\infty})$ *is (f)-minimal and (f)-good;*
(iv) $(\{x_t^*\}_{t=S}^{\infty}, \{u_t^*\}_{t=S}^{\infty})$ *is (f)-minimal and satisfies* $\lim_{t\to\infty} \rho_E(x_t^*, x_t^f) = 0$,
(v) $(\{x_t^*\}_{t=S}^{\infty}, \{u_t^*\}_{t=S}^{\infty})$ *is (f)-minimal and satisfies*

$$\liminf_{t\to\infty} \rho_E(x_t^*, x_t^f) = 0.$$

4.4 Perturbed Problems

In this section we suppose that the following assumption holds.

(A3) For each $\epsilon > 0$, there exists $\delta > 0$ such that for each $(T, z) \in \mathcal{A}$ satisfying $\rho_E(z, x_T^f) \le \delta$, there exist integers $\tau_1, \tau_2 \in (0, b_f]$,

$$(\{x_t^{(1)}\}_{t=T}^{T+\tau_1}, \{u_t^{(1)}\}_{t=T}^{T+\tau_1-1}) \in X(T, T+\tau_1),$$

$$(\{x_t^{(2)}\}_{t=T-\tau_2}^{T}, \{u_t^{(2)}\}_{t=T-\tau_2}^{T-1}) \in X(T-\tau_2, T)$$

such that

$$x_T^{(1)} = z,\ x_{T+\tau_1}^{(1)} = x_{T+\tau_1}^f,$$

$$\sum_{t=T}^{T+\tau_1-1} f(t, x_t^{(1)}, u_t^{(1)}) \le \sum_{t=T}^{T+\tau_1-1} f(t, x_t^f, u_t^f) + \epsilon,$$

$$\rho_E(x_t^{(1)}, x_t^f) \le \epsilon,\ t = T, \dots, T+\tau_1,$$

$$x_T^{(2)} = z,\ x_{T-\tau_2}^{(2)} = x_{T-\tau_2}^f,$$

$$\sum_{t=T-\tau_2}^{T-1} f(t, x_t^{(2)}, u_t^{(2)}) \le \sum_{t=T-\tau_2}^{T-1} f(t, x_t^f, u_t^f) + \epsilon,$$

$$\rho_E(x_t^{(2)}, x_t^f) \le \epsilon,\ t = T-\tau_2, \dots, T.$$

Clearly, (A3) implies (A2). Assume that f has LSC property and that $\phi : E \times E \to [0, 1]$ is a continuous function satisfying $\phi(x, x) = 0$ for all $x \in E$ and such that the following property holds:

(i) for each $\epsilon > 0$, there exists $\delta > 0$ such that if $x, y \in E$ and $\phi(x, y) \le \delta$, then $\rho_E(x, y) \le \epsilon$.

For each $r \in (0, 1)$ set

$$f_r(t, x, u) = f(t, x, u) + r\phi(x, x_t^f),\ (t, x, u) \in \mathcal{M}. \tag{4.13}$$

Clearly, for any $r \in (0, 1)$, (4.4), (A1), (A3), LSC property hold for f_r with $(x_t^{f_r}, u_t^{f_r}) = (x_t^f, u_t^f)$, $t = 0, 1, \dots$. The next result is proved in Section 4.13.

Theorem 4.16. *Let $r \in (0, 1)$ and $(\{x_t^f\}_{t=-\infty}^{\infty}, \{u_t^f\}_{t=-\infty}^{\infty})$ be (f)-minimal. Then f_r has STP.*

4.5 Auxiliary Results for Theorems 4.1 and 4.2

(A2) easily implies the following result.

Proposition 4.17. *Let $\gamma > 0$. Then there exists $\delta > 0$ such that if $(T, z_1) \in \mathcal{A}$, $(T + 2b_f, z_2) \in \mathcal{A}$ satisfy*

$$\rho_E(z_1, x_T^f) \le \delta, \ \rho_E(z_2, x_{T+2b_f}^f) \le \delta,$$

then there exists

$$(\{x_t\}_{t=T}^{T+2b_f}, \{u_t\}_{t=T}^{T+2b_f-1}) \in X(T, T + 2b_f)$$

such that

$$x_T = z_1, \ x_{T+2b_f} = z_2, \ \sum_{t=T}^{T+2b_f-1} f(t, x_t, u_t) \le \sum_{t=T}^{T+2b_f-1} f(t, x_t^f, u_t^f) + \gamma.$$

Lemma 4.18. *There exists $S > 0$ such that for each pair of integers $T_2 > T_1$ and each $(\{x_t\}_{t=T_1}^{T_2}, \{u_t\}_{t=T_1}^{T_2-1}) \in X(T_1, T_2)$,*

$$\sum_{t=T_1}^{T_2-1} f(t, x_t^f, u_t^f) \le \sum_{t=T_1}^{T_2-1} f(t, x_t, u_t) + S. \tag{4.14}$$

Proof. In view of (4.3), there exists $S_1 > 0$ such that

$$\psi(S_1) > a_0 + 1 + |\Delta_f|. \tag{4.15}$$

By (A1), there exist $S_2 > 0$ and an integer $c_0 > 0$ such that the following property holds:

(i)

$$\sum_{t=T_1}^{T_2-1} f(t, x_t^f, u_t^f) \le \sum_{t=T_1}^{T_2-1} f(t, x_t, u_t) + S_2$$

for each pair of integers T_1, $T_2 \ge T_1 + c_0$, and each $(\{x_t\}_{t=T_1}^{T_2}, \{u_t\}_{t=T_1}^{T_2-1}) \in X(T_1, T_2)$ satisfying $\rho_E(\theta_0, x_j) \le S_1$, $j = T_1, T_2$.

Fix

$$S > S_2 + 2a_0 + 2(c_0 + 1)(a_0 + 1 + |\Delta_f|) + 2. \tag{4.16}$$

Assume that $T_2 > T_1$ are integers and that $(\{x_t\}_{t=T_1}^{T_2}, \{u_t\}_{t=T_1}^{T_2-1}) \in X(T_1, T_2)$. We show that (4.14) is true. In view of (4.4), (4.10), and (4.16), if $T_2 \le T_1 + c_0$, then (4.14) holds. Therefore we may assume without loss of generality that

$$T_2 > T_1 + c_0. \tag{4.17}$$

Assume that

$$\rho_E(\theta_0, x_t) \ge S_1, \ t = T_1, \ldots, T_2 - 1. \tag{4.18}$$

By (4.4), (4.15), and (4.18), for all $t = T_1, \ldots, T_2 - 1$,

$$f(t, x_t, u_t) \ge -a_0 + \psi(\rho_E(\theta_0, x_t)) \ge -a_0 + \psi(S_1). \tag{4.19}$$

It follows from (4.10), (4.15), and (4.19),

$$\sum_{t=T_1}^{T_2-1} f(t, x_t, u_t) \ge (T_2 - T_1)(\psi(S_1) - a_0)$$

$$\ge (T_2 - T_1)|\Delta_f| \ge \sum_{t=T_1}^{T_2-1} f(t, x_t^f, u_t^f)$$

and (4.14) holds. Assume that

$$\min\{\rho_E(\theta_0, x_t) : \ t = T_1, \ldots, T_2 - 1\} < S_1. \tag{4.20}$$

Set

$$\tau_1 = \min\{t \in \{T_1, \ldots, T_2 - 1\} : \ \rho_E(\theta_0, x_t) \le S_1\}, \tag{4.21}$$

$$\tau_2 = \max\{t \in \{T_1, \ldots, T_2 - 1\} : \ \rho_E(\theta_0, x_t) \le S_1\}. \tag{4.22}$$

There are two cases:

$$\tau_2 - \tau_1 \ge c_0; \tag{4.23}$$

$$\tau_2 - \tau_1 < c_0. \tag{4.24}$$

Assume that (4.13) holds. It follows from property (i) and (4.21)–(4.23) that

$$\sum_{t=\tau_1}^{\tau_2-1} f(t, x_t^f, u_t^f) \le \sum_{t=\tau_1}^{\tau_2-1} f(t, x_t, u_t) + S_2. \tag{4.25}$$

By (4.4), (4.15), (4.21), and (4.22), for each integer

$$t \in ([T_1, \tau_1] \setminus \{\tau_1\}) \cup ([\tau_2, T_2 - 1] \setminus \{\tau_2\}),$$

we have

$$\rho_E(\theta_0, x_t) > S_1, \ f(t, x_t, u_t) \geq -a_0 + \psi(S_1) > \Delta_f. \tag{4.26}$$

It follows from (4.20) and (4.26) that

$$\sum\{f(t, x_t, u_t) : t \in (\{T_1, \dots, \tau_1\} \setminus \{\tau_1\}) \cup (\{\tau_2, \dots, T_2 - 1\})\} + a_0 + \Delta_f$$
$$\geq \sum\{f(t, x_t^f, u_t^f) : t \in (\{T_1, \dots, \tau_1\} \setminus \{\tau_1\}) \cup (\{\tau_2, \dots, T_2 - 1\})\}. \tag{4.27}$$

In view of (4.16), (4.25), and (4.27),

$$\sum_{t=T_1}^{T_2-1} f(t, x_t^f, u_t^f) \leq \sum_{t=T_1}^{T_2-1} f(t, x_t, u_t) + S_2 + \Delta_f + a_0 \leq \sum_{t=T_1}^{T_2-1} f(t, x_t, u_t) + S.$$

Assume that (4.24) holds. By (4.4), (4.15), (4.21), and (4.22), for all $t \in (\{T_1, \dots, \tau_1\} \setminus \{\tau_1\}) \cup (\{\tau_2, \dots, T_2 - 1\} \setminus \{\tau_2\})$, (4.26) holds. It follows from (4.4), (4.10), and (4.26) that (4.27) is true. In view of (4.4), (4.10), and (4.24),

$$\sum_{t=\tau_1}^{\tau_2-1} f(t, x_t^f, u_t^f) - \sum_{t=\tau_1}^{\tau_2-1} f(t, x_t, u_t) \leq (\tau_2 - \tau_1)(|\Delta_f| + a_0) \leq c_0(|\Delta_f| + a_0). \tag{4.28}$$

Equations (4.27) and (4.28) imply that

$$\sum_{t=T_1}^{T_2-1} f(t, x_t, u_t) + (c_0 + 1)(|\Delta_f| + a_0) \geq \sum_{t=T_1}^{T_2-1} f(t, x_t^f, u_t^f).$$

Lemma 4.18 is proved.

The next auxiliary result easily follows from (4.3) and (4.4).

Lemma 4.19. *Let $M_1 > 0$ and τ_0 be a natural number. Then there exists $M_2 > M_1$ such that for each integer T_1, each $T_2 \in (T_1, T_1 + \tau_0]$, and each $(\{x_t\}_{t=T_1}^{T_2}, \{u_t\}_{t=T_1}^{T_2-1}) \in X(T_1, T_2)$ which satisfies $\sum_{t=T_1}^{T_2-1} f(t, x_t, u_t) \leq M_1$ the inequality $\rho_E(x_t, \theta_0) \leq M_2$ holds for all $t = T_1, \dots, T_2 - 1$.*

Lemma 4.20. *Let $\Delta > 0$. Then there exists $\delta > 0$ such that for each $(T_i, z_i) \in \mathcal{A}$, $i = 1, 2$ satisfying*

$$T_2 \geq T_1 + 2b_f, \ \rho_E(z_i, x_{T_i}^f) \leq \delta, \ i = 1, 2 \tag{4.29}$$

the following inequality holds:

$$U^f(T_1, T_2, z_1, z_2) \leq \sum_{t=T_1}^{T_2-1} f(t, x_t^f, u_t^f) + \Delta.$$

Proof. Let $\delta > 0$ be as guaranteed by (A2) with $\epsilon = \Delta/4$ and let $(T_i, z_i) \in \mathcal{A}$, $i = 1, 2$ satisfy (4.29). By (A2) with $\epsilon = \Delta/4$ and (4.29), there exists $(\{y_t\}_{t=T_1}^{T_2}, \{v_t\}_{t=T_1}^{T_2-1}) \in X(T_1, T_2)$ such that

$$y_{T_1} = z_1,\ y_{T_2} = z_2,\ y_t = x_t^f,\ t = T_1 + b_f, \dots, T_2 - b_f,$$

$$v_t = u_t^f,\ t \in \{T_1 + b_f, \dots, T_2 - b_f\} \setminus \{T_2 - b_f\},$$

$$\sum_{t=T_1}^{T_1+b_f-1} f(t, y_t, v_t) \le \sum_{t=T_1}^{T_1+b_f-1} f(t, x_t^f, u_t^f) + \Delta/4,$$

$$\sum_{t=T_2-b_f}^{T_2-1} f(t, y_t, v_t) \le \sum_{t=T_2-b_f}^{T_2-1} f(t, x_t^f, u_t^f) + \Delta/4.$$

By the relations above

$$U^f(T_1, T_2, z_1, z_2) \le \sum_{t=T_1}^{T_2-1} f(t, y_t, v_t) \le \sum_{t=T_1}^{T_2-1} f(t, x_t^f, u_t^f) + \Delta/2.$$

Lemma 4.20 is proved.

4.6 Proof of Theorems 4.1 and 4.2

Proof of Theorem 4.1. Assertion 1 of Theorem 4.1 follows from Lemma 4.18. Let us prove assertion 2. Assume that there exists a sequence of natural numbers $\{T_k\}_{k=1}^{\infty}$ such that

$$T_k \to \infty \text{ as } k \to \infty,$$

$$\sum_{t=-T_k}^{T_k-1} f(t, x_t, u_t) - \sum_{t=-T_k}^{T_k-1} f(t, x_t^f, u_t^f) \to \infty \text{ as } k \to \infty. \tag{4.30}$$

Let a number $S > 0$ be as guaranteed by Assertion 1. Let $k \ge 1$ be an integer and let $\tau > T_k$ be an integer. Then

$$\sum_{t=-\tau}^{\tau-1} f(t, x_t, u_t) - \sum_{t=-\tau}^{\tau-1} f(t, x_t^f, u_t^f)$$

$$= \sum_{t=-T_k}^{T_k-1} f(t, x_t, u_t) - \sum_{t=-T_k}^{T_k-1} f(t, x_t^f, u_t^f) + \sum_{t=-\tau}^{-T_k-1} f(t, x_t, u_t) - \sum_{t=-\tau}^{-T_k-1} f(t, x_t^f, u_t^f)$$

$$+ \sum_{t=T_k}^{\tau-1} f(t, x_t, u_t) - \sum_{t=T_k}^{\tau-1} f(t, x_t^f, u_t^f)$$

$$\geq \sum_{t=-T_k}^{T_k-1} f(t, x_t, u_t) - \sum_{t=-T_k}^{T_k-1} f(t, x_t^f, u_t^f) - 2S \to \infty \text{ as } k \to \infty.$$

Assertion 2 is proved.

Assume that

$$\sup\{|\sum_{t=-T}^{T-1} f(t, x_t, u_t) - \sum_{t=-T}^{T-1} f(t, x_t^f, u_t^f)| : T = 1, 2, \dots\} < \infty.$$

By the relation above and (4.10),

$$\sup\{f(t, x_t, u_t) : t \in \mathbf{Z}\} < \infty.$$

Together with Lemma 4.19 the inequality above implies that $\{x_t : t \in \mathbf{Z}\}$ is bounded. Theorem 4.1 is proved.

Proof of Theorem 4.2. By Theorem 4.1, there exists $S_0 > 0$ such that the following property holds:

(i) for each pair of integers $T_2 > T_1$ and each $(\{x_t\}_{t=T_1}^{T_2}, \{u_t\}_{t=T_1}^{T_2-1}) \in X(T_1, T_2)$,

$$\sum_{t=T_1}^{T_2-1} f(t, x_t, u_t) + S_0 \geq \sum_{t=T_1}^{T_2-1} f(t, x_t^f, u_t^f).$$

In view of (4.3), there exists $M_1 > 0$ such that

$$\psi(M_1) > a_0 + |\Delta_f| + 3S_0 + M_0. \tag{4.31}$$

Let $T_2 > T_1$ be integers and $(\{x_t\}_{t=T_1}^{T_2}, \{u_t\}_{t=T_1}^{T_2-1}) \in X(T_1, T_2)$ satisfy

$$\sum_{t=T_1}^{T_2-1} f(t, x_t, u_t) \leq \sum_{t=T_1}^{T_2-1} f(t, x_t^f, u_t^f) + M_0. \tag{4.32}$$

We show that

$$\rho_E(\theta_0, x_t) \leq M_1,\ t = T_1, \ldots, T_2 - 1.$$

Assume the contrary. Then there exists $j_0 \in \{T_1, \ldots, T_2 - 1\}$ such that

$$\rho_E(\theta_0, x_{j_0}) > M_1. \tag{4.33}$$

Property (i), (4.4), (4.10), (4.32), and (4.33) imply that

$$\sum_{t=T_1}^{T_2-1} f(t, x_t^f, u_t^f) + M_0 \geq \sum_{t=T_1}^{T_2-1} f(t, x_t, u_t)$$

$$\geq \sum \{f(t, x_t, u_t) : t \in \{T_1, \ldots, j_0\} \setminus \{j_0\}\}$$

$$+ f(j_0, x_{j_0}, u_{j_0}) + \sum \{f(t, x_t, u_t) : t \in \{j_0, \ldots, T_2 - 1\} \setminus \{j_0\}\}$$

$$\geq \psi(M_1) - a_0 + \sum \{f(t, x_t^f, u_t^f) : t \in \{T_1, \ldots, j_0\} \setminus \{j_0\}\} - S_0$$

$$- \sum \{f(t, x_t^f, u_t^f) : t \in \{j_0, \ldots, T_2 - 1\} \setminus \{j_0\}\} - S_0$$

$$\geq \psi(M_1) - a_0 - 2S_0 - |\Delta_f| + \sum_{t=T_1}^{T_2-1} f(t, x_t^f, u_t^f),$$

$$\psi(M_1) \leq a_0 + 2S_0 + |\Delta_f| + M_0.$$

This contradicts (4.31). The contradiction we have reached proves Theorem 4.2.

4.7 Proofs of Propositions 4.10, 4.12 and Theorems 4.3–4.5

Proof of Proposition 4.10. Let $S > 0$ be as guaranteed by Theorem 4.1. Let $(T_1, z_1) \in \mathcal{A}_{L_0}$, $(T_2, z_2) \in \widehat{\mathcal{A}}_{L_0}$, $T_2 \geq T_1 + 2L_0$. There exist integers $\tau_1 \in (0, L_0]$, $\tau_2 \in (0, L_0]$ and $(\{y_t\}_{t=T_1}^{T_2}, \{v_t\}_{t=T_1}^{T_2-1}) \in X(T_1, T_2)$ such that

$$y_{T_1} = z_1,\ y_{T_2} = z_2,\ y_t = x_t^f,\ t = T_1 + \tau_1, \ldots, T_2 - \tau_2,$$

$$v_t = u_t^f,\ t \in \{T_1 + \tau_1, \ldots, T_2 - \tau_2\} \setminus \{T_2 - \tau_2\},$$

$$\sum_{t=T_1}^{T_1+\tau_1-1} f(t, y_t, v_t) \leq L_0,\ \sum_{t=T_2-\tau_2}^{T_2-1} f(t, y_t, v_t) \leq L_0.$$

In view of the relations above, (4.4) and the choice of S_0,

$$U^f(T_1, T_2, z_1, z_2) \le \sum_{t=T_1}^{T_2-1} f(t, y_t, v_t)$$

$$\le 2L + \sum\{f(t, x_t^f, u_t^f) : t \in \{T_1 + \tau_1, \ldots, T_2 - \tau_2\} \setminus \{T_2 - \tau_2\}\}$$

$$\le 2L + 2a_0 L + \sum_{t=T_1}^{T_2-1} f(t, x_t^f, u_t^f) \le 4L(1 + a_0) + \sigma^f(T_1, T_2) + S.$$

Proposition 4.10 is proved.

Theorem 4.3 follows from Theorem 4.2 and Proposition 4.10.

Proof of Proposition 4.12. Let $S > 0$ be as guaranteed by Theorem 4.1. Let $(T_1, z_1) \in \mathcal{A}_{L_0}$, $T_2 \ge T_1 + L_0$ be an integer. Clearly, there exist an integer $\tau \in (0, L_0]$ and $(\{y_t\}_{t=T_1}^{T_2}, \{v_t\}_{t=T_1}^{T_2-1}) \in X(T_1, T_2)$ such that

$$y_{T_1} = z_1,\ y_t = x_t^f,\ t = T_1 + \tau, \ldots, T_2,\ v_t = u_t^f,\ t \in \{T_1 + \tau, \ldots, T_2\} \setminus \{T_2\},$$

$$\sum_{t=T_1}^{T_1+\tau-1} f(t, y_t, v_t) \le L.$$

In view of the relations above, the choice of S and (4.32),

$$\sigma^f(T_1, T_2, z_1) \le \sum_{t=T_1}^{T_2-1} f(t, y_t, v_t)$$

$$\le L + \sum\{f(t, y_t, v_t) : t \in \{T_1 + \tau, \ldots, T_2\} \setminus \{T_2\}\}$$

$$\le L + \sum\{f(t, x_t^f, u_t^f) : t \in \{T_1 + \tau, \ldots, T_2\} \setminus \{T_2\}\}$$

$$\le L + a_0 L + \sum_{t=T_1}^{T_2-1} f(t, x_t^f, u_t^f) \le \sigma^f(T_1, T_2) + S + L(1 + a_0).$$

Proposition 4.12 is proved.

Theorem 4.4 follows from Theorem 4.2 and Proposition 4.12. Theorem 4.5 follows from Theorem 4.2 and the next result.

Proposition 4.21. *Let $L > 0$ be an integer, $T_2 > T_1$ be integers, $(T_2, z) \in \widehat{\mathcal{A}}_L$, $T_2 \geq T_1 + L$. Then*

$$\widehat{\sigma}^f(T_1, T_2, z) \leq \sum_{t=T_1}^{T_2-1} f(t, x_t^f, u_t^f) + a_0(1 + L).$$

Proof. By the definition of $\widehat{\mathcal{A}}_L$, there exist an integer $\tau \in (0, L]$ and

$$(\{y_t\}_{t=T_1}^{T_2}, \{v_t\}_{t=T_1}^{T_2-1}) \in X(T_1, T_2)$$

such that

$$y_{T_2} = z,\ y_t = x_t^f,\ t = T_1, \dots, T_2 - \tau,\ v_t = u_t^f,\ t \in \{T_1, \dots, T_2 - \tau\} \setminus \{T_2 - \tau\},$$

$$\sum_{t=T_2-\tau}^{T_2-1} f(t, y_t, v_t) \leq L.$$

In view of the relations above,

$$\widehat{\sigma}^f(T_1, T_2, z) \leq \sum_{t=T_1}^{T_2-1} f(t, y_t, v_t)$$

$$\leq L + \sum \{f(t, x_t^f, u_t^f) : t \in \{T_1, \dots, T_2 - \tau\} \setminus \{T_2 - \tau\}\}$$

$$\leq L + a_0 L + \sum_{t=T_1}^{T_2-1} f(t, x_t^f, u_t^f).$$

Proposition 4.21 is proved.

4.8 Auxiliary Results for Theorem 4.7

Lemma 4.22. *Let $\epsilon > 0$. Then there exists an integer $L > 0$ such that the following properties hold:*

(i) for each pair of integers $T_2 > T_1 \geq L$,

$$\sum_{t=T_1}^{T_2-1} f(t, x_t^f, u_t^f) \leq U^f(T_1, T_2, x_{T_1}^f, x_{T_2}^f) + \epsilon; \tag{4.34}$$

(ii) for each pair of integers $T_1 < T_2 \leq -L$, (4.34) is true.

Proof. Assume the contrary. Then there exists a sequence of closed intervals $[a_i, b_i], i = 1, 2, \dots$ such that $a_i < b_i, i = 1, 2, \dots$ are integers,

$$[a_i, b_i] \cap [a_j, b_j] = \emptyset \text{ for each pair of natural numbers } j > i, \tag{4.35}$$

$$\sum_{t=a_i}^{b_i-1} f(t, x_t, u_t) > U^f(a_i, b_i, x^f_{a_i}, x^f_{b_i}) + \epsilon,\ i = 1, 2, \dots. \tag{4.36}$$

By (4.35) and (4.36), there exists $(\{y_t\}_{t=-\infty}^{\infty}, \{v_t\}_{t=-\infty}^{\infty}) \in X(-\infty, \infty)$ such that

$$y_t = x_t^f \text{ for all integers } t \notin \cup_{p=1}^{\infty}(a_p, b_p), \tag{4.37}$$

$$v_t = u_t^f \text{ for all integers } t \notin \cup_{p=1}^{\infty}[a_p, b_p - 1], \tag{4.38}$$

and for each natural number i,

$$\sum_{t=a_i}^{b_i-1} f(t, x_t^f, u_t^f) > \sum_{t=a_i}^{b_i-1} f(t, y_t, v_t) + \epsilon. \tag{4.39}$$

Let $q \geq 1$ be an integer. Then for every integer $T > 0$ such that

$$[a_i, b_i] \subset [-T, T],\ i = 1, \dots, q,$$

it follows from (4.37)–(4.39) that

$$\sum_{t=-T}^{T-1} f(t, x_t^f, u_t^f) - \sum_{t=-T}^{T-1} f(t, y_t, v_t)$$

$$\geq \sum_{i=1}^{q}\left(\sum_{t=a_i}^{b_i-1} f(t, x_t^f, u_t^f) - \sum_{t=a_i}^{b_i-1} f(t, y_t, v_t)\right) \geq q\epsilon,$$

$$\sum_{t=-T}^{T-1} f(t, y_t, v_t) \leq \sum_{t=-T}^{T-1} f(t, x_t^f, u_t^f) - q\epsilon.$$

Since q is an arbitrary natural number, we conclude that

$$\sum_{t=-T}^{T-1} f(t, y_t, v_t) - \sum_{t=-T}^{T-1} f(t, x_t^f, u_t^f) \to -\infty \text{ as } T \to \infty.$$

This contradicts Theorem 4.1. The contradiction we have reached proves Lemma 4.22.

Lemma 4.23. *Assume that (P1) holds and that* $\epsilon > 0$. *Then there exists* $\delta > 0$ *and an integer* $T_\epsilon > 0$ *such that the following properties hold:*

(iii) *for each pair of integers* $T_2 > T_1 \geq T_\epsilon$ *and each* $(\{y_t\}_{t=T_1}^{T_2}, \{v_t\}_{t=T_1}^{T_2-1}) \in X(T_1, T_2)$ *which satisfies*

$$\rho_E(y_{T_i}, x_{T_i}^f) \leq \delta, \; i = 1, 2, \tag{4.40}$$

we have

$$\sum_{t=T_1}^{T_2-1} f(t, y_t, v_t) \geq \sum_{t=T_1}^{T_2-1} f(t, x_t^f, u_t^f) - \epsilon; \tag{4.41}$$

(iv) *for each pair of integers* $T_1 < T_2 \leq -T_\epsilon$ *and each*

$$(\{y_t\}_{t=T_1}^{T_2}, \{v_t\}_{t=T_1}^{T_2-1}) \in X(T_1, T_2)$$

which satisfies (4.40) *inequality* (4.41) *holds.*

Proof. By (A2), there exists $\delta \in (0, \epsilon/4)$ such that the following property holds:

(v) for each $(T, z) \in \mathcal{A}$ satisfying $\rho_E(z, x_T^f) \leq \delta$ there exist integers $\tau_1, \tau_2 \in (0, b_f]$,

$$(\{\tilde{x}_t^{(1)}\}_{t=T}^{T+\tau_1}, \{\tilde{u}_t^{(1)}\}_{t=T}^{T+\tau_1-1}) \in X(T, T + \tau_1),$$

$$(\{\tilde{x}_t^{(2)}\}_{t=T-\tau_2}^{T}, \{\tilde{u}_t^{(2)}\}_{t=T-\tau_2}^{T-1}) \in X(T - \tau_2, T)$$

such that

$$\tilde{x}_{T_1}^{(1)} = z, \; \tilde{x}_{T+\tau_1}^{(1)} = x_{T+\tau_1}^f,$$

$$\sum_{t=T}^{T+\tau_1-1} f(t, \tilde{x}_t^{(1)}, \tilde{u}_t^{(1)}) \leq \sum_{t=T}^{T+\tau_1-1} f(t, x_t^f, u_t^f) + \epsilon/4,$$

$$\tilde{x}_T^{(2)} = z, \; \tilde{x}_{T-\tau_2}^{(2)} = x_{T-\tau_2}^f,$$

$$\sum_{t=T-\tau_2}^{T-1} f(t, \tilde{x}_t^{(2)}, \tilde{u}_t^{(2)}) \leq \sum_{t=T-\tau_2}^{T-1} f(t, x_t^f, u_t^f) + \epsilon/4.$$

Lemma 4.22 implies that there exists an integer $L > b_f$ such that the following properties hold:

(vi) for each pair of integers $T_2 > T_1 \geq L$,

$$\sum_{t=T_1}^{T_2-1} f(t, x_t^f, u_t^f) \leq U^f(T_1, T_2, x_{T_1}^f, x_{T_2}^f) + \epsilon/4; \tag{4.42}$$

(vii) for each pair of integers $T_1 < T_2 \leq -L$, (4.42) is true.

Set

$$T_\epsilon = L + b_f. \tag{4.43}$$

Assume that $T_2 > T_1$ are integers,

$$\text{either } T_1 \geq L + b_f \text{ or } T_2 \leq -L - b_f, \tag{4.44}$$

$$(\{y_t\}_{t=T_1}^{T_2}, \{v_t\}_{t=T_1}^{T_2-1}) \in X(T_1, T_2),$$

$$\rho_E(y_{T_i}, x_{T_i}^f) \leq \delta,\ i = 1, 2. \tag{4.45}$$

We show that (4.41) holds. Assume the contrary. Then

$$\sum_{t=T_1}^{T_2-1} f(t, y_t, v_t) < \sum_{t=T_1}^{T_2-1} f(t, x_t^f, u_t^f) - \epsilon. \tag{4.46}$$

Property (v) and (4.45) imply that there exist integers $\tau_1, \tau_2 \in (0, b_f]$,

$$(\{x_t\}_{t=T_1-\tau_1}^{T_2+\tau_2}, \{u_t\}_{t=T_1-\tau_1}^{T_2+\tau_2-1}) \in X(T_1 - \tau_1, T_2 + \tau_2)$$

such that

$$x_t = y_t,\ t = T_1, \ldots, T_2,\ u_t = v_t,\ t \in \{T_1, \ldots, T_2\} \setminus \{T_2\}, \tag{4.47}$$

$$x_{T_1-\tau_1} = x_{T_1-\tau_1}^f,\ x_{T_2+\tau_2} = x_{T_2+\tau_2}^f, \tag{4.48}$$

$$\sum_{t=T_1-\tau_1}^{T_1-1} f(t, x_t, u_t) \leq \sum_{t=T_1-\tau_1}^{T_1-1} f(t, x_t^f, u_t^f) + \epsilon/4, \tag{4.49}$$

$$\sum_{t=T_2}^{T_2+\tau_2-1} f(t, x_t, u_t) \leq \sum_{t=T_2}^{T_2+\tau_2-1} f(t, x_t^f, u_t^f) + \epsilon/4. \tag{4.50}$$

By (4.46)–(4.50),

$$U^f(T_1-\tau_1, T_2+\tau_2, x^f_{T_1-\tau_1}, x^f_{T_2+\tau_2}) \le \sum_{t=T_1-\tau_1}^{T_2+\tau_2-1} f(t, x_t, u_t)$$

$$= \sum_{t=T_1-\tau_1}^{T_1-1} f(t, x_t, u_t) + \sum_{t=T_1}^{T_2-1} f(t, y_t, v_t) + \sum_{t=T_2}^{T_2+\tau_2-1} f(t, x_t, u_t)$$

$$\le \sum_{t=T_1-\tau_1}^{T_1-1} f(t, x^f_t, u^f_t) + \epsilon/4$$

$$+ \sum_{t=T_1}^{T_2-1} f(t, x^f_t, u^f_t) - \epsilon + \sum_{t=T_2}^{T_2+\tau_2-1} f(t, x^f_t, u^f_t) + \epsilon/4$$

$$= \sum_{t=T_1-\tau_1}^{T_2+\tau_2-1} f(t, x^f_t, u^f_t) - \epsilon/2. \tag{4.51}$$

On the other hand, properties (vi) and (vii), (4.42), and (4.44) imply that either $T_1 - \tau_1 \ge L$ or $T_2 + \tau_2 \le -L$,

$$\sum_{t=T_1-\tau_1}^{T_2+\tau_2-1} f(t, x^f_t, u^f_t) \le U^f(T_1-\tau_1, T_2+\tau_2, x^f_{T_1-\tau_1}, x^f_{T_2+\tau_2}) + \epsilon/4.$$

This contradicts (4.51). The contradiction we have reached proves (4.41) and Lemma 4.23 itself.

4.9 STP Implies (P1)–(P3)

Let f have STP. Then (P2) follows from Theorem 4.1 and STP. We show that (P1) holds. Assume that $(\{x_t\}_{t=-\infty}^{\infty}, \{u_t\}_{t=-\infty}^{\infty}) \in X(-\infty, \infty)$ is (f)-good. Theorem 4.1 implies that there exists $M_0 > 0$ such that

$$\rho_E(x_t, \theta_0) \le M_0 \text{ for all integers } t, \tag{4.52}$$

for each integer $T > 0$,

$$\left|\sum_{t=-T}^{T-1} f(t, x_t, u_t) - \sum_{t=-T}^{T-1} f(t, x^f_t, u^f_t)\right| \le M_0. \tag{4.53}$$

By Theorem 4.1, there exists $M_1 > 0$ such that the following property holds:

(i) for each pair of integers $T_2 > T_1$ and each $(\{y_t\}_{t=T_1}^{T_2}, \{v_t\}_{t=T_1}^{T_2-1}) \in X(T_1, T_2)$,

$$\sum_{t=T_1}^{T_2-1} f(t, y_t, v_t) + M_1 \geq \sum_{t=T_1}^{T_2-1} f(t, x_t^f, u_t^f).$$

Let $S_1 < S_2$ be integers. Choose a natural number T such that

$$-T < S_1 < S_2 < T. \tag{4.54}$$

Property (i), (4.53), and (4.54) imply that

$$\sum_{t=S_1}^{S_2-1} f(t, x_t, u_t) = -\sum_{t=-T}^{S_1-1} f(t, x_t, u_t) - \sum_{t=S_2}^{T_2-1} f(t, x_t, u_t) + \sum_{t=-T}^{T-1} f(t, x_t, u_t)$$

$$\leq \sum_{t=-T}^{T-1} f(t, x_t^f, u_t^f) + M_0 - \sum_{t=-T}^{S_1-1} f(t, x_t^f, u_t^f) + M_1$$

$$-\sum_{t=S_2}^{T-1} f(t, x_t^f, u_t^f) + M_1 \leq \sum_{t=S_1}^{S_2-1} f(t, x_t^f, u_t^f) + M_0 + 2M_1.$$

Thus for each pair of integers $S_2 > S_1$,

$$\sum_{t=S_1}^{S_2-1} f(t, x_t, u_t) \leq \sum_{t=S_1}^{S_2-1} f(t, x_t^f, u_t^f) + M_0 + 2M_1. \tag{4.55}$$

Let $\gamma > 0$. We show that there exists an integer $T_\gamma > 0$ such that the following properties hold:

(ii) for each pair of integers $S_1 < S_2 \leq -T_\gamma$,

$$\sum_{t=S_1}^{S_2-1} f(t, x_t, u_t) \leq U^f(S_1, S_2, x_{S_1}, x_{S_2}) + \gamma; \tag{4.56}$$

(iii) for each pair of integers $S_2 > S_1 \geq T_\gamma$, (4.56) is true.

Assume the contrary. Then there exists a sequence of closed intervals $[a_i, b_i]$, $i = 1, 2, \ldots$ such that $a_i < b_i$, $i = 1, 2, \ldots$ are integers,

$$[a_i, b_i] \cap [a_j, b_j] = \emptyset \text{ for each pair of natural numbers } j > i, \tag{4.57}$$

$$\sum_{t=a_i}^{b_i-1} f(t, x_t, u_t) > U^f(a_i, b_i, x^f_{a_i}, x^f_{b_i}) + \gamma, \ i = 1, 2, \dots. \tag{4.58}$$

By (4.57) and (4.58), there exists $(\{y_t\}_{t=-\infty}^{\infty}, \{v_t\}_{t=-\infty}^{\infty}) \in X(-\infty, \infty)$ such that

$$y_t = x_t \text{ for all integers } t \notin \cup_{i=1}^{\infty}(a_i, b_i), \tag{4.59}$$

$$v_t = u_t \text{ for all integers } t \notin \cup_{i=1}^{\infty}[a_i, b_i), \tag{4.60}$$

and for each natural number i,

$$\sum_{t=a_i}^{b_i-1} f(t, y_t, v_t) < \sum_{t=a_i}^{b_i-1} f(t, x_t, u_t) - \gamma. \tag{4.61}$$

Fix a natural number q. There exists a natural number T such that

$$[a_i, b_i] \subset [-T, T], \ i = 1, \dots, q. \tag{4.62}$$

it follows from (4.53) and (4.59)–(4.61) that

$$\sum_{t=-T}^{T-1} f(t, x_t, u_t) - \sum_{t=-T}^{T-1} f(t, y_t, v_t)$$

$$\geq \sum_{i=1}^{q} (\sum_{t=a_i}^{b_i-1} f(t, x_t, u_t) - \sum_{t=a_i}^{b_i-1} f(t, y_t, v_t)) \geq \gamma q,$$

$$\sum_{t=-T}^{T-1} f(t, y_t, v_t) \leq \sum_{t=-T}^{T-1} f(t, x_t, u_t) - q\gamma \leq \sum_{t=-T}^{T-1} f(t, x^f_t, u^f_t) + M_0 - \gamma q.$$

Thus for all natural numbers T satisfying (4.62),

$$\sum_{t=-T}^{T-1} f(t, y_t, v_t) - \sum_{t=-T}^{T-1} f(t, x^f_t, u^f_t) \leq M_0 - \gamma q.$$

Since q is an arbitrary natural number, we conclude that

$$\lim_{T\to\infty} [\sum_{t=-T}^{T-1} f(t, y_t, v_t) - \sum_{t=-T}^{T-1} f(t, x^f_t, u^f_t)] = -\infty.$$

This contradicts Theorem 4.1. The contradiction we have reached proves that for each $\gamma > 0$ there exists an integer $T_\gamma > 0$ such that properties (ii) and (iii) hold.

Let $\epsilon > 0$. By STP, there exist $\delta > 0$ and an integer $L > 0$ such that the following property holds:

(iv) for each pair of integers T_1, $T_2 \geq T_1 + 2L$ and each

$$(\{y_t\}_{t=T_1}^{T_2}, \{v_t\}_{t=T_1}^{T_2-1}) \in X(T_1, T_2)$$

which satisfies

$$\sum_{t=T_1}^{T_2-1} f(t, y_t, v_t) \leq \min\{\sigma^f(T_1, T_2) + 2M_0 + 3M_1, U^f(T_1, T_2, x_{T_1}, x_{T_2}) + \delta\}$$

we have

$$\rho_E(y_t, x_t^f) \leq \epsilon, \ t = T_1 + L, \dots, T_2 - L.$$

Let a natural number T_δ be as guaranteed by properties (ii), (iii) with $\gamma = \delta$ and let $k \geq 1$ be an integer. By properties (ii), (iii) with $\gamma = \delta$, property (i), and (4.55),

$$\sum_{t=T_\delta}^{T_\delta+2L+k-1} f(t, x_t, u_t) \leq U^f(T_\delta, T_\delta + 2L + k, x_{T_\delta}, x_{T_\delta+2L+k}) + \delta,$$

$$\sum_{t=T_\delta}^{T_\delta+2L+k-1} f(t, x_t, u_t) \leq \sum_{t=T_\delta}^{T_\delta+2L+k-1} f(t, x_t^f, u_t^f) + M_0 + 2M_1$$

$$\leq \sigma^f(T_\delta, T_\delta + 2L + k) + M_0 + 3M_1,$$

$$\sum_{t=-T_\delta-2L-k}^{-T_\delta-1} f(t, x_t, u_t) \leq U^f(-T_\delta - 2L - k, -T_\delta, x_{-T_\delta-2L-k}, x_{-T_\delta}) + \delta,$$

$$\sum_{t=-T_\delta-2L-k}^{-T_\delta-1} f(t, x_t, u_t) \leq \sum_{t=-T_\delta-2L-k}^{-T_\delta-1} f(t, x_t^f, u_t^f) + M_0 + 2M_1$$

$$\leq \sigma^f(-T_\delta - 2L - k, -T_\delta) + M_0 + 3M_1.$$

The relations above and property (iv) imply that

$$\rho_E(x_t, x_t^f) \leq \epsilon \text{ for all integers } t \in [T_\delta + L, T_\delta + L + k] \cup [-T_\delta - L - k, -T_\delta - L].$$

Since k is an arbitrary natural number, we conclude that

$$\rho_E(x_t, x_t^f) \le \epsilon \text{ for all integers } t \in [T_\delta + L, \infty) \cup (-\infty, -T_\delta - L].$$

Since ϵ is any positive number, (P1) holds.

We show that (P3) holds. In view of LSC property, for each integer $k \ge 1$ there exists $(\{x_t^{(k)}\}_{t=-k}^k, \{u_t^{(k)}\}_{t=-k}^{k-1}) \in X(-k, k)$ such that

$$x_{-k}^{(k)} = x_{-k}^f,\ x_k^{(k)} = x_k^f,\ \sum_{t=-k}^{k-1} f(t, x_t^{(k)}, u_t^{(k)}) = U^f(-k, k, x_{-k}^f, x_k^f). \tag{4.63}$$

It follows from (A2) that for each integer $k \ge 1$ there exists $\delta_k \in (0, 2^{-k})$ such that the following property holds:

(v) for each $(T, z) \in \mathcal{A}$ satisfying $\rho_E(z, x_T^f) \le \delta_k$ there exist integers $\tau_1, \tau_2 \in (0, b_f]$,

$$(\{y_t^{(1)}\}_{t=T_1}^{T+\tau_1}, \{v_t^{(1)}\}_{t=T}^{T+\tau_1-1}) \in X(T, T+\tau_1),$$

$$(\{y_t^{(2)}\}_{t=T-\tau_2}^{T}, \{v_t^{(2)}\}_{t=T-\tau_2}^{T-1}) \in X(T-\tau_2, T_2)$$

such that

$$y_T^{(1)} = z,\ y_{T+\tau_1}^{(1)} = x_{T+\tau_1}^f,$$

$$\sum_{t=T}^{T+\tau_1-1} f(t, y_t^{(1)}, v_t^{(1)}) \le \sum_{t=T}^{T+\tau_1-1} f(t, x_t^f u_t^f) + k^{-1},$$

$$y_T^{(2)} = z,\ y_{T-\tau_2}^{(2)} = x_{T-\tau_2}^f,$$

$$\sum_{t=T-\tau_2}^{T-1} f(t, y_t^{(2)}, v_t^{(2)}) \le \sum_{t=T-\tau_2}^{T-1} f(t, x_t^f u_t^f) + k^{-1}.$$

Theorem 4.1 implies that there exists $M_0 > 0$ such that for each pair of integers $S_2 > S_1$ and each $(\{y_t\}_{t=S_1}^{S_2}, \{v_t\}_{t=S_1}^{S_2-1}) \in X(S_1, S_2)$,

$$\sum_{t=S_1}^{S_2-1} f(t, y_t, v_t) \ge \sum_{t=S_1}^{S_2-1} f(t, x_t^f, u_t^f) - M_0. \tag{4.64}$$

By (4.63) and (4.64), for each pair of integers $S_2 > S_1$ and each natural number $k > \max\{S_2, -S_1\}$,

$$\sum_{t=S_1}^{S_2-1} f(t, x_t^{(k)}, u_t^{(k)}) = \sum_{t=-k}^{k-1} f(t, x_t^{(k)}, u_t^{(k)})$$

$$-\sum_{t=-k}^{S_1-1} f(t, x_t^{(k)}, u_t^{(k)}) - \sum_{t=S_2}^{k-1} f(t, x_t^{(k)}, u_t^{(k)})$$

$$\leq \sum_{t=-k}^{k-1} f(t, x_t^f, u_t^f) - \sum_{t=-k}^{S_1-1} f(t, x_t^f, u_t^f) - \sum_{t=S_2}^{k-1} f(t, x_t^f, u_t^f) + 2M_0$$

$$= \sum_{t=S_1}^{S_2-1} f(t, x_t^f, u_t^f) + 2M_0. \tag{4.65}$$

It follows from STP, (4.63), and (4.65) that the following property holds:

(vi) for each $\epsilon > 0$ there exists an integer $L > 0$ such that for each integer $k > L$,

$$\rho_E(x_t^{(k)}, x_t^f) \leq \epsilon, \; t = -k + L, \ldots, k - L. \tag{4.66}$$

In view of (4.65), for each integer i and each integer $k > |i| + 1$,

$$f(i, x_i^{(k)}, u_i^{(k)}) \leq f(i, x_i^f, u_i^f) + 2M_0. \tag{4.67}$$

By LSC property and (4.67), extracting subsequences and re-indexing, we obtain that there exists a strictly increasing sequence of natural numbers $\{k_p\}_{p=1}^{\infty}$ and there exists $(\{\tilde{x}_t\}_{t=-\infty}^{\infty}, \{\tilde{u}_t\}_{t=-\infty}^{\infty}) \in X(-\infty, \infty)$ such that for each integer t,

$$f(t, \tilde{x}_t, \tilde{u}_t) \leq \liminf_{p\to\infty} f(t, x_t^{(k_p)}, u_t^{(k_p)}), \tag{4.68}$$

$$x_t^{(k_p)} \to \tilde{x}_t \text{ as } p \to \infty. \tag{4.69}$$

It follows from (4.66) and (4.69) that that for all integers t,

$$\tilde{x}_t = x_t^f. \tag{4.70}$$

Let S be a natural number. By (4.65), (4.68), and (4.70),

$$\sum_{t=-S}^{S-1} f(t, \tilde{x}_t, \tilde{u}_t) \leq \liminf_{p\to\infty} \sum_{t=-S}^{S-1} f(t, x_t^{(k_p)}, u_t^{(k_p)}) \leq \sum_{t=-S}^{S-1} f(t, x_t^f, u_t^f) + 2M_0. \tag{4.71}$$

Theorem 4.1 and (4.71) imply that $(\{x_t^f\}_{t=-\infty}^{\infty}, \{\tilde{u}_t^f\}_{t=-\infty}^{\infty})$ is (f)-good. We show that it is (f)-minimal.

Assume the contrary. Then there exist $\gamma > 0$ and an integer $S_0 \geq 1$ such that

$$\sum_{t=-S_0}^{S_0-1} f(t, x_t^f, u_t^f) > U^f(-S_0, S_0, x_{-S_0}^f, x_{S_0}^f) + \gamma. \tag{4.72}$$

In view of (4.72), there exists $(\{y_t^*\}_{t=-S_0}^{S_0}, \{v_t^*\}_{t=-S_0}^{S_0-1}) \in X(-S_0, S_0)$ such that

$$\sum_{t=-S_0}^{S_0-1} f(t, x_t^f, \tilde{u}_t^f) > \sum_{t=-S_0}^{S_0-1} f(t, y_t^*, v_t^*) + \gamma, \tag{4.73}$$

$$y_{-S_0}^* = x_{-S_0}^f,\ y_{S_0}^* = x_{S_0}^f, \tag{4.74}$$

Lemma 4.23 implies that there exist $\delta_* > 0$ and an integer $T_* > 0$ such that the following properties hold:

(vii) for each pair of integers $T_2 > T_1 \geq T_*$ and each $(\{y_t\}_{t=T_1}^{T_2}, \{v_t\}_{t=T_1}^{T_2-1}) \in X(T_1, T_2)$ which satisfies $\rho_E(y_{T_i}, x_{T_i}^f) \leq \delta_*$, $i = 1, 2$ we have:

$$\sum_{t=T_1}^{T_2-1} f(t, y_t, v_t) \geq \sum_{t=T_1}^{T_2-1} f(t, x_t^f, u_t^f) - \gamma/16; \tag{4.75}$$

(viii) for each pair of integers $T_1 < T_2 \leq -T_*$ and each

$$(\{y_t\}_{t=T_1}^{T_2}, \{v_t\}_{t=T_1}^{T_2-1}) \in X(T_1, T_2)$$

which satisfies $\rho_E(y_{T_i}, x_{T_i}^f) \leq \delta_*$, $i = 1, 2$, (4.75) holds.

Choose a natural number

$$k_0 > 8\gamma^{-1} + S_0 + 8. \tag{4.76}$$

It follows from (4.68)–(4.70) that there exists a natural number q such that

$$k_q > k_0 + T_* + b_f + 4, \tag{4.77}$$

for each integer $t \in [-k_0 - T_* - b_f - 4, k_0 + T_* + b_f + 4]$,

$$\rho_E(x_t^{(k_q)}, x_t^f) \leq \min\{\delta_{k_0}, \delta_*\}, \tag{4.78}$$

$$\sum_{t=-k_0-T_*}^{k_0+T_*-1} f(t, x_t^f, \tilde{u}_t^f) \le \sum_{t=-k_0-T_*}^{k_0+T_*-1} f(t, x_t^{(k_q)}, u_t^{(k_q)}) + \gamma/16. \tag{4.79}$$

For all integers $t \in [-k_0 - T_*, k_0 + T_*] \setminus [-S_0, S_0]$ set

$$y_t^* = x_t^f, \tag{4.80}$$

and for all integers $t \in [-k_0 - T_*, k_0 + T_* - 1] \setminus [-S_0, S_0 - 1]$ set

$$v_t^* = \tilde{u}_t^f. \tag{4.81}$$

In view of (4.73), (4.74), (4.80), and (4.81), $(\{y_t^*\}_{t=-k_0-T_*}^{k_0+T_*}, \{v_t^*\}_{t=-k_0-T_*}^{k_0+T_*-1}) \in X(-k_0 - T_*, k_0 + T_*)$,

$$y^*_{-k_0-T_*} = x^f_{-k_0-T_*},\ y^*_{k_0+T_*} = x^f_{k_0+T_*}, \tag{4.82}$$

$$\sum_{t=-k_0-T_*}^{k_0+T_*-1} f(t, x_t^f, \tilde{u}_t^f) > \sum_{t=-k_0-T_*}^{k_0+T_*-1} f(t, y_t^*, v_t^*) + \gamma. \tag{4.83}$$

Property (v) and (4.78) imply that there exist integers $\tau_1, \tau_2 \in (0, b_f]$,

$$(\{y_t^{(1)}\}_{t=k_0+T_*+b_f-\tau_1}^{k_0+T_*+b_f}, \{v_t^{(1)}\}_{t=k_0+T_*+b_f-\tau_1}^{k_0+T_*+b_f-1}) \in X(k_0 + T_* + b_f - \tau_1, k_0 + T_* + b_f),$$

$$(\{y_t^{(2)}\}_{t=-k_0-T_*-b_f}^{-k_0-T_*-b_f+\tau_2}, \{v_t^{(2)}\}_{t=-k_0-T_*-b_f}^{-k_0-T_*-b_f+\tau_2-1})$$

$$\in X(-k_0 - T_* - b_f, -k_0 - T_* - b_f + \tau_2)$$

such that

$$y^{(1)}_{k_0+T_*+b_f-\tau_1} = x^f_{k_0+T_*+b_f-\tau_1},\ y^{(1)}_{k_0+T_*+b_f} = x^{k_q}_{k_0+T_*+b_f}, \tag{4.84}$$

$$\sum_{t=k_0+T_*+b_f-\tau_1}^{k_0+T_*+b_f-1} f(t, y_t^{(1)}, v_t^{(1)}) \le \sum_{t=k_0+T_*+b_f-\tau_1}^{k_0+T_*+b_f-1} f(t, x_t^f, u_t^f) + k_0^{-1}, \tag{4.85}$$

$$y^{(2)}_{-k_0-T_*-b_f} = x^{k_q}_{-k_0-T_*-b_f},\ y^{(2)}_{-k_0-T_*-b_f+\tau_2} = x^f_{-k_0-T_*-b_f+\tau_2}, \tag{4.86}$$

$$\sum_{t=-k_0-T_*-b_f}^{-k_0-T_*-b_f+\tau_2-1} f(t, y_t^{(2)}, v_t^{(2)}) \le \sum_{t=-k_0-T_*-b_f}^{-k_0-T_*-b_f+\tau_2-1} f(t, x_t^f, u_t^f) + k_0^{-1}. \tag{4.87}$$

Define

$$\widehat{y}_t = y_t^{(2)},\ \widehat{v}_t = v_t^{(2)} \text{ for all integers } t \in [-k_0 - T_* - b_f, -k_0 - T_* - b_f + \tau_2), \tag{4.88}$$

$$\widehat{y}_t = x_t^f,\ \widehat{v}_t = u_t^f \text{ for all integers } t \in [-k_0 - T_* - b_f + \tau_2, -k_0 - T_*] \setminus \{-k_0 - T_*\}, \tag{4.89}$$

$$\widehat{y}_t = y_t^*,\ \widehat{v}_t = v_t^* \text{ for all integers } t \in [-k_0 - T_*, k_0 + T_*), \tag{4.90}$$

$$\widehat{y}_t = x_t^f \text{ for all integers } t = k_0 + T_*, \dots, k_0 + T_* + b_f - \tau_1, \tag{4.91}$$

$$\widehat{v}_t = u_t^f \text{ for all integers } t \in [k_0 + T_*, k_0 + T_* + b_f - \tau_1] \setminus \{k_0 + T_* + b_f - \tau_1\}, \tag{4.92}$$

$$\widehat{y}_t = y_t^{(1)} \text{ for all integers } t = k_0 + T_* + b_f - \tau_1, \dots, k_0 + T_* + b_f, \tag{4.93}$$

$$\widehat{v}_t = v_t^{(1)} \text{ for all integers } t \in [k_0 + T_* + b_f - \tau_1, k_0 + T_* + b_f). \tag{4.94}$$

By (4.82), (4.84), (4.86), and (4.88)–(4.94),

$$(\{\widehat{y}_t\}_{t=-k_0-T_*-b_f}^{k_0+T_*+b_f}, \{\widehat{v}_t\}_{t=-k_0-T_*-b_f}^{k_0+T_*+b_f-1}) \in X(-k_0 - T_* - b_f, k_0 + T_* + b_f),$$

$$\widehat{y}_{-k_0-T_*-b_f} = x^{(k_q)}_{-k_0-T_*-b_f},\ \widehat{y}_{k_0+T_*+b_f} = x^{(k_q)}_{k_0+T_*+b_f}. \tag{4.95}$$

It follows from (4.83), (4.85), and (4.87–4.94) that

$$\sum_{t=-k_0-T_*-b_f}^{k_0+T_*+b_f-1} f(t, \widehat{y}_t, \widehat{v}_t) = \sum_{t=-k_0-T_*-b_f}^{-k_0-T_*-b_f+\tau_2-1} f(t, \widehat{y}_t, \widehat{v}_t)$$

$$+\sum\{f(t, \widehat{y}_t, \widehat{v}_t) :\ t \in \{-k_0 - T_* - b_f + \tau_2, \dots, -k_0 - T_*\} \setminus \{-k_0 - T_*\}\}$$

$$+ \sum_{t=-k_0-T_*}^{k_0+T_*-1} f(t, y_t^*, v_t^*)$$

$$+\sum\{f(t, \widehat{y}_t, \widehat{v}_t) :\ t \in \{k_0 + T_*, \dots, k_0 + T_* + b_f - \tau_1\} \setminus \{k_0 + T^* + b_f - \tau_1\}\}$$

$$+ \sum_{t=k_0+T_*+b_f-\tau_1}^{k_0+T_*+b_f-1} f(t, y_t^{(1)}, v_t^{(1)})$$

$$\leq \sum_{t=-k_0-T_*-b_f}^{-k_0-T_*-b_f+\tau_2-1} f(t, x_t^f, u_t^f) + k_0^{-1}$$

$$+\sum\{f(t,x_t^f,u_t^f):\ t\in\{-k_0-T_*-b_f+\tau_2,\dots,-k_0-T_*\}\setminus\{-k_0-T_*\}\}$$

$$+\sum_{t=-k_0-T_*}^{k_0+T_*-1} f(t,y_t^*,v_t^*)$$

$$+\sum\{f(t,x_t^f,u_t^f):\ t\in\{k_0+T_*,\dots,k_0+T_*+b_f-\tau_1\}\setminus\{k_0+T_*+b_f-\tau_1\}\}$$

$$+\sum_{t=k_0+T_*+b_f-\tau_1}^{k_0+T_*+b_f-1} f(t,x_t^f,u_t^f)+k_0^{-1}$$

$$\le\sum_{t=-k_0-T_*-b_f}^{-k_0-T_*-1} f(t,x_t^f,u_t^f)+\sum_{t=-k_0-T_*}^{k_0+T_*-1} f(t,x_t^f,\tilde{u}_t^f)-\gamma$$

$$+\sum_{t=k_0+T_*}^{k_0+T_*+b_f-1} f(t,x_t^f,u_t^f)+2k_0^{-1}. \tag{4.96}$$

In view of (4.79) and (4.96),

$$\sum_{t=-k_0-T_*-b_f}^{k_0+T_*+b_f-1} f(t,\widehat{y}_t,\widehat{v}_t)\le 2k_0^{-1}-\gamma$$

$$+\sum_{t=-k_0-T_*-b_f}^{-k_0-T_*-1} f(t,x_t^f,u_t^f)+\sum_{t=-k_0-T_*}^{k_0+T_*-1} f(t,x_t^{(k_q)},u_t^{(k_q)})$$

$$+\sum_{t=k_0+T_*}^{k_0+T_*+b_f-1} f(t,x_t^f,u_t^f)+\gamma/16. \tag{4.97}$$

Properties (vii) and (viii) and (4.75) and (4.78) imply that

$$\sum_{t=-k_0-T_*-b_f}^{-k_0-T_*-1} f(t,x_t^{(k_q)},u_t^{(k_q)})\ge\sum_{t=-k_0-T_*-b_f}^{-k_0-T_*-1} f(t,x_t^f,u_t^f)-\gamma/16, \tag{4.98}$$

$$\sum_{t=k_0+T_*}^{k_0+T_*+b_f-1} f(t,x_t^{(k_q)},u_t^{(k_q)})\ge\sum_{t=k_0+T_*}^{k_0+T_*+b_f-1} f(t,x_t^f,u_t^f)-\gamma/16. \tag{4.99}$$

By (4.76) and (4.97)–(4.99),

$$\sum_{t=-k_0-T_*-b_f}^{k_0+T_*+b_f-1} f(t, \widehat{y}_t, \widehat{v}_t) - \sum_{t=-k_0-T_*-b_f}^{k_0+T_*+b_f-1} f(t, x_t^{(k_q)}, u_t^{(k_q)})$$

$$\leq 2k_0^{-1} - \gamma + \gamma/16 + \gamma/8 \leq -\gamma/4.$$

This contradicts (4.63) and (4.95). The contradiction we have reached proves that $(\{x_t^f\}_{t=-\infty}^{\infty}, \{\tilde{u}_t^f\}_{t=-\infty}^{\infty})$ is (f)-minimal.

Assume that $(\{x_t\}_{t=-\infty}^{\infty}, \{u_t\}_{t=-\infty}^{\infty}) \in X(-\infty, \infty)$ is (f)-good and (f)-minimal. Then

$$\sup\{|\sum_{t=-T}^{T-1} f(t, x_t, u_t) - \sum_{t=-T}^{T-1} f(t, x_t^f, u_t^f)| : T \in \{1, 2, \dots\}\} < \infty.$$

Property (P1) implies that

$$\lim_{t\to\infty} \rho_E(x_t, x_t^f) = 0, \quad \lim_{t\to-\infty} \rho_E(x_t, x_t^f) = 0.$$

STP and the relations above imply that $x_t = x_t^f$ for all integers t. Thus (P3) holds.

4.10 (P1)–(P3) Imply STP

Lemma 4.24. *Assume that (P1) holds and that $\epsilon > 0$. Then there exists $\delta > 0$ and an integer $L > 0$ such that the following properties hold:*

(i) for each integer $T_1 \geq L$, each integer $T_2 \geq T_1 + 2b_f$, and each $(\{x_t\}_{t=T_1}^{T_2}, \{u_t\}_{t=T_1}^{T_2-1}) \in X(T_1, T_2)$ which satisfies

$$\rho_E(x_{T_i}, x_{T_i}^f) \leq \delta, \ i = 1, 2, \tag{4.100}$$

$$\sum_{t=T_1}^{T_2-1} f(t, x_t, u_t) \leq U^f(T_1, T_2, x_{T_1}, x_{T_2}) + \delta \tag{4.101}$$

the inequality

$$\rho_E(x_t, x_t^f) \leq \epsilon, \ t = T_1, \dots, T_2 \tag{4.102}$$

is true;

(ii) *for each integer* $T_2 \le -L$, *each integer* $T_1 \le T_2 - 2b_f$, *and each* $(\{x_t\}_{t=T_1}^{T_2}, \{u_t\}_{t=T_1}^{T_2-1}) \in X(T_1, T_2)$ *which satisfies* (4.100) *and* (4.101) *the relation* (4.102) *is true.*

Proof. By (A2), for each integer $k \ge 1$, there exists $\delta_k \in (0, 2^{-k})$ such that the following property holds:

(iii) for each $(T, z) \in \mathcal{A}$ satisfying $\rho_E(z, x_T^f) \le \delta_k$ there exist integers $\tau_1, \tau_2 \in (0, b_f]$,

$$(\{x_t^{(1)}\}_{t=T}^{T+\tau_1}, \{u_t^{(1)}\}_{t=T}^{T+\tau_1-1}) \in X(T, T+\tau_1),$$

$$(\{x_t^{(2)}\}_{t=T-\tau_2}^{T}, \{u_t^{(2)}\}_{t=T-\tau_2}^{T-1}) \in X(T-\tau_2, T)$$

such that

$$x_T^{(1)} = z,\ x_{T+\tau_1}^{(1)} = x_{T+\tau_1}^f,\ x_T^{(2)} = z,\ x_{T-\tau_2}^{(2)} = x_{T-\tau_2}^f,$$

$$\sum_{t=T}^{T+\tau_1-1} f(t, x_t^{(1)}, u_t^{(1)}) \le \sum_{t=T}^{T+\tau_1-1} f(t, x_t^f, u_t^f) + 2^{-k},$$

$$\sum_{t=T-\tau_2}^{T-1} f(t, x_t^{(2)}, u_t^{(2)}) \le \sum_{t=T-\tau_2}^{T-1} f(t, x_t^f, u_t^f) + 2^{-k}.$$

Assume that the lemma does not hold. Then for each integer $k \ge 1$, there exist integers $T_{k,1}$, $T_{k,2} \ge T_{k,1} + 2b_f$, and $(\{x_t^{(k)}\}_{t=T_{k,1}}^{T_{k,2}}, \{u_t^{(k)}\}_{t=T_{k,1}}^{T_{k,2}-1}) \in X(T_{k,1}, T_{k,2})$ such that

$$\rho_E(x_{T_{k,i}}^{(k)}, x_{T_{k,i}}^f) \le \delta_k,\ i = 1, 2, \tag{4.103}$$

$$\sum_{t=T_{k,1}}^{T_{k,2}-1} f(t, x_t^{(k)}, u_t^{(k)}) \le U^f(T_{k,1}, T_{k,2}, x_{T_{k,1}}^{(k)}, x_{T_{k,2}}^{(k)}) + \delta_k, \tag{4.104}$$

$$\max\{\rho_E(x_t^{(k)}, x_t^f) : t \in \{T_{k,1}, \dots, T_{k,2}\}\} > \epsilon \tag{4.105}$$

and that either for all integers $k \ge 1$,

$$T_{k,1} \ge k + b_f, \tag{4.106}$$

or for all integers $k \ge 1$,

$$T_{k,2} \le -k - b_f. \tag{4.107}$$

Extracting a subsequence and re-indexing, we may assume without loss of generality that for each integer $k \geq 1$,

$$\min\{|T_{k+1,1}|, |T_{k+1,2}|\} \geq \max\{|T_{k,1}|, |T_{k,2}|\} + 4b_f. \tag{4.108}$$

Let $k \geq 1$ be an integer. Property (iii) and (4.103) imply that there exist integers $\tau_{k,1}, \tau_{k,2} \in (0, b_f]$ and $(\{\tilde{x}_t^{(k)}\}_{t=T_{k,1}-\tau_{k,1}}^{T_{k,2}+\tau_{k,2}}, \{\tilde{u}_t^{(k)}\}_{t=T_{k,1}-\tau_{k,1}}^{T_{k,2}+\tau_{k,2}-1}) \in X(T_{k,1} - \tau_{k,1}, T_{k,2} + \tau_{k,2})$ such that

$$\tilde{x}_t^{(k)} = x_t^{(k)},\ t = T_{k,1}, \dots, T_{k,2},\ \tilde{u}_t^{(k)} = u_t^{(k)},\ t = T_{k,1}, \dots, T_{k,2} - 1, \tag{4.109}$$

$$\tilde{x}_{T_{k,1}-\tau_{k,1}}^{(k)} = x_{T_{k,1}-\tau_{k,1}}^f,\ \tilde{x}_{T_{k,2}+\tau_{k,2}}^{(k)} = x_{T_{k,2}+\tau_{k,2}}^f, \tag{4.110}$$

$$\sum_{t=T_{k,1}-\tau_{k,1}}^{T_{k,1}-1} f(t, \tilde{x}_t^{(k)}, \tilde{u}_t^{(k)}) \leq \sum_{t=T_{k,1}-\tau_{k,1}}^{T_{k,1}-1} f(t, x_t^f, u_t^f) + 2^{-k}, \tag{4.111}$$

$$\sum_{t=T_{k,2}}^{T_{k,2}+\tau_{k,2}-1} f(t, \tilde{x}_t^{(k)}, \tilde{u}_t^{(k)}) \leq \sum_{t=T_{k,2}}^{T_{k,2}+\tau_{k,2}-1} f(t, x_t^f, u_t^f) + 2^{-k}. \tag{4.112}$$

Property (iii) and (4.103) imply that there exist integers $\tau_{k,3}, \tau_{k,4} \in (0, b_f]$ and $(\{\widehat{x}_t^{(k)}\}_{t=T_{k,1}}^{T_{k,2}}, \{\widehat{u}_t^{(k)}\}_{t=T_{k,1}}^{T_{k,2}-1}) \in X(T_{k,1}, T_{k,2})$ such that

$$\widehat{x}_{T_{k,1}}^{(k)} = x_{T_{k,1}}^{(k)},\ \widehat{x}_{T_{k,2}}^{(k)} = x_{T_{k,2}}^{(k)}, \tag{4.113}$$

$$\widehat{x}_t^{(k)} = x_t^f,\ t = T_{k,1} + \tau_{k,3}, \dots, T_{k,2} - \tau_{k,4}, \tag{4.114}$$

$$\widehat{u}_t^{(k)} = u_t^f,\ t \in \{T_{k,1} + \tau_{k,3}, \dots, T_{k,2} - \tau_{k,4}\} \setminus \{T_{k,2} - \tau_{k,4}\}, \tag{4.115}$$

$$\sum_{t=T_{k,1}}^{T_{k,1}+\tau_{k,3}-1} f(t, \widehat{x}_t^{(k)}, \widehat{u}_t^{(k)}) \leq \sum_{t=T_{k,1}}^{T_{k,1}+\tau_{k,3}-1} f(t, x_t^f, u_t^f) + 2^{-k}, \tag{4.116}$$

$$\sum_{t=T_{k,2}-\tau_{k,4}}^{T_{k,2}-1} f(t, \widehat{x}_t^{(k)}, \widehat{u}_t^{(k)}) \leq \sum_{t=T_{k,2}-\tau_{k,4}}^{T_{k,2}-1} f(t, x_t^f, u_t^f) + 2^{-k}. \tag{4.117}$$

By (4.104) and (4.113)–(4.117),

$$U^f(T_{k,1}, T_{k,2}, x_{T_{k,1}}^{(k)}, x_{T_{k,2}}^{(k)}) \leq \sum_{t=T_{k,1}}^{T_{k,2}-1} f(t, x_t^{(k)}, u_t^{(k)})$$

$$\leq U^f(T_{k,1}, T_{k,2}, x^{(k)}_{T_{k,1}}, x^{(k)}_{T_{k,2}}) + \delta_k \leq \sum_{t=T_{k,1}}^{T_{k,2}-1} f(t, \widehat{x}_t^{(k)}, \widehat{u}_t^{(k)}) + \delta_k$$

$$\leq \sum_{t=T_{k,1}}^{T_{k,2}-1} f(t, x_t^f, u_t^f) + 3 \cdot 2^{-k}. \tag{4.118}$$

By (4.109)–(4.112) and (4.118),

$$\sum_{t=T_{k,1}-\tau_{k,1}}^{T_{k,2}+\tau_{k,2}-1} f(t, \tilde{x}_t^{(k)}, \tilde{u}_t^{(k)}) \leq \sum_{t=T_{k,1}-\tau_{k,1}}^{T_{k,1}-1} f(t, x_t^f, u_t^f) + \sum_{t=T_{k,1}}^{T_{k,2}-1} f(t, x_t^{(k)}, u_t^{(k)})$$

$$+ \sum_{t=T_{k,2}}^{T_{k,2}+\tau_{k,2}-1} f(t, x_t^f, u_t^f) + 2^{-k+1} \leq \sum_{t=T_{k,1}-\tau_{k,1}}^{T_{k,2}+\tau_{k,2}-1} f(t, x_t^f, u_t^f) + 2^{-k+3}. \tag{4.119}$$

By (4.108) and (4.110), there exists $(\{x_t\}_{t=-\infty}^{\infty}, \{u_t\}_{t=-\infty}^{\infty}) \in X(-\infty, \infty)$ such that for every integer $k \geq 1$,

$$x_t = \tilde{x}_t^{(k)},\ t = T_{k,1} - \tau_{k,1}, \dots, T_{k,2} + \tau_{k,2},\ u_t = \tilde{u}_t^{(k)},$$

$$t = T_{k,1} - \tau_{k,1}, \dots, T_{k,2} + \tau_{k,2} - 1, \tag{4.120}$$

$$x_t = x_t^f,\ u_t = u_t^f \text{ for all integers } t \notin \cup_{k=1}^{\infty} (T_{k,1} - \tau_{k,1}, T_{k,2} + \tau_{k,2}). \tag{4.121}$$

It follows from (4.119)–(4.121) that for each integer $k \geq 1$, in the case of (4.106),

$$\sum \{f(t, x_t, u_t) :\ t = 0, \dots, T_{k,2} + \tau_{k,2} - 1\}$$

$$- \sum \{f(t, x_t^f, u_t^f) :\ t = 0, \dots, T_{k,2} + \tau_{k,2} - 1\}$$

$$= \sum_{i=1}^{k} \Big(\sum_{t=T_{i,1}-\tau_{i,1}}^{T_{i,2}+\tau_{i,2}-1} f(t, \tilde{x}_t^{(i)}, \tilde{u}_t^{(i)}) - \sum_{t=T_{i,1}-\tau_{i,1}}^{T_{i,2}+\tau_{i,2}-1} f(t, x_t^f, u_t^f) \Big) \leq \sum_{i=1}^{k} 2^{-i+3} \leq 8 \tag{4.122}$$

and in the case of (4.107),

$$\sum \{f(t, x_t, u_t) :\ t = T_{k,1} - \tau_{k,1}, \dots, -1\}$$

$$- \sum \{f(t, x_t^f, u_t^f) :\ t = T_{k,1} - \tau_{k,1}, \dots, -1\}$$

$$= \sum_{i=1}^{k} \Big(\sum_{t=T_{i,1}-\tau_{i,1}}^{T_{i,2}+\tau_{i,2}-1} f(t, \tilde{x}_t^{(i)}, \tilde{u}_t^{(i)}) - \sum_{t=T_{i,1}-\tau_{i,1}}^{T_{i,2}+\tau_{i,2}-1} f(t, x_t^f, u_t^f) \Big) \le \sum_{i=1}^{k} 2^{-i+3} \le 8. \tag{4.123}$$

Theorem 4.1 and (4.120)–(4.123) imply that in the both cases

$$(\{x_t\}_{t=-\infty}^{\infty}, \{u_t\}_{t=-\infty}^{\infty})$$

is (f)-good. In view of (P1), $\lim_{t\to\infty} \rho_E(x_t, x_t^f) = 0$ and $\lim_{t\to-\infty} \rho_E(x_t, x_t^f) = 0$. On the other hand in view of (4.105), (4.109), (4.120), and (4.121), at least one of the following relations holds: $\limsup_{t\to\infty} \rho_E(x_t, x_t^f) \ge \epsilon$; $\limsup_{t\to-\infty} \rho_E(x_t, x_t^f) \ge \epsilon$. The contradiction we have reached completes the proof of Lemma 4.24.

Lemma 4.25. *Assume that (P1), (P2), (P3) hold,*

$$\tilde{u}_t = u_t^f \text{ for all integers } t \tag{4.124}$$

and that $\epsilon \in (0, 1)$. Then there exists $\delta > 0$ such that for each pair of integers T_1, $T_2 \ge T_1 + 3b_f$ and each $(\{x_t\}_{t=T_1}^{T_2}, \{u_t\}_{t=T_1}^{T_2-1}) \in X(T_1, T_2)$ satisfying

$$\rho_E(x_{T_i}, x_{T_i}^f) \le \delta,\ i = 1, 2$$

$$\sum_{t=T_1}^{T_2-1} f(t, x_t, u_t) \le U^f(T_1, T_2, x_{T_1}, x_{T_2}) + \delta$$

the inequality $\rho_E(x_t, x_t^f) \le \epsilon$ for all $t = T_1, \ldots, T_2$.

Proof. Lemma 4.24 and (A2) imply that there exist $\delta_0 \in (0, \epsilon/4)$ and an integer $L_0 > 0$ such that:

(iv) (A2) holds with $\epsilon = 1$ and $\delta = \delta_0$;
(v) for each pair of integers T_1, $T_2 \ge T_1 + 2b_f$ such that either $T_1 \ge L_0$ or $T_2 \le -L_0$ and each $(\{x_t\}_{t=T_1}^{T_2}, \{u_t\}_{t=T_1}^{T_2-1}) \in X(T_1, T_2)$ satisfying

$$\rho_E(x_{T_i}, x_{T_i}^f) \le \delta_0,\ i = 1, 2$$

$$\sum_{t=T_1}^{T_2-1} f(t, x_t, u_t) \le U^f(T_1, T_2, x_{T_1}, x_{T_2}) + \delta_0$$

the inequality $\rho_E(x_t, x_t^f) \le \epsilon$ holds for all $t = T_1, \ldots, T_2$.

Theorem 4.1 implies that there exists $S_0 > 0$ such that for each pair of integers $T_2 > T_1$ and each $(\{x_t\}_{t=T_1}^{T_2}, \{u_t\}_{t=T_1}^{T_2-1}) \in X(T_1, T_2)$,

$$\sum_{t=T_1}^{T_2-1} f(t, x_t, u_t) + S_0 \geq \sum_{t=T_1}^{T_2-1} f(t, x_t^f, u_t^f). \tag{4.125}$$

It follows from (P2) that there exist $\delta_1 \in (0, \delta_0)$ and a natural number L_1 such that the following property holds:

(vi) for each integer T and each $(\{x_t\}_{t=T}^{T+L_1}, \{u_t\}_{t=T}^{T+L_1-1}) \in X(T, T+L_1)$ satisfying

$$\sum_{t=T}^{T+L_1-1} f(t, x_t, u_t) \leq \min\{ \sum_{t=T}^{T+L_1-1} f(t, x_t^f, u_t^f) + 2S_0 + 3,$$

$$U^f(T, T + L_1, x_T, x_{T+L_1}) + \delta_1\}$$

we have

$$\min\{\rho_E(x_t, x_t^f) : t = T, \ldots, T + L_1\} \leq \delta_0.$$

By (A2), there exists a sequence $\{\delta_i\}_{i=1}^{\infty} \subset (0, 1)$ such that

$$\delta_i < 2^{-1}\delta_{i-1}, \; i = 2, 3, \ldots. \tag{4.126}$$

and that for each integer $i \geq 1$ the following property holds:

(vii) for each $(T, z) \in \mathcal{A}$ satisfying $\rho_E(z, x_T^f) \leq \delta_i$ there exist

$$(\{y_t^{(1)}\}_{t=T}^{T+b_f}, \{v_t^{(1)}\}_{t=T}^{T+b_f-1}) \in X(T, T + b_f),$$

$$(\{y_t^{(2)}\}_{t=T-b_f}^{T}, \{v_t^{(2)}\}_{t=T-b_f}^{T-1}) \in X(T - b_f, T)$$

such that

$$y_T^{(1)} = z, \; y_{T+b_f}^{(1)} = x_{T+b_f}^f, \; y_T^{(2)} = z, \; y_{T-b_f}^{(2)} = x_{T-b_f}^f,$$

$$\sum_{t=T}^{T+b_f-1} f(t, y_t^{(1)}, v_t^{(1)}) \leq \sum_{t=T}^{T+b_f-1} f(t, x_t^f, u_t^f) + 1/i,$$

$$\sum_{t=T-b_f}^{T-1} f(t, y_t^{(2)}, v_t^{(2)}) \leq \sum_{t=T-b_f}^{T-1} f(t, x_t^f, u_t^f) + 1/i.$$

Assume that the lemma does not hold. Then for each integer $i \geq 1$ there exist integers $T_{i,1}$, $T_{i,2} \geq T_{i,1} + 3b_f$ and $(\{x_t^{(i)}\}_{t=T_{i,1}}^{T_{i,2}}, \{u_t^{(i)}\}_{t=T_{i,1}}^{T_{i,2}-1}) \in X(T_{i,1}, T_{i,2})$ such that

$$\rho_E(x_{T_{i,j}}^{(i)}, x_{T_{i,j}}^f) \leq \delta_i, \; j = 1, 2, \tag{4.127}$$

$$\sum_{t=T_{i,1}}^{T_{i,2}-1} f(t, x_t^{(i)}, u_t^{(i)}) \leq U^f(T_{i,1}, T_{i,2}, x_{T_{i,1}}^{(i)}, x_{T_{i,2}}^{(i)}) + \delta_i \tag{4.128}$$

and $t_i \in \{T_{i,1}, \ldots, T_{i,2}\}$ for which

$$\rho_E(x_{t_i}^{(i)}, x_{t_i}^f) > \epsilon. \tag{4.129}$$

Let i be a natural number. Property (v) and (4.126)–(4.129) imply that

$$T_{i,1} < L_0, \; T_{i,2} > -L_0. \tag{4.130}$$

We show that

$$-2b_f - 1 - L_1 - L_0 \leq t_i \leq 2b_f + 1 + L_1 + L_0. \tag{4.131}$$

Property (iv), (A2) with $\epsilon = 1$, $\delta = \delta_0$, (4.126) and (4.127) imply that there exists $(\{y_t^{(i)}\}_{t=T_{i,1}}^{T_{i,2}}, \{v_t^{(i)}\}_{t=T_{i,1}}^{T_{i,2}-1}) \in X(T_{i,1}, T_{i,2})$ such that

$$y_{T_{i,1}}^{(i)} = x_{T_{i,1}}^{(i)}, \; y_{T_{i,2}}^{(i)} = x_{T_{i,2}}^{(i)}, \tag{4.132}$$

$$y_t^{(i)} = x_t^f, \; t \in \{T_{i,1} + b_f, \ldots, T_{i,2} - b_f\},$$

$$v_t^{(i)} = u_t^f, \; t \in \{T_{i,1} + b_f, \ldots, T_{i,2} - b_f - 1\}, \tag{4.133}$$

$$\sum_{t=T_{i,1}}^{T_{i,1}+b_f-1} f(t, y_t^{(i)}, v_t^{(i)}) \leq \sum_{t=T_{i,1}}^{T_{i,1}+b_f-1} f(t, x_t^f, u_t^f) + 1, \tag{4.134}$$

$$\sum_{t=T_{i,2}-b_f}^{T_{i,2}-1} f(t, y_t^{(i)}, v_t^{(i)}) \leq \sum_{t=T_{i,2}-b_f}^{T_{i,2}-1} f(t, x_t^f, u_t^f) + 1. \tag{4.135}$$

It follows from (4.128) and (4.132)–(4.135) that

$$\sum_{t=T_{i,1}}^{T_{i,2}-1} f(t, x_t^{(i)}, u_t^{(i)}) \le \sum_{t=T_{i,1}}^{T_{i,2}-1} f(t, y_t^{(i)}, v_t^{(i)}) + 1$$

$$\le \sum_{t=T_{i,1}}^{T_{i,2}-1} f(t, x_t^f, u_t^f) + 3. \tag{4.136}$$

In view of (4.125), (4.136), for each pair of integers $S_1, S_2 \in [T_{i,1}, T_{i,2}]$ satisfying $S_1 < S_2$,

$$\sum_{t=S_1}^{S_2-1} f(t, x_t^{(i)}, u_t^{(i)}) = \sum_{t=T_{i,1}}^{T_{i,2}-1} f(t, x_t^{(i)}, u_t^{(i)})$$

$$-\sum\{f(t, x_t^{(i)}, u_t^{(i)}) : t \in \{T_{i,1}, \dots, S_1\} \setminus \{S_1\}\}$$

$$-\sum\{f(t, x_t^{(i)}, u_t^{(i)}) : t \in \{S_2, \dots, T_{i,2}\} \setminus \{T_{i,2}\}\}$$

$$\le \sum_{t=T_{i,1}}^{T_{i,2}-1} f(t, x_t^f, u_t^f) + 3 - \sum\{f(t, x_t^f, u_t^f) : t \in \{T_{i,1}, \dots, S_1\} \setminus \{S_1\}\} + S_0$$

$$-\sum\{f(t, x_t^f, u_t^f) : t \in \{S_2, \dots, T_{i,2}\} \setminus \{T_{i,2}\}\} + S_0$$

$$\le \sum_{t=S_1}^{S_2-1} f(t, x_t^f, u_t^f) + 3 + 2S_0. \tag{4.137}$$

Assume that

$$t_i > 2b_f + 1 + L_1 + L_0. \tag{4.138}$$

In view of (4.138),

$$[t_i - 2b_f - 1 - L_1, t_i - 1 - 2b_f] \subset [L_0, \infty). \tag{4.139}$$

Consider the pair $(\{x_t^{(i)}\}_{t=t_i-L_1-2b_f-1}^{t_i-2b_f-1}, \{u_t^{(i)}\}_{t=t_i-2b_f-L_1-1}^{t_i-2b_f-2}) \in X(t_i - L_1 - 2b_f - 1, t_i - 2b_f - 1)$. Property (vi), (4.126), (4.128), and (4.137) imply that there exists

$$\tilde{t} \in \{t_i - L_1 - 2b_f - 1, \dots, t_i - 2b_f - 1\} \tag{4.140}$$

such that

$$\rho_E(x_{\tilde{t}}^{(i)}, x_{\tilde{t}}^f) \le \delta_0. \tag{4.141}$$

By property (v), (4.126)–(4.128) and (4.139)–(4.141),

$$\rho_E(x_t^{(i)}, x_t^f) \le \epsilon, \; t = \tilde{t}, \ldots, T_{i,2}$$

and in particular,

$$\rho_E(x_{t_i}^{(i)}, x_{t_i}^f) \le \epsilon. \tag{4.142}$$

This contradicts (4.129). The contradiction we have reached proves that

$$t_i \le 2b_f + 1 + L_1 + L_0. \tag{4.143}$$

Assume that

$$t_i < -2b_f - 1 - L_1 - L_0. \tag{4.144}$$

In view of (4.144),

$$[t_i + 2b_f + 1, t_i + 1 + 2b_f + L_1] \subset (-\infty, -L_0]. \tag{4.145}$$

Consider the pair $(\{x_t^{(i)}\}_{t=t_i+1+2b_f}^{t_i+2b_f+L_1+1}, \{u_t^{(i)}\}_{t=t_i+1+2b_f}^{t_i+2b_f+L_1}) \in X(t_i+1+2b_f, t_i+2b_f+1+L_1)$. Property (vi), (4.126), (4.128), (4.130), (4.137), and (4.145) imply that there exists

$$\tilde{t} \in \{t_i + 1 + 2b_f, \ldots, t_i + 2b_f + 1 + L_1\} \tag{4.146}$$

such that

$$\rho_E(x_{\tilde{t}}^{(i)}, x_{\tilde{t}}^f) \le \delta_0. \tag{4.147}$$

By property (v), (4.126)–(4.128), (4.130), (4.145), and (4.147),

$$\rho_E(x_t^{(i)}, x_t^f) \le \epsilon, \; t = T_{i,1} \ldots, \tilde{t},$$

and in particular,

$$\rho_E(x_{t_i}^{(i)}, x_{t_i}^f) \le \epsilon.$$

This contradicts (4.129). The contradiction we have reached proves that

$$t_i \ge -2b_f - 1 - L_1 - L_0.$$

Together with (4.143) this implies (4.131). In view of (4.10) and (4.137), there exists $M_1 > 0$ such that

$$f(t, x_t^{(i)}, u_t^{(i)}) \le M_1 \text{ for each integer } i \ge 1 \text{ each } t \in \{T_{i,1}, \dots, T_{i,2} - 1\}. \tag{4.148}$$

It follows from (4.3), (4.4), (4.9), (4.127), and (4.148) that there exists $M_2 > 0$ such that

$$\rho_E(x_t^{(i)}, \theta_0) \le M_2, \ i = 1, 2, \dots, \ t \in \{T_{i,1}, \dots, T_{i,2}\}. \tag{4.149}$$

Extracting a subsequence and re-indexing, we may assume without loss of generality that

$$\text{either } T_{i,1} = T_{1,1} \text{ for all integers } i \ge 1 \text{ or } \lim_{i\to\infty} T_{i,1} = -\infty, \tag{4.150}$$

$$\text{either } T_{i,2} = T_{1,2} \text{ for all integers } i \ge 1 \text{ or } \lim_{i\to\infty} T_{i,2} = \infty, \tag{4.151}$$

$$t_i = t_1 \text{ for all integers } i \ge 1. \tag{4.152}$$

There are four cases:

(a) $T_{i,1} = T_{1,1}$, $T_{i,2} = T_{1,2}$ for all integers $i \ge 1$;
(b) $T_{i,1} = T_{1,1}$ for all integers $i \ge 1$ and $\{T_{i,2}\}_{i=1}^{\infty}$ is a strictly increasing sequence;
(c) $T_{i,2} = T_{1,2}$ for all integers $i \ge 1$ and $\{T_{i,1}\}_{i=1}^{\infty}$ is a strictly decreasing sequence;
(d) $\{T_{i,1}\}_{i=1}^{\infty}$ is a strictly decreasing sequence and $\{T_{i,2}\}_{i=1}^{\infty}$ is a strictly increasing sequence.

Assume that the case (a) holds. In view of (4.48) and LSC property, extracting a subsequence and re-indexing we may assume without loss of generality that there exists $(\{\widehat{x}_t\}_{t=T_{1,1}}^{T_{1,2}}, \{\widehat{u}_t\}_{t=T_{1,1}}^{T_{1,2}-1}) \in X(T_{1,1}, T_{1,2})$ such that

$$x_t^{(i)} \to \widehat{x}_t \text{ as } i \to \infty \text{ for all } t \in \{T_{1,1}, \dots, T_{1,2}\} \tag{4.153}$$

$$f(t, \widehat{x}_t, \widehat{u}_t) \le \liminf_{i\to\infty} f(t, x_t^{(i)}, u_t^{(i)}), \ t \in \{T_{1,1}, \dots, T_{1,2} - 1\}. \tag{4.154}$$

It follows from (4.126), (4.127), (4.129), and (4.153) that

$$\widehat{x}_{T_{1,1}} = x^f_{T_{1,1}}, \ \widehat{x}_{T_{1,2}} = x^f_{T_{1,2}}, \ \rho_E(\widehat{x}_{t_1}, x^f_{t_1}) \ge \epsilon. \tag{4.155}$$

Property (P3), (4.124), and (4.155) imply that there exists $\gamma > 0$ such that

$$\sum_{t=T_{1,1}}^{T_{1,2}-1} f(t, x_t^f, u_t^f) < \sum_{t=T_{1,1}}^{T_{1,2}-1} f(t, \widehat{x}_t, \widehat{u}_t) - \gamma. \tag{4.156}$$

Lemma 4.20, (4.127), (4.128), (4.154), and (4.156) imply that

$$\sum_{t=T_{1,1}}^{T_{1,2}-1} f(t, x_t^f, u_t^f) + \gamma \le \liminf_{i\to\infty} \sum_{t=T_{1,1}}^{T_{1,2}-1} f(t, x_t^{(i)}, u_t^{(i)})$$

$$\le \liminf_{i\to\infty} U^f(T_{1,1}, T_{1,2}, x_{T_{1,1}}^{(i)}, x_{T_{1,2}}^{(i)}) \le \sum_{t=T_{1,1}}^{T_{1,2}-1} f(t, x_t^f, u_t^f),$$

a contradiction. The contradiction we have reached proves that the case (a) does not hold.

Assume that the case (b) holds. In view of (4.48) and LSC property, extracting a subsequence, using the diagonalization process and re-indexing, we may assume without loss of generality that there exists

$$(\{\widehat{x}_t\}_{t=T_{1,1}}^{\infty}, \{\widehat{u}_t\}_{t=T_{1,1}}^{\infty}) \in X(T_{1,1}, \infty)$$

such that for every integer $t \ge T_{1,1}$,

$$\widehat{x}_t = \lim_{i\to\infty} x_t^{(i)}, \tag{4.157}$$

$$f(t, \widehat{x}_t, \widehat{u}_t) \le \liminf_{i\to\infty} f(t, x_t^{(i)}, u_t^{(i)}). \tag{4.158}$$

By (4.137) and (4.158), for each pair of integers $S_2 > S_1 \ge T_{1,1}$,

$$\sum_{t=S_1}^{S_2-1} f(t, \widehat{x}_t, \widehat{u}_t) \le \sum_{t=S_1}^{S_2-1} f(t, x_t^f, u_t^f) + 3 + 2S_0. \tag{4.159}$$

Set

$$\widehat{x}_t = x_t^f,\ \widehat{u}_t = u_t^f \text{ for all integers } t < T_{1,1}. \tag{4.160}$$

Theorem 4.1, (4.127), (4.157), (4.159), and (4.160) imply that

$$(\{\widehat{x}_t\}_{t=-\infty}^{\infty}, \{\widehat{u}_t\}_{t=-\infty}^{\infty}) \in X(-\infty, \infty)$$

is (f)-good. (P1) implies that

$$\lim_{t\to\infty} \rho_E(\widehat{x}_t, x_t^f) = 0. \tag{4.161}$$

By (4.129) and (4.157),

$$\rho_E(x_{t_1}^f, \widehat{x}_{t_1}) \ge \epsilon. \tag{4.162}$$

Property (P3), (4.102), and (4.124) imply that $(\{\widehat{x}_t\}_{t=-\infty}^{\infty}, \{\widehat{u}_t\}_{t=-\infty}^{\infty})$ is not (f)-minimal. There exist $\Delta > 0$ and an integer $S_* > |T_{1,1}| + 1 + b_f$ such that

$$\sum_{t=-S_*}^{S_*-1} f(t, \widehat{x}_t, \widehat{u}_t) > U^f(-S_*, S_*, \widehat{x}(-S_*), \widehat{x}(S_*)) + 2\Delta. \tag{4.163}$$

Lemma 4.20 and (A2) imply that there exists $\delta(\Delta) \in (0, \Delta/8)$ such that the following properties hold:

(viii) for each $(T_i, z_i) \in \mathcal{A}$, $i = 1, 2$ satisfying

$$T_2 \geq T_1 + 2b_f, \ \rho_E(z_i, x^f_{T_i}) \leq \delta(\Delta), \ i = 1, 2$$

we have

$$U^f(T_1, T_2, z_1, z_2) \leq \sum_{t=T_1}^{T_2-1} f(t, x^f_t, u^f_t) + \Delta/8;$$

(ix) for each $(T, z) \in \mathcal{A}$ satisfying $\rho_E(z, x^f_T) \leq \delta(\Delta)$ there exist

$$(\{y^{(1)}_t\}_{t=T}^{T+b_f}, \{v^{(1)}_t\}_{t=T}^{T+b_f-1}) \in X(T, T + b_f),$$

$$(\{y^{(2)}_t\}_{t=T-b_f}^{T}, \{v^{(2)}_t\}_{t=T-b_f}^{T-1}) \in X(T - b_f, T)$$

such that

$$y^{(1)}_T = z, \ y^{(1)}_{T+b_f} = x^f_{T+b_f}, \ y^{(2)}_T = z, \ y^{(2)}_{T-b_f} = x^f_{T-b_f},$$

$$\sum_{t=T}^{T+b_f-1} f(t, y^{(1)}_t, v^{(1)}_t) \leq \sum_{t=T}^{T+b_f-1} f(t, x^f_t, u^f_t) + \Delta/16,$$

$$\sum_{t=T-b_f}^{T-1} f(t, y^{(2)}_t, v^{(2)}_t) \leq \sum_{t=T-b_f}^{T-1} f(t, x^f_t, u^f_t) + \Delta/16.$$

In view of (4.126), there exists an integer $k(\Delta) \geq 1$ such that

$$\delta_k < \delta(\Delta)/4 \text{ for all integers } k \geq k(\Delta). \tag{4.164}$$

Property (viii), (4.64), (4.127), and (4.128) imply that the following property holds:

(x) for each integer $k \geq k(\Delta)$,

$$\sum_{t=T_{1,1}}^{T_{k,2}-1} f(t, x_t^{(k)}, u_t^{(k)}) \leq \sum_{t=T_{1,1}}^{T_{k,2}-1} f(t, x_t^f, u_t^f) + \Delta/2.$$

Property (P1) implies that there exists an integer $\tau_0 > 1$ such that

$$\rho_E(\widehat{x}_t, x_t^f) \leq \delta(\Delta)/4 \text{ for all integers } t \geq \tau_0. \tag{4.165}$$

Let $q \geq 1$ be an integer satisfying

$$T_{q,2} > \tau_0 + S_*. \tag{4.166}$$

By (4.157), there exists an integer $k_1 \geq k(\Delta) + q$ such that for each integer $k \geq k_1$,

$$\rho_E(\widehat{x}_{T_{q,2}}, x_{T_{q,2}}^{(k)}) \leq \delta(\Delta)/4. \tag{4.167}$$

Assume that an integer $k \geq k_1$. Then (4.167) holds. By (4.128) and (4.164),

$$\sum_{t=T_{1,1}}^{T_{q,2}-1} f(t, x_t^{(k)}, u_t^{(k)}) \leq U^f(T_{1,1}, T_{q,2}, x_{T_{1,1}}^{(k)}, x_{T_{q,2}}^{(k)}) + \delta(\Delta)/4. \tag{4.168}$$

It follows from (4.127) and (4.164) that

$$\rho_E(x_{T_{1,1}}^{(k)}, x_{T_{1,1}}^f) \leq \delta_k \leq \delta(\Delta)/4. \tag{4.169}$$

In view of (4.165) and (4.167),

$$\rho_E(x_{T_{q,2}}^{(k)}, x_{T_{q,2}}^f) \leq \rho_E(x_{T_{q,2}}^{(k)}, \widehat{x}_{T_{q,2}}) + \rho_E(\widehat{x}_{T_{q,2}}, x_{T_{q,2}}^f) \leq \delta(\Delta)/4 + \delta(\Delta)/4. \tag{4.170}$$

Property (viii), (4.118), (4.164), (4.169), and (4.170) imply that

$$\sum_{t=T_{1,1}}^{T_{q,2}-1} f(t, x_t^{(k)}, u_t^{(k)}) \leq \sum_{t=T_{1,1}}^{T_{q,2}-1} f(t, x_t^f, u_t^f) + \Delta/2. \tag{4.171}$$

In view of (4.158), (4.170), and (4.171),

$$\sum_{t=T_{1,1}}^{T_{q,2}-1} f(t, \widehat{x}_t, \widehat{u}_t) \leq \sum_{t=T_{1,1}}^{T_{q,2}-1} f(t, x_t^f, u_t^f) + \Delta/2, \tag{4.172}$$

$$\rho_E(\widehat{x}_{T_{q,2}}, x_{T_{q,2}}^f) \leq \delta(\Delta)/2 \tag{4.173}$$

for each integer $q \geq 1$ satisfying (4.166).

Let an integer $q \geq 1$ satisfy (4.166) and

$$q^{-1} < \Delta/4. \tag{4.174}$$

Property (ix), (4.163), and (4.173) imply that there exists

$$(\{\xi_t\}_{t=-\infty}^{\infty}, \{\eta_t\}_{t=-\infty}^{\infty}) \in X(-\infty, \infty)$$

such that

$$\xi_t = \widehat{x}_t,\ \eta_t = \widehat{u}_t \text{ for all integers } t < -S_*,\ \xi_{-S_*} = \widehat{x}_{-S_*},\ \xi_{S_*} = \widehat{x}_{S_*}, \tag{4.175}$$

$$\sum_{t=-S_*}^{S_*-1} f(t, \xi_t, \eta_t) < \sum_{t=-S_*}^{S_*-1} f(t, \widehat{x}_t, \widehat{u}_t) - 2\Delta, \tag{4.176}$$

$$\xi_t = \widehat{x}_t \text{ for all integers } t \in (S_*, T_{q,2}], \tag{4.177}$$

$$\eta_t = \widehat{u}_t, \text{ for all integers } t \in (S_*, T_{q,2}), \tag{4.178}$$

$$\xi_{T_{q,2}+b_f} = x^f_{T_{q,2}+b_f}, \tag{4.179}$$

$$\sum_{t=T_{q,2}}^{T_{q,2}+b_f-1} f(t, \xi_t, \eta_t) \leq \sum_{t=T_{q,2}}^{T_{q,2}+b_f-1} f(t, x^f_t, u^f_t) + \Delta/16. \tag{4.180}$$

It follows from the choice of S_*, (4.160), (4.168), and (4.175)–(4.180) that

$$x^f_{-S_*} = \widehat{x}_{-S_*} = \xi_{-S_*},\ \xi_{T_{q,2}+b_f} = x^f_{T_{q,2}+b_f}. \tag{4.181}$$

Property (P3), (4.124), and (4.181) imply that

$$0 \leq \sum_{t=-S_*}^{T_{q,2}+b_f-1} f(t, \xi_t, \eta_t) - \sum_{t=-S_*}^{T_{q,2}+b_f-1} f(t, x^f_t, u^f_t)$$

$$= \sum_{t=-S_*}^{S_*-1} f(t, \xi_t, \eta_t) + \sum_{t=S_*}^{T_{q,2}-1} f(t, \widehat{x}_t, \widehat{u}_t)$$

$$+ \sum_{t=T_{q,2}}^{T_{q,2}+b_f-1} f(t, \xi_t, \eta_t) - \sum_{t=-S_*}^{T_{q,2}+b_f-1} f(t, x^f_t, u^f_t)$$

$$\leq \sum_{t=-S_*}^{S_*-1} f(t, \widehat{x}_t, \widehat{u}_t) - 2\Delta + \sum_{t=S_*}^{T_{q,2}-1} f(t, \widehat{x}_t, \widehat{u}_t)$$

$$+ \sum_{t=T_{q,2}}^{T_{q,2}+b_f-1} f(t, x_t^f, u_t^f) + \Delta/16 - \sum_{t=-S_*}^{T_{q,2}+b_f-1} f(t, x_t^f, u_t^f)$$

$$= \sum_{t=-S_*}^{T_{q,2}-1} f(t, \widehat{x}_t, \widehat{u}_t) - \sum_{t=-S_*}^{T_{q,2}-1} f(t, x_t^f, u_t^f) - 2\Delta + \Delta/16$$

$$= \sum_{t=T_{1,1}}^{T_{q,2}-1} f(t, \widehat{x}_t, \widehat{u}_t) - \sum_{t=T_{1,1}}^{T_{q,2}-1} f(t, x_t^f, u_t^f) - 2\Delta + \Delta/16 < -\Delta,$$

a contradiction. The contradiction we have reached proves that the case (b) does not hold.

Assume that the case (c) holds. In view of (4.148) and LSC property, extracting a subsequence and re-indexing, we may assume without loss of generality that there exists $(\{\widehat{x}_t\}_{t=-\infty}^{T_{1,2}}, \{\widehat{u}_t\}_{t=-\infty}^{T_{1,2}-1}) \in X(-\infty, T_{1,2})$ such that for every integer $t \leq T_{1,2}$,

$$\widehat{x}_t = \lim_{i\to\infty} x_t^{(i)}, \tag{4.182}$$

$$f(t, \widehat{x}_t, \widehat{u}_t) \leq \liminf_{i\to\infty} f(t, x_t^{(i)}, u_t^{(i)}) \text{ for all integers } t < T_{1,2}. \tag{4.183}$$

By (4.137) and (4.182), for each pair of integers $S_1 < S_2 \leq T_{1,2}$,

$$\sum_{t=S_1}^{S_2-1} f(t, \widehat{x}_t, \widehat{u}_t) \leq \sum_{t=S_1}^{S_2-1} f(t, x_t^f, u_t^f) + 3 + 2S_0. \tag{4.184}$$

Set

$$\widehat{x}_t = x_t^f,\ \widehat{u}_t = u_t^f \text{ for all integers } t > T_{1,2}. \tag{4.185}$$

It follows from (4.127), (4.182), (4.184), and (4.185) that

$$(\{\widehat{x}_t\}_{t=-\infty}^{\infty}, \{\widehat{u}_t\}_{t=-\infty}^{\infty}) \in X(-\infty, \infty)$$

is (f)-good. (P1) implies that

$$\lim_{t\to-\infty} \rho_E(\widehat{x}_t, x_t^f) = 0. \tag{4.186}$$

By (4.129), (4.152), and (4.182),

$$\rho_E(x^f_{t_1}, \widehat{x}_{t_1}) \geq \epsilon. \tag{4.187}$$

Property (P3), (4.124), and (4.187) imply that $(\{\widehat{x}_t\}_{t=-\infty}^{\infty}, \{\widehat{u}_t\}_{t=-\infty}^{\infty})$ is not (f)-minimal. Therefore there exist $\Delta > 0$ and an integer $S_* > 0$ such that

$$S_* > |T_{1,2}| + 1 + b_f, \tag{4.188}$$

$$\sum_{t=-S_*}^{S_*-1} f(t, \widehat{x}_t, \widehat{u}_t) > U^f(-S_*, S_*, \widehat{x}(-S_*), \widehat{x}(S_*)) + 2\Delta. \tag{4.189}$$

Lemma 4.20 and (A2) imply that there exists $\delta(\Delta) \in (0, \Delta/8)$ such that the following properties hold:

(xi) for each $(T_i, z_i) \in \mathcal{A}$, $i = 1, 2$ satisfying

$$T_2 \geq T_1 + 2b_f,\ \rho_E(z_i, x^f_{T_i}) \leq \delta(\Delta),\ i = 1, 2$$

we have

$$U^f(T_1, T_2, z_1, z_2) \leq \sum_{t=T_1}^{T_2-1} f(t, x^f_t, u^f_t) + \Delta/8;$$

(xii) for each $(T, z) \in \mathcal{A}$ satisfying $\rho_E(z, x^f_T) \leq \delta(\Delta)$ there exist

$$(\{y^{(1)}_t\}_{t=T}^{T+b_f}, \{v^{(1)}_t\}_{t=T}^{T+b_f-1}) \in X(T, T + b_f),$$

$$(\{y^{(2)}_t\}_{t=T-b_f}^{T}, \{v^{(2)}_t\}_{t=T-b_f}^{T-1}) \in X(T - b_f, T)$$

such that

$$y^{(1)}_T = z,\ y^{(1)}_{T+b_f} = x^f_{T+b_f},\ y^{(2)}_T = z,\ y^{(2)}_{T-b_f} = x^f_{T-b_f},$$

$$\sum_{t=T}^{T+b_f-1} f(t, y^{(1)}_t, v^{(1)}_t) \leq \sum_{t=T}^{T+b_f-1} f(t, x^f_t, u^f_t) + \Delta/16,$$

$$\sum_{t=T-b_f}^{T-1} f(t, y^{(2)}_t, v^{(2)}_t) \leq \sum_{t=T-b_f}^{T-1} f(t, x^f_t, u^f_t) + \Delta/16.$$

In view of (4.126), there exists an integer $k(\Delta) \geq 1$ such that

$$\delta_k < \delta(\Delta)/4 \text{ for all integers } k \geq k(\Delta). \tag{4.190}$$

Property (xi), (4.127), (4.128), and (4.190) imply that for each integer $k \geq k(\Delta)$,

$$\sum_{t=T_{1,1}}^{T_{k,2}-1} f(t, x_t^{(k)}, u_t^{(k)}) \leq \sum_{t=T_{1,1}}^{T_{k,2}-1} f(t, x_t^f, u_t^f) + \Delta/2. \tag{4.191}$$

By (4.186), there exists an integer $\tau_0 < -1$ such that

$$\rho_E(\widehat{x}_t, x_t^f) \leq \delta(\Delta)/4 \text{ for all integers } t \leq \tau_0. \tag{4.192}$$

Let $q \geq 1$ be an integer satisfying

$$T_{q,1} < -\tau_0 - S_*. \tag{4.193}$$

By (4.182), there exists an integer $k_1 \geq k(\Delta) + q$ such that for each integer $k \geq k_1$,

$$\rho_E(\widehat{x}_{T_{q,1}}, x_{T_{q,1}}^{(k)}) \leq \delta(\Delta)/4. \tag{4.194}$$

Assume that an integer $k \geq k_1$. Then (4.194) holds. By (4.127), (4.128), and (4.190),

$$\sum_{t=T_{q,1}}^{T_{1,2}-1} f(t, x_t^{(k)}, u_t^{(k)}) \leq U^f(T_{q,1}, T_{1,2}, x_{T_{q,1}}^{(k)}, x_{T_{1,2}}^{(k)}) + \delta(\Delta)/4, \tag{4.195}$$

$$\rho_E(x_{T_{1,2}}^{(k)}, x_{T_{1,2}}^f) \leq \delta_k \leq \delta(\Delta)/4. \tag{4.196}$$

In view of (4.192)–(4.194),

$$\rho_E(x_{T_{q,1}}^{(k)}, x_{T_{q,1}}^f) \leq \rho_E(x_{T_{q,1}}^{(k)}, \widehat{x}_{T_{q,1}}) + \rho_E(\widehat{x}_{T_{q,1}}, x_{T_{q,1}}^f) \leq \delta(\Delta)/4 + \delta(\Delta)/4. \tag{4.197}$$

Property (xi), (4.128), (4.196), and (4.197) imply that

$$\sum_{t=T_{q,1}}^{T_{1,2}-1} f(t, x_t^{(k)}, u_t^{(k)}) \leq \sum_{t=T_{q,1}}^{T_{1,2}-1} f(t, x_t^f, u_t^f) + \Delta/2. \tag{4.198}$$

In view of (4.182), (4.183), (4.197), and (4.198),

$$\sum_{t=T_{q,1}}^{T_{1,2}-1} f(t, \widehat{x}_t, \widehat{u}_t) \leq \sum_{t=T_{q,1}}^{T_{1,2}-1} f(t, x_t^f, u_t^f) + \Delta/2, \tag{4.199}$$

$$\rho_E(\widehat{x}_{T_{q,1}}, x^f_{T_{q,1}}) \le \delta(\Delta)/2 \tag{4.200}$$

for each integer $q \ge 1$ satisfying (4.193).

Let an integer $q \ge 1$ satisfy (4.193) and

$$q^{-1} < \Delta/4. \tag{4.201}$$

Property (xii), (4.189), and (4.200) imply that there exists

$$(\{\xi_t\}_{t=T_{q,1}-b_f}^{S_*}, \{\eta_t\}_{t=T_{q,1}-b_f}^{S_*-1}) \in X(T_{q,1} - b_f, S_*)$$

such that

$$\xi_{-S_*} = \widehat{x}_{-S_*},\ \xi_{S_*} = \widehat{x}_{S_*}, \tag{4.202}$$

$$\sum_{t=-S_*}^{S_*-1} f(t, \xi_t, \eta_t) < \sum_{t=-S_*}^{S_*-1} f(t, \widehat{x}_t, \widehat{u}_t) - 2\Delta, \tag{4.203}$$

$$\xi_t = \widehat{x}_t \text{ for all integers } t \in [T_{q,1}, -S_*], \tag{4.204}$$

$$\eta_t = \widehat{u}_t \text{ for all integers } t \in [T_{q,1}, -S_*), \tag{4.205}$$

$$\xi_{T_{q,1}-b_f} = x^f_{T_{q,1}-b_f}, \tag{4.206}$$

$$\sum_{t=T_{q,1}-b_f}^{T_{q,1}-1} f(t, \xi_t, \eta_t) \le \sum_{t=T_{q,1}-b_f}^{T_{q,1}-1} f(t, x^f_t, u^f_t) + \Delta/16. \tag{4.207}$$

It follows from (4.185), (4.188), (4.202) that

$$x^f_{S_*} = \widehat{x}_{S_*} = \xi_{S_*}. \tag{4.208}$$

Property (P3), (4.24) (4.193), and (4.203)–(4.208) imply that

$$0 \le \sum_{t=T_{q,1}-b_f}^{S_*-1} f(t, \xi_t, \eta_t) - \sum_{t=T_{q,1}-b_f}^{S_*-1} f(t, x^f_t, u^f_t)$$

$$= \sum_{t=T_{q,1}-b_f}^{T_{q,1}-1} f(t, \xi_t, \eta_t) + \sum_{t=T_{q,1}}^{-S_*-1} f(t, \xi_t, \eta_t)$$

$$+\sum_{t=-S_*}^{S_*-1} f(t,\xi_t,\eta_t) - \sum_{t=T_{q,1}-b_f}^{S_*-1} f(t,x_t^f,u_t^f)$$

$$\leq \sum_{t=T_{q,1}-b_f}^{T_{q,1}-1} f(t,x_t^f,u_t^f) + \Delta/16 + \sum_{t=T_{q,1}}^{-S_*-1} f(t,\widehat{x}_t,\widehat{u}_t)$$

$$+\sum_{t=-S_*}^{S_*-1} f(t,\widehat{x}_t,\widehat{u}_t) - 2\Delta - \sum_{t=T_{q,1}-b_f}^{S_*-1} f(t,x_t^f,u_t^f)$$

$$= \sum_{t=T_{q,1}}^{S_*-1} f(t,\widehat{x}_t,\widehat{u}_t) - \sum_{t=T_{q,1}}^{S_*-1} f(t,x_t^f,u_t^f) - 2\Delta + \Delta/16 \leq -\Delta,$$

a contradiction. The contradiction we have reached proves that the case (c) does not hold.

Assume that the case (d) holds. In view of (4.148) and LSC property, extracting a subsequence and re-indexing, we may assume without loss of generality that there exists $(\{\widehat{x}_t\}_{t=-\infty}^{\infty}, \{\widehat{u}_t\}_{t=-\infty}^{\infty}) \in X(-\infty,\infty)$ such that for every integer t,

$$\widehat{x}_t = \lim_{i\to\infty} x_t^{(i)}, \tag{4.209}$$

$$f(t,\widehat{x}_t,\widehat{u}_t) \leq \liminf_{i\to\infty} f(t,x_t^{(i)},u_t^{(i)}). \tag{4.210}$$

By (4.137) and (4.210), for each pair of integers $S_1 < S_2$,

$$\sum_{t=S_1}^{S_2-1} f(t,\widehat{x}_t,\widehat{u}_t) \leq \sum_{t=S_1}^{S_2-1} f(t,x_t^f,u_t^f) + 3 + 2S_0. \tag{4.211}$$

It follows from Theorem 4.1 and (4.211) that $(\{\widehat{x}_t\}_{t=-\infty}^{\infty}, \{\widehat{u}_t\}_{t=-\infty}^{\infty}) \in X(-\infty,\infty)$ is (f)-good. (P1) implies that

$$\lim_{t\to-\infty} \rho_E(\widehat{x}_t,x_t^f) = 0, \ \lim_{t\to\infty} \rho_E(\widehat{x}_t,x_t^f) = 0. \tag{4.212}$$

By (4.129), (4.152), and (4.209),

$$\rho_E(x_{t_1}^f,\widehat{x}_{t_1}) \geq \epsilon. \tag{4.213}$$

Property (P3) and (4.213) imply that $(\{\widehat{x}_t\}_{t=-\infty}^{\infty}, \{\widehat{u}_t\}_{t=-\infty}^{\infty})$ is not (f)-minimal. Therefore there exist $\Delta > 0$ and an integer S_* such that

$$S_* > 1 + b_f, \tag{4.214}$$

$$\sum_{t=-S_*}^{S_*-1} f(t, \widehat{x}_t, \widehat{u}_t) > U^f(-S_*, S_*, \widehat{x}(-S_*), \widehat{x}(S_*)) + 2\Delta. \tag{4.215}$$

Lemma 4.20 and (A2) imply that there exists $\delta(\Delta) \in (0, \Delta/8)$ such that the following properties hold:

(xiii) for each $(T_i, z_i) \in \mathcal{A}$, $i = 1, 2$ satisfying

$$T_2 \geq T_1 + 2b_f,\ \rho_E(z_i, x^f_{T_i}) \leq \delta(\Delta),\ i = 1, 2$$

we have

$$U^f(T_1, T_2, z_1, z_2) \leq \sum_{t=T_1}^{T_2-1} f(t, x^f_t, u^f_t) + \Delta/8;$$

(xiv) for each $(T, z) \in \mathcal{A}$ satisfying $\rho_E(z, x^f_T) \leq \delta(\Delta)$ there exist

$$(\{y^{(1)}_t\}_{t=T}^{T+b_f}, \{v^{(1)}_t\}_{t=T}^{T+b_f-1}) \in X(T, T + b_f),$$

$$(\{y^{(2)}_t\}_{t=T-b_f}^{T}, \{v^{(2)}_t\}_{t=T-b_f}^{T-1}) \in X(T - b_f, T)$$

such that

$$y^{(1)}_T = z,\ y^{(1)}_{T+b_f} = x^f_{T+b_f},\ y^{(2)}_T = z,\ y^{(2)}_{T-b_f} = x^f_{T-b_f},$$

$$\sum_{t=T}^{T+b_f-1} f(t, y^{(1)}_t, v^{(1)}_t) \leq \sum_{t=T}^{T+b_f-1} f(t, x^f_t, u^f_t) + \Delta/16,$$

$$\sum_{t=T-b_f}^{T-1} f(t, y^{(2)}_t, v^{(2)}_t) \leq \sum_{t=T-b_f}^{T-1} f(t, x^f_t, u^f_t) + \Delta/16.$$

In view of (4.126), there exists an integer $k(\Delta) \geq 1$ such that

$$\delta_k < \delta(\Delta)/4 \text{ for all integers } k \geq k(\Delta). \tag{4.216}$$

Property (xiii), (4.127), (4.128), and (4.216) imply that for each integer $k \geq k(\Delta)$,

$$\sum_{t=T_{k,1}}^{T_{k,2}-1} f(t, x^{(k)}_t, u^{(k)}_t) \leq \sum_{t=T_{k,1}}^{T_{k,2}-1} f(t, x^f_t, u^f_t) + \Delta/2. \tag{4.217}$$

By (4.212), there exists an integer $\tau_0 > 1$ such that

$$\rho_E(\widehat{x}_t, x_t^f) \le \delta(\Delta)/4 \text{ for all integers } t \in [\tau_0, \infty) \cup (-\infty, -\tau_0]. \tag{4.218}$$

Let $q \ge 1$ be an integer satisfying

$$T_{q,1} < -\tau_0 - S_*, \ T_{q,2} > S_* + \tau_0. \tag{4.219}$$

By (4.209), there exists an integer $k_1 \ge k(\Delta) + q$ such that for each integer $k \ge k_1$,

$$\rho_E(\widehat{x}_{T_{q,1}}, x_{T_{q,1}}^{(k)}) \le \delta(\Delta)/4, \ \rho_E(\widehat{x}_{T_{q,2}}, x_{T_{q,2}}^{(k)}) \le \delta(\Delta)/4. \tag{4.220}$$

Assume that an integer $k \ge k_1$. Then (4.220) holds. By (4.128) and (4.216),

$$\sum_{t=T_{q,1}}^{T_{q,2}-1} f(t, x_t^{(k)}, u_t^{(k)}) \le U^f(T_{q,1}, T_{q,2}, x_{T_{q,1}}^{(k)}, x_{T_{q,2}}^{(k)}) + \delta(\Delta)/4. \tag{4.221}$$

In view of (4.218)–(4.220), for $j = 1, 2$,

$$\rho_E(x_{T_{q,j}}^{(k)}, x_{T_{q,j}}^f) \le \rho_E(x_{T_{q,j}}^{(k)}, \widehat{x}_{T_{q,j}}) + \rho_E(\widehat{x}_{T_{q,j}}, x_{T_{q,j}}^f) \le \delta(\Delta)/4 + \delta(\Delta)/4. \tag{4.222}$$

Property (xiii), (4.221), and (4.222) imply that

$$\sum_{t=T_{q,1}}^{T_{q,2}-1} f(t, x_t^{(k)}, u_t^{(k)}) \le \sum_{t=T_{q,1}}^{T_{q,2}-1} f(t, x_t^f, u_t^f) + \Delta/2. \tag{4.223}$$

In view of (4.210) and (4.223),

$$\sum_{t=T_{q,1}}^{T_{q,2}-1} f(t, \widehat{x}_t, \widehat{u}_t) \le \sum_{t=T_{q,1}}^{T_{q,2}-1} f(t, x_t^f, u_t^f) + \Delta/2. \tag{4.224}$$

By (4.209) and (4.222),

$$\rho_E(\widehat{x}_{T_{q,j}}, x_{T_{q,j}}^f) \le \delta(\Delta)/4, \ j = 1, 2. \tag{4.225}$$

Let an integer $q \ge 1$ satisfy (4.219). Property (xiv), (4.215), (4.219), and (4.225) imply that there exists $(\{\xi_t\}_{t=T_{q,1}-b_f}^{T_{q,2}+b_f}, \{\eta_t\}_{t=T_{q,1}-b_f}^{T_{q,2}+b_f-1}) \in X(T_{q,1}-b_f, T_{q,2}+b_f)$ such that

$$\xi_{-S_*} = \widehat{x}_{-S_*}, \ \xi_{S_*} = \widehat{x}_{S_*}, \tag{4.226}$$

$$\sum_{t=-S_*}^{S_*-1} f(t,\xi_t,\eta_t) < \sum_{t=-S_*}^{S_*-1} f(t,\widehat{x}_t,\widehat{u}_t) - 2\Delta, \tag{4.227}$$

$$\xi_t = \widehat{x}_t \text{ for all integers } t \in [T_{q,1}, -S_*] \cup [S_*, T_{q,2}], \tag{4.228}$$

$$\eta_t = \widehat{u}_t \text{ for all integers } t \in [T_{q,1}, -S_*) \cup [S_*, T_{q,2}), \tag{4.229}$$

$$\xi_{T_{q,1}-b_f} = x^f_{T_{q,1}-b_f},\ \xi_{T_{q,2}+b_f} = x^f_{T_{q,2}+b_f}, \tag{4.230}$$

$$\sum_{t=T_{q,1}-b_f}^{T_{q,1}-1} f(t,\xi_t,\eta_t) \le \sum_{t=T_{q,1}-b_f}^{T_{q,1}-1} f(t,x^f_t,u^f_t) + \Delta/16, \tag{4.231}$$

$$\sum_{t=T_{q,2}}^{T_{q,2}+b_f-1} f(t,\xi_t,\eta_t) \le \sum_{t=T_{q,2}}^{T_{q,2}+b_f-1} f(t,x^f_t,u^f_t) + \Delta/16. \tag{4.232}$$

Property (P3), (4.124), (4.219), (4.224), and (4.227)–(4.232) imply that

$$0 \le \sum_{t=T_{q,1}-b_f}^{T_{q,2}+b_f-1} f(t,\xi_t,\eta_t) - \sum_{t=T_{q,1}-b_f}^{T_{q,2}+b_f-1} f(t,x^f_t,u^f_t)$$

$$\le \sum_{t=T_{q,1}-b_f}^{T_{q,1}-1} f(t,\xi_t,\eta_t) + \sum_{t=T_{q,1}}^{-S_*-1} f(t,\widehat{x}_t,\widehat{u}_t) + \sum_{t=-S_*}^{S_*-1} f(t,\widehat{x}_t,\widehat{u}_t) - 2\Delta$$

$$+ \sum_{t=S_*}^{T_{q,2}-1} f(t,\widehat{x}_t,\widehat{u}_t) + \sum_{t=T_{q,2}}^{T_{q,2}+b_f-1} f(t,\xi_t,\eta_t) - \sum_{t=T_{q,1}-b_f}^{T_{q,2}+b_f-1} f(t,x^f_t,u^f_t)$$

$$\le \sum_{t=T_{q,1}-b_f}^{T_{q,1}-1} f(t,x^f_t,u^f_t) + \Delta/16 + \sum_{t=T_{q,1}}^{T_{q,2}-1} f(t,\widehat{x}_t,\widehat{u}_t) + \sum_{t=T_{q,2}}^{T_{q,2}+b_f-1} f(t,x^f_t,u^f_t) + \Delta/16$$

$$-2\Delta - \sum_{t=T_{q,1}-b_f}^{T_{q,2}+b_f-1} f(t,x^f_t,u^f_t)$$

$$\le \sum_{t=T_{q,1}}^{T_{q,2}-1} f(t,\widehat{x}_t,\widehat{u}_t) - \sum_{t=T_{q,1}}^{T_{q,2}-1} f(t,x^f_t,u^f_t) - 2\Delta + \Delta/8 \le -\Delta,$$

a contradiction. The contradiction we have reached proves that the case (d) does not hold too. Therefore all the cases (a)-((d) do not hold. The contradiction we have reached completes the proof of Lemma 4.25.

Completion of the Proof of Theorem 4.7. Assume that properties (P1), (P2), and (P3) hold and $\tilde{u}_t = u_t$ for all integers t. Let $\epsilon, M > 0$. By Lemma 4.25, there exist $\delta_0 \in (0, \epsilon)$ such that the following property holds:

(xv) for each integer T_1, each integer $T_2 \geq T_1 + 3b_f$ and each

$$(\{x_t\}_{t=T_1}^{T_2}, \{u_t\}_{t=T_1}^{T_2-1}) \in X(T_1, T_2)$$

which satisfies

$$\rho_E(x_{T_i}, x_{T_i}^f) \leq \delta_0, \ i = 1, 2,$$

$$\sum_{t=T_1}^{T_2-1} f(t, x_t, u_t) \leq U^f(T_1, T_2, x_{T_1}, x_{T_2}) + \delta_0$$

the inequality $\rho_E(x_t, x_t^f) \leq \epsilon$ holds for all $t = T_1, \dots, T_2$.

By Theorem 4.1, there exists $S_0 > 0$ such that for each pair of integers $T_2 > T_1$ and each $(\{y_t\}_{t=T_1}^{T_2}, \{v_t\}_{t=T_1}^{T_2-1}) \in X(T_1, T_2)$, we have

$$\sum_{t=T_1}^{T_2-1} f(t, y_t, v_t) + S_0 \geq \sum_{t=T_1}^{T_2-1} f(t, x_t^f, u_t^f). \tag{4.233}$$

In view of (P2), there exist $\delta \in (0, \delta_0)$ and an integer $L_0 > 0$ such that the following property holds:

(xvi) for each integer T and each $(\{y_t\}_{t=T}^{T+L_0}, \{v_t\}_{t=T}^{T+L_0-1}) \in X(T, T + L_0)$ which satisfies

$$\sum_{t=T}^{T+L_0-1} f(t, y_t, v_t)$$

$$\leq \min\{U^f(T, T + L_0, y_T, y_{T+L_0}) + \delta, \ \sum_{t=T}^{T+L_0-1} f(t, x_t^f, u_t^f) + 2S_0 + M\}$$

there exists an integer $s \in [T, T + L_0]$ such that $\rho_E(y_s, x_s^f) \leq \delta_0$.

Set

$$L = 2L_0 + 2b_f + 2. \tag{4.234}$$

Assume that T_1 and $T_2 \geq T_1 + 2L$ are integers and that $(\{x_t\}_{t=T_1}^{T_2}, \{u_t\}_{t=T_1}^{T_2-1}) \in X(T_1, T_2)$ satisfies

$$\sum_{t=T_1}^{T_2-1} f(t, x_t, u_t) \leq \sigma^f(T_1, T_2) + M, \tag{4.235}$$

$$\sum_{t=T_1}^{T_2-1} f(t, x_t, u_t) \leq U^f(T_1, T_2, x_{T_1}, x_{T_2}) + \delta. \tag{4.236}$$

In view of (4.236),

$$\sum_{t=T_1}^{T_1+L_0-1} f(t, x_t, u_t) \leq U^f(T_1, T_1 + L_0, x_{T_1}, x_{T_1+L_0}) + \delta, \tag{4.237}$$

$$\sum_{t=T_2-L_0}^{T_2-1} f(t, x_t, u_t) \leq U^f(T_2 - L_0, T_2, x_{T_2-L_0}, x_{T_2}) + \delta. \tag{4.238}$$

By (4.235),

$$\sum_{t=T_1}^{T_2-1} f(t, x_t, u_t) \leq \sum_{t=T_1}^{T_2-1} f(t, x_t^f, u_t^f) + M. \tag{4.239}$$

It follows from (4.233) and (4.239) that for each pair of integers $S_1, S_2 \in [T_1, T_2]$ satisfying $S_1 < S_2$,

$$\sum_{t=S_1}^{S_2-1} f(t, x_t, u_t) = \sum_{t=T_1}^{T_2-1} f(t, x_t, u_t) - \sum\{f(t, x_t, u_t) : t \in \{T_1, \ldots, S_1\} \setminus \{S_1\}\}$$

$$-\sum\{f(t, x_t, u_t) : t \in \{S_2, \ldots, T_2\} \setminus \{T_2\}\}$$

$$\leq \sum_{t=T_1}^{T_2-1} f(t, x_t^f, u_t^f) + M - \sum\{f(t, x_t^f, u_t^f) : t \in \{T_1, \ldots, S_1\} \setminus \{S_1\}\} + S_0$$

$$-\sum\{f(t, x_t^f, u_t^f) : t \in \{S_2, \ldots, T_2\} \setminus \{T_2\}\} + S_0 = \sum_{t=S_1}^{S_2-1} f(t, x_t^f, u_t^f) + M + 2S_0. \tag{4.240}$$

It follows from (4.237), (4.238), (4.240), and property (xvi) that there exist integers $\tau_1 \in [T_1, T_1 + L_0]$, $\tau_2 \in [T_2 - L_0, T_2]$ such that

$$\rho_E(x_{\tau_i}, x^f_{\tau_i}) \le \delta_0, \ i = 1, 2. \tag{4.241}$$

If $\rho_E(x_{T_1}, x^f_{T_1}) \le \delta$, then we may assume that $\tau_1 = T_1$ and if $\rho_E(x_{T_2}, x^f_{T_2}) \le \delta$, then we may assume that $\tau_2 = T_2$. Property (xv), (4.234), (4.236), and (4.241) imply that

$$\rho_E(x_t, x^f_t) \le \epsilon, \ t = \tau_1, \dots, \tau_2.$$

Theorem 4.7 is proved.

4.11 Proof of Theorem 4.8

Lemma 4.26. *Let* $(\{x_t\}_{t=-\infty}^{\infty}, \{u_t\}_{t=-\infty}^{\infty}) \in X(-\infty, \infty)$ *be* (f)*-good and* $\epsilon > 0$. *Then there exists an integer* $L > 0$ *such that for each pair of integers* $T_2 > T_1 \ge L$,

$$\sum_{t=T_1}^{T_2-1} f(t, x_t, u_t) \le U^f(T_1, T_2, x_{T_1}, x_{T_2}) + \epsilon$$

for each pair of integers $T_1 < T_2 \le -L$,

$$\sum_{t=T_1}^{T_2-1} f(t, x_t, u_t) \le U^f(T_1, T_2, x_{T_1}, x_{T_2}) + \epsilon.$$

Proof. There exists $M_0 > 0$ such that for each integer $T > 0$,

$$|\sum_{t=-T}^{T-1} f(t, x_t, u_t) - \sum_{t=-T}^{T-1} f(t, x^f_t, u^f_t)| < M_0. \tag{4.242}$$

Theorem 4.1 implies that there exists $M_1 > 0$ such that for each pair of integers $S_2 > S_1$ and each $(\{y_t\}_{t=S_1}^{S_2}, \{v_t\}_{t=S_1}^{S_2-1}) \in X(S_1, S_2)$

$$M_1 + \sum_{t=S_1}^{S_2-1} f(t, y_t, v_t) \ge \sum_{t=S_1}^{S_2-1} f(t, x^f_t, u^f_t). \tag{4.243}$$

Assume that the lemma does not hold. Then there exists a sequence of intervals $[a_i, b_i]$, $i = 1, 2, \dots$ such that for each integer $i \ge 1$, $a_i < b_i$ are integers, for each pair of integers $i < j$,

$$[a_i, b_i] \cap [a_j, b_j] = \emptyset, \tag{4.244}$$

$$\sum_{t=a_i}^{b_i-1} f(t, x_t, u_t) > U^f(a_i, b_i, x_{a_i}, x_{b_i}) + \epsilon, \ i = 1, 2, \ldots. \tag{4.245}$$

By (4.244) and (4.245), there exists $(\{y_t\}_{t=-\infty}^{\infty}, \{v_t\}_{t=-\infty}^{\infty}) \in X(-\infty, \infty)$ such that

$$y_t = x_t \text{ for all integers } t \notin \cup_{i=1}^{\infty}(a_i, b_i), \tag{4.246}$$

$$v_t = u_t \text{ for all integers } t \notin \cup_{i=1}^{\infty}[a_i, b_i), \tag{4.247}$$

for all integer $i \geq 1$,

$$\sum_{t=a_i}^{b_i-1} f(t, y_t, v_t) < \sum_{t=a_i}^{b_i-1} f(t, x_t, u_t) - \epsilon. \tag{4.248}$$

Let $q \geq 1$ be an integer and

$$q > (M_0 + M_1)\epsilon^{-1}. \tag{4.249}$$

Choose an integer $T > 0$ such that

$$[a_i, b_i] \subset [-T, T], \ i = 1, \ldots, q. \tag{4.250}$$

By (4.246)–(4.248) and (4.250),

$$\sum_{t=-T}^{T-1} f(t, y_t, v_t) - \sum_{t=-T}^{T-1} f(t, x_t, u_t)$$

$$\leq \sum_{i=1}^{q}\Big(\sum_{t=a_i}^{b_i-1} f(t, y_t, v_t) - \sum_{t=a_i}^{b_i-1} f(t, x_t, u_t)\Big) \leq -q\epsilon. \tag{4.251}$$

In view of (4.242), (4.243), and (4.251),

$$\sum_{t=-T}^{T-1} f(t, x_t^f, u_t^f) - M_1 \leq \sum_{t=-T}^{T-1} f(t, y_t, v_t) \leq \sum_{t=-T}^{T-1} f(t, x_t, u_t) - q\epsilon$$

$$\leq \sum_{t=-T}^{T-1} f(t, x_t^f, u_t^f) + M_0 - q\epsilon,$$

$$q\epsilon \leq M_0 + M_1.$$

This contradicts (4.249). The contradiction we have reached proves Lemma 4.26.

Proof of Theorem 4.8. Assume that (P4) holds. Clearly, (P4) implies (P2) and (P1) follows from (P4) and Lemma 4.26.

Assume that (P1) and (P2) hold. Then (P4) follows from (P2) and Lemma 4.24. Theorem 4.8 is proved.

4.12 Proof of Theorem 4.13

Clearly, WTP implies (P1) and (P2). Assume that f has (P1) and (P2). We show that WTP holds. Let $\epsilon, M > 0$. By Theorem 4.1, there exists $S_0 > 0$ such that the following property holds:

(i) for each pair of integers $\tau_2 > \tau_1$ and each $(\{y_t\}_{t=\tau_1}^{\tau_2}, \{v_t\}_{t=\tau_1}^{\tau_2-1}) \in X(\tau_1, \tau_2)$,

$$\sum_{t=\tau_1}^{\tau_2-1} f(t, y_t, v_t) + S_0 \geq \sum_{t=\tau_1}^{\tau_2-1} f(t, x_t^f, u_t^f).$$

Lemma 4.24 implies that there exist $\delta_0 > 0$ and an integer $L_0 > 2b_f$ such that the following properties hold:

(ii) for each integer $T_1 \geq L_0$, each integer $T_2 \geq T_1 + 2b_f$, and each $(\{y_t\}_{t=T_1}^{T_2}, \{v_t\}_{t=T_1}^{T_2-1}) \in X(T_1, T_2)$ which satisfies

$$\rho_E(y_{T_i}, x_{T_i}^f) \leq \delta_0, \ i = 1, 2, \tag{4.252}$$

$$\sum_{t=T_1}^{T_2-1} f(t, y_t, v_t) \leq U^f(T_1, T_2, y_{T_1}, y_{T_2}) + \delta_0 \tag{4.253}$$

we have

$$\rho_E(y_t, x_t^f) \leq \epsilon, \ t = T_1, \ldots, T_2; \tag{4.254}$$

(iii) for each integer $T_2 \leq -L_0$, each integer $T_1 \leq T_2 - 2b_f$, and each $(\{y_t\}_{t=T_1}^{T_2}, \{v_t\}_{t=T_1}^{T_2-1}) \in X(T_1, T_2)$ which satisfies (4.252) and (4.253) relation (4.254) holds.

Property (P2) implies that there exist $\delta \in (0, \delta_0)$ and an integer $L_1 > 0$ such that the following property holds:

(iv) for each integer T and each $(\{y_t\}_{t=T}^{T+L_1}, \{v_t\}_{t=T}^{T+L_1-1}) \in X(T, T + L_1)$ which satisfies

$$\sum_{t=T}^{T+L_1-1} f(t, y_t, v_t)$$

$$\leq \min\{U^f(T, T+L_1, y_T, y_{T+L_1}) + \delta, \sum_{t=T}^{T+L_1-1} f(t, x_t^f, u_t^f) + M + 2S_0\}$$

there exists an integer $s \in [T, T+L_1]$ such that $\rho_E(x_s, x_s^f) \leq \delta_0$.

Set

$$l = 8L_0 + 8L_1 + 8 + 8b_f. \tag{4.255}$$

Choose a natural number

$$Q \geq 3(2 + 2(M + S_0)\delta^{-1}). \tag{4.256}$$

Assume that T_1, $T_2 \geq T_1 + lQ$ are integers and $(\{x_t\}_{t=T_1}^{T_2}, \{u_t\}_{t=T_1}^{T_2-1}) \in X(T_1, T_2)$ satisfies

$$\sum_{t=T_1}^{T_2-1} f(t, x_t, u_t) \leq \sum_{t=T_1}^{T_2-1} f(t, x_t^f, u_t^f) + M. \tag{4.257}$$

Property (i) and (4.257) imply that for each pair of integers $\tau_1, \tau_2 \in [T_1, T_2]$ satisfying $\tau_1 < \tau_2$,

$$\sum_{t=\tau_1}^{\tau_2-1} f(t, x_t, u_t) = \sum_{t=T_1}^{T_2-1} f(t, x_t, u_t) - \sum\{f(t, x_t, u_t) : t \in \{T_1, \dots, \tau_1\} \setminus \{\tau_1\}\}$$

$$- \sum\{f(t, x_t, u_t) : t \in \{\tau_2, \dots, T_2\} \setminus \{T_2\}\}$$

$$\leq \sum_{t=T_1}^{T_2-1} f(t, x_t^f, u_t^f) + M - \sum\{f(t, x_t^f, u_t^f) : t \in \{T_1, \dots, \tau_1\} \setminus \{\tau_1\}\} + S_0$$

$$- \sum\{f(t, x_t^f, u_t^f) : t \in \{\tau_2, \dots, T_2\} \setminus \{T_2\}\} + S_0$$

$$= \sum_{t=\tau_1}^{\tau_2-1} f(t, x_t^f, u_t^f) + M + 2S_0. \tag{4.258}$$

Set

$$t_0 = T_1. \tag{4.259}$$

If

$$\sum_{t=T_1}^{T_2-1} f(t, x_t, u_t) \le U^f(T_1, T_2, x_{T_1}, x_{T_2}) + \delta,$$

then we set $t_1 = T_2$. Assume that

$$\sum_{t=T_1}^{T_2-1} f(t, x_t, u_t) > U^f(T_1, T_2, x_{T_1}, x_{T_2}) + \delta. \tag{4.260}$$

Set

$$t_1 = \min\{t \in \{T_1 + 1, \dots, T_2\} : \sum_{\tau=T_1}^{t-1} f(\tau, x_\tau, u_\tau) > U^f(T_1, t, x_{T_1}, x_t) + \delta\}. \tag{4.261}$$

Assume that $k \ge 1$ is an integer and we defined a sequence of integers $\{t_i\}_{i=0}^k \subset [T_1, T_2]$ such that

$$t_0 < t_1 \cdots < t_k, \tag{4.262}$$

for each integer $i = 1, \dots, k$, if $t_i < T_2$, then

$$\sum_{t=t_{i-1}}^{t_i-1} f(t, x_t, u_t) - U^f(t_{i-1}, t_i, x_{t_{i-1}}, x_{t_i}) > \delta, \tag{4.263}$$

and if $t_i - t_{i-1} \ge 2$, then

$$\sum_{t=t_{i-1}}^{t_i-2} f(t, x_t, u_t) - U^f(t_{i-1}, t_i - 1, x_{t_{i-1}}, x_{t_i-1}) \le \delta. \tag{4.264}$$

(Note that in view of (4.261), our assumption holds for $k = 1$.)

By (4.262)–(4.264), there exists $(\{y_t\}_{t=T_1}^{T_2}, \{v_t\}_{t=T_1}^{T_2-1}) \in X(T_1, T_2)$ such that

$$y_{t_i} = x_{t_i}, \ i = 1, \dots, k, \tag{4.265}$$

$$\sum_{t=t_{i-1}}^{t_i-1} f(t, x_t, u_t) - \sum_{t=t_{i-1}}^{t_i-1} f(t, y_t, v_t) > \delta, \ i \in \{1, \dots, k\} \setminus \{k\}, \tag{4.266}$$

$$y_t = x_t, \ t \in \{t_k, \dots, T_2\}, \ v_t = u_t, \ t \in \{t_k, \dots, T_2\} \setminus \{T_2\}. \tag{4.267}$$

Property (i), (4.257), and (4.265)–(4.267) imply that

$$M + \sum_{t=T_1}^{T_2-1} f(t, x_t^f, u_t^f) \geq \sum_{t=T_1}^{T_2-1} f(t, x_t, u_t) \geq \sum_{t=T_1}^{T_2-1} f(t, y_t, v_t) + \delta(k-1)$$

$$\geq \sum_{t=T_1}^{T_2-1} f(t, x_t^f, u_t^f) - S_0 + \delta(k-1),$$

$$k \leq 1 + \delta^{-1}(M + S_0). \tag{4.268}$$

If $t_k = T_2$, then the construction of the sequence is completed. Assume that $t_k < T_2$. If

$$\sum_{t=t_k}^{T_2-1} f(t, x_t, u_t) \leq U^f(t_k, T_2, x_{t_k}, x_{T_2}) + \delta,$$

then we set $t_{k+1} = T_2$ and the construction of the sequence is completed. Assume that

$$\sum_{t=t_k}^{T_2-1} f(t, x_t, u_t) > U^f(t_k, T_2, x_{t_k}, x_{T_2}) + \delta.$$

Set

$$t_{k+1} = \min\{t \in \{t_k + 1, \dots, T_2\} : \sum_{\tau=t_k}^{t-1} f(\tau, x_\tau, u_\tau) - U^f(t_k, t, x_{t_k}, x_t) > \delta\}.$$

Clearly, t_{k+1} is well-defined, and the assumption made for k also holds for $k+1$. By induction and by (4.263), (4.264), and (4.268), we constructed a finite sequence of integers $\{t_i\}_{i=0}^q \subset [T_1, T_2]$ such that $T_1 = t_0 < t_1 \cdots < t_q = T_2$,

$$q \leq 2 + \delta^{-1}(M + S_0), \tag{4.269}$$

for each integer $i = 1, \dots, q$ if $t_i < T_2$, then

$$\sum_{t=t_{i-1}}^{t_i-1} f(t, x_t, u_t) - U^f(t_{i-1}, t_i, x_{t_{i-1}}, x_{t_i}) > \delta,$$

and if $t_i - t_{i-1} \geq 2$, then

$$\sum_{t=t_{i-1}}^{t_i-2} f(t, x_t, u_t) - U^f(t_{i-1}, t_i - 1, x_{t_{i-1}}, x_{t_i-1}) \leq \delta. \tag{4.270}$$

Let

$$i \in \{0, \dots, q-1\},\ t_{i+1} - t_i \geq 8 + 8L_1 + 8L_0 + 8b_f. \tag{4.271}$$

By (4.270) and (4.271),

$$\sum_{t=t_i}^{t_{i+1}-2} f(t, x_t, u_t) \leq U^f(t_i, t_{i+1} - 1, x_{t_i}, x_{t_{i+1}-1}) + \delta. \tag{4.272}$$

Assume that

$$t_i \geq -2L_0 - 2L_1. \tag{4.273}$$

Property (iv), (4.258), (4,271), and (4.272) imply that there exist

$$t_{i,1} \in \{t_i + 3L_0 + 2L_1, \dots, t_i + 3L_0 + 3L_1\} \tag{4.274}$$

such that

$$\rho_E(x_{t_{i,1}}, x^f_{t_{i,1}}) \leq \delta_0 \tag{4.275}$$

and

$$t_{i,2} \in \{t_{i+1} - L_1 - 1, \dots, t_{i+1} - 1\} \tag{4.276}$$

such that

$$\rho_E(x_{t_{i,2}}, x^f_{t_{i,2}}) \leq \delta_0. \tag{4.277}$$

It follows from (4.271), (4.272), (4.274)–(4.277), and property (ii) that

$$\rho_E(x_t, x^f_t) \leq \epsilon,\ t = t_{i,1}, \dots, t_{i,2}.$$

Together with (4.274) and (4.276), this implies that

$$\rho_E(x_t, x^f_t) \leq \epsilon,\ t = t_i + 3L_0 + 3L_1, \dots, t_{i+1} - L_1 - 1. \tag{4.278}$$

Assume that

$$t_{i+1} \leq 2L_0 + 2L_1 + 1. \tag{4.279}$$

Property (iv), (4.258), (4,271), and (4.272) imply that there exist

$$t_{i,1} \in \{t_i, \dots, t_i + L_1\} \tag{4.280}$$

such that

$$\rho_E(x_{t_{i,1}}, x^f_{t_{i,1}}) \le \delta_0 \tag{4.281}$$

and

$$t_{i,2} \in \{t_{i+1} - 1 - 3L_0 - 3L_1, \dots, t_{i+1} - 1 - 3L_0 - 2L_1\} \tag{4.282}$$

such that

$$\rho_E(x_{t_{i,2}}, x^f_{t_{i,2}}) \le \delta_0. \tag{4.283}$$

It follows from (4.271), (4.272), (4.281), (4.283), and property (ii) that

$$\rho_E(x_t, x^f_t) \le \epsilon,\ t = t_{i,1}, \dots, t_{i,2}.$$

Together with (4.280) and (4.282), this implies that

$$\rho_E(x_t, x^f_t) \le \epsilon,\ t = t_i + L_1, \dots, t_{i+1} - 1 - 3L_1 - 3L_0.$$

Therefore (see (4.278)) in both cases

$$\rho_E(x_t, x^f_t) \le \epsilon,\ t = t_i + 3L_0 + 3L_1, \dots, t_{i+1} - 2 - 3L_1 - 3L_0. \tag{4.284}$$

Assume that

$$t_i < -2L_0 - 2L_1,\ t_{i+1} > 2L_0 + 2L_1 + 1. \tag{4.285}$$

Property (iv), (4.258), (4.271), and (4.272) imply that there exist

$$t_{i,1} \in \{L_0, \dots, L_0 + L_1\} \tag{4.286}$$

such that

$$\rho_E(x_{t_{i,1}}, x^f_{t_{i,1}}) \le \delta_0 \tag{4.287}$$

and

$$t_{i,2} \in \{t_{i+1} - 1 - L_1, \dots, t_{i+1} - 1] \tag{4.288}$$

such that

$$\rho_E(x_{t_{i,2}}, x^f_{t_{i,2}}) \le \delta_0, \tag{4.289}$$

$$t_{i,3} \in \{t_i, \dots, t_i + L_1\} \tag{4.290}$$

such that

$$\rho_E(x_{t_{i,3}}, x^f_{t_{i,3}}) \le \delta_0 \tag{4.291}$$

and

$$t_{i,4} \in \{-L_0 - L_1, \dots, -L_0\} \tag{4.292}$$

such that

$$\rho_E(x_{t_{i,4}}, x^f_{t_{i,4}}) \le \delta_0. \tag{4.293}$$

It follows from property (ii), (4.272), and (4.285)–(4.293) that

$$\rho_E(x_t, x^f_t) \le \epsilon,\ t \in \{t_{i,1}, \dots, t_{i,2}\} \cup \{t_{i,3} \dots, t_{i,4}\},$$

$$\rho_E(x_t, x^f_t) \le \epsilon,\ t \in \{L_0 + L_1, \dots, t_{i+1} - 1 - L_1\} \cup \{t_i + L_1, \dots, -L_1 - L_0.\}. \tag{4.294}$$

In view of (4.294),

$$\rho_E(x_t, x^f_t) \le \epsilon,\ t \in \{t_i + L_1, \dots, t_{i+1} - 1 - L_1\} \setminus (-L_0 - L_1, L_0 + L_1). \tag{4.295}$$

In view of (4.284) and (4.295),

$$\{t \in \{T_1, \dots, T_2\} :\ \rho_E(x_t, x^f_t) \le \epsilon\}$$

$$\supset \cup\{\{t_i + 3L_0 + 3L_1, \dots, t_{i+1} - 2 - 3L_0 - 3L_1\} :$$

$$i \in \{0, \dots, q-1\} \text{ and } t_{i+1} - t_i \ge 8L_0 + 8L_1 + 8b_f + 8,$$

$$t_i \ge -2L_0 - 2L_1 \text{ or } t_{i+1} \le 2L_0 + 2L_1 + 1\}$$

$$\cup\{\{t_i + L_1, \dots, t_{i+1} - 1 - L_1\} \setminus (-L_0 - L_1, L_0 + L_1) :\ i \in \{0, \dots, q-1\}$$

$$\text{and } t_{i+1} - t_i \ge 8L_0 + 8L_1 + 8b_f + 8,$$

$$t_i < -2L_0 - 2L_1,\ t_{i+1} > 2L_0 + 2L_1 + 1\}. \tag{4.296}$$

By (4.296),

$$\{t \in \{T_1, \dots, T_2\} :\ \rho_E(x_t, x^f_t) > \epsilon\}$$

$$\subset \cup\{\{t_i, \dots t_{i+1}\}: \ i \in \{0, \dots, q-1\}, \ t_{i+1} - t_i < 8L_0 + 8L_1 + 8b_f + 8\}$$

$$\cup\{\{t_i, \dots, t_i + 3L_0 + 3L_1\} \cup \{t_{i+1} - 3L_0 - 3L_1 - 2, \dots t_{i+1}\}:$$

$$i \in \{0, \dots, q-1\}, \ t_{i+1} - t_i > 8L_1 + 8L_0 + 8 + 8b_f\}$$

$$\cup\{\{t_i, \dots, t_i + L_1\} \cup \{t_{i+1} - 1 - L_1, \dots, t_{i+1}\} \cup \{-L_0 - L_1, \dots, L_0 + L_1\}:$$

$$i \in \{0, \dots, q-1\}: \ t_i < -2L_0 - 2L_1, \ t_{i+1} > 2L_0 + 2L_1 + 1\}. \tag{4.297}$$

The right-hand side of equation (4.297) is the union of intervals. By (4.256) and (4.269), the number of the intervals does not exceed

$$3q \leq 3(M + S_0)\delta^{-1} + 6 < Q. \tag{4.298}$$

In view of (4.255), the maximal length of these intervals does not exceed l. Together with (4.298), this completes the proof of Theorem 4.13.

4.13 Proof of Theorem 4.16

By Theorem 4.7, it is sufficient to show that f_r possesses (P1), (P2), and (P3) with $\tilde{u}_t = u_t$ for all integers t. Theorem 4.1 implies that there exists $M_0 > 0$ such that for each pair of integers $S_2 > S_1$ and each $(\{x_t\}_{t=S_1}^{S_2}, \{u_t\}_{t=S_1}^{S_2-1}) \in X(S_1, S_2)$, we have

$$\sum_{t=S_1}^{S_2-1} f(t, x_t, u_t) + M_0 \geq \sum_{t=S_1}^{S_2-1} f(t, x_t^f, u_t^f). \tag{4.299}$$

We show that (P2) holds. Let $\epsilon, M > 0$. Property (i) implies that there exists $\delta > 0$ such that the following property holds:

(a) for each $x_1, x_2 \in E$ satisfying $\phi(x_1, x_2) \leq \delta$, we have $\rho_E(x_1, x_2) \leq \epsilon$.

Choose a natural number

$$L > (r\delta)^{-1}(M + M_0). \tag{4.300}$$

Assume that T in a integer and $(\{x_t\}_{t=T}^{T+L}, \{u_t\}_{t=T}^{T+L-1}) \in X(T, T+L)$ satisfy

$$\sum_{t=T}^{T+L-1} f_r(t, x_t, u_t) \leq \sum_{t=T}^{T+L-1} f_r(t, x_t^f, u_t^f) + M. \tag{4.301}$$

It follows from (4.13) and (4.299)–(4.301) that

$$\sum_{t=T}^{T+L-1} f(t, x_t, u_t) + r \sum_{t=T}^{T+L-1} \phi(x_t, x_t^f) = \sum_{t=T}^{T+L-1} f_r(t, x_t, u_t)$$

$$\leq \sum_{t=T}^{T+L-1} f_r(t, x_t^f, u_t^f) + M \leq \sum_{t=T}^{T+L-1} f(t, x_t, u_t) + M + M_0,$$

$$\sum_{t=T}^{T+L-1} \phi(x_t, x_t^f) \leq (M_0 + M)r^{-1},$$

$$\min\{\phi(x_t, x_t^f) : t \in \{T, \dots, T + L - 1\}\} \leq (Lr)^{-1}(M_0 + M) < \delta$$

and there exists an integer $\tau \in [T, T + L - 1]$ such that $\phi(x_\tau, x_\tau^f) < \delta$. Property (a) implies that $\rho_E(x_\tau, x_\tau^f) \leq \epsilon$. Therefore f_r has (P2).

We show that f_r has (P1). Let $(\{x_t\}_{t=-\infty}^{\infty}, \{u_t\}_{t=-\infty}^{\infty}) \in X(-\infty, \infty)$ be (f_r)-good. There exists $M_1 > 0$ such that for each integer $T > 0$,

$$\left|\sum_{t=-T}^{T-1} f_r(t, x_t, u_t) - \sum_{t=-T}^{T-1} f(t, x_t^f, u_t^f)\right| < M_1. \tag{4.302}$$

We claim that

$$\lim_{t\to\infty} \rho_E(x_t, x_t^f) = 0, \quad \lim_{t\to-\infty} \rho_E(x_t, x_t^f) = 0.$$

Assume the contrary. Then there exist $\epsilon > 0$ such that at least one of the following properties hold:

(b) there exists a sequence of natural numbers $\{t_k\}_{k=1}^{\infty}$ such that for each integer $k \geq 1$,

$$t_{k+1} \geq t_k + 8, \ \rho_E(x_{t_k}, x_{t_k}^f) > \epsilon;$$

(c) there exists a sequence of negative integers $\{t_k\}_{k=1}^{\infty}$ such that for each integer $k \geq 1$,

$$t_{k+1} < t_k - 8, \ \rho_E(x_{t_k}, x_{t_k}^f) > \epsilon.$$

In view of (4.13), (4.299), and (4.302), for each integer $T > 0$,

$$M_1 \geq \sum_{t=-T}^{T-1} f_r(t, x_t, u_t) - \sum_{t=-T}^{T-1} f(t, x_t^f, u_t^f)$$

$$= \sum_{t=-T}^{T-1} f(t, x_t, u_t) - \sum_{t=-T}^{T-1} f(t, x_t^f, u_t^f) + r \sum_{t=-T}^{T-1} \phi(x_t, x_t^f)$$

$$\geq r \sum_{t=-T}^{T-1} \phi(x_t, x_t^f) - M_0,$$

$$\Delta := \sum_{t=-\infty}^{\infty} \phi(x_t, x_t^f) = \lim_{T\to\infty} \sum_{t=-T}^{T-1} \phi(x_t, x_t^f) < (M_0 + M_1) r^{-1}.$$

Properties (a), (b), and (c) imply that for each integer $k \geq 1$, $\phi(x_{t_k}, x_{t_k}^f) > \delta$,

$$\Delta \geq \sum_{k=1}^{\infty} \phi(x_{t_k}, x_{t_k}^f) = \infty,$$

a contradiction. The contradiction we have reached shows that (P1) holds.

Let us show that (P3) holds. It is clear that $(\{x_t^f\}_{t=-\infty}^{\infty}, \{u_t^f\}_{t=-\infty}^{\infty})$ is (f_r)-minimal. Assume that $(\{x_t\}_{t=-\infty}^{\infty}, \{u_t\}_{t=-\infty}^{\infty}) \in X(-\infty, \infty)$ is (f_r)-minimal and (f_r)-good. Property (P1) implies that

$$\lim_{t\to\infty} \rho_E(x_t, x_t^f) = 0, \ \lim_{t\to-\infty} \rho_E(x_t, x_t^f) = 0. \tag{4.303}$$

We show that $x_t = x_t^f$ for all integers t. Assume the contrary. Then there exists an integer t_0 such that

$$\gamma := \rho(x_{t_0}, x_{t_0}^f) > 0. \tag{4.304}$$

Lemma 4.23 and (A3) imply that there exist $\delta_1 > 0$ and an integer $L_1 > 0$ such that the following properties hold:

(d) for each $(S, z) \in \mathcal{A}$ satisfying $\rho_E(z, x_T^f) \leq \delta_1$ there exist an integer $\tau_1 \in (0, b_f]$ and $(\{y_t^{(1)}\}_{t=S}^{S+\tau_1}, \{v_t^{(1)}\}_{t=S}^{S+\tau_1-1}) \in X(S, S+\tau_1)$ satisfying

$$y_S^{(1)} = z\ ,\ y_{S+\tau_1}^{(1)} = x_{S+\tau_1}^f,$$

$$\sum_{t=S}^{S+\tau_1-1} f_r(t, y_t^{(1)}, v_t^{(1)}) \leq \sum_{t=S}^{S+\tau_1-1} f_r(t, x_t^f, u_t^f) + \delta_0 r/8$$

and an integer $\tau_2 \in (0, b_f]$ and $(\{y_t^{(2)}\}_{t=S-\tau_2}^{S}, \{v_t^{(2)}\}_{t=S-\tau_2}^{S-1}) \in X(S-\tau_2, S)$ satisfying

$$y^{(2)}_{S-\tau_2} = x^f_{S-\tau_2}, \; y^{(2)}_S = z,$$

$$\sum_{t=S-\tau_2}^{S-1} f_r(t, y^{(2)}_t, v^{(2)}_t) \le \sum_{t=S-\tau_2}^{S-1} f_r(t, x^f_t, u^f_t) + \delta_0 r/8,$$

$$\rho_E(y^{(1)}_t, x^f_t) \le \delta_0 r/8, \; t = S, \dots, S + \tau_1,$$

$$\rho_E(y^{(2)}_t, x^f_t) \le \delta_0 r/8, \; t = S - \tau_2, \dots, S;$$

(e) for each pair of integers $T_2 > T_1$ such that either $T_2 \le -L_1$ or $T_1 \ge L_1$ and each $(\{y_t\}_{t=T_1}^{T_2}, \{v_t\}_{t=T_1}^{T_2-1}) \in X(T_1, T_2)$ which satisfies $\rho_E(y_{T_i}, x^f_{T_i}) \le \delta_1$, $i = 1, 2$, we have

$$\sum_{t=T_1}^{T_2-1} f_r(t, y_t, v_t) \ge \sum_{t=T_1}^{T_2-1} f_r(t, x^f_t, y^f_t) - \delta_0 r/8. \tag{4.305}$$

In view of (4.303), there exists an integer $T_0 > 0$ such that

$$[-L_1 - |t_0| - 2b_f - 2, |t_0| + 2b_f + 2 + L_1] \subset [-T_0, T_0], \tag{4.306}$$

$$\rho_E(x_t, x^f_t) \le \delta_1 \text{ for all integers } t \in (-\infty, -T_0] \cup [T_0, \infty). \tag{4.307}$$

Property (d) and (4.307) imply that there exists

$$(\{y_t\}_{t=-T_0-b_f}^{T_0+b_f}, \{v_t\}_{t=-T_0-b_f}^{T_0+b_f-1}) \in X(-T_0 - b_f, T_0 + b_f)$$

such that

$$y_{-T_0-b_f} = x_{-T_0-b_f}, \tag{4.308}$$

$$y_t = x^f_t, \; t = -T_0, \dots, T_0, \; v_t = u^f_t, \; t = -T_0, \dots, T_0 - 1, \tag{4.309}$$

$$\sum_{t=-T_0-b_f}^{-T_0-1} f_r(t, y_t, v_t) \le \sum_{t=-T_0-b_f}^{-T_0-1} f_r(t, x^f_t, u^f_t) + 8^{-1}\delta_0 r, \tag{4.310}$$

$$y_{T_0+b_f} = x_{T_0+b_f}, \tag{4.311}$$

$$\sum_{t=T_0}^{T_0+b_f-1} f_r(t, y_t, v_t) \le \sum_{t=T_0}^{T_0+b_f-1} f_r(t, x^f_t u^f_t) + 8^{-1}\delta_0 r. \tag{4.312}$$

In view of (4.308) and (4.311),

$$\sum_{t=-T_0-b_f}^{T_0+b_f-1} f_r(t, y_t, v_t) \geq \sum_{t=-T_0-b_f}^{T_0+b_f-1} f_r(t, x_t, u_t). \tag{4.313}$$

It follows from (4.306), (4.309), (4.310), (4.312), (4.313), and property (e) that

$$\sum_{t=-T_0-b_f}^{-T_0-1} f_r(t, x_t, u_t) - \delta_0 r/8 + \sum_{t=-T_0}^{T_0-1} f_r(t, x_t, u_t) + \sum_{t=T_0}^{T_0+b_f-1} f_r(t, x_t^f, u_t^f) - \delta_0 r/8$$

$$\leq \sum_{t=-T_0-b_f}^{-T_0-1} f_r(t, x_t, u_t) + \sum_{t=-T_0}^{T_0-1} f_r(t, x_t, u_t) + \sum_{t=T_0}^{T_0+b_f-1} f_r(t, x_t, u_t)$$

$$\leq \sum_{t=-T_0-b_f}^{T_0+b_f-1} f_r(t, y_t, v_t) \leq \sum_{t=-T_0-b_f}^{-T_0-1} f(t, x_t^f, u_t^f) + \delta_0 r/8$$

$$+ \sum_{t=-T_0}^{T_0-1} f(t, x_t^f, u_t^f) + \sum_{t=T_0}^{T_0+b_f-1} f(t, x_t^f, u_t^f) + \delta_0 r/8,$$

$$\sum_{t=-T_0}^{T_0-1} f_r(t, x_t, u_t) \leq \sum_{t=-T_0}^{T_0-1} f(t, x_t^f, u_t^f) + \delta_0 r/2. \tag{4.314}$$

By property (d), (4.306), and (4.307), there exists

$$(\{\xi_t\}_{t=-T_0-b_f}^{T_0+b_f}, \{\eta_t\}_{t=-T_0-b_f}^{T_0+b_f-1}) \in X(-T_0 - b_f, T_0 + b_f)$$

such that

$$\xi_{-T_0-b_f} = x_{-T_0-b_f}^f,\ \xi_t = x_t,\ t = -T_0, \dots, T_0,\ \eta_t = u_t,\ t = -T_0, \dots, T_0-1, \tag{4.315}$$

$$\xi_{T_0+b_f} = x_{T_0+b_f}^f, \tag{4.316}$$

$$\sum_{t=-T_0-b_f}^{-T_0-1} f_r(t, \xi_t, \eta_t) \leq \sum_{t=-T_0-b_f}^{-T_0-1} f_r(t, x_t^f, u_t^f) + 8^{-1}\delta_0, \tag{4.317}$$

$$\sum_{t=T_0}^{T_0+b_f-1} f_r(t,\xi_t,\eta_t) \le \sum_{t=T_0}^{T_0+b_f-1} f_r(t,x_t^f,u_t^f) + 8^{-1}\delta_0 r. \tag{4.318}$$

It follows from (4.13), and (4.315)–(4.318) that

$$\sum_{t=-T_0-b_f}^{T_0+b_f-1} f(t,x_t^f,u_t^f) \le \sum_{t=-T_0-b_f}^{T_0+b_f-1} f(t,\xi_t,\eta_t)$$

$$= \sum_{t=-T_0-b_f}^{-T_0-1} f(t,\xi_t,\eta_t) + \sum_{t=-T_0}^{T_0-1} f(t,x_t,u_t) + \sum_{t=T_0}^{T_0+b_f-1} f(t,\xi_t,\eta_t)$$

$$\le \sum_{t=-T_0}^{T_0-1} f(t,x_t,u_t) + \sum_{t=-T_0-b_f}^{-T_0-1} f(t,x_t^f,u_t^f) + \sum_{t=T_0}^{T_0+b_f-1} f(t,x_t^f,u_t^f) + \delta_0 r/4,$$

$$\sum_{t=-T_0}^{T_0-1} f(t,x_t^f,u_t^f) \le \sum_{t=-T_0}^{T_0-1} f(t,x_t,u_t) + \delta_0 r/4.$$

By (4.13), (4.19), (4.305), and (4.314),

$$\sum_{t=-T_0}^{T_0-1} f(t,x_t^f,u_t^f) + \delta_0 r/2 \ge \sum_{t=-T_0}^{T_0-1} f_r(t,x_t,u_t)$$

$$\ge \sum_{t=-T_0}^{T_0-1} f(t,x_t,u_t) + r\phi(x_{t_0},x_{t_0}^f) \ge \sum_{t=-T_0}^{T_0-1} f(t,x_t^f,u_t^f) + r\delta_0 - r\delta_0/4,$$

a contradiction. The contradiction we have reached proves (P3). This completes the proof of Theorem 4.16.

Chapter 5
Continuous-Time Autonomous Problems

In this chapter we establish sufficient and necessary conditions for the turnpike phenomenon for continuous-time optimal control problems in infinite dimensional spaces. For these optimal control problems the turnpike is a singleton. We also study the structure of approximate solutions on large intervals in the regions close to the endpoints and the existence of solutions of the corresponding infinite horizon optimal control problems. The results of this chapter will be obtained for two large classes of problems which will be treated simultaneously.

5.1 Preliminaries

We begin with the description of the first class of problems. Let $(E, \|\cdot\|)$ be a Banach space and (F, ρ_F) be a metric space. We suppose that $\mathcal{A}$ is a nonempty subset of E, $\mathcal{U} : \mathcal{A} \to 2^F$ is a point to set mapping with a graph

$$\mathcal{M} = \{(x, u) : \ x \in \mathcal{A}, \ u \in \mathcal{U}(x)\}. \tag{5.1}$$

We suppose that $\mathcal{M}$ is a Borel measurable subset of $E \times F$, $G : \mathcal{M} \to E$ is a Borelian function and a linear operator $A : \mathcal{D}(A) \to E$ generates a C_0 semigroup e^{At} on E. Let $f : \mathcal{M} \to R^1$ be a bounded from below Borelian function.

Let $0 \le T_1 < T_2$. We consider the following equation:

$$x'(t) = Ax(t) + G(x(t), u(t)), \ t \in [T_1, T_2]. \tag{5.2}$$

A pair of functions $x : [T_1, T_2] \to E$, $u : [T_1, T_2] \to F$ is called a (mild) solution of (5.2) if $x : [T_1, T_2] \to E$ is a continuous function, $u : [T_1, T_2] \to F$ is a Lebesgue measurable function,

A. J. Zaslavski, *Turnpike Conditions in Infinite Dimensional Optimal Control*, Springer Optimization and Its Applications 148,
https://doi.org/10.1007/978-3-030-20178-4_5

$$x(t) \in \mathcal{A},\ t \in [T_1, T_2], \tag{5.3}$$

$$u(t) \in \mathcal{U}(x(t)),\ t \in [T_1, T_2] \text{ almost everywhere (a.e.)}, \tag{5.4}$$

$G(x(s), u(s)),\ s \in [T_1, T_2]$ is Bochner integrable, and

$$x(t) = e^{A(t-T_1)}x(T_1) + \int_{T_1}^{t} e^{A(t-s)}G(x(s), u(s))ds,\ t \in [T_1, T_2]. \tag{5.5}$$

The set of all pairs (x, u) which are solutions of (5.2) is denoted by

$$X(T_1, T_2, A, G).$$

In the sequel for simplicity, we use the notation $X(T_1, T_2) = X(T_1, T_2, A, G)$ if the pair (A, G) is understood.

Let $0 \le T_1 < T_2$, $x : [T_1, T_2] \to E$, $u : [T_1, T_2] \to F$. Set

$$y(t) = x(t + T_1),\ v(t) = u(t + T_1),\ t \in [0, T_2 - T_1].$$

Assume that $(y, v) \in X(0, T_2 - T_1)$. We show that $(x, u) \in X(T_1, T_2)$. Indeed, we have

$$y(t) = e^{At}y(0) + \int_0^t e^{A(t-s)}G(y(s), v(s))ds,\ t \in [0, T_2 - T_1].$$

In view of the relations above, for all $t \in [T_1, T_2]$,

$$x(t) = y(t - T_1) = e^{A(t-T_1)}x(T_1) + \int_0^{t-T_1} e^{A(t-T_1-s)}G(y(s), v(s))ds$$

$$= e^{A(t-T_1)}x(T_1) + \int_0^{t-T_1} e^{A(t-T_1-s)}G(x(s + T_1), u(s + T_1))ds$$

$$= e^{A(t-T_1)}x(T_1) + \int_{T_1}^{t} e^{A(t-s)}G(x(s), u(s))ds$$

and $(x, u) \in X(T_1, T_2)$.

Assume that $(x, u) \in X(T_1, T_2)$. We show that $(x, u) \in X(\tau, T_2)$ for all $\tau \in (T_1, T_2]$. Let $\tau \in (T_1, T_2)$. By (5.5), for all $t \in [\tau, T_2]$,

$$x(t) = e^{A(t-T_1)}x(T_1) + \int_{T_1}^{t} e^{A(t-s)}G(x(s), u(s))ds$$

$$= e^{A(t-\tau)}e^{A(\tau-T_1)}x(T_1) + \int_{T_1}^{\tau} e^{A(t-\tau)}e^{A(\tau-s)}G(x(s), u(s))ds$$

$$+\int_{\tau}^{t} e^{A(t-\tau)}e^{A(\tau-s)}G(x(s),u(s))ds$$

$$= e^{A(t-\tau)}(e^{A(\tau-T_1)}x(T_1) + \int_{T_1}^{\tau} e^{A(\tau-s)}G(x(s),u(s))ds)$$

$$+\int_{\tau}^{t} e^{A(t-s)}G(x(s),u(s))ds$$

$$= e^{A(t-\tau)}x(\tau) + \int_{\tau}^{T_2} e^{A(t-s)}G(x(s),u(s))ds.$$

Thus $(x,u) \in X(\tau, T_2)$ for all $\tau \in (T_1, T_2)$.

Let $0 \le T_1 < T_2 < T_3$, $(x_1,u_1) \in X(T_1,T_2)$, $(x_2,u_2) \in X(T_2,T_3)$,

$$x_1(T_2) = x_2(T_2).$$

For each $t \in [T_1, T_2]$ set

$$x(t) = x_1(t),\ u(t) = u_1(t)$$

and for each $t \in (T_2, T_3]$ set

$$x(t) = x_2(t),\ u(t) = u_2(t).$$

We show that $(x,u) \in X(T_1,T_3)$. It is not difficult to see that for each $t \in [T_1, T_2]$,

$$e^{A(t-s)}G(x(s),u(s)) = e^{A(t-s)}G(x_1(s),u_1(s)),\ \ s \in [T_1,t],$$

$$x(t) = x_1(t) = e^{A(t-T_1)}x_1(T_1) + \int_{T_1}^{t} e^{A(t-s)}G(x_1(s),u_1(s))ds$$

$$= e^{A(t-T_1)}x(T_1) + \int_{T_1}^{t} e^{A(t-s)}G(x(s),u(s))ds.$$

Let $t \in (T_2, T_3]$. For every $s \in [T_1, t]$, if $s \le T_2$, then

$$e^{A(t-s)}G(x(s),u(s)) = e^{A(t-T_2)}e^{A(T_2-s)}G(x_1(s),u_1(s))$$

and if $s > T_2$, then

$$e^{A(t-s)}G(x(s),u(s)) = e^{A(t-s)}G(x_2(s),u_2(s)),$$

$$e^{A(t-T_1)}x(T_1) + \int_{T_1}^{t} e^{A(t-s)}G(x(s),u(s))ds = e^{A(t-T_2)}e^{A(T_2-T_1)}x(T_1)$$

$$+\int_{T_1}^{T_2} e^{A(t-T_2)}e^{A(T_2-s)}G(x(s),u(s))ds+\int_{T_2}^{t} e^{A(t-s)}G(x(s),u(s))ds$$

$$=e^{A(t-T_2)}[e^{A(T_2-T_1)}x(T_1)+\int_{T_1}^{T_2} e^{A(T_2-s)}G(x_1(s),u_2(s))ds]$$

$$+\int_{T_2}^{t} e^{A(t-s)}G(x_2(s),u_2(s))ds$$

$$=e^{A(t-T_2)}x_2(T_1)+\int_{T_2}^{t} e^{A(t-s)}G(x_2(s),u_2(s))ds=x_2(t)=x(t).$$

Thus $(x,u)\in X(T_1,T_3)$.

Now we describe the second class of problems.

Let $(E,\langle\cdot,\cdot\rangle)_E$ be a Hilbert space equipped with an inner product $\langle\cdot,\cdot\rangle_E$ which induces the norm $\|\cdot\|_E$, and let $(F,\langle\cdot,\cdot\rangle_F)$ be a Hilbert space equipped with an inner product $\langle\cdot,\cdot\rangle_F$ which induces the norm $\|\cdot\|_F$. For simplicity, we set $\langle\cdot,\cdot\rangle_E=\langle\cdot,\cdot\rangle$, $\|\cdot\|_E=\|\cdot\|$, $\langle\cdot,\cdot\rangle_F=\langle\cdot,\cdot\rangle$, $\|\cdot\|_F=\|\cdot\|$, if E,F are understood. We suppose that $\mathcal{A}$ is a nonempty subset of E, $\mathcal{U}:\mathcal{A}\to 2^F$ is a point to set mapping with a graph

$$\mathcal{M}=\{(x,u):\ x\in\mathcal{A},\ u\in\mathcal{U}(x)\}.$$

We suppose that $\mathcal{M}$ is a Borel measurable subset of $E\times F$ and that a linear operator $A:\mathcal{D}(A)\to E$ generates a C_0 semigroup $S(t)=e^{At}$, $t\ge 0$ on E. As usual, we denote by $S(t)^*$ the adjoint operator of $S(t)$. Then $S(t)^*$, $t\in[0,\infty)$ is C_0 semigroup, and its generator is the adjoint operator A^* of A. The domain $\mathcal{D}(A^*)$ is a Hilbert space equipped with the graph norm $\|\cdot\|_{\mathcal{D}(A^*)}$ of the operator A^*:

$$\|z\|^2_{\mathcal{D}(A^*)}:=\|z\|^2_E+\|A^*z\|^2_E,\ z\in\mathcal{D}(A^*).$$

Let $\mathcal{D}(A^*)'$ be the dual space of $\mathcal{D}(A^*)$ with the pivot space E. In particular,

$$E_1^d:=\mathcal{D}(A^*)\subset E\subset\mathcal{D}(A^*)'=E_{-1}.$$

(Here we use the notation of Section 1.7.)

Let $G:\mathcal{M}\to\mathcal{D}(A^*)'=E_{-1}$ and $f:\mathcal{M}\to R^1$ be Borelian functions, $B\in\mathcal{L}(F,E_{-1})$ is an admissible control operator for e^{At}, $t\ge 0$,

$$G(x,u)=Bu,\ (x,u)\in\mathcal{M}.$$

Let $0\le T_1<T_2$. We consider the following equation

$$x'(t)=Ax(t)+Bu(t),\ t\in[T_1,T_2]\text{ a. e. .}\tag{5.6}$$

A pair of functions $x : [T_1, T_2] \to E$, $u : [T_1, T_2] \to F$ is called a (mild) solution of (5.6) if $x : [T_1, T_2] \to E$ is a continuous function, $u : [T_1, T_2] \to F$ is a Lebesgue measurable function, $u \in L^2([T_1, T_2]; F)$,

$$x(t) \in \mathcal{A},\ t \in [T_1, T_2],$$

$$u(t) \in \mathcal{U}(x(t)),\ t \in [T_1, T_2] \text{ a.e.},$$

and for each $t \in [T_1, T_2]$,

$$x(t) = e^{A(t-T_1)}x(T_1) + \int_{T_1}^{t} e^{A(t-s)}Bu(s)ds \tag{5.7}$$

in E_{-1}. The set of all pairs (x, u) which are solutions of (5.6) is denoted by $X(T_1, T_2, A, G)$. In the sequel for simplicity, we use the notation $X(T_1, T_2) = X(T_1, T_2, A, G)$ if the pair (A, G) is understood.

From now these two problems will be treated simultaneously.

Let $T_1 \geq 0$. A pair of functions $x : [T_1, \infty) \to E$, $u : [T_1, \infty) \to F$ is called a (mild) solution of the system

$$x'(t) = Ax(t) + G(x(t), u(t)),\ t \in [T_1, \infty)$$

if for every $T_2 > T_1$, $x : [T_1, T_2] \to E$, $u : [T_1, T_2] \to F$ is a solution of (5.2) for the first problem and is a solution of (5.6) for the second problem. The set of all such pairs (x, u) which are solutions of the equation above is denoted by $X(T_1, \infty)$.

A function $x : I \to E$, where I is either $[T_1, T_2]$ or $[T_1, \infty)$ $(0 \leq T_1 < T_2)$ is called a trajectory if there exists a Lebesgue measurable function $u : I \to F$ (referred to as a control) such that $(x, u) \in X(T_1, T_2)$ or $(x, u) \in X(T_1, \infty)$, respectively).

Let $T_2 > T_1 \geq 0$, $(x, u) \in X(T_1, T_2)$. Define

$$I^f(T_1, T_2, x, u) = \int_{T_1}^{T_2} f(x(t), u(t))dt$$

which is well-defined but can be ∞. Proposition 1.3 implies that there exist $M_* \geq 1$, $\omega_* \in R^1$ such that

$$\|e^{At}\| \leq M_* e^{\omega_* t},\ t \in [0, \infty). \tag{5.8}$$

Let $a_0, K_0 > 0$ and let $\psi : [0, \infty) \to [0, \infty)$ be an increasing function such that

$$\psi(t) \to \infty \text{ as } t \to \infty. \tag{5.9}$$

We suppose that for the first problem the function f satisfies

$$f(x,u) \geq -a_0 + \max\{\psi(\|x\|),\ \psi((\|G(x,u)\| - a_0\|x\|)_+)(\|G(x,u)\| - a_0\|x\|)_+\} \tag{5.10}$$

for each $(x,u) \in \mathcal{M}$ and for the second problem f satisfies

$$f(x,u) \geq -a_0 + \max\{\psi(\|x\|),\ K_0\|u\|^2\},\ (x,u) \in \mathcal{M}. \tag{5.11}$$

Let $T_2 > T_1 \geq 0$ and $y, z \in \mathcal{A}$. We consider the following problems:

$$I^f(T_1, T_2, x, u) \to \min,\ (x,u) \in X(T_1, T_2),\ x(T_1) = y,\ x(T_2) = z, \tag{P_1}$$

$$I^f(T_1, T_2, x, u) \to \min,\ (x,u) \in X(T_1, T_2),\ x(T_1) = y, \tag{P_2}$$

$$I^f(T_1, T_2, x, u) \to \min,\ (x,u) \in X(T_1, T_2). \tag{P_3}$$

We consider functionals of the form $I^f(T_1, T_2, x, u)$, where $0 \leq T_1 < T_2$ and $(x,u) \in X(T_1, T_2)$. For each pair of numbers $T_2 > T_1 \geq 0$ and each pair of points $y, z \in \mathcal{A}$, we define

$$U^f(T_1, T_2, y, z) = \inf\{I^f(T_1, T_2, x, u) :$$

$$(x,u) \in X(T_1, T_2),\ x(T_1) = y,\ x(T_2) = z\}, \tag{5.12}$$

$$\sigma^f(T_1, T_2, y) = \inf\{U^f(T_1, T_2, y, h) : h \in \mathcal{A}\}, \tag{5.13}$$

$$\widehat{\sigma}^f(T_1, T_2, y) = \inf\{U^f(T_1, T_2, h, y) : h \in \mathcal{A}\}, \tag{5.14}$$

$$\sigma^f(T_1, T_2) = \inf\{U^f(T_1, T_2, h, y) : h, y \in \mathcal{A}\}. \tag{5.15}$$

We suppose that

$$x_f \in E,\ u_f \in F,\ (x_f, u_f) \in X(0, \infty). \tag{5.16}$$

This means that $\bar{x}(t) = x_f, \bar{u}(t) = u_f, t \geq 0$ is a solution of the equation

$$x'(t) = Ax(t) + G(x(t), u(t)),\ t \in [0, \infty).$$

For the first class of problems, Proposition 1.6 implies that this is equivalent to the following property:

for every $x^* \in \mathcal{D}(A^*)$, $\langle \bar{x}(t), x^* \rangle$ is an absolutely continuous (a. c.) function on any bounded subinterval of $[0, \infty)$ and for a. e. $t \in [0, \infty)$,

$$0 = (d/dt)\langle \bar{x}(t), x^* \rangle = \langle x_f, A^* x^* \rangle + \langle G(x_f, u_f), x^* \rangle. \tag{5.17}$$

Thus for the first class of problems, (5.16) holds if and only if for every $x^* \in \mathcal{D}(A^*)$,

$$0 = \langle x_f, A^*x^* \rangle + \langle G(x_f, u_f), x^* \rangle. \tag{5.18}$$

It is not difficult to see that for the second class of problems, (5.16) holds if and only if (5.18) is true for every $x^* \in \mathcal{D}(A^*)$ (see Section 1.7).

We suppose that there exists a number $b_f > 0$ and the following assumptions hold.

(A1) For each $S_1 > 0$ there exist $S_2 > 0$ and an integer $c > 0$ such that

$$(T_2 - T_1) f(x_f, u_f) \leq I^f(T_1, T_2, x, u) + S_2$$

for each $T_1 \geq 0$, each $T_2 \geq T_1 + c$ and each $(x, u) \in X(T_1, T_2)$ satisfying $\|x(T_j)\| \leq S_1$, $j = 1, 2$.

(A2) For each $\epsilon > 0$ there exists $\delta > 0$ such that for each $z_i \in \mathcal{A}$, $i = 1, 2$ satisfying $\|z_i - x_f\| \leq \delta$, $i = 1, 2$ there exist $\tau \in (0, b_f]$ and $(x, u) \in X(0, \tau)$ which satisfies $x(0) = z_1$, $x(\tau) = z_2$ and

$$I^f(0, \tau, x, u) \leq \tau f(x_f, u_f) + \epsilon.$$

Section 5.34 contains examples of optimal control problems satisfying assumptions (A1) and (A2). Many examples can also be found in [106–108, 118, 124, 125, 134].

5.2 Boundedness Results

The following result is proved in Section 5.8.

Theorem 5.1.

1. *There exists $S > 0$ such that for each pair of numbers $T_2 > T_1 \geq 0$ and each $(x, u) \in X(T_1, T_2)$,*

$$I^f(T_1, T_2, x, u) + S \geq (T_2 - T_1) f(x_f, u_f).$$

2. *For each $(x, u) \in X(0, \infty)$ either*

$$I^f(0, t, x, u) - t f(x_f, u_f) \to \infty \text{ as } t \to \infty$$

or

$$\sup\{|I^f(0, t, x, u) - t f(x_f, u_f)| : t \in (0, \infty)\} < \infty. \tag{5.19}$$

Moreover, if (5.19) *holds, then* $\sup\{\|x(t)\| : t \in (0, \infty)\} < \infty$.

We say that $(x, u) \in X(0, \infty)$ is (f, A, G)-good (or (f)-good if A, G are understood) if (5.19) holds [29, 104, 116, 120, 129, 134]. The next boundedness result is proved in Section 5.8.

Theorem 5.2. *Let $M_0 > 0$, $c > 0$, $c_0 \in (0, c)$. Then there exists $M_1 > 0$ such that for each $T_1 \geq 0$, each $T_2 \geq T_1 + c$, and each $(x, u) \in X(T_1, T_2)$ satisfying*

$$I^f(T_1, T_2, x, u) \leq (T_2 - T_1) f(x_f, u_f) + M_0,$$

the inequality $\|x(t)\| \leq M_1$ holds for all $t \in [T_1 + c_0, T_2]$.

Let $L > 0$. Denote by $\mathcal{A}_L$ the set of all $z \in \mathcal{A}$ for which there exist $\tau \in (0, L]$ and $(x, u) \in X(0, \tau)$ such that

$$x(0) = z, \ x(\tau) = x_f, \ I^f(0, \tau, x, u) \leq L.$$

Denote by $\widehat{\mathcal{A}}_L$ the set of all $z \in \mathcal{A}$ for which there exist $\tau \in (0, L]$ and $(x, u) \in X(0, \tau)$ such that

$$x(0) = x_f, \ x(\tau) = z, \ I^f(0, \tau, x, u) \leq L.$$

The following Theorems 5.3–5.5 are also boundedness results. They are proved in Section 5.8.

Theorem 5.3. *Let $L > 0$, $M_0 > 0$, and $c \in (0, L)$. Then there exists $M_1 > 0$ such that for each $T_1 \geq 0$, each $T_2 \geq T_1 + 2L$, and each $(x, u) \in X(T_1, T_2)$ satisfying*

$$x(T_1) \in \mathcal{A}_L, \ x(T_2) \in \widehat{\mathcal{A}}_L,$$

$$I^f(T_1, T_2, x, u) \leq U^f(T_1, T_2, x(T_1), x(T_2)) + M_0,$$

the inequality $\|x(t)| \leq M_1$ holds for all $t \in [T_1 + c, T_2]$.

Theorem 5.4. *Let $L > 0$, $M_0 > 0$, and $c \in (0, L)$. Then there exists $M_1 > 0$ such that for each $T_1 \geq 0$, each $T_2 \geq T_1 + L$, and each $(x, u) \in X(T_1, T_2)$ satisfying*

$$x(T_1) \in \mathcal{A}_L, \ I^f(T_1, T_2, x, u) \leq \sigma^f(T_1, T_2, x(T_1)) + M_0,$$

the inequality $\|x(t)\| \leq M_1$ holds for all $t \in [T_1 + c, T_2]$.

Theorem 5.5. *Let $L > 0$, $M_0 > 0$, and $c \in (0, L)$. Then there exists $M_1 > 0$ such that for each $T_1 \geq 0$, each $T_2 \geq T_1 + L$, and each $(x, u) \in X(T_1, T_2)$ satisfying*

$$x(T_2) \in \widehat{\mathcal{A}}_L, \ I^f(T_1, T_2, x, u) \leq \widehat{\sigma}^f(T_1, T_2, x_{T_2}) + M_0,$$

the inequality $\|x(t)\| \leq M_1$ holds for all $t \in [c + T_1, T_2]$.

The following Theorems 5.6–5.9 are proved in Section 5.9.

Theorem 5.6. *Let $M_0 > 0$ and $c > 0$. Then there exists $M_1 > 0$ such that for each $T_1 \geq 0$, each $T_2 \geq T_1 + c$, and each $(x, u) \in X(T_1, T_2)$ satisfying*

$$\|x(T_1)\| \leq M_0,\ I^f(T_1, T_2, x, u) \leq (T_2 - T_1) f(x_f, u_f) + M_0,$$

the inequality $\|x(t)| \leq M_1$ holds for all $t \in [T_1, T_2]$.

Theorem 5.7. *Let $L > 0$ and $M_0 > 0$. Then there exists $M_1 > 0$ such that for each $T_1 \geq 0$, each $T_2 \geq T_1 + 2L$, and each $(x, u) \in X(T_1, T_2)$ satisfying*

$$\|x(T_1)\| \leq M_0,\ x(T_1) \in \mathcal{A}_L,\ \ x(T_2) \in \widehat{\mathcal{A}}_L,$$

$$I^f(T_1, T_2, x, u) \leq U^f(T_1, T_2, x(T_1), x(T_2)) + M_0,$$

the inequality $\|x(t)\| \leq M_1$ holds for all $t \in [T_1, T_2]$.

Theorem 5.8. *Let $L > 0$ and $M_0 > 0$. Then there exists $M_1 > 0$ such that for each $T_1 \geq 0$, each $T_2 \geq T_1 + L$, and each $(x, u) \in X(T_1, T_2)$ satisfying*

$$\|x(T_1)\| \leq M_0,\ x(T_1) \in \mathcal{A}_L,\ I^f(T_1, T_2, x, u) \leq \sigma^f(T_1, T_2, x(T_1)) + M_0,$$

the inequality $\|x(t)\| \leq M_1$ holds for all $t \in [T_1, T_2]$.

Theorem 5.9. *Let $L > 0$ and $M_0 > 0$. Then there exists $M_1 > 0$ such that for each $T_1 \geq 0$, each $T_2 \geq T_1 + L$, and each $(x, u) \in X(T_1, T_2)$ satisfying*

$$\|x(T_1)\| \leq M_0,\ x(T_2) \in \widehat{\mathcal{A}}_L,\ I^f(T_1, T_2, x, u) \leq \widehat{\sigma}^f(T_1, T_2, x_{T_2}) + M_0,$$

the inequality $\|x(t)\| \leq M_1$ holds for all $t \in [T_1, T_2]$.

5.3 Turnpike Results

We say that the triplet (f, A, G) (or f if the pair (A, G) is understood) possesses the turnpike property (or TP for short) if for each $\epsilon > 0$ and each $M > 0$ there exist $\delta > 0$ and $L > 0$ such that for each $T_1 \geq 0$, each $T_2 \geq T_1 + 2L$, and each $(x, u) \in X(T_1, T_2)$ which satisfies

$$I^f(T_1, T_2, x, u) \leq \min\{\sigma^f(T_1, T_2) + M, U^f(T_1, T_2, x(T_1), x(T_2)) + \delta\},$$

we have

$$\|x(t) - x_f\| \leq \epsilon \text{ for all } t \in [T_1 + L, T_2 - L].$$

Moreover, if $\|x(T_1) - x_f\| \le \delta$ then $\|x(t) - x_f\| \le \epsilon$ for all $t \in [T_1, T_2 - L]$ and if $\|x(T_2) - x_f\| \le \delta$ then $\|x(t) - x_f\| \le \epsilon$ for all $t \in [T_1 + L, T_2]$.

Theorem 5.1 implies the following result.

Proposition 5.10. *Assume that f has TP and that $\epsilon, M > 0$. Then there exist $\delta > 0$ and $L > 0$ such that for each $T_1 \ge 0$, each $T_2 \ge T_1 + 2L$, and each $(x, u) \in X(T_1, T_2)$ which satisfies*

$$I^f(T_1, T_2, x, u) \le \min\{(T_2 - T_1) f(x_f, u_f) + M, U^f(T_1, T_2, x(T_1), x(T_2)) + \delta\},$$

there exist $\tau_1 \in [T_1, T_1 + L]$, $\tau_2 \in [T_2 - L, T_2]$ such that

$$\|x(t) - x_f\| \le \epsilon \text{ for all } t \in [\tau_1, \tau_2].$$

Moreover, if $\|x(T_1) - x_f\| \le \delta$ then $\tau_1 = T_1$ and if $\|x(T_2) - x_f\| \le \delta$ then $\tau_2 = T_2$.

The next two results are proved in Section 5.10.

Theorem 5.11. *Assume that f has TP and that $\epsilon, L > 0$. Then there exist $\delta > 0$ and $L_0 > L$ such that for each $T_1 \ge 0$, each $T_2 \ge T_1 + 2L_0$, and each $(x, u) \in X(T_1, T_2)$ which satisfies*

$$x(T_1) \in \mathcal{A}_L, \ x(T_2) \in \widehat{\mathcal{A}}_L,$$

$$I^f(T_1, T_2, x, u) \le U^f(T_1, T_2, x(T_1), x(T_2)) + \delta,$$

there exist $\tau_1 \in [T_1, T_1 + L_0]$, $\tau_2 \in [T_2 - L_0, T_2]$ such that

$$\|x(t) - x_f\| \le \epsilon \text{ for all } t \in [\tau_1, \tau_2].$$

Moreover, if $\|x(T_1) - x_f\| \le \delta$ then $\tau_1 = T_1$ and if $\|x(T_2) - x_f\|) \le \delta$ then $\tau_2 = T_2$.

Theorem 5.12. *Assume that f has TP and that $\epsilon, L, M > 0$. Then there exist $\delta > 0$ and $L_0 > L$ such that for each $T_1 \ge 0$, each $T_2 \ge T_1 + 2L_0$, and each $(x, u) \in X(T_1, T_2)$ which satisfies*

$$x(T_1) \in \mathcal{A}_L,$$

$$I^f(T_1, T_2, x, u) \le \min\{\sigma^f(T_1, T_2, x(T_1)) + M, \ U^f(T_1, T_2, x(T_1), x(T_2)) + \delta\},$$

there exist $\tau_1 \in [T_1, T_1 + L_0]$, $\tau_2 \in [T_2 - L_0, T_2]$ such that

$$\|x(t) - x_f\| \le \epsilon \text{ for all } t \in [\tau_1, \tau_2].$$

Moreover, if $\|x(T_1) - x_f\| \le \delta$ then $\tau_1 = T_1$ and if $\|x(T_2) - x_f\| \le \delta$ then $\tau_2 = T_2$.

The next theorem is the main result of this chapter. It is proved in Section 5.11.

Theorem 5.13. *f has TP if and only if the following properties hold:*

(P1) for each (f)-good pair $(x, u) \in X(0, \infty)$,

$$\lim_{t\to\infty} \|x(t) - x_f\| = 0;$$

(P2) for each $\epsilon > 0$ and each $M > 0$, there exist $\delta > 0$ and $L > 0$ such that for each $(x, u) \in X(0, L)$ which satisfies

$$I^f(0, L, x, u) \le \min\{U^f(0, L, x(0), x(L)) + \delta,\ Lf(x_f, u_f) + M\},$$

there exists $s \in [0, L]$ such that $\|x(s) - x_f\| \le \epsilon$.

The next result is proved in Section 5.13.

Theorem 5.14. *Assume that f has properties (P1) and (P2). Let $\epsilon, M > 0$. Then there exist a natural number Q and $l > 0$ such that for each $T_1 \ge 0$, each $T_2 \ge T_1 + lQ$, and each $(x, u) \in X(T_1, T_2)$ which satisfies*

$$I^f(T_1, T_2, x, u) \le (T_2 - T_1) f(x_f, u_f) + M,$$

there exist finite sequences $\{a_i\}_{i=1}^q$, $\{b_i\}_{i=1}^q \subset [T_1, T_2]$ such that an integer $q \le Q$,

$$0 \le b_i - a_i \le l,\ i = 1, \dots, q,$$

$$b_i \le a_{i+1} \text{ for all integers } i \text{ satisfying } 1 \le i < q,$$

$$\|x(t) - x_f\| \le \epsilon \text{ for all } t \in [T_1, T_2] \setminus \cup_{i=1}^q [a_i, b_i].$$

Theorem 5.14 leads to the following definition.

We say that the triplet (f, A, G) (or f if the pair (A, G) is understood) possesses the weak turnpike property (or WTP for short) if for each $\epsilon > 0$ and each $M > 0$, there exist a natural number Q and $l > 0$ such that for each $T_1 \ge 0$, each $T_2 \ge T_1 + lQ$, and each $(x, u) \in X(T_1, T_2)$ which satisfies

$$I^f(T_1, T_2, x, u) \le (T_2 - T_1) f(x_f, u_f) + M,$$

there exist finite sequences $\{a_i\}_{i=1}^q$, $\{b_i\}_{i=1}^q \subset [T_1, T_2]$ such that an integer $q \le Q$,

$$0 \le b_i - a_i \le l,\ i = 1, \dots, q,$$

$$b_i \le a_{i+1} \text{ for all integers } i \text{ satisfying } 1 \le i < q,$$

$$\|x(t) - x_f\| \le \epsilon \text{ for all } t \in [T_1, T_2] \setminus \cup_{i=1}^q [a_i, b_i].$$

WTP was studied in [95, 96] for discrete-time unconstrained problems, in [97] for variational problems, and in [114] for discrete-time constrained problems.

Theorem 5.15. *f has WTP if and only if f has (P1) and (P2).*

In view of Theorem 5.14, (P1) and (P2) imply WTP. Clearly, WTP implies (P2). Therefore Theorem 5.15 follows from the next result which is proved in Section 5.14.

Proposition 5.16. *Let f have WTP. Then f has (P1).*

5.4 Lower Semicontinuity Property

We say that (f, A, G) (or f if the pair (A, G) is understood) possesses lower semicontinuity property (LSC property for short) if for each $T_2 > T_1 \geq 0$ and each sequence $(x_j, u_j) \in X(T_1, T_2)$, $j = 1, 2, \dots$ which satisfies

$$\sup\{I^f(T_1, T_2, x_j, u_j) : \ j = 1, 2, \dots\} < \infty,$$

there exist a subsequence $\{(x_{j_k}, u_{j_k})\}_{k=1}^{\infty}$ and $(x, u) \in X(T_1, T_2)$ such that,

$$x_{j_k}(t) \to x(t) \text{ as } k \to \infty \text{ for every } t \in [T_1, T_2],$$

$$I^f(T_1, T_2, x, u) \leq \liminf_{j\to\infty} I^f(T_1, T_2, x_j, u_j).$$

LSC property plays an important role in the calculus of variations and optimal control theory [32].

Theorem 5.17. *Assume that f possesses LSC property. Then f has TP if and only if f has (P1).*

Theorem 5.17 follows from Theorem 5.13 and the next proposition which is proved in Section 5.15.

Proposition 5.18. *Assume that f has LSC property and (P1). Then f has (P2).*

A pair $(x, u) \in X(0, \infty)$ is called (f, A, G)-overtaking optimal (or (f)-overtaking optimal if the pair (A, G) is understood) [29, 104, 120] if for every $(y, v) \in X(0, \infty)$ satisfying $x(0) = y(0)$,

$$\limsup_{T\to\infty}[I^f(0, T, x, u) - I^f(0, T, y, v)] \leq 0.$$

A pair $(x, u) \in X(0, \infty)$ is called (f, A, G)-weakly optimal (or (f)-weakly optimal if the pair (A, G) is understood) [29, 104, 120] if for every $(y, v) \in X(0, \infty)$ satisfying $x(0) = y(0)$,

$$\liminf_{T\to\infty}[I^f(0, T, x, u) - I^f(0, T, y, v)] \leq 0.$$

A pair $(x, u) \in X(0, \infty)$ is called (f, A, G)-minimal (or (f)-minimal optimal if the pair (A, G) is understood) [12, 94] if for every $T > 0$

$$I^f(0, T, x, u) = U^f(0, T, x(0), x(T)).$$

In infinite horizon optimal control the main goal is to show the existence of solutions using the optimality criterions above.

The next result is proved in Section 5.17.

Theorem 5.19. *Assume that f has (P1) and LSC property and that $(x, u) \in X(0, \infty)$ is (f)-good. Then there is an (f)-overtaking optimal pair $(x_*, u_*) \in X(0, \infty)$ such that $x_*(0) = x(0)$.*

The following theorem is also proved in Section 5.17.

Theorem 5.20. *Assume that f has (P1) and LSC property, $(\tilde{x}, \tilde{u}) \in X(0, \infty)$ is (f)-good and that $(x_*, u_*) \in X(0, \infty)$ satisfies $x_*(0) = \tilde{x}(0)$. Then the following conditions are equivalent:*

(i) (x_*, u_*) is (f)-overtaking optimal.
(ii) (x_*, u_*) is (f)-weakly optimal.
(iii) (x_*, u_*) is (f)-minimal and (f)-good.
(iv) (x_*, u_*) is (f)-minimal and satisfies $\lim_{t\to\infty} x_*(t) = x_f$.
(v) (x_*, u_*) is (f)-minimal and satisfies $\liminf_{t\to\infty} \|x_*(t) - x_f\| = 0$.

The following three results are proved in Section 5.18.

Theorem 5.21. *Assume that f has (P1) and LSC property and $\epsilon > 0$. Then there exists $\delta > 0$ such that the following assertions hold.*

(i) For every $z \in \mathcal{A}$ satisfying $\|z - x_f\| \le \delta$, there exists an (f)-overtaking optimal pair $(x, u) \in X(0, \infty)$ which satisfies $x(0) = z$.
(ii) If an (f)-overtaking optimal pair $(x, u) \in X(0, \infty)$ satisfies $\|x(0) - x_f\| \le \delta$, then $\|x(t) - x_f\| \le \epsilon$ for all $t \ge 0$.

Theorem 5.22. *Assume that f has (P1) and LSC property, $\epsilon > 0$, and $L > 0$. Then there exists $\tau_0 > 0$ such that for every (f)-overtaking optimal pair $(x, u) \in X(0, \infty)$ which satisfies $x(0) \in \mathcal{A}_L$, the inequality $\|x(t) - x_f\| \le \epsilon$ holds for all $t \ge \tau_0$.*

Theorem 5.23. *Assume that f has (P1) and LSC property, $(x, u) \in X(0, \infty)$ is an (f)-overtaking optimal and (f)-good pair, and numbers $t_2 > t_1 \ge 0$ satisfy $x(t_1) = x(t_2)$. Then $x(t) = x_f$ for all $t \ge t_1$.*

In the sequel we use the following result which easily follows the definitions.

Proposition 5.24. *Let f have (P1) and $z \in \mathcal{A}$. There exists an (f)-good pair $(x, u) \in X(0, \infty)$ satisfying $x(0) = z$ if and only if $z \in \cup\{\mathcal{A}_L : L > 0\}$.*

LSC property implies the following result.

Proposition 5.25. *Let f have LSC property, $T_2 > T_2 \geq 0$, and $L > 0$. Then*

1. *There exists $(x, u) \in X(T_1, T_2)$ such that*

$$I^f(T_1, T_2, x, u) \leq I^f(T_1, T_2, y, v) \text{ for all } (y, v) \in X(T_1, T_2).$$

2. *If $T_2 - T_1 \geq 2L$, $z_1 \in \mathcal{A}_L$, $z_2 \in \widehat{\mathcal{A}}_L$, then there exists $(x, u) \in X(T_1, T_2)$ such that $x(T_i) = z_i$, $i = 1, 2$ and*

$$I^f(T_1, T_2, x, u) = U^f(T_1, T_2, z_1, z_2).$$

3. *If $T_2 - T_1 \geq L$, $z \in \mathcal{A}_L$, then there exists $(x, u) \in X(T_1, T_2)$ such that $x(T_1) = z$ and*

$$I^f(T_1, T_2, x, u) = \sigma^f(T_1, T_2, z).$$

4. *If $T_2 - T_1 \geq L$, $z \in \widehat{\mathcal{A}}_L$, then there exists $(x, u) \in X(T_1, T_2)$ such that $x(T_2) = z$ and*

$$I^f(T_1, T_2, x, u) = \widehat{\sigma}^f(T_1, T_2, z).$$

Theorem 5.26. *f has (P1) and (P2) if and only if the following property holds*

(P3) for each $\epsilon > 0$ and each $M > 0$, there exists $L > 0$ such that for each $T_1 \geq 0$, each $T_2 \geq T_1 + L$, and each $(x, u) \in X(T_1, T_2)$ which satisfies

$$I^f(T_1, T_2, x, u) \leq (T_2 - T_1) f(x_f, u_f) + M$$

the inequality

$$mes(\{t \in [T_1, T_2] : \|x(t) - x_f\| > \epsilon\}) \leq L$$

holds.

In view of Theorem 5.14, (P1) and (P2) imply (P3). Clearly, (P3) implies (P2). Therefore Theorem 5.26 follows from the next result which is proved in Section 5.19.

Proposition 5.27. *Let f have (P3). Then f has (P1).*

5.5 Perturbed Problems

In this section we use the following assumption.

(A3) For each $\epsilon > 0$, there exists $\delta > 0$ such that for each $z_i \in \mathcal{A}$, $i = 1, 2$ satisfying $\|z_i - x_f\| \leq \delta$, $i = 1, 2$, there exist $\tau \in (0, b_f]$ and $(x, u) \in X(0, \tau)$ which satisfies $x(0) = z_1$, $x(\tau) = z_2$ and

$$\|x(t) - x_f\| \le \epsilon, \ t \in [0, \tau],$$

$$I^f(0, \tau, x, u) \le \tau f(x_f, u_f) + \epsilon.$$

Clearly, (A3) implies (A2). Assume that $\phi : E \to [0, 1]$ is a continuous function satisfying $\phi(0) = 0$ and such that the following property holds:

(i) for each $\epsilon > 0$ there exists $\delta > 0$ such that for each $x \in E$ satisfying $\phi(x) \le \delta$ we have $\|x\| \le \epsilon$.

For each $r \in (0, 1)$ set,

$$f_r(x, u) = f(x, u) + r\phi(x - x_f), \ (x, u) \in \mathcal{M}.$$

Clearly, for any $r \in (0, 1)$, f_r is a Borelian function, if (A3) holds, then (A1) and (A3) hold for f_r with $(x_{f_r}, u_{f_r}) = (x_f, u_f)$.

Theorem 5.28. *Let (A3) hold and $r \in (0, 1)$. Then f_r has (P1) and (P2).*

Theorem 5.28 is proved in Section 5.20.

5.6 The Triplet $(f, -A, -G)$

We use the notation, definitions, and assumptions introduced in Sections 5.1–5.4. We assume that $-A : \mathcal{D}(A) \to E$ also generates a C_0 semigroup e^{-At}, $t \ge 0$ on E. Clearly, $\mathcal{D}(-A) = \mathcal{D}(A)$, $(-A)^* = -A^*$.

Let $0 \le T_1 < T_2$. We consider the following equation

$$x'(t) = -Ax(t) - G(x(t), u(t)), \ t \in [T_1, T_2]. \tag{5.20}$$

Let us consider our first class of problems.

A pair of functions $x : [T_1, T_2] \to E$, $u : [T_1, T_2] \to F$ is called a (mild) solution of (5.20) if $x : [T_1, T_2] \to E$ is a continuous function, $u : [T_1, T_2] \to F$ is a Lebesgue measurable function,

$$x(t) \in \mathcal{A}, \ t \in [T_1, T_2], \tag{5.21}$$

$$u(t) \in \mathcal{U}(x(t)), \ t \in [T_1, T_2] \text{ a. e.}, \tag{5.22}$$

$G(x(s), u(s))$, $s \in [T_1, T_2]$ is Bochner integrable, and

$$x(t) = e^{-A(t-T_1)}x(T_1) - \int_{T_1}^{t} e^{-A(t-s)}G(x(s), u(s))ds, \ t \in [T_1, T_2]. \tag{5.23}$$

For the second class of problems, a pair of functions $x : [T_1, T_2] \to E$, $u : [T_1, T_2] \to F$ is a solution of (5.20) if $x : [T_1, T_2] \to E$ is a continuous function, $u : [T_1, T_2] \to F$ is a Lebesgue measurable function, $u \in L^2(T_1, T_2; F)$, (5.21) and (5.22) hold and for each $t \in [T_1, T_2]$,

$$x(t) = e^{-A(t-T_1)}x(T_1) + \int_{T_1}^{t} e^{-A(t-s)}(-Bu(s))ds \tag{5.24}$$

in E_{-1}. It is not difficult to show that $-B$ is an admissible control operator for e^{-At}, $t \geq 0$.

The set of all pairs (x, u) which are solutions of (5.20) is denoted by $X(T_1, T_2, -A, -G)$.

Let $T_1 \geq 0$. A pair of functions $x : [T_1, \infty) \to E$, $u : [T_1, \infty) \to F$ is called a (mild) solution of the system

$$x'(t) = -Ax(t) - G(x(t), u(t)),\ t \in [T_1, \infty)$$

if for every $T_2 > T_1$, $x : [T_1, T_2] \to E$, $u : [T_1, T_2] \to F$ is a solution of (5.20). The set of all such pairs (x, u), which are solutions of the equation above, is denoted by $X(T_1, \infty, -A - G)$.

A function $x : I \to E$, where I is either $[T_1, T_2]$ or $[T_1, \infty)$ $(0 \leq T_1 < T_2)$, is called a trajectory if there exists a Lebesgue measurable function $u : I \to F$ (referred to as a control) such that $(x, u) \in X(T_1, T_2, -A, -G)$ or $(x, u) \in X(T_1, \infty, -A, -G)$, respectively.

Let $T_2 > T_1 \geq 0$ and $(x, u) \in X(T_1, T_2, -A, -G)$. Define

$$I^f(T_1, T_2, x, u) = \int_{T_1}^{T_2} f(x(t), u(t))dt \tag{5.25}$$

which is well-defined but can be ∞. Proposition 1.3 implies that there exists $\bar{M}_* \geq 1$, $\bar{\omega}_* \in R^1$ such that

$$\|e^{-At}\| \leq \bar{M}_* e^{\bar{\omega}_* t},\ t \in [0, \infty). \tag{5.26}$$

We consider functionals of the form $I^f(T_1, T_2, x, u)$, where $0 \leq T_1 < T_2$ and $(x, u) \in X(T_1, T_2, -A, -G)$. For each pair of numbers $T_2 > T_1 \geq 0$ and each pair of points $y, z \in \mathcal{A}$, we define

$$U_-^f(T_1, T_2, y, z) = \inf\{I^f(T_1, T_2, x, u) :$$

$$(x, u) \in X(T_1, T_2, -A, -G),\ x(T_1) = y,\ x(T_2) = z\}, \tag{5.27}$$

$$\sigma_-^f(T_1, T_2, y) = \inf\{U_-^f(T_1, T_2, y, h) : h \in \mathcal{A}\}, \tag{5.28}$$

$$\widehat{\sigma}_-^f(T_1, T_2, y) = \inf\{U_-^f(T_1, T_2, h, y) : h \in \mathcal{A}\}, \tag{5.29}$$

$$\sigma_-^f(T_1, T_2) = \inf\{U_-^f(T_1, T_2, h, y) : h, y \in \mathcal{A}\}. \tag{5.30}$$

Assume that $S_2 > S_1 \geq 0$, $(x, u) \in X(S_1, S_2, A, G)$. For every $t \in [T_1, T_2]$ set

$$\bar{x}(t) = x(S_2 - t + S_1),\ \bar{u}(t) = u(S_2 - t + S_1). \tag{5.31}$$

We show that $(\bar{x}, \bar{u}) \in X(S_1, S_2, -A, -G)$. First we consider the first class of problems. Proposition 1.6 implies that for every $x^* \in \mathcal{D}(A^*)$, $\langle x(\cdot), x^*\rangle$ is an a. c. function on $[S_1, S_2]$ and for all $t \in [S_1, S_2]$,

$$\langle x(t), x^*\rangle = \langle x(S_1), x^*\rangle + \int_{S_1}^t [\langle x(s), A^*x^*\rangle + \langle G(x(s), u(s)), x^*\rangle]ds. \tag{5.32}$$

Let $x^* \in \mathcal{D}(A^*)$. By (5.31) and (5.32), for every $t \in [S_1, S_2]$,

$$\langle \bar{x}(t), x^*\rangle = \langle x(S_2 - t + S_1), x^*\rangle$$

$$= \langle x(S_1), x^*\rangle + \int_{S_1}^{S_2-t+S_1} [\langle x(s), A^*x^*\rangle + \langle G(x(s), u(s)), x^*\rangle]ds$$

$$= \langle x(S_2), x^*\rangle - \int_{S_1}^{S_2} [\langle x(s), A^*x^*\rangle + \langle G(x(s), u(s)), x^*\rangle]ds$$

$$+ \int_{S_1}^{S_2-t+S_1} [\langle x(s), A^*x^*\rangle + \langle G(x(s), u(s)), x^*\rangle]ds$$

$$= \langle \bar{x}(S_1), x^*\rangle - \int_{S_2-t+S_1}^{S_2} [\langle x(s), A^*x^*\rangle + \langle G(x(s), u(s)), x^*\rangle]ds$$

$$= \langle \bar{x}(S_1), x^*\rangle$$

$$- \int_{S_1}^t [\langle x(S_2 - s + S_1), A^*x^*\rangle + \langle G(x(S_2 - s + S_1), u(S_2 - s + S_1)), x^*\rangle]ds$$

$$= \langle \bar{x}(S_1), x^*\rangle + \int_{S_1}^t [\langle \bar{x}(s), -A^*x^*\rangle + \langle -G(\bar{x}(s), \bar{u}(s)), x^*\rangle]ds$$

and for every $t \in [S_1, S_2]$,

$$\langle \bar{x}(t), x^* \rangle = \langle \bar{x}(S_1), x^* \rangle + \int_{S_1}^{t} [\langle \bar{x}(s), -A^* x^* \rangle + \langle -G(\bar{x}(s), \bar{u}(s)), x^* \rangle] ds. \tag{5.33}$$

Proposition 1.6 and (5.33) imply that $(\bar{x}, \bar{u})$ is a solution of (5.20) on $[S_1, S_2]$ and satisfies (5.23). This implies that $(\bar{x}, \bar{u}) \in X(S_1, S_2, -A, -G)$. For the second class of problems $(x, u) \in X(S_1, S_2, -A, -G)$ because the second problem can be considered as a special case of the first class of problems when E is replaced by E_{-1}.

By (5.31),

$$I^f(S_1, S_2, \bar{x}, \bar{u}) = \int_{S_1}^{S_2} f(\bar{x}(t), \bar{u}(t)) dt$$

$$= \int_{S_1}^{S_2} f(x(S_2 - t + S_1), u(S_2 - t + S_1)) dt = \int_{S_1}^{S_2} f(x(t), u(t)) dt. \tag{5.34}$$

Thus we showed that for each $S_2 > S_1 \geq 0$ and each $(x, u) \in X(S_1, S_2, A, G)$,

$$(\bar{x}, \bar{u}) \in X(S_1, S_2, -A, -G), \ I^f(S_1, S_2, \bar{x}, \bar{u}) = I^f(S_1, S_2, x, u), \tag{5.35}$$

and moreover, if $(x, u) \in X(S_1, S_2, -A, -G)$, then $(\bar{x}, \bar{u}) \in X(S_1, S_2, A, G)$.

This implies the following result.

Proposition 5.29. *Let $S_2 > S_1 \geq 0$, $M \geq 0$ and $(x_i, u_i) \in X(S_1, S_2, A, G)$, $i = 1, 2$. Then*

$$I^f(S_1, S_2, x_1, u_1) \geq I^f(S_1, S_2, x_2, u_2) - M \tag{5.36}$$

$$\text{if and only if } I^f(S_1, S_2, \bar{x}_1, \bar{u}_1) \geq I^f(S_1, S_2, \bar{x}_2, \bar{u}_2) - M. \tag{5.37}$$

Proposition 5.29 implies the following result.

Proposition 5.30. *Let $S_2 > S_1 \geq 0$, $M \geq 0$ and $(x, u) \in X(S_1, S_2, A, G)$. Then the following assertions hold:*

$$I^f(S_1, S_2, x, u) \leq \sigma^f(S_1, S_2) + M \text{ if and only if}$$

$$I^f(S_1, S_2, \bar{x}, \bar{u}) \leq \sigma_-^f(S_1, S_2) + M;$$

$$I^f(S_1, S_2, x, u) \leq U^f(S_1, S_2, x(S_1), x(S_2)) + M$$

$$\text{if and only if } I^f(S_1, S_2, \bar{x}, \bar{u}) \leq U_-^f(S_1, S_2, \bar{x}(S_1), \bar{x}(S_2)) + M;$$

$$I^f(S_1, S_2, x, u) \leq \sigma^f(S_1, S_2, x(S_1)) + M$$

if and only if $I^f(S_1, S_2, \bar{x}, \bar{u}) \leq \widehat{\sigma}_-^f(S_1, S_2, \bar{x}(S_2)) + M$;

$$I^f(S_1, S_2, x, u) \leq \widehat{\sigma}^f(S_1, S_2, x(S_2)) + M$$

if and only if $I^f(S_1, S_2, \bar{x}, \bar{u}) \leq \sigma_-^f(S_1, S_2, \bar{x}(S_1) + M$.

Clearly,

$$(x_f, u_f) \in X(0, \infty, -A, -G). \tag{5.38}$$

In view of (5.35), (A1) and (A2) hold for the triplet $(f, -A, -G)$, and if (A3) holds for (f, A, G), then (A3) holds for $(f, -A, -G)$ too. Therefore the results obtained for (f, A, G) also hold for the triplet $(f, -A, -G)$.

Theorem 5.31. *Assume that (f, A, G) has properties (P1) and (P2). Then $(f, -A, -G)$ possesses properties (P1) and (P2).*

Proof. By Theorem 5.13, (f, A, G) has TP. Proposition 5.30, Theorems 5.13 and (5.35) imply that $(f, -A, -G)$ has TP, (P1), and (P2).

We can now improve our boundedness results. Proposition 5.30 and Theorems 5.2 and (5.35) for (f, A, G) and $(f, -A, -G)$ imply the following result.

Theorem 5.32. *Let $M_0 > 0$ and $c > 0$. Then there exists $M_1 > 0$ such that for each $T_1 \geq 0$, each $T_2 \geq T_1 + c$, and each $(x, u) \in X(T_1, T_2)$ satisfying*

$$I^f(T_1, T_2, x, u) \leq (T_2 - T_1)f(x_f, u_f) + M_0,$$

the inequality $\|x(t)\| \leq M_1$ holds for all $t \in [T_1, T_2]$.

Proposition 5.30, (5.35) and Theorem 5.3 for (f, A, G) and $(f, -A, -G)$ imply the following result.

Theorem 5.33. *Let $L > 0$ and $M_0 > 0$. Then there exists $M_1 > 0$ such that for each $T_1 \geq 0$, each $T_2 \geq T_1 + 2L$, and each $(x, u) \in X(T_1, T_2)$ satisfying*

$$x(T_1) \in \mathcal{A}_L, \ x(T_2) \in \widehat{\mathcal{A}}_L,$$

$$I^f(T_1, T_2, x, u) \leq U^f(T_1, T_2, x(T_1), x(T_2)) + M_0,$$

the inequality $\|x(t)| \leq M_1$ holds for all $t \in [T_1, T_2]$.

Propositions 5.30 and (5.35) and Theorem 5.4 for (f, A, G) and Theorem 5.5 for $(f, -A, -G)$ imply the following result.

Theorem 5.34. *Let $L > 0$ and $M_0 > 0$. Then there exists $M_1 > 0$ such that for each $T_1 \geq 0$, each $T_2 \geq T_1 + L$, and each $(x, u) \in X(T_1, T_2)$ satisfying*

$$x(T_1) \in \mathcal{A}_L, \; I^f(T_1, T_2, x, u) \le \sigma^f(T_1, T_2, x(T_1)) + M_0,$$

the inequality $\|x(t)\| \le M_1$ *holds for all* $t \in [T_1, T_2]$.

Proposition 5.30, (5.35), Theorem 5.5 for (f, A, G), and Theorem 5.4 $(f, -A, -G)$ imply the following result.

Theorem 5.35. *Let* $L > 0$ *and* $M_0 > 0$. *Then there exists* $M_1 > 0$ *such that for each* $T_1 \ge 0$, *each* $T_2 \ge T_1 + L$, *and each* $(x, u) \in X(T_1, T_2)$ *satisfying*

$$x(T_2) \in \widehat{\mathcal{A}}_L, \; I^f(T_1, T_2, x, u) \le \widehat{\sigma}^f(T_1, T_2, x_{T_2}) + M_0,$$

the inequality $\|x(t)\| \le M_1$ *holds for all* $t \in [T_1, T_2]$.

5.7 Auxiliary Results for Theorem 5.1

Lemma 5.36. *There exist numbers* $S > 0$ *and* $c_0 \ge 1$ *such that for each* $T_1 \ge 0$, *each* $T_2 \ge T_1 + c_0$, *and each* $(x, u) \in X(T_1, T_2)$,

$$(T_2 - T_1) f(x_f, u_f) \le I^f(T_1, T_2, x, u) + S_2. \tag{5.39}$$

Proof. By (5.9), there exists $S_1 > 0$ such that

$$\psi(S_1) > |f(x_f, u_f)| + a_0 + 1. \tag{5.40}$$

By (A1), there exist $S_2 > 0$, $c_0 \ge 1$ such that

$$(T_2 - T_1) f(x_f, u_f) \le I^f(T_1, T_2, x, u) + S_2 \tag{5.41}$$

for each $T_1 \ge 0$, each $T_2 \ge T_1 + c_0$, and each $(x, u) \in X(T_1, T_2)$ satisfying

$$\|x(T_1)\|, \|x(T_2)\| \le S_1. \tag{5.42}$$

Set

$$S = S_2 + c_0(|f(x_f, u_f)| + a_0 + 1). \tag{5.43}$$

Assume that

$$T_1 \ge 0, \; T_2 \ge T_1 + c_0, \; (x, u) \in X(T_1, T_2). \tag{5.44}$$

We show that (5.39) is true. Assume that

$$\|x(t)\| \geq S_1, \ t \in [T_1, T_2]. \tag{5.45}$$

By (5.10), (5.11), (5.40), and (5.45),

$$f(x(t), u(t)) \geq -a_0 + \psi(\|x(t)\|) \geq -a_0 + \psi(S_1) > |f(x_f, u_f)| + 1. \tag{5.46}$$

In view of (5.46),

$$I^f(T_1, T_2, x, u) \geq (T_2 - T_1)(|f(x_f, u_f)| + 1) \geq (T_2 - T_1) f(x_f, u_f)$$

and (5.39) holds. Assume that

$$\inf\{\|x(t)\| : \ t \in [T_1, T_2]\} < S_1. \tag{5.47}$$

Set

$$\tau_1 = \inf\{t \in [T_1, T_2] : \ \|x(t)\| \leq S_1\}, \tag{5.48}$$

$$\tau_2 = \sup\{t \in [T_1, T_2] : \ \|x(t)\| \leq S_1\}. \tag{5.49}$$

By (5.47)–(5.49), τ_1, τ_2 are well-defined, $\tau_1 \leq \tau_2$ and

$$\|x(\tau_i)\| \leq S_1, \ i = 1, 2. \tag{5.50}$$

It follows from (5.10), (5.11), (5.40), (5.48), and (5.49) that for all $t \in [T_1, \tau_1] \cup [\tau_2, T_2]$, (5.46) is true. There are two cases:

$$\tau_2 - \tau_1 \geq c_0; \tag{5.51}$$

$$\tau_2 - \tau_1 < c_0. \tag{5.52}$$

Assume that (5.51) holds. It follows from (5.41), (5.50), and (5.51) and the choice of S_2 and c_0 that

$$(\tau_2 - \tau_1) f(x_f, u_f) \leq I^f(\tau_1, \tau_2, x, u) + S_2. \tag{5.53}$$

By (5.43), (5.46) holding for each $t \in [T_1, \tau_1] \cup [\tau_2, T_2]$, (5.48), (5.49), and (5.53),

$$(T_2 - T_1) f(x_f, u_f) \leq (\tau_1 - T_1) f(x_f, u_f) + (\tau_2 - \tau_1) f(x_f, u_f) + (T_2 - \tau_2) f(x_f, u_f)$$

$$\leq I^f(T_1, \tau_1, x, u) + I^f(\tau_1, \tau_2, x, u) + S_2 + I^f(\tau_2, T_2, x, u)$$

$$\leq I^f(T_1, T_2, x, u) + S_2 \leq I^f(T_1, T_2, x, u) + S$$

and (5.39) holds.

Assume that (5.52) holds. By (5.46) holding for each $t \in [T_1, \tau_1] \cup [\tau_2, T_2]$, (5.10), (5.11), (5.43), and (5.52),

$$I^f(T_1, T_2, x, u) = I^f(T_1, \tau_1, x, u) + I^f(\tau_1, \tau_2, x, u) + I^f(\tau_2, T_2, x, u)$$
$$\geq (|f(x_f, u_f)| + 1)(\tau_1 - T_1) - a_0(\tau_2 - \tau_1) + (|f(x_f, u_f)| + 1)(T_2 - \tau_2)$$
$$\geq (|f(x_f, u_f)| + 1)(T_2 - T_1) - (a_0 + |f(x_f, u_f)| + 1)(\tau_2 - \tau_1)$$
$$\geq (|f(x_f, u_f)| + 1)(T_2 - T_1) - c_0(|f(x_f, u_f)| + 1 + a_0),$$

$$(T_2 - T_1)f(x_f, u_f) \leq I^f(T_1, T_2, x, u)$$
$$+c_0(|f(x_f, u_f)| + 1 + a_0) \leq I^f(T_1, T_2, x, u) + S.$$

Lemma 5.36 is proved.

Lemma 5.37. *Let $M_0, M_1, \tau_0 > 0$. Then there exists $M_2 > M_1$ such that for each $T_1 \geq 0$, each $T_2 \in (T_1, T_1 + \tau_0]$, and each $(x, u) \in X(T_1, T_2)$ satisfying*

$$\|x(T_1)\| \leq M_1, \tag{5.54}$$

$$I^f(T_1, T_2, x, u) \leq M_0, \tag{5.55}$$

the following inequality holds:

$$\|x(t)\| \leq M_2 \text{ for all } t \in [T_1, T_2]. \tag{5.56}$$

Proof. Let us consider the first class of problems. Fix a positive number

$$\delta < 2^{-1}\min\{1, 16^{-1}\tau_0, 8^{-1}(a_0 + 1)^{-1}e^{-|\omega_*|\delta}(a_0 + 1)^{-1}(M_* + 1)^{-1}\}. \tag{5.57}$$

By (5.9) and (5.10), there exist $h_0 > M_1 + 1$ and $\gamma_0 > 0$ such that

$$f(x, u) \geq 4(M_0 + a_0\tau_0)\delta^{-1} \text{ for each } (x, u) \in \mathcal{M} \text{ satisfying } \|x\| \geq h_0, \tag{5.58}$$

$$f(x, u) \geq 8(\|G(x, u)\| - a_0\|x\|)_+$$
$$\text{for each } (x, u) \in \mathcal{M} \text{ satisfying } \|G(x, u)\| - a_0\|x\| \geq \gamma_0. \tag{5.59}$$

Choose a number

$$M_2 > (2(h_0 + \gamma_0\delta + M_0 + a_0\tau_0) + 8M_1)(M_* + 1)e^{|\omega_*|}\delta^{-1}. \tag{5.60}$$

Let

$$T_1 \geq 0,\ T_2 \in (T_1, T_1 + \tau_0],\ (x, u) \in X(T_1, T_2),$$

(5.54) and (5.55) hold. We show that (5.56) hold. Assume the contrary. Then there exists $t_0 \in [T_1, T_2]$ such that

$$\|x(t_0)\| > M_2. \tag{5.61}$$

By the choice of h_0, (5.54), (5.55), and (5.58), there exists $t_1 \in [T_1, t_0]$ such that

$$\|x(t_1)\| \leq h_0,\ |t_1 - t_0| \leq \delta. \tag{5.62}$$

There exists $t_2 \in [t_1, t_0]$ such that

$$\|x(t_2)\| \geq \|x(t)\| \text{ for all } t \in [t_1, t_0]. \tag{5.63}$$

In view of (5.61)–(5.63),

$$t_2 > t_1. \tag{5.64}$$

By the definition of $X(T_1, T_2)$,

$$\|x(t_2) - e^{A(t_2-t_1)}x(t_1)\| = \|\int_{t_1}^{t_2} e^{A(t_2-s)}G(x(s), u(s))ds\|. \tag{5.65}$$

It follows from (5.8), (5.62), (5.64), and (5.65) that

$$\|x(t_2)\| - M_* e^{|\omega_*|\delta} h_0 \leq \|x(t_2)\| - \|e^{A(t_2-t_1)}\| \|x(t_1)\|$$

$$\leq M_* e^{|\omega_*|\delta} \int_{t_1}^{t_2} \|G(x(s), u(s))\| ds. \tag{5.66}$$

Define

$$E_1 = \{t \in [t_1, t_2] :\ \|G(x(t), u(t))\| \geq a_0 \|x(t)\| + \gamma_0\}, \tag{5.67}$$

$$E_2 = [t_1, t_2] \setminus E_1. \tag{5.68}$$

By (5.10), (5.55), (5.59), (5.62), (5.63), (5.66)–(5.68),

$$\|x(t_2)\| M_*^{-1} e^{-|\omega_*|\delta} - h_0 \leq \int_{t_1}^{t_2} \|G(x(s), u(s))\| ds$$

$$\leq a_0 \int_{t_1}^{t_2} \|x(t)\| dt + \int_{t_1}^{t_2} (\|G(x(s), u(s))\| - a_0 \|x(s)\|)_+ ds$$

$$\leq a_0 \int_{t_1}^{t_2} \|x(t)\| dt + \int_{E_1} (\|G(x(t), u(t))\| - a_0 \|x(t)\|)_+ dt$$

$$+ \int_{E_2} (\|G(x(t), u(t))\| - a_0 \|x(t)\|)_+ dt$$

$$\leq a_0 \|x(t_2)\| \delta + \gamma_0 \delta + \int_{E_1} (\|G(x(t), u(t))\| - a_0 \|x(t)\|)_+ dt$$

$$\leq a_0 \|x(t_2)\| \delta + \gamma_0 \delta + 8^{-1} \int_{E_1} f(x(t), u(t)) dt$$

$$\leq a_0 \|x(t_2)\| \delta + \gamma_0 \delta + 8^{-1}(M_0 + a_0 \tau_0). \tag{5.69}$$

It follows from (5.57), (5.61), (5.63) and (5.69) that

$$2^{-1} M_*^{-1} e^{-|\omega_*|\delta} M_2 \leq M_2 (M_*^{-1} e^{-|\omega_*|\delta} - a_0 \delta)$$

$$\leq \|x(t_2)\| (M_*^{-1} e^{-|\omega_*|\delta} - a_0 \delta) \leq h_0 + \gamma_0 \delta + 8^{-1}(M_0 + a_0 \tau_0).$$

This contradicts (5.60). The contradiction we have reached proves (5.56) and Lemma 5.37 for the first class of problems.

Consider the second class of problems. Recall (see (5.8)) that

$$\|e^{At}\| \leq M_* e^{\omega_* t}, \ t \in [0, \infty) \tag{5.70}$$

and that for every $\tau \geq 0$,

$$\Phi_\tau u = \int_0^\tau e^{A(\tau - s)} B u(s) ds, \tag{5.71}$$

where $\Phi_\tau \in \mathcal{L}(L^2(0, \tau; U), F)$ (see (1.9)). Choose a number

$$\delta \in (0, 1). \tag{5.72}$$

By (5.11), there exists

$$h_0 > M_1 + 1 \tag{5.73}$$

and $\gamma > 1$ such that

$$f(x, u) \geq 4(M_0 + a_0 \tau_0) \delta^{-1} \text{ for each } (x, u) \in \mathcal{M} \text{ satisfying } \|x\| \geq h_0, \tag{5.74}$$

$$f(x, u) \geq K_0 \|u\|^2 / 2 \text{ for each } u \in F \text{ satisfying } \|u\| \geq \gamma. \tag{5.75}$$

Choose

$$M_2 > h_0 M_* e^{|\omega_*|} + \|\Phi_1\|\gamma + \|\Phi_1\|(2K_0^{-1}(M_0 + a_0\tau_0) + 1) + 1. \tag{5.76}$$

Let $T_1 \geq 0,\ T_2 \in (T_1, T_1 + \tau_0]$, $(x, u) \in X(T_1, T_2)$, (5.54) and (5.55) hold. We show that (5.56) holds. Assume the contrary. Then there exists $t_0 \in [T_1, T_2]$ such that (5.61) holds. By the choice of h_0, (5.54), (5.55), and (5.74), there exists

$$t_1 \in [T_1, t_0] \tag{5.77}$$

satisfying (5.62). There exists

$$t_2 \in [t_1, t_0] \tag{5.78}$$

satisfying (5.63) and (5.64). Clearly,

$$(x, u) \in X(t_1, t_2), \tag{5.79}$$

$$x(t_2) = e^{A(t_2 - t_1)}x(t_1) + \int_{t_1}^{t_2} e^{A(t_2 - s)}Bu(s)ds \tag{5.80}$$

in $E_{-1} = \mathcal{D}(A^*)'$. Proposition 1.18, (5.62), (5.70)–(5.72), (5.78), and (5.80) imply that

$$\|x(t_2)\| \leq \|x(t_1)\|M_* e^{|\omega_*|\delta} + \|\int_{t_1}^{t_2} e^{A(t_2 - s)}Bu(s)ds\|$$

$$\leq M_* e^{|\omega_*|\delta}h_0 + \|\int_0^{t_2 - t_1} e^{A(t_2 - t_1 - s)}Bu(t_1 + s)ds\|$$

$$\leq M_* e^{|\omega_*|\delta}h_0 + \|\Phi_{t_2 - t_1}\|\|u(t_1 + \cdot)\|_{L^2(0, t_2 - t_1; F)}$$

$$\leq M_* e^{|\omega_*|\delta}h_0 + \|\Phi_1\|(\int_{t_1}^{t_2} \|u(s)\|^2 ds)^{1/2}. \tag{5.81}$$

Define

$$\Omega_1 = \{t \in [t_1, t_2] :\ \|u(t)\| \geq \gamma\}, \tag{5.82}$$

$$\Omega_2 = [t_1, t_2] \setminus \Omega_1. \tag{5.83}$$

By (5.55), (5.63), (5.75), (5.76), (5.81), and (5.82),

$$\|x(t_0)\| \leq \|x(t_2)\| \leq M_* e^{|\omega_*|\delta}h_0$$

$$+\|\Phi_1\|[(\int_{\Omega_1} \|u(s)\|^2 ds)^{1/2} + (\int_{\Omega_2} \|u(s)\|^2 ds)^{1/2}]$$

$$\leq M_* e^{|\omega_*|\delta} h_0 + \|\Phi_1\|(2K_0^{-1} \int_{\Omega_1} f(x(s), u(s)) ds)^{1/2} + \|\Phi_1\|\gamma\delta^{1/2}$$

$$\leq M_* e^{|\omega_*|} h_0 + \|\Phi_1\|\gamma + \|\Phi_1\|(2K_0^{-1}(M_0 + a_0\tau_0))^{1/2} < M_2. \quad (5.84)$$

This contradicts (5.61). The contradiction we have reached proves (5.56) and Lemma 5.37 for the second class of problems.

Lemma 5.37, (5.9) and (5.10) imply the following result.

Lemma 5.38. *Let $M_1 > 0, 0 < \tau < \tau_0 < \tau_1$. Then there exists $M_2 > 0$ such that for each $T_1 \geq 0$, each $T_2 \in [T_1 + \tau_0, T_1 + \tau_1]$, and each $(x, u) \in X(T_1, T_2)$ satisfying*

$$I^f(T_1, T_2, x, u) \leq M_1,$$

the inequality

$$\|x(t)\| \leq M_2 \text{ for all } t \in [T_1 + \tau, T_2]$$

holds.

5.8 Proofs of Theorems 5.1–5.4

Proof of Theorem 5.1. Assertion 1 follows from Lemma 5.36, (5.10), and (5.11). We will prove Assertion 2. Let $(x, u) \in X(0, \infty)$. Assume that there exists a sequence of numbers $\{t_k\}_{k=1}^{\infty}$ such that

$$t_k \to \infty \text{ as } k \to \infty, \quad (5.85)$$

$$I^f(0, t_k, x, u) - t_k f(x_f, u_f) \to \infty \text{ as } k \to \infty. \quad (5.86)$$

We show that

$$I^f(0, t, x, u) - t f(x_f, u_f) \to \infty \text{ as } t \to \infty. \quad (5.87)$$

In view of (5.9), there exists $S_0 > 0$ such that

$$S_0 > 2\|x(0)\| + 8, \quad (5.88)$$

$$\psi(S_0 - 4) > a_0 + 8 + 8|f(x_f, u_f)|. \tag{5.89}$$

Let $S > 0$ be as guaranteed by Assertion 1. By (5.10), (5.11), and (5.89), we may assume that

$$\liminf_{t\to\infty} \|x(t)\| \le S_0 - 1.$$

For each integer $k \ge 1$ set,

$$\tau_k = \inf\{t \in [t_k, \infty) : \|x(t)\| \le S_0\}. \tag{5.90}$$

Let $k \ge 1$ be an integer and $t \ge \tau_k$. It follows from (5.10), (5.11), (5.86), (5.89), and (5.90) the choice of S and Assertion 1 that

$$I^f(0, t, x, u) - tf(x_f, u_f) \ge I^f(0, t_k, x, u) - t_k f(x_f, u_f)$$

$$+ I^f(t_k, \tau_k, x, u) - (\tau_k - t_k) f(x_f, u_f) - S$$

$$\ge I^f(0, t_k, x, u) - t_k f(x_f, u_f) - S \to \infty \text{ as } k \to \infty.$$

Thus Assertion 2 holds.

Assume that

$$\sup\{|I^f(0, t, x, u) - tf(x_f, u_f)| : t \in [0, \infty)\} < \infty.$$

Therefore there exists $S_1 > 0$ such that for each $T_2 > T_1 \ge 0$,

$$|I^f(T_1, T_2, x, u) - (T_2 - T_1) f(x_f, u_f)| \le S_1. \tag{5.91}$$

Let $S_0 > 0$ be such that

$$\psi(S_0) > |f(x_f, u_f)| + 2 + a_0. \tag{5.92}$$

It follows from (5.10), (5.11), (5.91) and (5.92) that for each $T \ge 0$,

$$\min\{\|x(t)\| : t \in [T, T + S_1]\} \le S_0.$$

By the relation above, there exists a sequence $\{T_i\}_{i=1}^{\infty} \subset [0, \infty)$ such that for each integer $i \ge 1$,

$$T_{i+1} - T_i \in [1, 1 + S_1],$$

$$\|x(T_i)\| \le S_0.$$

This relation, Lemma 5.37 and (5.91) imply that

$$\sup\{\|x(t)\| : t \in [0, \infty)\} < \infty.$$

Theorem 5.1 is proved.

Proof of Theorem 5.2. We may assume that $c_0 < c/4$. By Theorem 5.1, there exists $S_0 > 0$ such that the following property holds:

(i) for each $T_2 > T_1 \geq 0$ and each $(x, u) \in X(T_1, T_2)$,

$$I^f(T_1, T_2, x, u) + S_0 \geq (T_2 - T_1) f(x_f, u_f).$$

Lemma 5.38 implies that there exists $M_1 > 0$ such that the following property holds:

(ii) for each $T \geq 0$ and each $(x, u) \in X(T, T + c/2)$ satisfying

$$I^f(T, T + c/2, x, u) \leq M_0 + 2S_0 + 1 + c|f(x_f, u_f)|,$$

we have

$$\|x(t)\| \leq M_1 \text{ for all } t \in [c_0 + T, T + c/2].$$

Assume that $T_1 \geq 0$, $T_2 \geq T_1 + c$, $(x, u) \in X(T_1, T_2)$ and

$$I^f(T_1, T_2, x, u) \leq (T_2 - T_1) f(x_f, u_f) + M_0. \tag{5.93}$$

In order to prove the theorem, it is sufficient to show that

$$\|x(t)\| \leq M_1, \ t \in [T_1 + c_0, T_2].$$

Assume the contrary. Then there exists

$$t_0 \in [T_1 + c_0, T_2] \tag{5.94}$$

such that

$$\|x(t_0)\| > M_1. \tag{5.95}$$

By (5.94), there exists a closed interval $[a, b] \subset [T_1, T_2]$ such that

$$t_0 \in [a + c_0, b], \ b - a = 2^{-1}c. \tag{5.96}$$

Property (ii), (5.95) and (5.96) imply that

$$I^f(a,b,x,u) > M_0 + 2S_0 + 1 + c|f(x_f,u_f)|. \tag{5.97}$$

Property (i), (5.96) and (5.97) imply that

$$I^f(T_1,T_2,x,u) = I^f(T_1,a,x,u) + I^f(a,b,x,u) + I^f(b,T_2,x,u)$$

$$\geq (a-T_1)f(x_f,u_f) - S_0 + c|f(x_f,u_f)| + M_0 + 2S_0 + 1 + (T_2-b)f(x_f,u_f) - S_0$$

$$> (T_2-T_1)f(x_f,u_f) + M_0 + 1.$$

This contradicts (5.93). The contradiction we have reached proves Theorem 5.2.

Proof of Theorem 5.3. Theorem 5.2 implies that there exists $M_1 > 0$ such that the following property holds:

(iii) for each $T_1 \geq 0$, each $T_2 \geq T_1 + 2L$, and each $(x,u) \in X(T_1,T_2)$ satisfying

$$I^f(T_1,T_2,x,u) \leq (T_2-T_1)f(x_f,u_f) + 2L + M_0 + 2L|f(x_f,u_f)|,$$

the inequality $\|x(t)\| \leq M_1$ holds for all $t \in [T_1+c, T_2]$.

Assume that $T_1 \geq 0$, $T_2 \geq T_1 + 2L$, and $(x,u) \in X(T_1,T_2)$ satisfy

$$x(T_1) \in \mathcal{A}_L,\ x(T_2) \in \widehat{\mathcal{A}}_L, \tag{5.98}$$

$$I^f(T_1,T_2,x,u) \leq U^f(T_1,T_2,x(T_1),x(T_2)) + M_0. \tag{5.99}$$

In view of (5.98), there exist $\tau_1 \in (0,L]$, $\tau_2 \in (0,L]$, and $(y,v) \in X(T_1,T_2)$ such that

$$y(T_1) = x(T_1),\ y(T_2) = x(T_2), \tag{5.100}$$

$$(y(t),v(t)) = (x_f,u_f),\ t \in [\tau_1+T_1, T_2-\tau_2], \tag{5.101}$$

$$I^f(T_1,T_1+\tau_1,y,v) \leq L,\ I^f(T_2-\tau_2,T_2,y,v) \leq L. \tag{5.102}$$

By (5.99)–(5.102),

$$I^f(T_1,T_2,x,u) \leq M_0 + I^f(T_1,T_2,y,v)$$

$$= I^f(T_1,T_1+\tau_1,y,v) + I^f(T_1+\tau_1,T_2-\tau_2,y,v) + I^f(T_2-\tau_2,T_2,y,v) + M_0$$

$$\leq 2L + M_0 + (T_2-T_1-\tau_2-\tau_1)f(x_f,u_f)$$

$$\leq (T_2-T_1)f(x_f,u_f) + 2L + M_0 + 2L|f(x_f,u_f)|.$$

Together with property (iii), this implies that $\|x(t)\| \le M_1$ for all $t \in [T_1 + c, T_2]$. Theorem 5.3 is proved.

Proof of Theorem 5.4. Theorem 5.2 implies that there exists $M_1 > 0$ such that the following property holds:

(iv) for each $T_1 \ge 0$, each $T_2 \ge T_1 + L$, and each $(x, u) \in X(T_1, T_2)$ satisfying

$$I^f(T_1, T_2, x, u) \le (T_2 - T_1) f(x_f, u_f) + L + M_0 + L|f(x_f, u_f)|,$$

the inequality $\|x(t)\| \le M_1$ holds for all $t \in [T_1 + c, T_2]$.

Assume that $T_1 \ge 0$, $T_2 \ge T_1 + L$, and $(x, u) \in X(T_1, T_2)$ satisfy

$$x(T_1) \in \mathcal{A}_L, \tag{5.103}$$

$$I^f(T_1, T_2, x, u) \le \sigma^f(T_1, T_2, x(T_1)) + M_0. \tag{5.104}$$

In view of (5.103), there exist $\tau \in (0, L]$ and $(y, v) \in X(T_1, T_2)$ such that

$$y(T_1) = x(T_1),\ (y(t), v(t)) = (x_f, u_f),\ t \in [\tau + T_1, T_2], \tag{5.105}$$

$$I^f(T_1, T_1 + \tau, y, v) \le L. \tag{5.106}$$

By (5.104)–(5.106),

$$\begin{aligned} I^f(T_1, T_2, x, u) &\le M_0 + I^f(T_1, T_2, y, v) \\ &= I^f(T_1, T_1 + \tau, y, v) + I^f(T_1 + \tau, T_2, y, v) + M_0 \\ &\le L + M_0 + (T_2 - T_1 - \tau) f(x_f, u_f) \\ &\le (T_2 - T_1) f(x_f, u_f) + L + M_0 + L|f(x_f, u_f)|. \end{aligned}$$

Together with property (iv), this implies that $\|x(t)\| \le M_1$ for all $t \in [T_1 + c, T_2]$. Theorem 5.4 is proved.

Proof of Theorem 5.5. Theorem 5.2 implies that there exists $M_1 > 0$ such that the following property holds:

(v) for each $T_1 \ge 0$, each $T_2 \ge T_1 + L$, and each $(x, u) \in X(T_1, T_2)$ satisfying

$$I^f(T_1, T_2, x, u) \le (T_2 - T_1) f(x_f, u_f) + L + M_0 + L|f(x_f, u_f)|,$$

the inequality $\|x(t)\| \le M_1$ holds for all $t \in [T_1 + c, T_2]$.

Assume that $T_1 \ge 0$, $T_2 \ge T_1 + L$, and $(x, u) \in X(T_1, T_2)$ satisfy

$$x(T_2) \in \widehat{\mathcal{A}}_L, \tag{5.107}$$

$$I^f(T_1, T_2, x, u) \le \widehat{\sigma}^f(T_1, T_2, x(T_2)) + M_0. \tag{5.108}$$

In view of (5.107), there exist $\tau \in (0, L]$ and $(y, v) \in X(T_1, T_2)$ such that

$$(y(t), v(t)) = (x_f, u_f),\ t \in [T_1, T_2 - \tau],\ y(T_2) = x(T_2), \tag{5.109}$$

$$I^f(T_2 - \tau, T_2, y, v) \le L. \tag{5.110}$$

By (5.108)–(5.110),

$$\begin{aligned} I^f(T_1, T_2, x, u) &\le M_0 + I^f(T_1, T_2, y, v) \\ &= I^f(T_1, T_2 - \tau, y, v) + I^f(T_2 - \tau, T_2, y, v) + M_0 \\ &\le L + M_0 + (T_2 - T_1 - \tau) f(x_f, u_f) \\ &\le (T_2 - T_1) f(x_f, u_f) + L + M_0 + L|f(x_f, u_f)|. \end{aligned}$$

Together with property (v), this implies that $\|x(t)\| \le M_1$ for all $t \in [T_1 + c, T_2]$. Theorem 5.5 is proved.

5.9 Proofs of Theorems 5.6–5.9

Proof of Theorem 5.6. Fix $c_0 = c/4$. Theorem 5.2 implies that there exists $M_2 > 0$ such that the following property holds:

(i) for each $T_1 \ge 0$, each $T_2 \ge T_1 + c$, and each $(x, u) \in X(T_1, T_2)$ satisfying

$$I^f(T_1, T_2, x, u) \le (T_2 - T_1) f(x_f, u_f) + M_0,$$

the inequality $\|x(t)\| \le M_2$ holds for all $t \in [T_1 + c_0, T_2]$.

By Theorem 5.1, there exists $S_0 > 0$ such that the following property holds:

(ii) for each $T_2 > T_1 \ge 0$, and each $(x, u) \in X(T_1, T_2)$,

$$I^f(T_1, T_2, x, u) + S_0 \ge (T_2 - T_1) f(x_f, u_f).$$

Lemma 5.37 implies that there exists $M_1 > M_2$ such that the following property holds:

(iii) for each $T \geq 0$ and each $(x, u) \in X(T, T + c_0)$ satisfying

$$\|x(T_1)\| \leq M_0,\ I^f(T, T + c_0, x, u) \leq M_0 + S_0 + 2c_0|f(x_f, u_f)|,$$

we have

$$\|x(t)\| \leq M_0 \text{ for all } t \in [T, T + c_0].$$

Assume that $T_1 \geq 0$, $T_2 \geq T_1 + c$, $(x, u) \in X(T_1, T_2)$,

$$\|x(T_1)\| \leq M_0, \tag{5.111}$$

$$I^f(T_1, T_2, x, u) \leq (T_2 - T_1)f(x_f, u_f) + M_0. \tag{5.112}$$

Property (i) and (5.112) imply that

$$\|x(t)\| \leq M_2 < M_1,\ t \in [T_1 + c_0, T_2]. \tag{5.113}$$

It follows from (5.112) and property (ii) that

$$I^f(T_1, T_1 + c_0, x, u) = I^f(T_1, T_2, x, u) - I^f(T_1 + c_0, T_2, x, u)$$

$$\leq (T_2 - T_1)f(x_f, u_f) + M_0 - (T_2 - T_1 - c_0)f(x_f, u_f) + S_0 = c_0 f(x_f, u_f) + M_0 + S_0.$$

By the relation above, (5.111) and property (iii),

$$\|x(t)\| \leq M_1,\ t \in [T_1, T_1 + c_0].$$

Together with (5.113), this completes the proof of Theorem 5.6.

Proof of Theorem 5.7. Theorem 5.6 implies that there exists $M_1 > 0$ such that the following property holds:

(iv) for each $T_1 \geq 0$, each $T_2 \geq T_1 + 2L$, and each $(x, u) \in X(T_1, T_2)$ satisfying

$$\|x(T_1)\| \leq M_0,$$

$$I^f(T_1, T_2, x, u) \leq (T_2 - T_1)f(x_f, u_f) + 2L + 2M_0 + 2L|f(x_f, u_f)|,$$

the inequality $\|x(t)\| \leq M_1$ holds for all $t \in [T_1, T_2]$.

By Theorem 5.1, there exists $S_0 > 0$ such that the following property holds:

(v) for each $T_2 > T_1 \geq 0$ and each $(x, u) \in X(T_1, T_2)$,

$$I^f(T_1, T_2, x, u) + S_0 \geq (T_2 - T_1)f(x_f, u_f).$$

Assume that $T_1 \geq 0$, $T_2 \geq T_1 + 2L$, $(x, u) \in X(T_1, T_2)$,

$$\|x(T_1)\| \leq M_0, \tag{5.114}$$

$$x(T_1) \in \mathcal{A}_L, \; x(T_2) \in \widehat{\mathcal{A}}_L, \tag{5.115}$$

$$I^f(T_1, T_2, x, u) \leq U^f(T_1, T_2, x(T_1), x(T_2)) + M_0. \tag{5.116}$$

In view of (5.115), there exist $\tau_1 \in (0, L]$, $\tau_2 \in (0, L]$ and $(y, v) \in X(T_1, T_2)$ such that

$$y(T_1) = x(T_1), \; y(T_2) = x(T_2), \tag{5.117}$$

$$(y(t), v(t)) = (x_f, u_f), \; t \in [\tau_1 + T_1, T_2 - \tau_2], \tag{5.118}$$

$$I^f(T_1, T_1 + \tau_1, y, v) \leq L, \; I^f(T_2 - \tau_2, T_2, y, v) \leq L. \tag{5.119}$$

By (5.116)–(5.119),

$$I^f(T_1, T_2, x, u) \leq M_0 + I^f(T_1, T_2, y, v)$$

$$= I^f(T_1, T_1 + \tau_1, y, v) + I^f(T_1 + \tau_1, T_2 - \tau_2, y, v) + I^f(T_2 - \tau_2, T_2, y, v) + M_0$$

$$\leq 2L + M_0 + (T_2 - T_1 - \tau_2 - \tau_1) f(x_f, u_f)$$

$$\leq (T_2 - T_1) f(x_f, u_f) + 2L + M_0 + 2L|f(x_f, u_f)|. \tag{5.120}$$

Property (iv), (5.114) and (5.120) imply that $\|x(t)\| \leq M_1$ for all $t \in [T_1, T_2]$. Theorem 5.7 is proved.

Proof of Theorem 5.8. Theorem 5.6 implies that there exists $M_1 > 0$ such that the following property holds:

(vi) for each $T_1 \geq 0$, each $T_2 \geq T_1 + L$, and each $(x, u) \in X(T_1, T_2)$ satisfying

$$\|x(T_1)\| \leq M_0,$$

$$I^f(T_1, T_2, x, u) \leq (T_2 - T_1) f(x_f, u_f) + L + 2M_0 + L|f(x_f, u_f)|,$$

the inequality $\|x(t)\| \leq M_1$ holds for all $t \in [T_1, T_2]$.

By Theorem 5.1, there exists $S_0 > 0$ such that the following property holds:

(vii) for each $T_2 > T_1 \geq 0$ and each $(x, u) \in X(T_1, T_2)$,

$$I^f(T_1, T_2, x, u) + S_0 \geq (T_2 - T_1) f(x_f, u_f).$$

Assume that $T_1 \geq 0$, $T_2 \geq T_1 + L$, $(x, u) \in X(T_1, T_2)$,

$$\|x(T_1)\| \leq M_0, \ x(T_1) \in \mathcal{A}_L, \tag{5.121}$$

$$I^f(T_1, T_2, x, u) \leq \sigma^f(T_1, T_2, x(T_1)) + M_0. \tag{5.122}$$

In view of (5.121), there exist $\tau \in (0, L]$, $(y, v) \in X(T_1, T_2)$ such that

$$y(T_1) = x(T_1), \ (y(t), v(t)) = (x_f, u_f), \ t \in [\tau + T_1, T_2], \tag{5.123}$$

$$I^f(T_1, T_1 + \tau, y, v) \leq L. \tag{5.124}$$

By (5.122)–(5.124),

$$\begin{aligned} I^f(T_1, T_2, x, u) &\leq M_0 + I^f(T_1, T_2, y, v) \\ &= I^f(T_1, T_1 + \tau, y, v) + I^f(T_1 + \tau, T_2, y, v) + M_0 \\ &\leq L + M_0 + (T_2 - T_1 - \tau) f(x_f, u_f) \\ &\leq (T_2 - T_1) f(x_f, u_f) + L + M_0 + L|f(x_f, u_f)|. \end{aligned} \tag{5.125}$$

Together with property (vi), (5.121) and (5.125), this implies that $\|x(t)\| \leq M_1$ for all $t \in [T_1, T_2]$. Theorem 5.8 is proved.

Proof of Theorem 5.9. Theorem 5.6 implies that there exists $M_1 > 0$ such that the following property holds:

(viii) for each $T_1 \geq 0$, each $T_2 \geq T_1 + L$, and each $(x, u) \in X(T_1, T_2)$ satisfying

$$\|x(T_1)\| \leq M_0,$$

$$I^f(T_1, T_2, x, u) \leq (T_2 - T_1) f(x_f, u_f) + L + 2M_0 + L|f(x_f, u_f)|,$$

the inequality $\|x(t)\| \leq M_1$ holds for all $t \in [T_1, T_2]$.

By Theorem 5.1, there exists $S_0 > 0$ such that the following property holds:

(ix) for each $T_2 > T_1 \geq 0$ and each $(x, u) \in X(T_1, T_2)$,

$$I^f(T_1, T_2, x, u) + S_0 \geq (T_2 - T_1) f(x_f, u_f).$$

Assume that $T_1 \geq 0$, $T_2 \geq T_1 + L$, $(x, u) \in X(T_1, T_2)$,

$$\|x(T_1)\| \leq M_0, \ x(T_2) \in \widehat{\mathcal{A}}_L, \tag{5.126}$$

$$I^f(T_1, T_2, x, u) \leq \widehat{\sigma}^f(T_1, T_2, x(T_2)) + M_0. \tag{5.127}$$

In view of (5.126), there exist $\tau \in (0, L]$ and $(y, v) \in X(T_1, T_2)$ such that

$$(y(t), v(t)) = (x_f, u_f),\ t \in [T_1, T_2 - \tau], \tag{5.128}$$

$$y(T_2) = x(T_2), \tag{5.129}$$

$$I^f(T_2 - \tau, T_2, y, v) \le L. \tag{5.130}$$

By (5.127)–(5.130),

$$I^f(T_1, T_2, x, u) \le M_0 + I^f(T_1, T_2, y, v)$$

$$\le L + M_0 + (T_2 - T_1 - \tau) f(x_f, u_f) \le (T_2 - T_1) f(x_f, u_f) + L + M_0 + L|f(x_f, u_f)|. \tag{5.131}$$

Property (viii), (5.126) and (5.131) imply that $\|x(t)\| \le M_1$ for all $t \in [T_1, T_2]$. Theorem 5.9 is proved.

5.10 Proofs of Theorems 5.11 and 5.12

Theorem 5.11 follows from Proposition 5.10 and the following result.

Proposition 5.39. *Let $L > 0$. Then for each $T_1 \ge 0$, each $T_2 \ge T_1 + 2L$, and each pair of points $y \in \mathcal{A}_L$, $z \in \widehat{\mathcal{A}}_L$,*

$$U^f(T_1, T_2, y, z) \le (T_2 - T_1) f(x_f, u_f) + 2L(1 + |f(x_f, u_f)|).$$

Proof. Let $T_1 \ge 0$, $T_2 \ge T_1 + 2L$, $y \in \mathcal{A}_L$, $z \in \widehat{\mathcal{A}}_L$. By the definition of $\mathcal{A}_L$, $\widehat{\mathcal{A}}_L$, there exist $\tau_1 \in (0, L]$, $\tau_2 \in (0, L]$, and $(x, u) \in X(T_1, T_2)$ such that

$$x(T_1) = y,\ (x(t), u(t)) = (x_f, u_f),\ t \in [T_1 + \tau_1, T_2 - \tau_2],\ x(T_2) = z_2,$$

$$I^f(T_1, T_1 + \tau_1, x, u) \le L,\ I^f(T_2 - \tau_2, T_2, x, u) \le L.$$

In view of the relations above,

$$U^f(T_1, T_2, y, z) \le I^f(T_1, T_2, x, u)$$

$$= I^f(T_1, T_1 + \tau_1, x, u) + I^f(T_1 + \tau_1, T_2 - \tau_2, x, u) + I^f(T_2 - \tau_2, T_2, x, u)$$

$$\le 2L + (T_2 - T_1 - \tau_2 - \tau_1) f(x_f, u_f) \le 2L + (T_2 - T_1) f(x_f, u_f) + 2L|f(x_f, u_f)|.$$

Proposition 5.39 is proved.

Theorem 5.12 follows from Proposition 5.10 and the following result.

Proposition 5.40. *Let $L > 0$. Then for each $T_1 \geq 0$, each $T_2 \geq T_1 + L$, and each $y \in \mathcal{A}_L$,*

$$\sigma^f(T_1, T_2, y) \leq (T_2 - T_1)f(x_f, u_f) + L(1 + |f(x_f, u_f)|).$$

Proof. Let $T_1 \geq 0$, $T_2 \geq T_1 + L$, $y \in \mathcal{A}_L$. By the definition of $\mathcal{A}_L$, there exist $\tau \in (0, L]$ and $(x, u) \in X(T_1, T_2)$ such that

$$x(T_1) = y, \ (x(t), u(t)) = (x_f, u_f), \ t \in [T_1 + \tau, T_2],$$

$$I^f(T_1, T_1 + \tau, x, u) \leq L.$$

In view of the relations above,

$$\sigma^f(T_1, T_2, y) \leq I^f(T_1, T_2, x, u)$$

$$= I^f(T_1, T_1 + \tau, x, u) + I^f(T_1 + \tau, T_2, x, u)$$

$$\leq L + (T_2 - T_1 - \tau)f(x_f, u_f) \leq (T_2 - T_1)f(x_f, u_f) + L(1 + |f(x_f, u_f)|).$$

Proposition 5.40 is proved.

5.11 Proof of Theorem 5.13

First we show that TP implies (P1) and (P2). In view of Theorem 5.1, TP implies (P2). We show that TP implies (P1). Assume that TP holds, $(x, u) \in X(0, \infty)$ is (f)-good. There exists $S > 0$ such that for each $T > 0$,

$$|I^f(0, T, x, u) - Tf(x_f, u_f)| < S.$$

This implies that for each pair of numbers $T_2 > T_1 \geq 0$,

$$|I^f(T_1, T_2, x, u) - (T_2 - T_1)f(x_f, u_f)| < 2S. \tag{5.132}$$

Let $\delta > 0$. We show that there exists $T_\delta > 0$ such that for each $T > T_\delta$,

$$I^f(T_\delta, T, x, u) \leq U^f(T_\delta, T, x(T_\delta), x(T)) + \delta.$$

Assume the contrary. Then for each $T \geq 0$, there exists $S > T$ such that

$$I^f(T, S, x, u) > U^f(T, S, x(T), x(S)) + \delta.$$

This implies that there exists a strictly increasing sequence of numbers $\{T_i\}_{i=0}^{\infty}$ such that

$$T_0 = 0$$

and for every integer $i \geq 0$,

$$I^f(T_i, T_{i+1}, x, u) > U^f(T_i, T_{i+1}, x(T_i), x(T_{i+1})) + \delta. \tag{5.133}$$

By (5.133), there exists $(y, v) \in X(0, \infty)$ such that for every integer $i \geq 0$,

$$y(T_i) = x(T_i), \tag{5.134}$$

$$I^f(T_i, T_{i+1}, x, u) > I^f(T_i, T_{i+1}, y, v) + \delta. \tag{5.135}$$

In view of (5.132), (5.134) and (5.135), for each integer $k \geq 1$,

$$I^f(0, T_k, y, v) - T_k f(x_f, u_f)$$

$$= I^f(0, T_k, y, v) - I^f(0, T_k, x, u) + I^f(0, T_k, x, u)$$

$$-T_k f(x_f, u_f) \leq -k\delta + 2S \to -\infty \text{ as } k \to \infty.$$

This contradicts Theorem 5.1. The contradiction we have reached proves that the following property holds:

(i) for each $\delta > 0$ there exists $T_\delta > 0$ such that for each integer $T > T_\delta$,

$$I^f(T_\delta, T, x, u) \leq U^f(T_\delta, T, x(T_\delta), x(T)) + \delta.$$

Let $\epsilon > 0$. Proposition 5.10 implies that there exist $\delta > 0$, $L > 0$ such that the following property holds:

(ii) for each $S_1 \geq 0$, each $S_2 \geq S_1 + 2L$, and each $(z, \xi) \in X(S_1, S_2)$ which satisfies

$$I^f(S_1, S_2, z, \xi) \leq \min\{(S_2 - S_1) f(x_f, u_f) + 2S, U^f(S_1, S_2, z(S_1), z(S_2)) + \delta\},$$

we have

$$\|z(t) - x_f\| \leq \epsilon, \; t \in [S_1 + L, S_2 - L].$$

Let $T_\delta > 0$ be as guaranteed by (i). Properties (i) and (ii) (5.132) and the choice of T_δ imply that for each $T \geq T_\delta + 2L$,

$$\|x(t) - x_f\| \leq \epsilon, \; t \in [T_\delta + L, T - L].$$

Therefore

$$\|x(t) - x_f\| \le \epsilon \text{ for all } t \ge T_\delta + L$$

and (P1) holds. Thus TP implies (P1).

Lemma 5.41. *Assume that (P1) holds and that $\epsilon > 0$. Then there exists $\delta > 0$ such that for each $T_1 \ge 0$, each $T_2 \ge T_1 + 2b_f$, and each $(x, u) \in X(T_1, T_2)$ which satisfies*

$$\|x(T_i) - x_f\| \le \delta,\ i = 1, 2,$$

$$I^f(T_1, T_2, x, u) \le U^f(T_1, T_2, x(T_1), x(T_2)) + \delta,$$

the inequality $\|x(t) - x_f\| \le \epsilon$ holds for all $t \in [T_1, T_2]$.

Proof. By (A2), for each integer $k \ge 1$, there exists $\delta_k \in (0, 2^{-k})$ such that the following property holds:

(iii) for each $z \in \mathcal{A}$ satisfying $\|z - x_f\| \le \delta_k$, there exist $\tau_1, \tau_2 \in (0, b_f]$ and $(x_1, u_1) \in X(0, \tau_1)$, $(x_2, u_2) \in X(0, \tau_2)$ which satisfy

$$x_1(0) = z,\ x_1(\tau_1) = x_f,\ x_2(0) = x_f,\ x_2(\tau_2) = z,$$

$$I^f(0, \tau_i, x_i, u_i) \le \tau_i f(x_f, u_f) + 2^{-k},\ i = 1, 2.$$

Assume that the lemma does not hold. Then for each integer $k \ge 1$, there exist $T_k \ge 2b_f$ and $(x_k, u_k) \in X(0, T_k)$ such that

$$\|x_k(0) - x_f\| \le \delta_k,\ \|x_k(T_k) - x_f\| \le \delta_k, \tag{5.136}$$

$$I^f(0, T_k, x_k, u_k) \le U^f(0, T_k, x_k(0), x_k(T_k)) + \delta_k, \tag{5.137}$$

$$\sup\{\|x_k(t) - x_f\| : t \in [0, T_k]\} > \epsilon. \tag{5.138}$$

Let $k \ge 1$ be an integer. Property (iii) and (5.136) imply that there exist $\tau_{k,1}, \tau_{k,2} \in (0, b_f]$ and $(\tilde{x}_k, \tilde{u}_k) \in X(0, T_k + \tau_{k,1} + \tau_{k,2})$ such that

$$\tilde{x}_k(0) = x_f,\ \tilde{x}_k(\tau_{k,1}) = x_k(0), \tag{5.139}$$

$$I^f(0, \tau_{k,1}, \tilde{x}_k, \tilde{u}_k) \le \tau_{k,1} f(x_f, u_f) + 2^{-k}, \tag{5.140}$$

$$(\tilde{x}_k(\tau_{k,1} + t), \tilde{u}_k(\tau_{k,1} + t)) = (x_k(t), u_k(t)),\ t \in [0, T_k], \tag{5.141}$$

$$\tilde{x}_k(T_k + \tau_{k,1} + \tau_{k,2}) = x_f, \tag{5.142}$$

$$I^f(T_k+\tau_{k,1}, T_k+\tau_{k,1}+\tau_{k,2}, \tilde{x}_k, \tilde{u}_k) \le \tau_{k,2} f(x_f, u_f) + 2^{-k}. \tag{5.143}$$

By (5.137), (5.140), (5.141), and (5.143),

$$I^f(0, T_k+\tau_{k,1}+\tau_{k,2}, \tilde{x}_k, \tilde{u}_k) \le (\tau_{k,1}+\tau_{k,2}) f(x_f, u_f)+2^{-k+1}+I^f(0, T_k, x_k, u_k)$$
$$\le (\tau_{k,1}+\tau_{k,2}) f(x_f, u_f) + U^f(0, T_k, x_k(0), x_k(T_k)) + \delta_k + 2^{-k+1}. \tag{5.144}$$

Property (iii) and (5.136) imply that there exist $\tau_{k,3}, \tau_{k,4} \in (0, b_f]$ and $(\widehat{x}_k, \widehat{u}_k) \in X(0, T_k)$ such that

$$\widehat{x}_k(0) = x_k(0), \ \widehat{x}_k(T_k) = x_k(T_k), \tag{5.145}$$

$$(\widehat{x}_k(t), \widehat{u}_k(t)) = (x_f, u_f), \ t \in [\tau_{k,3}, T_k - \tau_{k,4}], \tag{5.146}$$

$$I^f(0, \tau_{k,3}, \widehat{x}_k, \widehat{u}_k) \le \tau_{k,3} f(x_f, u_f) + 2^{-k}, \tag{5.147}$$

$$I^f(T_k - \tau_{k,4}, T_k, \widehat{x}_k, \widehat{u}_k) \le \tau_{k,4} f(x_f, u_f) + 2^{-k}. \tag{5.148}$$

In view of (5.145)–(5.148),

$$U^f(0, T_k, x_k(0), x_k(T_k)) \le I^f(0, T_k, \widehat{x}_k, \widehat{u}_k) \le T_k f(x_f, u_f) + 2^{-k+1}. \tag{5.149}$$

By (5.144) and (5.149),

$$I^f(0, T_k+\tau_{k,1}+\tau_{k,2}, \tilde{x}_k, \tilde{u}_k) \le (T_k+\tau_{k,1}+\tau_{k,2}) f(x_f, u_f)+3\cdot 2^{-k+1}. \tag{5.150}$$

By (5.139) and (5.142), there exists $(x, u) \in X(0, \infty)$ such that

$$(x(t), u(t)) = (\tilde{x}_1(t), \tilde{u}_1(t)), \ t = [0, T_1 + \tau_{1,1} + \tau_{1,2}], \tag{5.151}$$

and for each integer $k \ge 1$ and each $t \in [0, T_{k+1} + \tau_{k+1,1} + \tau_{k+1,2}]$,

$$(x(\sum_{i=1}^{k}(T_i + \tau_{i,1} + \tau_{i,2} + t)), u(\sum_{i=1}^{k}(T_i + \tau_{i,1} + \tau_{i,2} + t))) = (\tilde{x}_{k+1}(t), \tilde{u}_{k+1}(t)). \tag{5.152}$$

In view of (5.150)–(5.152), for each integer $k \ge 1$,

$$I^f(0, \sum_{i=1}^{k}(T_i + \tau_{i,1} + \tau_{i,2}), \tilde{x}, \tilde{u}) = \sum_{i=1}^{k} I^f(0, T_i + \tau_{i,1} + \tau_{i,2}, \tilde{x}_i, \tilde{u}_i)$$

$$\leq \sum_{i=1}^{k}(T_i+\tau_{i,1}+\tau_{i,2})f(x_f,u_f)+3\sum_{i=1}^{k}2^{-i-1} \leq \sum_{i=1}^{k}(T_i+\tau_{i,1}+\tau_{i,2})f(x_f,u_f)+6.$$

Theorem 5.1 implies that $(\tilde{x},\tilde{u})$ is (f)-good. Together with (P1), this implies that $\lim_{t\to\infty}\|\tilde{x}(t)-x_f\| = 0$. On the other hand, in view of (5.138), (5.141), and (5.152), $\limsup_{t\to\infty}\|\tilde{x}(t)-x_f\| > \epsilon$. The contradiction we have reached completes the proof of Lemma 5.41.

Completion of the Proof of Theorem 5.13. Assume that properties (P1) and (P2) hold. Let $\epsilon, M > 0$. By Lemma 5.41, there exists $\delta_0 > 0$ such that the following property holds:

(iv) for each $T_1 \geq 0$, each $T_2 \geq T_1 + 2b_f$, and each $(x,u) \in X(T_1,T_2)$ which satisfies

$$\|x(T_i)-x_f\| \leq \delta_0,\ i=1,2,$$

$$I^f(T_1,T_2,x,u) \leq U^f(T_1,T_2,x(T_1),x(T_2))+\delta_0,$$

the inequality $\|x(t)-x_f\| \leq \epsilon$ holds for all $t \in [T_1,T_2]$.

By Theorem 5.1, there exists $S_0 > 0$ such that the following property holds:

(v) for each $T_2 > T_1 \geq 0$ and each $(x,u) \in X(T_1,T_2)$, we have

$$I^f(T_1,T_2,x,u)+S_0 \geq (T_2-T_1)f(x_f,u_f).$$

Set

$$S_1 = 2S_0 + M. \tag{5.153}$$

In view of (P2), there exist $\delta \in (0,\delta_0)$ and $L_0 > 0$ such that the following property holds:

(vi) for each $(x,u) \in X(0,L_0)$ which satisfies

$$I^f(0,L_0,x,u) \leq \min\{U^f(0,L_0,x(0),x(L_0))+\delta,\ L_0 f(x_f,u_f)+S_1\},$$

there exists $t_0 \in [0,L_0]$ such that $\|x(t_0)-x_f\| \leq \delta_0$.

Set

$$L = L_0 + b_f. \tag{5.154}$$

Assume that $T_1 \geq 0$ and $T_2 \geq T_1 + 2L$ and that $(x,u) \in X(T_1,T_2)$ satisfy

$$I^f(T_1,T_2,x,u) \leq \min\{\sigma^f(T_1,T_2)+M, U^f(T_1,T_2,x(T_1),x(T_2))+\delta\}. \tag{5.155}$$

Property (v) and (5.155) imply that for each pair of numbers $Q_1, Q_2 \in [T_1, T_2]$ satisfying $Q_1 < Q_2$,

$$I^f(Q_1, Q_2, x, u) = I^f(T_1, T_2, x, u) - I^f(T_1, Q_1, x, u) - I^f(Q_2, T_2, x, u)$$

$$\leq (T_2 - T_1)f(x_f, u_f) + M - (Q_1 - T_1)f(x_f, u_f) + S_0 - (T_2 - Q_2)f(x_f, u_f) + S_0$$

$$= (Q_2 - Q_1)f(x_f, u_f) + M + 2S_0. \tag{5.156}$$

It follows from (5.156) that

$$I^f(T_1, T_1 + L_0, x, u) \leq L_0 f(x_f, u_f) + 2S_0 + M, \tag{5.157}$$

$$I^f(T_2 - L_0, T_2, x, u) \leq L_0 f(x_f, u_f) + 2S_0 + M. \tag{5.158}$$

By (5.153), (5.155), (5.157), (5.158), and property (vi), there exist

$$\tau_1 \in [T_1, T_1 + L_0], \ \tau_2 \in [T_2 - L_0, T_2] \tag{5.159}$$

such that

$$\|x(\tau_i) - x_f\| \leq \delta_0, \ i = 1, 2. \tag{5.160}$$

If $\|x(T_1) - x_f\| \leq \delta$, then we may assume that $\tau_1 = T_1$ and if $\|x(T_2) - x_f\| \leq \delta$, then we may assume that $\tau_2 = T_2$. In view of (5.155) and (5.159),

$$I^f(\tau_1, \tau_2, x, u) \leq U^f(\tau_1, \tau_2, x(\tau_1), x(\tau_2)) + \delta. \tag{5.161}$$

It follows from (5.154) and (5.159) that

$$\tau_2 - \tau_1 \geq T_2 - T_1 - 2L_0 \geq 2L - 2L_0 \geq 2b_f. \tag{5.162}$$

Property (iv) and (5.160)–(5.162) imply that

$$\|x(t) - x_f\| \leq \epsilon, \ t \in [\tau_1, \tau_2].$$

Theorem 5.13 is proved.

5.12 An Auxiliary Result

Lemma 5.42. *Assume that (P2) holds and that* $M, \epsilon > 0$. *Then there exists* $L > 0$ *such that for each* $(x, u) \in X(0, L)$ *which satisfies*

$$I^f(0, L, x, u) \le Lf(x_f, u_f) + M,$$

the following inequality holds:

$$\inf\{\|x(t) - x_f\| : t \in [0, L]\} \le \epsilon.$$

Proof. By Theorem 5.1, there exists $S_0 > 0$ such that the following property holds:

(i) for each $T_2 > T_1 \ge 0$ and each $(x, u) \in X(T_1, T_2)$,

$$I^f(T_1, T_2, x, u) + S_0 \ge (T_2 - T_1)f(x_f, u_f).$$

By (P2), there exist $\delta_0 \in (0, \epsilon)$ and $L_0 > 0$ such that the following property holds:

(ii) for each $(x, u) \in X(0, L_0)$ which satisfies

$$I^f(0, L_0, x, u) \le \min\{U^f(0, L_0, x(0), x(L_0)) + \delta_0,$$

$$L_0 f(x_f, u_f) + 2S_0 + M + |f(x_f, u_f)|\},$$

we have

$$\inf\{\|x(t) - x_f\| : t \in [0, L_0]\} \le \epsilon.$$

Choose an integer

$$q_0 > (M + S_0)\delta^{-1} \tag{5.163}$$

and set

$$L = q_0 L_0. \tag{5.164}$$

Let $(x, u) \in X(0, L)$ satisfy

$$I^f(0, L, x, u) \le Lf(x_f, u_f) + M. \tag{5.165}$$

We show that

$$\inf\{\|x(t) - x_f\| : t \in [0, L]\} \le \epsilon.$$

Assume the contrary. Then

$$\|x(t) - x_f\| > \epsilon,\ t \in [0, L]. \tag{5.166}$$

Let an integer $i \in \{0, \ldots, q_0 - 1\}$. Property (i) and (5.165) imply that

$$I^f(iL_0, (i+1)L_0, x, u) = I^f(0, L, x, u) - I^f(0, iL_0, x, u) - I^f((i+1)L_0, L, x, u)$$
$$\le Lf(x_f, u_f) + M - iL_0 f(x_f, u_f) + S_0 - (L - (i+1)L_0) f(x_f, u_f) + S_0$$
$$= L_0 f(x_f, u_f) + M + 2S_0. \quad (5.167)$$

By (5.166), (5.167), and property (ii),

$$I^f(iL_0, (i+1)L_0, x, u) > U^f(iL_0, (i+1)L_0, x(iL_0), x((i+1)L_0)) + \delta_0. \quad (5.168)$$

It follows from (5.168) that there exists $(y, v) \in X(0, L)$ such that

$$y(iL_0) = x(iL_0),\ i = 0, \ldots, q_0 \quad (5.169)$$

and for all $i = 0, \ldots, q_0 - 1$,

$$I^f(iL_0, (i+1)L_0, x, u) > I^f(iL_0, (i+1)L_0, y, v) + \delta_0. \quad (5.170)$$

In view of (5.165), (5.169), (5.170) and property (i),

$$Lf(x_f, u_f) + M \ge I^f(0, L, x, u) > I^f(0, L, y, v) + q_0\delta \ge Lf(x_f, u_f) - S_0 + q_0\delta,$$

$$q_0 \le (M + S_0)\delta^{-1}.$$

This contradicts (5.163). The contradiction we have reached proves Lemma 5.42.

5.13 Proof of Theorem 5.14

By Theorem 5.1, there exists $S_0 > 0$ such that the following property holds:

(i) for each $\tau_2 > \tau_1 \ge 0$ and each $(y, v) \in X(\tau_1, \tau_2)$,

$$I^f(\tau_1, \tau_2, y, v) + S_0 \ge (\tau_2 - \tau_1) f(x_f, u_f).$$

Theorem 5.13 and STP imply that there exist $\delta \in (0, \epsilon)$ and $L_0 > 0$ such that the following property holds:

(ii) for each $\tau_1 \ge 0$, each $\tau_2 \ge \tau_1 + 2L_0$, and each $(y, v) \in X(\tau_1, \tau_2)$ which satisfies

$$I^f(\tau_1, \tau_2, y, v) \le \min\{\sigma^f(\tau_1, \tau_2) + M + 3S_0,\ U^f(\tau_1, \tau_2, y(\tau_1), y(\tau_2)) + \delta\},$$

we have

$$\|y(t) - x_f\| \le \epsilon, \ t \in [\tau_1 + L_0, \tau_2 - L_0].$$

Set

$$l = 2L_0 + 1. \tag{5.171}$$

Choose a natural number

$$Q \ge 2 + 2(M + S_0)\delta^{-1}. \tag{5.172}$$

Assume that $T_1 \ge 0$, $T_2 \ge T_1 + lQ$ and $(x, u) \in X(T_1, T_2)$ satisfy

$$I^f(T_1, T_2, x, u) \le (T_2 - T_1) f(x_f, u_f) + M. \tag{5.173}$$

Property (i), (5.172), and (5.173) imply that for each pair of numbers $\tau_1, \tau_2 \in [T_1, T_2]$ satisfying $\tau_1 < \tau_2$,

$$I^f(\tau_1, \tau_2, x, u) = I^f(T_1, T_2, x, u) - I^f(T_1, \tau_1, x, u) - I^f(\tau_2, T_2, x, u)$$

$$\le M + (T_2 - T_1) f(x_f, u_f) - (\tau_1 - T_1) f(x_f, u_f) + S_0 - (T_2 - \tau_2) f(x_f, u_f) + S_0$$

$$\le M + 2S_0 + (\tau_2 - \tau_1) f(x_f, u_f). \tag{5.174}$$

Set

$$t_0 = T_1.$$

If

$$I^f(T_1, T_2, x, u) \le U^f(T_1, T_2, x(T_1), x(T_2)) + \delta,$$

then we set $t_1 = T_2$, $s_1 = T_2$. Assume that

$$I^f(T_1, T_2, x, u) > U^f(T_1, T_2, x(T_1), x(T_2)) + \delta. \tag{5.175}$$

It is easy to see that for each $t \in (T_1, T_2)$ such that $t - T_1$ is sufficiently small, we have

$$I^f(T_1, t, x, u) - U^f(T_1, t, x(T_1), x(t)) < \delta. \tag{5.176}$$

Set

$$\tilde{t}_1 = \inf\{t \in (T_1, T_2] : \ I^f(T_1, t, x, u) - U^f(T_1, t, x(T_1), x(t)) > \delta\}. \tag{5.177}$$

Clearly, $\tilde{t}_1 \in (T_1, T_2]$ is well-defined. By (5.175) and (5.176), there exist $s_1, t_1 \in [T_1, T_2]$ such that

$$t_0 < s_1 < \tilde{t}_1,\ s_1 \geq \tilde{t}_1 - 1/4, \tag{5.178}$$

$$I^f(t_0, s_1, x, u) - U^f(t_0, s_1, x(t_0), x(s_1)) \leq \delta, \tag{5.179}$$

$$\tilde{t}_1 \leq t_1 < \tilde{t}_1 + 1/4,\ t_1 \leq T_2, \tag{5.180}$$

$$I^f(t_0, t_1, x, u) - U^f(t_0, t_1, x(t_0), x(t_1)) > \delta. \tag{5.181}$$

Assume that $k \geq 1$ is an integer and we defined finite sequences of numbers $\{t_i\}_{i=0}^k \subset [T_1, T_2]$, $\{s_i\}_{i=1}^k \subset [T_1, T_2]$ such that

$$t_0 < t_1 \cdots < t_k, \tag{5.182}$$

for each integer $i = 1, \dots, k$

$$t_{i-1} < s_i \leq t_i,\ t_i - s_i \leq 2^{-1}, \tag{5.183}$$

$$I^f(t_{i-1}, s_i, x, u) - U^f(t_{i-1}, s_i, x(t_{i-1}), x(s_i)) \leq \delta, \tag{5.184}$$

and if $t_i < T_2$, then

$$I^f(t_{i-1}, t_i, x, u) - U^f(t_{i-1}, t_i, x(t_{i-1}), x(t_i)) > \delta. \tag{5.185}$$

(Note that in view of (5.175), (5.178)–(5.181), our assumption holds for $k = 1$.)

By (5.182) and (5.185), there exists $(y, v) \in X(T_1, T_2)$ such that

$$y(t_i) = x(t_i),\ i = 1, \dots, k, \tag{5.186}$$

$$I^f(t_{i-1}, t_i, x, u) - I^f(t_{i-1}, t_i, y, v) > \delta \text{ for all intergers } i \in [1, k]\setminus\{k\}, \tag{5.187}$$

$$(y(t), v(t)) = (x(t), u(t)),\ t \in [t_{k-1}, T_2]. \tag{5.188}$$

Property (i), (5.173), and (5.187) imply that

$$(T_2 - T_1)f(x_f, u_f) + M \geq I^f(T_1, T_2, x, u) \geq I^f(T_1, T_2, y, v) + \delta(k-1)$$

$$\geq (T_2 - T_1)f(x_f, u_f) - S_0 + \delta(k-1),$$

$$k \leq 1 + \delta^{-1}(M + S_0). \tag{5.189}$$

If $t_k = T_2$, then the construction of the sequence is completed. Assume that $t_k < T_2$. If

$$I^f(t_k, T_2, x, u) \le U^f(t_k, T_2, x(t_k), x(T_2)) + \delta,$$

then we set $t_{k+1} = T_2, s_{k+1} = T_2$, and the construction of the sequence is completed. Assume that

$$I^f(t_k, T_2, x, u) > U^f(t_k, T_2, x(t_k), x(T_2)) + \delta. \tag{5.190}$$

It is easy to see that for each $t \in (t_k, T_2)$ such that $t - t_k$ is sufficiently small, we have

$$I^f(t_k, t, x, u) - U^f(t_k, t, x(t_k), x(t)) < \delta. \tag{5.191}$$

Set

$$\tilde{t} = \inf\{t \in (t_k, T_2] : \ I^f(t_k, t, x, u) - U^f(t_k, t, x(t_k), x(t)) > \delta\}. \tag{5.192}$$

By (5.190) and (5.191), $\tilde{t}$ is well-defined and

$$\tilde{t} \in (t_k, T_2]. \tag{5.193}$$

There exist $s_{k+1}, t_{k+1} \in [T_1, T_2]$ such that

$$t_k < s_{k+1} < \tilde{t}, \ s_{k+1} > \tilde{t} - 1/4, \ \tilde{t} + 4^{-1} > t_{k+1} \ge \tilde{t},$$

$$I^f(t_k, s_{k+1}, x, u) - U^f(t_k, s_{k+1}, x(t_k), x(s_{k+1})) \le \delta,$$

$$I^f(t_k, t_{k+1}, x, u) - U^f(t_k, t_{k+1}, x(t_k), x(t_{k+1})) > \delta.$$

It is not difficult to see that the assumption made for k also holds for $k + 1$. In view of (5.189), by induction we constructed finite sequences $\{t_i\}_{i=0}^q \subset [T_1, T_2]$, $\{s_i\}_{i=0}^q \subset [T_1, T_2]$ such that

$$q \le 1 + \delta^{-1}(M + S_0), \tag{5.194}$$

$$T_1 = t_0 < t_1 \cdots < t_q = T_2,$$

for each integer $i = 1, \ldots, q$, (5.183) and (5.184) hold and if $t_i < T_2$, then (5.185) holds.

Assume that

$$i \in \{0, \ldots, q-1\}, \ t_{i+1} - t_i \ge 2L_0 + 1. \tag{5.195}$$

By (5.183) and (5.195),

$$s_{i+1} - t_i \geq 2L_0. \tag{5.196}$$

In view of (5.184),

$$I^f(t_i, s_{i+1}, x, u) - U^f(t_i, s_{i+1}, x(t_i), x(s_{i+1})) \leq \delta. \tag{5.197}$$

Property (i) and (5.174) imply that

$$I^f(t_i, s_{i+1}, x, u) \leq M + 2S_0 + (s_{i+1} - t_i) f(x_f, u_f) \leq M + 3S_0 + \sigma^f(t_i, s_{i+1}). \tag{5.198}$$

It follows from (5.196)–(5.198)) and property (ii) that

$$\|x(t) - x_f\| \leq \epsilon, \ t \in [t_i + L_0, s_{i+1} - L_0],$$

and in view of (5.183),

$$\|x(t) - x_f\| \leq \epsilon, \ t \in [t_i + L_0, t_{i+1} - L_0 - 1]. \tag{5.199}$$

By (5.199),

$$\{t \in [T_1, T_2] : \ \|x(t) - x_f\| > \epsilon\}$$

$$\subset \cup\{[t_i, t_i + L_0] \cup [t_{i+1} - 1 - L_0, t_{i+1}] : \ i \in \{0, \dots, q-1\} \text{ and } t_{i+1} - t_i \geq 2L_0 + 1\}$$

$$\cup \{[t_i, t_{i+1}] : \ i \in \{0, \dots, q-1\} \text{ and } t_{i+1} - t_i < 2L_0 + 1\}. \tag{5.200}$$

In view of (5.171) and (5.193), the maximal length of intervals in the right-hand side of (5.200) does not exceed l, and their number does not exceed $2q \leq 2(1 + \delta^{-1}(M + S_0)) \leq Q$. Theorem 5.14 is proved.

5.14 Proof of Proposition 5.16

Let us show that (P1) holds. Let $(x, u) \in X(0, \infty)$ be (f)-good. There exists $M_0 > 0$ such that for each $T > 0$,

$$|I^f(0, T, x, u) - Tf(x_f, u_f)| \leq M_0.$$

This implies that for each pair of numbers $T_2 > T_1 \geq 0$,

$$I^f(T_1, T_2, x, u) - (T_2 - T_1) f(x_f, u_f) = I^f(0, T_2, x, u) - T_2 f(x_f, u_f)$$

$$- I^f(0, T_1, x, u) + T_1 f(x_f, u_f) \le 2M_0. \tag{5.201}$$

Let $\epsilon > 0$, and let $l > 0$ and a natural number Q be as guaranteed by WTP with ϵ and $M = 2M_0$. Set

$$\Omega = \{t \in [0, \infty) : \|x(t) - x_f\| > \epsilon\}.$$

We show that Ω is bounded. Assume that contrary. Then there exists a sequence of positive numbers $\{t_i\}_{i=1}^{\infty}$ such that for all integers $i \ge 1$,

$$t_{i+1} - t_i \ge 2l + 2, \ \|x(t_i) - x_f\| > \epsilon. \tag{5.202}$$

Let $k \ge 1 + Q$ be an integer. It follows from (5.202) that

$$t_Q \ge Q(2l + 2). \tag{5.203}$$

By WTP and the choice of Q, l, there exist an integer $q \le Q$ and finite sequences $\{a_i\}_{i=1}^{q}$, $\{b_i\}_{i=1}^{q} \subset [0, t_k]$ such that

$$0 \le b_i - a_i \le l, \ i = 1, \dots, q, \tag{5.204}$$

$$b_i \le a_{i+1} \text{ for all integers } i \text{ satisfying } 1 \le i < q,$$

$$\|x(t) - x_f\| \le \epsilon \text{ for all } t \in [0, t_k] \setminus \cup_{i=1}^{q} [a_i, b_i]. \tag{5.205}$$

In view of (5.202) and (5.205),

$$\{t_i\}_{i=1}^{k} \subset \cup_{i=1}^{q} [a_i, b_i]. \tag{5.206}$$

By (5.202), (5.204), and (5.206), $k \le q \le Q$, a contradiction. The contradiction we have reached proves that Ω is bounded. Since ϵ is an arbitrary positive number, this completes the proof of Proposition 5.16.

5.15 Proof of Proposition 5.18

Assume that (P2) does not hold. Then there exist $\epsilon, M > 0$ such that for each natural number k, there exists an integer $n_k \ge k$ and $(x_k, u_k) \in X(0, n_k)$ which satisfies

$$I^f(0, n_k, x_k, u_k) \le n_k f(x_f, u_f) + M, \tag{5.207}$$

$$\|x_k(t) - x_f\| > \epsilon, \ t \in [0, n_k]. \tag{5.208}$$

Let $S_0 > 0$ be as guaranteed by Theorem 5.1. By the choice of S_0, Theorem 5.1 and (5.207), for each integer $k \geq 1$ and each pair of numbers $\tau_1, \tau_2 \in [0, k]$ satisfying $\tau_1 < \tau_2$,

$$I^f(\tau_1, \tau_2, x_k, u_k) - (\tau_2 - \tau_1) f(x_f, u_f) = I^f(0, n_k, x_k, u_k) - n_k f(x_f, u_f)$$

$$-I^f(0, \tau_1, x_k, u_k) + \tau_1 f(x_f, u_f) - I^f(\tau_2, n_k, x_k, u_k)$$

$$+ (n_k - \tau_2) f(x_f, u_f) \leq M + 2S_0. \tag{5.209}$$

By the relation above and LSC property, extracting subsequences, using the diagonalization process and re-indexing, we may assume that for each integer $i \geq 1$, there exists $\lim_{k\to\infty} I^f(i-1, i, x_k, u_k)$, and there exists $(y_i, v_i) \in X(i-1, i)$ such that

$$x_k(t) \to y_i(t) \text{ as } k \to \infty \text{ for every } t \in [i-1, i], \tag{5.210}$$

$$I^f(i-1, i, y_i, v_i) \leq \lim_{k\to\infty} I^f(i-1, i, x_k, u_k). \tag{5.211}$$

In view of (5.210), there exists $(y, v) \in X(0, \infty)$ such that for each integer $i \geq 1$,

$$(y(t), v(t)) = (y_i(t), v_i(t)),\ t \in [i-1, i].$$

It follows from the relation above, (5.209) and (5.211), that for every integer $m \geq 1$

$$I^f(0, m, y, v) = \sum_{i=1}^{m} I^f(i-1, i, y, v) = \sum_{i=1}^{m} I^f(i-1, i, y_i, v_i)$$

$$\leq \sum_{i=1}^{m} (\lim_{k\to\infty} I^f(i-1, i, x_k, u_k)) = \lim_{k\to\infty} I^f(0, m, x_k, u_k) \leq m f(x_f, u_f) + M + 2S_0.$$

Together with Theorem 5.1, this implies that (y, v) is (f)-good. Property (P1) implies that there exists an integer $\tau_0 > 0$ such that

$$\|y(t) - x_f\| \leq \epsilon/4 \text{ for all } t \geq \tau_0 - 1,$$

and in view of the definition of (y, v),

$$\|y_{\tau_0}(t) - x_f\| \leq \epsilon/4 \text{ for all } t \in [\tau_0 - 1, \tau_0]. \tag{5.212}$$

By (5.210), there exists an integer $k_0 > \tau_0$ such that for all integers $k \geq k_0$,

$$\|y_{\tau_0}(\tau_0) - x_k(\tau_0)\| \leq \epsilon/4.$$

In view of (5.212) and the relation above, for all integers $k \geq k_0$,

$$\|x_k(\tau_0) - x_f\| \leq \|x_k(\tau_0) - y_{\tau_0}(\tau_0)\| + \|y_{\tau_0}(\tau_0) - x_f\| \leq \epsilon/2.$$

This contradicts (5.208). The contradiction we have reached proves Proposition 5.18.

5.16 Auxiliary Results for Theorems 5.19 and 5.20

Lemma 5.43. *Let $\epsilon > 0$. Then there exists $\delta > 0$ such that for each $T \geq b_f$ and each $(x, u) \in X(0, T)$ which satisfies*

$$\|x(0) - x_f\| \leq \delta, \ \|x(T) - x_f\| \leq \delta,$$

the following inequality holds:

$$I^f(0, T, x, u) \geq Tf(x_f, u_f) - \epsilon.$$

Proof. By (A2), there exists $\delta > 0$ such that the following property holds:

(i) for each $z \in \mathcal{A}$ satisfying $\|z - x_f\| \leq \delta$, there exist $\tau_1 \in (0, b_f]$, $\tau_2 \in (0, b_f]$ and $(\tilde{x}_1, \tilde{u}_1) \in X(0, \tau_1)$, $(\tilde{x}_2, \tilde{u}_2) \in X(0, \tau_2)$ such that

$$\tilde{x}_1(0) = z \ , \tilde{x}_1(\tau_1) = x_f, \ I^f(0, \tau_1, \tilde{x}_1, \tilde{u}_1) \leq \tau_1 f(x_f, u_f) + \epsilon/4,$$

$$\tilde{x}_2(0) = x_f \ , \tilde{x}_2(\tau_2) = z, \ I^f(0, \tau_2, \tilde{x}_2, \tilde{u}_2) \leq \tau_2 f(x_f, u_f) + \epsilon/4.$$

Assume that $T \geq b_f$, $(x, u) \in X(0, T)$ and

$$\|x(0) - x_f\| \leq \delta, \ \|x(T) - x_f\| \leq \delta.$$

By the relation above and property (i), there exists $(y, v) \in X(0, \infty)$ such that

$$(y(t + 2b_f + T), v(t + 2b_f + T)) = (y(t), v(t)) \text{ for all } t \geq 0, \tag{5.213}$$

$$y(0) = x_f, \ y(b_f) = x(0), \ y(2b_f + T) = x_f, \tag{5.214}$$

$$I^f(0, b_f, y, v) \leq b_f f(x_f, u_f) + \epsilon/4, \tag{5.215}$$

$$(y(t), v(t)) = (x(t - b_f), u(t - b_f)), \ t \in [b_f, b_f + T], \tag{5.216}$$

$$I^f(b_f + T, 2b_f + T, y, v) \leq b_f f(x_f, u_f) + \epsilon/4. \tag{5.217}$$

Theorem 5.1, (5.213), (5.214), and (5.216) imply that

$$(T + 2b_f)f(x_f, u_f) \le I^f(0, T + 2b_f, y, v) = I^f(0, b_f, y, v)$$
$$+I^f(0, T, x, u) + I^f(T + b_f, T + 2b_f, y, v).$$

By the relation above, (5.215) and (5.217),

$$I^f(0, T, x, u) \ge (T+2b_f)f(x_f, u_f) - 2b_f f(x_f, u_f) + \epsilon/4 \ge Tf(x_f, u_f) - \epsilon/2.$$

Lemma 5.43 is proved.

Assume that f has LSC property and (P1). Theorem 5.13 and Proposition 5.18 imply that f has (P2) and TP.

Proposition 5.44. *Assume that* $z \in \cup\{\mathcal{A}_L : L > 0\}$. *Then there exists an* (f)*-good and* (f)*-minimal pair* $(x_*, u_*) \in X(0, \infty)$ *such that* $x_*(0) = z$.

Proof. There exists $L_0 > 0$ such that

$$z \in \mathcal{A}_{L_0}. \tag{5.218}$$

It follows from Theorem 5.1 that there exists $S_0 > 0$ such that for each $T_2 > T_1 \ge 0$ and each $(x, u) \in X(T_1, T_2)$,

$$I^f(T_1, T_2, x, u) + S_0 \ge (T_2 - T_1)f(x_f, u_f).$$

Fix an integer $k_0 \ge L_0$. LSC property implies that for each integer $k \ge k_0$, there exists $(x_k, u_k) \in X(0, k)$ satisfying

$$x_k(0) = z,\ I^f(0, k, x_k, u_k) = \sigma^f(0, k, z). \tag{5.219}$$

In view of (5.218), for each integer $k \ge k_0$,

$$\sigma^f(0, k, z) \le L_0 + f(x_f, u_f)(k - L_0) + L_0|f(x_f, u_f)|. \tag{5.220}$$

By (5.219), (5.220) and the choice of S_0, for each integer $k \ge k_0$ and each pair of numbers $T_1, T_2 \in [0, k]$ satisfying $T_1 < T_2$,

$$I^f(T_1, T_2, x_k, u_k) = I^f(0, k, x_k, u_k) - I^f(0, T_1, x_k, u_k) - I^f(T_2, k, x_k, u_k)$$
$$\le kf(x_f, u_f) + L_0(1+2|f(x_f, u_f)|) - T_1 f(x_f, u_f) + S_0 - (k - T_2)f(x_f, u_f) + S_0$$
$$\le (T_2 - T_1)f(x_f, u_f) + L_0(1 + 2|f(x_f, u_f)|) + 2S_0. \tag{5.221}$$

By (5.221) and LSC property, extracting subsequences, using the diagonalization process and re-indexing, we obtain that there exists a strictly increasing sequence of natural numbers $\{k_p\}_{p=1}^{\infty}$ such that $k_1 \geq k_0$ and for each integer $i \geq 0$, there exists $\lim_{p\to\infty} I^f(i, i+1, x_{k_p}, u_{k_p})$, and there exists $(y_i, v_i) \in X(i, i+1)$ such that

$$x_{k_p}(t) \to y_i(t) \text{ as } p \to \infty \text{ for all } t \in [i, i+1], \tag{5.222}$$

$$I^f(i, i+1, y_i, v_i) \leq \lim_{p\to\infty} I^f(i, i+1, x_{k_p}, u_{k_p}). \tag{5.223}$$

In view of (5.222), there exists $(x_*, u_*) \in X(0, \infty)$ such that for each integer $i \geq 0$,

$$(x_*(t), u_*(t)) = (y_i(t), v_i(t)), \ t \in [i, i+1].$$

It follows from (5.221), (5.223), and the definition of (x_*, u_*) that for every integer $q \geq 1$,

$$I^f(0, q, x_*, u_*) \leq \lim_{p\to\infty} I^f(0, q, x_{k_p}, u_{k_p})$$

$$\leq q f(x_f, u_f) + L_0(1 + 2|f(x_f, u_f)|) + 2S_0. \tag{5.224}$$

Theorem 5.1, the definition of (x_*, u_*), (5.219), (5.222), and (5.224), imply that (x_*, u_*) is (f)-good and

$$x_*(0) = z. \tag{5.225}$$

In order to complete the proof of the proposition, it is sufficient to show that (x_*, u_*) is (f)-minimal. Assume the contrary. Then there exist $\Delta > 0$, an integer $\tau_0 \geq 1$ and $(y, v) \in X(0, \tau_0)$ such that

$$y(0) = x_*(0), \ y(\tau_0) = x_*(\tau_0),$$

$$I^f(0, \tau_0, x_*, u_*) > I^f(0, \tau_0, y, v) + 2\Delta. \tag{5.226}$$

By (A2) and Lemma 5.43, there exists $\delta > 0$ such that the following properties hold:

(ii) for each $\xi \in \mathcal{A}$ satisfying $\|\xi - x_f\| \leq \delta$, there exist $\tau_1, \tau_2 \in (0, b_f]$ and $(\tilde{x}_1, \tilde{u}_1) \in X(0, \tau_1)$, $(\tilde{x}_2, \tilde{u}_2) \in X(0, \tau_2)$ such that

$$\tilde{x}_1(0) = \xi, \ \tilde{x}_1(\tau_1) = x_f, \ I^f(0, \tau_1, \tilde{x}_1, \tilde{u}_1) \leq \tau_1 f(x_f, u_f) + \Delta/8,$$

$$\tilde{x}_2(0) = x_f, \ \tilde{x}_2(\tau_2) = \xi, \ I^f(0, \tau_2, \tilde{x}_2, \tilde{u}_2) \leq \tau_2 f(x_f, u_f) + \Delta/8;$$

(iii) for each $(x, u) \in X(0, b_f)$ which satisfies $\|x(0)-x_f\| \le \delta$, $\|x(b_f)-x_f\| \le \delta$, we have

$$I^f(0, b_f, x, u) \ge b_f f(x_f, u_f) - \Delta/8.$$

Theorem 5.1, (TP), (5.219) and (5.220) imply that there exists an integer $L_1 > L_0$ such that for each integer $k \ge k_0 + 2L_1$,

$$\|x_k(t) - x_f\| \le \delta, \; t \in [L_1, k - L_1]. \tag{5.227}$$

In view of (5.222), (5.227) and the definition of x_*,

$$\|x_*(t) - x_f\| \le \delta \text{ for all } t \ge L_1. \tag{5.228}$$

By (5.223) and the definition of (x_*, u_*), there exists a natural number p_0 such that

$$k_{p_0} > k_0 + 2L_1 + 2\tau_0 + 2 + 2b_f, \tag{5.229}$$

$$I^f(0, \tau_0 + L_1, x_*, u_*) \le I^f(0, \tau_0 + L_1, x_{k_{p_0}}, u_{k_{p_0}}) + \Delta/2. \tag{5.230}$$

Property (ii) and (5.226)–(5.229) imply that there exists $(x, u) \in X(0, k_{p_0})$ such that

$$(x(t), u(t)) = (y(t), v(t)), \; t \in [0, \tau_0],$$

$$(x(t), u(t)) = (x_*(t), u_*(t)), \; t \in (\tau_0, \tau_0 + L_1],$$

$$x(\tau_0+L_1+b_f) = x_f, \; I^f(\tau_0+L_1, \tau_0+L_1+b_f, x, u) \le b_f f(x_f, u_f)+\Delta/8,$$

$$(x(t), u(t)) = (x_{k_{p_0}}(t), u_{k_{p_0}}(t)), \; t \in [\tau_0 + L_1 + 2b_f, k_{p_0}],$$

$$I^f(\tau_0 + L_1 + b_f, \tau_0 + L_1 + 2b_f, x, u) \le b_f f(x_f, u_f) + \Delta/8.$$

It follows from the relations above, (5.219), (5.225), (5.226), and property (iii) that

$$I^f(0, k_{p_0}, x, u) \ge I^f(0, k_{p_0}, x_{k_{p_0}}, u_{k_{p_0}}).$$

By the relation above, (5.226), (5.227), (5.229), (5.230), property (iii), and the choice of (x, u),

$$0 \le I^f(0, k_{p_0}, x, u) - I^f(0, k_{p_0}, x_{k_{p_0}}, u_{k_{p_0}})$$

$$= I^f(0, \tau_0 + L_1 + 2b_f, x, u) - I^f(0, \tau_0 + L_1 + 2b_f, x_{k_{p_0}}, u_{k_{p_0}})$$

$$= I^f(0, \tau_0, y, v) + I^f(\tau_0, \tau_0 + L_1, x_*, u_*) + 2b_f f(x_f, u_f) + \Delta/4$$

$$-I^f(0, \tau_0 + L_1, x_{k_{p_0}}, u_{k_{p_0}}) - I^f(\tau_0 + L_1, \tau_0 + L_1 + b_f, x_{k_{p_0}}, u_{k_{p_0}})$$

$$-I^f(\tau_0 + L_1 + b_f, \tau_0 + L_1 + 2b_f, x_{k_{p_0}}, u_{k_{p_0}})$$

$$\le I^f(0, \tau_0, y, v) + I^f(\tau_0, \tau_0 + L_1, x_*, u_*) + 2b_f f(x_f, u_f) + \Delta/4$$

$$-I^f(0, \tau_0 + L_1, x_{k_{p_0}}, u_{k_{p_0}}) - 2b_f f(x_f, u_f) + \Delta/4$$

$$\le I^f(0, \tau_0, x_*, u_*) - 2\Delta + I^f(\tau_0, \tau_0 + L_1, x_*, u_*)$$

$$+\Delta/2 - I^f(0, \tau_0 + L_1, x_{k_{p_0}}, u_{k_{p_0}}) \le -\Delta,$$

a contradiction. The contradiction we have reached completes the proof of Proposition 5.44.

5.17 Proofs of Theorems 5.19 and 5.20

Proof of Theorem 5.20. Clearly, (i) implies (ii). In view of Theorem 5.1, (ii) implies (iii). By (P1), (iii) implies (iv). Evidently (iv) implies (v). We show that (v) implies (iii). Assume that $(x_*, u_*) \in X(0, \infty)$ is (f)-minimal and satisfies

$$\liminf_{t \to \infty} \|x_*(t) - x_f\| = 0. \tag{5.231}$$

(P1) implies that

$$\lim_{t \to \infty} \|\tilde{x}(t) - x_f\| = 0. \tag{5.232}$$

In view of Theorem 5.1, there exists $S_0 > 0$ such that for all numbers $T > 0$,

$$|I^f(0, T, \tilde{x}, \tilde{u}) - Tf(x_f, u_f)| \le S_0. \tag{5.233}$$

By (A2), there exists $\delta > 0$ such that the following property holds:

(a) for each $z \in \mathcal{A}$ satisfying $\|z - x_f\| \le \delta$, there exist $\tau_1, \tau_2 \in (0, b_f]$ and $(x_i, u_i) \in X(0, \tau_i)$, $i = 1, 2$ which satisfy

$$x_1(0) = z,\ x_1(\tau_1) = x_f,\ x_2(0) = x_f,\ x_2(\tau_2) = z,$$

$$I^f(0, \tau_i, x_i, u_i) \le \tau_i f(x_f, u_f) + 1,\ i = 1, 2.$$

In view of (5.231) and (5.232), there exists an increasing sequence $\{t_k\}_{k=1}^{\infty} \subset (0, \infty)$ such that

$$\lim_{k\to\infty} t_k = \infty,\ \|x_*(t_k + 2b_f) - x_f\| < \delta \text{ for all integers } k \ge 1, \tag{5.234}$$

$$\|\tilde{x}(t) - x_f\| \le \delta \text{ for all } t \ge t_0. \tag{5.235}$$

Let $k \ge 1$ be an integer. By (a), (5.234) and (5.235), there exists $(y, v) \in X(0, t_k + 2b_f)$ such that

$$(y(t), v(t)) = (\tilde{x}(t), \tilde{u}(t)),\ t \in [0, t_k],$$

$$y(t_k + b_f) = x_f,$$

$$I^f(t_k, t_k + b_f, y, v) \le b_f f(x_f, u_f) + 1,$$

$$y(t_k + 2b_f) = x_*(t_k + 2b_f),\ I^f(t_k + b_f, t_k + 2b_f, y, v) \le b_f f(x_f, u_f) + 1.$$

The relations above and (5.233) imply that

$$I^f(0, t_k + 2b_f, x_*, u_*) \le I^f(0, t_k + 2b_f, y, v)$$

$$\le I^f(0, t_k, \tilde{x}, \tilde{u}) + 2b_f f(x_f, u_f) + 2$$

$$\le t_k f(x_f, u_f) + S_0 + 2b_f f(x_f, u_f) + 2 = (t_k + 2b_f) f(x_f, u_f) + 2S_0.$$

Together with Theorem 5.1, this implies that (x_*, u_*) is (f)-good and (iii) holds.

We show that (iii) implies (i). Assume that $(x_*, u_*) \in X(0, \infty)$ is (f)-minimal and (f)-good. (P1) implies that

$$\lim_{t\to\infty} \|x_*(t) - x_f\| = 0. \tag{5.236}$$

Clearly, there exists $S_0 > 0$ such that

$$|I^f(0, T, x_*, u_*) - T f(x_f, u_f)| \le S_0 \text{ for all } T > 0. \tag{5.237}$$

Let $(x, u) \in X(0, \infty)$ satisfy

$$x(0) = x_*(0). \tag{5.238}$$

We show that

$$\limsup_{T\to\infty}[I^f(0,T,x_*,u_*)-I^f(0,T,x,u)]\leq 0. \tag{5.239}$$

In view of Theorem 5.1, we may assume that (x, u) is (f)-good. (P1) implies that

$$\lim_{t\to\infty}\|x(t)-x_f\|=0. \tag{5.240}$$

Let $\epsilon > 0$. By (A2) and Lemma 5.43, there exists $\delta \in (0, \epsilon)$ such that the following properties hold:

(b) for each $z \in \mathcal{A}$ satisfying $\|z - x_f\| \leq \delta$, there exist $\tau_1, \tau_2 \in (0, b_f]$ and $(x_i, u_i) \in X(0, \tau_i)$, $i = 1, 2$ satisfying

$$x_1(0)=z\,, x_1(\tau_1)=x_f,\ x_2(0)=x_f\,, x_2(\tau_2)=z,$$

$$I^f(0,\tau_i,x_i,u_i)\leq\tau_i f(x_f,u_f)+\epsilon/8,\ i=1,2;$$

(c) for each $(y, v) \in X(0, b_f)$ which satisfies $\|y(0)-x_f\| \leq \delta$, $\|y(b_f)-x_f\| \leq \delta$, we have

$$I^f(0,b_f,y,v)\geq b_f f(x_f,u_f)-\epsilon/8.$$

It follows from (5.236) and (5.240) that there exists $\tau_0 > 0$ such that

$$\|x(t)-x_f\|\leq\delta,\ \|x_*(t)-x_f\|\leq\delta \text{ for all } t\geq\tau_0. \tag{5.241}$$

Let $T \geq \tau_0$. Property (b) and (5.241) imply that there exists $(y, v) \in X(0, T + 2b_f)$ which satisfies

$$(y(t),v(t))=(x(t),u(t)),\ t\in[0,T], \tag{5.242}$$

$$y(T+b_f)=x_f, \tag{5.243}$$

$$I^f(T,T+b_f,y,v)\leq b_f f(x_f,u_f)+\epsilon/8, \tag{5.244}$$

$$y(T+2b_f)=x_*(T+2b_f), \tag{5.245}$$

$$I^f(T+b_f,T+2b_f,y,v)\leq b_f f(x_f,u_f)+\epsilon/8. \tag{5.246}$$

By property (c) and (5.241),

$$I^f(T,T+b_f,x_*,u_*),\ I^f(T+b_f,T+2b_f,x_*,u_*)\geq b_f f(x_f,u_f)-\epsilon/8. \tag{5.247}$$

It follows from (5.238), (5.242), and (5.244)–(5.246) that

$$I^f(0, T, x_*, u_*) + 2b_f f(x_f, u_f) - \epsilon/4 \le I^f(0, T + 2b_f, x_*, u_*)$$

$$\le I^f(0, T + 2b_f, y, v) \le I^f(0, T, x, u) + 2b_f f(x_f, u_f) + \epsilon/4,$$

$$I^f(0, T, x_*, u_*) \le I^f(0, T, x, u) + \epsilon/2 \text{ for all } T \ge \tau_0.$$

Since ϵ is any positive number, we conclude that (5.239) holds and (x_*, u_*) is (f)-overtaking optimal. Thus (iii) implies (i) and Theorem 5.20 is proved.

Theorem 5.20 and Proposition 5.44 imply Theorem 5.19.

5.18 Proofs of Theorems 5.21–5.23

Proof of Theorem 5.21. Proposition 5.18, Lemma 5.41, TP, and (A2) imply that there exists $\delta \in (0, \epsilon)$ such that the following properties hold:

(a) for each $z \in \mathcal{A}$ satisfying $\|z - x_f\| \le \delta$, there exist $\tau_0 \in (0, b_f]$ and $(x_0, u_0) \in X(0, \tau_0)$ such that

$$x_0(0) = z,\ x_0(\tau_0) = x_f,\ I^f(0, \tau_0, x_0, u_0) \le \tau_0 f(x_f, u_f) + 1;$$

(b) for each $T_1 \ge 0$, each $T_2 \ge T_1 + 2b_f$, and each $(y, v) \in X(T_1, T_2)$ which satisfies

$$\|y(T_i) - x_f\| \le \delta,\ i = 1, 2,$$

$$I^f(T_1, T_2, y, v) \le U^f(T_1, T_2, y(T_1), y(T_2)) + \delta,$$

we have

$$\|y(t) - x_f\| \le \epsilon,\ t \in [T_1, T_2].$$

Assertion (i) follows from (a) and Theorem 5.19. We prove assertion (ii). Assume that $(x, u) \in X(0, \infty)$ is (f)-overtaking optimal and

$$\|x(0) - x_f\| \le \delta. \tag{5.248}$$

Properties (P1) and (a) and (5.248) imply that

$$\lim_{t \to \infty} \|x(t) - x_f\| = 0. \tag{5.249}$$

By (5.249), there exists $\tau_0 > 0$ such that for each $t \geq \tau_0$,

$$\|x(t) - x_f\| \leq \delta. \tag{5.250}$$

Let $T \geq \tau_0 + 2b_f$. It follows from property (b), (5.248), and (5.250) that

$$\|x(t) - x_f\| \leq \epsilon, \; t \in [0, T].$$

Theorem 5.21 is proved.

Proof of Theorem 5.22. By Proposition 5.18, f has (P2). In view of Theorem 5.13, TP holds. Together with Proposition 5.10, this implies that there exists $\tau_0 > 0$ such that the following property holds:

(c) for each $T \geq 2\tau_0$ and each $(x, u) \in X(0, T)$ which satisfies

$$I^f(0, T, x, u) = U^f(0, T, x(0), x(T)),$$

$$I^f(0, T, x, u) \leq Tf(x_f, u_f) + L(1 + |f(x_f, u_f)|) + 2$$

we have

$$\|x(t) - x_f\| \leq \epsilon, \; t \in [\tau_0, T - \tau_0].$$

Let $(x, u) \in X(0, \infty)$ be (f)-overtaking optimal and

$$x(0) \in \mathcal{A}_L. \tag{5.251}$$

Property (P1) and (5.251) imply that

$$\lim_{t \to \infty} \|x(t) - x_f\| = 0. \tag{5.252}$$

In view of (5.251), there exist $S_0 \in (0, L]$ and $(y, v) \in X(0, \infty)$ such that

$$I^f(0, S_0, y, v) \leq L, \; y(0) = x(0), \; y(S_0) = x_f,$$

$$y(t) = x_f, \; t \in [S_0, \infty), \; v(t) = u_f, \; t \in (S_0, \infty).$$

By the relations above, for all sufficiently large $T > L$,

$$I^f(0, T, x, u) \leq I^f(0, T, y, v) + 1 \leq 1 + L + I^f(S_0, T, y, v)$$

$$\leq 1 + L + (T - S_0)f(x_f, u_f) \leq Tf(x_f, u_f) + 1 + L(1 + |f(x_f, u_f)|). \tag{5.253}$$

It follows from (5.253) and property (c) that for all sufficiently large $T > L + 2\tau_0$,

$$\|x(t) - x_f\| \le \epsilon, \ t \in [\tau_0, T - \tau_0].$$

Since the relation above holds for all sufficiently large numbers T, we conclude that

$$\|x(t) - x_f\| \le \epsilon, \ t \in [\tau_0, \infty).$$

This completes the proof of Theorem 5.22.

Proof of Theorem 5.23. We may assume without loss of generality that $t_1 = 0$. Clearly, there exists $(y, v) \in X(0, \infty)$ such that

$$(y(t), v(t)) = (x(t), u(t)), \ t \in [0, t_2], \tag{5.254}$$

$$(y(t + t_2), v(t + t_2)) = (y(t), v(t)) \text{ for all } t \ge 0. \tag{5.255}$$

Theorem 5.1, (5.254) and (5.255) imply that

$$I^f(0, t_2, y, v) \ge t_2 f(x_f, u_f). \tag{5.256}$$

We show that

$$I^f(0, t_2, y, v) = t_2 f(x_f, u_f).$$

Assume the contrary. Then in view of (5.254),

$$I^f(0, t_2, x, u) = I^f(0, t_2, y, v) > t_2 f(x_f, u_f). \tag{5.257}$$

Set

$$\Delta = I^f(0, t_2, x, u) - t_2 f(x_f, u_f) > 0. \tag{5.258}$$

By (A2) and Lemma 5.43, there exists $\delta \in (0, 1)$ such that the following properties hold:

(d) for each $z \in \mathcal{A}$ satisfying $\|z - x_f\| \le \delta$, there exist $\tau_1, \tau_2 \in (0, b_f]$, and $(x_i, u_i) \in X(0, \tau_i)$, $i = 1, 2$ satisfying

$$x_1(0) = z, \ x_1(\tau_1) = x_f, \ x_2(0) = x_f, x_2(\tau_2) = z,$$

$$I^f(0, \tau_i, x_i, u_i) \le \tau_i f(x_f, u_f) + \Delta/8, \ i = 1, 2;$$

(e) for each $(z, \xi) \in X(0, b_f)$ which satisfies $\|z(0) - x_f\| \le \delta$, $\|z(b_f) - x_f\| \le \delta$, we have

$$I^f(0, b_f, z, \xi) \ge b_f f(x_f, u_f) - \Delta/8.$$

It follows from (P1) that there exists $T_0 > 0$ such that

$$\|x(t) - x_f\| \le \delta \text{ for all } t \ge \tau_0. \tag{5.259}$$

Choose

$$T_1 > 2T_0 + 2t_2 + 2b_f + 4. \tag{5.260}$$

By (d), (5.259), and (5.260), there exists $(\widehat{x}, \widehat{u}) \in X(0, \infty)$ such that

$$(\widehat{x}(t), \widehat{u}(t)) = (x(t + t_2), u(t + t_2)),\ t \in [0, T_0 + t_2 + 2], \tag{5.261}$$

$$I^f(T_0 + t_2 + 2, T_0 + t_2 + 2 + b_f, \widehat{x}, \widehat{u}) \le b_f f(x_f, u_f) + \Delta/8, \tag{5.262}$$

$$\widehat{x}(T_0 + t_2 + b_f + 2) = x_f, \tag{5.263}$$

$$(\widehat{x}(t), \widehat{u}(t)) = (x_f, u_f),\ t \in [T_0+t_2+2+b_f, T_0+2t_2+2+b_f], \tag{5.264}$$

$$\widehat{x}(T_0 + 2t_2 + 2b_f + 2) = x(T_0 + 2t_2 + 2b_f + 2), \tag{5.265}$$

$$I^f(T_0 + 2t_2 + 2 + b_f, T_0 + 2t_2 + 2 + 2b_f, \widehat{x}, \widehat{u}) \le b_f f(x_f, u_f) + \Delta/8, \tag{5.266}$$

$$(\widehat{x}(t), \widehat{u}(t)) = (x(t), u(t)) \text{ for all } t \ge T_0 + 2t_2 + 2 + 2b_f. \tag{5.267}$$

In view of (5.261) and (5.265),

$$I^f(0, T_0+2t_2+2+2b_f, x, u) \le I^f(0, T_0+2t_2+2+2b_f, \widehat{x}, \widehat{u}). \tag{5.268}$$

It follows from (5.261), (5.262), (5.264), and (5.266) that

$$I^f(0, T_0+2t_2+2+2b_f, \widehat{x}, \widehat{u}) \le I^f(t_2, T_0+2t_2+2, x, u) + t_2 f(x_f, u_f) + \Delta/8$$

$$+ f(x_f, u_f) b_f + b_f f(x_f, u_f) + \Delta/8$$

$$= I^f(t_2, T_0 + 2t_2 + 2, x, u) + (t_2 + 2b_f) f(x_f, u_f) + \Delta/4. \tag{5.269}$$

Property (e), (5.258), and (5.259) imply that

$$I^f(0, T_0 + 2t_2 + 2 + 2b_f, x, u) = I^f(0, t_2, x, u) + I^f(t_2, T_0 + 2t_2 + 2, x, u)$$

$$+I^f(T_0+2t_2+2, T_0+2t_2+2+b_f, x, u)+I^f(T_0+2t_2+2+b_f, T_0+2t_2+2+2b_f, x, u)$$

$$\geq t_2 f(x_f, u_f) + \Delta + I^f(t_2, T_0 + 2t_2 + 2, x, u) + 2b_f f(x_f, u_f) - \Delta/4.$$

By (5.269) and the relations above,

$$I^f(0, T_0+2t_2+2b_f+2, \widehat{x}, \widehat{u})-I^f(0, T_0+2t_2+2b_f+2, x, u) \leq \Delta/4-\Delta+\Delta/4.$$

This contradicts (5.268). The contradiction we have reached proves that

$$I^f(0, t_2, y, v) = t_2 f(x_f, u_f).$$

Theorem 5.1, (5.255), and the relation above imply that (y, v) is (f)-good. By (P1),

$$\lim_{t\to\infty} y(t) = x_f.$$

Together with (5.254) and (5.255), this implies that $y(t) = x_f$ for all $t \geq 0$ and that $x(t) = x_f$ for all $t \in [0, t_2]$. Combined with Theorem 5.21, this implies that $x(t) = x_f$ for all $t \geq 0$. Theorem 5.23 is proved.

5.19 Proof of Proposition 5.27

Assume that $(x, u) \in X(0, \infty)$ is (f)-good. Theorem 5.1 implies that there exists $M_1 > 0$ such that

$$\|x(t)\| \leq M_1, \ t \in [0, \infty), \tag{5.270}$$

$$|I^f(0, t, x, u) - tf(x_f, u_f)| < M_1, \ t \in (0, \infty).$$

By the relation above, for all $S > T \geq 0$,

$$|I^f(T, S, x, u) - (S - T)f(x_f, u_f)| \leq 2M_1. \tag{5.271}$$

Proposition 1.3 imply that there exist $M \geq 1$, $\omega \in R^1$ such that

$$\|e^{At}\| \leq Me^{\omega t}, \ t \in [0, \infty)$$

and in the case of the second class of problems $\|e^{A^*t}\| \leq Me^{\omega t}, \ t \in [0, \infty)$. (5.272)

Set

$$\epsilon_1 = \epsilon(Me^{|\omega|})^{-1}. \tag{5.273}$$

We show that

$$\lim_{t\to\infty} x(t) = x_f.$$

Assume the contrary. Then there exists $\epsilon \in (0, 1)$ and a sequence of numbers $\{t_k\}_{k=1}^{\infty}$ such that

$$t_1 \geq 8,\ t_{k+1} \geq t_k + 8 \text{ for all integers } k \geq 1, \tag{5.274}$$

$$\|x(t_k) - x_f\| > \epsilon \text{ for all integers } k \geq 1. \tag{5.275}$$

From now we consider the two classes of problems separately and begin with the first class of problems.

By (5.9) and (5.10), there exists $h_0 > 0$ such that

$$f(x, u) \geq 4\epsilon_1^{-1}(M_1 + a_0 + 8)(\|G(x, u)\| - a_0\|x\|_+) \tag{5.276}$$

for each $(x, u) \in \mathcal{M}$ satisfying $\|G(x, u)\| - a_0\|x\| \geq h_0$. There exists $\delta > 0$ such that

$$\delta(|f(x_f, u_f)| + a_0 + 1) < 1, \tag{5.277}$$

$$\delta \in (0, \epsilon_1(4a_0 + 4h_0 + 4 + 4a_0M_1)^{-1}), \tag{5.278}$$

$$Me^{|\omega|}\delta < \epsilon_1/16, \tag{5.279}$$

$$\|e^{At}x_f - x_f\| < \epsilon_1/16 \text{ for all } t \in [0, \delta]. \tag{5.280}$$

Let $k \geq 1$ be an integer. We show that for all $t \in [t_k - \delta, t_k]$,

$$\|x(t) - x_f\| \geq \delta. \tag{5.281}$$

Assume the contrary. Then there exists

$$\tau \in [t_k - \delta, t_k] \tag{5.282}$$

such that

$$\|x(\tau) - x_f\| < \delta. \tag{5.283}$$

By (5.5),

$$x(t_k) = e^{A(t_k-\tau)}x(\tau) + \int_{\tau}^{t_k} e^{A(t_k-s)}G(x(s), u(s))ds. \tag{5.284}$$

It follows from (5.272), (5.279), (5.280), and (5.282)–(5.284) that

$$\|x(t_k) - x_f\| = \|e^{A(t_k-\tau)}x(\tau) + \int_{\tau}^{t_k} e^{A(t_k-s)}G(x(s), u(s))ds - x_f\|$$

$$\leq \|e^{A(t_k-\tau)}x(\tau) - e^{A(t_k-\tau)}x_f\| + \|e^{A(t_k-\tau)}x_f - x_f\|$$

$$+Me^{|\omega|\delta} \int_{\tau}^{t_k} \|G(x(s), u(s))\|ds$$

$$\leq Me^{|\omega|\delta}\|x(\tau) - x_f\| + \|e^{A(t_k-\tau)}x_f - x_f\| + Me^{|\omega|\delta} \int_{\tau}^{t_k} \|G(x(s), u(s))\|ds$$

$$\leq Me^{|\omega|}\delta + \|e^{A(t_k-\tau)}x_f - x_f\| + Me^{|\omega|} \int_{\tau}^{t_k} \|G(x(s), u(s))\|ds$$

$$< \epsilon/16 + \epsilon/16 + Me^{|\omega|} \int_{\tau}^{t_k} \|G(x(s), u(s))\|ds. \tag{5.285}$$

Set

$$\Omega_1 = \{t \in [\tau, t_k] : \|G(x(t), u(t))\| - a_0\|x(t)\| \geq h_0\}, \tag{5.286}$$

$$\Omega_2 = [\tau, t_k] \setminus \Omega_1. \tag{5.287}$$

By (5.270), (5.271), (5.276)–(5.278), (5.282), (5.286) and (5.287),

$$\int_{\tau}^{t_k} \|G(x(s), u(s))\|ds \leq a_0 \int_{\tau}^{t_k} \|x(s)\|ds$$

$$+ \int_{\tau}^{t_k} (\|G(x(s), u(s))\| - a_0\|x(s)\|)_+ ds$$

$$\leq a_0\delta M_1 + h_0\delta + \int_{\Omega_1} (\|G(x(s), u(s))\| - a_0\|x(s)\|)ds$$

$$\leq a_0\delta M_1 + h_0\delta + 4^{-1}\epsilon_1(M_1 + a_0 + 8)^{-1} \int_{\Omega_1} f(x(s), u(s)ds$$

$$\le a_0\delta M_1 + h_0\delta + (I^f(\tau, t_k, x, u) + a_0\delta)(4^{-1}\epsilon_1(M_1 + a_0 + 8)^{-1})$$

$$\le a_0\delta M_1 + h_0\delta + (4^{-1}\epsilon_1(M_1 + a_0 + 8)^{-1})(2M_1 + \delta|f(x_f, u_f)| + a_0\delta)$$

$$\le \delta(a_0M_1 + h_0) + 4^{-1}\epsilon_1(M_1 + a_0 + 8)^{-1}(2M_1 + 2) \le \epsilon_1/4 + \epsilon_1/2. \quad (5.288)$$

In view of (5.279), (5.285), and (5.288),

$$\|x(t_k) - x_f\| < \epsilon/16 + \epsilon/16 + Me^{|\omega|}3\epsilon_1/4 \le \epsilon/8 + 3\epsilon/4 < \epsilon.$$

This contradicts (5.275). The contradiction we have reached proves that (5.281) holds for all $t \in [t_k - \delta, t_k]$. This implies that

$$\mathrm{mes}(\{t \in [0, T] : \|x(t) - x_f\| \ge \delta\}) \to \infty \text{ as } T \to \infty.$$

On the other hand, property (P3) and (5.271) imply that there exists $L > 0$ such that for all $T > 0$,

$$\mathrm{mes}(\{t \in [0, T] : \|x(t) - x_f\| \ge \delta\}) \le L.$$

The contradiction we have reached proves that $\lim_{t\to\infty} x(t) = x_f$ and (P1) holds. Therefore for the first class of problems Proposition 5.27 is proved.

Consider the second class of problems. Recall that for each $\tau \ge 0$, $\Phi_\tau \in \mathcal{L}(L^2(0, \tau; F), E)$ is defined by

$$\Phi_\tau u = \int_0^\tau e^{A(\tau-s)}Bu(s)ds, \ u \in L^2(0, \tau; F). \quad (5.289)$$

Lemma 5.43 and (A2) imply that there exists $\delta \in (0, 1)$ such that

$$M_* e^{|\omega_*|}\delta < \epsilon/8, \quad (5.290)$$

$$32(a_0 + |f(x_f, u_f)| + 1)\delta < K_0\epsilon^2(\|\Phi_1\| + 1)^{-1}, \quad (5.291)$$

$$\|e^{As}x_f - x_f\| < \epsilon/8 \text{ for all } s \in [0, \delta], \quad (5.292)$$

and the following properties hold:

(a) for each $z_i \in \mathcal{A}$, $i = 1, 2$ satisfying $\|z_i - x_f\| \le \delta$, $i = 1, 2$ there exist $\tau \in (0, b_f]$ and $(x, u) \in X(0, \tau)$ satisfying

$$x(0) = z_1, \ x(\tau) = z_2,$$

$$I^f(0, \tau, x, u) \le \tau f(x_f, u_f) + 32^{-1}\epsilon^2 K_0(\|\Phi_1\| + 1)^{-2};$$

(b) for each $T \geq b_f$ and each $(y, v) \in X(0, T)$ which satisfies $\|y(0) - x_f\| \leq \delta$, $\|y(T) - x_f\| \leq \delta$, we have

$$I^f(0, T, y, v) \geq Tf(x_f, u_f) - 32^{-1}\epsilon^2 K_0(\|\Phi_1\| + 1)^{-2}$$

(recall K_0 in (5.11)).

Property (P3) and (5.271) imply that there exists $L > 0$ such that for all $T > 0$,

$$\text{mes}(\{t \in [0, T] : \|x(t) - x_f\| \geq \delta\}) \leq L.$$

Therefore there exists $T_0 > 0$ such that for each $S > T_0$,

$$\text{mes}(\{t \in [T_0, S] : \|x(t) - x_f\| > \delta\}) \leq \delta/2. \tag{5.293}$$

Since the pair (x, u) is (f)-good, it follows from Theorem 5.1 that there exists $T_1 > 0$ such that for each $S > T_1$,

$$I^f(T_1, S, x, u) \leq U^f(T_1, S, x(T_1), x(S)) + 16^{-1}\epsilon^2 K_0(\|\Phi_1\| + 1)^{-2}. \tag{5.294}$$

Let $k \geq 1$ be an integer such that

$$t_k > T_0 + T_1 + 1. \tag{5.295}$$

Assume that

$$\inf\{\|x(t) - x_f\| : t \in [t_k - \delta, t_k]\} < \delta. \tag{5.296}$$

Therefore there exists

$$\tau \in [t_k - \delta, t_k] \tag{5.297}$$

such that

$$\|x(\tau) - x_f\| < \delta. \tag{5.298}$$

It is clear that $(x, u) \in X(\tau, t_k)$ and in view of (5.7),

$$x(t_k) = e^{A(t_k - \tau)}x(\tau) + \int_\tau^{t_k} e^{A(t_k - s)}Bu(s)ds$$

$$= e^{A(t_k - \tau)}x(\tau) + \Phi_{t_k - \tau}u(\tau + \cdot) \tag{5.299}$$

in $E_{-1} = \mathcal{D}(A^*)'$. It follows from (5.272), (5.275), (5.290), (5.292), (5.297)–(5.299), and Proposition 1.18 that

$$\epsilon \le \|x(t_k) - x_f\| = \|e^{A(t_k-\tau)}x(\tau) + \Phi_{t_k-\tau}u(\tau+\cdot) - x_f\|$$

$$\le \|e^{A(t_k-\tau)}x(\tau) - e^{A(t_k-\tau)}x_f\| + \|e^{A(t_k-\tau)}x_f - x_f\| + \|\Phi_{t_k-\tau}u(\tau+\cdot)\|$$

$$\le \|e^{A(t_k-\tau)}\|\|x(\tau) - x_f\| + \|e^{A(t_k-\tau)}x_f - x_f\| + \|\Phi_{t_k-\tau}\|\|u(\tau+\cdot)\|_{L^2(0,t_k-\tau;F)}$$

$$\le M_* e^{|\omega_*|}\delta + \epsilon/8 + \|\Phi_1\|\|u(\tau+\cdot)\|_{L^2(0,t_k-\tau;F)}$$

$$\le \epsilon/4 + \|\Phi_1\|(\int_\tau^{t_k} \|u(s)\|^2 ds)^{1/2}. \tag{5.300}$$

In view of (5.300),

$$\int_\tau^{t_k} \|u(s)\|^2 ds \ge (2^{-1}\epsilon(\|\Phi_1\| + 1)^{-1})^2. \tag{5.301}$$

By (5.11), (5.297), and (5.301),

$$\int_\tau^{t_k} f(x(s), u(s))ds + a_0(t_k - \tau) \ge K_0 \int_\tau^{t_k} \|u(s)\|^2 ds \ge 4^{-1}\epsilon^2 K_0(\|\Phi_1\| + 1)^{-2},$$

$$\int_\tau^{t_k} f(x(s), u(s))ds \ge 4^{-1}K_0\epsilon^2(\|\Phi_1\| + 1)^{-2} - a_0\delta. \tag{5.302}$$

In view of (5.293) and (5.295), for each $S > t_k$,

$$\text{mes}(\{t \in [t_k, S] : \|x(t) - x_f\| > \delta\}) \le \delta/2. \tag{5.303}$$

Inequality (5.303) implies that there exists

$$S_1 \in [t_k, t_k + \delta] \tag{5.304}$$

such that

$$\|x(S_1) - x_f\| \le \delta. \tag{5.305}$$

By (5.303) and (5.304), there exists

$$S_2 \in [S_1 + 2b_f, S_1 + 2b_f + \delta] \tag{5.306}$$

such that

$$\|x(S_2) - x_f\| \le \delta. \tag{5.307}$$

It follows from (5.11), (5.297), (5.302), and (5.304) that

$$I^f(\tau, S_1, x, u) = I^f(\tau, t_k, x, u) + I^f(t_k, S_1, x, u)$$
$$\geq 4^{-1}K_0\epsilon^2(\|\Phi_1\| + 1)^{-2} - a_0\delta - a_0\delta. \tag{5.308}$$

Property (b) and (5.305)–(5.307) imply that

$$I^f(S_1, S_2, x, u) \geq (S_2 - S_1)f(x_f, u_f) - 32^{-1}K_0\epsilon^2(\|\Phi_1\| + 1)^{-2}. \tag{5.309}$$

By (5.304), (5.306), (5.308), and (5.309),

$$I^f(\tau, S_2, x, u) \geq (S_2 - S_1)f(x_f, u_f) - 32^{-1}K_0\epsilon^2(\|\Phi_1\| + 1)^{-2}$$
$$+4^{-1}K_0\epsilon^2(\|\Phi_1\| + 1)^{-2} - 2a_0\delta$$
$$\geq K_0\epsilon^2(\|\Phi_1\|+1)^{-2}(4^{-1} - 32^{-1}) - 2a_0\delta + (S_2 - \tau)f(x_f, u_f) - (S_1 - \tau)|f(x_f, u_f)|$$
$$\geq K_0\epsilon^2(\|\Phi_1\| + 1)^{-2}(4^{-1} - 32^{-1}) + (S_2 - \tau)f(x_f, u_f) - 2\delta(a_0 + |f(x_f, u_f)|). \tag{5.310}$$

Property (a), (5.298), (5.304), (5.306), and (5.307) imply that that there exists $(y, v) \in X(\tau, S_2)$ such that

$$y(\tau) = x(\tau),\ y(\tau + b_f) = x_f,$$

$$I^f(\tau, \tau + b_f, y, v) \leq b_f f(x_f, u_f) + 32^{-1}\epsilon^2 K_0(\|\Phi_1\| + 1)^{-2}, \tag{5.311}$$

$$y(S_2) = x(S_2),\ y(S_2 - b_f) = x_f,$$

$$I^f(S_2 - b_f, S_2, y, v) \leq b_f f(x_f, u_f) + 32^{-1}\epsilon^2 K_0(\|\Phi_1\| + 1)^{-2}, \tag{5.312}$$

$$(y(t), v(t)) = (x_f, u_f),\ t \in [\tau + b_f, S_2 - b_f]. \tag{5.313}$$

In view of (5.311)–(5.313),

$$I^f(\tau, S_2, y, v) \leq (S_2 - \tau)f(x_f, u_f) + 16^{-1}\epsilon^2 K_0(\|\Phi_1\| + 1)^{-2}.$$

Together with (5.291) and (5.310), the relation above implies that

$$I^f(\tau, S_2, x, u) - I^f(\tau, S_2, y, v)$$
$$\geq \epsilon^2 K_0(\|\Phi_1\| + 1)^{-2}(4^{-1} - 16^{-1} - 32^{-1}) - 2\delta(a_0 + |f(x_f, u_f)|)$$
$$\geq \epsilon^2 K_0(\|\Phi_1\| + 1)^{-2}(4^{-1} - 8^{-1}) = 8^{-1}\epsilon^2 K_0(\|\Phi_1\| + 1)^{-2}. \tag{5.314}$$

Since $t_k > T_1 + 1$ it follows from (5.295) and (5.297) that $\tau > T_1$. In view of the choice of T_1 (see (5.294)),

$$I^f(\tau, S_2, x, u) \le U^f(\tau, S_2, x(\tau), x(S_2)) + 16^{-1}\epsilon^2 K_0(\|\Phi_1\| + 1)^{-2}.$$

This contradicts (5.311), (5.312), and (5.314). The contradiction we have reached proves that (5.296) is not true and

$$\inf\{\|x(t) - x_f\| : t \in [t_k - \delta, t_k]\} \ge \delta$$

for each natural number k satisfying $t_k \ge T_0 + T_1 + 1$. This implies that

$$\text{mes}(\{t \in [0, T] : \|x(t) - x_f\| \ge \delta\}) \to \infty \text{ as } T \to \infty.$$

This contradicts (5.293). The contradiction we have reached proves that $\lim_{t\to\infty} x(t) = x_f$ and (P1) holds for the second class of problems too. Proposition 5.27 is proved.

5.20 Proof of Theorem 5.28

In view of Theorem 5.26, it is sufficient to show that f_r has (P3). Let $\epsilon, M > 0$. Theorem 5.1 implies that there exists $M_0 > 0$ such that for each $S_2 > S_1 \ge 0$ and each $(x, u) \in X(S_1, S_2)$,

$$I^f(S_1, S_2, x, u) \ge (S_2 - S_1) f(x_f, u_f) - M_0. \tag{5.315}$$

There exists $\delta > 0$ such that

$$\text{if } z \in E \text{ satisfies } \phi(z) \le \delta, \text{ then } \|z\| \le \epsilon. \tag{5.316}$$

Set

$$L = \delta^{-1} r^{-1}(M_0 + M). \tag{5.317}$$

Assume that $T_1 \ge 0$, $T_2 \ge T_1 + L$, $(x, u) \in X(T_1, T_2)$ satisfies

$$I^{f_r}(T_1, T_2, x, u) \le (T_2 - T_1) f_r(x_f, u_f) + M. \tag{5.318}$$

By (5.318) and the definition of f_r,

$$I^f(T_1, T_2, x, u) + r \int_{T_1}^{T_2} \phi(x(t) - x_f))dt \le (T_2 - T_1) f(x_f, u_f) + M. \tag{5.319}$$

In view of (5.315) and (5.319),

$$(T_2 - T_1)f(x_f, u_f) - M_0 + r\int_{T_1}^{T_2} \phi(x(t) - x_f)dt \le (T_2 - T_1)f(x_f, u_f) + M,$$

$$\int_{T_1}^{T_2} \phi(x(t) - x_f)dt \le r^{-1}(M_0 + M). \tag{5.320}$$

It follows from the choice of δ, (5.316), (5.317), and (5.320) that

$$\mathrm{mes}(\{t \in [T_1, T_2] : \ \|x(t) - x_f\| \ge \epsilon\}) \le \mathrm{mes}(\{t \in [T_1, T_2] : \ \phi(x(t) - x_f) > \delta\})$$

$$\le \delta^{-1}\int_{T_1}^{T_2} \phi(x(t) - x_f)dt \le \delta^{-1}r^{-1}(M_0 + M) \le L.$$

Thus f_r has (P3). Theorem 5.28 is proved.

5.21 Auxiliary Results

We suppose that all the assumptions used in Section 5.6 hold and that (P1) and (P2) hold. By Theorem 5.1, there exists $S_* > 0$ such that for each pair of numbers $T_2 > T_1 \ge 0$ and each $(x, u) \in X(T_1, T_2)$,

$$I^f(T_1, T_2, x, u) \ge (T_2 - T_1)f(x_f, u_f) - S_*. \tag{5.321}$$

For all $z \in \mathcal{A} \setminus \cup\{\mathcal{A}_L : \ L \in (0, \infty)\}$, set

$$\pi^f(z) = \infty. \tag{5.322}$$

Let

$$z \in \cup\{\mathcal{A}_L : \ L \in (0, \infty)\}. \tag{5.323}$$

Define

$$\pi^f(z) = \inf\{\liminf_{T\to\infty}(I^f(0, T, x, u) - Tf(x_f, u_f)) : \ (x, u) \in X(0, \infty), \ x(0) = z\}. \tag{5.324}$$

By (5.321), (5.323) and (5.324),

$$- S_* \le \pi^f(z) < \infty. \tag{5.325}$$

There exists a $L > 0$ such that

$$z \in \mathcal{A}_L. \tag{5.326}$$

In view of (5.326), there exist $(\tilde{x}, \tilde{u}) \in X(0, \infty)$ and $\tau \in (0, L]$ such that

$$\tilde{x}(0) = z, \ \tilde{x}(\tau) = x_f, \ I^f(0, \tau, \tilde{x}, \tilde{u}) \leq L,$$

$$(\tilde{x}(t), \tilde{u}(t)) = (x_f, u_f) \text{ for all } t \geq \tau.$$

By the relations above, for all $T \geq \tau$,

$$I^f(0, T, \tilde{x}, \tilde{u}) - Tf(x_f, u_f)$$

$$\leq \tau + (T - \tau)f(x_f, u_f) - Tf(x_f, u_f) \leq L(1 + |f(x_f, u_f)|)$$

and

$$\pi^f(z) \leq L(1 + |f(x_f, u_f)|) \text{ for all } z \in \mathcal{A}_L. \tag{5.327}$$

Definition (5.324) implies the following result.

Proposition 5.45.

1. *Let $S > T \geq 0$ and $(x, u) \in X(T, S)$ satisfy*

$$\pi^f(x(T)), \ \pi^f(x(S)) < \infty.$$

   Then

$$\pi^f(x(T)) \leq I^f(T, S, x, u) - (S - T)f(x_f, u_f) + \pi^f(x(S)). \tag{5.328}$$

2. *Let $(x, u) \in X(0, \infty)$ be an (f)-good pair. Then for each pair of numbers $S > T > 0$, (5.328) holds.*

Proposition 5.46. $\pi^f(x_f) = 0$.

Proof. In view of (5.324),

$$\pi^f(x_f) \leq 0. \tag{5.329}$$

Assume that

$$\pi^f(x_f) < 0. \tag{5.330}$$

There exists $(x_0, u_0) \in X(0, \infty)$ such that

$$x_0(0) = x_f, \tag{5.331}$$

$$\liminf_{T\to\infty}[I^f(0,T,x_0,u_0) - Tf(x_f,u_f)] < 2^{-1}\pi^f(x_f). \tag{5.332}$$

It follows from (5.332) that there exists a strictly increasing sequence of positive numbers $\{T_k\}_{k=1}^{\infty}$ such that

$$T_k \to \infty \text{ as } k \to \infty,$$

$$I^f(0,T_k,x_0,u_0) - T_k f(x_f,u_f) < 2^{-1}\pi^f(x_f),\ k = 1,2,\dots. \tag{5.333}$$

By (A2), there exists $\delta \in (0,1)$ such that the following property holds:

(i) for each $z \in \mathcal{A}$, satisfying $\|z - x_f\| \le \delta$, there exist $\tau_1, \tau_2 \in (0, b_f]$ and $(y_i, v_i) \in X(0,\tau_i)$, $i = 1,2$ which satisfy

$$y_1(0) = z,\ y_1(\tau_1) = x_f,\ I^f(0,\tau_1,y_1,v_1) \le \tau_1 f(x_f,u_f) - \pi^f(x_f)/8,$$

$$y_2(0) = x_f,\ y_2(\tau_2) = z,\ I^f(0,\tau_2,y_2,v_2) \le \tau_2 f(x_f,u_f) - \pi^f(x_f)/8.$$

Theorem 5.1 and (5.333) imply that (x_0,u_0) is (f)-good. In view of (P1),

$$\lim_{t\to\infty} \|x_0(t) - x_f\| = 0.$$

Choose a natural number k such that

$$\|x_0(T_k) - x_f\| < \delta. \tag{5.334}$$

Property (i), (5.331), and (5.334) imply that there exists $(y,v) \in X(0,\infty)$ such that

$$(y(t+T_k+b_f), v(t+T_k+b_f)) = (y(t), v(t)) \text{ for all numbers } t \ge 0, \tag{5.335}$$

$$(y(t), v(t)) = (x_0(t), u_0(t)),\ t \in [0,T_k], \tag{5.336}$$

$$y(T_k+b_f) = x_f, \tag{5.337}$$

$$I^f(T_k, T_k+b_f, y, v) \le b_f f(x_f,u_f) - \pi^f(x_f)/8. \tag{5.338}$$

By (5.333), (5.336), and (5.338),

$$\begin{aligned} I^f(0,T_k+b_f,y,v) &= I^f(0,T_k,y,v) + I^f(T_k,T_k+b_f,y,v) \\ &< T_k f(x_f,u_f) + 2^{-1}\pi^f(x_f) + b_f f(x_f,u_f) - \pi^f(x_f)/8 \\ &\le (T_k+b_f) f(x_f,u_f) + 4^{-1}\pi^f(x_f). \end{aligned}$$

Combined with (5.335) this implies that

$$\liminf_{T\to\infty}[I^f(0,T,y,v)-Tf(x_f,u_f)]=-\infty,$$

a contradiction which proves that $\pi^f(x_f)=0$ and Proposition 5.46 itself.

Proposition 5.47. *The function* $\pi^f:\mathcal{A}\to R^1\cup\{\infty\}$ *is finite in a neighborhood of* x_f *and continuous at* x_f.

Proof. Let $\epsilon>0$. By (A2), there exists $\delta>0$ such that the following property holds:

(ii) for each $z\in\mathcal{A}$ satisfying $\|z-x_f\|\le\delta$ there exist $\tau_1,\tau_2\in(0,b_f]$ and $(x_i,u_i)\in X(0,\tau_i)$, $i=1,2$ which satisfy

$$x_1(0)=z,\ x_1(\tau_1)=x_f,\ x_2(0)=x_f,\ x_2(\tau_2)=z,$$

$$I^f(0,\tau_i,x_i,u_i)\le\tau_i f(x_f,u_f)+\epsilon/8,\ i=1,2.$$

Let $z\in\mathcal{A}$ satisfy

$$\|z-x_f\|\le\delta, \tag{5.339}$$

and let $\tau_1,\tau_2\in(0,b_f]$ and $(x_i,u_i)\in X(0,\tau_i)$, $i=1,2$ be as guaranteed by property (ii). Set

$$(\tilde{x}_1(t),\tilde{u}_1(t))=(x_1(t),u_1(t))\text{ for all }t\in[0,\tau_1], \tag{5.340}$$

$$(\tilde{x}_1(t),\tilde{u}_1(t))=(x_f,u_f)\text{ for all }t>\tau_1. \tag{5.341}$$

Clearly, $(\tilde{x}_1,\tilde{u}_1)\in X(0,\infty)$. Property (ii), (5.340), and (5.341) imply that

$$\pi^f(z)\le\liminf_{T\to\infty}[I^f(0,T,\tilde{x}_1,\tilde{u}_1)-Tf(x_f,u_f)]$$

$$=I^f(0,\tau_1,x_1,u_1)-\tau_1 f(x_f,u_f)\le\epsilon/8.$$

Propositions 5.45 and 5.46 and property (ii) imply that

$$0=\pi^f(x_f)\le I^f(0,\tau_2,x_2,u_2)-\tau_2 f(x_f,u_f)+\pi^f(z)\le\epsilon/8+\pi^f(z),$$

$$\pi^f(z)\ge-\epsilon/8.$$

Proposition 5.47 is proved.

Proposition 5.48. *For each $M > 0$ the set $\{z \in \mathcal{A} : \pi^f(z) \le M\}$ is bounded,*

Proof. Let $M > 0$ and suppose that the set $\{z \in \mathcal{A} : \pi^f(z) \le M\}$ is nonempty. Propositions 5.46 and 5.47 imply that there exists $\delta > 0$ such that for each $z \in \mathcal{A}$ satisfying $\|z - x_f\| \le \delta$, $\pi^f(z)$ is finite and

$$|\pi^f(z)| \le 1. \tag{5.342}$$

By Theorem 5.14, there exists $L_0 > 0$ such that the following property holds:

(iii) for each $T \ge L_0$, each $(x, u) \in X(0, T)$ which satisfies

$$I^f(0, T, x, u) \le T_f(x_f, u_f) + M + 4,$$

and each $S \in [0, T - L_0]$, we have

$$\min\{\|x(t) - x_f\| : t \in [S, S + L_0]\} \le \delta.$$

Theorem 5.32 implies that there exists $M_1 > 0$ such that the following property holds:

(iv) for each $S \in [1, L_0 + 1]$ and each $(x, u) \in X(0, S)$ which satisfies

$$I^f(0, S, x, u) \le (1 + L_0)|f(x_f, u_f)| + M + 4,$$

we have

$$\|x(t)\| \le M_1, \ t \in [0, S].$$

Assume that $z \in \mathcal{A}$ satisfies

$$\pi^f(z) \le M. \tag{5.343}$$

In view of (5.324) and (5.343), there exists $(x, u) \in X(0, \infty)$ such that

$$x(0) = z, \tag{5.344}$$

$$\liminf_{T \to \infty} [I^f(0, T, x, u) - Tf(x_f, u_f)] < \pi^f(z) + 1 \le M + 1. \tag{5.345}$$

By property (iii) and (5.345), there exists

$$t_0 \in [1, L_0 + 1]$$

such that

$$\|x(t_0) - x_f\| \le \delta. \tag{5.346}$$

It follows from (5.346) and the choice of δ (see (5.342)) that

$$|\pi^f(x(t_0))| \le 1. \tag{5.347}$$

Proposition 5.45, (5.343)–(5.345), and (5.347) imply that

$$\pi^f(x(0)) \le I^f(0, t_0, x, u) - t_0 f(x_f, x_f) + \pi^f(x(t_0))$$

$$\le \liminf_{T\to\infty}[I^f(0, T, x, u) - Tf(x_f, u_f)] < \pi^f(z) + 1 \le M + 1. \tag{5.348}$$

In view of (5.347) and (5.348),

$$I^f(0, t_0, x, u) \le t_0 f(x_f, u_f) + M + 2.$$

Together with property (iv), this implies that

$$\|x(t)\| \le M_1 \text{ for all } t \in [0, t_0]. \tag{5.349}$$

By (5.344) and (5.349), $\|z\| = \|x(0)\| \le M_1$. Proposition 5.48 is proved.

Corollary 5.49.

$$\pi^f(z) \to \infty \text{ as } z \in \mathcal{A} \text{ and } \|z\| \to \infty.$$

Let $T_2 > T_1 \ge 0$ and $(x, u) \in X(T_1, T_2)$ satisfy $\pi^f(x(T_1)) < \infty$. Define

$$\Gamma^f(T_1, T_2, x, u) = I^f(T_1, T_2, x, u) - (T_2 - T_1) f(x_f, u_f) - \pi^f(x(T_1)) + \pi^f(x(T_2)). \tag{5.350}$$

Proposition 5.45, (5.322), (5.325), and (5.350) implies that

$$0 \le \Gamma^f(T_1, T_2, x, u) \le \infty. \tag{5.351}$$

Proposition 5.50. *Let $(x, u) \in X(0, \infty)$ and $x(0) \in \cup\{\mathcal{A}_L : L \in (0, \infty)\}$. Then (x, u) is (f)-good if and only if*

$$\sup\{\Gamma^f(0, T, x, u) : T \in (0, \infty)\} = \lim_{T\to\infty} \Gamma^f(0, T, x, u) < \infty.$$

If (x, u) is (f)-good, then

$$\lim_{T\to\infty} [I^f(0, T, x, u) - Tf(x_f, u_f)] = \pi^f(x(0)) + \lim_{T\to\infty} \Gamma^f(0, T, x, u).$$

Proof. Assume that $(x, u) \in X(0, \infty)$ is (f)-good. Propositions 5.46, 5.47, (P1), (5.350), and (5.351) imply that

$$\lim_{t\to\infty} x(t) = x_f, \ \lim_{t\to\infty} \pi^f(x(t)) = 0, \tag{5.352}$$

$$\pi^f(x(t)) < \infty \text{ for all } t \geq 0 \tag{5.353}$$

and

$$\sup\{\Gamma^f(0, T, x, u) : T \in (0, \infty)\} = \lim_{T\to\infty} \Gamma^f(0, T, x, u)$$

$$\leq \lim_{T\to\infty} (I^f(0, T, x, u) - Tf(x_f, u_f) - \pi^f(x(0)) < \infty.$$

Assume that

$$\Delta := \lim_{T\to\infty} \Gamma^f(0, T, x, u) < \infty.$$

It follows from (5.325), (5.327), (5.350), (5.351), and (5.353) that for all $T > 0$, $\pi^f(x(T)) < \infty$ and

$$I^f(0, T, x, u) - Tf(x_f, u_f) = \Gamma^f(0, T, x, u)$$

$$+\pi^f(x(0)) - \pi^f(x(T)) \leq \Delta + \pi^f(x(0)) + S_*$$

and (x, u) is (f)-good.

Assume that (x, u) is (f)-good. By Propositions 5.46 and 5.47, (P1), (5.350) and (5.352), for all $T > 0$,

$$I^f(0, T, x, u) - Tf(x_f, u_f) = \Gamma^f(0, T, x, u) + \pi^f(x(0)) - \pi^f(x(T))$$

$$\to \lim_{T\to\infty} \Gamma^f(0, T, x, u) + \pi^f(x(0)).$$

Proposition 5.50 is proved.

Corollary 5.51. *Let $z \in \mathcal{A}$ satisfy $\pi^f(z) < \infty$ and $\epsilon > 0$. Then there exists an (f)-good $(x, u) \in X(0, \infty)$ such that $x(0) = z$ and*

$$\lim_{T\to\infty} \Gamma^f(0, T, x, u) \leq \epsilon.$$

Set

$$\inf(\pi^f) = \inf\{\pi^f(z) : \ z \in \mathcal{A}\}. \tag{5.354}$$

Proposition 5.46, (5.325), and (5.354) imply that $\inf(\pi^f)$ is finite. Set

$$\mathcal{A}_f = \{z \in \mathcal{A} : \ \pi^f(z) \leq \inf(\pi^f) + 1\}. \tag{5.355}$$

Proposition 5.52. *There exists $M_0 > 0$ such that $\mathcal{A}_f \subset \mathcal{A}_{M_0}$.*

Proof. Proposition 5.48, (5.354), and (5.355) imply that there exists $M_1 > 0$ such that

$$\mathcal{A}_f \subset \{z \in \mathcal{A} : \ \|z\| \le M_1\}. \tag{5.356}$$

By Theorem 5.6 and Lemma 5.37, there exists $M_2 > 0$ such that for each $T > 0$ and each $(x, u) \in X(0, T)$ satisfying

$$\|x(0)\| \le M_1, \ I^f(0, T, x, u) \le Tf(x_f, u_f) + |\inf(\pi^f)| + 3,$$

we have

$$\|x(t)\| \le M_2, \ t \in [0, T]. \tag{5.357}$$

Theorem 5.1 implies that there exists $c_1 > 0$ such that for each $T > 0$ and each $(x, u) \in X(0, T)$,

$$I^f(0, T, x, u) \ge Tf(x_f, u_f) - c_1. \tag{5.358}$$

In view of (A2), there exists $\delta \in (0, 1)$ such that the following property holds:

(v) for each $z \in \mathcal{A}$ satisfying $\|z - x_f\| \le \delta$, there exist $\tau_1 \in (0, b_f]$ and $(x_1, u_1) \in X(0, \tau_1)$ which satisfies

$$x_1(0) = z, \ x_1(\tau_1) = x_f, \ I^f(0, \tau_1, x_1, u_1) \le \tau_1 f(x_f, u_f) + 1.$$

By Theorem 5.14, there exists $L_0 > 0$ such that the following property holds:

(vi) for each $T_1 \ge 0$, each $T_2 \ge T_1 + L_0$, and each $(x, u) \in X(T_1, T_2)$ which satisfies

$$I^f(T_1, T_2, x, u) - (T_2 - T_1)f(x_f, u_f) \le |\inf(\pi^f)| + 3 + c_1,$$

there exists $t_0 \in [0, L_0]$ such that $\|x(t_0) - x_f\| \le \delta$.

Set

$$M_0 = (L_0 + b_f)(1 + |f(x_f, u_f)|) + |\inf(\pi^f)| + 4 + c_1. \tag{5.359}$$

Let

$$z \in \mathcal{A}_f.$$

In view of (5.355) and (5.356),

$$\pi^f(z) \le \inf(\pi^f) + 1, \ \|z\| \le M_1, \tag{5.360}$$

and by Proposition 5.50, there exists $(x, u) \in X(0, \infty)$ such that

$$x(0) = z, \tag{5.361}$$

$$\lim_{T\to\infty} [I^f(0, T, x, u) - Tf(x_f, u_f)]$$

$$= \lim_{T\to\infty} \Gamma^f(0, T, x, u) + \pi^f(z) \le \pi^f(z) + 1 \le \inf(\pi^f) + 2. \tag{5.362}$$

By the relation above, there exists a number T_0 such that for all $T \ge T_0$,

$$I^f(0, T, x, u) - Tf(x_f, u_f) \le \inf(\pi^f) + 3. \tag{5.363}$$

It follows from the choice of M_2 (see (5.357)), (5.360), (5.361), and (5.363) that

$$\|x(t)\| \le M_1 \text{ for all } t \ge 0. \tag{5.364}$$

It follows from (5.358) and (5.363) that for any $T \in (0, T_0)$,

$$I^f(0, T, x, u) - Tf(x_f, u_f) = I^f(0, T_0, x, u) - T_0 f(x_f, u_f)$$

$$-(I^f(T, T_0, x, u) - (T_0 - T)f(x_f, u_f)) \le \inf(\pi^f) + 3 + c_1.$$

Therefore for all $T > 0$,

$$I^f(0, T, x, u) - Tf(x_f, u_f) \le \inf(\pi^f) + 3 + c_1. \tag{5.365}$$

Property (vi) and (5.365) imply that there exists a number t_0 such that

$$t_0 \in [0, L_0], \ \|(x(t_0) - x_f\| \le \delta. \tag{5.366}$$

It follows from property (v) and (5.366) that there exists $(y, v) \in X(0, t_0 + b_f)$ such that

$$(y(t), v(t)) = (x(t), u(t)), \ t \in [0, t_0], \tag{5.367}$$

$$y(t_0 + b_f) = x_f, \ I^f(t_0, t_0 + b_f, y, v) \le b_f f(x_f, u_f) + 1. \tag{5.368}$$

In view of (5.359) and (5.366),

$$t_0 + b_f \le L_0 + b_f \le M_0.$$

By (5.359) and (5.365)–(5.368),

$$I^f(0, t_0 + b_f, y, v) \le t_0 f(x_f, u_f) + \inf(\pi^f) + 3 + c_1 + b_f f(x_f, u_f) + 1$$

$$\le |f(x_f, u_f)|(L_0 + b_f) + |\inf(\pi^f)| + 4 + c_1 \le M_0.$$

Proposition 5.52 is proved.

Proposition 5.53. *Let $M_0 > 0$ be an integer. Then the set $\mathcal{A}_{M_0}$ is bounded.*

Proof. Let $z \in \mathcal{A}_{M_0}$. There exists $(x, u) \in X(0, \infty)$ such that

$$x(0) = z, \ x(t) = x_f, \ u(t) = u_f \text{ for all } t \ge M_0,$$

$$I^f(0, M_0, x, u) \le M_0 + M_0|f(x_f, u_f)|.$$

By the relations above,

$$\pi^f(z) \le \liminf_{T\to\infty}[I^f(0, T, x, u) - Tf(x_f, u_f)]$$

$$\le M_0(1 + |f(x_f, u_f)|) + M_0|f(x_f, u_f)|.$$

Thus $\mathcal{A}_{M_0} \subset \{z \in \mathcal{A} : \ \pi^f(z) \le 2M_0(1 + |f(x_f, u_f)|)\}$ which is bounded by Proposition 5.48. This completes the proof of Proposition 5.53.

Let

$$z \in \cup\{\mathcal{A}_L : \ L \in (0, \infty)\}.$$

Denote by $\Lambda(f, z)$ the set of all (f)-overtaking optimal pairs $(x, u) \in X(0, \infty)$ such that $x(0) = z$. If f has LSC property, then in view of Theorem 5.19, $\Lambda(f, z) \ne \emptyset$, and by Theorem 5.1, any element of $\Lambda(f, z)$ is (f)-good. Equation (5.324) implies the following result.

Proposition 5.54. *Let f have LSC property. Then for every $z \in \cup\{\mathcal{A}_L : \ L \in (0, \infty)\}$ and every $(x, u) \in \Lambda(f, z)$,*

$$\pi^f(z) = \liminf_{T\to\infty}[I^f(0, T, x, u) - Tf(x_f, u_f)].$$

The next result follows from Proposition 5.54.

Proposition 5.55. *Let f have LSC property and $(x, u) \in X(0, \infty)$ be (f)-overtaking optimal and (f)-good. Then for each pair of numbers $S > T \ge 0$,*

$$\pi^f(x(T)) = I^f(T, S, x, u) - (S - T)f(x_f, u_f) + \pi^f(x(S)).$$

Proposition 5.56. *Let f have LSC property and $(x, u) \in X(0, \infty)$ be (f)-overtaking optimal and (f)-good. Then*

$$\pi^f(x_0) = \lim_{T\to\infty} [I^f(0, T, x, u) - Tf(x_f, u_f)].$$

Proof. Propositions 5.46, 5.47, 5.55, and (P1) imply that for all $T > 0$,

$$I^f(0, T, x, u) - Tf(x_f, u_f) = \pi^f(x(0)) - \pi^f(x(T)) \to \pi^f(x(0)) \text{ as } T \to \infty.$$

Proposition 5.56 is proved.

Proposition 5.57. *Let LSC property holds. Then $\pi^f : \mathcal{A} \to R^1 \cup \{\infty\}$ is a lower semicontinuous function on $\mathcal{A}$.*

Proof. Let $\{x_k\}_{k=1}^{\infty} \subset \mathcal{A}$, $x \in \mathcal{A}$ and

$$x = \lim_{k\to\infty} x_k.$$

We show that $\pi^f(x) \le \liminf_{k\to\infty} \pi^f(x_k)$. Extracting subsequences and re-indexing, we may assume without loss of generality that there exists

$$\lim_{k\to\infty} \pi^f(x_k) < \infty \tag{5.369}$$

and that $\pi^f(x_k) < \infty$ for all integers $k \ge 1$. By Theorem 5.19, for each integer $k \ge 1$, there exists (f, A, G)-overtaking optimal and (f, A, G)-good $(y_k, v_k) \in X(0, \infty, A, G)$ such that

$$y_k(0) = x_k. \tag{5.370}$$

Proposition 5.56 implies that for every integer $k \ge 1$,

$$\pi^f(x_k) = \lim_{T\to\infty} [I^f(0, T, y_k, v_k) - Tf(x_f, u_f)]. \tag{5.371}$$

It follows from Proposition 5.55, (5.325), and (5.369) that for every integer $T \ge 1$, the sequence $\{I^f(0, T, y_k, v_k)\}_{k=1}^{\infty}$ is bounded. By LSC property, extracting subsequences, using the diagonalization process and re-indexing, we may assume that there exists $(y, v) \in X(0, \infty, A, G)$, and a subsequence (y_{i_k}, v_{i_k}), $k = 1, 2, \dots$ such that for every integer $t \ge 0$,

$$y_k(t) \to y(t) \text{ as } k \to \infty, \tag{5.372}$$

and for every natural number T, there exists

$$\lim_{k\to\infty} I^f(T-1, T, y_{i_k}, v_{i_k})$$

and

$$I^f(T-1, T, y, v) \le \lim_{k\to\infty} I^f(T-1, T, y_{i_k}, v_{i_k}). \tag{5.373}$$

Let $\epsilon > 0$. By Propositions 5.46 and 5.47, there exists $\delta > 0$ such that for each $\xi \in \mathcal{A}$ satisfying $\|\xi - x_f\| \le \delta$, we have

$$|\pi^f(\xi)| \le \epsilon/2. \tag{5.374}$$

Since the pairs (y_k, v_k), $k = 1, 2, \dots$ are (f, A, G)-overtaking optimal, it follows from (5.325), (5.369), (5.371), and Proposition 5.10 that there exists $L_0 > 0$ such that for each integer $k \ge 1$ and each $t \ge L_0$,

$$\|y_k(t) - x_f\| \le \delta. \tag{5.375}$$

It follows from (5.375) and the choice of δ (see (5.374)) that for each integer $k \ge 1$ and each integer $T \ge L_0$,

$$|\pi^f(y_k(T))| \le \epsilon/2 \tag{5.376}$$

and in view of Proposition 5.55,

$$\begin{aligned} I^f(0, T, y_k, v_k) &= Tf(x_f, u_f) + \pi^f(y_k(0)) - \pi^f(y_k(T)) \\ &\le Tf(x_f, u_f) + \pi^f(y_k(0)) + \epsilon/2. \end{aligned} \tag{5.377}$$

In view of (5.369), (5.370), (5.372), (5.373), and (5.377), for each integer $T \ge L_0$,

$$I^f(0, T, y, v) \le Tf(x_f, u_f) + \lim_{k\to\infty} \pi^f(y_k(0)) + \epsilon/2,$$

$$I^f(0, T, y, v) - Tf(x_f, u_f) \le \lim_{k\to\infty} \pi^f(x_k) + \epsilon/2. \tag{5.378}$$

By (5.370), (5.372), and (5.378), $y(0) = x$,

$$\pi^f(x) \le \liminf_{T\to\infty}[I^f(0, T, y, v) - Tf(x_f, u_f)] \le \lim_{k\to\infty} \pi^f(x_k) + \epsilon/2.$$

Since ϵ is any positive number, this completes the proof of Proposition 5.57.

5.22 Structure of Solutions in the Regions Close to the Endpoints

We suppose that all the assumptions of Sections 5.6 and 5.21 hold and use all the notations and definitions introduced in these sections. As we have already mentioned in Section 5.6, all the results obtained for the triplet (f, A, G) also hold for the triplet $(f, -A, -G)$. Assume that (f, A, G) has properties (P1) and (P2). By Theorem 5.31, $(f, -A - G)$ has (P1) and (P2) too.

For all $z \in \mathcal{A} \setminus \cup\{\widehat{\mathcal{A}}_L : L \in (0, \infty)\}$, set

$$\pi_-^f(z) = \infty,$$

and for all $z \in \cup\{\widehat{\mathcal{A}}_L : L \in (0, \infty)\}$, define

$$\pi_-^f(z) = \inf\{\liminf_{T\to\infty}(I^f(0, T, x, u) - Tf(x_f, u_f)) :$$

$$(x, u) \in X(0, \infty, -A, -G),\ x(0) = z\}. \tag{5.379}$$

Let $T_2 > T_1 \geq 0$ and $(x, u) \in X(T_1, T_2, -A, -G)$ satisfy $\pi_-^f(x(T_1)) < \infty$. Define

$$\Gamma_-^f(T_1, T_2, x, u) = I^f(T_1, T_2, x, u) - (T_2 - T_1)f(x_f, u_f) - \pi_-^f(x(T_1)) + \pi_-^f(x(T_2)). \tag{5.380}$$

Analogously to (5.351) we have,

$$0 \leq \Gamma_-^f(T_1, T_2, x, u) \leq \infty. \tag{5.381}$$

In this section we state the results which describe the asymptotic behavior of approximate solutions in the regions close to the endpoints.

The following two results are proved in Section 5.24.

Theorem 5.58. *Let $L_0, M > 0$ and $\epsilon \in (0, 1)$. Then there exist $\delta > 0$ and $L_1 > L_0$ such that for each $T \geq L_1$ and each $(x, u) \in X(0, T, A, G)$ which satisfies*

$$x(0) \in \mathcal{A}_M,\ I^f(0, T, x, u) \leq \sigma^f(0, T, x(0)) + \delta,$$

the inequalities

$$\pi_-^f(x(T)) \leq \inf(\pi_-^f) + \epsilon,\ \Gamma_-^f(0, L_0, \bar{x}, \bar{u}) \leq \epsilon$$

hold where $\bar{x}(t) = x(T - t)$, $\bar{u}(t) = u(T - t)$, $t \in [0, L_0]$.

Theorem 5.59. *Let $L_0 > 0$ and $\epsilon \in (0, 1)$. Then there exist $\delta > 0$ and $L_1 > L_0$ such that for each $T \geq L_1$ and each $(x, u) \in X(0, T, A, G)$ which satisfies*

$$I^f(0, T, x, u) \leq \sigma^f(0, T) + \delta,$$

the inequalities

$$\pi^f(x(0)) \leq \inf(\pi^f) + \epsilon,\ \Gamma^f(0, L_0, x, u) \leq \epsilon,$$

$$\pi_-^f(x(T)) \le \inf(\pi_-^f) + \epsilon,\ \Gamma_-^f(0, L_0, \bar{x}, \bar{u}) \le \epsilon$$

hold where $\bar{x}(t) = x(T-t)$, $\bar{u}(t) = u(T-t)$, $t \in [0, T]$.

The following two results are proved in Section 5.25.

Theorem 5.60. *Let* (f, A, G) *have LSC property,* $L_0, M > 0$ *and* $\epsilon \in (0, 1)$. *Then there exist* $\delta > 0$ *and* $L_1 > L_0$ *such that for each* $T \ge L_1$ *and each* $(x, u) \in X(0, T, A, G)$ *which satisfies*

$$x(0) \in \mathcal{A}_M,\ I^f(0, T, x, u) \le \sigma^f(0, T, x(0)) + \delta,$$

there exists an $(f, -A, -G)$*-overtaking optimal* $(x_*, u_*) \in X(0, \infty, -A, -G)$ *such that*

$$\pi_-^f(x_*(0)) = \inf(\pi_-^f),\ \|x_*(t) - x(T-t)\| \le \epsilon,\ t \in [0, L_0].$$

Theorem 5.61. *Let* (f, A, G) *have LSC property,* $L_0 > 0$ *and* $\epsilon \in (0, 1)$. *Then there exist* $\delta > 0$ *and* $L_1 > L_0$ *such that for each* $T \ge L_1$ *and each* $(x, u) \in X(0, T, A, G)$ *which satisfies*

$$I^f(0, T, x, u) \le \sigma^f(0, T) + \delta,$$

there exist an (f, A, G)*-overtaking optimal* $(x_{*,1}, u_{*,1}) \in X(0, \infty, A, G)$ *and an* $(f, -A, -G)$*-overtaking optimal* $(x_{*,2}, u_{*,2}) \in X(0, \infty, -A, -G)$ *such that*

$$\pi^f(x_{*,1}(0)) = \inf(\pi^f),\ \pi_-^f(x_{*,2}(0)) = \inf(\pi_-^f),$$

$$\|x(t) - x_{*,1}(t)\| \le \epsilon,\ \|x(T-t) - x_{*,2}(t)\| \le \epsilon,\ t \in [0, L_0].$$

5.23 Auxiliary Results for Theorems 5.58–5.61

Since the results obtained for the triplet (f, A, G) also hold for the triplet $(f, -A, -G)$, in view of Theorem 5.12, we have the following result.

Proposition 5.62. *Assume that* $L, M > 0$ *and that* $\epsilon > 0$. *Then there exist* $\delta > 0$ *and* $L_0 > L$ *such that for each* $T_1 \ge 0$, *each* $T_2 \ge T_1 + 2L_0$, *and each* $(x, u) \in X(T_1, T_2, -A, -G)$ *which satisfies*

$$x(T_1) \in \widehat{\mathcal{A}}_L,$$

$$I^f(T_1, T_2, x, u) \le \min\{\sigma_-^f(T_1, T_2, x(T_1)) + M,\ U_-^f(T_1, T_2, x(T_1), x(T_2)) + \delta\},$$

there exist $\tau_1 \in [T_1, T_1 + L_0]$, $\tau_2 \in [T_2 - L_0, T_2]$ *such that*

$$\|x(t) - x_f\| \le \epsilon, \ t \in [\tau_1, \tau_2].$$

Moreover, if $\|x(T_1) - x_f\| \le \delta$ *then* $\tau_1 = T_1$ *and if* $\|x(T_2) - x_f\| \le \delta$ *then* $\tau_2 = T_2$.

Propositions 5.30 and 5.62 imply the following result.

Proposition 5.63. *Assume that* $L, M > 0$ *and that* $\epsilon > 0$. *Then there exist* $\delta > 0$ *and* $L_0 > 0$ *such that for each* $T_1 \ge 0$, *each* $T_2 \ge T_1 + 2L_0$, *and each* $(x, u) \in X(T_1, T_2, A, G)$ *which satisfies*

$$x(T_2) \in \widehat{\mathcal{A}}_L,$$

$$I^f(T_1, T_2, x, u) \le \min\{(\widehat{\sigma}^f(T_1, T_2, x(T_2)) + M, \ U^f(T_1, T_2, x(T_1), x(T_2)) + \delta\},$$

there exist $\tau_1 \in [T_1, T_1 + L_0]$, $\tau_2 \in [T_2 - L_0, T_2]$ *such that*

$$\|x(t) - x_f\| \le \epsilon, \ t \in [\tau_1, \tau_2].$$

Moreover, if $\|x(T_1) - x_f\| \le \delta$ *then* $\tau_1 = T_1$ *and if* $\|x(T_2) - x_f\| \le \delta$, *then* $\tau_2 = T_2$.

5.24 Proofs of Theorems 5.58 and 5.59

We prove the following result.

Theorem 5.64. *Let* $L_0, M > 0$ *and* $\epsilon \in (0, 1)$. *Then there exist* $\delta > 0$ *and* $L_1 > L_0$ *such that for each* $T \ge L_1$ *and each* $(x, u) \in X(0, T, A, G)$ *which satisfies*

$$x(T) \in \widehat{\mathcal{A}}_M, \ I^f(0, T, x, u) \le \widehat{\sigma}^f(0, T, x(T)) + \delta,$$

the inequalities

$$\pi^f(x(0)) \le \inf(\pi^f) + \epsilon, \ \Gamma^f(0, L_0, x, u) \le \epsilon$$

hold.

Applying Theorem 5.64 with the triplet $(f, -A, -G)$ and using Proposition 5.30, we obtain Theorem 5.58. Theorems 5.58 and 5.64 and TP imply Theorem 5.59.

Proof of Theorem 5.64. By Propositions 5.46, 5.47 and Lemma 5.43 and (A2), there exists $\delta_1 \in (0, \epsilon/4)$ such that:

(i) for each $z \in \mathcal{A}$ satisfying $\|z - x_f\| \le 2\delta_1$, we have $|\pi^f(z)| \le \epsilon/16$;
(ii) for each $(x, u) \in X(0, b_f, A, G)$ which satisfies $\|x(0) - x_f\| \le 2\delta_1$, $\|x(b_f) - x_f\| \le 2\delta_1$, we have

$$I^f(0, b_f, x, u) \ge b_f f(x_f, u_f) - \epsilon/16;$$

(iii) for each $z_i \in \mathcal{A}$, $i = 1, 2$ satisfying $\|z_i - x_f\| \le 2\delta_1$, $i = 1, 2$, there exist $\tau \in (0, b_f]$ and $(x, u) \in X(0, \tau, A, G)$ which satisfies $x(0) = z_1$, $x(\tau) = z_2$ and

$$I^f(0, \tau, x, u) \le \tau f(x_f, u_f) + \epsilon/16.$$

By Proposition 5.63, there exist $\delta_2 \in (0, \delta_1/8)$ and $l_0 > 0$ such that the following property holds:

(iv) for each $T_1 \ge 0$, each $T_2 \ge T_1 + 2l_0$, and each $(x, u) \in X(T_1, T_2, A, G)$ which satisfies

$$x(T_2) \in \widehat{\mathcal{A}}_M,$$

$$I^f(T_1, T_2, x, u) \le \widehat{\sigma}^f(T_1, T_2, x(T_2)) + \delta_2,$$

we have

$$\|x(t) - x_f\| \le \delta_1,\ t \in [T_1 + l_0, T_2 - l_0].$$

Clearly, there exists $z_* \in \mathcal{A}$ such that

$$\pi^f(z_*) \le \inf(\pi^f) + \delta_2/8. \tag{5.382}$$

Corollary 5.51 and (5.382) imply that there exists an (f, A, G)-good $(x_*, u_*) \in X(0, \infty, A, G)$ for which

$$x_*(0) = z_*, \tag{5.383}$$

$$\lim_{T\to\infty} \Gamma^f(0, T, x_*, u_*) \le \delta_2/8. \tag{5.384}$$

By (P1), there exists $l_1 > 0$ such that

$$\|x_*(t) - x_f\| \le \delta_2/8 \text{ for all numbers } t \ge l_1. \tag{5.385}$$

Choose

$$\delta \in (0, \delta_2/4) \tag{5.386}$$

and

$$L_1 > 2L_0 + 2l_0 + 2l_1 + 2b_f + 8. \tag{5.387}$$

Assume that $T \ge L_1$ and that $(x, u) \in X(0, T, A, G)$ satisfies

$$x(T) \in \widehat{\mathcal{A}}_M, \tag{5.388}$$

$$I^f(0, T, x, u) \le \widehat{\sigma}^f(0, T, x(T)) + \delta. \tag{5.389}$$

Property (iv), (5.382), (5.386), (5.387), and (5.389) imply that

$$\|x(t) - x_f\| \le \delta_1, \ t \in [l_0, T - l_0]. \tag{5.390}$$

In view of (5.387),

$$[l_0 + l_1 + L_0, l_0 + l_1 + L_0 + 2b_f + 8] \subset [l_0, T - l_0 - l_1 - L_0]. \tag{5.391}$$

It follows from (5.390) and (5.391) that

$$\|x(t) - x_f\| \le \delta_1, \ t \in [l_0 + l_1 + L_0, l_0 + l_1 + L_0 + 2b_f + 8]. \tag{5.392}$$

Property (iii), (5.385), and (5.392) imply that there exist $\tau_1, \tau_2 \in (0, b_f]$ and $(x_1, u_1) \in X(0, T, A, G)$ such that

$$x_1(t) = x_*(t), \ u_1(t) = u_*(t), \ t \in [0, l_0 + l_1 + L_0 + 4], \tag{5.393}$$

$$x_1(l_0 + l_1 + L_0 + 4 + \tau_1) = x_f, \tag{5.394}$$

$$I^f(l_0 + l_1 + L_0 + 4, l_0 + l_1 + L_0 + 4 + \tau_1, x_1, u_1) \le \tau_1 f(x_f, u_f) + \epsilon/16, \tag{5.395}$$

$$x_1(l_0 + l_1 + L_0 + 4 + 2b_f) = x(l_0 + l_1 + L_0 + 4 + 2b_f), \tag{5.396}$$

$$x_1(l_0 + l_1 + L_0 + 4 + 2b_f - \tau_2) = x_f, \tag{5.397}$$

$$I^f(l_0+l_1+L_0+4+2b_f-\tau_2, l_0+l_1+L_0+4+2b_f, x_1, u_1) \le \tau_2 f(x_f, u_f)+\epsilon/16, \tag{5.398}$$

$$x_1(t) = x_f, \ u_1(t) = u_f, \ t \in [l_0+l_1+L_0+4+\tau_1, l_0+l_1+L_0+4+2b_f-\tau_2], \tag{5.399}$$

$$x_1(t) = x(t), u_1(t) = u(t), \ t \in [l_0 + l_1 + L_0 + 4 + 2b_f, T]. \tag{5.400}$$

It follows from (5.389) and (5.400) that

$$I^f(0, T, x, u) \le I^f(0, T, x_1, u_1) + \delta. \tag{5.401}$$

By (5.393), (5.395), and (5.398)–(5.401),

$$\delta \ge I^f(0, T, x, u) - I^f(0, T, x_1, u_1)$$

$$= I^f(0, l_0 + l_1 + L_0 + 4 + 2b_f, x, u) - I^f(0, l_0 + l_1 + L_0 + 4 + 2b_f, x_1, u_1)$$

$$\geq I^f(0, l_0+l_1+L_0+4, x, u) + I^f(l_0+l_1+L_0+4, l_0+l_1+L_0+4+b_f, x, u)$$

$$+I^f(l_0+l_1+L_0+4+b_f, l_0+l_1+L_0+4+2b_f, x, u) - I^f(0, l_0+l_1+L_0+4, x_*, u_*)$$

$$-\tau_1 f(x_f, u_f) - \epsilon/16 - \tau_2 f(x_f, u_f) - \epsilon/16 - f(x_f, u_f)(2b_f - \tau_1 - \tau_2). \quad (5.402)$$

Property (ii) and (5.392) imply that

$$I^f(l_0+l_1+L_0+4, l_0+l_1+L_0+4+b_f, x, u),$$

$$I^f(l_0+l_1+L_0+4+b_f, l_0+l_1+L_0+4+2b_f, x, u) \geq b_f f(x_f, u_f) - \epsilon/16. \quad (5.403)$$

In view of (5.402) and (5.403),

$$\delta \geq I^f(0, l_0+l_1+L_0+4, x, u) + 2b_f f(x_f, u_f) - \epsilon/8$$

$$- I^f(0, l_0+l_1+L_0+4, x_*, u_*) - 2b_f f(x_f, u_f) - \epsilon/8. \quad (5.404)$$

By property (i), (5.350), (5.382)–(5.386), (5.392), and (5.404),

$$I^f(0, l_0+l_1+L_0+4, x, u) \leq \epsilon/4 + \epsilon/16 + I^f(0, l_0+l_1+L_0+4, x_*, u_*)$$

$$= 5\epsilon/16 + \Gamma^f(0, l_0+l_1+L_0+4, x_*, u_*) + (l_0+l_1+L_0+4) f(x_f, u_f)$$

$$+\pi^f(x_*(0)) - \pi^f(x_*(l_0+l_1+L_0+4))$$

$$\leq \Gamma^f(0, l_0+l_1+L_0+4, x_*, u_*) + (l_0+l_1+L_0+4) f(x_f, u_f) + \pi^f(x_*(0)) + 6\epsilon/16$$

$$\leq \delta_2/8 + (l_0+l_1+L_0+4) f(x_f, u_f) + \pi^f(x_*(0)) + 6\epsilon/16$$

$$\leq \delta_2/8 + (l_0+l_1+L_0+4) f(x_f, u_f) + \inf(\pi^f) + \delta_2/8 + 6\epsilon/16. \quad (5.405)$$

It follows from (5.350), (5.392), (5.405), and property (i) that

$$\inf(\pi^f) + (l_0+l_1+L_0+4) f(x_f, u_f) \geq -3\epsilon/8 - \delta_2/4 + I^f(0, l_0+l_1+L_0+4, x, u)$$

$$\geq -7\epsilon/16 + \Gamma^f(0, l_0+l_1+L_0+4, x, u) + (l_0+l_1+L_0+4) f(x_f, u_f)$$

$$+\pi^f(x(0)) + \pi^f(x(l_0+l_1+L_0+4))$$

$$\geq -7\epsilon/16 + \Gamma^f(0, l_0+l_1+L_0+4, x, u) + (l_0+l_1+L_0+4) f(x_f, u_f) + \pi^f(x_0) - \epsilon/16. \quad (5.406)$$

By (5.351) and (5.406),

$$\pi^f(x_0) + \Gamma^f(0, l_0+l_1+L_0+4, x, u) \leq \inf(\pi^f) + \epsilon/2.$$

The inequality above implies that

$$\pi^f(x(0)) \le \inf(\pi^f) + \epsilon, \ \Gamma^f(0, l_0 + l_1 + L_0 + 4, x, u) \le \epsilon.$$

Theorem 5.64 is proved.

5.25 Proofs of Theorems 5.60 and 5.61

We prove the following result.

Theorem 5.65. *Let (f, A, G) have LSC property, $L_0, M > 0$ and $\epsilon \in (0, 1)$. Then there exist $\delta > 0$ and $L_1 > L_0$ such that for each $T \ge L_1$ and each $(x, u) \in X(0, T, A, G)$ which satisfies*

$$x(T) \in \widehat{\mathcal{A}}_M, \ I^f(0, T, x, u) \le \widehat{\sigma}^f(0, T, x(T)) + \delta,$$

there exists an (f, A, G)-overtaking optimal $(x_*, u_*) \in X(0, \infty, A, G)$ such that

$$\pi^f(x_*(0)) = \inf(\pi^f), \ \|x(t) - x_*(t)\| \le \epsilon, \ t \in [0, L_0].$$

Note that Theorem 5.60 follows from Theorem 5.65, applied for the triplet $(f, -A, -G)$ and Proposition 5.30. Theorem 5.61 follows from Theorems 5.60, 5.65 and TP. Theorem 5.65 follows from Theorem 5.64 and the following result.

Proposition 5.66. *Let (f, A, G) have LSC property, $L_0 > 0$ and $\epsilon \in (0, 1)$. Then there exist $\delta \in (0, \epsilon)$ such that for each $(x, u) \in X(0, L_0, A, G)$ which satisfies*

$$\pi^f(x(0)) \le \inf(\pi^f) + \delta, \ \Gamma^f(0, L_0, x, u) \le \delta,$$

there exists an (f, A, G)-overtaking optimal $(x_*, u_*) \in X(0, \infty, A, G)$ such that

$$\pi^f(x_*(0)) = \inf(\pi^f), \ \|x_*(t) - x(t)\| \le \epsilon, \ t \in [0, L_0].$$

Proof. Assume that the proposition does not hold. Then there exist a sequence $\{\delta_k\}_{k=1}^{\infty} \subset (0, 1]$ and a sequence $\{(x_k, u_k)\}_{k=1}^{\infty} \subset X(0, L_0, A, G)$ such that

$$\lim_{k\to\infty} \delta_k = 0 \tag{5.407}$$

and that for all integers $k \ge 1$,

$$\pi^f(x_k(0)) \le \inf(\pi^f) + \delta_k, \tag{5.408}$$

$$\Gamma^f(0, L_0, x_k, u_k) \le \delta_k, \tag{5.409}$$

and the following property holds:

(i) for each (f, A, G)-overtaking optimal pair $(y, v) \in X(0, \infty, A, G)$ satisfying $\pi^f(y(0)) = \inf(\pi^f)$, we have $\max\{\|x_k(t) - y(t)\| : t \in [0, L_0]\} > \epsilon$.

In view of (5.407)–(5.409) and the boundedness from below of the function π^f, the sequence $\{I^f(0, L_0, x_k, u_k)\}_{k=1}^{\infty}$ is bounded. By LSC property, extracting a subsequence and re-indexing if necessary, we may assume without loss of generality that there exists $(x, u) \in X(0, L_0, A, G)$ such that

$$x_k(t) \to x(t) \text{ as } k \to \infty \text{ for all } t \in [0, L_0], \tag{5.410}$$

$$I^f(0, L_0, x, u) \leq \liminf_{k\to\infty} I^f(0, L_0, x_k, u_k). \tag{5.411}$$

The lower semicontinuity of π^f, (5.408) and (5.410) implies that

$$\pi^f(x(0)) \leq \liminf_{k\to\infty} \pi^f(x_k(0)) = \inf(\pi^f),$$

$$\pi^f(x(0)) = \inf(\pi^f). \tag{5.412}$$

By the lower semicontinuity of π^f and (5.410),

$$\pi^f(x(L_0)) \leq \liminf_{k\to\infty} \pi^f(x_k(L_0)). \tag{5.413}$$

It follows from (5.350), (5.407)–(5.409), and (5.411)–(5.413) that

$$\begin{aligned} &I^f(0, L_0, x, u) - L_0 f(x_f, u_f) - \pi^f(x(0)) + \pi^f(x(L_0)) \\ &\leq \liminf_{k\to\infty}[I^f(0, L_0, x_k, u_k) - L_0 f(x_f, u_f)] \\ &\quad - \lim_{k\to\infty} \pi^f(x_k(0)) + \liminf_{k\to\infty} \pi^f(x_k(L_0)) \\ &\leq \liminf_{k\to\infty}[I^f(0, L_0, x_k, u_k) - L_0 f(x_f, u_f) - \pi^f(x_k(0)) + \pi^f(x_k(L_0))] \leq 0. \end{aligned} \tag{5.414}$$

In view of (5.350), (5.351), and (5.414),

$$I^f(0, L_0, x, u) - L_0 f(x_f, u_f) - \pi^f(x(0)) + \pi^f(x(L_0)) = 0. \tag{5.415}$$

Theorem 5.19, (5.412), and (5.415) imply that there exists an (f, A, G)-overtaking optimal pair $(\tilde{x}, \tilde{u}) \in X(0, \infty, A, G)$ such that

$$\tilde{x}(0) = x(L_0). \tag{5.416}$$

For all numbers $t > L_0$, set

$$x(t) = \tilde{x}(t - L_0),\ u(t) = \tilde{u}(t - L_0). \tag{5.417}$$

It is clear that $(x, u) \in X(0, \infty, A, G)$ is (f, A, G)-good. By Proposition 5.55, (5.350), (5.351), and (5.415), for all $S > 0$,

$$I^f(0, S, x, u) - Sf(x_f, u_f) - \pi^f(x(0)) + \pi^f(x(S)) = 0. \tag{5.418}$$

In view of (5.412), (5.418), and Theorem 5.20, (x, u) is (f, A, G)-overtaking optimal pair satisfying

$$\pi^f(x(0)) = \inf(\pi^f).$$

By (5.410), for all sufficiently large natural numbers k,

$$\|x_k(t) - x(t)\| \leq \epsilon/2,\ t \in [0, L_0].$$

This contradicts property (i). The contradiction we have reached proves Proposition 5.66.

5.26 The First Bolza Problem

We use the notation, definitions, and assumptions introduced and used in Sections 5.6, 5.21, and 5.22. Let $a_1 > 0$. Denote by $\mathfrak{A}(E))$ the set of all lower semicontinuous functions $h : E \to R^1$ which are bounded on bounded subsets of E and such that

$$h(z) \geq -a_1 \text{ for all } z \in E. \tag{5.419}$$

The set $\mathfrak{A}(E)$ is equipped with the uniformity which is determined by the base

$$\mathcal{E}(N, \epsilon) = \{(h_1, h_2) \in \mathfrak{A}(E) \times \mathfrak{A}(E) : |h_1(z) - h_2(z)| \leq \epsilon$$

$$\text{for all } z \in E \text{ satisfying } \|z\| \leq N\}, \tag{5.420}$$

where $N, \epsilon > 0$. It is not difficult to see that the uniform space $\mathfrak{A}(E)$ is metrizable and complete.

For each pair of numbers $T_2 > T_1 \geq 0$, each $y \in \mathcal{A}$, and each $h \in \mathfrak{A}(E)$, we define

$$\sigma^{f,h}(T_1, T_2, y) = \inf\{I^f(T_1, T_2, x, u) + h(x(T_2)) :$$

$$(x, u) \in X(T_1, T_2, A, G),\ x(T_1) = y\}, \tag{5.421}$$

$$\widehat{\sigma}^{f,h}(T_1, T_2, y) = \inf\{I^f(T_1, T_2, x, u) + h(x(T_1)) :$$

$$(x, u) \in X(T_1, T_2, A, G),\ x(T_2) = y\}, \tag{5.422}$$

$$\sigma_{-}^{f,h}(T_1, T_2, y) = \inf\{I^f(T_1, T_2, x, u) + h(x(T_2)) :$$

$$(x, u) \in X(T_1, T_2, -A, -G),\ x(T_1) = y\}, \tag{5.423}$$

$$\widehat{\sigma}_{-}^{f,h}(T_1, T_2, y) = \inf\{I^f(T_1, T_2, x, u) + h(x(T_1)) :$$

$$(x, u) \in X(T_1, T_2, -A - G),\ x(T_2) = y\}. \tag{5.424}$$

The next result follows from Proposition 5.29 and (5.35).

Proposition 5.67. *Let $S_2 > S_1 \geq 0$, $M \geq 0$, $h \in \mathfrak{A}(E)$ and $(x, u) \in X(S_1, S_2, A, G)$. Then the following assertions hold:*

$$I^f(S_1, S_2, x, u) + h(x(S_2)) \leq \sigma^{f,h}(S_1, S_2, x(S_1)) + M \text{ if and only if}$$

$$I^f(S_1, S_2, \bar{x}, \bar{u}) + h(\bar{x}(S_1)) \leq \widehat{\sigma}_{-}^{f,h}(S_1, S_2, \bar{x}(S_2)) + M;$$

$$I^f(S_1, S_2, x, u) + h(x(S_1)) \leq \widehat{\sigma}^{f,h}(S_1, S_2, x(S_2)) + M \text{ if and only if}$$

$$I^f(S_1, S_2, \bar{x}, \bar{u}) + h(\bar{x}(S_2)) \leq \sigma_{-}^{f,h}(S_1, S_2, \bar{x}(S_1)) + M.$$

The next result is proved in Section 5.27.

Theorem 5.68. *Assume that (f, A, G) has TP, $M, L > 0$ and that $\epsilon > 0$. Then there exist $\delta > 0$ and $L_0 > L$ such that for each $T_1 \geq 0$, each $T_2 \geq T_1 + 2L_0$, each $h \in \mathfrak{A}(E)$ satisfying $|h(x_f)| \leq M$ and each $(x, u) \in X(T_1, T_2, A, G)$ which satisfies*

$$x(T_1) \in \mathcal{A}_L,$$

$$I^f(T_1, T_2, x, u) + h(x(T_2)) \leq \sigma^{f,h}(T_1, T_2, x(T_1)) + \delta$$

there exist $\tau_1 \in [T_1, T_1 + L_0]$, $\tau_2 \in [T_2 - L_0, T_2]$ such that

$$\|x(t) - x_f\| \leq \epsilon,\ t \in [\tau_1, \tau_2].$$

Moreover, if $\|x(T_1) - x_f\| \leq \delta$ then $\tau_1 = T_1$ and if $\|x(T_2) - x_f\| \leq \delta$ then $\tau_2 = T_2$.

Note that Theorem 5.68 is valid for the $(f, h, -A, -G)$ too and combined with Proposition 5.67 this implies the following result.

Theorem 5.69. *Assume that (f, A, G) has TP, $M, L > 0$, and $\epsilon > 0$. Then there exist $\delta > 0$ and $L_0 > L$ such that for each $T_1 \geq 0$, each $T_2 \geq T_1 + 2L_0$, each $h \in \mathfrak{A}(E)$ satisfying $|h(x_f)| \leq M$, and each $(x, u) \in X(T_1, T_2, A, G)$ which satisfies*

$$x(T_2) \in \widehat{\mathcal{A}}_L,$$

$$I^f(T_1, T_2, x, u) + h(x(T_1)) \leq \widehat{\sigma}^{f,h}(T_1, T_2, x(T_2)) + \delta,$$

there exist $\tau_1 \in [T_1, T_1 + L_0]$, $\tau_2 \in [T_2 - L_0, T_2]$ such that

$$\|x(t) - x_f\| \leq \epsilon,\ t \in [\tau_1, \tau_2].$$

Moreover, if $\|x(T_1) - x_f\| \leq \delta$, then $\tau_1 = T_1$, and if $\|x(T_2) - x_f\| \leq \delta$, then $\tau_2 = T_2$.

Let $g \in \mathfrak{A}(E)$ be given. By Corollary 5.49, (5.325), and (5.419), $\pi^f + g : \mathcal{A} \to R^1 \cup \{\infty\}$ is bounded from below and satisfies

$$(\pi^f + g)(z) \to \infty \text{ as } z \in \mathcal{A},\ \|z\| \to \infty. \tag{5.425}$$

We prove the following results which describe the asymptotic behavior of approximate solutions in the regions close to the endpoints.

Theorem 5.70. *Let (f, A, G) have TP, $L_0, M > 0$ and $\epsilon \in (0, 1)$. Then there exist $\delta > 0$, $L_1 > L_0$, and a neighborhood $\mathfrak{U}$ of g in $\mathfrak{A}(E)$ such that for each $T \geq L_1$, each $h \in \mathfrak{U}$, and each $(x, u) \in X(0, T, A, G)$ which satisfies*

$$x(0) \in \mathcal{A}_M,\ I^f(0, T, x, u) + h(x(T)) \leq \sigma^{f,h}(0, T, x(0)) + \delta,$$

the inequalities

$$(g + \pi_-^f)(x(T)) \leq \inf(g + \pi_-^f) + \epsilon,\ \Gamma_-^f(0, L_0, \bar{x}, \bar{u}) \leq \epsilon$$

hold where $\bar{x}(t) = x(T - t)$, $\bar{u}(t) = u(T - t)$, $t \in [0, T]$.

Theorem 5.71. *Let (f, A, G) have TP, $L_0, M > 0$ and $\epsilon \in (0, 1)$. Then there exist $\delta > 0$, $L_1 > L_0$, and a neighborhood $\mathfrak{U}$ of g in $\mathfrak{A}(E)$ such that for each $T \geq L_1$, each $h \in \mathfrak{U}$, and each $(x, u) \in X(0, T, A, G)$ which satisfies*

$$x(T) \in \widehat{\mathcal{A}}_M,\ I^f(0, T, x, u) + h(x(0)) \leq \widehat{\sigma}^{f,h}(0, T, x(T)) + \delta,$$

the inequalities

$$(g + \pi^f)(x(0)) \leq \inf(g + \pi^f) + \epsilon,\ \Gamma^f(0, L_0, x, u) \leq \epsilon.$$

Theorem 5.71 is proved in Section 5.28. It is also valid for $(f, -A, -G)$. Together with Proposition 5.67, this implies Theorem 5.70.

Theorem 5.72. *Let (f, A, G) have TP and LSC property, $L_0, M > 0$ and $\epsilon \in (0, 1)$. Then there exist $\delta > 0$, $L_1 > L_0$, and a neighborhood $\mathfrak{U}$ of g in $\mathfrak{A}(E)$ such that for each $T \geq L_1$, each $h \in \mathfrak{U}$, and each $(x, u) \in X(0, T, A, G)$ which satisfies*

$$x(0) \in \mathcal{A}_M, \; I^f(0, T, x, u) + h(x(T)) \leq \sigma^{f,h}(0, T, x(0)) + \delta,$$

there exists an $(f, -A, -G)$-overtaking optimal pair

$$(x_*, u_*) \in X(0, \infty, -A, -G)$$

such that

$$(\pi_-^f + g)(x_*(0)) = \inf(\pi_-^f + g), \; \|x_*(t) - x(T - t)\| \leq \epsilon, \; t \in [0, L_0].$$

Theorem 5.73. *Let (f, A, G) have TP and LSC property, $L_0, M > 0$ and $\epsilon \in (0, 1)$. Then there exist $\delta > 0$, $L_1 > L_0$, and a neighborhood $\mathfrak{U}$ of g in $\mathfrak{A}(E)$ such that for each $T \geq L_1$, each $h \in \mathfrak{U}$, and each $(x, u) \in X(0, T, A, G)$ which satisfies*

$$x(T) \in \widehat{\mathcal{A}}_M, \; I^f(0, T, x, u) + h(x(0)) \leq \widehat{\sigma}^{f,h}(0, T, x(T)) + \delta,$$

there exists an (f, A, G)-overtaking optimal pair $(x_*, u_*) \in X(0, \infty, A, G)$ such that

$$(\pi^f + g)(x_*(0)) = \inf(\pi^f + g), \; \|x_*(t) - x(t)\| \leq \epsilon, \; t \in [0, L_0].$$

We prove Theorem 5.73. It is also valid for $(f, -A, -G)$. Together with Proposition 5.67, this implies Theorem 5.72. Theorem 5.73 follows from Theorem 5.71 and the following result which is proved in Section 5.29.

Proposition 5.74. *Let (f, A, G) have TP and LSC property, $L_0 > 0$ and $\epsilon \in (0, 1)$. Then there exist $\delta \in (0, \epsilon)$ such that for each and each $(x, u) \in X(0, L_0, A, G)$ which satisfies*

$$(g + \pi^f)(x(0)) \leq \inf(\pi^f + g) + \delta,$$

$$\Gamma^f(0, L_0, x, u) \leq \delta,$$

there exists an (f, A, G)-overtaking optimal pair $(x_*, u_*) \in X(0, \infty, A, G)$ such that

$$(\pi^f + g)(x_*(0)) = \inf(\pi^f + g), \; \|x_*(t) - x(t)\| \leq \epsilon, \; t \in [0, L_0].$$

5.27 Proof of Theorem 5.68

By Proposition 5.10, there exist $\delta \in (0, 1)$ and $L_0 > L$ such that the following property holds:

(i) for each $T_1 \ge 0$, each $T_2 \ge T_1 + 2L_0$, and each $(x, u) \in X(T_1, T_2, A, G)$ which satisfies

$$I^f(T_1, T_2, x, u) \le \min\{(T_2-T_1)f(x_f, u_f)+a_1+M+a_0+2+2L(1+|f(x_f, u_f)|),$$
$$U^f(T_1, T_2, x(T_1), x(T_2)) + \delta\},$$

there exist $\tau_1 \in [T_1, T_1 + L_0]$, $\tau_2 \in [T_2 - L_0, T_2]$ such that

$$\|x(t) - x_f\| \le \epsilon, \ t \in [\tau_1, \tau_2];$$

moreover, if $\|x(T_1) - x_f\| \le \delta$ then $\tau_1 = T_1$, and if $\|x(T_2) - x_f\| \le \delta$, then $\tau_2 = T_2$.

Assume that $T_1 \ge 0$, $T_2 \ge T_1 + 2L_0$, $h \in \mathfrak{A}(E)$ satisfies

$$|h(x_f)| \le M, \tag{5.426}$$

and $(x, u) \in X(T_1, T_2, A, G)$ satisfies

$$x(T_1) \in \mathcal{A}_L, \tag{5.427}$$

$$I^f(T_1, T_2, x, u) + h(x(T_2)) \le \sigma^{f,h}(T_1, T_2, x(T_1)) + \delta. \tag{5.428}$$

By (5.428),

$$I^f(T_1, T_2, x, u) \le U^f(T_1, T_2, x(T_1), x(T_2)) + \delta. \tag{5.429}$$

In view of (5.427), there exists $(y, v) \in X(T_1, T_2, A, G)$ such that

$$y(T_1) = x(T_1), \tag{5.430}$$

$$y(t) = x_f, \ v(t) = u_f, \ t \in [T_1 + L, T_2], \tag{5.431}$$

$$I^f(T_1, T_1 + L, y, v) \le L + L|f(x_f, u_f)|. \tag{5.432}$$

It follows from (5.29), (5.419), (5.428), (5.430), and (5.432) that

$$-a_1+I^f(T_1, T_2, x, u) \le I^f(T_1, T_2, x, u)+h(x(T_2)) \le 1+I^f(T_1, T_2, y, v)+h(x_f)$$
$$\le h(x_f) + 1 + L(1 + |f(x_f, u_f)|) + (T_2 - T_1 - L)f(x_f, u_f),$$

and together with (5.426), this implies that

$$I^f(T_1, T_2, x, u) \le a_1 + 1 + M + 2L(1 + |f(x_f, u_f)|) + (T_2 - T_1) f(x_f, u_f). \tag{5.433}$$

In view of (5.429), (5.433), and property (i), there exist $\tau_1 \in [T_1, T_1 + L_0]$, $\tau_2 \in [T_2 - L_0, T_2]$ such that

$$\|x(t) - x_f\| \le \epsilon, \ t \in [\tau_1, \tau_2].$$

Moreover, if $\|x(T_1) - x_f\| \le \delta$, then $\tau_1 = T_1$, and if $\|x(T_2) - x_f\| \le \delta$, then $\tau_2 = T_2$. Theorem 5.68 is proved.

5.28 Proof of Theorem 5.71

By Propositions 5.46, 5.47 and Lemma 5.43 and (A2), there exists $\delta_1 \in (0, \epsilon/4)$ such that:

(i) for each $z \in \mathcal{A}$ satisfying $\|z - x_f\| \le 2\delta_1$, we have $|\pi^f(z)| \le \epsilon/16$;

(ii) for each $(x, u) \in X(0, b_f, A, G)$ which satisfies $\|x(0) - x_f\| \le 2\delta_1$, $\|x(b_f) - x_f\| \le 2\delta_1$, we have

$$I^f(0, b_f, x, u) \ge b_f f(x_f, u_f) - \epsilon/16;$$

(iii) for each $z_i \in \mathcal{A}$, $i = 1, 2$ satisfying $\|z_i - x_f\| \le 2\delta_1$, $i = 1, 2$, there exist $\tau \in (0, b_f]$ and $(x, u) \in X(0, \tau, A, G)$ which satisfies $x(0) = z_1$, $x(\tau) = z_2$ and

$$I^f(0, \tau, x, u) \le \tau f(x_f, u_f) + \epsilon/16.$$

By (5.419), (5.420), and Corollary 5.49, there exists a neighborhood $\mathfrak{U}_1$ of g in $\mathfrak{A}(E)$ such that for each $h \in \mathfrak{U}_1$,

$$|\inf(\pi^f + g) - \inf(\pi^f + h)| \le \delta_1/16. \tag{5.434}$$

Theorem 5.69 implies that there exist $\delta_2 \in (0, \delta_1/8)$, $l_0 > M$, and a neighborhood $\mathfrak{U}_2$ of g in $\mathfrak{A}(E)$ such that the following property holds:

(iv) for each $T_1 \ge 0$, each $T_2 \ge T_1 + 2l_0$, each $h \in \mathfrak{U}_2$, and each $(x, u) \in X(T_1, T_2, A, G)$ which satisfies

$$x(T_2) \in \widehat{\mathcal{A}}_M,$$

$$I^f(T_1, T_2, x, u) + h(x(T_1)) \le \widehat{\sigma}^{f,h}(T_1, T_2, x(T_2)) + \delta_2,$$

we have

$$\|x(t) - x_f\| \le \delta_1, \ t \in [T_1 + l_0, T_2 - l_0].$$

Clearly, there exists $z_* \in \mathcal{A}$ such that

$$(\pi^f + g)(z_*) \le \inf(\pi^f + g) + \delta_2/8. \tag{5.435}$$

Corollary 5.51 implies that there exists an (f, A, G)-good

$$(x_*, u_*) \in X(0, \infty, A, G)$$

for which

$$x_*(0) = z_*, \ \lim_{T\to\infty} \Gamma^f(0, T, x_*, u_*) \le \delta_2/8. \tag{5.436}$$

Property (P1) implies that

$$\lim_{t\to\infty} \|x_*(t) - x_f\| = 0. \tag{5.437}$$

By (5.437), there exists $l_1 > 0$ such that

$$\|x_*(t) - x_f\| \le \delta_2/8 \text{ for all numbers } t \ge l_1. \tag{5.438}$$

Theorem 5.32 implies that there exists $\tilde{M} > 0$ such that the following property holds:

(v) for each $(x, u) \in X(0, l_0 + l_1 + L_0 + 4, A, G)$ satisfying

$$I^f(0, l_0+l_1+L_0+4, x, u) \le (l_0+l_1+L_0+4) f(x_f, u_f) + a_1 + 2 + a_0 + \inf(\pi^f + g),$$

we have

$$\|x(t)\| \le \tilde{M}, \ t \in [0, l_0 + l_1 + L_0 + 4].$$

Proposition 5.53 implies that the set $\mathcal{A}_M$ is bounded. In view of (5.420), there exists a neighborhood $\mathfrak{U}$ of g in $\mathfrak{A}(E)$ such that

$$\mathfrak{U} \subset \mathfrak{U}_1 \cap \mathfrak{U}_2, \tag{5.439}$$

$$|h(x_*(0)) - g(x_*(0))| \le \delta_1/16 \text{ for all } h \in \mathfrak{U}, \tag{5.440}$$

$$|h(z) - g(z)| \le \delta_1/16 \text{ for all } h \in \mathfrak{U} \text{ and all } z \in \mathcal{A}_M, \tag{5.441}$$

$$|h(z) - g(z)| \le \delta_1/16 \text{ for all } h \in \mathfrak{U} \text{ and all } z \in E \text{ satisfying } \|z\| \le \tilde{M}. \tag{5.442}$$

Choose

$$\delta \in (0, \delta_2/4), \tag{5.443}$$

$$L_1 > 2L_0 + 2l_0 + 2l_1 + 2b_f + 8. \tag{5.444}$$

Assume that

$$T \geq L_1, \ h \in \mathfrak{U}, \tag{5.445}$$

$$(x, u) \in X(0, T, A, G), \tag{5.446}$$

$$x(T) \in \widehat{\mathcal{A}}_M, \tag{5.447}$$

$$I^f(0, T, x, u) + h(x(0)) \leq \sigma^{f,h}(0, T, x(T)) + \delta. \tag{5.448}$$

Property (iv), (5.439), (5.443)–(5.446), and (5.448) imply that

$$\|x(t) - x_f\| \leq \delta_1, \ t \in [l_0, T - l_0]. \tag{5.449}$$

In view of (5.444) and (5.445),

$$[l_0 + l_1 + L_0, l_0 + l_1 + L_0 + 2b_f + 8] \subset [l_0, T - l_0 - l_1 - L_0]. \tag{5.450}$$

It follows from (5.449) and (5.450) that

$$\|x(t) - x_f\| \leq \delta_1, \ t \in [l_0 + l_1 + L_0, l_0 + l_1 + L_0 + 2b_f + 8]. \tag{5.451}$$

Property (iii), (5.438), (5.443), and (5.444), (5.451) imply that there exist $\tau_1, \tau_2 \in (0, b_f]$ and $(x_1, u_1) \in X(0, T, A, G)$ such that

$$x_1(t) = x_*(t), \ u_1(t) = u_*(t), \ t \in [0, l_0 + l_1 + L_0 + 4], \tag{5.452}$$

$$x_1(l_0 + l_1 + L_0 + 4 + \tau_1) = x_f, \tag{5.453}$$

$$I^f(l_0+l_1+L_0+4, l_0+l_1+L_0+4+\tau_1, x_1, u_1) \leq \tau_1 f(x_f, u_f)+\epsilon/16, \tag{5.454}$$

$$x_1(l_0 + l_1 + L_0 + 4 + 2b_f) = x(l_0 + l_1 + L_0 + 4 + 2b_f), \tag{5.455}$$

$$x_1(l_0 + l_1 + L_0 + 4 + 2b_f - \tau_2) = x_f, \tag{5.456}$$

$$I^f(l_0+l_1+L_0+4+2b_f-\tau_2, l_0+l_1+L_0+4+2b_f, x_1, u_1) \leq \tau_2 f(x_f, u_f)+\epsilon/16, \tag{5.457}$$

$$x_1(t) = x_f,\ u_1(t) = u_f,\ t \in [l_0 + l_1 + L_0 + 4 + \tau_1, l_0 + l_1 + L_0 + 4 + 2b_f - \tau_2], \tag{5.458}$$

$$x_1(t) = x(t), u_1(t) = u(t),\ t \in [l_0 + l_1 + L_0 + 4 + 2b_f, T]. \tag{5.459}$$

It follows from (5.448) and (5.459) that

$$I^f(0, T, x, u) + h(x(0)) \le I^f(0, T, x_1, u_1) + h(x_1(0)) + \delta. \tag{5.460}$$

By (5.458)–(5.460),

$$\delta \ge I^f(0, T, x, u) + h(x(0)) - I^f(0, T, x_1, u_1) - h(x_1(0))$$

$$= I^f(0, l_0 + l_1 + L_0 + 4 + 2b_f, x, u) + h(x(0))$$

$$-I^f(0, l_0 + l_1 + L_0 + 4 + 2b_f, x_1, u_1) - h(x_1(0))$$

$$= I^f(0, l_0+l_1+L_0+4, x, u)+I^f(l_0+l_1+L_0+4, l_0+l_1+L_0+4+2b_f, x, u)+h(x(0))$$

$$-I^f(0, l_0+l_1+L_0+4, x_1, u_1) - I^f(l_0+l_1+L_0+4, l_0+l_1+L_0+4+\tau_1, x_1, u_1)$$

$$-f(x_f, u_f)(2b_f - \tau_1 - \tau_2)$$

$$-I^f(l_0+l_1+L_0+4+2b_f-\tau_2, l_0+l_1+L_0+4+2b_f, x_1, u_1)-h(x_1(0)). \tag{5.461}$$

Property (ii) and (5.451) imply that

$$I^f(l_0+l_1+L_0+4, l_0+l_1+L_0+4+2b_f, x, u) \ge 2b_f f(x_f, u_f)-\epsilon/8. \tag{5.462}$$

By property (i), (5.350), (5.435), (5.436), (5.438), (5.452), (5.454), (5.457), (5.461), and (5.462),

$$I^f(0, l_0 + l_1 + L_0 + 4, x, u) + h(x(0))$$

$$\le \delta + \epsilon/4 + I^f(0, l_0 + l_1 + L_0 + 4, x_1, u_1) + h(x_1(0))$$

$$= \delta + \epsilon/4 + I^f(0, l_0 + l_1 + L_0 + 4, x_*, u_*) + h(x_*(0))$$

$$= \delta + \epsilon/4 + \Gamma^f(0, l_0 + l_1 + L_0 + 4, x_*, u_*) + (l_0 + l_1 + L_0 + 4)f(x_f, u_f)$$

$$+\pi^f(x_*(0)) - \pi^f(x_*(l_0 + l_1 + L_0 + 4)) + h(x_*(0))$$

$$\le \Gamma^f(0, l_0 + l_1 + L_0 + 4, x_*, u_*) + (l_0 + l_1 + L_0 + 4)f(x_f, u_f)$$

$$+\pi^f(x_*(0)) + \epsilon/16 + h(x_*(0)) + \delta + \epsilon/4$$

$$\le (l_0 + l_1 + L_0 + 4) f(x_f, u_f) + \delta_2/8 + \delta + \epsilon/4 + (\pi^f + h)(x_*(0)) + \epsilon/16$$

$$\le (l_0 + l_1 + L_0 + 4) f(x_f, u_f) + \epsilon/64 + \epsilon/64 + \epsilon/4 + (\pi^f + g)(x_*(0)) + \epsilon/16$$

$$\le (l_0 + l_1 + L_0 + 4) f(x_f, u_f) + \inf(\pi^f + g) + \delta_2/8 + \epsilon/16 + \epsilon/4 + \epsilon/32$$

$$\le (l_0 + l_1 + L_0 + 4) f(x_f, u_f) + \inf(\pi^f + g) + \epsilon/4 + \epsilon/16 + \epsilon/16. \quad (5.463)$$

In view of (5.419) and (5.463),

$$I^f(0, l_0 + l_1 + L_0 + 4, x, u) \le (l_0 + l_1 + L_0 + 4) f(x_f, u_f) + \inf(\pi^f + g) + a_1 + 1. \quad (5.464)$$

Property (v) and (5.464) imply that

$$\|x(0)\| \le \tilde{M}.$$

Together with (5.442) and (5.444), this implies that

$$|h(x(0)) - g(x(0))| \le \delta_1/16 \le \epsilon/16. \quad (5.465)$$

It follows from (5.340), (5.463), and (5.465) that

$$I^f(0, l_0 + l_1 + L_0 + 4, x, u) + g(x(0))$$

$$\le (l_0 + l_1 + L_0 + 4) f(x_f, u_f) + \inf(\pi^f + g) + \epsilon/4 + \epsilon/8 + \epsilon/16. \quad (5.466)$$

By (5.350) and (5.466),

$$\Gamma^f(0, l_0 + l_1 + L_0 + 4, x, u) + \pi^f(x(0)) + \pi^f(x(l_0 + l_1 + L_0 + 4)) + g(x(0))$$

$$\le \inf(\pi^f + g) + \epsilon/4 + \epsilon/4. \quad (5.467)$$

It follows from (5.451) and property (i) that

$$|\pi^f(x(l_0 + l_1 + L_0 + 4))| \le \epsilon/16. \quad (5.468)$$

By (5.467) and (5.468),

$$\Gamma^f(0, l_0 + l_1 + L_0 + 4, x, u) + (\pi^f + g)(x(0)) \le \inf(\pi^f + g) + \epsilon/2 + \epsilon/16.$$

This implies that

$$\Gamma^f(0, l_0 + l_1 + L_0 + 4, x, u) \leq \epsilon,$$

$$(\pi^f + g)(x(0)) \leq \inf(\pi^f + g) + 3\epsilon/4.$$

Theorem 5.71 is proved.

5.29 Proof of Proposition 5.74

Assume that the proposition does not hold. Then there exist a sequence $\{\delta_k\}_{k=1}^\infty \subset (0, 1]$ and a sequence $\{(x_k, u_k)\}_{k=1}^\infty \subset X(0, L_0, A, G)$ such that

$$\lim_{k\to\infty} \delta_k = 0 \tag{5.469}$$

and that for all integers $k \geq 1$,

$$(\pi^f + g)(x_k(0)) \leq \inf(\pi^f + g) + \delta_k, \tag{5.470}$$

$$\Gamma^f(0, L_0, x_k, u_k) \leq \delta_k \tag{5.471}$$

and the following property holds:

(i) for each (f, A, G)-overtaking optimal pair $(y, v) \in X(0, \infty, A, G)$ satisfying $(\pi^f + g)(y(0)) = \inf(\pi^f + g)$, we have $\sup\{\|x_k(t) - y(t)\| : t \in [0, L_0]\} > \epsilon$.

In view of (5.419), (5.469), and (5.470), the sequence $\{\pi^f(x_k(0))\}_{k=1}^\infty$ is bounded. Together with Corollary 5.49, this implies that the sequence $\{x_k(0)\}_{k=1}^\infty$ is bounded. By (5.325), (5.350), (5.419), (5.469), and (5.471), the sequence $\{I^f(0, L_0, x_k, u_k)\}_{k=1}^\infty$ is bounded. By LSC property, extracting a subsequence and re-indexing if necessary, we may assume without loss of generality that there exists $(x, u) \in X(0, L_0, A, G)$ such that

$$x_k(t) \to x(t) \text{ as } k \to \infty \text{ for all } t \in [0, L_0], \tag{5.472}$$

$$I^f(0, L_0, x, u) \leq \liminf_{k\to\infty} I^f(0, L_0, x_k, u_k). \tag{5.473}$$

Proposition 5.57, (5.470), and (5.472) imply that the function $\pi^f + g$ is lower semicontinuous and that

$$\pi^f(x(0)) \leq \liminf_{k\to\infty} \pi^f(x_k(0)), \tag{5.474}$$

$$g(x(0)) \leq \liminf_{k\to\infty} g(x_k(0)), \tag{5.475}$$

$$(\pi^f + g)(x(0)) \le \liminf_{k\to\infty}(\pi^f + g)(x_k(0)) \le \inf(\pi^f + g),$$

$$(\pi^f + g)(x(0)) = \inf(\pi^f + g) = \lim_{k\to\infty} (\pi^f + g)(x_k(0)). \tag{5.476}$$

In view of (5.474)–(5.476),

$$g(x(0)) = \lim_{k\to\infty} g(x_k(0)), \tag{5.477}$$

$$\pi^f(x(0)) = \lim_{k\to\infty} \pi^f(x_k(0)). \tag{5.478}$$

By the lower semicontinuity of π^f and (5.472),

$$\pi^f(x(L_0)) \le \liminf_{k\to\infty} \pi^f(x_k(L_0)). \tag{5.479}$$

It follows from (5.350), (5.351), (5.469), (5.471), (5.473), (5.478), and (5.479) that

$$I^f(0, L_0, x, u) - L_0 f(x_f, u_f) - \pi^f(x(0)) + \pi^f(x(L_0))$$

$$\le \liminf_{k\to\infty}[I^f(0, L_0, x_k, u_k) - L_0 f(x_f, u_f)] - \lim_{k\to\infty} \pi^f(x_k(0)) + \liminf_{k\to\infty} \pi^f(x_k(L_0))$$

$$\le \liminf_{k\to\infty}[I^f(0, L_0, x_k, u_k) - L_0 f(x_f, u_f) - \pi^f(x_k(0)) + \pi^f(x_k(L_0))] \le 0.$$

In view of the relation above and (5.351),

$$I^f(0, L_0, x, u) - L_0 f(x_f, u_f) - \pi^f(x(0)) + \pi^f(x(L_0)) = 0. \tag{5.480}$$

Theorem 5.19 and (5.480) imply that there exists an (f, A, G)-overtaking optimal pair $(\tilde{x}, \tilde{u}) \in X(0, \infty, A, G)$ such that

$$\tilde{x}(0) = x(L_0). \tag{5.481}$$

For all numbers $t > L_0$, set

$$x(t) = \tilde{x}(t - L_0),\ u(t) = \tilde{u}(t - L_0). \tag{5.482}$$

It is clear that $(x, u) \in X(0, \infty, A, G)$ is (f, A, G)-good. By Propositions 5.55, (5.350), (5.351), and (5.480), for all $S > 0$,

$$I^f(0, S, x, u) - S f(x_f, u_f) - \pi^f(x(0)) + \pi^f(x(S)) = 0.$$

In view of the relation above and Theorem 5.20, (x, u) is (f, A, G)-overtaking optimal pair satisfying

$$(\pi^f + g)(x(0)) = \inf(\pi^f + g).$$

By (5.472), for all sufficiently large natural numbers k,

$$\|x_k(t) - x(t)\| \le \epsilon/2,\ t \in [0, L_0].$$

This contradicts property (i). The contradiction we have reached proves Proposition 5.74.

5.30 The Second Bolza Problem

We use the notation, definitions, and assumptions introduced and used in Sections 5.6, 5.21, and 5.22. Let $a_1 > 0$. Denote by $\mathfrak{A}(E \times E)$ the set of all lower semicontinuous functions $h : E \times E \to R^1$ which are bounded on bounded subsets of $E \times E$ and such that

$$h(z_1, z_2) \ge -a_1 \text{ for all } z_1, z_2 \in E. \tag{5.483}$$

The set $\mathfrak{A}(E \times E)$ is equipped with the uniformity which is determined by the base

$$\mathcal{E}(N, \epsilon) = \{(h_1, h_2) \in \mathfrak{A}(E \times E) \times \mathfrak{A}(E \times E) : |h_1(z_1, z_2) - h_2(z_1, z_2)| \le \epsilon$$

$$\text{for all } z_1, z_2 \in E \text{ satisfying } \|z_i\| \le N,\ i = 1, 2\}, \tag{5.484}$$

where $N, \epsilon > 0$. It is not difficult to see that the uniform space $\mathfrak{A}(E \times E)$ is metrizable and complete.

For each pair of numbers $T_2 > T_1 \ge 0$ and each $h \in \mathfrak{A}(E \times E)$, we define

$$\sigma^{f,h}(T_1, T_2) = \inf\{I^f(T_1, T_2, x, u) + h(x(T_1), x(T_2)) : (x, u) \in X(T_1, T_2, A, G)\}. \tag{5.485}$$

The next result is proved in Section 5.31.

Theorem 5.75. *Assume that (f, A, G) has TP, $M, \epsilon > 0$. Then there exist $\delta > 0$ and $L > 0$ such that for each $T_1 \ge 0$, each $T_2 \ge T_1 + 2L$, each $h \in \mathfrak{A}(E \times E)$ satisfying $|h(x_f, x_f)| \le M$, and each $(x, u) \in X(T_1, T_2, A, G)$ which satisfies*

$$I^f(T_1, T_2, x, u) + h(x(T_1), x(T_2)) \le \sigma^{f,h}(T_1, T_2) + \delta,$$

there exist $\tau_1 \in [T_1, T_1 + L]$, $\tau_2 \in [T_2 - L, T_2]$ such that

$$\|x(t) - x_f\| \le \epsilon, \ t \in [\tau_1, \tau_2].$$

Moreover, if $\|x(T_1) - x_f\| \le \delta$ *then* $\tau_1 = T_1$ *and if* $\|x(T_2) - x_f\| \le \delta$ *then* $\tau_2 = T_2$.

Let $h \in \mathfrak{A}(E \times E)$. Define

$$\psi_h(z_1, z_2) = \pi^f(z_1) + \pi_-^f(z_2) + h(z_1, z_2), \ z_1, z_2 \in \mathcal{A}. \tag{5.486}$$

By (5.325), (5.483), and Corollary 5.49, $\psi_h : \mathcal{A} \times \mathcal{A} \to R^1 \cup \{\infty\}$ is bounded from below and satisfies

$$\psi_h(z_1, z_2) \to \infty \text{ as } z_1, z_2 \in \mathcal{A}, \ \|z_1\| + \|z_2\| \to \infty. \tag{5.487}$$

Fix $g \in \mathfrak{A}(E \times E)$.

In this section we state the following results which describe the asymptotic behavior of approximate solutions in the regions close to the endpoints.

The next result is proved in Section 5.32.

Theorem 5.76. *Let* (f, A, G) *have TP,* $L_0 > 0$ *and* $\epsilon \in (0, 1)$. *Then there exist* $\delta > 0$, $L_1 > L_0$, *and a neighborhood* $\mathfrak{U}$ *of* g *in* $\mathfrak{A}(E \times E)$ *such that for each* $T \ge L_1$, *each* $h \in \mathfrak{U}$, *and each* $(x, u) \in X(0, T, A, G)$ *which satisfies*

$$I^f(0, T, x, u) + h(x(0), x(T)) \le \sigma^{f,h}(0, T) + \delta,$$

the inequalities

$$\psi_g(x(0), x(T)) \le \inf(\psi_g) + \epsilon,$$

$$\Gamma^f(0, L_0, x, u) \le \epsilon, \ \Gamma_-^f(0, L_0, \bar{x}, \bar{u}) \le \epsilon$$

hold where $\bar{x}(t) = x(T - t)$, $\bar{u}(t) = u(T - t)$, $t \in [0, T]$.

Theorem 5.77. *Let* (f, A, G) *have TP and LSC property,* $L_0 > 0$ *and* $\epsilon > 0$. *Then there exist* $\delta > 0$, $L_1 > L_0$, *and a neighborhood* $\mathfrak{U}$ *of* g *in* $\mathfrak{A}(E \times E)$ *such that for each* $T \ge L_1$, *each* $h \in \mathfrak{U}$, *and each* $(x, u) \in X(0, T, A, G)$ *which satisfies*

$$I^f(0, T, x, u) + h(x(0), x(T))) \le \sigma^{f,h}(0, T) + \delta,$$

there exists an (f, A, G)*-overtaking optimal pair* $(x_{*,1}, u_{*,1}) \in X(0, \infty, A, G)$ *and an* $(f, -A, -G)$*-overtaking optimal pair* $(x_{*,2}, u_{*,2}) \in X(0, \infty, -A, -G)$ *such that*

$$\psi_g(x_{*,1}(0), x_{*,2}(0)) = \inf(\psi_g),$$

$$\|x_{*,1}(t) - x(t))\| \le \epsilon, \ t \in [0, L_0],$$

$$\|x_{*,2}(t) - x(T - t)\| \le \epsilon, \ t \in [0, L_0].$$

Theorem 5.77 follows from Theorem 5.76 and the following result which is proved in Section 5.33.

Proposition 5.78. *Let (f, A, G) have TP and LSC property, $L_0 > 0$ and $\epsilon \in (0, 1)$. Then there exist $\delta > 0$ such that for each $(x_1, u_1) \in X(0, L_0, A, G)$ and each $(x_2, u_2) \in X(0, L_0, -A, -G)$ which satisfy*

$$\psi_g(x_1(0), x_2(0)) \leq \inf(\psi_g) + \delta,$$

$$\Gamma^f(0, L_0, x_1, u_1) \leq \delta, \ \Gamma_-^f(0, L_0, x_2, u_2) \leq \delta,$$

there exists an (f, A, G)-overtaking optimal pair $(y_1, v_1) \in X(0, \infty, A, G)$ and an $(f, -A, -G)$-overtaking optimal pair $(y_2, v_2) \in X(0, \infty, -A, -G)$ such that

$$\psi_g(y_1(0), y_2(0)) = \inf(\psi_g),$$

$$\|x_i(t) - y_i(t)\| \leq \epsilon, \ t \in [0, L_0], \ i = 1, 2.$$

5.31 Proof of Theorem 5.75

By Proposition 5.10, there exist $\delta \in (0, 1)$ and $L > 0$ such that the following property holds:

(i) for each $T_1 \geq 0$, each $T_2 \geq T_1 + 2L$, and each $(x, u) \in X(T_1, T_2, A, G)$ which satisfies

$$I^f(T_1, T_2, x, u) \leq \min\{(T_2 - T_1) f(x_f, u_f) + a_1 + M + a_0 + 1,$$

$$U^f(T_1, T_2, x(T_1), x(T_2)) + \delta\},$$

there exist $\tau_1 \in [T_1, T_1 + L]$, $\tau_2 \in [T_2 - L, T_2]$ such that

$$\|x(t) - x_f\| \leq \epsilon, \ t \in [\tau_1, \tau_2];$$

moreover, if $\|x(T_1) - x_f\| \leq \delta$, then $\tau_1 = T_1$, and if $\|x(T_2) - x_f\| \leq \delta$, then $\tau_2 = T_2$.

Assume that $T_1 \geq 0$, $T_2 \geq T_1 + 2L$, $h \in \mathfrak{A}(E \times E)$ satisfies

$$|h(x_f, x_f)| \leq M, \tag{5.488}$$

and $(x, u) \in X(T_1, T_2, A, G)$ satisfies

$$I^f(T_1, T_2, x, u) + h(x(T_1), x(T_2)) \leq \sigma^{f,h}(T_1, T_2) + \delta. \tag{5.489}$$

By (5.489),

$$I^f(T_1, T_2, x, u) \le U^f(T_1, T_2, x(T_1), x(T_2)) + \delta. \tag{5.490}$$

It follows from (5.483), (5.488), and (5.489) that

$$-a_1 + I^f(T_1, T_2, x, u) \le I^f(T_1, T_2, x, u) + h(x(T_1), x(T_2))$$

$$\le 1 + (T_2 - T_1) f(x_f, u_f) + h(x_f, x_f),$$

$$I^f(T_1, T_2, x, u) \le a_1 + 1 + M + (T_2 - T_1) f(x_f, u_f). \tag{5.491}$$

In view of (5.490), (5.491), and property (i), there exist $\tau_1 \in [T_1, T_1 + L]$, $\tau_2 \in [T_2 - L, T_2]$ such that

$$\|x(t) - x_f\| \le \epsilon,\ t \in [\tau_1, \tau_2].$$

Moreover, if $\|x(T_1) - x_f\| \le \delta$, then $\tau_1 = T_1$, and if $\|x(T_2) - x_f\| \le \delta$, then $\tau_2 = T_2$. Theorem 5.75. is proved.

5.32 Proof of Theorem 5.76

By Propositions 5.46 and 5.47, Lemma 5.43, and (A2), there exists $\delta_1 \in (0, \epsilon/4)$ such that:

(i) for each $z \in \mathcal{A}$ satisfying $\|z - x_f\| \le 2\delta_1$, we have

$$|\pi^f(z)| \le \epsilon/16,\ |\pi_-^f(z)| \le \epsilon/16;$$

(ii) for each $(x, u) \in X(0, b_f, A, G)$ which satisfies $\|x(0) - x_f\| \le 2\delta_1$, $\|x(b_f) - x_f\| \le 2\delta_1$, we have

$$I^f(0, b_f, x, u) \ge b_f f(x_f, u_f) - \epsilon/16;$$

(iii) for each $z_i \in E$, $i = 1, 2$ satisfying $\|z_i - x_f\| \le 2\delta_1$, $i = 1, 2$, there exist $\tau \in (0, b_f]$ and $(x, u) \in X(0, \tau, A, G)$ which satisfies $x(0) = z_1$, $x(\tau) = z_2$ and

$$I^f(0, \tau, x, u) \le \tau f(x_f, u_f) + \epsilon/16.$$

By (5.483), (5.484), and Corollary 5.49, there exists a neighborhood $\mathfrak{U}_1$ of g in $\mathfrak{A}(E \times E)$ such that for each $h \in \mathfrak{U}_1$,

$$|\inf(\psi_g) - \inf(\psi_h)| \le \delta_1/16. \tag{5.492}$$

Theorem 5.75 implies that there exist $\delta_2 \in (0, \delta_1/8)$, $l_0 > 0$, and a neighborhood $\mathfrak{U}_2$ of g in $\mathfrak{A}(E \times E)$ such that the following property holds:

(iv) for each $T_1 \ge 0$, each $T_2 \ge T_1 + 2l_0$, each $h \in \mathfrak{U}_2$ and each $(x, u) \in X(T_1, T_2, A, G)$ which satisfies

$$I^f(T_1, T_2, x, u) + h(x(T_1), x(T_2)) \le \sigma^{f,h}(T_1, T_2) + \delta_2,$$

we have

$$\|x(t) - x_f\| \le \delta_1, \; t \in [T_1 + l_0, T_2 - l_0].$$

Clearly, there exists $z_{*,1}, z_{*,2} \in \mathcal{A}$ such that

$$\psi_g(z_{*,1}, z_{*,2}) \le \inf(\psi_g) + \delta_2/8. \tag{5.493}$$

Corollary 5.51 and (5.493) imply that there exist an (f, A, G)-good

$$(x_{*,1}, u_{*,1}) \in X(0, \infty, A, G)$$

and an $(f, -A, -G)$-good $(x_{*,2}, u_{*,2}) \in X(0, \infty, -A, -G)$ for which

$$x_{*,1}(0) = z_{*,1}, \; x_{*,2}(0) = z_{*,2}, \tag{5.494}$$

$$\lim_{T \to \infty} \Gamma^f(0, T, x_{*,1}, u_{*,1}) \le \delta_2/8, \tag{5.495}$$

$$\lim_{T \to \infty} \Gamma^f_-(0, T, x_{*,2}, u_{*,2}) \le \delta_2/8, \tag{5.496}$$

Property (P1), (5.495), and (5.496) imply that

$$\lim_{t \to \infty} \|x_{*,i}(t) - x_f\| = 0, \; i = 1, 2. \tag{5.497}$$

By (5.497), there exists $l_1 > 0$ such that

$$\|x_{*,i}(t) - x_f\| \le \delta_2/8, \; i = 1, 2 \text{ for all numbers } t \ge l_1. \tag{5.498}$$

It follows from Corollary 5.49, (5.483), (5.484), and (5.487) that there exists a neighborhood $\mathfrak{U}$ of g in $\mathfrak{A}(E \times E)$ such that

$$\mathfrak{U} \subset \mathfrak{U}_1 \cap \mathfrak{U}_2, \tag{5.499}$$

$$|h(x_{*,1}(0), x_{*,2}(0)) - g(x_{*,1}(0), x_{*,2}(0))| \le \delta_1/16 \text{ for all } h \in \mathcal{U}, \tag{5.500}$$

and that for all $h \in \mathcal{U}$ and all $(z_1, z_2) \in \mathcal{A} \times \mathcal{A}$ satisfying

$$\psi_h(z_1, z_2) \le \inf(\psi_g) + 4,$$

we have

$$|\psi_h(z_1, z_2) - \psi_g(z_1, z_2)| \le \epsilon/64. \tag{5.501}$$

Choose

$$\delta \in (0, \delta_2/8), \tag{5.502}$$

$$L_1 > 4L_0 + 4l_0 + 4l_1 + 4b_f + 16. \tag{5.503}$$

Assume that

$$T \ge L_1,\ h \in \mathcal{U}, \tag{5.504}$$

$$(x, u) \in X(0, T, A, G),$$

$$I^f(0, T, x, u) + h(x(0), x(T)) \le \sigma^{f,h}(0, T) + \delta. \tag{5.505}$$

Property (iv), (5.499), (5.502)–(5.505), and (5.511) imply that

$$\|x(t) - x_f\| \le \delta_1,\ t \in [T_1 + l_0, T_2 - l_0]. \tag{5.506}$$

Set

$$\bar{x}(t) = x(T - t),\ \bar{u}(t) = u(T - t),\ t \in [0, T]. \tag{5.507}$$

Property (iii), (5.498), (5.503), (5.504), and (5.506) imply that there exist $\tau_1, \tau_2, \tau_3, \tau_4 \in (0, b_f]$ and $(x_1, u_1) \in X(0, T, A, G)$ such that

$$x_1(t) = x_{*,1}(t),\ u_1(t) = u_{*,1}(t),\ t \in [0, l_0 + l_1 + L_0 + 4], \tag{5.508}$$

$$x_1(l_0 + l_1 + L_0 + 4 + \tau_1) = x_f, \tag{5.509}$$

$$I^f(l_0+l_1+L_0+4, l_0+l_1+L_0+4+\tau_1, x_1, u_1) \le \tau_1 f(x_f, u_f)+\epsilon/16, \tag{5.510}$$

$$x_1(l_0 + l_1 + L_0 + 4 + 2b_f) = x(l_0 + l_1 + L_0 + 4 + 2b_f), \tag{5.511}$$

$$x_1(l_0 + l_1 + L_0 + 4 + 2b_f - \tau_2) = x_f, \tag{5.512}$$

$$I^f(l_0+l_1+L_0+4+2b_f-\tau_2, l_0+l_1+L_0+4+2b_f, x_1, u_1) \le \tau_2 f(x_f, u_f)+\epsilon/16, \quad (5.513)$$

$$x_1(t) = x_f,\ u_1(t) = u_f,\ t \in [l_0+l_1+L_0+4+\tau_1, l_0+l_1+L_0+4+2b_f-\tau_2], \quad (5.514)$$

$$x_1(T-t) = x_{*,2}(t),\ u_1(T-t) = u_{*,2}(t),\ t \in [0, l_0+l_1+L_0+4], \quad (5.515)$$

$$x_1(T-l_0-l_1-L_0-4-\tau_3) = x_f, \quad (5.516)$$

$$I^f(T-l_0-l_1-L_0-4-\tau_3, T-l_0-l_1-L_0-4, x_1, u_1) \le \tau_3 f(x_f, u_f)+\epsilon/16, \quad (5.517)$$

$$x_1(T-l_0-l_1-L_0-4-2b_f) = x(T-l_0-l_1-L_0-4-2b_f), \quad (5.518)$$

$$x_1(T-l_0-l_1-L_0-4-2b_f+\tau_4) = x_f, \quad (5.519)$$

$$I^f(T-l_0-l_1-L_0-4-2b_f, T-l_0-l_1-L_0-4-2b_f+\tau_4, x_1, u_1)$$

$$\le \tau_4 f(x_f, u_f)+\epsilon/16, \quad (5.520)$$

$$x_1(t) = x_f,\ u_1(t) = u_f,\ t \in [T-l_0-l_1-L_0-4-2b_f+\tau_4, T-l_0-l_1-L_0-4-\tau_3], \quad (5.521)$$

$$x_1(t) = x(t), u_1(t) = u(t),\ t \in [l_0+l_1+L_0+4+2b_f, T-l_0-l_1-L_0-4-2b_f]. \quad (5.522)$$

It follows from (5.505) that

$$I^f(0, T, x, u) + h(x(0), x(T)) \le I^f(0, T, x_1, u_1) + h(x_1(0), x_1(T)) + \delta. \quad (5.523)$$

By (5.522) and (5.523),

$$\delta \ge I^f(0, T, x, u) + h(x(0), x(T)) - I^f(0, T, x_1, u_1) - h(x_1(0), x_1(T))$$

$$= I^f(0, l_0+l_1+L_0+4, x, u) + I^f(l_0+l_1+L_0+4, l_0+l_1+L_0+4+b_f, x, u)$$

$$+I^f(l_0+l_1+L_0+4+b_f, l_0+l_1+L_0+4+2b_f, x, u)$$

$$+I^f(T-l_0-l_1-L_0-4-2b_f, T-l_0-l_1-L_0-4-b_f, x, u)$$

$$+I^f(T-l_0-l_1-L_0-4-b_f, T-l_0-l_1-L_0-4, x, u)$$

$$+I^f(T-l_0-l_1-L_0-4, T, x, u) + h(x(0), x(T))$$

$$-I^f(0, l_0+l_1+L_0+4, x_1, u_1) - I^f(l_0+l_1+L_0+4, l_0+l_1+L_0+4+\tau_1, x_1, u_1)$$

$$-I^f(l_0+l_1+L_0+4+\tau_1, l_0+l_1+L_0+4+2b_f-\tau_2, x_1, u_1)$$

$$-I^f(l_0+l_1+L_0+4+2b_f-\tau_2, l_0+l_1+L_0+4+2b_f, x_1, u_1)$$

$$-I^f(T-l_0-l_1-L_0-4-2b_f, T-l_0-l_1-L_0-4-2b_f+\tau_4, x_1, u_1)$$

$$-I^f(T-l_0-l_1-L_0-4-2b_f+\tau_4, T-l_0-l_1-L_0-4-\tau_3, x_1, u_1)$$

$$-I^f(T-l_0-l_1-L_0-4-\tau_3, T-l_0-l_1-L_0-4, x_1, u_1)$$

$$-I^f(T-l_0-l_1-L_0-4, T, x_1, u_1)-h(x_1(0), x_1(T)). \tag{5.524}$$

Proposition 5.28, property (ii), (5.503), (5.504), (5.506), (5.508), (5.510), (5.513)–(5.515), (5.517), (5.520), (5.521), and (5.524) imply that

$$\delta \geq I^f(0, l_0+l_1+L_0+4, x, u)+2b_f f(x_f, u_f)-\epsilon/8+2b_f f(x_f, u_f)-\epsilon/8$$

$$+I^f(0, l_0+l_1+L_0+4, \bar{x}, \bar{u})+h(x(0), \bar{x}(0))$$

$$-I^f(0, l_0+l_1+L_0+4, x_{*,1}, u_{*,1})-\tau_1 f(x_f, u_f)-\epsilon/16-(2b_f-\tau_1-\tau_2)f(x_f, u_f)$$

$$-\tau_2 f(x_f, u_f)-\epsilon/16-\tau_4 f(x_f, u_f)-\epsilon/16-(2b_f-\tau_3-\tau_4)f(x_f, u_f)$$

$$-\tau_3 f(x_f, u_f)-\epsilon/16-I^f(0, l_0+l_1+L_0+4, x_{*,2}, u_{*,2})-h(x_{*,1}(0), x_{*,2}(0))$$

$$\geq I^f(0, l_0+l_1+L_0+4, x, u)+I^f(0, l_0+l_1+L_0+4, \bar{x}, \bar{u})+h(x(0), \bar{x}(0))$$

$$-I^f(0, l_0+l_1+L_0+4, x_{*,1}, u_{*,1})-I^f(0, l_0+l_1+L_0+4, x_{*,2}, u_{*,2})$$

$$-h(x_{*,1}(0), x_{*,2}(0))-\epsilon/2. \tag{5.525}$$

Property (i), (5.350), (5.351), (5.408), (5.410), (5.495)–(5.498), (5.500), (5.503), (5.504), (5.506), (5.517), and (5.525) imply that

$$\delta+\epsilon/2 \geq \Gamma^f(0, l_0+l_1+L_0+4, x, u)$$

$$+(l_0+l_1+L_0+4)f(x_f, u_f)+\pi^f(x(0))-\pi^f(x(l_0+l_1+L_0+4))$$

$$+\Gamma^f_-(0, l_0+l_1+L_0+4, \bar{x}, \bar{u})+(l_0+l_1+L_0+4)f(x_f, u_f)$$

$$+\pi^f_-(\bar{x}(0))-\pi^f_-(\bar{x}(l_0+l_1+L_0+4))+h(x(0), \bar{x}(0))$$

$$-\Gamma^f(0, l_0+l_1+L_0+4, x_{*,1}, u_{*,1})-(l_0+l_1+L_0+4)f(x_f, u_f)$$

$$-\pi^f(x_{*,1}(0))+\pi^f(x_{*,1}(l_0+l_1+L_0+4))-h(x_{*,1}(0), x_{*,2}(0))$$

$$-\Gamma_-^f(0, l_0 + l_1 + L_0 + 4, x_{*,2}, u_{*,2}) - (l_0 + l_1 + L_0 + 4) f(x_f, u_f)$$

$$-\pi_-^f(x_{*,2}(0)) + \pi_-^f(x_{*,2}(l_0 + l_1 + L_0 + 4))$$

$$\geq \Gamma^f(0, l_0 + l_1 + L_0 + 4, x, u) + \pi^f(x(0))$$

$$+\Gamma_-^f(0, l_0 + l_1 + L_0 + 4, \bar{x}, \bar{u}) + \pi_-^f(\bar{x}(0)) + h(x(0), \bar{x}(0)) - \epsilon/8$$

$$-\Gamma^f(0, l_0 + l_1 + L_0 + 4, x_{*,1}, u_{*,1}) - \pi^f(x_{*,1}(0))$$

$$-\Gamma_-^f(0, l_0 + l_1 + L_0 + 4, x_{*,2}, u_{*,2}) - \pi_-^f(x_{*,2}(0)) - h(x_{*,1}(0), x_{*,2}(0)) - \epsilon/8$$

$$\geq \Gamma^f(0, l_0 + l_1 + L_0 + 4, x, u) + \Gamma_-^f(0, l_0 + l_1 + L_0 + 4, \bar{x}, \bar{u}) + \psi_h(x(0), \bar{x}(0))$$

$$-\delta_2/4 - \psi_h(x_{*,1}(0) x_{*,2}(0)) - \epsilon/4$$

$$\geq \Gamma^f(0, l_0 + l_1 + L_0 + 4, x, u) + \Gamma_-^f(0, l_0 + l_1 + L_0 + 4, \bar{x}, \bar{u}) + \psi_h(x(0), \bar{x}(0))$$

$$-\delta_2/4 - \epsilon/4 - \psi_g(x_{*,1}(0), x_{*,2}(0)) - \delta_1/16. \tag{5.526}$$

By (5.493), (5.494), (5.502), and (5.526),

$$\Gamma^f(0, l_0 + l_1 + L_0 + 4, x, u) + \Gamma_-^f(0, l_0 + l_1 + L_0 + 4, \bar{x}, \bar{u}) + \psi_h(x(0), \bar{x}(0))$$

$$\leq \delta + \epsilon/2 + \epsilon/4 + \delta_2/4 + \delta_1/16 + \psi_g(x_{*,1}(0) x_{*,2}(0))$$

$$\leq 3\epsilon/4 + 3\epsilon/64 + \inf(\psi_g) + \epsilon/64 \leq 3\epsilon/4 + \epsilon/8 + \inf(\psi_g). \tag{5.527}$$

In view of (5.492), (5.504), and (5.527),

$$\Gamma^f(0, l_0 + l_1 + L_0 + 4, x, u) + \Gamma_-^f(0, l_0 + l_1 + L_0 + 4, \bar{x}, \bar{u}) + \psi_h(x(0), \bar{x}(0))$$

$$\leq 3\epsilon/4 + \epsilon/8 + \inf(\psi_h) + \delta_1/16$$

$$\leq \inf(\psi_h) + 3\epsilon/4 + \epsilon/8 + \epsilon/64.$$

The relation above, (5.492), and (5.504) imply that

$$\Gamma^f(0, l_0 + l_1 + L_0 + 4, x, u),\ \Gamma_-^f(0, l_0 + l_1 + L_0 + 4, \bar{x}, \bar{u}) \leq \epsilon,$$

$$\psi_h(x(0), \bar{x}(0)) \leq \inf(\psi_h) + 57\epsilon/64 \leq \inf(\psi_g) + 58\epsilon/64. \tag{5.528}$$

By (5.501), (5.504), and (5.528),

$$\psi_g(x(0), \bar{x}(0)) \le \psi_h(x(0), \bar{x}(0)) + \epsilon/64 \le \inf(\psi_g) + (59/64)\epsilon.$$

Theorem 5.76 is proved.

5.33 Proof of Proposition 5.78

Assume that the proposition does not hold. Then there exist a sequence $\{\delta_k\}_{k=1}^{\infty} \subset (0, 1]$ and sequences

$$\{(x_{k,1}, u_{k,1})\}_{k=1}^{\infty} \subset X(0, L_0, A, G),\ \{(x_{k,2}, u_{k,2})\}_{k=1}^{\infty} \subset X(0, L_0, -A, -G)$$

such that

$$\lim_{k\to\infty} \delta_k = 0 \tag{5.529}$$

and that for all integers $k \ge 1$,

$$\Gamma^f(0, L_0, x_{k,1}, u_{k,1}) \le \delta_k, \tag{5.530}$$

$$\Gamma_-^f(0, L_0, x_{k,2}, u_{k,2}) \le \delta_k, \tag{5.531}$$

$$\psi_g(x_{k,1}(0), x_{k,2}(0)) \le \inf(\psi_g) + \delta_k, \tag{5.532}$$

and the following property holds:

(i) for each (f, A, G)-overtaking optimal pair $(y_1, v_1) \in X(0, \infty, A, G)$ and each $(f, -A, -G)$-overtaking optimal pair $(y_2, v_2) \in X(0, \infty, -A, -G)$ satisfying

$$\psi_g(y_1(0), y_2(0)) = \inf(\psi_g),$$

we have

$$\sup\{\|x_{k,1}(t) - y_1(t)\| + \|x_{k,2}(t) - y_2(t)\| : t \in [0, L_0]\} > \epsilon.$$

In view of (5.325), (5.529), (5.531) , and (5.532), the sequences

$$\{\pi^f(x_{k,1}(0))\}_{k=1}^{\infty},\ \{\pi_-^f(x_{k,2}(0))\}_{k=1}^{\infty} \text{ are bounded.} \tag{5.533}$$

By (5.350), (5.530), and (5.531), the sequences

$$\{\pi^f(x_{k,1}(L_0))\}_{k=1}^{\infty},\ \{\pi_-^f(x_{k,2}(L_0))\}_{k=1}^{\infty} \text{ are bounded.} \tag{5.534}$$

It follows from (5.350) and (5.529)–(5.534) that the sequences

$$\{I^f(0, L_0, x_{k,1}, u_{k,1})\}_{k=1}^{\infty}, \ \{I^f(0, L_0, x_{k,2}, u_{k,1})\}_{k=1}^{\infty}$$

are bounded. By LSC property, extracting a subsequence and re-indexing if necessary, we may assume without loss of generality that there exist $(y_1, v_1) \in X(0, L_0, A, G)$, $(y_2, v_2) \in X(0, L_0, -A, -G)$ such that for $i = 1, 2$,

$$x_{k,i}(t) \to y_i(t) \text{ as } k \to \infty \text{ for all } t \in [0, L_0], \tag{5.535}$$

$$I^f(0, L_0, y_i, v_i) \le \liminf_{k\to\infty} I^f(0, L_0, x_{k,i}, u_{k,i}). \tag{5.536}$$

Proposition 5.57, (5.529), (5.532), and (5.535) imply that

$$\psi_g(y_1(0), y_2(0)) \le \liminf_{k\to\infty} \psi_g(x_{k,1}(0), x_{k,2}(0)) = \inf(\psi_g), \tag{5.537}$$

$$\psi_g(y_1(0), y_2(0)) = \inf(\psi_g) = \lim_{k\to\infty} \psi_g(x_{k,1}(0), x_{k,2}(0)). \tag{5.538}$$

Proposition 5.57 and (5.535) imply that

$$\pi^f(y_1(0)) \le \liminf_{k\to\infty} \pi^f(x_{k,1}(0)), \tag{5.539}$$

$$\pi_-^f(y_2(0)) \le \liminf_{k\to\infty} \pi_-^f(x_{k,2}(0)), \tag{5.540}$$

$$g(y_1(0), y_2(0)) \le \liminf_{k\to\infty} g(x_{k,1}(0), x_{k,2}(0)). \tag{5.541}$$

By (5.537)–(5.541),

$$\pi^f(y_1(0)) = \lim_{k\to\infty} \pi^f(x_{k,1}(0)), \ \pi_-^f(y_2(0)) = \lim_{k\to\infty} \pi_-^f(x_{k,2}(0)), \tag{5.542}$$

$$g(y_1(0), y_2(0)) = \lim_{k\to\infty} g(x_{k,1}(0), x_{k,2}(0)). \tag{5.543}$$

Proposition 5.57 and (5.535) imply that

$$\pi^f(y_1(L_0)) \le \liminf_{k\to\infty} \pi^f(x_{k,1}(L_0)), \tag{5.544}$$

$$\pi_-^f(y_2(L_0)) \le \liminf_{k\to\infty} \pi_-^f(x_{k,2}(L_0)). \tag{5.545}$$

It follows from (5.529), (5.530), (5.536), (5.542), and (5.544) that

$$I^f(0, L_0, y_1, v_1) - L_0 f(x_f, u_f) - \pi^f(y_1(0)) + \pi^f(y_1(L_0))$$

$$\leq \liminf_{k\to\infty} I^f(0, L_0, x_{k,1}, u_{k,1}) - L_0 f(x_f, u_f)$$

$$- \lim_{k\to\infty} \pi^f(x_{k,1}(0)) + \liminf_{k\to\infty} \pi^f(x_{k,1}(L_0))$$

$$\leq \liminf_{k\to\infty} [I^f(0, L_0, x_{k,1}, u_{k,1}) - L_0 f(x_f, u_f) - \pi^f(x_{k,1}(0)) + \pi^f(x_{k,1}(L_0))] \leq 0. \tag{5.546}$$

It follows from (5.529), (5.531), (5.536), (5.542), and (5.545) that

$$I^f(0, L_0, y_2, v_2) - L_0 f(x_f, u_f) - \pi_-^f(y_2(0)) + \pi_-^f(y_2(L_0))$$

$$\leq \liminf_{k\to\infty} I^f(0, L_0, x_{k,2}, u_{k,2}) - L_0 f(x_f, u_f)$$

$$- \lim_{k\to\infty} \pi_-^f(x_{k,2}(0)) + \liminf_{k\to\infty} \pi_-^f(x_{k,2}(L_0))$$

$$\leq \liminf_{k\to\infty} [I^f(0, L_0, x_{k,2}, u_{k,2}) - L_0 f(x_f, u_f) - \pi_-^f(x_{k,2}(0)) + \pi_-^f(x_{k,2}(L_0))] \leq 0. \tag{5.547}$$

In view of (5.351), (5.546), and (5.547),

$$I^f(0, L_0, y_1, v_1) - L_0 f(x_f, u_f) - \pi^f(y_1(0)) + \pi^f(y_1(L_0)) = 0, \tag{5.548}$$

$$I^f(0, L_0, y_2, v_2) - L_0 f(x_f, u_f) - \pi_-^f(y_2(0)) + \pi_-^f(y_2(L_0)) = 0. \tag{5.549}$$

By (5.548) and (5.549), $\pi^f(y_1(L_0))$, $\pi_-^f(y_2(L_0))$ are finite. Theorem 5.19 implies that there exist an (f, A, G)-overtaking optimal pair

$$(\tilde{y}_1, \tilde{v}_1) \in X(0, \infty, A, G)$$

and an $(f, -A, -G)$-overtaking optimal pair $(\tilde{y}_2, \tilde{v}_2) \in X(0, \infty, -A, -G)$ such that

$$\tilde{y}_i(0) = y_i(L_0), \ i = 1, 2. \tag{5.550}$$

For $i = 1, 2$ and all numbers $t > L_0$, set

$$y_i(t) = \tilde{y}_i(t - L_0), \ v_i(t) = \tilde{v}_i(t - L_0). \tag{5.551}$$

It is clear that $(y_1, v_1) \in X(0, \infty, A, G)$ is (f, A, G)-good and $(y_2, v_2) \in X(0, \infty, -A, -G)$ is $(f, -A, -G)$-good. Proposition 5.55, (5.350), (5.351), (5.548), and (5.549) imply that for all $T_2 > T_1 \geq 0$,

$$I^f(T_1, T_2, y_1, v_1) - (T_2 - T_1) f(x_f, u_f) - \pi^f(y_1(T_1)) + \pi^f(y_1(T_2)) = 0,$$

$$I^f(T_1, T_2, y_2, v_2) - (T_2 - T_1) f(x_f, u_f) - \pi_-^f(y_2(T_1)) + \pi_-^f(y_2(T_2)) = 0.$$

In view of the relations above and Theorem 5.20, (y_1, v_1) is (f, A, G)-overtaking optimal pair and (y_2, v_2) is $(f, -A, -G)$-overtaking optimal pair. By (5.545), for all sufficiently large natural numbers k and $i = 1, 2$,

$$\|x_{k,i}(t) - y_i(t)\| \le \epsilon/4, \ t \in [0, L_0].$$

This contradicts property (i). The contradiction we have reached proves Proposition 5.78.

5.34 Examples

In this section we present a family of problems which belong to the second class of problems and for which the results of this chapter hold.

Let $(E, \langle\cdot,\cdot\rangle_E)$ be a Hilbert space equipped with an inner product $\langle\cdot,\cdot\rangle_E$ which induces the norm $\|\cdot\|_E$, and let $(F, \langle\cdot,\cdot\rangle_F)$ be a Hilbert space equipped with an inner product $\langle\cdot,\cdot\rangle_F$ which induces the norm $\|\cdot\|_F$. For simplicity, we set $\langle\cdot,\cdot\rangle_E = \langle\cdot,\cdot\rangle$, $\|\cdot\|_E = \|\cdot\|$, $\langle\cdot,\cdot\rangle_F = \langle\cdot,\cdot\rangle$, $\|\cdot\|_F = \|\cdot\|$, if E *and* F are understood. We suppose that $\mathcal{A}$ is a nonempty subset of E, and $\mathcal{U} : \mathcal{A} \to 2^F$ is a point to set mapping with a graph

$$\mathcal{M} = \{(x, u) : x \in \mathcal{A}, \ u \in \mathcal{U}(x)\}.$$

We suppose that $\mathcal{M}$ is a Borel measurable subset of $E \times F$ and a linear operator $A : \mathcal{D}(A) \to E$ generates a C_0 semigroup $S(t) = e^{At}$, $t \ge 0$ on E. Let $B \in \mathcal{L}(F, E_{-1})$ be an admissible control operator for e^{At}, $t \ge 0$, $f : \mathcal{M} \to R^1$ be a borelian function. For $\tau \ge 0$,

$$\Phi_\tau u = \int_0^\tau e^{A(\tau-s)} Bu(s) ds, \ u \in L^2(0, \tau; F). \tag{5.552}$$

Since B is an admissible control operator for e^{At}, $t \ge 0$,

$$\Phi_\tau \in \mathcal{L}(L^2(0, \tau; F), E), \ \tau \ge 0 \tag{5.553}$$

(see Proposition 1.18). We consider the control system

$$x' = Ax + Bu. \tag{5.554}$$

Assume that there exists $T_f > 0$ such that

$$\mathrm{Ran}\Phi_{T_f} = E. \tag{5.555}$$

In other words the pair (A, B) is exactly controllable. We suppose that $x_f \in E$, $u_f \in F$, the pair $\bar{x}(t) = x_f, \bar{u}(t) = u_f, t \geq 0$ is a solution of (5.554) and that (x_f, u_f) is an interior point of $\mathcal{M}$ in $E \times F$. Assume that $L : E \times F \to [0, \infty)$ is a borelian function which is upper semicontinuous at (x_f, u_f),

$$L(x_f, u_f) = 0, \tag{5.556}$$

$\psi_0 : [0, \infty) \to [0, \infty)$ is an increasing function, $K_1, a_1 > 0$ such that

$$\lim_{t\to\infty} \psi_0(t) = \infty,$$

$$L(x, u) \geq -a_1 + \max\{\psi_0(\|x\|)\|x\|,\ K_1\|u\|^2\} \tag{5.557}$$

for all $(x, u) \in E \times F, \mu \in R^1, \bar{p} \in \mathcal{D}(A^*)$. Let for all $(x, u) \in E \times F$

$$f(x, u) = L(x, u) + \mu + \langle x, A^*\bar{p}\rangle + \langle Bu, \bar{p}\rangle. \tag{5.558}$$

Clearly, f is a Borelian function. It is not difficult to see that there exist $a_0, K_0 > 0$ and an increasing function $\psi : [0, \infty) \to [0, \infty)$ such that $\lim_{t\to\infty} \psi(t) = \infty$,

$$f(x, u) \geq -a_0 + \max\{\psi(x)\|, K_0\|u\|^2\},\ x \in E,\ u \in F,$$

f is upper semicontinuous at (x_f, u_f) and (5.21) holds. It follows from (5.555), Theorem 1.23, and the upper semicontinuity of f at (x_f, u_f) that (A3) holds.

Let $0 \leq T_1 < T_2$, $(x, u) \in X(T_1, T_2)$. It is not difficult to see that

$$I^f(T_1, T_2, x, u) = \int_{T_1}^{T_2} f(x(t), u(t))dt = \int_{T_1}^{T_2} L(x(t), u(t))dt + \mu(T_2 - T_1)$$

$$+ \int_{T_1}^{T_2} \langle x(t), A^*\bar{p}\rangle dt + \int_{T_1}^{T_2} \langle Bu(t), \bar{p}\rangle dt \geq \mu(T_2 - T_1) + \langle x(T_2) - x(T_1).\bar{p}\rangle. \tag{5.559}$$

This implies that (A1) holds.

Assume now that the following property holds:

(a) for each $M, \epsilon > 0$ there exists $\delta > 0$ such that for each $(x, u) \in E \times F$ which satisfies

$$\|x\| \leq M \text{ and } L(x, u) \leq \delta,$$

we have $\|x - x_f\| \leq \epsilon$.

We claim that f has TP. In view of Theorems 5.13 and 5.26, it is sufficient to show that property (P3) of Theorem 5.26 holds.

Let $\epsilon, M > 0$ and $M_1 > 0$ be as guaranteed by Theorem 5.2 with $M_0 = M$, $c_0 = 1, c = 2$. Let $S > 0$ be as guaranteed by Theorem 5.1, and let $\delta \in (0, 1)$ be as guaranteed by property (a) with $M = M_1$. Choose

$$L > 3 + \delta^{-1}(M + S + 2M_1\|\bar{p}\|).$$

Assume that $T_1 \geq 0$, $T_2 \geq T_1 + L$ and that $(x, u) \in X(T_1, T_2)$ satisfies

$$I^f(T_1, T_2, x, u) \leq (T_2 - T_1)f(x_f, u_f) + M.$$

Combined with Theorem 5.2 and the choice of M_1, this implies that

$$\|x(t)\| \leq M_1, \; t \in [T_1 + 1, T_2].$$

It is not difficult to see that, in view of the choice of S and Theorem 5.1,

$$I^f(T_1 + 1, T_2, x, u) = I^f(T_1, T_2, x, u) - I^f(T_1, T_1 + 1, x, u)$$

$$\leq (T_2 - T_1)f(x_f, u_f) + M - f(x_f, u_f) + S = (T_2 - T_1 - 1)f(x_f, u_f) + M + S.$$

Combined with (5.559), this implies that

$$\int_{T_1+1}^{T_2} L(x(t), u(t))dt$$

$$= I^f(T_1 + 1, T_2, x, u) - \mu(T_2 - T_1 - 1) - \langle x(T_2) - x(T_1 + 1), \bar{p}\rangle$$

$$\leq M + S + 2M_1\|\bar{p}|.$$

Together with the choice of δ and property (a) with $M = M_1$, this implies that

$$M + S + 2M_1\|\bar{p}\| \geq \int_{T_1+1}^{T_2} L(x(t), u(t))dt$$

$$\geq \delta \text{mes}(\{t \in [T_1 + 1, T_2] : \; L(x(t), u(t)) > \delta\})$$

$$\geq \delta \text{mes}(\{t \in [T_1 + 1, T_2] : \; \|x(t) - x_f\| > \epsilon\})$$

and

$$\text{mes}(\{t \in [T_1, T_2] : \|x(t) - x_f\| > \epsilon\})$$

$$\leq 1 + \delta^{-1}(M + S + 2M_1\|\bar{p}\|) < L.$$

Thus (P3) holds and f has TP.

We can easily obtain particular cases of this example with the pairs (A, B) considered in Section 1.8.

5.35 Exercises for Chapter 5

Exercise 5.79. Use the control problem considered in Example 1.24, and construct an optimal control problem which is a special case of the example analyzed in Section 5.34.

Exercise 5.80. Use the control problem considered in Example 1.25, and construct an optimal control problem which is a special case of the example analyzed in Section 5.34.

Exercise 5.81. Use the control problem considered in Example 1.26, and construct an optimal control problem which is a special case of the example analyzed in Section 5.34.

Exercise 5.82. Use the control problem considered in Example 1.27, and construct an optimal control problem which is a special case of the example analyzed in Section 5.34.

Exercise 5.83. Use the control problem considered in Example 1.28, and construct an optimal control problem which is a special case of the example analyzed in Section 5.34.

Exercise 5.84. Use the control problem considered in Example 1.29, and construct an optimal control problem which is a special case of the example analyzed in Section 5.34.

Exercise 5.85. Use the control problem considered in Example 1.30, and construct an optimal control problem which is a special case of the example analyzed in Section 5.34.

Exercise 5.86. Use the control problem considered in Example 1.31, and construct an optimal control problem which is a special case of the example analyzed in Section 5.34.

Chapter 6
Continuous-Time Nonautonomous Problems on the Half-Axis

In this chapter we establish sufficient and necessary conditions for the turnpike phenomenon for continuous-time optimal control problems on subintervals of half-axis in infinite dimensional spaces. For these optimal control problems the turnpike is not a singleton. We also study the existence of solutions of the corresponding infinite horizon optimal control problems. The results of this chapter will be obtained for two large classes of problems which will be treated simultaneously.

6.1 Preliminaries

We begin with the description of the first class of problems. Let $(E, \|\cdot\|)$ be a Banach space and E^* be its dual. Let $\{A(t) : t \in [0, \infty)\}$ be the family of closed densely defined linear operators with the domain and range in the Banach space E. Let (F, ρ_F) be a metric space. We suppose that $\mathcal{A}$ is a nonempty subset of $[0, \infty) \times E$, $\mathcal{U} : \mathcal{A} \to 2^F$ is a point to set mapping with a graph

$$\mathcal{M} = \{(t, x, u) : (t, x) \in \mathcal{A},\ u \in \mathcal{U}(t, x)\}.$$

We suppose that $\mathcal{M}$ is a Borel measurable subset of $[0, \infty) \times E \times F$, $G : \mathcal{M} \to E$ is a Borelian function. Let $f : \mathcal{M} \to R^1$ be a Borelian function bounded from below. We consider the homogeneous Cauchy problem

$$x'(t) = A(t)x(t),\ t \in [0, \infty). \tag{6.1}$$

We assume that there exists a function $U : \{(t, s) \in R^2 : 0 \le s \le t < \infty\} \to L(E)$ which has the following properties [18]:

A. J. Zaslavski, *Turnpike Conditions in Infinite Dimensional Optimal Control*, Springer Optimization and Its Applications 148,
https://doi.org/10.1007/978-3-030-20178-4_6

(i) for each $x_0 \in E$ the function $(t, s) \to U(t, s)x_0$ is continuous on the set $\{(t, s) \in R^2 : 0 \le s \le t < \infty\}$;
(ii) $U(s, s) = Id$ for all $s \in [0, \infty)$, where Id is the identity operator;
(iii) $U(t, s)U(s, \tau) = U(t, \tau)$ for all numbers $t \ge s \ge \tau \ge 0$;
(iv) for each $s \ge 0$ there exists a densely linear subspace E_s of E such that for each $x_0 \in E_s$ the function $t \to U(t, s)x_0$ is continuously differentiable on $[s, \infty)$ and

$$(\partial/\partial t)U(t, s)x_0 = A(t)U(t, s)x_0, \ t \in [s, \infty); \tag{6.2}$$

(v) there exists an increasing function $\tau \to \Delta_\tau > 0$, $\tau > 0$ such that for each $\tau > 0$, each $s \ge 0$ and each $t \in [s, s + \tau]$,

$$\|U(t, s)\| \le \Delta_\tau.$$

In this case problem (6.1) is called well-posed [18].

Let $0 \le T_1 < T_2$ and consider the following equation

$$x'(t) = A(t)x(t) + f(t), \ t \in [T_1, T_2], \ x(0) = x_{T_1}, \tag{6.3}$$

where $x_{T_1} \in E$ and $f \in L^1(T_1, T_2; E)$.

A continuous function $x : [T_1, T_2] \to E$ is a solution of (6.3) if

$$x(t) = U(t, T_1)x(T_1) + \int_{T_1}^{t} U(t, s)f(s)ds, \ t \in [T_1, T_2]. \tag{6.4}$$

Assume that (6.4) holds and $\tau \in [T_1, T_2]$. In view of (6.4),

$$x(\tau) = U(\tau, T_1)x(T_1) + \int_{T_1}^{\tau} U(\tau, s)f(s)ds. \tag{6.5}$$

Property (iii), (6.4), and (6.5) imply that for all $t \in [\tau, T_2]$,

$$x(t) = U(t, \tau)U(\tau, T_1)x(T_1) + \int_{T_1}^{\tau} U(t, s)f(s)ds + \int_{\tau}^{t} U(t, s)f(s)ds$$

$$= U(t, \tau)U(\tau, T_1)x(T_1) + U(t, \tau)\int_{T_1}^{\tau} U(\tau, s)f(s)ds + \int_{\tau}^{t} U(t, s)f(s)ds$$

$$= U(t, \tau)x(\tau) + \int_{\tau}^{t} U(t, s)f(s)ds.$$

Thus $x : [\tau, T_2] \to E$ is a solution of the equation

$$y'(t) = A(t)y(t) + f(t), \ t \in [\tau, T_2], \ y(\tau) = x(\tau).$$

Let $0 \le T_1 < T_2 < T_3$, $z_0 \in E$, $f \in L^1(T_1, T_3; E)$, a continuous function $x_1 : [T_1, T_2] \to E$ is a solution of the equation

$$x'(t) = A(t)x(t) + f(t),\ t \in [T_1, T_2],\ x(T_1) = z_0,$$

a continuous function $x_2 : [T_2, T_3] \to E$ is a solution of the equation

$$x'(t) = A(t)x(t) + f(t),\ t \in [T_2, T_3],\ x(T_2) = x_1(T_2).$$

Set

$$x(t) = x_1(t),\ t \in [T_1, T_2],\ x(t) = x_2(t),\ t \in [T_2, T_3]. \tag{6.6}$$

Clearly, the function $x : [T_1, T_3] \to E$ is continuous. In view of (6.4) and (6.6), for all $t \in [T_1, T_2]$,

$$x(t) = x_1(t) = U(t, T_1)z_0 + \int_{T_1}^{t} U(t, s)f(s)ds. \tag{6.7}$$

Property (iii), (6.4), (6.6), and (6.7) imply that for all $t \in [T_2, T_3]$,

$$x(t) = x_2(t) = U(t, T_2)x_1(T_2) + \int_{T_2}^{t} U(t, s)f(s)ds$$

$$= U(t, T_2)(U(T_2, T_1)z_0 + \int_{T_1}^{T_2} U(T_2, s)f(s)ds) + \int_{T_2}^{t} U(t, s)f(s)ds$$

$$= U(t, T_1)z_0 + \int_{T_1}^{T_2} U(t, s)f(s)ds + \int_{T_2}^{t} U(t, s)f(s)ds$$

$$= U(t, T_1)z_0 + \int_{T_1}^{t} U(t, s)f(s)ds.$$

Thus $x(\cdot)$ is a solution of the equation

$$x'(t) = A(t)x(t) + f(t),\ t \in [T_1, T_3],\ x(T_1) = z_0.$$

Let $0 \le T_1 < T_2$. We consider the following equation

$$x'(t) = A(t)x(t) + G(t, x(t), u(t)),\ t \in [T_1, T_2]. \tag{6.8}$$

A pair of functions $x : [T_1, T_2] \to E$, $u : [T_1, T_2] \to F$ is called a (mild) solution of (6.8) if $x : [T_1, T_2] \to E$ is a continuous function, $u : [T_1, T_2] \to F$ is a Lebesgue measurable function,

$$(t, x(t)) \in \mathcal{A},\ t \in [T_1, T_2], \tag{6.9}$$

$$u(t) \in \mathcal{U}(t, x(t)),\ t \in [T_1, T_2] \text{ almost everywhere (a.e.)}, \tag{6.10}$$

$G(s, x(s), u(s)),\ s \in [T_1, T_2]$ is Bochner integrable and for every $t \in [T_1, T_2]$,

$$x(t) = U(t, T_1)x(T_1) + \int_{T_1}^{t} U(t, s)G(s, x(s), u(s))ds. \tag{6.11}$$

The set of all pairs (x, u) which are solutions of (6.8) is denoted by $X(T_1, T_2)$.

Let $T_1 \geq 0$. A pair of functions $x : [T_1, \infty) \to E$, $u : [T_1, \infty) \to F$ is called a (mild) solution of the system

$$x'(t) = A(t)x(t) + G(t, x(t), u(t)),\ t \in [T_1, \infty) \tag{6.12}$$

if for every $T_2 > T_1$, $x : [T_1, T_2] \to E$, $u : [T_1, T_2] \to F$ is a solution of (6.8). The set of all such pairs (x, u) which are solutions of the equation above, is denoted by $X(T_1, \infty)$.

A function $x : I \to E$, where I is either $[T_1, T_2]$ or $[T_1, \infty)$ $(0 \leq T_1 < T_2)$ is called a trajectory if there exists a Lebesgue measurable function $u : I \to F$ (referred to as a control) such that $(x, u) \in X(T_1, T_2)$ or $(x, u) \in X(T_1, \infty)$, respectively).

Now we describe the second class of problems.

Let $(E, \langle\cdot,\cdot\rangle)_E$ be a Hilbert space equipped with an inner product $\langle\cdot,\cdot\rangle_E$ which induces the norm $\|\cdot\|_E$, and let $(F, \langle\cdot,\cdot\rangle_F)$ be a Hilbert space equipped with an inner product $\langle\cdot,\cdot\rangle_F$ which induces the norm $\|\cdot\|_F$. For simplicity, we set $\langle\cdot,\cdot\rangle_E = \langle\cdot,\cdot\rangle$, $\|\cdot\|_E = \|\cdot\|$, $\langle\cdot,\cdot\rangle_F = \langle\cdot,\cdot\rangle$, $\|\cdot\|_F = \|\cdot\|$, if E, F are understood. We suppose that $\mathcal{A}_0$ is a nonempty subset of E, $\mathcal{U}_0 : \mathcal{A}_0 \to 2^F$ is a point to set mapping with a graph

$$\mathcal{M}_0 = \{(x, u) :\ x \in \mathcal{A}_0,\ u \in \mathcal{U}_0(x)\}.$$

We suppose that $\mathcal{M}_0$ is a Borel measurable subset of $E \times F$. Define

$$\mathcal{A} = [0, \infty) \times \mathcal{A}_0,$$

$\mathcal{U} : \mathcal{A} \to 2^F$ by

$$\mathcal{U}(t, x) = \mathcal{U}_0(x),\ (t, x) \in \mathcal{A},$$

$$\mathcal{M} = [0, \infty) \times \mathcal{M}_0.$$

Let a linear operator $A : \mathcal{D}(A) \to E$ generates a C_0 semigroup $S(t) = e^{At}$, $t \in [0, \infty)$ on E. As usual, we denote by $S(t)^*$ the adjoint of $S(t)$. Then $S(t)^*,\ t \in$

$[0, \infty)$ is C_0 semigroup and its generator is the adjoint A^* of A. The domain $\mathcal{D}(A^*)$ is a Hilbert space equipped with the graph norm $\|\cdot\|_{\mathcal{D}(A^*)}$:

$$\|z\|^2_{\mathcal{D}(A^*)} = \|z\|^2_E + \|A^*z\|^2_E, \ z \in \mathcal{D}(A^*). \tag{6.13}$$

Let $\mathcal{D}(A^*)'$ be the dual of $\mathcal{D}(A^*)$ with the pivot space E. In particular,

$$E_1^d := \mathcal{D}(A^*) \subset E \subset \mathcal{D}(A^*)' = E_{-1}.$$

(Here we use the notation of Section 1.7.)

Let $G : \mathcal{M} \to \mathcal{D}(A^*)' = E_{-1}$ and $f : \mathcal{M} \to R^1$ be Borelian functions, $B \in \mathcal{L}(F, E_{-1})$ is an admissible control operator for e^{At}, $t \geq 0$, for all $(t, x, u) \in \mathcal{M}$,

$$G(t, x, u) = Bu.$$

Let $0 \leq T_1 < T_2$. We consider the following equation

$$x'(t) = Ax(t) + Bu(t), \ t \in [T_1, T_2] \text{ a. e. .} \tag{6.14}$$

A pair of functions $x : [T_1, T_2] \to E$, $u : [T_1, T_2] \to F$ is called a (mild) solution of (6.14) if $x : [T_1, T_2] \to E$ is a continuous function, $u : [T_1, T_2] \to F$ is a Lebesgue measurable function, $u \in L^2(T_1, T_2; F)$,

$$(t, x(t)) \in \mathcal{A}, \ t \in [T_1, T_2], \tag{6.15}$$

$$u(t) \in \mathcal{U}(t, x(t)), \ t \in [T_1, T_2] \text{ a.e.,} \tag{6.16}$$

and for each $t \in [T_1, T_2]$,

$$x(t) = e^{A(t-T_1)}x(T_1) + \int_{T_1}^t e^{A(t-s)}Bu(s)ds \tag{6.17}$$

in E_{-1}. The set of all pairs (x, u) which are solutions of (6.14) is denoted by $X(T_1, T_2, A, G)$. In the sequel for simplicity, we use the notation $X(T_1, T_2) = X(T_1, T_2, A, G)$ if the pair (A, G) is understood.

Let $T_1 \geq 0$. A pair of functions $x : [T_1, \infty) \to E$, $u : [T_1, \infty) \to F$ is called a (mild) solution of the system

$$x'(t) = Ax(t) + Bu(t), \ t \in [T_1, \infty) \tag{6.18}$$

if for every $T_2 > T_1$, $x : [T_1, T_2] \to E$, $u : [T_1, T_2] \to F$ is a solution of (6.14). The set of all such pairs (x, u) which are solutions of (6.18) is denoted by $X(T_1, \infty)$.

A function $x : I \to E$, where I is either $[T_1, T_2]$ or $[T_1, \infty)$ $(0 \leq T_1 < T_2)$ is called a trajectory if there exists a Lebesgue measurable function $u : I \to F$

(referred to as a control) such that $(x, u) \in X(T_1, T_2)$ or $(x, u) \in X(T_1, \infty)$, respectively.

We treat the two problems simultaneously.

Let $T_2 > T_1 \geq 0$, $(x, u) \in X(T_1, T_2)$. Define

$$I^f(T_1, T_2, x, u) = \int_{T_1}^{T_2} f(t, x(t), u(t))dt$$

which is well-defined but can be ∞.

Let $a_0 > 0$ and let $\psi : [0, \infty) \to [0, \infty)$ be an increasing function such that

$$\psi(t) \to \infty \text{ as } t \to \infty. \tag{6.19}$$

We suppose that for the first class of problems the function f satisfies

$$f(t, x, u) \geq -a_0 + \max\{\psi(\|x\|),$$

$$\psi((\|G(t, x, u)\| - a_0\|x\|)_+)(\|G(t, x, u)\| - a_0\|x\|)_+\} \tag{6.20}$$

for each $(t, x, u) \in \mathcal{M}$ and for the second class of problems f satisfies

$$f(t, x, u) \geq -a_0 + \max\{\psi(\|x\|),\ K_0\|u\|^2\},\ (t, x, u) \in \mathcal{M}. \tag{6.21}$$

We consider functionals of the form $I^f(T_1, T_2, x, u)$, where $0 \leq T_1 < T_2$ and $(x, u) \in X(T_1, T_2)$. For each pair of numbers $T_2 > T_1 \geq 0$ and each pair of points $(T_1, y), (T_2, z) \in \mathcal{A}$ we define

$$U^f(T_1, T_2, y, z) = \inf\{I^f(T_1, T_2, x, u) :$$

$$(x, u) \in X(T_1, T_2),\ x(T_1) = y,\ x(T_2) = z\},$$

$$\sigma^f(T_1, T_2, y) = \inf\{I^f(T_1, T_2, x, u) : (x, u) \in X(T_1, T_2),\ x(T_1) = y\},$$

$$\widehat{\sigma}^f(T_1, T_2, z) = \inf\{I^f(T_1, T_2, x, u) : (x, u) \in X(T_1, T_2),\ x(T_2) = z\},$$

$$\sigma^f(T_1, T_2) = \inf\{I^f(T_1, T_2, x, u) : (x, u) \in X(T_1, T_2)\}. \tag{6.22}$$

We suppose that $(x_f, u_f) \in X(0, \infty)$ satisfies

$$\sup\{\|x_f(t)\| :\ t \in [0, \infty)\} < \infty, \tag{6.23}$$

$$\Delta_f := \sup\{I^f(j, j+1, x_f, u_f) :\ j = 0, 1, \dots\} < \infty. \tag{6.24}$$

We suppose that there exists a number $b_f > 0$ and the following assumptions hold.

(A1) For each $S_1 > 0$ there exist $S_2 > 0$ and $c > 0$ such that

$$I^f(T_1, T_2, x_f, u_f) \le I^f(T_1, T_2, x, u) + S_2$$

for each $T_1 \ge 0$, each $T_2 \ge T_1 + c$, and each $(x, u) \in X(T_1, T_2)$ satisfying $\|x(T_j)\| \le S_1$, $j = 1, 2$.

(A2) For each $\epsilon > 0$ there exists $\delta > 0$ such that for each $(T_i, z_i) \in \mathcal{A}$, $i = 1, 2$ satisfying $\|z_i - x_f(T_i)\| \le \delta$, $i = 1, 2$ and $T_2 \ge b_f$ there exist $\tau_1, \tau_2 \in (0, b_f]$ and $(x_1, u_1) \in X(T_1, T_1 + \tau_1)$, $(x_2, u_2) \in X(T_2 - \tau_2, T_2)$ which satisfies

$$x_1(T_1) = z_1, \; x_1(T_1 + \tau_1) = x_f(T_1 + \tau_1),$$

$$I^f(T_1, T_1 + \tau_1, x_1, u_1) \le I^f(T_1, T_1 + \tau_1, x_f, u_f) + \epsilon,$$

$$x_2(T_2) = z_2, \; x_2(T_2 - \tau_2) = x_f(T_2 - \tau_2),$$

$$I^f(T_2 - \tau_2, T_2, x_2, u_2) \le I^f(T_2 - \tau_2, T_2, x_f, u_f) + \epsilon.$$

Relations (6.20), (6.21), and (6.24) imply the following result.

Lemma 6.1. *Let $c > 0$. Then*

$$\Delta_f(c) = \sup\{I^f(T_1, T_2, x_f, u_f) : \; T_1 \ge 0, \; T_2 \in (T_1, T_1 + c]\} < \infty.$$

Section 6.19 contains examples of optimal control problems satisfying assumptions (A1) and (A2). Many examples can also be found in [106–108, 118, 124, 125, 134].

6.2 Boundedness Results

The following result is proved in Section 6.7.

Theorem 6.2.

1. *There exists $S > 0$ such that for each pair of numbers $T_2 > T_1 \ge 0$ and each $(x, u) \in X(T_1, T_2)$,*

$$I^f(T_1, T_2, x, u) + S \ge I^f(T_1, T_2, x_f, u_f).$$

2. *For each $(x, u) \in X(0, \infty)$ either*

$$I^f(0, T, x, u) - I^f(0, T, x_f, u_f) \to \infty \text{ as } T \to \infty$$

or

$$\sup\{|I^f(0,T,x,u) - I^f(0,T,x_f,u_f)| : \ T \in (0,\infty)\} < \infty. \tag{6.25}$$

Moreover, if (6.25) holds, then $\sup\{\|x(t)\| : \ t \in (0,\infty)\} < \infty$.

We say that $(x,u) \in X(0,\infty)$ is (f)-good if (6.25) holds. The next boundedness result is proved in Section 6.7.

Theorem 6.3. *Let $M_0 > 0$, $c > 0$, $c_0 \in (0,c)$. Then there exists $M_1 > 0$ such that for each $T_1 \geq 0$, each $T_2 \geq T_1 + c$, and each $(x,u) \in X(T_1,T_2)$ satisfying*

$$I^f(T_1,T_2,x,u) \leq I^f(T_1,T_2,x_f,u_f) + M_0$$

the inequality $\|x(t)\| \leq M_1$ holds for all $t \in [T_1 + c_0, T_2]$.

Let $L > 0$. Denote by $\mathcal{A}_L$ the set of all $(s,z) \in \mathcal{A}$ for which there exist $\tau \in (0,L]$ and $(x,u) \in X(s,s+\tau)$ such that

$$x(s) = z, \ x(s+\tau) = x_f(s+\tau), \ I^f(s,s+\tau,x,u) \leq L.$$

Denote by $\widehat{\mathcal{A}}_L$ the set of all $(s,z) \in \mathcal{A}$ for which $s \geq L$ and there exist $\tau \in (0,L]$ and $(x,u) \in X(s-\tau,s)$ such that

$$x(s-\tau) = x_f(s-\tau), \ x(s) = z, \ I^f(s-\tau,s,x,u) \leq L.$$

Theorems 6.4–6.10 stated below are also boundedness results. They are proved in Section 6.8.

Theorem 6.4. *Let $L > 0$, $M_0 > 0$ and $c \in (0,L)$. Then there exists $M_1 > 0$ such that for each $T_1 \geq 0$, each $T_2 \geq T_1 + 2L$, and each $(x,u) \in X(T_1,T_2)$ satisfying*

$$(T_1, x(T_1)) \in \mathcal{A}_L, \ (T_2, x(T_2)) \in \widehat{\mathcal{A}}_L,$$

$$I^f(T_1,T_2,x,u) \leq U^f(T_1,T_2,x(T_1),x(T_2)) + M_0$$

the inequality $\|x(t)\| \leq M_1$ holds for all $t \in [T_1 + c, T_2]$.

Theorem 6.5. *Let $L > 0$, $M_0 > 0$ and $c \in (0,L)$. Then there exists $M_1 > 0$ such that for each $T_1 \geq 0$, each $T_2 \geq T_1 + L$, and each $(x,u) \in X(T_1,T_2)$ satisfying*

$$(T_1, x(T_1)) \in \mathcal{A}_L, \ I^f(T_1,T_2,x,u) \leq \sigma^f(T_1,T_2,x(T_1)) + M_0$$

the inequality $\|x(t)\| \leq M_1$ holds for all $t \in [T_1 + c, T_2]$.

Theorem 6.6. *Let $L > 0$, $M_0 > 0$ and $c \in (0,L)$. Then there exists $M_1 > 0$ such that for each $T_1 \geq 0$, each $T_2 \geq T_1 + L$, and each $(x,u) \in X(T_1,T_2)$ satisfying*

$$(T_2, x(T_2)) \in \widehat{\mathcal{A}}_L,\ I^f(T_1, T_2, x, u) \le \widehat{\sigma}^f(T_1, T_2, x(T_2)) + M_0$$

the inequality $\|x(t)\| \le M_1$ *holds for all* $t \in [c + T_1, T_2]$.

Theorem 6.7. *Let* $M_0 > 0$ *and* $c > 0$. *Then there exists* $M_1 > 0$ *such that for each* $T_1 \ge 0$, *each* $T_2 \ge T_1 + c$, *and each* $(x, u) \in X(T_1, T_2)$ *satisfying*

$$\|x(T_1)\| \le M_0,\ I^f(T_1, T_2, x, u) \le I^f(T_1, T_2, x_f, u_f) + M_0$$

the inequality $\|x(t)| \le M_1$ *holds for all* $t \in [T_1, T_2]$.

Theorem 6.8. *Let* $L > 0$, $M_0 > 0$. *Then there exists* $M_1 > 0$ *such that for each* $T_1 \ge 0$, *each* $T_2 \ge T_1 + 2L$, *and each* $(x, u) \in X(T_1, T_2)$ *satisfying*

$$\|x(T_1)\| \le M_0,\ (T_1, x(T_1)) \in \mathcal{A}_L,\ \ (T_2, x(T_2)) \in \widehat{\mathcal{A}}_L,$$

$$I^f(T_1, T_2, x, u) \le U^f(T_1, T_2, x(T_1), x(T_2)) + M_0$$

the inequality $\|x(t)\| \le M_1$ *holds for all* $t \in [T_1, T_2]$.

Theorem 6.9. *Let* $L > 0$, $M_0 > 0$. *Then there exists* $M_1 > 0$ *such that for each* $T_1 \ge 0$, *each* $T_2 \ge T_1 + L$, *and each* $(x, u) \in X(T_1, T_2)$ *satisfying*

$$\|x(T_1)\| \le M_0,\ (T_1, x(T_1)) \in \mathcal{A}_L,\ I^f(T_1, T_2, x, u) \le \sigma^f(T_1, T_2, x(T_1)) + M_0$$

the inequality $\|x(t)\| \le M_1$ *holds for all* $t \in [T_1, T_2]$.

Theorem 6.10. *Let* $L > 0$, $M_0 > 0$. *Then there exists* $M_1 > 0$ *such that for each* $T_1 \ge 0$, *each* $T_2 \ge T_1 + L$, *and each* $(x, u) \in X(T_1, T_2)$ *satisfying*

$$\|x(T_1)\| \le M_0,\ (T_2, x(T_2)) \in \widehat{\mathcal{A}}_L,\ I^f(T_1, T_2, x, u) \le \widehat{\sigma}^f(T_1, T_2, x(T_2)) + M_0$$

the inequality $\|x(t)\| \le M_1$ *holds for all* $t \in [T_1, T_2]$.

6.3 Turnpike Results

We say that the integrand f possesses the turnpike property (or TP for short) if for each $\epsilon > 0$ and each $M > 0$, there exist $\delta > 0$ and $L > 0$ such that for each $T_1 \ge 0$, each $T_2 \ge T_1 + 2L$, and each $(x, u) \in X(T_1, T_2)$ which satisfies

$$I^f(T_1, T_2, x, u) \le \min\{\sigma^f(T_1, T_2) + M,\ U^f(T_1, T_2, x(T_1), x(T_2)) + \delta\}$$

there exist $\tau_1, \tau_2 \in [0, L]$ such that

$$\|x(t) - x_f(t)\| \le \epsilon \text{ for all } t \in [T_1 + \tau_1, T_2 - \tau_2].$$

Moreover, if $\|x(T_2) - x_f(T_2)\| \le \delta$ then $\tau_2 = 0$ and if $T_1 \ge L$ and $\|x(T_1) - x_f(T_1)\| \le \delta$, then $\tau_1 = 0$.

We say that the integrand f possesses the strong turnpike property (or STP for short) if for each $\epsilon > 0$ and each $M > 0$ there exist $\delta > 0$ and $L > 0$ such that for each $T_1 \ge 0$, each $T_2 \ge T_1 + 2L$, and each $(x, u) \in X(T_1, T_2)$ which satisfies

$$I^f(T_1, T_2, x, u) \le \min\{\sigma^f(T_1, T_2) + M,\ U^f(T_1, T_2, x(T_1), x(T_2)) + \delta\}$$

there exist $\tau_1, \tau_2 \in [0, L]$ such that

$$\|x(t) - x_f(t)\| \le \epsilon \text{ for all } t \in [T_1 + \tau_1, T_2 - \tau_2].$$

Moreover, if $\|x(T_1) - x_f(T_1)\| \le \delta$ then $\tau_1 = 0$ and if $\|x(T_2) - x_f(T_2)\| \le \delta$, then $\tau_2 = 0$.

Theorem 6.2 implies the following result.

Theorem 6.11. *Assume that f has TP and that $\epsilon, M > 0$. Then there exist $\delta > 0$ and $L > 0$ such that for each $T_1 \ge 0$, each $T_2 \ge T_1 + 2L$, and each $(x, u) \in X(T_1, T_2)$ which satisfies*

$$I^f(T_1, T_2, x, u) \le \min\{I^f(T_1, T_2, x_f, u_f) + M,\ U^f(T_1, T_2, x(T_1), x(T_2)) + \delta\},$$

there exist $\tau_1 \in [T_1, T_1 + L]$, $\tau_2 \in [T_2 - L, T_2]$ such that

$$\|x(t) - x_f(t)\| \le \epsilon \textit{ for all } t \in [\tau_1, \tau_2].$$

Moreover, if $\|x(T_2) - x_f(T_2)\| \le \delta$, then $\tau_2 = T_2$, and if $T_1 \ge L$ and $\|x(T_1) - x_f(T_1)\| \le \delta$, then $\tau_1 = T_1$.

Theorem 6.12. *Assume that f has STP and that $\epsilon, M > 0$. Then there exist $\delta > 0$ and $L > 0$ such that for each $T_1 \ge 0$, each $T_2 \ge T_1 + 2L$, and each $(x, u) \in X(T_1, T_2)$ which satisfies*

$$I^f(T_1, T_2, x, u) \le \min\{I^f(T_1, T_2, x_f, u_f) + M,\ U^f(T_1, T_2, x(T_1), x(T_2)) + \delta\},$$

there exist $\tau_1 \in [T_1, T_1 + L]$, $\tau_2 \in [T_2 - L, T_2]$ such that

$$\|x(t) - x_f(t)\| \le \epsilon \textit{ for all } t \in [\tau_1, \tau_2].$$

Moreover, for $i = 1, 2$, if $\|x(T_i) - x_f(T_i)\| \le \delta$, then $\tau_i = T_i$.

Proposition 6.13. *Let $L > 0$. Then for each $T_1 \ge 0$, each $T_2 \ge T_1 + 2L$, and each pair of points $(T_1, z_1) \in \mathcal{A}_L$, $(T_2, z_2) \in \widehat{\mathcal{A}}_L$,*

$$U^f(T_1, T_2, z_1, z_2) \le I^f(T_1, T_2, x_f, u_f) + 2L(1 + a_0).$$

Proof. Let $T_1 \geq 0$, $T_2 \geq T_1 + 2L$, $(T_1, z_1) \in \mathcal{A}_L$, $(T_2, z_2) \in \widehat{\mathcal{A}}_L$. By the definition of $\mathcal{A}_L, \widehat{\mathcal{A}}_L$, there exist $\tau_1 \in (0, L]$, $\tau_2 \in (0, L]$ and $(y, v) \in X(T_1, T_2)$ such that

$$y(T_1) = z_1,\ y(T_2) = z_2,\ (y(t), v(t)) = (x_f(t), u_f(t)),\ t \in [T_1 + \tau_1, T_2 - \tau_2],$$

$$I^f(T_1, T_1 + \tau_1, y, v) \leq L,\ I^f(T_2 - \tau_2, T_2, y, v) \leq L.$$

In view of the relations above, (6.20) and (6.21),

$$U^f(T_1, T_2, z_1, z_2) \leq I^f(T_1, T_2, y, v)$$

$$\leq 2L + I^f(T_1 + \tau_1, T_2 - \tau_2, x_f, u_f) \leq 2L + 2a_0 L + I^f(T_1, T_2, x_f, u_f).$$

Proposition 6.13 is proved.

Proposition 6.13 and Theorems 6.11 and 6.12 imply the following two results.

Theorem 6.14. *Assume that f has TP and that $\epsilon, L_0 > 0$. Then there exist $\delta > 0$ and $L > L_0$ such that for each $T_1 \geq 0$, each $T_2 \geq T_1 + 2L$, and each $(x, u) \in X(T_1, T_2)$ which satisfies*

$$(T_1, x(T_1)) \in \mathcal{A}_{L_0},\ (T_2, x(T_2)) \in \widehat{\mathcal{A}}_{L_0},$$

$$I^f(T_1, T_2, x, u) \leq U^f(T_1, T_2, x(T_1), x(T_2)) + \delta$$

there exist $\tau_1 \in [T_1, T_1 + L]$, $\tau_2 \in [T_2 - L, T_2]$ such that

$$\|x(t) - x_f(t)\| \leq \epsilon \text{ for all } t \in [\tau_1, \tau_2].$$

Moreover, if $\|x(T_2) - x_f(T_2)\| \leq \delta$ then $\tau_2 = T_2$ and if $T_1 \geq L$ and $\|x(T_1) - x_f(T_1)\| \leq \delta$ then $\tau_1 = T_1$.

Theorem 6.15. *Assume that f has STP and that $\epsilon, L_0 > 0$. Then there exist $\delta > 0$ and $L > L_0$ such that for each $T_1 \geq 0$, each $T_2 \geq T_1 + 2L$, and each $(x, u) \in X(T_1, T_2)$ which satisfies*

$$(T_1, x(T_1)) \in \mathcal{A}_{L_0},\ (T_2, x(T_2)) \in \widehat{\mathcal{A}}_{L_0},$$

$$I^f(T_1, T_2, x, u) \leq U^f(T_1, T_2, x(T_1), x(T_2)) + \delta$$

there exist $\tau_1 \in [T_1, T_1 + L]$, $\tau_2 \in [T_2 - L, T_2]$ such that

$$\|x(t) - x_f(t)\| \leq \epsilon \text{ for all } t \in [\tau_1, \tau_2].$$

Moreover, if $\|x(T_1) - x_f(T_1)\| \leq \delta$, then $\tau_1 = T_1$, and if $\|x(T_2) - x_f(T_2)\| \leq \delta$, then $\tau_2 = T_2$.

Proposition 6.16. *Let $L > 0$. Then for each $(T_1, z) \in \mathcal{A}_L$, each $T_2 \geq T_1 + L$,*

$$\sigma^f(T_1, T_2, z) \leq I^f(T_1, T_2, x_f, u_f) + L(1 + a_0).$$

Proof. Let $(T, z) \in \mathcal{A}_L$, $T_2 \geq T_1 + L$. By the definition of $\mathcal{A}_L$, there exist $\tau \in (0, L]$ and $(y, v) \in X(T_1, T_2)$ such that

$$y(T_1) = z, \ (y(t), v(t)) = (x_f(t), u_f(t)), \ t \in [T_1 + \tau, T_2],$$

$$I^f(T_1, T_1 + \tau, y, v) \leq L.$$

In view of the relations above, (6.20) and (6.21),

$$\sigma^f(T_1, T_2, z) \leq I^f(T_1, T_2, y, v)$$

$$\leq L + I^f(T_1 + \tau, T_2, x_f, u_f) \leq L + a_0 L + I^f(T_1, T_2, x_f, u_f).$$

Proposition 6.16 is proved.

Proposition 6.16 and Theorems 6.11 and 6.12 imply the following two results.

Theorem 6.17. *Assume that f has TP and that $\epsilon, L_0 > 0$. Then there exist $\delta > 0$ and $L > L_0$ such that for each $T_1 \geq 0$, each $T_2 \geq T_1 + 2L$, and each $(x, u) \in X(T_1, T_2)$ which satisfies*

$$(T_1, x(T_1)) \in \mathcal{A}_{L_0},$$

$$I^f(T_1, T_2, x, u) \leq \sigma^f(T_1, T_2, x(T_1)) + \delta$$

there exist $\tau_1 \in [T_1, T_1 + L]$, $\tau_2 \in [T_2 - L, T_2]$ such that

$$\|x(t) - x_f(t)\| \leq \epsilon \text{ for all } t \in [\tau_1, \tau_2].$$

Moreover, if $\|x(T_2) - x_f(T_2)\| \leq \delta$, then $\tau_2 = T_2$, and if $T_1 \geq L$ and $\|x(T_1) - x_f(T_1)\| \leq \delta$, then $\tau_1 = T_1$.

Theorem 6.18. *Assume that f has STP and that $\epsilon, L_0 > 0$. Then there exist $\delta > 0$ and $L > L_0$ such that for each $T_1 \geq 0$, each $T_2 \geq T_1 + 2L$, and each $(x, u) \in X(T_1, T_2)$ which satisfies*

$$(T_1, x(T_1)) \in \mathcal{A}_{L_0},$$

$$I^f(T_1, T_2, x, u) \leq \sigma^f(T_1, T_2, x(T_1)) + \delta,$$

there exist $\tau_1 \in [T_1, T_1 + L]$, $\tau_2 \in [T_2 - L, T_2]$ such that

$$\|x(t) - x_f(t)\| \le \epsilon \text{ for all } t \in [\tau_1, \tau_2].$$

Moreover, if $\|x(T_1) - x_f(T_1)\| \le \delta$, *then* $\tau_1 = T_1$, *and if* $\|x(T_2) - x_f(T_2)\| \le \delta$, *then* $\tau_2 = T_2$.

Proposition 6.19. *Let* $L > 0$. *Then for each* $T_1 \ge 0$ *and each* $(T_2, z) \in \widehat{\mathcal{A}_L}$ *satisfying* $T_2 \ge T_1 + L$,

$$\widehat{\sigma}^f(T_1, T_2, z) \le I^f(T_1, T_2, x_f, u_f) + L(1 + a_0).$$

Proof. Let $T_1 \ge 0$, $(T_2, z) \in \widehat{\mathcal{A}_L}$, $T_2 \ge T_1 + L$. Then there exist $\tau \in (0, L]$ and $(y, v) \in X(T_1, T_2)$ such that

$$y(T_2) = z, \ (y(t), v(t)) = (x_f(t), u_f(t)), \ t \in [T_1, T_2 - \tau],$$

$$I^f(T_2 - \tau, T_2, y, v) \le L.$$

In view of the relations above, (6.21) and (6.26),

$$\widehat{\sigma}^f(T_1, T_2, z) \le I^f(T_1, T_2, y, v)$$

$$\le L + I^f(T_1, T_2 - \tau, x_f, u_f) \le L + a_0 L + I^f(T_1, T_2, x_f, u_f).$$

Proposition 6.19 is proved.

Proposition 6.19 and Theorems 6.11 and 6.12 imply the following two results.

Theorem 6.20. *Assume that* f *has TP and that* $\epsilon, L_0 > 0$. *Then there exist* $\delta > 0$ *and* $L > L_0$ *such that for each* $T_1 \ge 0$, *each* $T_2 \ge T_1 + 2L$, *and each* $(x, u) \in X(T_1, T_2)$ *which satisfies*

$$(T_2, x(T_2)) \in \widehat{\mathcal{A}_{L_0}},$$

$$I^f(T_1, T_2, x, u) \le \widehat{\sigma}^f(T_1, T_2, x(T_2)) + \delta,$$

there exist $\tau_1 \in [T_1, T_1 + L]$, $\tau_2 \in [T_2 - L, T_2]$ *such that*

$$\|x(t) - x_f(t)\| \le \epsilon \text{ for all } t \in [\tau_1, \tau_2].$$

Moreover, if $\|x(T_2) - x_f(T_2)\| \le \delta$, *then* $\tau_2 = T_2$, *and if* $T_1 \ge L$ *and* $\|x(T_1) - x_f(T_1)\| \le \delta$, *then* $\tau_1 = T_1$.

Theorem 6.21. *Assume that* f *has STP and that* $\epsilon, L_0 > 0$. *Then there exist* $\delta > 0$ *and* $L > L_0$ *such that for each* $T_1 \ge 0$, *each* $T_2 \ge T_1 + 2L$, *and each* $(x, u) \in X(T_1, T_2)$ *which satisfies*

$$(T_2, x(T_2)) \in \widehat{\mathcal{A}}_{L_0},$$

$$I^f(T_1, T_2, x, u) \le \widehat{\sigma}^f(T_1, T_2, x(T_2)) + \delta,$$

there exist $\tau_1 \in [T_1, T_1 + L]$, $\tau_2 \in [T_2 - L, T_2]$ *such that*

$$\|x(t) - x_f(t)\| \le \epsilon \text{ for all } t \in [\tau_1, \tau_2].$$

Moreover, if $\|x(T_1) - x_f(T_1)\| \le \delta$, *then* $\tau_1 = T_1$, *and if* $\|x(T_2) - x_f(T_2)\| \le \delta$, *then* $\tau_2 = T_2$.

The next theorem is the main result of this chapter. It is proved in Section 6.9.

Theorem 6.22. *f has TP if and only if the following properties hold:*

(P1) for each (f)-good pair $(x, u) \in X(0, \infty)$,

$$\lim_{t\to\infty} \|x(t) - x_f(t)\| = 0;$$

(P2) for each $\epsilon > 0$ and each $M > 0$, there exist $\delta > 0$ and $L > 0$ such that for each $T \ge 0$ and each $(x, u) \in X(T, T + L)$ which satisfies

$$I^f(T, T + L, x, u) \le \min\{U^f(T, T + L, x(T), x(T + L)) + \delta,$$

$$I^f(T, T + L, x_f, u_f) + M\}$$

there exists $s \in [T, T + L]$ such that $\|x(s) - x_f(s)\| \le \epsilon$.

The next result is proved in Section 6.11.

Theorem 6.23. *Assume that f has properties (P1) and (P2). Let $\epsilon, M > 0$. Then there exist a natural number Q and $l > 0$ such that for each $T_1 \ge 0$, each $T_2 \ge T_1 + lQ$, and each $(x, u) \in X(T_1, T_2)$ which satisfies*

$$I^f(T_1, T_2, x, u) \le I^f(T_1, T_2, x_f, u_f) + M,$$

there exist finite sequences $\{a_i\}_{i=1}^q$, $\{b_i\}_{i=1}^q \subset [T_1, T_2]$ such that an integer $q \le Q$,

$$0 \le b_i - a_i \le l, \ i = 1, \dots, q,$$

$$b_i \le a_{i+1} \text{ for all integers } i \text{ satisfying } 1 \le i < q,$$

$$\|x(t) - x_f(t)\| \le \epsilon \text{ for all } t \in [T_1, T_2] \setminus \cup_{i=1}^q [a_i, b_i].$$

Theorem 6.23 leads to the following definition.

We say that f possesses the weak turnpike property (or WTP for short) if for each $\epsilon > 0$ and each $M > 0$ there exist a natural number Q and $l > 0$ such that for each $T_1 \geq 0$, each $T_2 \geq T_1 + lQ$, and each $(x, u) \in X(T_1, T_2)$ which satisfies

$$I^f(T_1, T_2, x, u) \leq I^f(T_1, T_2, x_f, u_f) + M,$$

there exist finite sequences $\{a_i\}_{i=1}^q$, $\{b_i\}_{i=1}^q \subset [T_1, T_2]$ such that an integer $q \leq Q$,

$$0 \leq b_i - a_i \leq l, \ i = 1, \dots, q,$$

$$b_i \leq a_{i+1} \text{ for all integers } i \text{ satisfying } 1 \leq i < q,$$

$$\|x(t) - x_f(t)\| \leq \epsilon \text{ for all } t \in [T_1, T_2] \setminus \cup_{i=1}^q [a_i, b_i].$$

Theorem 6.24. *f has WTP if and only if f has (P1) and (P2).*

In view of Theorem 6.23, (P1) and (P2) imply WTP. Clearly, WTP implies (P2). Therefore Theorem 6.24 follows from the next result which can be easily proved.

Proposition 6.25. *Let f have WTP. Then f has (P1).*

Theorem 6.26. *For the second class of problems, we assume that the following assumption holds:*

(A3) for each $\epsilon > 0$ there exist $\delta, L > 0$ such that for each $S_1 \geq L$ and each $S_2 \in [S_1, S_1 + \delta]$ the inequality $I^f(S_1, S_2, x_f, u_f) \leq \epsilon$.

The integrand f has (P1) and (P2) if and only if the following property holds:

(P3) for each $\epsilon > 0$ and each $M > 0$ there exists $L > 0$ such that for each $T_1 \geq 0$, each $T_2 \geq T_1 + L$, and each $(x, u) \in X(T_1, T_2)$ which satisfies

$$I^f(T_1, T_2, x, u) \leq I^f(T_1, T_2, x_f, u_f) + M$$

the inequality

$$mes(\{t \in [T_1, T_2] : \ \|x(t) - x_f(t)\| > \epsilon\}) \leq L$$

is valid.

In view of Theorem 6.23, (P1) and (P2) imply (P3). Clearly, (P3) implies (P2). Therefore Theorem 6.26 follows from the next result which is proved in Section 6.13.

Proposition 6.27. *For the second class of problems, we assume that assumption (A3) holds. Let f have (P3). Then f has (P1).*

6.4 Lower Semicontinuity Property

We say that f possesses lower semicontinuity property (or LSC property for short) if for each $T_2 > T_1 \geq 0$ and each sequence $(x_j, u_j) \in X(T_1, T_2)$, $j = 1, 2, \ldots$ which satisfies

$$\sup\{I^f(T_1, T_2, x_j, u_j) : \ j = 1, 2, \ldots\} < \infty,$$

there exist a subsequence $\{(x_{j_k}, u_{j_k})\}_{k=1}^{\infty}$ and $(x, u) \in X(T_1, T_2)$ such that

$$x_{j_k}(t) \to x(t) \text{ as } k \to \infty \text{ for every } t \in [T_1, T_2],$$

$$I^f(T_1, T_2, x, u) \leq \liminf_{j\to\infty} I^f(T_1, T_2, x_j, u_j).$$

LSC property plays an important role in the calculus of variations and optimal control theory [32].

Let $S \geq 0$. A pair $(x, u) \in X(S, \infty)$ is called (f)-overtaking optimal if for every $(y, v) \in X(S, \infty)$ satisfying $x(S) = y(S)$,

$$\limsup_{T\to\infty} [I^f(S, T, x, u) - I^f(S, T, y, v)] \leq 0.$$

A pair $(x, u) \in X(S, \infty)$ is called f-weakly optimal if for every $(y, v) \in X(S, \infty)$ satisfying $x(S) = y(S)$,

$$\liminf_{T\to\infty} [I^f(S, T, x, u) - I^f(S, T, y, v)] \leq 0.$$

A pair $(x, u) \in X(S, \infty)$ is called (f)-minimal if for every $T > S$

$$I^f(S, T, x, u) = U^f(S, T, x(S), x(T)).$$

The next result is proved in Section 6.15.

Theorem 6.28. *Assume that f has (P1), (P2), and (LSC) property, $S \geq 0$, and that $(x, u) \in X(S, \infty)$ is (f)-good. Then there exists an (f)-overtaking optimal pair $(x_*, u_*) \in X(S, \infty)$ such that $x_*(S) = x(S)$.*

The following theorem is also proved in Section 6.15.

Theorem 6.29. *Assume that f has (P1), (P2), and LSC property, $S \geq 0$, $(\tilde{x}, \tilde{u}) \in X(S, \infty)$ is (f)-good, and that $(x_*, u_*) \in X(S, \infty)$ satisfies $x_*(S) = \tilde{x}(S)$. Then the following conditions are equivalent:*

(i) (x_*, u_*) *is (f)-overtaking optimal;*
(ii) (x_*, u_*) *is (f)-weakly optimal;*
(iii) (x_*, u_*) *is (f)-minimal and (f)-good;*

(iv) (x_*, u_*) *is* (f)*-minimal and satisfies* $\lim_{t\to\infty}(x_*(t) - x_f(t)) = 0$;
(v) (x_*, u_*) *is* (f)*-minimal and satisfies* $\liminf_{t\to\infty} \|x_*(t) - x_f(t)\| = 0$.

The next result easily follows from Theorems 6.14 and 6.23.

Theorem 6.30. *Assume that* f *has (P1), (P2), and LSC property and* $\epsilon, L > 0$. *Then there exists* $\tau_0 > 0$ *such that for every* $T_0 \geq 0$ *and every* (f)*-overtaking optimal pair* $(x, u) \in X(T_0, \infty)$ *which satisfies* $(T_0, x(T_0)) \in \mathcal{A}_L$ *the inequality* $\|x(t) - x_f(t)\| \leq \epsilon$ *holds for all* $t \geq T_0 + \tau_0$.

In the sequel we use the following result which can be easily proved.

Proposition 6.31. *Let* f *have (P1) and* $(T_0, z_0) \in \mathcal{A}$. *There exists an* (f)*-good pair* $(x, u) \in X(T_0, \infty)$ *satisfying* $x(T_0) = z_0$ *if and only if* $(T_0, z_0) \in \cup\{\mathcal{A}_L :\ L > 0\}$.

LSC property implies the following result.

Proposition 6.32. *Let* f *have LSC property,* $T_2 > T_2 \geq 0$, $L > 0$. *Then the following assertions hold.*

1. *There exists* $(x, u) \in X(T_1, T_2)$ *such that*

$$I^f(T_1, T_2, x, u) \leq I^f(T_1, T_2, y, v) \text{ for all } (y, v) \in X(T_1, T_2).$$

2. *If* $T_2 - T_1 \geq 2L$, $(T_1, z_1) \in \mathcal{A}_L$, $(T_2, z_2) \in \widehat{\mathcal{A}}_L$, *then there exists* $(x, u) \in X(T_1, T_2)$ *such that* $x(T_i) = z_i$, $i = 1, 2$ *and*

$$I^f(T_1, T_2, x, u) = U^f(T_1, T_2, z_1, z_2).$$

3. *If* $T_2 - T_1 \geq L$, $(T_1, z) \in \mathcal{A}_L$, *then there exists* $(x, u) \in X(T_1, T_2)$ *such that* $x(T_1) = z$ *and*

$$I^f(T_1, T_2, x, u) = \sigma^f(T_1, T_2, z).$$

4. *If* $T_2 - T_1 \geq L$, $(T_2, z) \in \widehat{\mathcal{A}}_L$, *then there exists* $(x, u) \in X(T_1, T_2)$ *such that* $x(T_2) = z$ *and*

$$I^f(T_1, T_2, x, u) = \widehat{\sigma}^f(T_1, T_2, z).$$

The next result is proved in Section 6.17.

Theorem 6.33. *For the first class of problems, assume that* $A(t) = 0$ *for all* $t \geq 0$, *and for the second class of problems, assume that* $A = 0$. *Let* f *have LSC property and* (x_f, u_f) *be* (f)*-overtaking optimal. Then* f *has STP if and only if (P1) and (P2) hold and the following property holds:*

(P3) for each (f)*-overtaking optimal pair of sequences* $(y, v) \in X(0, \infty)$ *satisfying* $y(0) = x_f(0)$ *the equality* $y(t) = x_f(t)$ *holds for all* $t \geq 0$.

The next result follows from (A2), (P1), and STP.

Theorem 6.34. *Assume that f has STP and $\epsilon > 0$. Then there exists $\delta > 0$ such that for every $T_1 \geq 0$ and every (f)-overtaking optimal pair $(x, u) \in X(T_1, \infty)$ satisfying $\|x(T_1) - x_f(T_1)\| \leq \delta$, the inequality $\|x(t) - x_f(t)\| \leq \epsilon$ holds for all $t \geq T_1$.*

6.5 Perturbed Problems

In this section we use the following assumption.

(A4) For each $\epsilon > 0$ there exists $\delta > 0$ such that for each $(T_i, z_i) \in \mathcal{A}$, $i = 1, 2$ satisfying $\|z_i - x_f(T_i)\| \leq \delta$, $i = 1, 2$ and $T_2 \geq b_f$, there exist $\tau_1, \tau_2 \in (0, b_f]$ and $(x_1, u_1) \in X(T_1, T_1 + \tau_1)$, $(x_2, u_2) \in X(T_2 - \tau_2, T_2)$ which satisfy

$$x_1(T_1) = z_1, \; x_1(T_1 + \tau_1) = x_f(T_1 + \tau_1),$$

$$I^f(T_1, T_1 + \tau_1, x_1, u_1) \leq I^f(T_1, T_1 + \tau_1, x_f, u_f) + \epsilon,$$

$$\|x_1(t) - x_f(t)\| \leq \epsilon, \; t \in [T_1, T_1 + \tau_1],$$

$$x_2(T_2) = z_2, \; x_2(T_2 - \tau_2) = x_f(T_2 - \tau_2),$$

$$I^f(T_2 - \tau_2, T_2, x_2, u_2) \leq I^f(T_2 - \tau_2, T_2, x_f, u_f) + \epsilon,$$

$$\|x_2(t) - x_f(t)\| \leq \epsilon, \; t \in [T_2 - \tau_2, T_2].$$

Clearly, (A4) implies (A2). Assume that $\phi : E \to [0, 1]$ is a continuous function satisfying $\phi(0) = 0$ and such that the following property holds:

(i) for each $\epsilon > 0$ there exists $\delta > 0$ such that for each $x \in E$ satisfying $\phi(x) \leq \delta$ we have $\|x\| \leq \epsilon$.

For each $r \in (0, 1)$ set

$$f_r(t, x, u) = f(t, x, u) + r\phi(x - x_f(t)), \; (t, x, u) \in \mathcal{M}.$$

Clearly, for any $r \in (0, 1)$, f_r is a Borealian function, if (A4) holds, then (A1) and (A4) hold for f_r with $(x_{f_r}, u_{f_r}) = (x_f, u_f)$.

Theorem 6.35. *Let (A4) hold and $r \in (0, 1)$. Then f_r has (P1) and (P2). Moreover, if (x_f, u_f) is (f)-minimal, then f_r has (P4) and (x_f, u_f) is (f_r)-overtaking optimal.*

Theorem 6.35 is proved in Section 6.18.

6.6 Auxiliary Results for Theorems 6.2 and 6.3

Proposition 6.36. *Let $\gamma > 0$. Then there exists $\delta > 0$ such that for each $(T, z_1), (T + 2b_f, z_2) \in \mathcal{A}$ satisfying*

$$\|z_1 - x_f(T)\| \le \delta, \ \|z_2 - x_f(T + 2b_f)\| \le \delta, \tag{6.26}$$

there exists $(x, u) \in X(T, T + 2b_f)$ which satisfies

$$x(T) = z_1, \ x(T + 2b_f) = z_2,$$

$$I^f(T, T + 2b_f, x, u) \le I^f(T, T + 2b_f, x_f, u_f) + \gamma.$$

Proof. Set $\epsilon = \gamma/2$ and let $\delta > 0$ be as guaranteed by (A2). Let $(T, z_1), (T + 2b_f, z_2) \in \mathcal{A}$ satisfy (6.26). By (A2), there exist $\tau_1, \tau_2 \in (0, b_f]$ and $(x_1, u_1) \in X(T, T + \tau_1)$, $(x_2, u_2) \in X(T + 2b_f - \tau_2, T + 2b_f)$ which satisfy

$$x_1(T) = z_1, \ x_1(T + \tau_1) = x_f(T + \tau_1),$$

$$I^f(T, T + \tau_1, x_1, u_1) \le I^f(T, T + \tau_1, x_f, u_f) + \gamma/2,$$

$$x_2(T + 2b_f) = z_2, \ x_2(T + 2b_f - \tau_2) = x_f(T + 2b_f - \tau_2),$$

$$I^f(T + 2b_f - \tau_2, T + 2b_f, x_2, u_2) \le I^f(T + 2b_f - \tau_2, T + 2b_f, x_f, u_f) + \gamma/2.$$

Define

$$x(t) = x_1(t), \ u(t) = u_1(t), \ t \in [T, T + \tau_1],$$

$$x(t) = x_2(t), \ u(t) = u_2(t), \ t \in (T + 2b_f - \tau_2, T + 2b_f],$$

$$x(t) = x_f(t), \ u(t) = u_f(t), \ t \in (T + \tau_1, T + 2b_f - \tau_2].$$

It is clear that $(x, u) \in X(T, T + 2b_f)$, $x(T) = z_1$, $x(T + 2b_f) = z_2$,

$$I^f(T, T + 2b_f, x, u) \le I^f(T, T + 2b_f, x_f, u_f) + \gamma.$$

Proposition 6.36 is proved.

Lemma 6.37. *There exist numbers $S > 0$, $c_0 \ge 1$ such that for each $T_1 \ge 0$, each $T_2 \ge T_1 + c_0$, and each $(x, u) \in X(T_1, T_2)$,*

$$I^f(T_1, T_2, x_f, u_f) \le I^f(T_1, T_2, x, u) + S. \tag{6.27}$$

Proof. In view of (6.19)–(6.21) and (6.24), there exists $S_1 > 0$ such that

$$\psi(S_1) > a_0 + 1 + \sup\{|I^f(j, j+1, x_f, u_f)| : j = 0, 1, \dots\}. \tag{6.28}$$

By (A1), there exist $S_2 > 0$, $c_0 > 1$ such that

$$I^f(T_1, T_2, x_f, u_f) \le I^f(T_1, T_2, x, u) + S_2 \tag{6.29}$$

for each $T_1 \ge 0$, each $T_2 \ge T_1 + c_0$, and each $(x, u) \in X(T_1, T_2)$ satisfying

$$\|x(T_1)\|, \|x(T_2)\| \le S_1.$$

Fix a number

$$S \ge S_2 + 2 + 2a_0(4 + 2c_0) + 4(c_0 + 1)\sup\{|I^f(j, j+1, x_f, u_f)| : j = 0, 1, \dots\}. \tag{6.30}$$

Assume that

$$T_1 \ge 0,\ T_2 \ge T_1 + c_0,\ (x, u) \in X(T_1, T_2).$$

We show that (6.27) is true. Assume that

$$\|x(t)\| \ge S_1,\ t \in [T_1, T_2]. \tag{6.31}$$

By (6.20), (6.21), (6.24), (6.28), (6.30), and (6.31), for all $t \in [T_1, T_2]$,

$$f(t, x(t), u(t)) \ge -a_0 + \psi(\|x(t)\|) \ge -a_0 + \psi(S_1)$$

and

$$I^f(T_1, T_2, x, u) \ge (T_2 - T_1)(\psi(S_1) - a_0)$$

$$\ge (T_2 - T_1)(\sup\{|I^f(j, j+1, x_f, u_f)| : j = 0, 1, \dots\} + 1)$$

$$\ge I^f(T_1, T_2, x_f, u_f) - 2\Delta_f - 2a_0 \ge I^f(T_1, T_2, x_f, u_f) - S$$

and (6.27) holds.

Assume that

$$\inf\{\|x(t)\| : t \in [T_1, T_2]\} < S_1. \tag{6.32}$$

Set

$$\tau_1 = \inf\{t \in [T_1, T_2] : \|x(t)\| \le S_1\}, \tag{6.33}$$

$$\tau_2 = \sup\{t \in [T_1, T_2] : \|x(t)\| \leq S_1\}. \tag{6.34}$$

Clearly, τ_1, τ_2 are well-defined, $\tau_1 \leq \tau_2$ and

$$\|x(\tau_i)\| \leq S_1, \ i = 1, 2. \tag{6.35}$$

There are two cases:

$$\tau_2 - \tau_1 \geq c_0; \tag{6.36}$$

$$\tau_2 - \tau_1 < c_0. \tag{6.37}$$

Assume that (6.36) holds. It follows from (6.29), (6.35), (6.36), and the choice of S_2 and c_0 that

$$I^f(\tau_1, \tau_2, x_f, u_f) \leq I^f(\tau_1, \tau_2, x, u) + S_2. \tag{6.38}$$

By (6.20), (6.21), (6.33), and (6.34), for each $t \in [T_1, \tau_1] \cup [\tau_2, T_2]$,

$$\|x(t)\| \geq S_1, \ f(t, x(t), u(t)) \geq -a_0 + \psi(\|x(t)\|) \geq -a_0 + \psi(S_1). \tag{6.39}$$

It follows from (6.24), (6.28), and (6.39) that

$$I^f(T_1, \tau_1, x, u) \geq (\tau_1 - T_1)(\psi(S_1) - a_0)$$

$$\geq (\tau_1 - T_1)(\Delta_f + 1) \geq I^f(T_1, \tau_1, x_f, u_f) - 2\Delta_f - 2a_0, \tag{6.40}$$

$$I^f(\tau_2, T_2, x, u) \geq (T_2 - \tau_2)(\psi(S_1) - a_0)$$

$$\geq (T_2 - \tau_2)(\Delta_f + 1) \geq I^f(\tau_2, T_2, x_f, u_f) - 2\Delta_f - 2a_0. \tag{6.41}$$

In view of (6.30), (6.38), (6.40), and (6.41),

$$I^f(T_1, T_2, x, u) = I^f(T_1, \tau_1, x, u) + I^f(\tau_1, \tau_2, x, u) + I^f(\tau_2, T_2, x, u)$$

$$\geq I^f(T_1, \tau_1, x_f, u_f) - 2\Delta_f - 2a_0 + I^f(\tau_1, \tau_2, x_f, u_f)$$

$$-S_2 + I^f(\tau_2, T_2, x_f, u_f) - 2\Delta_f - 2a_0$$

$$\geq I^f(T_1, T_2, x_f, u_f) - S.$$

Assume that (6.37) holds. By (6.20), (6.21), (6.33), and (6.34), for each $t \in [T_1, \tau_1] \cup [\tau_2, T_2]$, (6.39) is true. In view of (6.24), (6.28), and (6.39), (6.40) and (6.41) are true.

It follows from (6.20), (6.21), (6.30), (6.38), (6.40), and (6.41) that

$$I^f(T_1,T_2,x,u) = I^f(T_1,\tau_1,x,u) + I^f(\tau_1,\tau_2,x,u) + I^f(\tau_2,T_2,x,u)$$

$$\geq I^f(T_1,\tau_1,x_f,u_f) - 2\Delta_f - 2a_0 - a_0(\tau_2-\tau_1) + I^f(\tau_2,T_2,x_f,u_f) - 2\Delta_f - 2a_0$$

$$\geq I^f(T_1,T_2,x_f,u_f) - 4\Delta_f - 4a_0 - a_0c_0 - \Delta_f c_0 - 2\Delta_f - 2a_0$$

$$\geq I^f(T_1,T_2,x_f,u_f) - S.$$

Lemma 6.37 is proved.

Lemma 6.38. *Let $M_0, M_1, \tau_0 > 0$. Then there exists $M_2 > M_1$ such that for each $T_1 \geq 0$, each $T_2 \in (T_1, T_1+\tau_0]$, and each $(x,u) \in X(T_1,T_2)$ satisfying*

$$\|x(T_1)\| \leq M_1, \tag{6.42}$$

$$I^f(T_1,T_2,x,u) \leq M_0 \tag{6.43}$$

the following inequality holds:

$$\|x(t)\| \leq M_2 \text{ for all } t \in [T_1,T_2]. \tag{6.44}$$

Proof. Let us consider the first class of problems. Fix a positive number

$$\delta < \min\{16^{-1}\tau_0, 8^{-1}(a_0+1)^{-1}, 8^{-1}(a_0+1)^{-1}\Delta_\delta^{-1}\}. \tag{6.45}$$

By (6.19) and (6.20), there exist $h_0 > M_1+1$ and $\gamma_0 > 0$ such that

$$f(t,x,u) \geq 4(M_0+a_0\tau_0)\delta^{-1} \text{ for each } (t,x,u) \in \mathcal{M} \text{ satisfying } \|x\| \geq h_0, \tag{6.46}$$

$$f(t,x,u) \geq 8(\|G(t,x,u)\| - a_0\|x\|)_+ \text{ for each } (t,x,u) \in \mathcal{M} \text{ satisfying}$$

$$\|G(t,x,u)\| - a_0\|x\| \geq \gamma_0. \tag{6.47}$$

Choose a number

$$M_2 > (8M_1 + 8 + 2h_0 + 2\gamma_0\delta + M_0 + a_0\tau_0)(\Delta_\delta + 1). \tag{6.48}$$

Let

$$T_1 \geq 0,\ T_2 \in (T_1, T_1+\tau_0],\ (x,u) \in X(T_1,T_2),$$

(6.42) and (6.43) hold. We show that (6.44) hold. Assume the contrary. Then there exists $t_0 \in [T_1,T_2]$ such that

$$\|x(t_0)\| > M_2. \tag{6.49}$$

By the choice of h_0 (see (6.46)), there exists $t_1 \in [T_1, t_0]$ such that

$$\|x(t_1)\| \le h_0, \ |t_1 - t_0| \le \delta. \tag{6.50}$$

There exists $t_2 \in [t_1, t_0]$ such that

$$\|x(t_2)\| \ge \|x(t)\| \text{ for all } t \in [t_1, t_0]. \tag{6.51}$$

In view of (6.48)–(6.51),

$$t_2 > t_1. \tag{6.52}$$

By (6.11), (6.51), and (6.52),

$$\|x(t_2) - U(t_2, t_1)x(t_1)\| = \|\int_{t_1}^{t_2} U(t_2, s)G(x(s), u(s))ds\|. \tag{6.53}$$

It follows from property (i) from Section 6.1, (6.50), and (6.53) that

$$\|x(t_2)\| - \Delta_\delta h_0 \le \|x(t_2)\| - \Delta_\delta \|x(t_1)\| \le \Delta_\delta \int_{t_1}^{t_2} \|G(s, x(s), u(s))\| ds. \tag{6.54}$$

Define

$$E_1 = \{t \in [t_1, t_2] : \ \|G(t, x(t), u(t))\| \ge a_0\|x(t)\| + \gamma_0\}, \tag{6.55}$$

$$E_2 = [t_1, t_2] \setminus E_1. \tag{6.56}$$

By (6.43), (6.47), (6.50), (6.51) and (6.54)–(6.56),

$$\Delta_\delta^{-1}\|x(t_2)\| \le h_0 + \int_{t_1}^{t_2} \|G(s, x(s), u(s))\| ds$$

$$\le h_0 + a_0 \int_{t_1}^{t_2} \|x(t)|dt + \int_{t_1}^{t_2} (\|G(s, x(s), u(s))\| - a_0\|x(s)\|)_+ ds$$

$$\le h_0 + a_0\delta\|x(t_2)\| + \int_{E_1} (\|G(t, x(t), u(t))\| - a_0\|x(t)\|)_+ dt$$

$$+ \int_{E_2} (\|G(t, x(t), u(t))\| - a_0\|x(t)\|)_+ dt$$

$$\le h_0 + a_0\|x(t_2)\|\delta + \gamma_0\delta + \int_{E_1} (\|G(t, x(t), u(t))\| - a_0\|x(t)\|)_+ dt$$

$$\le h_0 + a_0\|x(t_2)\|\delta + \gamma_0\delta + 8^{-1} \int_{E_1} f(t, x(t), u(t))dt$$

$$\le h_0 + a_0\|x(t_2)\|\delta + \gamma_0\delta + 8^{-1}(M_0 + a_0\tau_0). \tag{6.57}$$

It follows from (6.45), (6.49), (6.51), and (6.57) that

$$2^{-1}M_2\Delta_\delta^{-1} \le 2^{-1}\Delta_\delta^{-1}\|x(t_2)\| \le h_0 + \gamma_0\delta + 8^{-1}(M_0 + a_0\tau_0)$$

and

$$M_2 \le (2h_0 + 2\gamma_0\delta + 4^{-1}(M_0 + a_0\tau_0))\Delta_\delta.$$

This contradicts (6.48). The contradiction we have reached proves Lemma 6.38 for the first class of problems.

Consider the second class of problems. Recall (see Proposition 1.3) that there exists $M_* \ge 1$ and $\omega_* \in R^1$ such that

$$\|e^{At}\| \le M_* e^{\omega_* t},\ t \in [0, \infty). \tag{6.58}$$

and that for every $\tau \ge 0$,

$$\Phi_\tau u = \int_0^\tau e^{A(\tau - s)}Bu(s)ds,\ u \in L^2(0, \tau; F), \tag{6.59}$$

where $\Phi_\tau \in \mathcal{L}(L^2(0, \tau; F), E)$ (see Proposition 1.18). Choose a number

$$\delta \in (0, 1).$$

By (6.19) and (6.21), there exists

$$h_0 > M_1 + 1 \tag{6.60}$$

and $\gamma > 1$ such that

$$f(t, x, u) \ge 4(M_0 + a_0\tau_0)\delta^{-1} \text{ for each } (t, x, u) \in \mathcal{M} \text{ satisfying } \|x\| \ge h_0, \tag{6.61}$$

$$f(t, x, u) \ge K_0\|u\|^2/2 \text{ for each } u \in F \text{ satisfying } \|u\| \ge \gamma. \tag{6.62}$$

Choose

$$M_2 > M_1 + M_* e^{|\omega_*|\delta} h_0 + \|\Phi_1\|\gamma\delta^{1/2} + \|\Phi_1\|(2K_0^{-1}(M_0 + a_0\tau_0))^{1/2}. \tag{6.63}$$

Let $T_1 \geq 0,\ T_2 \in (T_1, T_1 + \tau_0]$, $(x, u) \in X(T_1, T_2)$, (6.42) and (6.43) hold. We show that (6.44) hold. Assume the contrary. Then there exists $t_0 \in [T_1, T_2]$ such that (6.49) holds. By the choice of h_0, (6.60), (6.61), there exists

$$t_1 \in [T_1, t_0]$$

satisfying (6.50). There exists $t_2 \in [t_1, t_0]$ such that

$$\|x(t_2)\| \geq \|x(t)\| \text{ for all } t \in [t_1, t_0]. \tag{6.64}$$

Clearly, $(x, u) \in X(t_1, t_2)$,

$$x(t_2) = e^{A(t_2-t_1)}x(t_1) + \int_{t_1}^{t_2} e^{A(t_2-s)}Bu(s)ds \tag{6.65}$$

in $E_{-1} = \mathcal{D}(A^*)$. Proposition 1.18, (6.50), (6.58), (6.59), and (6.65) imply that

$$\|x(t_2)\| \leq \|x(t_1)\|M_* e^{|\omega_*|\delta} + \|\int_{t_1}^{t_2} e^{A(t_2-s)}Bu(s)ds\|$$

$$\leq M_* e^{|\omega_*|\delta} h_0 + \|\int_0^{t_2-t_1} e^{A(t_2-t_1-s)}Bu(t_1+s)ds\|$$

$$\leq M_* e^{|\omega_*|\delta} h_0 + \|\Phi_{t_2-t_1}\|\|u(t_1+\cdot)\|_{L^2(0,t_2-t_1;F)}$$

$$\leq M_* e^{|\omega_*|\delta} h_0 + \|\Phi_1\|(\int_{t_1}^{t_2} \|u(s)\|^2 ds)^{1/2}. \tag{6.66}$$

Define

$$\Omega_1 = \{t \in [t_1, t_2] :\ \|u(t)\| \geq \gamma\},\ \ \Omega_2 = [t_1, t_2] \setminus \Omega_1. \tag{6.67}$$

By (6.21), (6.43), (6.49), (6.50), (6.62)–(6.64), (6.60), and (6.67),

$$M_2 < \|x(t_0)\| \leq \|x(t_2)\| \leq M_* e^{|\omega_*|\delta} h_0$$

$$+\|\Phi_1\|[(\int_{\Omega_1} \|u(s)\|^2 ds)^{1/2} + (\int_{\Omega_2} \|u(s)\|^2 ds)^{1/2}]$$

$$\leq M_* e^{|\omega_*|\delta} h_0 + \|\Phi_1\|(2K_0^{-1} \int_{\Omega_1} f(s, x(s), u(s))ds)^{1/2} + \|\Phi_1\|\gamma\delta^{1/2}$$

$$\leq M_* e^{|\omega_*|\delta} h_0 + \|\Phi_1\|\gamma\delta^{1/2} + \|\Phi_1\|(2K_0^{-1}(M_0 + a_0\tau_0))^{1/2} < M_2.$$

The contradiction we have reached proves Lemma 6.38 for the second class of problems.

Lemma 6.38, (6.20), and (6.21) imply the following result.

Lemma 6.39. *Let* $M_1 > 0, 0 < \tau < \tau_0 < \tau_1$. *Then there exists* $M_2 > 0$ *such that for each* $T_1 \geq 0$, *each* $T_2 \in [T_1 + \tau_0, T_1 + \tau_1]$, *and each* $(x, u) \in X(T_1, T_2)$ *satisfying*

$$I^f(T_1, T_2, x, u) \leq M_1$$

the following inequality holds:

$$\|x(t)\| \leq M_2 \text{ for all } t \in [T_1 + \tau, T_2].$$

6.7 Proofs of Theorems 6.2 and 6.3

Proof of Theorem 6.2. Assertion 1 follows from Lemma 6.37, (6.20), (6.21), and (6.24).

We will prove Assertion 2. Let $(x, u) \in X(0, \infty)$. Assume that there exists a sequence of numbers $\{t_k\}_{k=1}^{\infty}$ such that

$$t_k \to \infty \text{ as } k \to \infty,$$

$$I^f(0, t_k, x, u) - I^f(0, t_k, x_f, u_f) \to \infty \text{ as } k \to \infty. \tag{6.68}$$

Let a number $S > 0$ be as guaranteed by Assertion 1. Let k be a natural number and $t \geq t_k$. In view of (6.68) and the choice of S,

$$I^f(0, t, x, u) - I^f(0, t, x_f, u_f)$$

$$= I^f(0, t_k, x, u) - I^f(0, t_k, x_f, u_f) + I^f(t_k, t, x, u) - I^f(t_k, t, x_f, u_f)$$

$$\geq I^f(0, t_k, x, u) - I^f(0, t_k, x_f, u_f) - S \to \infty \text{ as } k \to \infty.$$

Assume that

$$\sup\{|I^f(0, t, x, u) - I^f(0, t, x_f, u_f)| : t \in [0, \infty)\} < \infty.$$

Therefore there exists an integer $S_1 > 2 + a_0$ such that for each $T_2 > T_1 \geq 0$,

$$|I^f(T_1, T_2, x, u) - I^f(T_1, T_2, x_f, u_f)| \leq S_1. \tag{6.69}$$

Let $S_0 > 0$ be such that

$$\psi(S_0) > 8|\Delta_f| + 32 + 4a_0. \tag{6.70}$$

We show that for each $T \geq 0$,

$$\min\{\|x(t)\| : t \in [T, T + S_1]\} \leq S_0.$$

Assume the contrary. Then there exists $T \geq 0$ such that

$$\|x(t)\| > S_0, \; t \in [T, T + S_1]. \tag{6.71}$$

By (6.20), (6.21), (6.24), (6.70), and (6.71),

$$I^f(T, T + S_1, x, u) \geq -2a_0 + I^f(\lfloor T \rfloor + 1, \lfloor T \rfloor + S_1, x, u)$$

$$\geq -2a_0 + (\psi(S_0) - a_0)(S_1 - 1) \geq -2a_0 + 4^{-1}\psi(S_0)S_1. \tag{6.72}$$

In view of (6.20), (6.21), and (6.24),

$$I^f(T, T + S_1, x_f, u_f) \leq I^f(\lfloor T \rfloor, \lfloor T \rfloor + S_1 + 1, x_f, u_f) + 2a_0$$

$$\leq \Delta_f(S_1 + 1) + 2a_0 \leq 2|\Delta_f|S_1 + 2a_0. \tag{6.73}$$

It follows from (6.72) and (6.73) that

$$I^f(T, T + S_1, x, u) - I^f(T, T + S_1, x_f, u_f) \geq 4^{-1}\psi(S_0)S_1 - 4a_0 - 2|\Delta_f|S_1$$
$$\geq S_1(4^{-1}\psi(S_0) - 2|\Delta_f|) - 4a_0 \geq -4a_0 + 8S_1 \geq 4S_1.$$

This contradicts (6.69). The contradiction we have reached proves that for each $T \geq 0$,

$$\min\{\|x(t)\| : t \in [T, T + S_1]\} \leq S_0. \tag{6.74}$$

By (6.74), there exists a sequence $\{T_i\}_{i=1}^{\infty} \subset (0, \infty)$ such that for each integer $i \geq 1$,

$$T_{i+1} - T_i \in [1, 1 + S_1],$$
$$\|x(T_i)\| \leq S_0. \tag{6.75}$$

Lemma 6.38, (6.20), (6.21), (6.24), (6.69), and (6.75) imply that

$$\sup\{\|x(t)\| : t \in [0, \infty)\} < \infty.$$

Theorem 6.2 is proved.

Proof of Theorem 6.3. We may assume that $c < 1/2$ and $c_0 < c/4$. By Theorem 6.2, there exists $S_0 > 0$ such that the following property holds:

(i) for each $T_2 > T_1 \geq 0$ and each $(x, u) \in X(T_1, T_2)$,

$$I^f(T_1, T_2, x, u) + S_0 \geq I^f(T_1, T_2, x_f, u_f).$$

Lemma 6.39 implies that there exists $M_1 > 0$ such that the following property holds:

(ii) for each $T \geq 0$ and each $(x, u) \in X(T, T + c/2)$ satisfying

$$I^f(T, T + c/2, x, u) \leq M_0 + 2S_0 + 2 + 2a_0 + 2|\Delta_f|,$$

we have

$$\|x(t)\| \leq M_1 \text{ for all } t \in [c_0 + T, T + c/2].$$

Assume that $T_1 \geq 0$, $T_2 \geq T_1 + c$, $(x, u) \in X(T_1, T_2)$ and

$$I^f(T_1, T_2, x, u) \leq I^f(T_1, T_2, x_f, u_f) + M_0. \tag{6.76}$$

We show that

$$\|x(t)\| \leq M_1,\ t \in [T_1 + c_0, T_2].$$

Assume the contrary. Then there exists

$$t_0 \in [T_1 + c_0, T_2] \tag{6.77}$$

such that

$$\|x(t_0)\| > M_1. \tag{6.78}$$

By (6.77), there exists a closed interval $[a, b] \subset [T_1, T_2]$ such that

$$t_0 \in [a + c_0, b],\ b - a = 2^{-1}c. \tag{6.79}$$

Property (ii), (6.78), and (6.79) imply that

$$I^f(a, b, x, u) > M_0 + 2S_0 + 2 + 2a_0 + 2|\Delta_f|. \tag{6.80}$$

Property (i), (6.19)–(6.21), (6.24), (6.76), and (6.79) imply that

$$I^f(a, b, x, u) = I^f(T_1, T_2, x, u) - I^f(T_1, a, x, u) - I^f(b, T_2, x, u)$$

$$\leq I^f(T_1, T_2, x_f, u_f) + M_0 - I^f(T_1, a, x_f, u_f) + S_0 - I^f(b, T_2, x_f, u_f) + S_0$$

$$= I^f(a, b, x_f, u_f) + 2S_0 + M_0 \leq 2\Delta_f + 2a + 2S_0 + M_0.$$

This contradicts (6.86). The contradiction we have reached proves Theorem 6.3.

6.8 Proofs of Theorems 6.4–6.10

Proof of Theorem 6.4. Theorem 6.2 implies that there exists $M_1 > 0$ such that the following property holds:

(i) for each $T_1 \geq 0$, each $T_2 \geq T_1 + 2L$ and each $(x, u) \in X(T_1, T_2)$ satisfying

$$I^f(T_1, T_2, x, u) \leq I^f(T_1, T_2, x_f, u_f) + M_0 + 2L(1 + a_0)$$

the inequality $\|x(t)\| \leq M_1$ holds for all $t \in [T_1 + c, T_2]$.

Assume that $T_1 \geq 0$, $T_2 \geq T_1 + 2L$ and $(x, u) \in X(T_1, T_2)$ satisfies

$$(T_1, x(T_1)) \in \mathcal{A}_L, \ (T_2, x(T_2)) \in \widehat{\mathcal{A}}_L, \tag{6.81}$$

$$I^f(T_1, T_2, x, u) \leq U^f(T_1, T_2, x(T_1), x(T_2)) + M_0. \tag{6.82}$$

In view of (6.81), there exist $\tau_1 \in (0, L]$, $\tau_2 \in (0, L]$ and $(y, v) \in X(T_1, T_2)$ such that

$$y(T_1) = x(T_1), \ y(T_2) = x(T_2),$$

$$(y(t), v(t)) = (x_f(t), u_f(t)), \ t \in [\tau_1 + T_1, T_2 - \tau_2],$$

$$I^f(T_1, T_1 + \tau_1, y, v) \leq L, \ I^f(T_2 - \tau_2, T_2, y, v) \leq L.$$

By the relations above, (6.20), (6.21), and (6.82),

$$I^f(T_1, T_2, x, u) \leq M_0 + I^f(T_1, T_2, y, v)$$

$$\leq M_0 + I^f(T_1, T_1 + \tau_1, y, v) + I^f(T_1 + \tau_1, T_2 - \tau_2, y, v) + I^f(T_2 - \tau_2, T_2, y, v)$$

$$\leq 2L + M_0 + I^f(T_1 + \tau_1, T_2 - \tau_2, x_f, u_f)$$

$$\leq 2L + M_0 + I^f(T_1, T_2, x_f, u_f) + 2La_0. \tag{6.83}$$

Property (i) and (6.83) imply that $\|x(t)\| \le M_1$ for all $t \in [T_1+c, T_2]$. Theorem 6.4 is proved.

Proof of Theorem 6.5. Theorem 6.3 implies that there exists $M_1 > 0$ such that the following property holds:

(ii) for each $T_1 \ge 0$, each $T_2 \ge T_1 + L$, and each $(x, u) \in X(T_1, T_2)$ satisfying

$$I^f(T_1, T_2, x, u) \le I^f(T_1, T_2, x_f, u_f) + M_0 + L(1 + a_0)$$

the inequality $\|x(t)\| \le M_1$ holds for all $t \in [T_1 + c, T_2]$.

Assume that $T_1 \ge 0$, $T_2 \ge T_1 + L$ and $(x, u) \in X(T_1, T_2)$ satisfies

$$(T_1, x(T_1)) \in \mathcal{A}_L, \tag{6.84}$$

$$I^f(T_1, T_2, x, u) \le \sigma^f(T_1, T_2, x(T_1)) + M_0. \tag{6.85}$$

In view of (6.84), there exist $\tau \in (0, L]$ and $(y, v) \in X(T_1, T_2)$ such that

$$y(T_1) = x(T_1), \ (y(t), v(t)) = (x_f(t), u_f(t)), \ t \in [\tau + T_1, T_2],$$

$$I^f(T_1, T_1 + \tau, y, v) \le L.$$

By the relations above, (6.20), (6.21), and (6.85),

$$I^f(T_1, T_2, x, u) \le M_0 + I^f(T_1, T_2, y, v)$$

$$= I^f(T_1, T_1 + \tau, y, v) + I^f(T_1 + \tau, T_2, x_f, u_f) + M_0$$

$$\le L + M_0 + I^f(T_1, T_2, x_f, u_f) + La_0. \tag{6.86}$$

Property (ii) and (6.86) imply that $\|x(t)\| \le M_1$ for all $t \in [T_1+c, T_2]$. Theorem 6.5 is proved.

Proof of Theorem 6.6. Theorem 6.3 implies that there exists $M_1 > 0$ such that the following property holds:

(iii) for each $T_1 \ge 0$, each $T_2 \ge T_1 + L$, and each $(x, u) \in X(T_1, T_2)$ satisfying

$$I^f(T_1, T_2, x, u) \le I^f(T_1, T_2, x_f, u_f) + M_0 + L(1 + a_0)$$

the inequality $\|x(t)\| \le M_1$ holds for all $t \in [T_1 + c, T_2]$.

Assume that $T_1 \ge 0$, $T_2 \ge T_1 + L$ and $(x, u) \in X(T_1, T_2)$ satisfies

$$(T_2, x(T_2)) \in \widehat{\mathcal{A}}_L, \tag{6.87}$$

$$I^f(T_1, T_2, x, u) \le \widehat{\sigma}^f(T_1, T_2, x(T_2)) + M_0. \quad (6.88)$$

In view of (6.87), there exist $\tau \in (0, L]$ and $(y, v) \in X(T_1, T_2)$ such that

$$(y(t), v(t)) = (x_f(t), u_f(t)),\ t \in [T_1, T_2 - \tau],\ y(T_2) = x(T_2),$$

$$I^f(T_2 - \tau, T_2, y, v) \le L.$$

By the relations above, (6.20), (6.21), and (6.88),

$$I^f(T_1, T_2, x, u) \le M_0 + I^f(T_1, T_2, y, v)$$

$$= I^f(T_1, T_2 - \tau, y, v) + I^f(T_2 - \tau, T_2, y, v) + M_0$$

$$\le L + M_0 + I^f(T_1, T_2 - \tau, x_f, u_f)$$

$$\le L + M_0 + I^f(T_1, T_2, x_f, u_f) + La_0. \quad (6.89)$$

Property (iii) and (6.89) imply that $\|x(t)\| \le M_1$ for all $t \in [T_1+c, T_2]$. Theorem 6.6 is proved.

Proof of Theorem 6.7. We may assume that $c < 1$. Theorem 6.2 implies that there exists $S_0 > 0$ such that the following property holds:

(iv) for each $T_2 > T_1 \ge 0$ and each $(x, u) \in X(T_1, T_2)$, we have

$$I^f(T_1, T_2, x, u) + S_0 \ge I^f(T_1, T_2, x_f, u_f).$$

By Theorem 6.3, there exists $M_2 > 0$ such that the following property holds:

(v) for each $T_1 \ge 0$, each $T_2 \ge T_1 + c$, and each $(x, u) \in X(T_1, T_2)$, satisfying

$$I^f(T_1, T_2, x, u) \le I^f(T_1, T_2, x_f, u_f) + M_0$$

the inequality $\|x(t)\| \le M_2$ holds for all $t \in [T_1 + c/4, T_2]$.

Lemma 6.38 implies that there exists $M_1 > M_2$ such that the following property holds:

(vi) for each $T \ge 0$ and each $(x, u) \in X(T, T + c/4)$ satisfying

$$\|x(T)\| \le M_0,\ I^f(T, T + c/4, x, u) \le M_0 + S_0 + a_0c + 2|\Delta_f| + 2a_0,$$

we have

$$\|x(t)\| \le M_1 \text{ for all } t \in [T, T + c/4].$$

Assume that $T_1 \ge 0$, $T_2 \ge T_1 + c$, $(x, u) \in X(T_1, T_2)$,

$$\|x(T_1)\| \le M_0, \tag{6.90}$$

$$I^f(T_1, T_2, x, u) \le I^f(T_1, T_2, x_f, u_f) + M_0. \tag{6.91}$$

Property (v) and (6.91) imply that

$$\|x(t)\| \le M_2 < M_1,\ t \in [T_1 + c/4, T_2]. \tag{6.92}$$

It follows from (6.20), (6.21), (6.24), (6.91), and property (iv) that

$$I^f(T_1, T_1 + c/4, x, u) = I^f(T_1, T_2, x, u) - I^f(T_1 + c/4, T_2, x, u)$$

$$\le I^f(T_1, T_2, x_f, u_f) + M_0 - I^f(T_1 + c/4, T_2, x, u) + S_0$$

$$\le S_0 + M_0 + I^f(T_1, T_1 + c/4, x_f, u_f) \le M_0 + S_0 + 2|\Delta_f| + 2a_0.$$

By the relation above, (6.90) and property (vi),

$$\|x(t)\| \le M_1,\ t \in [T_1, T_1 + c/4].$$

Theorem 6.7 is proved.

Proof of Theorem 6.8. Theorem 6.7 implies that there exists $M_1 > 0$ such that the following property holds:

(vii) for each $T_1 \ge 0$, each $T_2 \ge T_1 + 2L$, and each $(x, u) \in X(T_1, T_2)$ satisfying

$$\|x(T_1)\| \le M_0,$$

$$I^f(T_1, T_2, x, u) \le I^f(T_1, T_2, x_f, u_f) + M_0 + 2L(1 + a_0)$$

the inequality $\|x(t)\| \le M_1$ holds for all $t \in [T_1, T_2]$.

Assume that $T_1 \ge 0$, $T_2 \ge T_1 + 2L$, $(x, u) \in X(T_1, T_2)$,

$$\|x(T_1)\| \le M_0, \tag{6.93}$$

$$(T_1, x(T_1)) \in \mathcal{A}_L,\ (T_2, x(T_2)) \in \widehat{\mathcal{A}}_L, \tag{6.94}$$

$$I^f(T_1, T_2, x, u) \le U^f(T_1, T_2, x(T_1), x(T_2)) + M_0. \tag{6.95}$$

In view of (6.94), there exist $\tau_1 \in (0, L]$, $\tau_2 \in (0, L]$ and $(y, v) \in X(T_1, T_2)$ such that

$$y(T_1) = x(T_1),\ y(T_2) = x(T_2),$$

$$(y(t), v(t)) = (x_f(t), u_f(t)),\ t \in [\tau_1 + T_1, T_2 - \tau_2],$$

$$I^f(T_1, T_1 + \tau_1, y, v) \le L,\ I^f(T_2 - \tau_2, T_2, y, v) \le L.$$

By the relations above, (6.20), (6.21), and (6.95),

$$I^f(T_1, T_2, x, u) \le M_0 + I^f(T_1, T_2, y, v)$$

$$= I^f(T_1, T_1 + \tau_1, y, v) + I^f(T_1 + \tau_1, T_2 - \tau_2, y, v) + I^f(T_2 - \tau_2, T_2, y, v) + M_0$$

$$\le 2L + M_0 + I^f(T_1 + \tau_1, T_2 - \tau_2, x_f, u_f)$$

$$\le I^f(T_1, T_2, x_f, u_f) + 2L + M_0 + 2La_0.$$

Property (vii), (6.93), and the relation above imply that $\|x(t)\| \le M_1$ for all $t \in [T_1, T_2]$. Theorem 6.8 is proved.

Proof of Theorem 6.9. Theorem 6.7 implies that there exists $M_1 > 0$ such that the following property holds:

(viii) for each $T_1 \ge 0$, each $T_2 \ge T_1 + L$, and each $(x, u) \in X(T_1, T_2)$ satisfying

$$\|x(T_1)\| \le M_0,$$

$$I^f(T_1, T_2, x, u) \le I^f(T_1, T_2, x_f, u_f) + M_0 + L(1 + a_0)$$

the inequality $\|x(t)\| \le M_1$ holds for all $t \in [T_1, T_2]$.

Assume that $T_1 \ge 0$, $T_2 \ge T_1 + L$, $(x, u) \in X(T_1, T_2)$,

$$\|x(T_1)\| \le M_0, \tag{6.96}$$

$$(T_1, x(T_1)) \in \mathcal{A}_L, \tag{6.97}$$

$$I^f(T_1, T_2, x, u) \le \sigma^f(T_1, T_2, x(T_1)) + M_0. \tag{6.98}$$

In view of (6.97), there exist $\tau \in (0, L]$, $(y, v) \in X(T_1, T_2)$ such that

$$y(T_1) = x(T_1),\ (y(t), v(t)) = (x_f(t), u_f(t)),\ t \in [\tau + T_1, T_2],$$

$$I^f(T_1, T_1 + \tau, y, v) \le L.$$

By the relations above, (6.20), (6.21), and (6.98),

$$I^f(T_1, T_2, x, u) \le M_0 + I^f(T_1, T_2, y, v)$$

$$= I^f(T_1, T_1 + \tau, y, v) + I^f(T_1 + \tau, T_2, x_f, u_f) + M_0$$

$$\le L + M_0 + I^f(T_1, T_2, x_f, u_f) + La_0.$$

The relations above, property (viii), and (6.96) imply that $\|x(t)\| \le M_1$ for all $t \in [T_1, T_2]$. Theorem 6.9 is proved.

Proof of Theorem 6.10. Theorem 6.7 implies that there exists $M_1 > 0$ such that the following property holds:

(ix) for each $T_1 \ge 0$, each $T_2 \ge T_1 + L$, and each $(x, u) \in X(T_1, T_2)$ satisfying

$$\|x(T_1)\| \le M_0,$$

$$I^f(T_1, T_2, x, u) \le I^f(T_1, T_2, x_f, u_f) + M_0 + L(1 + a_0)$$

the inequality $\|x(t)\| \le M_1$ holds for all $t \in [T_1, T_2]$.

Assume that $T_1 \ge 0$, $T_2 \ge T_1 + L$, $(x, u) \in X(T_1, T_2)$,

$$\|x(T_1)\| \le M_0, \tag{6.99}$$

$$(T_2, x(T_2)) \in \widehat{\mathcal{A}}_L, \tag{6.100}$$

$$I^f(T_1, T_2, x, u) \le \widehat{\sigma}^f(T_1, T_2, x(T_2)) + M_0. \tag{6.101}$$

In view of (6.100), there exist $\tau \in (0, L]$ and $(y, v) \in X(T_1, T_2)$ such that

$$(y(t), v(t)) = (x_f(t), u_f(t)),\ t \in [T_1, T_2 - \tau],\ y(T_2) = x(T_2),$$

$$I^f(T_2 - \tau, T_2, y, v) \le L.$$

By the relations above, (6.20), (6.21), and (6.101),

$$I^f(T_1, T_2, x, u) \le M_0 + I^f(T_1, T_2, y, v)$$

$$= I^f(T_1, T_2 - \tau, y, v) + I^f(T_2 - \tau, T_2, y, v) + M_0$$

$$\le L + M_0 + I^f(T_1, T_2 - \tau, x_f, u_f)$$

$$\le L + M_0 + La_0 + I^f(T_1, T_2, x_f, u_f).$$

The relations above, property (ix) and (6.99) imply that $\|x(t)\| \le M_1$ for all $t \in [T_1, T_2]$. Theorem 6.10 is proved.

6.9 Proof of Theorem 6.22

First we show that TP implies (P1) and (P2). In view of Theorem 6.11, TP implies (P2). We show that TP implies (P1). Assume that TP holds, $(x, u) \in X(0, \infty)$ is (f)-good. Let $\epsilon > 0$. There exists $S > 0$ such that for each $T > 0$,

$$|I^f(0, T, x, u) - I^f(0, T, x_f, u_f)| < S.$$

This implies that for each pair of numbers $T_2 > T_1 \geq 0$,

$$|I^f(T_1, T_2, x, u) - I^f(T_1, T_2, x_f, u_f)| < 2S. \tag{6.102}$$

Let $\delta > 0$. We show that there exists $T_\delta > 0$ such that for each $T > T_\delta$,

$$I^f(T_\delta, T, x, u) \leq U^f(T_\delta, T, x(T_\delta), x(T)) + \delta.$$

Assume the contrary. Then for each $T \geq 0$ there exists $S > T$ such that

$$I^f(T, S, x, u) > U^f(T, S, x(T), x(S)) + \delta.$$

This implies that there exists a strictly increasing sequence of numbers $\{T_i\}_{i=0}^{\infty}$ such that

$$T_0 = 0$$

and for every integer $i \geq 0$,

$$I^f(T_i, T_{i+1}, x, u) > U^f(T_i, T_{i+1}, x(T_i), x(T_{i+1})) + \delta.$$

By the relation above, there exists $(y, v) \in X(0, \infty)$ such that for every integer $i \geq 0$,

$$y(T_i) = x(T_i),$$

$$I^f(T_i, T_{i+1}, x, u) > I^f(T_i, T_{i+1}, y, v) + \delta.$$

Combined with (6.102) this implies that for each integer $k \geq 1$,

$$I^f(0, T_k, y, v) - I^f(0, T_k, x_f, u_f)$$

$$= I^f(0, T_k, y, v) - I^f(0, T_k, x, u) + I^f(0, T_k, x, u) - I^f(0, T_k, x_f, u_f)$$

$$\leq -k\delta + 2S \to -\infty \text{ as } k \to \infty.$$

This contradicts Theorem 6.1. The contradiction we have reached proves that the following property holds:

(i) for each $\delta > 0$ there exists $T_\delta > 0$ such that for each $T > T_\delta$,

$$I^f(T_\delta, T, x, u) \leq U^f(T_\delta, T, x(T_\delta), x(T)) + \delta.$$

Theorem 6.11 implies that there exist $\delta > 0$, $L > 0$ such that the following property holds:

(ii) for each $S_1 \geq 0$, each $S_2 \geq S_1 + 2L$, and each $(z, \xi) \in X(S_1, S_2)$ which satisfies

$$I^f(S_1, S_2, z, \xi) \leq \min\{I^f(S_1, S_2, x_f, u_f) + 2S,\ U^f(S_1, S_2, z(S_1), z(S_2)) + \delta\}$$

we have

$$\|z(t) - x_f(t)\| \leq \epsilon,\ t \in [S_1 + L, S_2 - L].$$

Let $T_\delta > 0$ be as guaranteed by (i). Properties (i) and (ii), (6.102), and the choice of T_δ imply that for each $T \geq T_\delta + 2L$,

$$\|x(t) - x_f(t)\| \leq \epsilon$$

and (P1) holds. Thus TP implies (P1).

Lemma 6.40. *Assume that (P1) holds and that $\epsilon > 0$. Then there exists $\delta, L > 0$ such that for each $T_1 \geq L$, each $T_2 \geq T_1 + 2b_f$, and each $(x, u) \in X(T_1, T_2)$ which satisfies*

$$\|x(T_i) - x_f(T_i)\| \leq \delta,\ i = 1, 2,$$

$$I^f(T_1, T_2, x, u) \leq U^f(T_1, T_2, x(T_1), x(T_2)) + \delta$$

the inequality $\|x(t) - x_f(t)\| \leq \epsilon$ holds for all $t \in [T_1, T_2]$.

Proof. By (A2), for each integer $k \geq 1$ there exists $\delta_k \in (0, 2^{-k})$ such that the following property holds:

(iii) for each $(T_i, z_i) \in \mathcal{A}$, $i = 1, 2$ satisfying $\|z_i - x_f(T_i)\| \leq \delta_k$, $i = 1, 2$ and $T_2 \geq b_f$, there exist $\tau_1, \tau_2 \in (0, b_f]$ and $(x_1, u_1) \in X(T_1, T_1 + \tau_1)$, $(x_2, u_2) \in X(T_2 - \tau_2, T_2)$ which satisfy

$$x_1(T_1) = z_1,\ x_1(T_1 + \tau_1) = x_f(T_1 + \tau_1),\ x_2(T_2) = z_2,\ x_2(T_2 - \tau_2) = x_f(T_2 - \tau_2),$$

$$I^f(T_1, T_1 + \tau_1, x_1, u_1) \leq I^f(T_1, T_1 + \tau_1, x_f, u_f) + 2^{-k},$$

$$I^f(T_2 - \tau_2, T_2, x_2, u_2) \leq I^f(T_2 - \tau_2, T_2, x_f, u_f) + 2^{-k}.$$

Assume that the lemma does not hold. Then for each integer $k \geq 1$ there exist

$$T_{k,1} \geq k + b_f, \ T_{k,2} \geq T_{k,1} + 2b_f, \tag{6.103}$$

and $(x_k, u_k) \in X(T_{k,1}, T_{k,2})$ such that

$$\|x_k(T_{k,i}) - x_f(T_{k,i})\| \leq \delta_k, \ i = 1, 2, \tag{6.104}$$

$$I^f(T_{k,1}, T_{k,2}, x_k, u_k) \leq U^f(T_{k,1}, T_{k,2}, x_k(T_{k,1}), x_k(T_{k,2})) + \delta_k, \tag{6.105}$$

$$\sup\{\|x_k(t) - x_f(t)\| : t \in [T_{k,1}, T_{k,2}]\} > \epsilon. \tag{6.106}$$

Extracting a subsequence and re-indexing, we may assume without loss of generality that for each integer $k \geq 1$,

$$T_{k+1,1} \geq T_{k,2} + 4b_f. \tag{6.107}$$

Let $k \geq 1$ be an integer. Property (iii), (6.103), and (6.104) imply that there exist $\tau_{k,1}, \tau_{k,2} \in (0, b_f]$ and $(\tilde{x}_k, \tilde{u}_k) \in X(T_{k,1} - \tau_{k,1}, T_{k,2} + \tau_{k,2})$ such that

$$(\tilde{x}_k(t), \tilde{u}_k(t)) = (x_k(t), u_k(t)), \ t \in [T_{k,1}, T_{k,2}], \tag{6.108}$$

$$\tilde{x}_k(T_{k,1} - \tau_{k,1}) = x_f(T_{k,1} - \tau_{k,1}), \ \tilde{x}_k(T_{k,2} + \tau_{k,2}) = x_f(T_{k,2} + \tau_{k,2}), \tag{6.109}$$

$$I^f(T_{k,1} - \tau_{k,1}, T_{k,1}, \tilde{x}_k, \tilde{u}_k) \leq I^f(T_{k,1} - \tau_{k,1}, T_{k,1}, x_f, u_f) + 2^{-k}, \tag{6.110}$$

$$I^f(T_{k,2}, T_{k,2} + \tau_{k,2}, \tilde{x}_k, \tilde{u}_k) \leq I^f(T_{k,2}, T_{k,2} + \tau_{k,2}, x_f, u_f) + 2^{-k}. \tag{6.111}$$

Property (iii), (6.103), and (6.104) imply that there exist $\tau_{k,3}, \tau_{k,4} \in (0, b_f]$ and $(\widehat{x}_k, \widehat{u}_k) \in X(T_{k,1}, T_{k,2})$ such that

$$\widehat{x}_k(T_{k,1}) = x_k(T_{k,1}), \ \widehat{x}_k(T_{k,2}) = x_k(T_{k,2}), \tag{6.112}$$

$$(\widehat{x}_k(t), \widehat{u}_k(t)) = (x_f(t), u_f(t)), \ t \in [T_{k,1} + \tau_{k,3}, T_{k,2} - \tau_{k,4}], \tag{6.113}$$

$$I^f(T_{k,1}, T_{k,1} + \tau_{k,3}, \widehat{x}_k, \widehat{u}_k) \leq I^f(T_{k,1}, T_{k,1} + \tau_{k,3}, x_f, u_f) + 2^{-k}, \tag{6.114}$$

$$I^f(T_{k,2} - \tau_{k,4}, T_{k,2}, \widehat{x}_k, \widehat{u}_k) \leq I^f(T_{k,2} - \tau_{k,4}, T_{k,2}, x_f, u_f) + 2^{-k}. \tag{6.115}$$

In view of (6.112),

$$\begin{aligned} U^f(T_{k,1}, T_{k,2}, x_k(T_{k,1}), x_k(T_{k,2})) &\leq I^f(T_{k,1}, T_{k,2}, \widehat{x}_k, \widehat{u}_k) \\ &\leq I^f(T_{k,1}, T_{k,2}, x_f, u_f) + 2^{-k+1}. \end{aligned} \tag{6.116}$$

By ((6.105) and (6.116),

$$I^f(T_{k,1}, T_{k,2}, x_k, u_k) \le I^f(T_{k,1}, T_{k,2}, x_f, u_f) + 2^{-k+2}. \tag{6.117}$$

It follows from (6.108), (6.110), (6.111), and (6.117) that

$$\begin{aligned} & I^f(T_{k,1} - \tau_{k,1}, T_{k,2} + \tau_{k,2}, \widehat{x}_k, \widehat{u}_k) \\ & \le I^f(T_{k,1} - \tau_{k,1}, T_{k,1}, x_f, u_f) \\ & + I^f(T_{k,1}, T_{k,2}, x_k, u_k) + I^f(T_{k,2}, T_{k,2} + \tau_{k,2}, x_f, u_f) + 2^{-k+1} \\ & \le I^f(T_{k,1} - \tau_{k,1}, T_{k,2} + \tau_{k,2}, x_f, u_f) + 2^{-k+3}. \end{aligned} \tag{6.118}$$

By (6.107) and (6.109), there exists $(x, u) \in X(0, \infty)$ such that for each integer $k \ge 1$,

$$(x(t), u(t)) = (\tilde{x}_k(t), \tilde{u}_k(t)),\ t \in [T_{k,1} - \tau_{k,1}, T_{k,2} + \tau_{k,2}], \tag{6.119}$$

$$(x(t), u(t)) = (x_f(t), u_f(t)),\ t \in [0, \infty) \setminus \cup_{k=1}^{\infty} [T_{k,1} - \tau_{k,1}, T_{k,2} + \tau_{k,2}]. \tag{6.120}$$

It follows from (6.118)–(6.120) that for each integer $k \ge 1$,

$$\begin{aligned} & I^f(0, T_{k,2} + \tau_{k,2}, x, u) - I^f(0, T_{k,2} + \tau_{k,2}, x_f, u_f) \\ & = \sum_{i=1}^{k} (I^f(T_{i,1} - \tau_{i,1}, T_{i,2} + \tau_{i,2}, \tilde{x}_i, \tilde{u}_i) - I^f(T_{i,1} - \tau_{i,1}, T_{i,2} + \tau_{i,2}, x_f, u_f)) \\ & \le \sum_{i=1}^{k} 2^{-i+3} \le 8. \end{aligned}$$

Theorem 6.1 implies that (x, u) is (f)-good. Together with (P1) this implies that $\lim_{t\to\infty} \|x(t) - x_f(t)\| = 0$. On the other hand in view of (6.106), (6.108), (6.117), and (6.119), $\limsup_{t\to\infty} \|x(t) - x_f(t)\| \ge \epsilon$. The contradiction we have reached completes the proof of Lemma 6.40.

Completion of the Proof of Theorem 6.22. Assume that properties (P1) and (P2) hold. Let $\epsilon, M > 0$. By Lemma 6.40, there exist $\delta_0, L_0 > 0$ such that the following property holds:

(iv) for each $T_1 \ge L_0$, each $T_2 \ge T_1 + 2b_f$, and each $(x, u) \in X(T_1, T_2)$ which satisfies

$$\|x(T_i) - x_f(T_i)\| \le \delta_0, \ i = 1, 2,$$

$$I^f(T_1, T_2, x, u) \le U^f(T_1, T_2, x(T_1), x(T_2)) + \delta_0$$

the inequality $\|x(t) - x_f(t)\| \le \epsilon$ holds for all $t \in [T_1, T_2]$.

By Theorem 6.1, there exists $S_0 > 0$ such that the following property holds:

(v) for each $T_2 > T_1 \ge 0$ and each $(x, u) \in X(T_1, T_2)$, we have

$$I^f(T_1, T_2, x, u) + S_0 \ge I^f(T_1, T_2, x_f, u_f).$$

In view of (P2), there exist $\delta \in (0, \delta_0)$ and $L_1 > 0$ such that the following property holds:

(vi) for each $T \ge 0$ and each $(x, u) \in X(T, T + L_1)$ which satisfies

$$I^f(T, T + L_1, x, u) \le \min\{U^f(T, T + L_1, x(T), x(T + L_1)) + \delta,$$

$$I^f(T, T + L_1, x_f, u_f) + 2S_0 + M\},$$

there exists $S_0 \in [0, L_1]$ such that $\|x(T + S_0) - x_f(T + S_0)\| \le \delta_0$.

Set

$$L = L_0 + L_1 + b_f. \tag{6.121}$$

Assume that $T_1 \ge 0$, $T_2 \ge T_1 + 2L$ and that $(x, u) \in X(T_1, T_2)$ satisfies

$$I^f(T_1, T_2, x, u) \le \sigma^f(T_1, T_2) + M, \tag{6.122}$$

$$I^f(T_1, T_2, x, u) \le U^f(T_1, T_2, x(T_1), x(T_2)) + \delta. \tag{6.123}$$

By (6.122),

$$I^f(T_1, T_2, x, u) \le I^f(T_1, T_2, x_f, u_f) + M. \tag{6.124}$$

Property (v) and (6.124) imply that for each pair of numbers $Q_1, Q_2 \in [T_1, T_2]$ satisfying $Q_1 < Q_2$,

$$I^f(Q_1, Q_2, x, u) = I^f(T_1, T_2, x, u) - I^f(T_1, Q_1, x, u) - I^f(Q_2, T_2, x, u)$$

$$\le I^f(T_1, T_2, x_f, u_f) + M - I^f(T_1, Q_1, x_f, u_f) + S_0 - I^f(Q_2, T_2, x_f, u_f) + S_0$$

$$= I^f(Q_1, Q_2, x_f, u_f) + M + 2S_0. \tag{6.125}$$

It follows from (6.125) that

$$I^f(\max\{T_1, L_0\}, \max\{T_1, L_0\} + L_1, x, u)$$
$$\leq I^f(\max\{T_1, L_0\}, \max\{T_1, L_0\} + L_1, x_f, u_f) + 2S_0 + M, \tag{6.126}$$

$$I^f(T_2 - L_1, T_2, x, u) \leq I^f(T_2 - L_1, T_2, x_f, u_f) + 2S_0 + M. \tag{6.127}$$

By (6.123), (6.126), (6.127), and property (v), there exist

$$\tau_1 \in [\max\{T_1, L_0\}, \max\{T_1, L_0\} + L_1], \ \tau_2 \in [T_2 - L_1, T_2] \tag{6.128}$$

such that

$$\|x(\tau_i) - x_f(\tau_i)\| \leq \delta_0, \ i = 1, 2. \tag{6.129}$$

If $\|x(T_2) - x_f(T_2)\| \leq \delta$, then we may assume that $\tau_2 = T_2$, and if $T_1 \geq L_0$ and $\|x(T_1) - x_f(T_1)\| \leq \delta$, then we may assume that $\tau_1 = T_1$. In view of (6.121) and (6.128),

$$\tau_2 - \tau_1 \geq T_2 - T_1 - L_1 - L_1 - L_0 \geq L_0 + 2b_f. \tag{6.130}$$

Property (iv), (6.123), and (6.128)–(6.130) imply that

$$\|x(t) - x_f(t)\| \leq \epsilon, \ t \in [\tau_1, \tau_2].$$

Thus TP holds and Theorem 6.22 is proved.

6.10 An Auxiliary Result for Theorem 6.23

Lemma 6.41. *Assume that (P2) holds and that $M, \epsilon > 0$. Then there exists $L > 0$ such that for each $T \geq 0$ and each $(x, u) \in X(T, T + L)$ which satisfies*

$$I^f(T, T + L, x, u) \leq I^f(T, T + L, x_f, u_f) + M$$

the following inequality holds:

$$\inf\{\|x(t) - x_f(t)\| : t \in [T, T + L]\} \leq \epsilon.$$

Proof. By Theorem 6.1, there exists $S_0 > 0$ such that the following property holds:

(i) for each $T_2 > T_1 \geq 0$ and each $(x, u) \in X(T_1, T_2)$,

$$I^f(T_1, T_2, x, u) + S_0 \geq I^f(T_1, T_2, x_f, u_f).$$

By (P2), there exist $\delta_0 \in (0, \epsilon)$ and $L_0 > 0$ such that the following property holds:

(ii) for each $T \geq 0$ and each $(x, u) \in X(T, T + L_0)$ which satisfies

$$I^f(T, T + L_0, x, u) \leq \min\{U^f(T, T + L_0, x(T), x(T + L_0)) + \delta_0,$$
$$I^f(T, T + L_0, x_f, u_f) + 2S_0 + M\}$$

we have

$$\inf\{\|x(t) - x_f(t)\| : t \in [T, T + L_0]\} \leq \epsilon.$$

Choose an integer

$$q_0 > (M + S_0)\delta_0^{-1} \tag{6.131}$$

and set

$$L = q_0 L_0. \tag{6.132}$$

Let $T \geq 0$ and $(x, u) \in X(T, T + L)$ satisfy

$$I^f(T, T + L, x, u) \leq I^f(T, T + L, x_f, u_f) + M. \tag{6.133}$$

We show that

$$\inf\{\|x(t) - x_f(t)\| : t \in [T, T + L]\} \leq \epsilon.$$

Assume the contrary. Then

$$\|x(t) - x_f(t)\| > \epsilon, \ t \in [T, T + L]. \tag{6.134}$$

Let an integer $i \in \{0, \ldots, q_0 - 1\}$. Property (i), (6.132), and (6.133) imply that

$$I^f(T + iL_0, T + (i + 1)L_0, x, u) = I^f(T, T + L, x, u)$$
$$-I^f(T, T + iL_0, x, u) - I^f(T + (i + 1)L_0, T + L, x, u)$$
$$\leq I^f(T, T + L, x_f, u_f) + M - I^f(T, T + iL_0, x_f, u_f) + S_0$$

$$-I^f(T+(i+1)L_0, T+L, x_f, u_f) + S_0$$

$$= I^f(T+iL_0, T+(i+1)L_0, x_f, u_f) + M + 2S_0. \tag{6.135}$$

By (6.134), (6.135), and property (ii),

$$I^f(T+iL_0, T+(i+1)L_0, x, u)$$

$$> U^f(T+iL_0, T+(i+1)L_0, x(T+iL_0), x(T+(i+1)L_0)) + \delta_0. \tag{6.136}$$

It follows from (6.136) that there exists $(y, v) \in X(T, T+L)$ such that

$$y(T+iL_0) = x(T+iL_0),\ i = 0, \dots, q_0$$

and for all $i = 0, \dots, q_0 - 1$,

$$I^f(T+iL_0, T+(i+1)L_0, x, u) > I^f(T+iL_0, T+(i+1)L_0, y, v)+\delta_0. \tag{6.137}$$

In view of (6.133), (6.137), and property (i),

$$I^f(T, T+L, x_f, u_f) + M \geq I^f(T, T+L, x, u)$$

$$\geq I^f(T, T+L, y, v) + q_0\delta_0 \geq I^f(T, T+L, x_f, u_f) - S_0 + q_0\delta_0$$

and

$$q_0 \leq (M + S_0)\delta_0^{-1}.$$

This contradicts (6.132). The contradiction we have reached proves Lemma 6.41.

6.11 Proof of Theorem 6.23

By Theorem 6.1, there exists $S_0 > 0$ such that the following property holds:

(i) for each $\tau_2 > \tau_1 \geq 0$ and each $(y, v) \in X(\tau_1, \tau_2)$,

$$I^f(\tau_1, \tau_2, y, v) + S_0 \geq I^f(\tau_1, \tau_2, x_f, u_f).$$

Theorem 6.22 implies that there exist $\delta \in (0, \epsilon)$ and $L_0 > 0$ such that the following property holds:

(ii) for each $\tau_1 \geq 0$, each $\tau_2 \geq \tau_1 + 2L_0$, and each $(y, v) \in X(\tau_1, \tau_2)$ which satisfies

$$I^f(\tau_1, \tau_2, y, v) \leq \min\{\sigma^f(\tau_1, \tau_2) + M + 3S_0, U^f(\tau_1, \tau_2, y(\tau_1), y(\tau_2)) + \delta\}$$

we have

$$\|y(t) - x_f(t)\| \leq \epsilon, \ t \in [\tau_1 + L_0, \tau_2 - L_0].$$

Set

$$l = 2L_0 + 1. \tag{6.138}$$

Choose a natural number

$$Q \geq 2 + 2(M + S_0)\delta^{-1}. \tag{6.139}$$

Assume that $T_1 \geq 0$, $T_2 \geq T_1 + lQ$ and $(x, u) \in X(T_1, T_2)$ satisfy

$$I^f(T_1, T_2, x, u) \leq I^f(T_1, T_2, x_f, u_f) + M. \tag{6.140}$$

Property (i) and (6.140) imply that for each pair of numbers $\tau_1, \tau_2 \in [T_1, T_2]$ satisfying $\tau_1 < \tau_2$,

$$I^f(\tau_1, \tau_2, x, u) = I^f(T_1, T_2, x, u) - I^f(T_1, \tau_1, x, u) - I^f(\tau_2, T_2, x, u)$$

$$\leq M + I^f(T_1, T_2, x_f, u_f) - I^f(T_1, \tau_1, x_f, u_f) + S_0 - I^f(\tau_2, T_2, x_f, u_f) + S_0$$

$$\leq I^f(\tau_1, \tau_2, x_f, u_f) + M + 2S_0. \tag{6.141}$$

Set

$$t_0 = T_1. \tag{6.142}$$

If

$$I^f(T_1, T_2, x, u) \leq U^f(T_1, T_2, x(T_1), x(T_2)) + \delta,$$

then we set

$$t_1 = T_2, \ s_1 = T_2. \tag{6.143}$$

Assume that

$$I^f(T_1, T_2, x, u) > U^f(T_1, T_2, x(T_1), x(T_2)) + \delta. \tag{6.144}$$

It is easy to see that for each $t \in (T_1, T_2)$ such that $t - T_1$ is sufficiently small, we have

$$I^f(T_1, t, x, u) - U^f(T_1, t, x(T_1), x(t)) < \delta.$$

Set

$$\tilde{t}_1 = \inf\{t \in (T_1, T_2] : I^f(T_1, t, x, u) - U^f(T_1, t, x(T_1), x(t)) > \delta\}.$$

Clearly, $\tilde{t}_1 \in (T_1, T_2]$ is well-defined. There exist $s_1, t_1 \in [T_1, T_2]$ such that

$$t_0 < s_1 < \tilde{t}_1, \ s_1 \geq \tilde{t}_1 - 1/4, \tag{6.145}$$

$$I^f(t_0, s_1, x, u) - U^f(t_0, s_1, x(t_0), x(s_1)) \leq \delta, \tag{6.146}$$

$$\tilde{t}_1 \leq t_1 < \tilde{t}_1 + 1/4, \ t_1 \leq T_2, \tag{6.147}$$

$$I^f(t_0, t_1, x, u) - U^f(t_0, t_1, x(t_0), x(t_1)) > \delta. \tag{6.148}$$

Assume that $k \geq 1$ is an integer and we defined finite sequences of numbers $\{t_i\}_{i=0}^k \subset [T_1, T_2]$, $\{s_i\}_{i=1}^k \subset [T_1, T_2]$ such that

$$T_1 = t_0 < t_1 \cdots < t_k, \tag{6.149}$$

for each integer $i = 1, \ldots, k$,

$$t_{i-1} < s_i \leq t_i, \ t_i - s_i \leq 2^{-1}, \tag{6.150}$$

$$I^f(t_{i-1}, s_i, x, u) - U^f(t_{i-1}, s_i, x(t_{i-1}), x(s_i)) \leq \delta, \tag{6.151}$$

and if $t_i < T_2$, then

$$I^f(t_{i-1}, t_i, x, u) - U^f(t_{i-1}, t_i, x(t_{i-1}), x(t_i)) > \delta. \tag{6.152}$$

(Note that in view of (6.142), (6.143), and (6.145)–(6.148), our assumption holds for $k = 1$.)

By (6.152), there exists $(y, v) \in X(T_1, T_2)$ such that

$$y(t_i) = x(t_i), \ i = 1, \ldots, k,$$

$$I^f(t_{i-1}, t_i, x, u) - I^f(t_{i-1}, t_i, y, v) > \delta \text{ for all integers } i \in [1, k] \setminus \{k\}, \tag{6.153}$$

$$(y(t), v(t)) = (x(t), u(t)), \ t \in [t_{k-1}, T_2]. \tag{6.154}$$

Property (i), (6.140), (6.153), and (6.154) imply that

$$I^f(T_1, T_2, x_f, u_f) + M \geq I^f(T_1, T_2, x, u) \geq I^f(T_1, T_2, y, v) + \delta(k-1)$$
$$\geq I^f(T_1, T_2, x_f, u_f) - S_0 + \delta(k-1),$$
$$k \leq 1 + \delta^{-1}(M + S_0). \tag{6.155}$$

If $t_k = T_2$, then the construction of the sequence is completed. Assume that $t_k < T_2$. If

$$I^f(t_k, T_2, x, u) \leq U^f(t_k, T_2, x(t_k), x(T_2)) + \delta,$$

then we set $t_{k+1} = T_2$, $s_{k+1} = T_2$ and the construction of the sequence is completed. Assume that

$$I^f(t_k, T_2, x, u) > U^f(t_k, T_2, x(t_k), x(T_2)) + \delta.$$

It is easy to see that for each $t \in (t_k, T_2)$ such that $t - t_k$ is sufficiently small, we have

$$I^f(t_k, t, x, u) - U^f(t_k, t, x(t_k), x(t)) < \delta.$$

Set

$$\tilde{t} = \inf\{t \in (t_k, T_2] : I^f(t_k, t, x, u) - U^f(t_k, t, x(t_k), x(t)) > \delta\}.$$

Clearly, $\tilde{t}$ is well-defined and $\tilde{t} \in (t_k, T_2]$. There exist $s_{k+1}, t_{k+1} \in [T_1, T_2]$ such that

$$t_k < s_{k+1} < \tilde{t}, \ s_{k+1} \geq \tilde{t} - 1/4, \ \tilde{t} + 4^{-1} > t_{k+1} \geq \tilde{t},$$
$$I^f(t_k, s_{k+1}, x, u) - U^f(t_k, s_{k+1}, x(t_k), x(s_{k+1})) \leq \delta,$$
$$I^f(t_k, t_{k+1}, x, u) - U^f(t_k, t_{k+1}, x(t_k), x(t_{k+1})) > \delta.$$

It is not difficult to see that the assumption made for k also holds for $k + 1$. In view of (6.155), by induction we constructed finite sequences $\{t_i\}_{i=0}^q \subset [T_1, T_2]$, $\{s_i\}_{i=1}^q \subset [T_1, T_2]$ such that

$$q \leq 1 + \delta^{-1}(M + S_0), \tag{6.156}$$

$$T_1 = t_0 < t_1 \cdots < t_q = T_2,$$

for each integer $i = 1, \dots, q$,

$$t_{i-1} < s_i \le t_i,\ t_i - s_i \le 2^{-1}, \tag{6.157}$$

$$I^f(t_{i-1}, s_i, x, u) - U^f(t_{i-1}, s_i, x(t_{i-1}), x(s_i)) \le \delta, \tag{6.158}$$

and if $t_i < T_2$, then

$$I^f(t_{i-1}, t_i, x, u) - U^f(t_{i-1}, t_i, x(t_{i-1}), x(t_i)) > \delta. \tag{6.159}$$

Assume that

$$i \in \{0, \dots, q-1\},\ t_{i+1} - t_i \ge 2L_0 + 1. \tag{6.160}$$

By (6.157) and (6.160),

$$s_{i+1} - t_i \ge 2L_0. \tag{6.161}$$

In view of (6.158),

$$I^f(t_i, s_{i+1}, x, u) - U^f(t_i, s_{i+1}, x(t_i), x(s_{i+1})) \le \delta. \tag{6.162}$$

Property (i) and (6.140) imply that

$$\begin{aligned} I^f(t_i, s_{i+1}, x, u) &= I^f(T_1, T_2, x, u) - I^f(T_1, t_i, x, u) - I^f(s_{i+1}, T_2, x, u) \\ &\le M + I^f(T_1, T_2, x_f, u_f) - I^f(T_1, t_i, x_f, u_f) + S_0 - I^f(s_{i+1}, T_2, x_f, u_f) + S_0 \\ &\le I^f(t_i, s_{i+1}, x_f, u_f) + M + 2S_0 \le M + 3S_0 + \sigma^f(t_i, s_{i+1}). \end{aligned} \tag{6.163}$$

It follows from (6.161)–(6.163) and property (ii) that

$$\|x(t) - x_f(t)\| \le \epsilon,\ t \in [t_i + L_0, s_{i+1} - L_0]$$

and together with (6.168) this implies that

$$\|x(t) - x_f(t)\| \le \epsilon,\ t \in [t_i + L_0, t_{i+1} - L_0 - 1]. \tag{6.164}$$

By (6.163),

$$\{t \in [T_1, T_2] :\ \|x(t) - x_f(t)\| > \epsilon\}$$

$$\subset \cup\{[t_i, t_i + L_0] \cup [t_{i+1} - 1 - L_0, t_{i+1}] :\ i \in \{0, \dots, q-1\} \text{ and } t_{i+1} - t_i \ge 2L_0 + 1\}$$

$$\cup\{[t_i, t_{i+1}] :\ i \in \{0, \dots, q-1\} \text{ and } t_{i+1} - t_i < 2L_0 + 1\}.$$

In view of (6.139) and (6.156), the maximal length of intervals in the right-hand side of the relation above does not exceed $2L_0 + 1 = l$ and their number does not exceed $2q \le Q$. Theorem 6.23 is proved.

6.12 An Auxiliary Result

Lemma 6.42. *Let $\epsilon > 0$. Then there exist $\delta, L > 0$ such that for each $T_1 \ge L$, each $T_2 > T_1$ and each $(x, u) \in X(T_1, T_2)$ which satisfies*

$$\|x(T_i) - x_f(T_i)\| \le \delta, \ i = 1, 2$$

the following inequality holds:

$$I^f(T_1, T_2, T, x, u) \ge I^f(T_1, T_2, T, x_f, u_f) - \epsilon.$$

Proof. By (A2), there exists $\delta > 0$ such that the following property holds:

(i) for each $(T, z) \in \mathcal{A}$ satisfying $\|z - x_f(T)\| \le \delta$, there exist $\tau_1 \in (0, b_f]$, $(\tilde{x}_1, \tilde{u}_1) \in X(T, T + \tau_1)$ such that

$$\tilde{x}_1(T) = z \ , \tilde{x}_1(T + \tau_1) = x_f(T + \tau_1),$$

$$I^f(T, T + \tau_1, \tilde{x}_1, \tilde{u}_1) \le I^f(T, T + \tau_1, x_f, u_f) + \epsilon/4,$$

and if $T \ge b_f$, then there exist $\tau_2 \in (0, b_f]$ and $(\tilde{x}_2, \tilde{u}_2) \in X(T - \tau_2, T)$ such that

$$\tilde{x}_2(T - \tau_2) = x_f(T - \tau_2) \ , \tilde{x}_2(T) = z,$$

$$I^f(T - \tau_2, T, \tilde{x}_2, \tilde{u}_2) \le I^f(T - \tau_2, T, x_f, u_f) + \epsilon/4.$$

Theorem 6.1 implies that there exists $L > 2b_f$ such that for each $T_1 \ge L - b_f$ and each $T_2 > T_1$

$$I^f(T_1, T_2, x_f, u_f) < U^f(T_1, T_2, x_f(T_1), x_f(T_2)) + \epsilon/4. \tag{6.165}$$

Assume that $T_1 \ge L$, $T_2 > T_1$, $(x, u) \in X(T_1, T_2)$, and

$$\|x(T_1) - x_f(T_1)\| \le \delta, \ \|x(T_2) - x_f(T_2)\| \le \delta. \tag{6.166}$$

By (6.166) and property (i), there exists $\tau_1, \tau_2 \in (0, b_f]$ and $(y, v) \in X(T_1 - \tau_1, T_2 + \tau_2)$ such that

$$y(T_1 - \tau_1) = x_f(T_1 - \tau_1), \tag{6.167}$$

$$(y(t), v(t)) = (x(t), u(t)) \text{ for all } t \in [T_1, T_2], \tag{6.168}$$

$$I^f(T_1 - \tau_1, T_1, y, v) \le I^f(T_1 - \tau_1, T_1, x_f, u_f) + \epsilon/4, \tag{6.169}$$

$$y(T_2 + \tau_2) = x_f(T_2 + \tau_2), \tag{6.170}$$

$$I^f(T_2, T_2 + \tau_2, y, v) \le I^f(T_2, T_2 + \tau_2, x_f, u_f) + \epsilon/4. \tag{6.171}$$

It follows from (6.165), (6.167), and (6.170) that

$$I^f(T_1 - \tau_1, T_2 + \tau_2, x_f, u_f) < I^f(T_1 - \tau_1, T_2 + \tau_2, y, v) + \epsilon/4. \tag{6.172}$$

Relations (6.168), (6.169), (6.171), and (6.172) imply that

$$\begin{aligned} I^f(T_1, T_2, x, u) &= I^f(T_1 - \tau_1, T_2 + \tau_2, y, v) \\ &\quad - I^f(T_1 - \tau_1, T_1, y, v) - I^f(T_2, T_2 + \tau_2, y, v) \\ &\ge I^f(T_1 - \tau_1, T_2 + \tau_2, x_f, u_f) - \epsilon/4 - I^f(T_1 - \tau_1, T_1, x_f, u_f) \\ &\quad - \epsilon/4 - I^f(T_2, T_2 + \tau_2, x_f, u_f) - \epsilon/4 \\ &= I^f(T_1, T_2, x_f, u_f) - 3\epsilon/4. \end{aligned}$$

Lemma 6.42 is proved.

6.13 Proof of Proposition 6.27

Proposition 1.3 implies that there exist $M_* \ge 1$, $\omega_* \in R^1$ such that

$$\|e^{At}\| \le M_* e^{\omega_* t}, \ t \in [0, \infty),$$

in the case of the second class of problems.

We show that (P1) holds.

Assume that $(x, u) \in X(0, \infty)$ is (f)-good. We show that $\lim_{t\to\infty} \|x(t) - x_f(t)\| = 0$. Assume the contrary. Then there exists $\epsilon > 0$ and a sequence of numbers $\{t_k\}_{k=1}^{\infty}$ such that

$$t_1 \geq 8, \; t_{k+1} \geq t_k + 8 \text{ for all integers } k \geq 1, \tag{6.173}$$

$$\|x(t_k) - x_f(t_k)\| > \epsilon \text{ for all integers } k \geq 1. \tag{6.174}$$

Theorem 6.2 implies that there exists $M_1 > 0$ such that

$$\|x_f(t)\| \leq M_1, \; \|x(t)\| \leq M_1, \; t \in [0, \infty), \tag{6.175}$$

$$|I^f(0, t, x, u) - I^f(0, t, x_f, u_f)| < M_1, \; t \in (0, \infty).$$

By the relation above, for all $S > T \geq 0$,

$$|I^f(T, S, x, u) - I^f(T, S, x_f, u_f)| \leq 2M_1. \tag{6.176}$$

Lemma 6.1 implies that there exists $M_2 > 0$ such that

$$I^f(S_1, S_2, x_f, u_f) \leq M_2 \text{ for all } S_1 \geq 0 \text{ and all } S_2 \in (S_1, S_1 + 1]. \tag{6.177}$$

Fix

$$\epsilon_1 = 4^{-1}\epsilon(2\Delta_1 + 1)^{-1}. \tag{6.178}$$

Form now we consider the two classes of problems separately and first consider the first class of problems. By (6.19) and (6.20), there exists $h_0 > 0$ such that

$$f(t, x, u) \geq 4\epsilon_1^{-1}(M_1 + M_2 + a_0 + 8)(\|G(t, x, u)\| - a_0\|x\|_+) \tag{6.179}$$

for each $(t, x, u) \in \mathcal{M}$ satisfying $\|G(t, x, u)\| - a_0\|x\| \geq h_0$.

There exists $\delta \in (0, 1)$ such that

$$\delta < \epsilon/8, \; \Delta_1\delta < \epsilon_1/32, \; \delta(a_0 M_1 + h_0) < \epsilon_1/4. \tag{6.180}$$

Let $k \geq 1$ be an integer. We show that for all $t \in [t_k - \delta, t_k]$,

$$\|x(t) - x_f(t)\| \geq \delta. \tag{6.181}$$

Assume the contrary. Then there exists

$$\tau \in [t_k - \delta, t_k] \tag{6.182}$$

such that

$$\|x(\tau) - x_f(\tau)\| < \delta. \tag{6.183}$$

By (6.11) and (6.182),

$$x(t_k) = U(t_k, \tau)x(\tau) + \int_\tau^{t_k} U(t_k, s)G(s, x(s), u(s))ds, \tag{6.184}$$

$$x_f(t_k) = U(t_k, \tau)x_f(\tau) + \int_\tau^{t_k} U(t_k, s)G(s, x_f(s), u_f(s))ds. \tag{6.185}$$

Property (v) (from Section 6.1), (6.181), (6.182), (6.184), and (6.185) imply that

$$\|x(t_k) - x_f(t_k)\| = \|U(t_k, \tau)\|\|x_f(\tau) - x(\tau)\|$$

$$+ \int_\tau^{t_k} \|U(t_k, s)\|\|G(s, x(s), u(s))\|ds$$

$$+ \int_\tau^{t_k} \|U(t_k, s)\|\|G(s, x_f(s), u_f(s))\|ds$$

$$\leq \Delta_\delta \|x_f(\tau) - x(\tau)\| + \Delta_\delta \int_\tau^{t_k} \|G(s, x(s), u(s))\|ds$$

$$+ \Delta_\delta \int_\tau^{t_k} \|G(s, x_f(s), u_g(s))\|ds$$

$$\leq \Delta_1 \delta + \Delta_\delta \int_\tau^{t_k} \|G(s, x(s), u(s))\|ds + \Delta_\delta \int_\tau^{t_k} \|G(s, x_f(s), u_f(s))\|ds. \tag{6.186}$$

Let

$$(y, v) \in \{(x, u),\ (x_f, u_f)\}. \tag{6.187}$$

Set

$$\Omega_1 = \{t \in [\tau, t_k] :\ \|G(t, y(t), v(t))\| - a_0\|y(t)\| \geq h_0\},\ \ \Omega_2 = [\tau, t_k] \setminus \Omega_1. \tag{6.188}$$

By (6.20), (6.175)–(6.177), (6.180), (6.182), (6.187), and (6.188),

$$\int_\tau^{t_k} \|G(s, y(s), v(s))\|ds \leq a_0 \int_\tau^{t_k} \|y(s)\|ds$$

$$+ \int_\tau^{t_k} (\|G(s, y(s), v(s))\| - a_0\|y(s)\|)_+ ds$$

$$\leq a_0\delta M_1 + h_0\delta + \int_{\Omega_1} (\|G(s, y(s), v(s))\| - a_0\|y(s)\|)_+ ds$$

$$\leq a_0\delta M_1 + h_0\delta + 4^{-1}\epsilon_1(M_1 + M_2 + a_0 + 8)^{-1} \int_{\Omega_1} f(s, y(s), v(s))ds$$

$$\leq a_0\delta M_1 + h_0\delta + (4^{-1}\epsilon_1(M_1 + M_2 + a_0 + 8)^{-1})(I^f(\tau, t_k, y, v) + a_0\delta)$$

$$\leq \delta(a_0 M_1 + h_0) + (4^{-1}\epsilon_1(M_1 + M_2 + a_0 + 8)^{-1})(I^f(\tau, t_k, x_f, u_f) + 2M_1 + a_0\delta)$$

$$\leq \delta(a_0 M_1 + h_0) + (4^{-1}\epsilon_1(M_1 + M_2 + a_0 + 8)^{-1})(M_2 + 2M_1 + a_0\delta) < \epsilon_1/4 + \epsilon_1/2.$$

In view of (6.174), (6.178), (6.180) and (6.186),

$$\epsilon \leq \|x(t_k) - x_f(t_k)\| \leq \Delta_1\delta + 2\Delta_\delta(\epsilon_1/4 + \epsilon_1/2)$$

$$\leq \epsilon_1/32 + 2\Delta_1\epsilon \leq \epsilon_1(2\Delta_1 + 1) = 4^{-1}\epsilon,$$

a contradiction. The contradiction we have reached proves that (6.181) holds for all $t \in [t_k - \delta, t_k]$. Together with (6.173) and (6.174), this implies that

$$\mathrm{mes}(\{t \in [0, T] : \|x(t) - x_f(t)\| \geq \delta\}) \to \infty \text{ as } T \to \infty.$$

On the other hand, property (P3) and (6.176) imply that there exists $L > 0$ such that for all $T > 0$,

$$\mathrm{mes}(\{t \in [0, T] : \|x(t) - x_f(t)\| \geq \delta\}) \leq L.$$

The contradiction we have reached proves that $\lim_{t\to\infty}(x(t) - x_f(t)) = 0$ and (P1) holds for the first class of problems.

Consider the second class of problems. Recall (see Proposition 1.18 and (1.9)) that for each $\tau \geq 0$, $\Phi_\tau \in \mathcal{L}(L^2(0, \tau; , F), E)$ is defined by

$$\Phi_\tau u = \int_0^\tau e^{A(\tau-s)} Bu(s)ds, \quad u \in L^2(0, \tau; F). \tag{6.189}$$

Lemma 6.42, (A2), and (A3) imply that there exist $\delta \in (0, 1)$, $\bar{L} > 0$ such that

$$M_* e^{|\omega_*|}\delta < \epsilon/8, \ \delta < b_f/16, \ 64a_0\delta < K_0\epsilon^2(\|\Phi_1\| + 1)^{-2} \tag{6.190}$$

and the following properties hold:

(i) for each $(T_i, z_i) \in \mathcal{A}$, $i = 1, 2$ satisfying $\|z_i - x_f(T_i)\| \leq \delta$, $i = 1, 2$ and $T_2 \geq b_f$ there exist $\tau_1, \tau_2 \in (0, b_f]$ and $(x_1, u_1) \in X(T_1, T_1 + \tau_1)$, $(x_2, u_2) \in X(T_2 - \tau_2, T_2)$ satisfying

$$x_1(T_1) = z_1 \,, x_1(T_1+\tau_1) = x_f(T_1+\tau_1), \; x_2(T_2) = z_2 \,, x_2(T_2-\tau_2) = x_f(T_2-\tau_2),$$

$$I^f(T_1, T_1+\tau_1, x_1, u_1) \le I^f(T_1, T_1+\tau_1, x_f, u_f) + 32^{-1}\epsilon^2 K_0(\|\Phi_1\|+1)^{-2},$$

$$I^f(T_2-\tau_2, T_2, x_2, u_2) \le I^f(T_2-\tau_2, T_2, x_f, u_f) + 32^{-1}\epsilon^2 K_0(\|\Phi_1\|+1)^{-2};$$

(ii) for each $t_1 \ge \bar{L}$, each $t_2 \in (t_1, t_1+4\delta]$,

$$I^f(t_1, t_2, x_f, u_f) \le 32^{-1}\epsilon^2 K_0(\|\Phi_1\|+1)^{-2};$$

(iii) for each $S_1 \ge \bar{L}$, each $S_2 > S_1$, and each $(\xi, \eta) \in X(S_1, S_2)$ which satisfies $\|\xi(S_i) - x_f(S_i)\| \le \delta$, $i = 1, 2$, we have

$$I^f(S_1, S_2, \xi, \eta) \ge I^f(S_1, S_2, x_f, u_f) - 32^{-1}\epsilon^2 K_0(\|\Phi_1\|+1)^{-2}.$$

Property (P3) and (6.176) imply that there exists $L > 0$ such that for all $T > 0$,

$$\mathrm{mes}(\{t \in [0, T] : \|x(t) - x_f(t)\| \ge \delta\}) \le L.$$

Therefore there exists $T_0 > 0$ such that for each $S > T_0$,

$$\mathrm{mes}(\{t \in [T_0, S] : \|x(t) - x_f(t)\| > \delta\}) \le \delta/2. \tag{6.191}$$

Since the pair (x, u) is (f)-good, it follows from Theorem 6.2 that there exists $T_1 > 0$ such that for each $S > T_1$,

$$I^f(T_1, S, x, u) \le U^f(T_1, S, x(T_1), x(S)) + 32^{-1}\epsilon^2 K_0(\|\Phi_1\|+1)^{-2}, \tag{6.192}$$

$$I^f(T_1, S, x_f, u_f) \le U^f(T_1, S, x_f(T_1), x_f(S)) + 32^{-1}\epsilon^2 K_0(\|\Phi_1\|+1)^{-2}. \tag{6.193}$$

Let $k \ge 1$ be an integer such that

$$t_k > T_0 + T_1 + 1 + \bar{L}. \tag{6.194}$$

Assume that

$$\inf\{\|x(t) - x_f(t)\| : t \in [t_k - \delta, t_k]\} < \delta. \tag{6.195}$$

Therefore there exists

$$\tau \in [t_k - \delta, t_k] \tag{6.196}$$

such that

$$\|x(\tau) - x_f(\tau)\| < \delta. \tag{6.197}$$

It is clear that $(x, u), (x_f, u_f) \in X(\tau, t_k)$. In view of (6.17), (6.189), and (6.196),

$$x(t_k) = e^{A(t_k-\tau)}x(\tau) + \int_\tau^{t_k} e^{A(t_k-s)}Bu(s)ds = e^{A(t_k-\tau)}x(\tau) + \Phi_{t_k-\tau}u(\tau + \cdot), \tag{6.198}$$

$$x_f(t_k) = e^{A(t_k-\tau)}x_f(\tau) + \int_\tau^{t_k} e^{A(t_k-s)}Bu_f(s)ds$$

$$= e^{A(t_k-\tau)}x_f(\tau) + \Phi_{t_k-\tau}u_f(\tau + \cdot), \tag{6.199}$$

Proposition 1.18, (6.174), (6.190), and (6.196)–(6.199) imply that

$$\epsilon \le \|x(t_k) - x_f(t_k)\|$$

$$= \|e^{A(t_k-\tau)}(x(\tau) - x_f(\tau))\| + \|\Phi_{t_k-\tau}u(\tau + \cdot)\| + \|\Phi_{t_k-\tau}u_f(\tau + \cdot)\|$$

$$\le M_* e^{|\omega_*|\delta}\|x(\tau) - x_f(\tau)\|$$

$$+\|\Phi_1\|(\int_\tau^{t_k} \|u(s)\|^2 ds)^{1/2} + \|\Phi_1\|(\int_\tau^{t_k} \|u_f(s)\|^2 ds)^{1/2}$$

$$< \epsilon/8 + \|\Phi_1\|(\int_\tau^{t_k} \|u(s)\|^2 ds)^{1/2} + \|\Phi_1\|(\int_\tau^{t_k} \|u_f(s)\|^2 ds)^{1/2}$$

and

$$(4^{-1}\epsilon(\|\Phi_1\| + 1)^{-1})^2 \le \max\{\int_\tau^{t_k} \|u(s)\|^2 ds, \int_\tau^{t_k} \|u_f(s)\|^2 ds\}. \tag{6.200}$$

Let

$$(y, v) \in \{(x, u), (x_f, u_f)\}, \tag{6.201}$$

$$\int_\tau^{t_k} \|v(s)\|^2 ds \ge \max\{\int_\tau^{t_k} \|u(s)\|^2 ds, \int_\tau^{t_k} \|u_f(s)\|^2 ds\}. \tag{6.202}$$

In view of (6.200)–(6.202),

$$\int_\tau^{t_k} \|v(s)\|^2 ds \ge (4^{-1}\epsilon)^2(\|\Phi_1\| + 1)^{-2}. \tag{6.203}$$

By (6.21) and (6.203),

$$\int_{\tau}^{t_k} f(s, y(s), v(s))ds + a_0(t_k - \tau) \geq K_0 \int_{\tau}^{t_k} \|v(s)\|^2 ds \geq 4^{-1}\epsilon^2 K_0(\|\Phi_1\| + 1)^{-2}. \tag{6.204}$$

It follows from (6.196) and (6.204) that

$$\int_{\tau}^{t_k} f(s, y(s), v(s))ds \geq 4^{-1} K_0 \epsilon^2 (\|\Phi_1\| + 1)^{-2} - a_0\delta. \tag{6.205}$$

In view of (6.191) and (6.194), for each $S > t_k$,

$$\mathrm{mes}(\{t \in [t_k, S] : \|y(t) - x_f(t)\| > \delta\}) \leq \delta/2. \tag{6.206}$$

Inequality (6.206) implies that there exist

$$S_1 \in [t_k, t_k + \delta] \tag{6.207}$$

such that

$$\|y(S_1) - x_f(S_1)\| \leq \delta \tag{6.208}$$

and

$$S_2 \in [S_1 + 2b_f, S_1 + 2b_f + \delta] \tag{6.209}$$

such that

$$\|y(S_2) - x_f(S_2)\| \leq \delta. \tag{6.210}$$

It follows from (6.205) and (6.207) that

$$\begin{aligned} I^f(\tau, S_1, y, v) &= I^f(\tau, t_k, y, v) + I^f(t_k, S_1, y, v) \\ &\geq 4^{-1} K_0 \epsilon^2 (\|\Phi_1\| + 1)^{-2} - a_0\delta - a_0\delta. \end{aligned} \tag{6.211}$$

Property (iii), (6.194), and (6.207)–(6.210) imply that

$$I^f(S_1, S_2, y, v) \geq I^f(S_1, S_2, x_f, u_f) - 32^{-1} K_0 \epsilon^2 (\|\Phi_1\| + 1)^{-2}. \tag{6.212}$$

By (6.196), (6.197), (6.207), (6.209), (6.211), (6.212), and property (ii),

$$\begin{aligned} I^f(\tau, S_2, y, v) &= I^f(\tau, S_1, y, v) + I^f(S_1, S_2, y, v) \\ &\geq 4^{-1} K_0 \epsilon^2 (\|\Phi_1\| + 1)^{-2} - 2a_0\delta \\ &+ I^f(S_1, S_2, x_f, u_f) - 32^{-1} K_0 \epsilon^2 (\|\Phi_1\| + 1)^{-2} \end{aligned}$$

$$\geq K_0\epsilon^2(\|\Phi_1\|+1)^{-2}(4^{-1}-32^{-1})$$

$$-2a_0\delta + I^f(\tau, S_2, x_f, u_f) - I^f(\tau, S_1, x_f, u_f)$$

$$\geq I^f(\tau, S_2, x_f, u_f) + K_0\epsilon^2(\|\Phi_1\|+1)^{-2}(4^{-1}-32^{-1})$$

$$-2a_0\delta - 16^{-1}K_0\epsilon^2(\|\Phi_1\|+1)^{-2}. \tag{6.213}$$

Property (i), (6.196), (6.197), (6.207), and (6.209) imply that that there exists $(\tilde{x}, \tilde{u}) \in X(\tau, S_2)$ such that

$$\tilde{x}(\tau) = y(\tau),\ \tilde{x}(\tau + b_f) = x_f(\tau + b_f), \tag{6.214}$$

$$I^f(\tau, \tau+b_f, \tilde{x}, \tilde{u}) \leq I^f(\tau, \tau+b_f, x_f, u_f) + 32^{-1}\epsilon^2 K_0(\|\Phi_1\|+1)^{-2}, \tag{6.215}$$

$$\tilde{x}(S_2) = y(S_2),\ \tilde{x}(S_2 - b_f) = x_f(S_2 - b_f), \tag{6.216}$$

$$I^f(S_2 - b_f, S_2, \tilde{x}, \tilde{u}) \leq I^f(S_2 - b_f, S_2, x_f, u_f) + 32^{-1}\epsilon^2 K_0(\|\Phi_1\|+1)^{-2}, \tag{6.217}$$

$$(\tilde{x}(t), \tilde{u}(t)) = (x_f(t), u_f(t)),\ t \in [\tau + b_f, S_2 - b_f]. \tag{6.218}$$

In view of (6.215), (6.217), and (6.218),

$$I^f(\tau, S_2, \tilde{x}, \tilde{u}) \leq I^f(\tau, S_2, x_f, u_f) + 16^{-1}\epsilon^2 K_0(\|\Phi_1\|+1)^{-2}. \tag{6.219}$$

It follows from (6.190), (6.213), and (6.219) that

$$I^f(\tau, S_2, y, v) - I^f(\tau, S_2, \tilde{x}, \tilde{u})$$

$$\geq \epsilon^2 K_0(\|\Phi_1\|+1)^{-2}(4^{-1} - 32^{-1} - 8^{-1}) - 2\delta a_0$$

$$\geq \epsilon^2 K_0(\|\Phi_1\|+1)^{-2}16^{-1}. \tag{6.220}$$

By (6.220),

$$I^f(\tau, S_2, y, v) \geq U^f(\tau, S_2, y(\tau), y(S_2)) + \epsilon^2 K_0(\|\Phi_1\|+1)^{-2}/16. \tag{6.221}$$

It follows from (6.192)–(6.194), (6.196), and (6.201) that $\tau > T_1$ and

$$I^f(\tau, S_2, y, v) \leq U^f(\tau, S_2, y(\tau), y(S_2)) + 32^{-1}\epsilon^2 K_0(\|\Phi_1\|+1)^{-2}.$$

This contradicts (6.221). The contradiction we have reached proves that (6.195) is not true and

$$\|x(t) - x_f(t)\| \geq \delta,\ t \in [t_k - \delta, t_k]$$

for each natural number k satisfying $t_k \geq T_0 + T_1 + 1 + \bar{L}$. Together with (6.173) and (6.174), this implies that

$$\text{mes}(\{t \in [0, T] : \|x(t) - x_f(t)\| \geq \delta\}) \to \infty \text{ as } T \to \infty.$$

This contradicts (6.191) holding for all $S > T_0$. The contradiction we have reached proves that $\lim_{t\to\infty}(x(t) - x_f(t) = 0$ and (P1) holds for the second class of problems too. Proposition 6.27 is proved.

6.14 Auxiliary Results for Theorems 6.28 and 6.29

Recall that for each $z \in R^1$, $\lfloor z \rfloor = \max\{i \in R^1 : i \text{ is an integer}, i \leq z\}$.

Proposition 6.43. *Assume that f has (P1), (P2), and LSC property, $T_0 \geq 0$, $z_0 \in E$, $(T_0, z_0) \in \cup\{\mathcal{A}_L : L \in (0, \infty)\}$. Then there exists an (f)-good and (f)-minimal pair $(x_*, u_*) \in X(T_0, \infty)$ such that $x_*(T_0) = z_0$.*

Proof. There exists $L_0 > 0$ such that

$$(T_0, z_0) \in \mathcal{A}_{L_0}. \tag{6.222}$$

It follows from Theorem 6.2 that there exists $S_0 > 0$ such that for each $T_2 > T_1 \geq 0$ and each $(x, u) \in X(T_1, T_2)$,

$$I^f(T_1, T_2, x, u) + S_0 \geq I^f(T_1, T_2, x_f, u_f). \tag{6.223}$$

Fix an integer $k_0 \geq L_0$. LSC property and (6.222) imply that for each integer $k \geq k_0$, there exists $(x_k, u_k) \in X(T_0, T_0 + k)$ satisfying

$$x_k(T_0) = z_0, \tag{6.224}$$

$$I^f(T_0, T_0 + k, x_k, u_k) = \sigma^f(T_0, T_0 + k, z_0). \tag{6.225}$$

In view of (6.20), (6.21), and (6.222), for each integer $k \geq k_0$,

$$\sigma^f(T_0, T_0 + k, z_0) \leq L_0 + I^f(T_0, T_0 + k, x_f, u_f) + a_0 L_0. \tag{6.226}$$

By (6.223), (6.225), and (6.226), for each integer $k \geq k_0$ and each pair of numbers $T_1, T_2 \in [T_0, T_0 + k]$ satisfying $T_1 < T_2$,

$$I^f(T_1, T_2, x_k, u_k) = I^f(T_0, T_0 + k, x_k, u_k)$$

$$-I^f(T_0, T_1, x_k, u_k) - I^f(T_0 + k, T_2, x_k, u_k)$$

$$\leq I^f(T_0, T_0+k, x_f, u_f) + L_0(1+a_0) - I^f(T_0, T_1, x_f, u_f)$$

$$+S_0 - I^f(T_0+k, T_2, x_k, u_k) + S_0$$

$$= I^f(T_1, T_2, x_f, u_f) + 2S_0 + L_0(1+a_0). \quad (6.227)$$

By (6.227) and LSC property, extracting subsequences, using the diagonalization process and re-indexing, we obtain that there exists a strictly increasing sequence of natural numbers $\{k_p\}_{p=1}^{\infty}$ such that $k_1 \geq k_0$ and for each integer $i \geq 0$ there exists $\lim_{p\to\infty} I^f(T_0+i, T_0+i+1, x_{k_p}, u_{k_p})$ and there exists $(y_i, v_i) \in X(T_0+i, T_0+i+1)$ such that

$$x_{k_p}(t) \to y_i(t) \text{ as } p \to \infty \text{ for all } t \in [T_0+i, T_0+i+1], \quad (6.228)$$

$$I^f(T_0+i, T_0+i+1, y_i, v_i) \leq \lim_{p\to\infty} I^f(T_0+i, T_0+i+1, x_{k_p}, u_{k_p}). \quad (6.229)$$

In view of (6.228), there exists $(x_*, u_*) \in X(T_0, \infty)$ such that for each integer $i \geq 0$,

$$(x_*(t), u_*(t)) = (y_i(t), v_i(t)), \; t \in [T_0+i, T_0+i+1]. \quad (6.230)$$

It follows from (6.227), (6.229), and (6.230) that for every integer $q \geq 1$,

$$I^f(T_0, T_0+q, x_*, u_*) \leq \lim_{p\to\infty} I^f(T_0, T_0+q, x_{k_p}, u_{k_p})$$

$$\leq I^f(T_0, T_0+q, x_f, u_f) + 2S_0 + L_0(1+a_0). \quad (6.231)$$

Theorem 6.2, (6.224), (6.228), (6.230), and (6.231) imply that (x_*, u_*) is (f)-good and

$$x_*(T_0) = z_0. \quad (6.232)$$

In order to complete the proof of the proposition, it is sufficient to show that (x_*, u_*) is (f)-minimal. Assume the contrary. Then there exist $\Delta > 0$, an integer $\tau_0 \geq 1$ and $(y, v) \in X(T_0, T_0+\tau_0)$ such that

$$y(T_0) = x_*(T_0), \; y(T_0+\tau_0) = x_*(T_0+\tau_0), \quad (6.233)$$

$$I^f(T_0, T_0+\tau_0, x_*, u_*) > I^f(T_0, T_0+\tau_0, y, v) + 2\Delta. \quad (6.234)$$

By (A2) and Lemma 6.42, there exist $L_1, \delta > 0$ such that the following properties hold:

(i) for each $(T, \xi) \in \mathcal{A}$ satisfying $\|\xi - x_f(T)\| \le \delta$, there exist $\tau_1 \in (0, b_f]$ and $(\tilde{x}_1, \tilde{u}_1) \in X(T, T + \tau_1)$ such that

$$\tilde{x}_1(T) = \xi \ , \tilde{x}_1(T + \tau_1) = x_f(T + \tau_1),$$

$$I^f(T, T + \tau_1, \tilde{x}_1, \tilde{u}_1) \le I^f(T, T + \tau_1, x_f, u_f) + \Delta/8$$

and if $T \ge b_f$, then there exist $\tau_2 \in (0, b_f]$ and $(\tilde{x}_2, \tilde{u}_2) \in X(T - \tau_2, T)$ such that

$$\tilde{x}_2(T - \tau_2) = x_f(T - \tau_2) \ , \tilde{x}_2(T) = \xi,$$

$$I^f(T - \tau_2, T, \tilde{x}_2, \tilde{u}_2) \le I^f(T - \tau_2, T, x_f, u_f) + \Delta/8;$$

(ii) if $T_2 > T_1 \ge L_1$, $(x, u) \in X(T_1, T_2)$, $\|x(T_i) - x_f(T_i)\| \le \delta$, $i = 1, 2$, we have

$$I^f(T_1, T_2, x, u) \ge I^f(T_1, T_2, x_f, u_f) - \Delta/8.$$

Theorems 6.17 and 6.22, (6.222), (6.224), and (6.225) imply that there exists an integer $L_2 > L_0 + L_1$ such that for each integer $k \ge k_0 + 2L_2$,

$$\|x_k(t) - x_f(t)\| \le \delta, \ t \in [T_0 + L_2, T_0 + k - L_2]. \tag{6.235}$$

In view of (6.228), (6.230), and (6.235),

$$\|x_*(t) - x_f(t)\| \le \delta \ \text{ for all } t \ge T_0 + L_2. \tag{6.236}$$

By (6.228) and (6.230), there exists a natural number p_0 such that

$$k_{p_0} > k_0 + 2L_1 + 2L_2 + 2\tau_0 + 2 + 2b_f + 2T_0, \tag{6.237}$$

$$I^f(T_0, T_0 + \tau_0 + L_2, x_*, u_*) \le I^f(T_0, T_0 + \tau_0 + L_2, x_{k_{p_0}}, u_{k_{p_0}}) + \Delta/2. \tag{6.238}$$

Property (i), (6.233), and (6.235)–(6.237) imply that there is $(x, u) \in X(T_0, T_0 + k_{p_0})$ such that

$$(x(t), u(t)) = (y(t), v(t)), \ t \in [T_0, T_0 + \tau_0], \tag{6.239}$$

$$(x(t), u(t)) = (x_*(t), u_*(t)), \ t \in (T_0 + \tau_0, T_0 + \tau_0 + L_2], \tag{6.240}$$

$$x(T_0 + \tau_0 + L_2 + b_f) = x_f(T_0 + \tau_0 + L_2 + b_f), \tag{6.241}$$

$$I^f(T_0 + \tau_0 + L_2, T_0 + \tau_0 + L_2 + b_f, x, u)$$

$$\leq I^f(T_0 + \tau_0 + L_2, T_0 + \tau_0 + L_2 + b_f, x_f, u_f) + \Delta/8, \tag{6.242}$$

$$(x(t), u(t)) = (x_{k_{p0}}(t), u_{k_{p0}}(t)),\ t \in [T_0 + \tau_0 + L_2 + 2b_f, k_{p0} + T_0], \tag{6.243}$$

$$I^f(T_0 + \tau_0 + L_2 + b_f, T_0 + \tau_0 + L_2 + 2b_f, x, u)$$

$$\leq I^f(T_0 + \tau_0 + L_2 + b_f, T_0 + \tau_0 + L_2 + 2b_f, x_f, u_f) + \Delta/8. \tag{6.244}$$

It follows from (6.224), (6.225), (6.232), (6.233), and (6.238) that

$$I^f(T_0, T_0 + k_{p0}, x, u) \geq I^f(T_0, T_0 + k_{p0}, x_{k_{p0}}, u_{k_{p0}}). \tag{6.245}$$

By (6.234), (6.235), (6.237)–(6.240), (6.242)–(6.245), and property (ii),

$$0 \leq I^f(T_0, T_0 + k_{p0}, x, u) - I^f(T_0, T_0 + k_{p0}, x_{k_{p0}}, u_{k_{p0}})$$

$$= I^f(T_0, T_0 + \tau_0, y, v) + I^f(T_0 + \tau_0, T_0 + \tau_0 + L_2, x_*, u_*)$$

$$+ I^f(T_0 + \tau_0 + L_2, T_0 + \tau_0 + L_2 + b_f, x_f, u_f) + \Delta/8$$

$$+ I^f(T_0 + \tau_0 + L_2 + b_f, T_0 + \tau_0 + L_2 + 2b_f, x_f, u_f) + \Delta/8$$

$$- I^f(T_0, T_0 + \tau_0 + L_2, x_{k_{p0}}, u_{k_{p0}})$$

$$- I^f(T_0 + \tau_0 + L_2, T_0 + \tau_0 + L_2 + b_f, x_{k_{p0}}, u_{k_{p0}})$$

$$- I^f(T_0 + \tau_0 + L_2 + b_f, T_0 + \tau_0 + L_2 + 2b_f, x_{k_{p0}}, u_{k_{p0}})$$

$$\leq I^f(T_0, T_0 + \tau_0, y, v) + I^f(T_0 + \tau_0, T_0 + \tau_0 + L_2, x_*, u_*)$$

$$+ I^f(T_0 + \tau_0 + L_2, T_0 + \tau_0 + L_2 + b_f, x_f, u_f) + \Delta/8$$

$$+ I^f(T_0 + \tau_0 + L_2 + b_f, T_0 + \tau_0 + L_2 + 2b_f, x_f, u_f) + \Delta/8$$

$$- I^f(T_0, T_0 + \tau_0 + L_2, x_{k_{p0}}, u_{k_{p0}})$$

$$- I^f(T_0 + \tau_0 + L_2, T_0 + \tau_0 + L_2 + 2b_f, x_{k_{p0}}, u_{k_{p0}}) + \Delta/8$$

$$= I^f(T_0, T_0 + \tau_0, y, v) + I^f(T_0 + \tau_0, T_0 + \tau_0 + L_2, x_*, u_*)$$

$$- I^f(T_0, T_0 + \tau_0 + L_2, x_{k_{p0}}, u_{k_{p0}}) + \Delta/2$$

$$< I^f(T_0, T_0 + \tau_0, x_*, u_*) - 2\Delta + I^f(T_0 + \tau_0, T_0 + \tau_0 + L_2, x_*, u_*)$$

$$-I^f(T_0, T_0 + \tau_0 + L_2, x_{k_{p0}}, u_{k_{p0}}) + \Delta/2 \le -2\Delta + \Delta/2 + \Delta/2,$$

a contradiction. The contradiction we have reached completes the proof of Proposition 6.43.

6.15 Proofs of Theorems 6.28 and 6.29

Proof of Theorem 6.29. Clearly, (i) implies (ii). In view of Theorem 6.2, (ii) implies (iii). By (P1), (iii) implies (iv). Evidently (iv) implies (v). We show that (v) implies (iii). Assume that (x_*, u_*) is (f)-minimal and satisfies

$$\liminf_{t\to\infty} \|\tilde{x}(t) - x_f(t)\| = 0. \tag{6.246}$$

Since $(\tilde{x}, \tilde{u})$ is (f)-good, there exists $S_0 > 0$ such that for all numbers $T > S$,

$$|I^f(S, T, \tilde{x}, \tilde{u}) - I^f(S, T, x_f, u_f)| \le S_0. \tag{6.247}$$

By (A2), there exists $\delta > 0$ such that the following property holds:

(a) for each $(T, z) \in \mathcal{A}$ satisfying $\|z - x_f(T)\| \le \delta$, there exist $\tau_1 \in (0, b_f]$ and $(x_1, u_1) \in X(T, T + \tau_1)$ which satisfy

$$x_1(T) = z, \ x_1(T + \tau_1) = x_f(T + \tau_1),$$

$$I^f(T, T + \tau_1, x_1, u_1) \le I^f(T, T + \tau_1, x_f, u_f) + 1,$$

and if $T \ge b_f$, then there exist $\tau_2 \in (0, b_f]$ and $(x_2, u_2) \in X(T - \tau_2, T)$ such that

$$x_2(T - \tau_2) = x_f(T - \tau_2), \ x_2(T) = z,$$

$$I^f(T - \tau_2, T, x_2, u_2) \le I^f(T - \tau_2, T, x_f, u_f) + 1.$$

In view of (6.246) and (P1), there exists an increasing sequence $\{t_k\}_{k=1}^{\infty} \subset (S, \infty)$ such that

$$\lim_{k\to\infty} t_k = \infty, \tag{6.248}$$

$$\|x_*(t_k + 2b_f) - x_f(t_k + 2b_f)\| \le \delta \text{ for all integers } k \ge 1, \tag{6.249}$$

$$\|\tilde{x}(t) - x_f(t)\| \le \delta \text{ for all } t \ge t_0. \tag{6.250}$$

Let $k \ge 1$ be an integer. By property (a), (6.249), and (6.250), there exists $(y, v) \in X(S, t_k + 2b_f)$ such that

$$(y(t), v(t)) = (\tilde{x}(t), \tilde{u}(t)),\ t \in [S, t_k],$$

$$y(t_k + b_f) = x_f(t_k + b_f),$$

$$I^f(t_k, t_k + b_f, y, v) \le I^f(t_k, t_k + b_f, x_f, u_f) + 1,$$

$$y(t_k + 2b_f) = x_*(t_k + 2b_f),$$

$$I^f(t_k + b_f, t_k + 2b_f, y, v) \le I^f(t_k + b_f, t_k + 2b_f, x_f, u_f) + 1.$$

The relations above and (6.247) imply that

$$I^f(S, t_k + 2b_f, x_*, u_*) \le I^f(S, t_k + 2b_f, y, v)$$

$$\le I^f(S, t_k, \tilde{x}, \tilde{u}) + I^f(t_k, t_k + 2b_f, x_f, u_f) + 2$$

$$\le I^f(S, t_k + 2b_f, x_f, u_f) + 2 + 2S_0.$$

Together with Theorem 6.2, this implies that (x_*, u_*) is (f)-good and (iii) holds.

We show that (iii) implies (i). Assume that (x_*, u_*) is (f)-minimal and (f)-good. (P1) implies that

$$\lim_{t\to\infty} \|x_*(t) - x_f(t)\| = 0. \tag{6.251}$$

Since (x_*, u_*) is (f)-good, there exists $S_0 > 0$ such that

$$|I^f(S, T, x_*, u_*) - I^f(S, T, x_f, u_f)| \le S_0 \text{ for all } T > S. \tag{6.252}$$

Let $(x, u) \in X(S, \infty)$ satisfy

$$x(S) = x_*(S). \tag{6.253}$$

We show that

$$\limsup_{T\to\infty} [I^f(S, T, x_*, u_*) - I^f(S, T, x, u)] \le 0.$$

In view of Theorem 6.2 and (6.252), we may assume that (x, u) is (f)-good. (P1) implies that

$$\lim_{t\to\infty} \|x(t) - x_f(t)\| = 0. \tag{6.254}$$

Let $\epsilon > 0$. By (A2) and Lemma 6.42, there exist $\delta \in (0, \epsilon)$ and $L_1 > 0$ such that the following properties hold:

(b) for each $(T, z) \in \mathcal{A}$ satisfying $\|z - x_f(T)\| \le \delta$, there exist $\tau_1 \in (0, b_f]$ and $(x_1, u_1) \in X(T, T + \tau_1)$ satisfying

$$x_1(T) = z \,, x_1(T + \tau_1) = x_f(T + \tau_1),$$

$$I^f(T, T + \tau_1, x_1, u_1) \le I^f(T, T + \tau_1, x_f, u_f) + \epsilon/8,$$

and if $T \ge b_f$, then there exist $\tau_2 \in (0, b_f]$ and $(x_2, u_2) \in X(T - \tau_2, T)$ satisfying

$$x_2(T - \tau_2) = x_f(T - \tau_2) \,, x_2(T) = z,$$

$$I^f(T - \tau_2, T, x_2, u_2) \le I^f(T - \tau_2, T, x_f, u_f) + \epsilon/8;$$

(c) for each $T_1 \ge L_1$, each $T_2 > T_1$, and each $(y, v) \in X(T_1, T_2)$ which satisfies $\|y(T_i) - x_f(T_i)\| \le \delta$, $i = 1, 2$, we have

$$I^f(T_1, T_2, y, v) \ge I^f(T_1, T_2, x_f, u_f) - \epsilon/8.$$

It follows from (6.251) and (6.254) that there exists $\tau_0 > 0$ such that

$$\|x(t) - x_f(t)\| \le \delta, \;\; \|x_*(t) - x_f(t)\| \le \delta \text{ for all } t \ge \tau_0. \tag{6.255}$$

Let

$$T \ge \tau_0 + L_1. \tag{6.256}$$

Property (b), (6.255), and (6.256) imply that there exists $(y, v) \in X(S, T + 2b_f)$ which satisfies

$$(y(t), v(t)) = (x(t), u(t)), \; t \in [S, T], \; y(T + b_f) = x_f(T + b_f), \tag{6.257}$$

$$I^f(T, T + b_f, y, v) \le I^f(T, T + b_f, x_f, u_f) + \epsilon/8, \tag{6.258}$$

$$y(T + 2b_f) = x_*(T + 2b_f), \tag{6.259}$$

$$I^f(T + b_f, T + 2b_f, y, v) \le I^f(T + b_f, T + 2b_f, x_f, u_f) + \epsilon/8. \tag{6.260}$$

By property (c) and (6.255),

$$I^f(T, T + 2b_f, x_*, u_*) \geq I^f(T, T + 2b_f, x_f, u_f) - \epsilon/8. \tag{6.261}$$

It follows from (6.253) and (6.257)–(6.261) that

$$I^f(S, T, x_*, u_*) + I^f(T, T + 2b_f, x_f, u_f) - \epsilon/8$$

$$\leq I^f(S, T + 2b_f, x_*, u_*) \leq I^f(S, T + 2b_f, y, v)$$

$$= I^f(S, T, x, u) + I^f(T, T + 2b_f, x_f, u_f) + \epsilon/4,$$

$$I^f(S, T, x_*, u_*) \leq I^f(S, T, x, u) + \epsilon \text{ for all } T \geq \tau_0 + L_1.$$

Since ϵ is any positive number, we conclude that

$$\limsup_{T \to \infty}[I^f(S, T, x_*, u_*) - I^f(S, T, x, u)] \leq 0,$$

(x_*, u_*) is (f)-overtaking optimal and (i) holds. Theorem 6.29 is proved.

Theorem 6.29 and Proposition 6.43 imply Theorem 6.28.

6.16 Auxiliary Results for Theorem 6.33

In the next lemma, we consider only the first class of problems with $A(t) = 0$ for all $t \geq 0$.

Lemma 6.44. *Let $M_0, M_1, \tau_0 > 0$. Then there exists $M_2 > M_1$ such that for each $T_1 \geq 0$, each $T_2 \in (T_1, T_1 + \tau_0]$, and each $(x, u) \in X(T_1, T_2)$ satisfying*

$$\inf\{\|x(t)\| : t \in [T_1, T_2]\} \leq M_1, \tag{6.262}$$

$$I^f(T_1, T_2, x, u) \leq M_0 \tag{6.263}$$

the following inequality holds:

$$\|x(t)\| \leq M_2 \text{ for all } t \in [T_1, T_2]. \tag{6.264}$$

Proof. Fix a positive number

$$\delta < \min\{8^{-1}\tau_0, 2^{-1}(a_0 + 1)^{-1}\}. \tag{6.265}$$

By (6.19) and (6.20), there exist $h_0 > M_1 + 1$ and $\gamma_0 > 0$ such that

$$f(t,x,u) \geq 4(M_0 + a_0\tau_0)\delta^{-1} \text{ for each } (t,x,u) \in \mathcal{M} \text{ satisfying } \|x\| \geq h_0, \tag{6.266}$$

$$f(t,x,u) \geq 8(\|G(t,x,u)\| - a_0\|x\|)_+$$

$$\text{for each } (t,x,u) \in \mathcal{M} \text{ satisfying } \|G(t,x,u)\| - a_0\|x\| \geq \gamma_0. \tag{6.267}$$

Choose a number

$$M_2 > 2h_0 + 2M_1 + 2\gamma_0\delta + M_0 + a_0\tau_0. \tag{6.268}$$

Let

$$T_1 \geq 0,\ T_2 \in (T_1, T_1 + \tau_0],\ (x,u) \in X(T_1, T_2),$$

(6.262) and (6.263) hold. We show that (6.264) hold. Assume the contrary. Then there exists $t_0 \in [T_1, T_2]$ such that

$$\|x(t_0)\| > M_2. \tag{6.269}$$

By the choice of h_0 (see (6.266)), there exists $t_1 \in [T_1, T_2]$ such that

$$\|x(t_1)\| \leq h_0,\ |t_1 - t_0| \leq \delta. \tag{6.270}$$

There exists

$$t_2 \in [\min\{t_1, t_0\}, \max\{t_1, t_0\}] \tag{6.271}$$

such that

$$\|x(t_2)\| \geq \|x(t)\| \text{ for all } t \in [\min\{t_1, t_0\}, \max\{t_1, t_0\}]. \tag{6.272}$$

In view of (6.268)–(6.270) and (6.272)

$$t_2 \neq t_1.$$

By (6.11) (with $U(t,s) = Id$),

$$\|x(t_2) - x(t_1)\| = \|\int_{t_1}^{t_2} G(x(s), u(s))ds\|. \tag{6.273}$$

It follows from (6.270) and (6.273) that

$$\|x(t_2)\| - h_0 \leq |\int_{t_1}^{t_2} \|G(s, x(s), u(s))\|ds|. \tag{6.274}$$

Define

$$E_1 = \{t \in [\min\{t_1, t_2\}, \max\{t_1, t_2\}] : \ \|G(t, x(t), u(t))\| \geq a_0\|x(t)\| + \gamma_0\}, \tag{6.275}$$

$$E_2 = [\min\{t_1, t_2\}, \max\{t_1, t_2\}] \setminus E_1. \tag{6.276}$$

By (6.263), (6.270), (6.272), and (6.274)–(6.276),

$$\|x(t_2)\| - h_0 \leq a_0 \int_{E_1 \cup E_2} \|x(t)\|dt + \int_{E_1 \cup E_2} [\|G(s, x(s), u(s))\| - a_0\|x(s)\|]_+ ds$$

$$\leq a_0\delta\|x(t_2)\| + \int_{E_1} (\|G(t, x(t), u(t))\| - a_0\|x(t)\|)_+ dt$$

$$+ \int_{E_2} (\|G(t, x(t), u(t))\| - a_0\|x(t)\|)_+ dt$$

$$\leq a_0\delta\|x(t_2)\| + \gamma_0\delta + \int_{E_1} (\|G(t, x(t), u(t))\| - a_0\|x(t)\|)_+ dt$$

$$\leq a_0\|x(t_2)\|\delta + \gamma_0\delta + 8^{-1} \int_{E_1} f(t, x(t), u(t))dt$$

$$\leq a_0\|x(t_2)\|\delta + \gamma_0\delta + 8^{-1}(M_0 + a_0\tau_0).$$

It follows from the relation above, (6.265), (6.269), and (6.272) that

$$2^{-1}M_2 \leq 2^{-1}\|x(t_2)\| \leq \|x(t_2)\|(1 - a_0\delta) \leq h_0 + \gamma_0\delta + 8^{-1}(M_0 + a_0\tau_0)$$

and

$$M_2 < 2h_0 + 2\gamma_0\delta + 4^{-1}(M_0 + a_0\tau_0).$$

This contradicts (6.268). The contradiction we have reached proves Lemma 6.44.

Lemma 6.44, (6.19), and (6.20) imply the following result.

Lemma 6.45. *Let $M_1 > 0, 0 < \tau_0 < \tau_1$. Then there exists $M_2 > 0$ such that for each $T_1 \geq 0$, each $T_2 \in [T_1 + \tau_0, T_1 + \tau_1]$, and each $(x, u) \in X(T_1, T_2)$ satisfying*

$$I^f(T_1, T_2, x, u) \leq M_1$$

the following inequality holds:

$$\|x(t)\| \leq M_2 \text{ for all } t \in [T_1, T_2].$$

Lemma 6.46. *Let $M_1 > 0$, $\epsilon \in (0, 1)$, $0 < \tau_0 < \tau_1$. Then there exists $\delta > 0$ such that for each $T_1 \ge 0$, each $T_2 \in [T_1 + \tau_0, T_1 + \tau_1]$, each $(x, u) \in X(T_1, T_2)$ satisfying*

$$I^f(T_1, T_2, x, u) \le M_0,$$

and each $t_1, t_2 \in [T_1, T_2]$ satisfying $|t_1 - t_2| \le \delta$ the inequality $\|x(t_1) - x(t_2)\| \le \epsilon$ holds.

Proof. Lemma 6.45 implies that there exists $M_2 > 0$ such that the following property holds:

(i) for each $T_1 \ge 0$, each $T_2 \in [T_1 + \tau_0, T_1 + \tau_1]$, and each $(x, u) \in X(T_1, T_2)$ satisfying

$$I^f(T_1, T_2, x, u) \le M_1,$$

we have

$$\|x(t)\| \le M_2 \text{ for all } t \in [T_1, T_2].$$

In view of (6.19) and (6.20), there exists $h_0 > 0$ such that

$$f(t, x, u) \ge 4\epsilon^{-1}(M_1 + a_0\tau_1 + 8)(\|G(t, x, u)\| - a_0\|x\|)_+ \tag{6.277}$$

$$\text{for each } (t, x, u) \in \mathcal{M} \text{ satisfying } \|G(t, x, u)\| - a_0\|x\| \ge h_0.$$

Choose a number

$$\delta \in (0, \epsilon(4a_0M_2 + 4h_0 + 4)^{-1}). \tag{6.278}$$

Let

$$T_1 \ge 0,\ T_2 \in [T_1 + \tau_0, T_1 + \tau_1],\ (x, u) \in X(T_1, T_2),$$

$$I^f(T_1, T_2, x, u) \le M_1, \tag{6.279}$$

$$t_1, t_2 \in [T_1, T_2],\ 0 < t_2 - t_2 \le \delta. \tag{6.280}$$

Property (i) and (6.279) imply that

$$\|x(t)\| \le M_2 \text{ for all } t \in [T_1, T_2]. \tag{6.281}$$

Define

$$\Omega_1 = \{t \in [t_1, t_2] : \|G(t, x(t), u(t))\| - a_0\|x(t)\| \geq h_0\}, \tag{6.282}$$

$$\Omega_2 = [t_1, t_2] \setminus \Omega_1. \tag{6.283}$$

By (6.11) and (6.277)–(6.283),

$$\|x(t_2) - x(t_1)\| = \| \int_{t_1}^{t_2} G(x(s), u(s))ds\|$$

$$\leq a_0 \int_{t_1}^{t_2} \|x(t)\|dt + \int_{t_1}^{t_2} (\|G(t, x(t), u(t))\| - a_0\|x(t)\|)_+ dt$$

$$\leq a_0\delta M_2 + \delta h_0 + \int_{E_1} (\|G(t, x(t), u(t)) - a_0\|x(t)\|)_+ dt$$

$$\leq a_0\delta M_2 + \delta h_0 + \epsilon(4(M_1 + a_0\tau_1 + 8))^{-1} \int_{E_1} f(t, x(t), u(t))dt$$

$$\leq a_0\delta M_2 + \delta h_0 + 4^{-1}\epsilon < \epsilon.$$

Lemma 6.46 is proved.

Lemma 6.47. *Let $\Delta > 0$. Then there exists $\delta > 0$ such that for each $(T_1, z_1), (T_2, z_2) \in \mathcal{A}$ satisfying*

$$T_2 \geq T_1 + 2b_f,\ \|z_i - x_f(T_i)\| \leq \delta,\ i = 1, 2 \tag{6.284}$$

the following inequality holds:

$$U^f(T_1, T_2, z_1, z_2) \leq I^f(T_1, T_2, x_f, u_f) + \Delta.$$

Proof. Let $\delta > 0$ be as guaranteed by (A2) with $\epsilon = \Delta/4$. Let

$$(T_1, z_1), (T_2, z_2) \in \mathcal{A}$$

satisfy (6.284). By (6.284), the choice of δ and (A2) with $\epsilon = \Delta/4$, there exist $(y, v) \in X(T_1, T_2)$ such that

$$y(T_1) = z_1,\ y(T_2) = z_2,$$

$$y(t) = x_f(t),\ v(t) = u_f(t),\ t \in [T_1 + b_f, T_2 - b_f],$$

$$I^f(T_1, T_1 + b_f, y, v) \leq I^f(T_1, T_1 + b_f, x_f, u_f) + \Delta/4,$$

$$I^f(T_2 - b_f, T_2, y, v) \leq I^f(T_2 - b_f, T_2, x_f, u_f) + \Delta/4.$$

By the relations above,

$$U^f(T_1, T_2, z_1, z_2) \le I^f(T_1, T_2, y, v) \le I^f(T_1, T_2, x_f, u_f) + \Delta/2.$$

Lemma 6.47 is proved.

Lemma 6.48. *Let $\Delta \in (0, 1)$. Then there exists $\delta > 0$ such that for each $T_1 \ge 0$, each $T_2 \ge T_1 + 3b_f$, each $(x, u) \in X(T_1, T_2)$ satisfying*

$$\|x(T_i) - x_f(T_i)\| \le \delta, \ i = 1, 2,$$

$$I^f(T_1, T_2, x, u) \le U^f(T_1, T_2, x(T_1), x(T_2)) + \delta,$$

and each $\tau_1 \in [T_1, T_1 + \delta]$ and each $\tau_2 \in [T_2 - \delta, T_2]$ the inequality

$$I^f(\tau_1, \tau_2, x, u) \le I^f(\tau_1, \tau_2, x_f, u_f) + \Delta$$

holds.

Proof. Lemma 6.47 implies that there exists

$$\delta_0 \in (0, \min\{4^{-1}\Delta, 2^{-1}b_f\})$$

such that the following property holds:

(ii) for each $(T_1, z_1), (T_2, z_2) \in \mathcal{A}$ satisfying

$$T_2 \ge T_1 + 2b_f, \ \|z_i - x_f(T_i)\| \le \delta_0, \ i = 1, 2$$

we have

$$U^f(T_1, T_2, z_1, z_2) \le I^f(T_1, T_2, x_f, u_f) + \Delta/4.$$

Theorem 6.2 implies that exists $M_0 > 0$ such that the following property holds:

(iii) for each pair of numbers $T_2 > T_1 \ge 0$ and each $(x, u) \in X(T_1, T_2)$,

$$I^f(T_1, T_2, x, u) + M_0 \ge I^f(T_1, T_2, x_f, u_f).$$

Lemma 6.46 implies that there exists $\delta_1 \in (0, \delta_0)$ such that the following property holds:

(iv) for each $S \ge 0$, each $(y, v) \in X(S, S + b_f)$ satisfying

$$I^f(S, S + b_f, y, v) \le |\Delta_f|(b_f + 2) + 3a_0 + 2M_0 + 1$$

and each $t_1, t_2 \in [S, S+b_f]$ satisfying $|t_1-t_2| \leq \delta_1$ we have $\|y(t_1)-y(t_2)\| \leq \delta_0/8$.

Set

$$\delta = \delta_1/8. \tag{6.285}$$

Assume that

$$T_1 \geq 0, \ T_2 \geq T_1 + 3b_f, \tag{6.286}$$

$(x, u) \in X(T_1, T_2)$,

$$\|x(T_i) - x_f(T_i)\| \leq \delta, \ i = 1, 2, \tag{6.287}$$

$$I^f(T_1, T_2, x, u) \leq U^f(T_1, T_2, x(T_1), x(T_2)) + \delta, \tag{6.288}$$

$$\tau_1 \in [T_1, T_1 + \delta], \ \tau_2 \in [T_2 - \delta, T_2]. \tag{6.289}$$

Property (ii) and (6.286)–(6.288) imply that

$$I^f(T_1, T_2, x, u) \leq I^f(T_1, T_2, x_f, u_f) + 1.$$

Property (iii) and the relation above imply that for each $\tau_1, \tau_2 \in [T_1, T_2]$ satisfying $\tau_1 < \tau_2$,

$$\begin{aligned} I^f(\tau_1, \tau_2, x, u) &= I^f(T_1, T_2, x, u) - I^f(T_1, \tau_1, x, u) - I^f(\tau_2, T_2, x, u) \\ &\leq I^f(T_1, T_2, x_f, u_f) + 1 - I^f(T_1, \tau_1, x_f, u_f) + M_0 - I^f(\tau_2, T_2, x_f, u_f) + M_0 \\ &= I^f(T_1, T_2, x_f, u_f) + 2M_0 + 1. \end{aligned} \tag{6.290}$$

In view of (6.20) and (6.24),

$$\begin{aligned} I^f(T_1, T_1 + b_f, x_f, u_f) &\leq I^f(\lfloor T_1 \rfloor, \lfloor T_1 \rfloor + \lfloor b_f \rfloor + 2, x_f, u_f) + 3a_0 \\ &\leq (\lfloor b_f \rfloor + 2)\Delta_f + 3a_0, \end{aligned} \tag{6.291}$$

$$\begin{aligned} I^f(T_2 - b_f, T_2, x_f, u_f) &\leq I^f(\lfloor T_2 \rfloor - \lfloor b_f \rfloor - 1, \lfloor T_2 \rfloor + 1, x_f, u_f) + 3a_0 \\ &\leq (\lfloor b_f \rfloor + 2)\Delta_f + 3a_0, \end{aligned} \tag{6.292}$$

It follows from (6.290)–(6.292) that

$$I^f(T_1, T_1 + b_f, x, u), \ I^f(T_2 - b_f, T_2, x, u) \leq \Delta_f(\lfloor b_f \rfloor + 2) + 3a_0 + 2M_0 + 1. \tag{6.293}$$

Property (iv), (6.285), (6.289), and (6.293) imply that

$$\|x_f(T_i) - x_f(\tau_i)\| \le \delta_0/8, \ i = 1, 2, \tag{6.294}$$

$$\|x(T_i) - x(\tau_i)\| \le \delta_0/8, \ i = 1, 2. \tag{6.295}$$

By (6.285), (6.287), (6.294), and (6.295), for $i = 1, 2$,

$$\|x_f(\tau_i) - x(\tau_i)\| \le \|x_f(\tau_i) - x_f(T_i)\| + \|x_f(T_i) - x(T_i)\| + \|x(T_i) - x(\tau_i)\|$$

$$\le \delta_0/8 + \delta_0/8 + \delta \le \delta_0/2. \tag{6.296}$$

Property (i), (6.286), (6.289), and (6.296) imply that

$$U^f(\tau_1, \tau_2, x(\tau_1), x(\tau_2)) \le I^f(\tau_1, \tau_2, x_f, u_f) + \Delta/4. \tag{6.297}$$

By (6.288), (6.289), and (6.297),

$$I^f(\tau_1, \tau_2, x_f, u_f) \le U^f(\tau_1, \tau_2, x(\tau_1), x(\tau_2)) + \delta$$

$$\le I^f(\tau_1, \tau_2, x_f, u_f) + \Delta/4 + \delta \le I^f(\tau_1, \tau_2, x_f, u_f) + \Delta.$$

Lemma 6.48 is proved.

6.17 Proof of Theorem 6.33

Assume that STP holds. Theorem 6.22 implies that (P1) and (P2) hold. Let us show that (P4) holds. Let $(\tilde{x}, \tilde{u}) \in X(0, \infty)$ be (f)-overtaking optimal and

$$\tilde{x}(0) = x_f(0). \tag{6.298}$$

(P1), STP, and (6.298) imply that $\tilde{x}(t) = x_f(t)$ for all $t \ge 0$ and (P4) holds.
Assume that (P1), (P2), and (P4) hold.

Lemma 6.49. *Let $\epsilon > 0$. Then there exists $\delta > 0$ such that for each pair of numbers $T_1 \ge 0$, $T_2 \ge T_1 + 3b_f$ and each $(x, u) \in X(T_1, T_2)$ satisfying*

$$\|x(T_i) - x_f(T_i)\| \le \delta, \ i = 1, 2$$

$$I^f(T_1, T_2, x, u) \le U^f(T_1, T_2, x(T_1), x(T_2)) + \delta$$

the inequality $\|x(t) - x_f(t)\| \le \epsilon$ holds for all $t \in [T_1, T_2]$.

Proof. Lemma 6.40 and (A2) imply that there exist $\delta_0 \in (0, \min\{1, \epsilon/4\})$ and $L_0 > 0$ such that the following properties hold:

(i) (A2) holds with $\epsilon = 1$ and $\delta = \delta_0$;
(ii) for each pair of numbers $T_1 \geq L_0$, $T_2 \geq T_1 + 2b_f$ and each $(x, u) \in X(T_1, T_2)$ satisfying

$$\|x(T_i) - x_f(T_i)\| \leq \delta_0, \ i = 1, 2$$

$$I^f(T_1, T_2, x, u) \leq U^f(T_1, T_2, x(T_1), x(T_2)) + \delta_0$$

we have

$$\|x(t) - x_f(t)\| \leq \epsilon, \ t \in [T_1, T_2].$$

Theorem 6.2 implies that there exists $S_0 > 0$ such that for each $T_2 > T_1 \geq 0$ and each $(x, u) \in X(T_1, T_2)$,

$$I^f(T_1, T_2, x, u) + S_0 \geq I^f(T_1, T_2, x_f, u_f). \tag{6.299}$$

It follows from Theorem 6.23; there exists $L_1 > 0$ such that the following property holds:

(iii) for each $T \geq 0$ and each $(x, u) \in X(T, T + L_1)$ satisfying

$$I^f(T_1, T_2, x, u) \leq I^f(T_1, T_2, x_f, u_f) + 2S_0 + 3$$

we have

$$\min\{\|x(t) - x_f(t)\| : t \in [T, T + L_1]\} \leq \delta_0.$$

Consider a sequence $\{\delta_i\}_{i=1}^{\infty} \subset (0, 1)$ such that

$$\delta_i < 2^{-1}\delta_{i-1}, \ i = 1, 2, \ldots. \tag{6.300}$$

Assume that the lemma does not hold. Then for each integer $i \geq 1$, there exist

$$T_{i,1} \geq 0, \ T_{i,2} \geq T_{i,1} + 3b_f \tag{6.301}$$

and $(x_i, u_i) \in X(T_{i,1}, T_{i,2})$ such that

$$\|x_i(T_{i,1}) - x_f(T_{i,1})\| \leq \delta_i, \ \|x_i(T_{i,2}) - x_f(T_{i,2})\| \leq \delta_i, \tag{6.302}$$

$$I^f(T_{i,1}, T_{i,2}, x_i, u_i) \leq U^f(T_{i,1}, T_{i,2}, x_i(T_{i,1}), x_i(T_{i,2})) + \delta_i \tag{6.303}$$

and $t_i \in [T_{i,1}, T_{i,2}]$ for which

$$\|x(t_i) - x_f(t_i)\| > \epsilon. \tag{6.304}$$

Let i be a natural number. Property (ii) and (6.300)–(6.304) imply that

$$T_{i,1} < L_0. \tag{6.305}$$

We show that

$$t_i \le 2b_f + 1 + L_1 + L_0. \tag{6.306}$$

Property (i), (A2) with $\epsilon = 1$, $\delta = \delta_0$, and (6.300)–(6.302) imply that there exists $(y_i, v_i) \in X(T_{i,1}, T_{i,2})$ such that

$$y_i(T_{i,1}) = x_i(T_{i,1}),\ y_i(T_{i,2}) = x_i(T_{i,2}), \tag{6.307}$$

$$y_i(t) = x_f(t),\ v_i(t) = u_f(t),\ t \in [T_{i,1} + b_f, T_{i,2} - b_f], \tag{6.308}$$

$$I^f(T_{i,1}, T_{i,1} + b_f, y_i, v_i) \le I^f(T_{i,1}, T_{i,1} + b_f, x_f, u_f) + 1, \tag{6.309}$$

$$I^f(T_{i,2} - b_f, T_{i,2}, y_i, v_i) \le I^f(T_{i,2} - b_f, T_{i,2}, x_f, u_f) + 1. \tag{6.310}$$

It follows from (6.303) and (6.307)–(6.310) that

$$I^f(T_{i,1}, T_{i,2}, x_i, u_i) \le 1 + I^f(T_{i,1}, T_{i,2}, y_i, v_i) \le I^f(T_{i,1}, T_{i,2}, x_f, u_f) + 3. \tag{6.311}$$

In view of (6.299) and (6.311), for each pair of integers $S_1, S_2 \in [T_{i,1}, T_{i,2}]$ satisfying $S_1 < S_2$,

$$\begin{aligned} I^f(S_1, S_2, x_i, u_i) &= I^f(T_{i,1}, T_{i,2}, x_i, u_i) - I^f(T_{i,1}, S_1, x_i, u_i) - I^f(S_2, T_{i,2}, x_i, u_i) \\ &\le I^f(T_{i,1}, T_{i,2}, x_f, u_f) + 3 - I^f(T_{i,1}, S_1, x_f, u_f) + S_0 - I^f(S_2, T_{i,2}, x_f, u_f) + S_0 \\ &= I^f(S_1, S_2, x_f, u_f) + 3 + 2S_0. \end{aligned} \tag{6.312}$$

Assume that (6.306) does not hold. Then

$$t_i > 2b_f + 1 + L_1 + L_0. \tag{6.313}$$

Consider the restriction of (x_i, u_i) to the interval

$$[t_i - 2b_f - 1 - L_1, t_i - 1 - 2b_f] \subset [L_0, \infty). \tag{6.314}$$

By property (iii) and (6.312), there exists

$$\tilde{t} \in [t_i - 2b_f - 1 - L_1, t_i - 1 - 2b_f] \tag{6.315}$$

such that

$$\|x_f(\tilde{t}) - x(\tilde{t})\| \le \delta_0. \tag{6.316}$$

Property (ii), (6.300), (6.302), (6.303), and (6.314)–(6.316) imply that

$$\|x(t) - x_f(t)\| \le \epsilon,\ t \in [\tilde{t}, T_{i,2}]$$

and in particular,

$$\|x(t_i) - x_f(t_i)\| \le \epsilon.$$

This contradicts (6.304). The contradiction we have reached proves (6.306).

In view of (6.20), (6.24), and (6.312), there exists $M_1 > 0$ such that the following property holds:

(iv) for each integer $i \ge 1$ and each $S_2 > S_1 \ge 0$ satisfying $S_1, S_2 \in [T_{i,1}, T_{i,2}]$, $S_2 - S_1 \le 2b_f$,

$$I^f(S_1, S_2, x_i, u_i) \le M_1.$$

Lemma 6.45 and property (iv) imply that there exists $M_2 > 0$ such that

$$\|x_i(t)\| \le M_2,\ t \in [T_{i,1}, T_{i,2}],\ i = 1, 2, \dots. \tag{6.317}$$

In view of (6.305) and (6.306), extracting a subsequence and re-indexing, we may assume without loss of generality that there exists

$$\tilde{T}_1 = \lim_{i\to\infty} T_{i,1} \in [0, L_0], \tag{6.318}$$

$$\tilde{t} = \lim_{i\to\infty} t_i \in [\tilde{T}_1, L_0 + L_1 + 2b_f + 1], \tag{6.319}$$

$$\tilde{T}_2 = \lim_{i\to\infty} T_{i,2} \in [\tilde{t}, \infty]. \tag{6.320}$$

There exist strictly decreasing sequence $\{T_1^{(k)}\}_{k=1}^{\infty}$ and a strictly increasing sequence $\{T_2^{(k)}\}_{k=1}^{\infty}$ such that

$$\tilde{T}_1 = \lim_{k\to\infty} T_1^{(k)},\ \tilde{T}_2 = \lim_{k\to\infty} T_2^{(k)}, \tag{6.321}$$

for all integers $k \geq 1$,

$$T_1^{(k)} < T_2^{(k)}. \tag{6.322}$$

If $\tilde{T}_2 = \infty$, then we may assume that

$$T_2^{(k)} = T_{k,2} \text{ for all integers } k \geq 1. \tag{6.323}$$

By LSC property and property (iv), extracting a subsequence and re-indexing we may assume without loss of generality that there exist $\widehat{x} : (\tilde{T}_1, \tilde{T}_2) \to E$, $\widehat{u} : (\tilde{T}_1, \tilde{T}_2) \to F$ such that for each $t \in (\tilde{T}_1, \tilde{T}_2)$,

$$\widehat{x}(t) = \lim_{i\to\infty} x_i(t), \tag{6.324}$$

for each integer $k \geq 1$,

$$(\widehat{x}, \widehat{u}) \in X(T_1^{(k)}, T_2^{(k)}), \tag{6.325}$$

$$I^f(T_1^{(1)}, T_2^{(1)}, \widehat{x}, \widehat{u}) \leq \liminf_{i\to\infty} I^f(T_1^{(1)}, T_2^{(1)}, x_i, u_i), \tag{6.326}$$

for each integer $k \geq 1$,

$$I^f(T_1^{(k+1)}, T_1^{(k)}, \widehat{x}, \widehat{u}) \leq \liminf_{i\to\infty} I^f(T_1^{(k+1)}, T_1^{(k)}, x_i, u_i), \tag{6.327}$$

$$I^f(T_2^{(k)}, T_2^{(k+1)}, \widehat{x}, \widehat{u}) \leq \liminf_{i\to\infty} I^f(T_2^{(k)}, T_2^{(k+1)}, x_i, u_i). \tag{6.328}$$

By (6.317) and (6.324),

$$\|\widehat{x}(t)\| \leq M_2,\ t \in (\tilde{T}_1, \tilde{T}_2). \tag{6.329}$$

We show that

$$x_f(\tilde{T}_1) = \lim_{t\to\tilde{T}_1^+} \widehat{x}(t). \tag{6.330}$$

Let $\gamma > 0$. Lemma 6.46, property (iv), and (6.24) imply that there exists $\delta > 0$ such that the following property holds:

(v) for each integer $i \geq 1$ and each $S_1, S_2 \in [T_{i,1}, T_{i,2}]$ satisfying $|S_2 - S_1| \leq 2\delta$, we have

$$\|x_i(S_1) - x_i(S_2)\| \leq \gamma/2;$$

(vi) for each $S_1, S_2 \in [0, \infty)$ satisfying $|S_2 - S_1| \le 2\delta$, we have

$$\|x_f(S_1) - x_f(S_2)\| \le \gamma/2.$$

Let

$$t \in (\tilde{T}_1, \tilde{T}_1 + 2\delta). \tag{6.331}$$

In view of property (v), (6.302), (6.321), and (6.331), for all sufficiently large natural numbers k,

$$t \in (T_{k,1}, T_{k,1} + 2\delta),$$

$$\|x_k(t) - x_k(T_{k,1})\| \le \gamma/2,$$

$$\|x_k(t) - x_f(T_{k,1})\| \le \|x_k(t) - x_k(T_{k,1})\| + \|x_k(T_{k,1}) - x_f(T_{k,1})\| \le \gamma/2 + \delta_k. \tag{6.332}$$

Property (vi) and (6.321) imply that for all sufficiently large natural numbers k,

$$\|x_f(T_{k,1}) - x_f(\tilde{T}_1)\| \le \gamma/2. \tag{6.333}$$

It follows from (6.332)–(6.334) that for all sufficiently large natural numbers k,

$$\|\widehat{x}(t) - x_f(\tilde{T}_1)\| = \lim_{k\to\infty} \|x_k(t) - x_f(\tilde{T}_1)\|$$

$$\le \limsup_{k\to\infty}(\|x_k(t) - x_f(T_{k,1})\| + \|x_f(T_{k,1}) - x_f(\tilde{T}_1)\|) \le \gamma.$$

Since γ is an arbitrary positive number, we conclude that

$$\lim_{t\to\tilde{T}_1^+} \|\widehat{x}(t) - x_f(\tilde{T}_1)\| = 0$$

and (6.330) holds. Analogously we can show that if $\tilde{T}_2 < \infty$, then

$$x_f(\tilde{T}_2) = \lim_{t\to\tilde{T}_2^-} \widehat{x}(t). \tag{6.334}$$

Set

$$\widehat{x}(\tilde{T}_1) = x_f(\tilde{T}_1),\ \widehat{u}(\tilde{T}_1) = u_f(\tilde{T}_1). \tag{6.335}$$

If $\tilde{T}_2 < \infty$, then we set

$$\widehat{x}(\tilde{T}_2) = x_f(\tilde{T}_2),\ \widehat{u}(\tilde{T}_2) = u_f(\tilde{T}_2). \tag{6.336}$$

We show that

$$\tilde{t} > \tilde{T}_1, \ \|\widehat{x}(\tilde{t}) - x_f(\tilde{t})\| \geq \epsilon. \tag{6.337}$$

Let $\gamma \in (0, \epsilon/4)$ and let $\delta > 0$ be such that properties (v) and (vi) hold. Let $k \geq 1$ be an integer. If $t_k - T_{k,1} \leq \delta$, then in view of properties (v), (vi), and the choice of δ,

$$\|x_f(t_k) - x_f(T_{k,1})\| \leq \gamma/2, \ \|x_k(t_k) - x_k(T_{k,1})\| \leq \gamma/2$$

and together with (6.302) and (6.304) these relations imply that

$$\epsilon < \|x_f(t_k) - x_k(t_k)\| \leq \|x_f(t_k) - x_f(T_{k,1})\| + \|x_f(T_{k,1}) - x_k(T_{k,1})\|$$

$$+\|x_k(T_{k,1}) - x_k(t_k)\| \leq \gamma/2 + \delta_k + \gamma/2 \leq \gamma + \epsilon/4 < \epsilon,$$

a contradiction. The contradiction we have reached proves that

$$t_k - T_{k,1} > \delta. \tag{6.338}$$

If $T_{k,2} - t_k \leq \delta$, then in view of properties (v), (vi), and the choice of δ,

$$\|x_f(t_k) - x_f(T_{k,2})\| \leq \gamma/2, \ \|x_k(t_k) - x_k(T_{k,2})\| \leq \gamma/2$$

and together with (6.302) and (6.304), these relations imply that

$$\epsilon < \|x_f(t_k) - x_k(t_k)\| \leq \|x_f(t_k) - x_f(T_{k,2})\| + \|x_f(T_{k,2}) - x_k(T_{k,2})\|$$

$$+\|x_k(T_{k,2}) - x_k(t_k)\| \leq \gamma/2 + \delta_k + \gamma/2 \leq \gamma + \epsilon/4 < \epsilon,$$

a contradiction. The contradiction we have reached proves that

$$T_{k,2} - t_k \geq \delta. \tag{6.339}$$

By (6.338) and (6.339), for all natural numbers k,

$$T_{k,2} - t_k \geq \delta, \ t_k - T_{k,1} \geq \delta. \tag{6.340}$$

It follows from (6.319) and (6.340) that

$$\tilde{T}_2 - \tilde{t} \geq \delta, \ \tilde{t} - \tilde{T}_1 \geq \delta. \tag{6.341}$$

Properties (v) and (vi), (6.306), (6.319), and (6.324) imply that

$$\|\widehat{x}(\tilde{t}) - x_f(\tilde{t})\| = \lim_{k\to\infty} \|x_k(\tilde{t}) - x_f(\tilde{t})\| = \lim_{k\to\infty} \|x_k(t_k) - x_f(t_k)\| \ge \epsilon$$

and (6.337) holds. In view of (6.325), for every $\tau \in (\tilde{T}_1, \tilde{T}_2)$, the function $G(t, \widehat{x}(t), \widehat{u}(t))$, $t \in [\tilde{T}_1, \tau]$ is strongly measurable.

Let k_0 be a natural number and

$$\tau = T_2^{(k_0)}. \tag{6.342}$$

We show that the function $G(t, \widehat{x}(t), \widehat{u}(t))$, $t \in [\tilde{T}_1, \tau]$ is Bochner integrable. By (6.20), there exists $\gamma_0 > 0$ such that

$$f(t, x, u) \ge 8(\|G(t, x, u)\| - a_0\|x\|)_+ \text{ for all } (t, x, u) \in \mathcal{M}$$
$$\text{satisfying } \|G(t, x, u)\| - a_0\|x\| \ge \gamma_0. \tag{6.343}$$

Let $k \ge 1$ be an integer and $T_1^{(k)} < \tau$. By (6.317),

$$\int_{T_1^{(k)}}^{\tau} \|G(t, \widehat{x}(t), \widehat{u}(t))\|dt \le \int_{T_1^{(k)}}^{\tau} a_0\|\widehat{x}(t)\|dt$$
$$+ \int_{T_1^{(k)}}^{\tau} (\|G(t, \widehat{x}(t), \widehat{u}(t)) - a_0\|\widehat{x}(t)\|)_+ dt$$
$$\le a_0 M_2(\tau - T_1^{(k)}) + \int_{T_1^{(k)}}^{\tau} (\|G(t, \widehat{x}(t), \widehat{u}(t))\| - a_0\|\widehat{x}(t)\|)_+ dt. \tag{6.344}$$

Set

$$E_1 = \{t \in [T_1^{(k)}, \tau] : \ \|G(t, \widehat{x}(t), \widehat{u}(t))\| \ge a_0\|\widehat{x}(t)\| + \gamma_0\}, \tag{6.345}$$

$$E_2 = [T_1^{(k)}, \tau] \setminus E_1. \tag{6.346}$$

It follows from (6.20), (6.312), (6.326), (6.328), and (6.342)–(6.346) that

$$\int_{T_1^{(k)}}^{\tau} \|G(t, \widehat{x}(t), \widehat{u}(t))\|dt \le a_0 M_2(\tau - T_1^{(k)})$$
$$+ \int_{E_1} (\|G(t, \widehat{x}(t), \widehat{u}(t))\| - a_0\|\widehat{x}(t)\|)_+ dt$$
$$+ \int_{E_2} (\|G(t, \widehat{x}(t), \widehat{u}(t))\| - a_0\|\widehat{x}(t)\|)_+ dt$$

$$\le a_0M_2(\tau - T_1^{(k)}) + \gamma_0(\tau - T_1^{(k)})$$

$$+ \int_{E_1} (\|G(t, \widehat{x}(t), \widehat{u}(t))\| - a_0\|\widehat{x}(t)\|)_+ dt$$

$$\le a_0M_2(\tau - T_1^{(k)}) + \gamma_0(\tau - T_1^{(k)}) + \int_{E_1} f(t, \widehat{x}(t), \widehat{u}(t))dt$$

$$\le a_0M_2(\tau - T_1^{(k)}) + \gamma_0(\tau - T_1^{(k)}) + I^f(T_1^{(k)}, \tau, \widehat{x}, \widehat{u}) + a_0(\tau - T_1^{(k)})$$

$$\le (a_0M_2 + \gamma_0 + a_0)(T_2^{(k_0)} - T_1^{(k)}) + I^f(T_1^{(k)}, T_2^{(k_0)}, \widehat{x}, \widehat{u})$$

$$\le (T_2^{(k_0)} - T_1^{(k)})(a_0M_2 + \gamma_0 + a_0) + \liminf_{i\to\infty} I^f(T_1^{(k)}, T_2^{(k_0)}, x_i, u_i)$$

$$\le (T_2^{(k_0)} - T_1^{(k)})(a_0M_2 + \gamma_0 + a_0) + I^f(T_1^{(k)}, T_2^{(k_0)}, x_f, u_f) + 3 + 2S_0. \quad (6.347)$$

By (6.321) and (6.347),

$$\lim_{k\to\infty} \int_{T_1^{(k)}}^{T_2^{(k_0)}} \|G(t, \widehat{x}(t), \widehat{u}(t))\|dt$$

$$\le (T_2^{(k_0)} - \tilde{T}_1)(a_0M_2 + \gamma_0 + a_0) + I^f(\tilde{T}_1, T_2^{(k_0)}, x_f, u_f) + 3 + 2S_0.$$

Fatou's lemma and (6.321) imply that

$$\int_{\tilde{T}_1}^{T_2^{(k_0)}} \|G(t, \widehat{x}(t), \widehat{u}(t))\|dt < \infty,$$

$G(t, \widehat{x}(t), \widehat{u}(t)),\ t \in [\tilde{T}_1, T_2^{(k_0)}]$ is Bochner integrable for all integers $k_0 \ge 1$ and

$$\int_{\tilde{T}_1}^{T_2^{(k_0)}} \|G(t, \widehat{x}(t), \widehat{u}(t))\|dt \le I^f(\tilde{T}_1, T_2^{(k_0)}, x_f, u_f) + 3 + 2S_0$$

$$+ (T_2^{(k_0)} - \tilde{T}_1)(a_0M_2 + \gamma_0 + a_0). \quad (6.348)$$

If $\tilde{T}_2 = \infty$, then by (6.348),

$G(t, \widehat{x}(\cdot), \widehat{u}(\cdot))$ is Bochner integrable on $[\tilde{T}_1, \tau]$ for any $\tau > \tilde{T}_1$. (6.349)

If $\tilde{T}_2 < \infty$, then by (6.321) and (6.348), $G(\cdot, \widehat{x}(\cdot), \widehat{u}(\cdot))$ is strongly measurable on $[\tilde{T}_1, \tilde{T}_2]$ and

$$\lim_{k\to\infty}\int_{\tilde{T}_1}^{T_2^{(k)}} \|G(t,\widehat{x}(t),\widehat{u}(t))\|dt \le I^f(\tilde{T}_1,\tilde{T}_2,x_f,u_f)+3+2S_0$$

$$+(\tilde{T}_2-\tilde{T}_1)(a_0M_2+\gamma_0+a_0)$$

and in view of Fatou's lemma

$$G(\cdot,\widehat{x}(\cdot),\widehat{u}(\cdot)) \text{ is Bochner integrable on } [\tilde{T}_1,\tilde{T}_2]. \tag{6.350}$$

Assume that $\tilde{T}_2=\infty$. We show that for every $\tau>\tilde{T}_1$, $(\widehat{x},\widehat{u})\in X(\tilde{T}_1,\tau)$. Let $\tau>\tilde{T}_1$. We claim that for every $t\in(\tilde{T}_1,\tau]$,

$$\widehat{x}(t)=\int_{\tilde{T}_1}^{t} G(s,\widehat{x}(s),\widehat{u}(s))ds.$$

Let

$$t\in(\tilde{T}_1,\tau],\ \bar{\epsilon}>0.$$

By (6.321), (6.330), (6.333), and (6.349), there exists an integer $k_0\ge 1$ such that

$$T_1^{(k_0)}<t\le\tau<T_2^{(k_0)}, \tag{6.351}$$

$$\|\widehat{x}(T_1^{(k)})-\widehat{x}(\tilde{T}_1)\|\le\bar{\epsilon}/4 \text{ for all integers } k\ge k_0, \tag{6.352}$$

$$\int_{\tilde{T}_1}^{T_1^{(k)}} \|G(s,\widehat{x}(s),\widehat{u}(s))\|dt<\bar{\epsilon}/4 \text{ for all integers } k\ge k_0. \tag{6.353}$$

Let $k>k_0$ be an integer. Since $(\widehat{x},\widehat{u})\in X(T_1^{(k)},T_2^{(k)})$ (see (6.335)), it follows from (6.351)–(6.353) that

$$\widehat{x}(t)=\widehat{x}(T_1^{(k)})+\int_{T_1^{(k)}}^{t} G(s,\widehat{x}(s),\widehat{u}(s))ds. \tag{6.354}$$

In view of (6.354),

$$\|\widehat{x}(t)-\widehat{x}(\tilde{T}_1)-\int_{\tilde{T}_1}^{t} G(s,\widehat{x}(s),\widehat{u}(s))ds\|$$

$$\le\|\widehat{x}(T_1^{(k)})-\widehat{x}(\tilde{T}_1)\|+\int_{\tilde{T}_1}^{T_1^{(k)}} \|G(s,\widehat{x}(s),\widehat{u}(s))ds\|\le\bar{\epsilon}.$$

Since $\bar{\epsilon}$ is any positive number, we conclude that

$$\widehat{x}(t) = \widehat{x}(\tilde{T}_1) + \int_{\tilde{T}_1}^{t} G(s, \widehat{x}(s), \widehat{u}(s))ds$$

for all $t \in (\tilde{T}_1, \tau]$ and

$$(\widehat{x}, \widehat{u}) \in X(\tilde{T}_1, \tau) \text{ for any } \tau > \tilde{T}_1. \tag{6.355}$$

Assume that $\tilde{T}_2 < \infty$ and let $t \in (\tilde{T}_1, \tilde{T}_2)$. We show that

$$\widehat{x}(t) = \widehat{x}(\tilde{T}_1) + \int_{\tilde{T}_1}^{t} G(s, \widehat{x}(s), \widehat{u}(s))ds.$$

Let $\bar{\epsilon} > 0$. By (6.320), (6.321), (6.330), and (6.335), there exists an integer $k_0 \geq 1$ such that

$$T_1^{(k)} < t < T_2^{(k)} \text{ for all } k \geq k_0. \tag{6.356}$$

$$\|\widehat{x}(T_1^{(k)}) - \widehat{x}(\tilde{T}_1)\| \leq \bar{\epsilon}/4 \text{ for all integers } k \geq k_0, \tag{6.357}$$

$$\int_{\tilde{T}_1}^{T_1^{(k)}} \|G(s, \widehat{x}(s), \widehat{u}(s))ds\| \leq \bar{\epsilon}/4 \text{ for all integers } k \geq k_0. \tag{6.358}$$

Let $k \geq k_0$ be an integer. Since $(\widehat{x}, \widehat{u}) \in X(T_1^{(k)}, T_2^{(k)})$ it follows from (6.324) and (6.356) that

$$\widehat{x}(t) = \widehat{x}(T_1^{(k)}) + \int_{T_1^{(k)}}^{t} G(s, \widehat{x}(s), \widehat{u}(s))ds. \tag{6.359}$$

In view of (6.359),

$$\|\widehat{x}(t) - \widehat{x}(\tilde{T}_1) - \int_{\tilde{T}_1}^{t} G(s, \widehat{x}(s), \widehat{u}(s))ds\|$$

$$= \|\widehat{x}(T_1^{(k)}) + \int_{T_1^{(k)}}^{t} G(s, \widehat{x}(s), \widehat{u}(s))ds - \widehat{x}(\tilde{T}_1) - \int_{\tilde{T}_1}^{t} G(s, \widehat{x}(s), \widehat{u}(s))ds\|$$

$$\leq \|\widehat{x}(T_1^{(k)}) - \widehat{x}(\tilde{T}_1)\| + \int_{\tilde{T}_1}^{T_1^{(k)}} \|G(s, \widehat{x}(s), \widehat{u}(s))\|ds < \bar{\epsilon}.$$

Since $\bar{\epsilon}$ is any positive number, we conclude that

$$\widehat{x}(t) = \widehat{x}(\tilde{T}_1) + \int_{\tilde{T}_1}^{t} G(s, \widehat{x}(s), \widehat{u}(s))ds \tag{6.360}$$

for all $t \in (\tilde{T}_1, \tilde{T}_2)$. By (6.334), (6.336), (6.350), and (6.360),

$$\widehat{x}(\tilde{T}_2) = \lim_{t \to \tilde{T}_2^-} \widehat{x}(t) = \widehat{x}(\tilde{T}_1) + \lim_{t \to \tilde{T}_2^-} \int_{\tilde{T}_1}^{t} G(s, \widehat{x}(s), \widehat{u}(s))ds$$

$$= \widehat{x}(\tilde{T}_1) + \int_{\tilde{T}_1}^{t} G(s, \widehat{x}(s), \widehat{u}(s))ds.$$

Together with (6.360) this implies that

$$(\widehat{x}, \widehat{u}) \in X(\tilde{T}_1, \tilde{T}_2). \tag{6.361}$$

It follows from (6.335)–(6.337) that

$$x_f(\tilde{T}_i) = \widehat{x}(\tilde{T}_i),\ i = 1, 2,\ \|\widehat{x}(\tilde{t}) - x_f(\tilde{t})\| \ge \epsilon. \tag{6.362}$$

Property (P4) and (6.362) imply that

$$I^f(\tilde{T}_1, \tilde{T}_2, x_f, u_f) < I^f(\tilde{T}_1, \tilde{T}_2, \widehat{x}, \widehat{u}). \tag{6.363}$$

Let $\Delta > 0$. Lemma 6.48 implies that there exists $\delta > 0$ such that the following property holds:

(vii) for each $\tau_1 \ge 0$, each $\tau_2 \ge \tau_1 + 3b_f$, each $(x, u) \in X(\tau_1, \tau_2)$ satisfying

$$\|x(\tau_i) - x_f(\tau_i)\| \le \delta,\ i = 1, 2,$$

$$I^f(\tau_1, \tau_2, x, u) \le U^f(\tau_1, \tau_2, x(\tau_1), x(\tau_2)) + \delta,$$

each $t_1 \in [\tau_1, \tau_1 + \delta]$, each $t_2 \in [\tau_2 - \delta, \tau_2]$,

$$I^f(t_1, t_2, x, u) \le I^f(t_1, t_2, x_f, u_f) + \Delta/2.$$

In view of (6.300), there exists an integer $k_0 \ge 1$ such that

$$\delta_k < \delta \text{ for all integers } k \ge k_0. \tag{6.364}$$

Let $q \ge 1$ be an integer such that for all integers $i \ge q$,

$$\|\tilde{T}_1 - T_1^{(i)}\| \le \delta/2,\ \|\tilde{T}_2 - T_2^{(i)}\| \le \delta/2. \tag{6.365}$$

By (6.318), (6.320), (6.321), and (6.365), there exists an integer $k_1 \geq k_0$ such that for each integer $k \geq k_1$,

$$T_{k,1} < T_1^{(q)} \leq T_{k,1} + \delta, \ T_{k,2} - \delta \leq T_2^{(q)} < T_{k,2}. \tag{6.366}$$

Property (vii), (6.301)–(6.303), (6.354), and (6.366) imply that for each integer $k \geq k_1$,

$$I^f(T_1^{(q)}, T_2^{(q)}, x_k, u_k) \leq I^f(T_1^{(q)}, T_2^{(q)}, x_f, u_f) + \Delta/2. \tag{6.367}$$

In view of (6.326)–(6.328) and (6.367),

$$\begin{aligned} I^f(T_1^{(q)}, T_2^{(q)}, \widehat{x}, \widehat{u}) &\leq \liminf_{k\to\infty} I^f(T_1^{(q)}, T_2^{(q)}, x_k, u_k) \\ &\leq I^f(T_1^{(q)}, T_2^{(q)}, x_f, u_f) + \Delta/2 \end{aligned} \tag{6.368}$$

for all integers $i \geq q$ and for every integer $q \geq 1$ satisfying (6.365). Fatou's lemma, (6.20), and (6.321) imply that

$$I^f(\tilde{T}_1, \tilde{T}_2, \widehat{x}, \widehat{u}) \leq I^f(\tilde{T}_1, \tilde{T}_2, x_f, u_f) + \Delta/2.$$

Since Δ is any positive number, we conclude that

$$I^f(\tilde{T}_1, \tilde{T}_2, \widehat{x}, \widehat{u}) \leq I^f(\tilde{T}_1, \tilde{T}_2, x_f, u_f).$$

This contradicts (6.363). The contradiction we have reached proves $\tilde{T}_2 = \infty$. Recall (see (6.335) and (6.337)) that

$$\widehat{x}(\tilde{T}_1) = x_f(\tilde{T}_1), \ \|\widehat{x}(\tilde{t}) - x_f(\tilde{t})\| \geq \epsilon. \tag{6.369}$$

Let $\Delta > 0$. Lemma 6.48 implies that there exists $\delta \in (0, b_f/4)$ such that the following property holds:

(viii) for each $\tau_1 \geq 0$, each $\tau_2 \geq \tau_1 + 3b_f$, each $(x, u) \in X(\tau_1, \tau_2)$ satisfying

$$\|x(\tau_i) - x_f(\tau_i)\| \leq \delta, \ i = 1, 2,$$

$$I^f(\tau_1, \tau_2, x, u) \leq U^f(\tau_1, \tau_2, x(\tau_1), x(\tau_2)) + \delta,$$

each $t_1 \in [\tau_1, \tau_1 + \delta]$, each $t_2 \in [\tau_2 - \delta, \tau_2]$,

$$I^f(t_1, t_2, x, u) \leq I^f(t_1, t_2, x_f, u_f) + \Delta/2.$$

Recall (see (6.323)) that for all integers $k \geq 1$,

$$T_2^{(k)} = T_{k,2}. \tag{6.370}$$

In view of (6.300), there exists an integer $k_0 \geq 1$ such that

$$\delta_k < \delta \text{ for all integers } k \geq k_0.$$

Property (viii), (6.301), (6.303), (6.368), and (6.370) imply that the following property holds:

(ix) for each integer $k \geq k_0$, each $\tau_1 \in [T_{k,1}, T_{k,1} + \delta]$, each $\tau_2 \in [T_{k,2} - \delta, T_{k,2}]$,

$$I^f(\tau_1, \tau_2, x_k, u_k) \leq I^f(\tau_1, \tau_2, x_f, u_f) + \Delta/2.$$

By (6.326)–(6.328) and (6.370), for each integer $k \geq 1$ and each integer $s \geq k$,

$$I^f(T_1^{(s)}, T_{k,2}, \widehat{x}, \widehat{u})$$

$$= I^f(T_1^{(s)}, T_2^{(k)}, \widehat{x}, \widehat{u}) \leq \liminf_{i \to \infty} I^f(T_1^{(s)}, T_2^{(k)}, x_i, u_i)$$

$$= \liminf_{i \to \infty} I^f(T_1^{(s)}, T_{k,2}, x_i, u_i). \tag{6.371}$$

In view of (6.312) and (6.371), for each integer $k \geq 1$ and each integer $s \geq k$,

$$I^f(T_1^{(s)}, T_{k,2}, \widehat{x}, \widehat{u}) \leq I^f(T_1^{(s)}, T_{k,2}, x_f, u_f) + 3 + 2S_0. \tag{6.372}$$

Fatou's lemma, (6.20), (6.321), and (6.372) imply that for all integers $k \geq 1$,

$$I^f(\tilde{T}_1, T_{k,2}, \widehat{x}, \widehat{u}) \leq I^f(\tilde{T}_1, T_{k,2}, x_f, u_f) + 4 + 2S_0.$$

Together with Theorem 6.2 and (6.369), this implies that $(\widehat{x}, \widehat{u})$ is (f)-good. Property (P1) implies that

$$\lim_{t \to \infty} \|\widehat{x}(t) - x_f(t)\| = 0.$$

Therefore there exists $\tau_0 > 1 + \tilde{T}_1$ such that

$$\|\widehat{x}(t) - x_f(t)\| \leq \delta/4 \text{ for all } t \geq \tau_0. \tag{6.373}$$

By (6.318)–(6.321) and (6.323), there exists an integer $q \geq 1$ such that

$$T_{q,2} > \tau_0, \ |I^f(\tilde{T}_1, T_1^{(q)}, \widehat{x}, \widehat{u})| < \Delta/8, \ |I^f(\tilde{T}_1, T_1^{(q)}, x_f, u_f)| < \Delta/8, \tag{6.374}$$

for all integers $i \geq q$,

$$|\tilde{T}_1 - T_1^{(i)}| \leq \delta/2. \tag{6.375}$$

It follows from (6.318), (6.321), (6.324), and (6.375) that there exists an integer $k_1 \geq k_0 + q$ such that for each integer $k \geq k_1$,

$$T_{k,1} < T_1^{(q)} \leq T_{k,1} + \delta, \ \|\widehat{x}(T_{q,2}) - x_k(T_{q,2})\| \leq \delta/4. \tag{6.376}$$

Assume that an integer $k \geq k_1$. Then (6.376) holds. By (6.301)–(6.303), (6.323), the choice of k_0, and (6.376),

$$I^f(T_1^q, T_{q,2}, x_k, u_k) \leq U^f(T_1^q, T_{q,2}, x_k(T_1^q), x_k(T_{q,2})) + \delta, \tag{6.377}$$

$$\|x_k(T_{k,1}) - x_f(T_{k,1})\| \leq \delta_k \leq \delta. \tag{6.378}$$

In view of (6.373)–(6.376),

$$\|x_k(T_{q,2}) - x_f(T_{q,2})\| \leq \|x_k(T_{q,2}) - \widehat{x}(T_{q,2})\| + \|\widehat{x}(T_{q,2}) - x_f(T_{q,2})\| \leq \delta/4 + \delta/4. \tag{6.379}$$

Properties (ix) and (vii) applied with $\tau_1 = T_{k,1}$, $\tau_2 = T_{q,2}$, $(x, u) = (x_k, u_k)$, $t_1 = T_1^{(q)}$, $t_2 = T_{q,2}$, (6.376)–(6.379) imply that

$$I^f(T_1^{(q)}, T_{q,2}, x_k, u_k) \leq I^f(T_1^{(q)}, T_{q,2}, x_f, u_f) + \Delta/2. \tag{6.380}$$

In view of (6.380),

$$\begin{aligned} I^f(T_1^{(q)}, T_{q,2}, \widehat{x}, \widehat{u}) &\leq \liminf_{k \to \infty} I^f(T_1^{(q)}, T_{q,2}, x_k, u_k) \\ &\leq I^f(T_1^{(q)}, T_{q,2}, x_f, u_f) + \Delta/2. \end{aligned} \tag{6.381}$$

By (6.375) and (6.381),

$$I^f(\tilde{T}_1, T_{q,2}, \widehat{x}, \widehat{u}) \leq I^f(\tilde{T}_1, T_{q,2}, x_f, u_f) + \Delta. \tag{6.382}$$

Since the relation above holds for any integer $q \geq 1$ satisfying (6.374) and (6.375), we have

$$\liminf_{T \to \infty} (I^f(\tilde{T}_1, T, \widehat{x}, \widehat{u}) - I^f(\tilde{T}_1, T, x_f, u_f)) \leq \Delta.$$

Since Δ is any positive, number we conclude that

$$\liminf_{T \to \infty} (I^f(\tilde{T}_1, T, \widehat{x}, \widehat{u}) - I^f(\tilde{T}_1, T, x_f, u_f)) \leq 0. \tag{6.383}$$

For all $t \in [0, \tilde{T}_1] \setminus \{\tilde{T}_1\}$, set

$$\widehat{x}(t) = x_f(t), \ \widehat{u}(t) = u_f(t).$$

In view of (6.383), $(\widehat{x}, \widehat{u}) \in X(0, \infty)$ is (f)-weakly optimal. Theorem 6.29 and (P4) imply that $(\widehat{x}, \widehat{u}) \in X(0, \infty)$ is (f)-overtaking optimal and that

$$\widehat{x}(t) = x_f(t), \ t \geq 0.$$

This contradicts (6.362). The contradiction we have reached completes the proof of Lemma 6.49.

Completion of Theorem 6.33. By Theorem 6.22, TP holds. Lemma 6.49 and TP imply STP. Theorem 6.33 is proved.

6.18 Proof of Theorem 6.35

In view of Theorem 6.26, in order to show that (P1), (P2) hold it is sufficient to show that f_r has (P3). Let $\epsilon, M > 0$. Theorem 6.2 implies that there exists $S_0 > 0$ such that for each $T_2 > T_1 \geq 0$ and each $(x, u) \in X(T_1, T_2)$,

$$I^f(T_1, T_2, x, u) + S_0 \geq I^f(T_1, T_2, x_f, u_f). \tag{6.384}$$

Property (i) of Section 6.5 implies there exists $\delta \in (0, \epsilon)$ such that

$$\text{if } z \in E \text{ satisfies } \phi(z) \leq \delta, \text{ then } \|z\| \leq \epsilon. \tag{6.385}$$

Set

$$L = \delta^{-1} r^{-1}(S_0 + M). \tag{6.386}$$

Assume that $T_1 \geq 0$, $T_2 \geq T_1 + L$, $(x, u) \in X(T_1, T_2)$ satisfies

$$I^{f_r}(T_1, T_2, x, u) \leq I^{f_r}(T_1, T_2, x_f, u_f) + M. \tag{6.387}$$

By (6.384), (6.385), the equality $\phi(0) = 0$, and the definition of f_r,

$$I^f(T_1, T_2, x_f, u_f) + M = M + I^{f_r}(T_1, T_2, x_f, u_f)$$

$$\geq I^f(T_1, T_2, x, u) + r \int_{T_1}^{T_2} \phi(x(t) - x_f(t))dt$$

$$\geq I^f(T_1, T_2, x_f, u_f) - S_0 + r\int_{T_1}^{T_2} \phi(x(t) - x_f(t))dt,$$

$$r^{-1}(S_0 + M) \geq \int_{T_1}^{T_2} \phi(x(t) - x_f(t))dt$$

$$\geq \delta \text{mes}(\{t \in [T_1, T_2] : \phi(x(t) - x_f(t)) \geq \delta\})$$

$$\geq \delta \text{mes}(\{t \in [T_1, T_2] : \|x(t) - x_f(t)\| > \epsilon\})$$

and

$$\text{mes}(\{t \in [T_1, T_2] : \|x(t) - x_f(t)\| > \epsilon\}) \leq \delta^{-1} r^{-1}(S_0 + M) \leq L.$$

Thus f_r has (P3) and (P1), (P2) too.

Assume that $(x_f, u_f) \in X(0, \infty)$ is (f)-minimal. It is not difficult to see that (x_f, u_f) is (f_r)-minimal too. Let $(x, u) \in X(0, \infty)$ be (f_r)-good,

$$x(0) = x_f(0). \tag{6.388}$$

Therefore

$$\lim_{t \to \infty} \|x(t) - x_f(t)\| = 0. \tag{6.389}$$

Let $\epsilon > 0$. By (A3), there exists $\delta > 0$ such that the following property holds:

(i) for each $(T, z) \in \mathcal{A}$ satisfying $\|z - x_f(T)\| \leq \delta$ there exist $\tau \in (0, b_f]$, $(y, v) \in X(T, T + \tau)$ such that $y(T) = z$, $y(T + \tau) = x_f(T + \tau)$,

$$I^{f_r}(T, T + \tau, y, v) \leq I^{f_r}(T, T + \tau, x_f, u_f) + \epsilon/8.$$

In view of (6.389), there exists $T_0 > 0$ such that

$$\|x(t) - x_f(t)\| \leq \delta \text{ for all } t \geq T_0. \tag{6.390}$$

Let $T \geq T_0$. Property (i) and (6.390) imply that there exist $\tau \in (0, b_f]$, $(y, v) \in X(T, T + \tau)$ such that

$$y(T) = x(T),\ \ y(T + \tau) = x_f(T + \tau),$$

$$I^{f_r}(T, T + \tau, y, v) \leq I^f(T, T + \tau, x_f, u_f) + \epsilon/8. \tag{6.391}$$

Set

$$y(t) = x(t),\ v(t) = u(t),\ t \in [0, T]. \tag{6.392}$$

Clearly, $(y, v) \in X(0, T + \tau)$. Since (x_f, u_f) is (f)-minimal, it follows from (6.388), (6.391), and (6.392) that

$$I^f(0, T + \tau, x_f, u_f) \le I^f(0, T + \tau, y, v). \tag{6.393}$$

By the definition of f_r and (6.391)–(6.393),

$$I^{f_r}(0, T, x_f, u_f) \le I^f(0, T, y, v) + I^f(T, T+\tau, y, v) - I^f(T, T + \tau, x_f, u_f)$$

$$= I^f(0, T, x, u) + I^{f_r}(T, T + \tau, y, v)$$

$$-r \int_T^{T+\tau} \phi(y(t) - x_f(t))dt - I^f(T, T + \tau, x_f, u_f)$$

$$\le I^f(0, T, x, u) - r \int_T^{T+\tau} \phi(y(t) - x_f(t))dt + \epsilon/8$$

$$= I^{f_r}(0, T, x, u) - r \int_0^T \phi(x(t) - x_f(t))dt - r \int_T^{T+\tau} \phi(y(t) - x_f(t))dt + \epsilon/8.$$

Since the relation above holds for any $T \ge T_0$, we obtain

$$\limsup_{T\to\infty}(I^{f_r}(0, T, x_f, u_f) - I^{f_r}(0, T, x, u)) \le -r \lim_{T\to\infty} \int_0^T \phi(x(t) - x_f(t))dt + \epsilon/8.$$

Since ϵ is any positive number, we have

$$\limsup_{T\to\infty}(I^{f_r}(0, T, x_f, u_f) - I^f(0, T, x, u)) \le -r \lim_{T\to\infty} \int_0^T \phi(x(t) - x_f(t))dt. \tag{6.394}$$

This implies that $(x_f, u_f) \in X(0, \infty)$ is (f_r)-overtaking optimal.

Assume that (x, u) is (f_r)-overtaking optimal. Then (6.394) implies that $x(t) = x_f(t)$ for all $t \ge 0$. Thus (P4) holds. Theorem 6.35 is proved.

6.19 Examples

In this section we present a family of problems which belong to the second class of problems and for which the results of this chapter hold. We use the notation introduced in Section 6.1.

Let $(E, \langle\cdot,\cdot\rangle)_E$ be a Hilbert space equipped with an inner product $\langle\cdot,\cdot\rangle_E$ which induces the norm $\|\cdot\|_E$, and let $(F, \langle\cdot,\cdot\rangle_F)$ be a Hilbert space equipped with an inner product $\langle\cdot,\cdot\rangle_F$ which induces the norm $\|\cdot\|_F$. For simplicity, we set $\langle\cdot,\cdot\rangle_E = \langle\cdot,\cdot\rangle$, $\|\cdot\|_E = \|\cdot\|$, $\langle\cdot,\cdot\rangle_F = \langle\cdot,\cdot\rangle$, $\|\cdot\|_F = \|\cdot\|$, if E, F are understood. We suppose that $\mathcal{A}_0$ is a nonempty subset of E, $\mathcal{U}_0 : \mathcal{A}_0 \to 2^F$ is a point to set mapping with a graph

$$\mathcal{M}_0 = \{(x,u) : x \in \mathcal{A}_0,\ u \in \mathcal{U}_0(x)\}.$$

We suppose that $\mathcal{M}_0$ is a Borel measurable subset of $E \times F$. Define

$$\mathcal{A} = [0,\infty) \times \mathcal{A}_0,$$

$\mathcal{U} : \mathcal{A} \to 2^F$ by

$$\mathcal{U}(t,x) = \mathcal{U}_0(x),\ (t,x) \in \mathcal{A},$$

$$\mathcal{M} = [0,\infty) \times \mathcal{M}_0.$$

Let a linear operator $A : \mathcal{D}(A) \to E$ generates a C_0 semigroup $S(t) = e^{At}$, $t \geq 0$ on E, $E_1^d = \mathcal{D}(A^*)$, $E_{-1} = \mathcal{D}(A^*)'$ and let $B \in \mathcal{L}(F, E_{-1})$ is an admissible control operator for e^{At}, $t \geq 0$.

For $T_2 > T_1 \geq 0$ we consider the following control system

$$x'(t) = Ax(t) + Bu(t),\ t \in [T_1, T_2] \text{ a. e.}.$$

Assume that $(x_f, u_f) \in X(0,\infty)$ and

$$\sup\{\|x_f(t)\| : t \in [0,\infty)\} < \infty.$$

Recall that for every $T > 0$,

$$\Phi_T \in \mathcal{L}(L^2(0,\infty; F), E)$$

is defined by

$$\Phi_T u = \int_0^T S(T-s)Bu(s)ds,\ u \in L^2(0,T;F). \tag{6.395}$$

Theorem 6.50. *Assume that $T_0 > 0$, $Ran(\Phi_{T_0}) = E$. Then there exists a constant $c > 0$ such that for each $T \geq 0$ and each $z^0, z^1 \in E$, there exist $u \in L^2(T, T + T_0; F)$ and $z \in C^0([T, T+T_0]); E)$ which is a solution of the initial value problem*

$$z'(t) = Az(t) + Bu(t),\ t \in [T, T+T_0] \text{ a. e.},\ z(T) = z^0$$

in E_{-1} and satisfies $z(T+T_0)=z^1$ and such that the following inequalities hold:

$$\|z(t)-x_f(t)\|,\ \|u(t)-u_f(t)\|$$

$$\leq c(\|z^1-x_f(T+T_0)\|+\|z^0-x_f(T)\|),\ t\in[T,T+T_0].$$

Proof. Let $L\in\mathcal{L}(E,L^2(0,T_0;F))$ be as guaranteed by Proposition 1.22. Therefore

$$\Phi_{T_0}Lx=x \text{ for all } x\in E. \tag{6.396}$$

Let $c_{T_0}>1$ be as guaranteed by Proposition 1.21 with $T=T_0$. Set

$$c=c_{T_0}+c_{T_0}^2\|L\|. \tag{6.397}$$

Let $z^0, z^1\in E$. Define

$$z_1(t)=S(t-T)(z^0-x_f(T))=e^{A(t-T)}(z^0-x_f(T)),\ t\in[T,T+T_0]. \tag{6.398}$$

In view of (6.398),

$$z_1'(t)=Az_1(t),\ t\in[T,T+T_0] \text{ a. e.}, \tag{6.399}$$

$$z_1(T)=z^0-x_f(T), \tag{6.400}$$

$(z_1+x_f,u_f)\in X(T,T+T_0)$,

$$(z_1+x_f)'(t)=A(z_1+x_f)(t)+Bu_f(t),\ t\in[T,T+T_0] \text{ a. e.} \tag{6.401}$$

in E_{-1},

$$(z_1+x_f)(T)=z^0. \tag{6.402}$$

By Proposition 1.19, there exists a unique $z_2\in C^0([0,T_0];E)$ such that

$$z_2'=Az_2+B(L(z^1-(z_1+x_f)(T+T_0))),\ t\in(0,T_0) \text{ a. e.}, \tag{6.403}$$

$$z_2(0)=0 \tag{6.404}$$

in E_{-1}. In view of (6.396), (6.403), and (6.404),

$$z_2(T_0)=S(T_0)z_2(0)+\Phi_{T_0}(L(z^1-(z_1+x_f)(T+T_0)))=z^1-(z_1+x_f)(T+T_0). \tag{6.405}$$

Set

$$z_3(t) = z_2(t - T),\ t \in [T, T + T_0], \tag{6.406}$$

$$z(t) = x_f(t) + z_1(t) + z_3(t), t \in [T, T + T_0]. \tag{6.407}$$

By (6.398) and (6.404)–(6.407),

$$z(T) = x_f(T) + z_1(T) + z_3(T) = x_f(T) + z^0 - x_f(T) + z_2(0) = z^0, \tag{6.408}$$

$$z(T + T_0) = x_f(T + T_0) + z_1(T + T_0) + z_2(T_0) = z^1. \tag{6.409}$$

It follows from (6.399), (6.403), (6.406), and (6.407) that for all $t \in [T, T + T_0]$,

$$z'(t) = x_f'(t) + z_1'(t) + z_3'(t) = Ax_f(t) + Bu_f(t) + Az_1(t) + z_2'(t - T)$$

$$= Ax_f(t) + Bu_f(t) + Az_1(t) + Az_3(t) + B(L(z^1 - (z_1 + x_f)(T + T_0)))$$

$$= Az(t) + B(u_f(t) + L(z^1 - (z_1 + x_f)(T + T_0))) \tag{6.410}$$

in E_{-1}. In view of (6.400),

$$\|L(z^1 - (z_1 + x_f)(T + T_0))\| \le \|L\|(\|z^1 - x_f(T + T_0)\| + \|z_1(T + T_0)\|). \tag{6.411}$$

Proposition 1.21, the choice of c_{T_0}, (6.399), (6.42), (6.404), and (6.406) imply that for all $t \in [T, T + T_0]$,

$$\|z_1(t)\| \le c_{T_0}\|z^0 - x_f(T)\|, \tag{6.412}$$

$$\|z_3(t)\| \le c_{T_0}\|L(z^1 - (z_1 + x_f)(T + T_0))\|$$

$$\le c_{T_0}\|L\|\|z^1 - (z_1 + x_f)(T + T_0)\|$$

$$\le c_{T_0}\|L\|(\|z^1 - x_f(T + T_0)\| + c_{T_0}\|z^0 - x_f(T)\|). \tag{6.413}$$

By (6.397) and (6.412),

$$\|L(z^1 - (z_1 + x_f)(T + T_0))\| \le \|L\|(\|z^1 - x_f(T + T_0)\| + c_{T_0}\|z^0 - x_f(T)\|)$$

$$\le c(\|z^1 - x_f(T + T_0)\| + \|z^0 - x_f(T)\|).$$

In view of (6.407), (6.412), and (6.413), for all $t \in [T, T + L_0]$,

$$\|z(t) - x_f(t)\| \le \|z_1\| + \|z_3(t)\|$$

$$\le c_{T_0}\|z^0 - x_f(T)\| + c_{T_0}\|L\|(\|z^1 - x_f(T + T_0)\| + c_{T_0}\|z^0 - x_f(T)\|)$$

$$\le c(\|z^0 - x_f(T)\| + \|z^1 - x_f(T + T_0)\|).$$

Theorem 6.50 is proved.

Assume that there exists $T_f > 0$ such that

$$\mathrm{Ran}\Phi_{T_f} = E.$$

In other words the pair (A, B) is exactly controllable. We suppose that there exists $r_* > 0$ such that for each $t \ge 0$,

$$\{(x, u) \in E \times F : \|x - x_f(t)\| \le r_*, \ \|u - u_f(t)\| \le r_*\} \subset \mathcal{M}_0.$$

Assume that $L : [0, \infty) \times E \times F \to [0, \infty)$ is a Borelian function,

$$L(t, x_f(t), u_f(t)) = 0, \ t \in [0, \infty),$$

$\psi_0 : [0, \infty) \to [0, \infty)$ is an increasing function, $K_1, a_1 > 0$ such that

$$\lim_{t\to\infty} \psi_0(t) = \infty,$$

$$L(t, x, u) \ge -a_1 + \max\{\psi_0(\|x\|)\|x\|, \ K_1\|u\|^2\}$$

for all $(t, x, u) \in \mathcal{M}$, $\mu \in R^1$, $\bar{p} \in \mathcal{D}(A^*)$. Let for all $(t, x, u) \in \mathcal{M}$,

$$f(t, x, u) = L(t, x, u) + \mu + \langle x, A^*\bar{p}\rangle + \langle Bu, \bar{p}\rangle_{E_{-1}, E_1^d}. \tag{6.414}$$

It is not difficult to see that f is a Borelian function; there exist a_0, $K_0 > 0$ and an increasing function $\psi : [0, \infty) \to [0, \infty)$ such that $\lim_{t\to\infty} \psi(t) = \infty$ and for all $(t, x, u) \in \mathcal{M}$,

$$f(t, x, u) \ge -a_0 + \max\{\psi(\|x\|), K_0\|u\|^2\}.$$

We suppose that the following property holds:

(a) for each $\epsilon > 0$ there exists $\delta > 0$ such that for each $(t, x, u) \in \mathcal{M}$ satisfying $\|x - x_f(t)\| + \|u - u_f(t)\| \le \delta$ we have

$$f(t, x, u) \le f(t, x_f(t), u_f(t)) + \epsilon.$$

It follows from property (a) and Theorem 6.50 that (A3) holds.

Let $0 \le T_1 < T_2$, $(x, u) \in X(T_1, T_2)$. It is not difficult to see that

$$I^f(T_1, T_2, x, u) - I^f(T_1, T_2, x_f, u_f)$$

$$= \int_{T_1}^{T_2} L(t, x(t), u(t))dt + \int_{T_1}^{T_2} \langle x(t), A^* \bar{p}\rangle dt + \int_{T_1}^{T_2} \langle Bu(t), \bar{p}\rangle_{E_{-1}, E_d} dt$$

$$- \int_{T_1}^{T_2} \langle x_f(t), A^* \bar{p}\rangle dt - \int_{T_1}^{T_2} \langle Bu_f(t), \bar{p}\rangle_{E_{-1}, E_d} dt$$

$$\ge \langle x(T_2) - x(T_1), \bar{p}\rangle - \langle x_f(T_2) - x_f(T_1), \bar{p}\rangle. \tag{6.415}$$

This implies that (A1) holds.

Assume now that x_f is uniformly continuous on $[0, \infty)$ and that the following property holds:

(a) for each $M, \epsilon > 0$ there exists $\delta > 0$ such that for each $(t, x, u) \in [0, \infty) \times E \times F$ which satisfies

$$\|x\| \le M \text{ and } L(t, x, u) \le \delta$$

we have $\|x - x_f(t)\| \le \epsilon$.

We claim that f has TP. In view of Theorems 6.22 and 6.26, it is sufficient to show that properties (A3) and (P3) of Theorem 6.26 hold. By (6.414), (A3) holds. Let us show that (P3) holds.

Let $\epsilon, M > 0$ and $M_1 > 0$ be as guaranteed by Theorem 6.3 with $M_0 = M$, $c_0 = 1$, $c = 2$. Let $S > 0$ be as guaranteed by Theorem 6.2, and let $\delta \in (0, 1)$ be as guaranteed by property (a) with $M = M_1$. Choose

$$L > 3 + \delta^{-1}(M + S + 4M_1\|\bar{p}\|).$$

Assume that $T_1 \ge 0$, $T_2 \ge T_1 + L$ and that $(x, u) \in X(T_1, T_2)$ satisfies

$$I^f(T_1, T_2, x, u) \le I^f(T_1, T_2, x_f, u_f) + M.$$

Combined with Theorem 6.3 and the choice of M_1, this implies that

$$\|x(t)\| \le M_1, \ t \in [T_1 + 1, T_2].$$

It is not difficult to see that, in view of the choice of S and Theorem 6.2,

$$I^f(T_1+1,T_2,x,u) = I^f(T_1,T_2,x,u) - I^f(T_1,T_1+1,x,u)$$

$$\leq I^f(T_1,T_2,x_f,u_f) + M - I^f(T_1,T_1+1,x_f,u_f) + S$$

$$= I^f(T_1+1,T_2,x_f,u_f) + M + S.$$

Combined with (6.415), this implies that

$$M + S \geq I^f(T_1+1,T_2,x,u) - I^f(T_1+1,T_2,x_f,u_f)$$

$$= \int_{T_1+1}^{T_2} L(t,x(t),u(t))dt$$

$$+ \int_{T_1+1}^{T_2} \langle x(t), A^*\bar{p}\rangle dt + \int_{T_1+1}^{T_2} \langle Bu(t), \bar{p}\rangle_{E_{-1},E_d} dt$$

$$- \int_{T_1+1}^{T_2} \langle x_f(t), A^*\bar{p}\rangle dt - \int_{T_1+1}^{T_2} \langle Bu_f(t), \bar{p}\rangle_{E_{-1},E_d} dt$$

$$= \int_{T_1+1}^{T_2} L(t,x(t),u(t))dt$$

$$+\langle x(T_1+1) - x(T_2), \bar{p}\rangle - \langle x_f(T_1+1) - x_f(T_2), \bar{p}\rangle$$

$$\geq \int_{T_1+1}^{T_2} L(t,x(t),u(t))dt - 4M_1\|\bar{p}\|.$$

Together with the choice of δ and property (a) with $M = M_1$, this implies that

$$M + S + 4M_1\|\bar{p}\| \geq \int_{T_1+1}^{T_2} L(t,x(t),u(t))dt$$

$$\geq \delta \text{mes}(\{t \in [T_1+1,T_2] : \ L(t,x(t),u(t)) > \delta\})$$

$$\geq \delta \text{mes}(\{t \in [T_1+1,T_2] : \ \|x(t) - x_f(t)\| > \epsilon\})$$

and

$$\text{mes}(\{t \in [T_1,T_2] : \ \|x(t) - x_f(t)\| > \epsilon\})$$

$$\leq 1 + \delta^{-1}(M + S + 4M_1\|\bar{p}\|) < L.$$

Thus (P3) holds and f has TP.

We can easily obtain particular cases of this example with the pairs (A, B) considered in Section 1.8.

6.20 Exercises for Chapter 6

Exercise 6.51. Use the control problem considered in Example 1.24, and construct an optimal control problem which is a special case of the example analyzed in Section 6.19.

Exercise 6.52. Use the control problem considered in Example 1.25, and construct an optimal control problem which is a special case of the example analyzed in Section 6.19.

Exercise 6.53. Use the control problem considered in Example 1.26, and construct an optimal control problem which is a special case of the example analyzed in Section 6.19.

Exercise 6.54. Use the control problem considered in Example 1.27, and construct an optimal control problem which is a special case of the example analyzed in Section 6.19.

Exercise 6.55. Use the control problem considered in Example 1.28, and construct an optimal control problem which is a special case of the example analyzed in Section 6.19.

Exercise 6.56. Use the control problem considered in Example 1.29, and construct an optimal control problem which is a special case of the example analyzed in Section 6.19.

Exercise 6.57. Use the control problem considered in Example 1.30, and construct an optimal control problem which is a special case of the example analyzed in Section 6.19.

Exercise 6.58. Use the control problem considered in Example 1.31, and construct an optimal control problem which is a special case of the example analyzed in Section 6.19.

Chapter 7
Continuous-Time Nonautonomous Problems on Axis

In this chapter we establish sufficient and necessary conditions for the turnpike phenomenon for continuous-time optimal control problems on subintervals of axis in infinite dimensional spaces. For these optimal control problems, the turnpike is not a singleton. We also study the existence of solutions of the corresponding infinite horizon optimal control problems.

7.1 Preliminaries

Let $(E, \|\cdot\|)$ be a Banach space and (F, ρ_F) be a metric space. We suppose that $\mathcal{A}$ is a nonempty subset of $R^1 \times E, \mathcal{U} : \mathcal{A} \to 2^F$ is a point to set mapping with a graph

$$\mathcal{M} = \{(t, x, u) : (t, x) \in \mathcal{A}, \ u \in \mathcal{U}(t, x)\}.$$

We suppose that $\mathcal{M}$ is a Borel measurable subset of $R^1 \times E \times F$, and $G : \mathcal{M} \to E$ is a Borelian function. Let $f : \mathcal{M} \to R^1$ be a Borelian function bounded from below.

Let $-\infty < T_1 < T_2 < \infty$ and consider the equation

$$x'(t) = G(t, x(t), u(t)), \ t \in [T_1, T_2]. \tag{7.1}$$

A pair of functions $x : [T_1, T_2] \to E$, $u : [T_1, T_2] \to F$ is called a solution of (7.1) if $x : [T_1, T_2] \to E$ is a continuous function, $u : [T_1, T_2] \to F$ is a Lebesgue measurable function,

$$(t, x(t)) \in \mathcal{A}, \ t \in [T_1, T_2], \tag{7.2}$$

$$u(t) \in \mathcal{U}(t, x(t)), \ t \in [T_1, T_2] \text{ almost everywhere (a.e.)}, \tag{7.3}$$

A. J. Zaslavski, *Turnpike Conditions in Infinite Dimensional Optimal Control*, Springer Optimization and Its Applications 148,
https://doi.org/10.1007/978-3-030-20178-4_7

$G(s, x(s), u(s))$, $s \in [T_1, T_2]$ is Bochner integrable and for every $t \in [T_1, T_2]$,

$$x(t) = x(T_1) + \int_{T_1}^{t} G(s, x(s), u(s))ds, \ t \in [T_1, T_2]. \tag{7.4}$$

The set of all pairs (x, u) which are solutions of (7.1) is denoted by $X(T_1, T_2)$.

Let $T_1 \in R^1$. A pair of functions $x : [T_1, \infty) \to E$, $u : [T_1, \infty) \to F$ is called a solution of the system

$$x'(t) = G(t, x(t), u(t)), \ t \in [T_1, \infty) \tag{7.5}$$

if for every $T_2 > T_1$, $x : [T_1, T_2] \to E$, $u : [T_1, T_2] \to F$ is a solution of (7.1). The set of all such pairs (x, u) which are solutions of the equation above is denoted by $X(T_1, \infty)$.

A pair of functions $x : R^1 \to E$, $u : R^1 \to F$ is called a solution of the system

$$x'(t) = G(t, x(t), u(t)), \ t \in R^1 \tag{7.6}$$

if for every pair $T_2 > T_1$, $x : [T_1, T_2] \to E$, $u : [T_1, T_2] \to F$ is a solution of (7.1). The set of all such pairs (x, u) which are solutions of (7.6) is denoted by $X(-\infty, \infty)$.

A function $x : I \to E$, where I is either $[T_1, T_2]$ or $[T_1, \infty)$ or R^1 $(T_1 < T_2)$, is called a trajectory if there exists a Lebesgue measurable function $u : I \to F$ (referred to as a control) such that $(x, u) \in X(T_1, T_2)$ or $(x, u) \in X(T_1, \infty)$ or $(x, u) \in X(-\infty, \infty)$, respectively.

Let $T_2 > T_1$, $(x, u) \in X(T_1, T_2)$. Define

$$I^f(T_1, T_2, x, u) = \int_{T_1}^{T_2} f(t, x(t), u(t))dt$$

which is well-defined but can be ∞.

Let $a_0 > 0$ and let $\psi : [0, \infty) \to [0, \infty)$ be an increasing function such that

$$\psi(t) \to \infty \text{ as } t \to \infty. \tag{7.7}$$

We suppose that the function f satisfies

$$f(t, x, u) \geq -a_0 + \max\{\psi(\|x\|),$$

$$\psi((\|G(t, x, u)\| - a_0\|x\|)_+)(\|G(t, x, u)\| - a_0\|x\|)_+\} \tag{7.8}$$

for each $(t, x, u) \in \mathcal{M}$, and we consider functionals of the form $I^f(T_1, T_2, x, u)$, where $T_1 < T_2$ and $(x, u) \in X(T_1, T_2)$. For each pair of points

$$(T_1, y), (T_2, z) \in \mathcal{A}$$

such that $T_1 < T_2$, we define

$$U^f(T_1, T_2, y, z) = \inf\{I^f(T_1, T_2, x, u) :$$

$$(x, u) \in X(T_1, T_2),\ x(T_1) = y,\ x(T_2) = z\}, \tag{7.9}$$

$$\sigma^f(T_1, T_2, y) = \inf\{I^f(T_1, T_2, x, u) : (x, u) \in X(T_1, T_2),\ x(T_1) = y\}, \tag{7.10}$$

$$\sigma^f(T_1, T_2) = \inf\{I^f(T_1, T_2, x, u) : (x, u) \in X(T_1, T_2)\}. \tag{7.11}$$

We suppose that $(x_f, u_f) \in X(-\infty, \infty)$ satisfies

$$\sup\{\|x_f(t)\| :\ t \in R^1\} < \infty, \tag{7.12}$$

$$\Delta_f := \sup\{I^f(j, j+1, x_f, u_f) :\ j \text{ is an integer }\} < \infty. \tag{7.13}$$

We suppose that there exists a number $b_f > 0$ and the following assumptions hold.

(A1) For each $S_1 > 0$, there exist $S_2 > 0$ and $c > 0$ such that

$$I^f(T_1, T_2, x_f, u_f) \le I^f(T_1, T_2, x, u) + S_2$$

for each $T_1 \in R^1$, each $T_2 \ge T_1 + c$, and each $(x, u) \in X(T_1, T_2)$ satisfying $\|x(T_j)\| \le S_1, j = 1, 2$.

(A2) For each $\epsilon > 0$, there exists $\delta > 0$ such that for each $(T_i, z_i) \in \mathcal{A}$, $i = 1, 2$ satisfying $\|z_i - x_f(T_i)\| \le \delta$, $i = 1, 2$ there exist $\tau_1, \tau_2 \in (0, b_f]$ and $(x_1, u_1) \in X(T_1, T_1 + \tau_1)$, $(x_2, u_2) \in X(T_2 - \tau_2, T_2)$ which satisfy

$$x_1(T_1) = z_1,\ x_1(T_1 + \tau_1) = x_f(T_1 + \tau_1),$$

$$I^f(T_1, T_1 + \tau_1, x_1, u_1) \le I^f(T_1, T_1 + \tau_1, x_f, u_f) + \epsilon,$$

$$x_2(T_2) = z_2,\ x_2(T_2 - \tau_2) = x_f(T_2 - \tau_2),$$

$$I^f(T_2 - \tau_2, T_2, x_2, u_2) \le I^f(T_2 - \tau_2, T_2, x_f, u_f) + \epsilon.$$

Relations (7.8) and (7.13) imply the following result.

Lemma 7.1. *Let $c > 0$. Then*

$$\Delta_f(c) := \sup\{I^f(T_1, T_2, x_f, u_f) :\ T_1 \in R^1,\ T_2 \in (T_1, T_1 + c]\} < \infty.$$

Section 7.18 contains examples of optimal control problems satisfying assumptions (A1) and (A2). Many examples can also be found in [106–108, 118, 124, 125, 134]

7.2 Boundedness Results

The following result is proved in Section 7.6.

Theorem 7.2.

1. *There exists $S > 0$ such that for each pair of numbers $T_2 > T_1$ and each $(x, u) \in X(T_1, T_2)$,*

$$I^f(T_1, T_2, x, u) + S \geq I^f(T_1, T_2, x_f, u_f).$$

2. *For each $(x, u) \in X(-\infty, \infty)$ either*

$$I^f(-T, T, x, u) - I^f(-T, T, x_f, u_f) \to \infty \text{ as } T \to \infty$$

or

$$\sup\{|I^f(-T, T, x, u) - I^f(-T, T, x_f, u_f)| : T \in (0, \infty)\} < \infty. \quad (7.14)$$

Moreover, if (7.14) holds, then $\sup\{\|x(t)\| : t \in R^1\} < \infty$.

We say that $(x, u) \in X(-\infty, \infty)$ is (f)-good if (7.14) holds.

The next boundedness result is proved in Section 7.6.

Theorem 7.3. *Let $M_0 > 0$, $c > 0$. Then there exists $M_1 > 0$ such that for each $T_1 \in R^1$, each $T_2 \geq T_1 + c$, and each $(x, u) \in X(T_1, T_2)$ satisfying*

$$I^f(T_1, T_2, x, u) \leq I^f(T_1, T_2, x_f, u_f) + M_0,$$

the inequality $\|x(t)\| \leq M_1$ holds for all $t \in [T_1, T_2]$.

Let $L > 0$. Denote by $\mathcal{A}_L$ the set of all $(s, z) \in \mathcal{A}$ for which there exist $\tau \in (0, L]$ and $(x, u) \in X(s, s + \tau)$ such that

$$x(s) = z, \ x(s + \tau) = x_f(s + \tau), \ I^f(s, s + \tau, x, u) \leq L.$$

Denote by $\widehat{\mathcal{A}}_L$ the set of all $(s, z) \in \mathcal{A}$ such that there exist $\tau \in (0, L]$ and $(x, u) \in X(s - \tau, s)$ such that

$$x(s - \tau) = x_f(s - \tau), \ x(s) = z, \ I^f(s - \tau, s, x, u) \leq L.$$

The following Theorems 7.4 and 7.5 are also boundedness results. They are proved in Section 7.6.

Theorem 7.4. *Let $L > 0$, $M_0 > 0$. Then there exists $M_1 > 0$ such that for each $T_1 \in R^1$, each $T_2 \geq T_1 + 2L$, and each $(x, u) \in X(T_1, T_2)$ satisfying*

$$(T_1, x(T_1)) \in \mathcal{A}_L, \ (T_2, x(T_2)) \in \widehat{\mathcal{A}}_L,$$

$$I^f(T_1, T_2, x, u) \le U^f(T_1, T_2, x(T_1), x(T_2)) + M_0,$$

the inequality $\|x(t)| \le M_1$ *holds for all* $t \in [T_1, T_2]$.

Theorem 7.5. *Let* $L > 0$, $M_0 > 0$. *Then there exists* $M_1 > 0$ *such that for each* $T_1 \ge 0$, *each* $T_2 \ge T_1 + L$, *and each* $(x, u) \in X(T_1, T_2)$ *satisfying*

$$(T_1, x(T_1)) \in \mathcal{A}_L, \ I^f(T_1, T_2, x, u) \le \sigma^f(T_1, T_2, x(T_1)) + M_0,$$

the inequality $\|x(t)\| \le M_1$ *holds for all* $t \in [T_1, T_2]$.

Theorem 7.6. *For each* $T_0 \in R^1$ *and each* $(x, u) \in X(T_0, \infty)$ *either*

$$I^f(T_0, T, x, u) - I^f(T_0, T, x_f, u_f) \to \infty \text{ as } T \to \infty$$

or

$$\sup\{|I^f(T_0, T, x, u) - I^f(T_0, T, x_f, u_f)| : T \in (T_0, \infty)\} < \infty. \tag{7.15}$$

Moreover, if (7.15) holds, then $\sup\{\|x(t)\| : t \in [T_0, \infty)\} < \infty$.

Theorem 7.6 easily follows from Assertion 1 of Theorem 7.2 and Theorem 7.3.

Let $T_0 \in R^1$. We say that $(x, u) \in X(T_0, \infty)$ is (f)-good if (7.15) holds.

Let $T_0 \in R^1$. We say that $(x, u) \in X(T_0, \infty)$ is (f)-minimal if for every $T > T_0$,

$$I^f(T_0, T, x, u) = U^f(T_0, T, x(T_0), x(T)) \tag{7.16}$$

and that $(x, u) \in X(-\infty, \infty)$ is called (f)-minimal if for every pair $T_2 > T_1$,

$$I^f(T_1, T_2, x, u) = U^f(T_1, T_2, x(T_1), x(T_2)). \tag{7.17}$$

7.3 Turnpike Results

We say that the integrand f possesses the strong turnpike property (or STP for short) if for each $\epsilon > 0$ and each $M > 0$, there exist $\delta > 0$ and $L > 0$ such that for each $T_1 \in R^1$, each $T_2 \ge T_1 + 2L$, and each $(x, u) \in X(T_1, T_2)$ which satisfies

$$I^f(T_1, T_2, x, u) \le \min\{\sigma^f(T_1, T_2) + M, \ U^f(T_1, T_2, x(T_1), x(T_2)) + \delta\},$$

there exist $\tau_1 \in [T_1, T_1 + L]$, $\tau_2 \in [T_2 - L, T_2]$ such that

$$\|x(t) - x_f(t)\| \le \epsilon \text{ for all } t \in [\tau_1, \tau_2].$$

Moreover, if $\|x(T_1) - x_f(T_1)\| \le \delta$, then $\tau_1 = T_1$, and if $\|x(T_2) - x_f(T_2)\| \le \delta$, then $\tau_2 = T_2$.

We say that f possesses lower semicontinuity property (or LSC property for short) if for each $T_2 > T_1$ and each sequence $(x_j, u_j) \in X(T_1, T_2)$, $j = 1, 2, \dots$ which satisfies

$$\sup\{I^f(T_1, T_2, x_j, u_j) : j = 1, 2, \dots\} < \infty,$$

there exist a subsequence $\{(x_{j_k}, u_{j_k})\}_{k=1}^{\infty}$ and $(x, u) \in X(T_1, T_2)$ such that

$$x_{j_k}(t) \to x(t) \text{ as } k \to \infty \text{ for every } t \in [T_1, T_2],$$

$$I^f(T_1, T_2, x, u) \le \liminf_{j\to\infty} I^f(T_1, T_2, x_j, u_j).$$

LSC property plays an important role in the calculus of variations and optimal control theory [32].

The next result follows from Lemma 7.21.

Proposition 7.7. *Let f have LSC property. Then for each $T_2 > T_1$ and each sequence $(x_j, u_j) \in X(T_1, T_2)$, $j = 1, 2, \dots$ which satisfies*

$$\sup\{I^f(T_1, T_2, x_j, u_j) : j = 1, 2, \dots\} < \infty$$

there exist a subsequence $\{(x_{j_k}, u_{j_k})\}_{k=1}^{\infty}$ and $(x, u) \in X(T_1, T_2)$ such that

$$x_{j_k}(\cdot) \to x(\cdot) \text{ as } k \to \infty \text{ uniformly on } [T_1, T_2],$$

$$I^f(T_1, T_2, x, u) \le \liminf_{j\to\infty} I^f(T_1, T_2, x_j, u_j).$$

The following result is proved in Section 7.11.

Theorem 7.8. *Let f have LSC property. If f has STP then the following three properties hold:*

(P1) for each (f)-good pair $(x, u) \in X(-\infty, \infty)$,

$$\lim_{t\to\infty} \|x(t) - x_f(t)\| = 0, \quad \lim_{t\to-\infty} \|x(t) - x_f(t)\| = 0;$$

(P2) for each $\epsilon > 0$ and each $M > 0$, there exist $\delta > 0$ and $L > 0$ such that for each $T \in R^1$ and each $(x, u) \in X(T, T + L)$ which satisfies

$$I^f(T, T + L, x, u) \le \min\{U^f(T, T + L, x(T), x(T + L)) + \delta,$$
$$I^f(T, T + L, x_f, u_f) + M\},$$

there exists $s \in [T, T + L]$ such that $\|x(s) - x_f(s)\| \le \epsilon$;

(P3) *there exists* $\tilde{u}_f : R^1 \to F$ *such that* $(x_f, \tilde{u}_f) \in X(-\infty, \infty)$ *is* (f)*-good and* (f)*-minimal and if* $(x, u) \in X(-\infty, \infty)$ *is* (f)*-good and* (f)*-minimal, then* $x(t) = x_f(t)$ *for all* $t \in R^1$.

If (P1)–(P3) hold and $\tilde{u}_f = u_f$*, then* f *has STP.*

The following result is proved in Section 7.14.

Theorem 7.9. *f has (P1) and (P2) if and only if the following property holds:*

(P4) *for each* $\epsilon > 0$ *and each* $M > 0$*, there exist* $\delta > 0$ *and* $L > 0$ *such that for each* $T_1 \in R^1$*, each* $T_2 > T_1$*, and each* $(x, u) \in X(T_1, T_2)$ *which satisfies*

$$I^f(T_1, T_2, x, u) \le \min\{\sigma^f(T_1, T_2) + M,\ U^f(T_1, T_2, x(T_1), x(T_2)) + \delta\}$$

if $T_2 \ge 2L + (T_1)_+$*, then*

$$\|x(t) - x_f(t)\| \le \epsilon \text{ for all } t \in [(T_1)_+ + L, T_2 - L]$$

and if $T_1 \le (-T_2)_+ - 2L$*, then*

$$\|x(t) - x_f(t)\| \le \epsilon \text{ for all } t \in [T_1 + L, -(-T_2)_+ - L].$$

Theorem 7.10. *Assume that f has STP and that $\epsilon, L_0 > 0$. Then there exist $\delta > 0$ and $L > L_0$ such that for each $T_1 \in R^1$, each $T_2 \ge T_1 + 2L$, and each $(x, u) \in X(T_1, T_2)$ which satisfies*

$$(T_1, x(T_1)) \in \mathcal{A}_{L_0},\ \ (T_2, x(T_2)) \in \widehat{\mathcal{A}}_{L_0},$$

$$I^f(T_1, T_2, x, u) \le U^f(T_1, T_2, x(T_1), x(T_2)) + \delta,$$

there exist $\tau_1 \in [T_1, T_1 + L]$, $\tau_2 \in [T_2 - L, T_2]$ *such that*

$$\|x(t) - x_f(t)\| \le \epsilon \text{ for all } t \in [\tau_1, \tau_2].$$

Moreover, if $\|x(T_2) - x_f(T_2)\| \le \delta$*, then* $\tau_2 = T_2$*, and if and* $\|x(T_1) - x_f(T_1)\| \le \delta$*, then* $\tau_1 = T_1$.

Theorem 7.10 follows from STP and the following result which is proved in Section 7.13.

Proposition 7.11. *Let* $L_0 > 0$*. Then there exists* $M > 0$ *such that for each pair of points* $(T_1, z_1) \in \mathcal{A}_{L_0}$*,* $(T_2, z_2) \in \widehat{\mathcal{A}}_{L_0}$ *satisfying* $T_2 \ge T_1 + 2L_0$*,*

$$U^f(T_1, T_2, z_1, z_2) \le \sigma^f(T_1, T_2) + M.$$

Theorem 7.12. *Assume that f has STP and that $\epsilon, L_0 > 0$. Then there exist $\delta > 0$ and $L > L_0$ such that for each $T_1 \in R^1$, each $T_2 \geq T_1 + 2L$, and each $(x, u) \in X(T_1, T_2)$ which satisfies*

$$(T_1, x(T_1)) \in \mathcal{A}_{L_0},$$

$$I^f(T_1, T_2, x, u) \leq \sigma^f(T_1, T_2, x(T_1)) + \delta,$$

there exist $\tau_1 \in [T_1, T_1 + L]$, $\tau_2 \in [T_2 - L, T_2]$ such that

$$\|x(t) - x_f(t)\| \leq \epsilon \text{ for all } t \in [\tau_1, \tau_2].$$

Moreover, if $\|x(T_2) - x_f(T_2)\| \leq \delta$, then $\tau_2 = T_2$, and if $\|x(T_1) - x_f(T_1)\| \leq \delta$, then $\tau_1 = T_1$.

Theorem 7.12 follows from STP and the following result which is proved in Section 7.13.

Proposition 7.13. *Let $L_0 > 0$. Then there exists $M > 0$ such that for each $(T_1, z_1) \in \mathcal{A}_{L_0}$ and each $T_2 \geq T_1 + L_0$,*

$$\sigma^f(T_1, T_2, z_1) \leq \sigma^f(T_1, T_2) + M.$$

We say that f possesses the weak turnpike property (or WTP for short) if for each $\epsilon > 0$ and each $M > 0$, there exist a natural number Q and $l > 0$ such that for each $T_1 \in R^1$, each $T_2 \geq T_1 + lQ$, and each $(x, u) \in X(T_1, T_2)$ which satisfies

$$I^f(T_1, T_2, x, u) \leq I^f(T_1, T_2, x_f, u_f) + M,$$

there exist finite sequences $\{a_i\}_{i=1}^q$, $\{b_i\}_{i=1}^q \subset [T_1, T_2]$ such that an integer $q \leq Q$,

$$0 \leq b_i - a_i \leq l, \; i = 1, \ldots, q,$$

$$b_i \leq a_{i+1} \text{ for all integers } i \text{ satisfying } 1 \leq i < q,$$

$$\|x(t) - x_f(t)\| \leq \epsilon \text{ for all } t \in [T_1, T_2] \setminus \cup_{i=1}^q [a_i, b_i].$$

The next result is proved in Section 7.14.

Theorem 7.14. *f has WTP if and only if f has (P1) and (P2).*

Let $T_0 \in R^1$. A pair $(x, u) \in X(T_0, \infty)$ is called (f)-overtaking optimal if (x, u) is (f)-good and if for every $(y, v) \in X(T_0, \infty)$ satisfying $x(T_0) = y(T_0)$,

$$\limsup_{T \to \infty} [I^f(T_0, T, x, u) - I^f(T_0, T, y, v)] \leq 0.$$

A pair $(x, u) \in X(T_0, \infty)$ is called f-weakly optimal if (x, u) is (f)-good and if for every $(y, v) \in X(T_0, \infty)$ satisfying $x(T_0) = y(T_0)$,

$$\liminf_{T\to\infty}[I^f(T_0, T, x, u) - I^f(T_0, T, y, v)] \le 0.$$

The next result is proved in Section 7.16.

Theorem 7.15. *Assume that f has (P1), (P2) and LSC property, $S \in R^1$, $(\tilde{x}, \tilde{u}) \in X(S, \infty)$ is (f)-good and that $(x_*, u_*) \in X(S, \infty)$ satisfies $x_*(S) = \tilde{x}(S)$. Then the following conditions are equivalent:*

(i) (x_*, u_*) is (f)-overtaking optimal;
(ii) (x_*, u_*) is (f)-weakly optimal;
(iii) (x_*, u_*) is (f)-minimal and (f)-good;
(iv) (x_*, u_*) is (f)-minimal and satisfies $\lim_{t\to\infty}(x_*(t) - x_f(t)) = 0$;
(v) (x_*, u_*) is (f)-minimal and satisfies $\liminf_{t\to\infty} \|x_*(t) - x_f(t)\| = 0$.

Moreover, there exists an (f)-overtaking optimal pair $(y, v) \in X(S, \infty)$ such that $\tilde{x}(S) = y(S)$.

7.4 Perturbed Problems

Let f has LSC property. In this section we suppose that the following assumption holds.

(A3) For each $\epsilon > 0$, there exists $\delta > 0$ such that for each $(T_i, z_i) \in \mathcal{A}$, $i = 1, 2$ satisfying $\|z_i - x_f(T_i)\| \le \delta$, $i = 1, 2$, there exist $\tau_1, \tau_2 \in (0, b_f]$ and $(x_1, u_1) \in X(T_1, T_1 + \tau_1)$, $(x_2, u_2) \in X(T_2 - \tau_2, T_2)$ which satisfy

$$x_1(T_1) = z_1,\ x_1(T_1 + \tau_1) = x_f(T_1 + \tau_1),$$

$$I^f(T_1, T_1 + \tau_1, x_1, u_1) \le I^f(T_1, T_1 + \tau_1, x_f, u_f) + \epsilon,$$

$$\|x_1(t) - x_f(t)\| \le \epsilon,\ t \in [T_1, T_1 + \tau_1],$$

$$x_2(T_2) = z_2,\ x_2(T_2 - \tau_2) = x_f(T_2 - \tau_2),$$

$$I^f(T_2 - \tau_2, T_2, x_2, u_2) \le I^f(T_2 - \tau_2, T_2, x_f, u_f) + \epsilon,$$

$$\|x_2(t) - x_f(t)\| \le \epsilon,\ t \in [T_2 - \tau_2, T_2].$$

Clearly, (A3) implies (A2). Assume that $\phi : E \to [0, 1]$ is a continuous function satisfying $\phi(0) = 0$ and such that the following property holds:

(i) for each $\epsilon > 0$ there exists $\delta > 0$ such that for each $x \in E$ satisfying $\phi(x) \le \delta$ we have $\|x\| \le \epsilon$.

For each $r \in (0, 1)$ set

$$f_r(t, x, u) = f(t, x, u) + r\phi(x - x_f(t)), \ (t, x, u) \in \mathcal{M}.$$

Clearly, for any $r \in (0, 1)$, f_r is a Borealian function and (A1) and (A3) hold for f_r with $(x_{f_r}, u_{f_r}) = (x_f, u_f)$.

Theorem 7.16. *Let $r \in (0, 1)$, (x_f, u_f) is (f)-minimal. Then f_r has STP.*

Theorem 7.16 is proved in Section 7.17.

7.5 Auxiliary Results

Assumption (A2) easily implies the following result.

Proposition 7.17. *Let $\gamma > 0$. Then there exists $\delta > 0$ such that for each $(T, z_1), (T + 2b_f, z_2) \in \mathcal{A}$ satisfying*

$$\|z_1 - x_f(T)\| \le \delta, \ \|z_2 - x_f(T + 2b_f)\| \le \delta$$

there exists $(x, u) \in X(T, T + 2b_f)$ which satisfies

$$x(T) = z_1, \ x(T + 2b_f) = z_2,$$

$$I^f(T, T + 2b_f, x, u) \le I^f(T, T + 2b_f, x_f, u_f) + \gamma.$$

Lemma 7.18. *There exist numbers $S > 0$, $c_0 \ge 1$ such that for each $T_1 \in R^1$, each $T_2 \ge T_1 + c_0$, and each $(x, u) \in X(T_1, T_2)$,*

$$I^f(T_1, T_2, x_f, u_f) \le I^f(T_1, T_2, x, u) + S. \tag{7.18}$$

Proof. In view of (7.7), there exists $S_1 > 0$ such that

$$\psi(S_1) > a_0 + 1 + |\Delta_f|. \tag{7.19}$$

By (A1), there exist $S_2 > 0$, $c_0 > 1$ such that

$$I^f(T_1, T_2, x_f, u_f) \le I^f(T_1, T_2, x, u) + S_2$$

for each $T_1 \in R^1$, each $T_2 \ge T_1 + c_0$, and each $(x, u) \in X(T_1, T_2)$ satisfying

$$\|x(T_1)\|, \|x(T_2)\| \le S_1.$$

Fix

$$S \geq S_2 + 2 + a_0(2 + 2c_0) + 8(c_0 + 1)|\Delta_f|. \tag{7.20}$$

Assume that

$$T_1 \in R^1,\ T_2 \geq T_1 + c_0,\ (x, u) \in X(T_1, T_2).$$

We show that (7.18) is true. Assume that

$$\|x(t)\| \geq S_1,\ t \in [T_1, T_2]. \tag{7.21}$$

By (7.8), (7.13), (7.19), and (7.21), for all $t \in [T_1, T_2]$,

$$f(t, x(t), u(t)) \geq -a_0 + \psi(\|x(t)\|) \geq -a_0 + \psi(S_1)$$

and

$$I^f(T_1, T_2, x, u) \geq (T_2 - T_1)(\psi(S_1) - a_0)$$

$$\geq (T_2 - T_1)(|\Delta_f| + 1) \geq I^f(T_1, T_2, x_f, u_f) - 2|\Delta_f| - 2a_0$$

$$\geq I^f(T_1, T_2, x_f, u_f) - S$$

and (7.18) holds.

Assume that

$$\inf\{\|x(t)\| :\ t \in [T_1, T_2]\} < S_1. \tag{7.22}$$

Set

$$\tau_1 = \inf\{t \in [T_1, T_2] :\ \|x(t)\| \leq S_1\}, \tag{7.23}$$

$$\tau_2 = \sup\{t \in [T_1, T_2] :\ \|x(t)\| \leq S_1\}. \tag{7.24}$$

There are two cases:

$$\tau_2 - \tau_1 \geq c_0; \tag{7.25}$$

$$\tau_2 - \tau_1 < c_0. \tag{7.26}$$

Assume that (7.25) holds. It follows from (7.23) to (7.25) and the choice of S_2 and c_0 that

$$I^f(\tau_1, \tau_2, x_f, u_f) \le I^f(\tau_1, \tau_2, x, u) + S_2. \tag{7.27}$$

By (7.8), (7.23), and (7.24), for each $t \in [T_1, \tau_1] \cup [\tau_2, T_2]$,

$$\|x(t)\| > S_1, \ f(t, x(t), u(t)) \ge -a_0 + \psi(\|x(t)\|) \ge -a_0 + \psi(S_1). \tag{7.28}$$

It follows from (7.13), (7.19), and (7.28) that

$$I^f(T_1, \tau_1, x, u) \ge (\tau_1 - T_1)(\psi(S_1) - a_0)$$

$$\ge (\tau_1 - T_1)(|\Delta_f| + 1) \ge I^f(T_1, \tau_1, x_f, u_f) - 2|\Delta_f| - 2a_0, \tag{7.29}$$

$$I^f(\tau_2, T_2, x, u) \ge (T_2 - \tau_2)(\psi(S_1) - a_0)$$

$$\ge (T_2 - \tau_2)(|\Delta_f| + 1) \ge I^f(\tau_2, T_2, x_f, u_f) - 2|\Delta_f| - 2a_0. \tag{7.30}$$

In view of (7.27), (7.29), and (7.30),

$$I^f(T_1, T_2, x, u) = I^f(T_1, \tau_1, x, u) + I^f(\tau_1, \tau_2, x, u) + I^f(\tau_2, T_2, x, u)$$

$$\ge I^f(T_1, \tau_1, x_f, u_f) - 2|\Delta_f| - 2a_0 + I^f(\tau_1, \tau_2, x_f, u_f)$$

$$-S_2 + I^f(\tau_2, T_2, x_f, u_f) - 2|\Delta_f| - 2a_0$$

$$\ge I^f(T_1, T_2, x_f, u_f) - S$$

and (7.18) holds.

Assume that (7.26) holds. By (7.23) and (7.24), for each $t \in [T_1, \tau_1] \cup [\tau_2, T_2]$, (7.28)–(7.30) are true.

It follows from (7.8), (7.13), (7.20), (7.29), and (7.30) that

$$I^f(T_1, T_2, x, u) = I^f(T_1, \tau_1, x, u) + I^f(\tau_1, \tau_2, x, u) + I^f(\tau_2, T_2, x, u)$$

$$\ge I^f(T_1, \tau_1, x_f, u_f) - 2|\Delta_f| - 2a_0 - a_0(\tau_2 - \tau_1) + I^f(\tau_2, T_2, x_f, u_f) - 2|\Delta_f| - 2a_0$$

$$\ge I^f(T_1, T_2, x_f, u_f) - a_0(c_0 + 2) - |\Delta_f|(c_0 + 2) - 4a_0 - 4|\Delta_f|$$

$$\ge I^f(T_1, T_2, x_f, u_f) - S.$$

Lemma 7.18 is proved.

Lemma 7.19. *Let $M_0, M_1, \tau_0 > 0$. Then there exists $M_2 > M_1$ such that for each $T_1 \in R^1$, each $T_2 \in (T_1, T_1 + \tau_0]$, and each $(x, u) \in X(T_1, T_2)$ satisfying*

$$\inf\{\|x(t)\| : t \in [T_1, T_2]\} \le M_1, \tag{7.31}$$

$$I^f(T_1, T_2, x, u) \le M_0, \tag{7.32}$$

the following inequality holds:

$$\|x(t)\| \le M_2 \text{ for all } t \in [T_1, T_2]. \tag{7.33}$$

Proof. Fix a positive number

$$\delta < 2^{-1}(a_0 + 1)^{-1}.$$

By (7.7) and (7.8), there exist $h_0 > M_1 + 1$ and $\gamma_0 > 0$ such that

$$f(t, x, u) \ge 4(M_0 + a_0\tau_0)\delta^{-1} \text{ for each } (t, x, u) \in \mathcal{M} \text{ satisfying } \|x\| \ge h_0, \tag{7.34}$$

$$f(t, x, u) \ge 8(\|G(t, x, u)\| - a_0\|x\|)_+$$

for each $(t, x, u) \in \mathcal{M}$ satisfying $\|G(t, x, u)\| - a_0\|x\| \ge \gamma_0$. (7.35)

Choose a number

$$M_2 > 2h_0 + 2M_1 + 2\gamma_0\delta + M_0 + a_0\tau_0. \tag{7.36}$$

Let

$$T_1 \in R^1,\ T_2 \in (T_1, T_1 + \tau_0],\ (x, u) \in X(T_1, T_2),$$

(7.31) and (7.32) hold. We show that (7.33) hold. Assume the contrary. Then there exists $t_0 \in [T_1, T_2]$ such that

$$\|x(t_0)\| > M_2. \tag{7.37}$$

By (7.31) and (7.34), there exists $t_1 \in [T_1, T_2]$ such that

$$\|x(t_1)\| \le h_0,\ |t_1 - t_0| \le \delta. \tag{7.38}$$

There exists

$$t_2 \in [\min\{t_1, t_0\}, \max\{t_1, t_0\}] \tag{7.39}$$

such that

$$\|x(t_2)\| \ge \|x(t)\| \text{ for all } t \in [\min\{t_1, t_0\}, \max\{t_1, t_0\}]. \tag{7.40}$$

In view of (7.36)–(7.38) and (7.40)

$$t_2 \neq t_1. \tag{7.41}$$

By (7.4),

$$\|x(t_2) - x(t_1)\| = \| \int_{t_1}^{t_2} G(x(s), u(s))ds \|. \tag{7.42}$$

It follows from (7.38) and (7.42) that

$$\|x(t_2)\| - h_0 \leq | \int_{t_1}^{t_2} \|G(s, x(s), u(s))\| ds|. \tag{7.43}$$

Define

$$E_1 = \{t \in [\min\{t_1, t_0\}, \max\{t_1, t_0\}] : \|G(t, x(t), u(t))\| \geq a_0\|x(t)\| + \gamma_0\},$$

$$E_2 = [\min\{t_1, t_0\}, \max\{t_1, t_0\}] \setminus E_1. \tag{7.44}$$

By (7.8), (7.32), (7.35), (7.38), (7.40), (7.43), and (7.44),

$$\begin{aligned}
\|x(t_2)\| - h_0 &\leq a_0 \int_{E_1 \cup E_2} \|x(t)\| dt + \int_{E_1 \cup E_2} [\|G(s, x(s), u(s))\| - a_0\|x(s)\|]_+ ds \\
&\leq a_0 \delta \|x(t_2)\| + \int_{E_1} [\|G(s, x(s), u(s))\| - a_0\|x(s)\|]_+ ds \\
&\quad + \int_{E_2} [\|G(s, x(s), u(s))\| - a_0\|x(s)\|]_+ ds \\
&\leq a_0\|x(t_2)\|\delta + \gamma_0\delta + 8^{-1} \int_{E_1} f(t, x(t), u(t)) dt \\
&\leq a_0\|x(t_2)\|\delta + \gamma_0\delta + 8^{-1}(M_0 + a_0\tau_0).
\end{aligned} \tag{7.45}$$

It follows from (7.36) to (7.38), (7.40), (7.45), and the choice of δ that

$$2^{-1}M_2 \leq 2^{-1}\|x(t_2)\| \leq \|x(t_2\|(1 - a_0\delta) \leq h_0 + \gamma_0\delta + 8^{-1}(M_0 + a_0\tau_0)$$

and

$$M_2 < 2h_0 + 2\gamma_0\delta + 4^{-1}(M_0 + a_0\tau_0).$$

This contradicts (7.36). The contradiction we have reached proves Lemma 7.19.

Lemma 7.19, (7.7) and (7.8) imply the following result.

Lemma 7.20. *Let $M_1 > 0, 0 < \tau_0 < \tau_1$. Then there exists $M_2 > 0$ such that for each $T_1 \in R^1$, each $T_2 \in [T_1 + \tau_0, T_1 + \tau_1]$, and each $(x, u) \in X(T_1, T_2)$ satisfying*

$$I^f(T_1, T_2, x, u) \le M_1,$$

the following inequality holds:

$$\|x(t)\| \le M_2 \text{ for all } t \in [T_1, T_2].$$

Lemma 7.21. *Let $M_1 > 0$, $\epsilon \in (0, 1)$, $0 < \tau_0 < \tau_1$. Then there exists $\delta > 0$ such that for each $T_1 \in R^1$, each $T_2 \in [T_1 + \tau_0, T_1 + \tau_1]$, and each $(x, u) \in X(T_1, T_2)$ satisfying*

$$I^f(T_1, T_2, x, u) \le M_1$$

and each $t_1, t_2 \in [T_1, T_2]$ satisfying $|t_1 - t_2| \le \delta$, the inequality $\|x(t_1) - x(t_2)\| \le \epsilon$ holds.

Proof. Lemma 7.20 implies that there exists $M_2 > 0$ such that the following property holds:

(i) for each $T_1 \ge 0$, each $T_2 \in [T_1 + \tau_0, T_1 + \tau_1]$, and each $(x, u) \in X(T_1, T_2)$ satisfying

$$I^f(T_1, T_2, x, u) \le M_1,$$

we have

$$\|x(t)\| \le M_2 \text{ for all } t \in [T_1, T_2].$$

In view of (7.7) and (7.8), there exists $h_0 > 0$ such that

$$f(t, x, u) \ge 4\epsilon^{-1}(M_1 + a_0\tau_1 + 8)(\|G(t, x, u)\| - a_0\|x\|)_+$$

$$\text{for each } (t, x, u) \in \mathcal{M} \text{ satisfying } \|G(t, x, u)\| - a_0\|x\| \ge h_0. \tag{7.46}$$

Choose a number

$$\delta \in (0, \epsilon(4a_0M_2 + 4h_0 + 4)^{-1}). \tag{7.47}$$

Let

$$T_1 \in R^1, \; T_2 \in [T_1 + \tau_0, T_1 + \tau_1], \; (x, u) \in X(T_1, T_2),$$

$$I^f(T_1, T_2, x, u) \le M_1, \tag{7.48}$$

$$t_1, t_2 \in [T_1, T_2],\ 0 < t_2 - t_1 \le \delta. \tag{7.49}$$

Property (i) and (7.48) imply that

$$\|x(t)\| \le M_2 \text{ for all } t \in [T_1, T_2]. \tag{7.50}$$

Define

$$\Omega_1 = \{t \in [t_1, t_2] : \|G(t, x(t), u(t))\| - a_0\|x(t)\| \ge h_0\}, \tag{7.51}$$

$$\Omega_2 = [t_1, t_2] \setminus \Omega_1. \tag{7.52}$$

By (7.4), (7.8), and (7.47)–(7.52),

$$\|x(t_2) - x(t_1)\| = \|\int_{t_1}^{t_2} G(s, x(s), u(s))ds\|$$

$$\le a_0 \int_{t_1}^{t_2} \|x(t)\|dt + \int_{t_1}^{t_2} (\|G(t, x(t), u(t))\| - a_0\|x(t)\|)_+ dt$$

$$\le a_0\delta M_2 + \delta h_0 + \int_{E_1} (\|G(t, x(t), u(t)) - a_0\|x(t)\|)_+ dt$$

$$\le a_0\delta M_2 + \delta h_0 + \epsilon(4(M_1 + a_0\tau_1 + 8))^{-1} \int_{E_1} f(t, x(t), u(t))dt$$

$$\le a_0\delta M_2 + \delta h_0 + 4^{-1}\epsilon < \epsilon.$$

Lemma 7.21 is proved.

Lemma 7.22. *Let $\Delta > 0$. Then there exists $\delta > 0$ such that for each $(T_1, z_1), (T_2, z_2) \in \mathcal{A}$ satisfying*

$$T_2 \ge T_1 + 2b_f,\ \|z_i - x_f(T_i)\| \le \delta,\ i = 1, 2 \tag{7.53}$$

the following inequality holds:

$$U^f(T_1, T_2, z_1, z_2) \le I^f(T_1, T_2, x_f, u_f) + \Delta.$$

Proof. Let $\delta > 0$ be as guaranteed by (A2) with $\epsilon = \Delta/4$. Let

$$(T_1, z_1), (T_2, z_2) \in \mathcal{A}$$

satisfy (7.53). By (7.53), the choice of δ and (A2) with $\epsilon = \Delta/4$, there exist $(y, v) \in X(T_1, T_2)$ such that

$$y(T_1) = z_1, \ y(T_2) = z_2,$$

$$y(t) = x_f(t), \ v(t) = u_f(t), \ t \in [T_1 + b_f, T_2 - b_f],$$

$$I^f(T_1, T_1 + b_f, y, v) \le I^f(T_1, T_1 + b_f, x_f, u_f) + \Delta/4,$$

$$I^f(T_2 - b_f, T_2, y, v) \le I^f(T_2 - b_f, T_2, x_f, u_f) + \Delta/4.$$

By the relations above,

$$U^f(T_1, T_2, z_1, z_2) \le I^f(T_1, T_2, y, v) \le I^f(T_1, T_2, x_f, u_f) + \Delta/2.$$

Lemma 7.22 is proved.

Lemma 7.23. *Let $\Delta \in (0, 1)$. Then there exists $\delta > 0$ such that for each $T_1 \in R^1$, each $T_2 \ge T_1 + 3b_f$, each $(x, u) \in X(T_1, T_2)$ satisfying*

$$\|x(T_i) - x_f(T_i)\| \le \delta, \ i = 1, 2,$$

$$I^f(T_1, T_2, x, u) \le U^f(T_1, T_2, x(T_1), x(T_2)) + \delta,$$

each $\tau_1 \in [T_1, T_1 + \delta]$ and each $\tau_2 \in [T_2 - \delta, T_2]$ the inequality

$$I^f(\tau_1, \tau_2, x, u) \le I^f(\tau_1, \tau_2, x_f, u_f) + \Delta$$

holds.

Proof. Lemma 7.22 implies that there exists

$$\delta_0 \in (0, \min\{4^{-1}\Delta, 2^{-1}b_f\})$$

such that the following property holds:

(ii) for each $(T_1, z_1), (T_2, z_2) \in \mathcal{A}$ satisfying

$$T_2 \ge T_1 + 2b_f, \ \|z_i - x_f(T_i)\| \le \delta_0, \ i = 1, 2$$

we have

$$U^f(T_1, T_2, z_1, z_2) \le I^f(T_1, T_2, x_f, u_f) + \Delta/4.$$

Lemmas 7.1 and 7.18, (7.8), and (7.13) imply that there exists $M_0 > 0$ such that the following property holds:

(iii) for each pair of numbers $T_2 > T_1$ and each $(x, u) \in X(T_1, T_2)$,

$$I^f(T_1, T_2, x, u) + M_0 \geq I^f(T_1, T_2, x_f, u_f).$$

Lemma 7.21 implies that there exists $\delta_1 \in (0, \delta_0)$ such that the following property holds:

(iv) for each $S \in R^1$ and each $(y, v) \in X(S, S + b_f)$ satisfying

$$I^f(S, S + b_f, y, v) \leq |\Delta_f|(b_f + 2) + 3a_0 + 2M_0 + 1$$

and each $t_1, t_2 \in [S, S+b_f]$ satisfying $|t_1 - t_2| \leq \delta_1$, we have $\|y(t_1) - y(t_2)\| \leq \delta_0/8$.

Set

$$\delta = \delta_1/8. \tag{7.54}$$

Assume that

$$T_1 \in R^1,\ T_2 \geq T_1 + 3b_f, \tag{7.55}$$

$(x, u) \in X(T_1, T_2)$,

$$\|x(T_i) - x_f(T_i)\| \leq \delta,\ i = 1, 2, \tag{7.56}$$

$$I^f(T_1, T_2, x, u) \leq U^f(T_1, T_2, x(T_1), x(T_2)) + \delta, \tag{7.57}$$

$$\tau_1 \in [T_1, T_1 + \delta],\ \tau_2 \in [T_2 - \delta, T_2]. \tag{7.58}$$

Property (ii), (7.54), and (7.55) imply that

$$I^f(T_1, T_2, x, u) \leq I^f(T_1, T_2, x_f, u_f) + 1. \tag{7.59}$$

Property (iii) and (7.59) imply that for each $t_1, t_2 \in [T_1, T_2]$ satisfying $t_1 < t_2$,

$$I^f(t_1, t_2, x, u) = I^f(T_1, T_2, x, u) - I^f(T_1, t_1, x, u) - I^f(t_2, T_2, x, u)$$

$$\leq I^f(T_1, T_2, x_f, u_f) + 1 - I^f(T_1, t_1, x_f, u_f) + M_0 - I^f(t_2, T_2, x_f, u_f) + M_0$$

$$= I^f(t_1, t_2, x_f, u_f) + 2M_0 + 1. \tag{7.60}$$

In view of (7.8), (7.13), and (7.55),

$$I^f(T_1, T_1 + b_f, x_f, u_f)$$

$$\leq I^f(\lfloor T_1, \rfloor T_1 + \lfloor b_f \rfloor + 2, x_f, u_f) + 3a_0 \leq (\lfloor b_f \rfloor + 2)\Delta_f + 3a_0, \tag{7.61}$$

$$I^f(T_2 - b_f, T_2, x_f, u_f)$$

$$\leq I^f(\lfloor T_2 \rfloor - \lfloor b_f \rfloor - 1, \lfloor T_2 \rfloor + 1, x_f, u_f) + 3a_0 \leq (\lfloor b_f \rfloor + 2)\Delta_f + 3a_0. \tag{7.62}$$

It follows from (7.60) and the relations above that

$$I^f(T_1, T_1 + b_f, x_f, u_f),\ I^f(T_2 - b_f, T_2, x_f, u_f) \leq \Delta_f(\lfloor b_f \rfloor + 2) + 3a_0 + 2M_0 + 1. \tag{7.63}$$

Property (iv), (7.54), (7.58), and (7.61)–(7.63) imply that

$$\|x_f(T_i) - x_f(\tau_i)\| \leq \delta_0/8,\ i = 1, 2, \tag{7.64}$$

$$\|x(T_i) - x(\tau_i)\| \leq \delta_0/8,\ i = 1, 2, \tag{7.65}$$

By (7.54), (7.56), (7.64), and (7.65), for $i = 1, 2$,

$$\|x_f(\tau_i) - x(\tau_i)\| \leq \|x_f(\tau_i) - x_f(T_i)\| + \|x_f(T_i) - x(T_i)\| + \|x(T_i) - x(\tau_i)\|$$

$$\leq \delta_0/8 + \delta_0/8 + \delta \leq \delta_0/2. \tag{7.66}$$

Property (ii), (7.55), and (7.66) imply that

$$U^f(\tau_1, \tau_2, x(\tau_1), x(\tau_2)) \leq I^f(\tau_1, \tau_2, x_f, u_f) + \Delta/4. \tag{7.67}$$

By (7.54), (7.57), (7.58), and (7.67),

$$I^f(\tau_1, \tau_2, x, u) \leq U^f(\tau_1, \tau_2, x(\tau_1), x(\tau_2)) + \delta$$

$$\leq I^f(\tau_1, \tau_2, x_f, u_f) + \Delta/4 + \delta \leq I^f(\tau_1, \tau_2, x_f, u_f) + \Delta.$$

Lemma 7.23 is proved.

7.6 Proofs of Theorems 7.2–7.5

Proof of Theorem 7.2. Assertion 1 of Theorem 7.2 follows from Lemma 7.18, (7.8), and (7.13).

We will prove Assertion 2. Let $(x, u) \in X(-\infty, \infty)$. Assume that there exists a sequence of positive numbers $\{T_k\}_{k=1}^{\infty}$ such that

$$T_k \to \infty \text{ as } k \to \infty, \tag{7.68}$$

$$I^f(-T_k, T_k, x, u) - I^f(-T_k, T_k, x_f, u_f) \to \infty \text{ as } k \to \infty. \quad (7.69)$$

Let a number $S > 0$ be as guaranteed by Assertion 1. Let k be a natural number and $T \geq T_k$. In view of (7.68), (7.69), Assertion 1, and the choice of S,

$$\begin{aligned} &I^f(-T, T, x, u) - I^f(-T, T, x_f, u_f) \\ &= I^f(-T_k, T_k, x, u) - I^f(-T_k, T_k, x_f, u_f) + I^f(-T, -T_k, x, u) \\ &\quad - I^f(-T, -T_k, x_f, u_f) \\ &\quad + I^f(T_k, T, x, u) - I^f(T_k, T, x_f, u_f) \\ &\geq I^f(-T_k, T_k, x, u) - I^f(-T_k, T_k, x_f, u_f) - 2S \to \infty \text{ as } k \to \infty. \end{aligned}$$

Thus Assertion 2 is proved.

Assume that

$$\sup\{|I^f(-T, T, x, u) - I^f(-T, T, x_f, u_f)| : T \in [0, \infty)\} < \infty.$$

Therefore there exists an integer $S_1 > 4 + a_0$ such that for each $T_1 < T_2$,

$$|I^f(T_1, T_2, x, u) - I^f(T_1, T_2, x_f, u_f)| \leq S_1. \quad (7.70)$$

Let $S_0 > 0$ be such that

$$\psi(S_0) > 4|\Delta_f| + 32. \quad (7.71)$$

We show that for each $T \in R^1$,

$$\min\{\|x(t)\| : t \in [T, T + S_1]\} \leq S_0. \quad (7.72)$$

Assume the contrary. Then there exists $T \in R^1$ such that

$$\|x(t)\| > S_0, \ t \in [T, T + S_1]. \quad (7.73)$$

By (7.8) and (7.73),

$$\begin{aligned} I^f(T, T + S_1, x, u) &\geq -2a_0 + I^f(\lfloor T \rfloor + 1, \lfloor T \rfloor + S_1, x, u) \\ &\geq -2a_0 + \psi(S_0)(S_1 - 2) \geq -2a_0 + 2^{-1}\psi(S_0)S_1. \end{aligned} \quad (7.74)$$

In view of (7.7) and (7.13),

$$I^f(T, T+S_1, x_f, u_f) \le I^f(\lfloor T \rfloor, \lfloor T \rfloor + S_1 + 1, x_f, u_f) + 2a_0$$

$$\le \Delta_f(S_1+1) + 2a_0 \le 2|\Delta_f|S_1 + 2a_0. \tag{7.75}$$

It follows from (7.71), (7.74), and (7.75) that

$$I^f(T, T+S_1, x, u) - I^f(T, T+S_1, x_f, u_f) \ge 2^{-1}\psi(S_0)S_1 - 4a_0 - 2|\Delta_f|S_1$$

$$\ge S_1(2^{-1}\psi(S_0) - 2|\Delta_f|) - 4a_0 \ge -4a_0 + 8S_1 \ge 4S_1.$$

This contradicts (7.70). The contradiction we have reached proves that for each $T \in R^1$, (7.72) is true. By (7.72), there exists a sequence $\{T_i\}_{i=-\infty}^{\infty} \subset R^1$ such that for each integer $i \ge 1$,

$$T_{i+1} - T_i \in [1, 1+S_1],$$

$$\|x(T_i)\| \le S_0. \tag{7.76}$$

Lemma 7.19, (7.7), (7.13), (7.70), and (7.76) imply that

$$\sup\{\|x(t)\| : t \in R^1\}.$$

Theorem 7.2 is proved.

Proof of Theorem 7.3. We may assume that $c < 1/2$. By Theorem 7.2, there exists $S_0 > 0$ such that the following property holds:

(i) for each $T_2 > T_1$ and each $(x, u) \in X(T_1, T_2)$,

$$I^f(T_1, T_2, x, u) + S_0 \ge I^f(T_1, T_2, x_f, u_f).$$

Lemma 7.20 implies that there exists $M_1 > 0$ such that the following property holds:

(ii) for each $T \in R^1$ and each $(x, u) \in X(T, T + c/2)$ satisfying

$$I^f(T, T + c/2, x, u) \le M_0 + 2S_0 + 2 + 2a_0 + 2|\Delta_f|$$

we have

$$\|x(t)\| \le M_1 \text{ for all } t \in [T, T + c/2].$$

Assume that $T_1 \in R^1$, $T_2 \ge T_1 + c$, $(x, u) \in X(T_1, T_2)$ and

$$I^f(T_1, T_2, x, u) \le I^f(T_1, T_2, x_f, u_f) + M_0. \tag{7.77}$$

We show that

$$\|x(t)\| \le M_1,\ t \in [T_1, T_2].$$

Assume the contrary. Then there exists

$$t_0 \in [T_1, T_2]$$

such that

$$\|x(t_0)\| > M_1. \tag{7.78}$$

Clearly, there exists a closed interval $[a, b] \subset [T_1, T_2]$ such that

$$t_0 \in [a, b],\ b - a = 2^{-1}c. \tag{7.79}$$

Property (ii), (7.78), and (7.79) imply that

$$I^f(a, b, x, u) > M_0 + 2S_0 + 2 + 2a_0 + 2|\Delta_f|. \tag{7.80}$$

Property (i), (7.7), (7.13), (7.77), and (7.79) imply that

$$I^f(a, b, x, u) = I^f(T_1, T_2, x, u) - I^f(T_1, a, x, u) - I^f(b, T_2, x, u)$$

$$\le I^f(T_1, T_2, x_f, u_f) + M_0 - I^f(T_1, a, x_f, u_f) + S_0 - I^f(b, T_2, x_f, u_f) + S_0$$

$$= I^f(a, b, x_f, u_f) + 2S_0 + M_0 \le 2|\Delta_f| + 2a_0 + 2S_0 + M_0.$$

This contradicts (7.80). The contradiction we have reached proves Theorem 7.3.

Proof of Theorem 7.4. Theorem 7.3 implies that there exists $M_1 > 0$ such that the following property holds:

(iii) for each $T_1 \in R^1$, each $T_2 \ge T_1 + 2L$, and each $(x, u) \in X(T_1, T_2)$ satisfying

$$I^f(T_1, T_2, x, u) \le I^f(T_1, T_2, x_f, u_f) + M_0 + 2L(1 + a_0),$$

the inequality $\|x(t)\| \le M_1$ holds for all $t \in [T_1, T_2]$.

Assume that $T_1 \in R^1$, $T_2 \ge T_1 + 2L$, and $(x, u) \in X(T_1, T_2)$ satisfy

$$(T_1, x(T_1)) \in \mathcal{A}_L,\ (T_2, x(T_2)) \in \widehat{\mathcal{A}}_L, \tag{7.81}$$

$$I^f(T_1, T_2, x, u) \le U^f(T_1, T_2, x(T_1), x(T_2)) + M_0. \tag{7.82}$$

In view of (7.81), there exist $\tau_1 \in (0, L]$, $\tau_2 \in (0, L]$, and $(y, v) \in X(T_1, T_2)$ such that

$$y(T_1) = x(T_1), \ y(T_2) = x(T_2), \tag{7.83}$$

$$(y(t), v(t)) = (x_f(t), u_f(t)), \ t \in [\tau_1 + T_1, T_2 - \tau_2], \tag{7.84}$$

$$I^f(T_1, T_1 + \tau_1, y, v) \leq L, \ I^f(T_2 - \tau_2, T_2, y, v) \leq L. \tag{7.85}$$

By (7.7) and (7.82)–(7.85),

$$I^f(T_1, T_2, x, u) \leq M_0 + I^f(T_1, T_2, y, v)$$

$$\leq M_0 + I^f(T_1, T_1 + \tau_1, y, v) + I^f(T_1 + \tau_1, T_2 - \tau_2, y, v) + I^f(T_2 - \tau_2, T_2, y, v)$$

$$\leq M_0 + 2L + I^f(T_1 + \tau_1, T_2 - \tau_2, x_f, u_f)$$

$$\leq 2L + M_0 + I^f(T_1, T_2, x_f, u_f) + 2La_0. \tag{7.86}$$

Property (iii), (7.85), and (7.86) imply that $\|x(t)\| \leq M_1$ for all $t \in [T_1, T_2]$. Theorem 7.4 is proved.

Proof of Theorem 7.5. Theorem 7.3 implies that there exists $M_1 > 0$ such that the following property holds:

(iv) for each $T_1 \in R^1$, each $T_2 \geq T_1 + L$, and each $(x, u) \in X(T_1, T_2)$ satisfying

$$I^f(T_1, T_2, x, u) \leq I^f(T_1, T_2, x_f, u_f) + M_0 + L(1 + a_0),$$

the inequality $\|x(t)\| \leq M_1$ holds for all $t \in [T_1, T_2]$.

Assume that $T_1 \in R^1$, $T_2 \geq T_1 + L$ and $(x, u) \in X(T_1, T_2)$ satisfies

$$(T_1, x(T_1)) \in \mathcal{A}_L, \tag{7.87}$$

$$I^f(T_1, T_2, x, u) \leq \sigma^f(T_1, T_2, x(T_1)) + M_0. \tag{7.88}$$

In view of (7.87), there exist $\tau \in (0, L]$ and $(y, v) \in X(T_1, T_2)$ such that

$$y(T_1) = x(T_1), \ (y(t), v(t)) = (x_f(t), u_f(t)), \ t \in [\tau + T_1, T_2], \tag{7.89}$$

$$I^f(T_1, T_1 + \tau, y, v) \leq L. \tag{7.90}$$

By (7.7) and (7.88)–(7.90),

$$I^f(T_1, T_2, x, u) \leq M_0 + I^f(T_1, T_2, y, v)$$

$$= I^f(T_1, T_1 + \tau, y, v) + I^f(T_1 + \tau, T_2, x_f, u_f) + M_0$$

$$\leq L + M_0 + I^f(T_1, T_2, x_f, u_f) + La_0. \tag{7.91}$$

Property (iv) and (7.91) imply that $\|x(t)\| \leq M_1$ for all $t \in [T_1, T_2]$. Theorem 7.5 is proved.

7.7 Auxiliary Results

Lemma 7.24. *Let $\epsilon > 0$. Then there exist $L > 0$ such that the following properties hold:*

(i) for each $T_2 \geq T_1 \geq L$,

$$I^f(T_1, T_2, x_f, u_f) \leq U^f(T_1, T_2, x_f(T_1), x_f(T_2)) + \epsilon;$$

(ii) for each $T_1 < T_2 \leq -L$,

$$I^f(T_1, T_2, x_f, u_f) \leq U^f(T_1, T_2, x_f(T_1), x_f(T_2)) + \epsilon.$$

Proof. Assume the contrary. Then there exists a sequence of closed intervals $\{[a_i, b_i]\}_{i=1}^{\infty}$ such that for each integer $i \geq 1$ and each integer $j > i$,

$$[a_i, b_i] \cap [a_j, b_j] = \emptyset,$$

$$I^f(a_i, b_i, x_f, u_f) > U^f(a_i, b_i, x_f(a_i), x_f(b_i)) + \epsilon.$$

By the relations above, there exists $(y, v) \in X(-\infty, \infty)$ such that

$$y(t) = x_f(t), \ v(t) = u_f(t), \ t \in R^1 \setminus \cup_{i=1}^{\infty}[a_i, b_i], \tag{7.92}$$

for every integer $i \geq 1$,

$$I^f(a_i, b_i, x_f, u_f) > I^f(a_i, b_i, y, v) + \epsilon.$$

Let $q \geq 1$ be an integer. It follows from (7.92) and the relation above that for each $T > 0$ satisfying

$$[a_i, b_i] \subset [-T, T], \ i = 1, \dots, q,$$

we have

$$I^f(-T, T, x_f, u_f) - I^f(-T, T, y, v)$$

$$\geq \sum_{i=1}^{q} (I^f(a_i, b_i, x_f, u_f) - I^f(a_i, b_i, y, v)) \geq q\epsilon,$$

$$I^f(-T, T, y, v) \leq I^f(-T, T, x_f, u_f) - q\epsilon.$$

Since q is any natural number, this implies that

$$I^f(-T, T, y, v) - I^f(-T, T, x_f, u_f) \to -\infty \text{ as } k \to \infty.$$

This contradicts Theorem 7.2. The contradiction we have reached proves Lemma 7.24.

Lemma 7.25. *Let (P1) hold and $\epsilon > 0$. Then there exist $\delta, T_\epsilon > 0$ such that the following properties hold:*

(iii) for each $T_1 \geq T_\epsilon$, each $T_2 > T_1$, and each $(y, v) \in X(T_1, T_2)$ which satisfies

$$\|y(T_i) - x_f(T_i)\| \leq \delta, \ i = 1, 2,$$

the inequality

$$I^f(T_1, T_2, y, v) \geq I^f(T_1, T_2, x_f, u_f) - \epsilon$$

holds;

(iv) for each $T_1 < T_2 \leq -T_\epsilon$ and each $(y, v) \in X(T_1, T_2)$ which satisfies

$$\|y(T_i) - x_f(T_i)\| \leq \delta, \ i = 1, 2,$$

the inequality

$$I^f(T_1, T_2, y, v) \geq I^f(T_1, T_2, x_f, u_f) - \epsilon$$

holds.

Proof. By (A2), there exists $\delta \in (0, \epsilon/4)$ such that the following property holds:

(v) for each $(T, z) \in \mathcal{A}$ satisfying $\|z - x_f(T)\| \leq \delta$, there exist $\tau_1, \tau_2 \in (0, b_f]$, $(\tilde{x}_1, \tilde{u}_1) \in X(T, T + \tau_1)$, $(\tilde{x}_2, \tilde{u}_2) \in X(T - \tau_2, T)$ such that

$$\tilde{x}_1(T) = z \ , \tilde{x}_1(T + \tau_1) = x_f(T + \tau_1),$$

$$I^f(T, T + \tau_1, \tilde{x}_1, \tilde{u}_1) \leq I^f(T, T + \tau_1, x_f, u_f) + \epsilon/4,$$

$$\tilde{x}_2(T - \tau_2) = x_f(T - \tau_2) \ , \tilde{x}_2(T) = z,$$

$$I^f(T - \tau_2, T, \tilde{x}_2, \tilde{u}_2) \leq I^f(T - \tau_2, T, x_f, u_f) + \epsilon/4.$$

Lemma 7.24 implies that there exists $L > b_f$ such that the following property holds:

(vi) for each $T_2 > T_1 \geq L$,

$$I^f(T_1, T_2, x_f, u_f) < U^f(T_1, T_2, x_f(T_1), x_f(T_2)) + \epsilon/4$$

and for each $T_1 < T_2 \leq -L$,

$$I^f(T_1, T_2, x_f, u_f) \leq U^f(T_1, T_2, x_f(T_1), x_f(T_2)) + \epsilon/4.$$

Set

$$T_\epsilon = L + b_f.$$

Let $T_2 > T_1$,

$$\text{either } T_1 > L + b_f \text{ or } T_2 < -L - b_f, \tag{7.93}$$

$(y, v) \in X(T_1, T_2)$ and

$$\|y(T_i) - x_f(T_i)\| \leq \delta, \; i = 1, 2. \tag{7.94}$$

We show that

$$I^f(T_1, T_2, y, v) \geq I^f(T_1, T_2, T, x_f, u_f) - \epsilon$$

Assume the contrary. Then

$$I^f(T_1, T_2, y, v) < I^f(T_1, T_2, x_f, u_f) - \epsilon. \tag{7.95}$$

By (7.94) and property (v), there exist $\tau_1, \tau_2 \in (0, b_f]$ and $(x, u) \in X(T_1 - \tau_1, T_2 + \tau_2)$ such that

$$(x(t), u(t)) = (y(t), v(t)) \text{ for all } t \in [T_1, T_2], \tag{7.96}$$

$$x(T_1 - \tau_1) = x_f(T_1 - \tau_1), \; x(T_2 + \tau_2) = x_f(T_2 + \tau_2), \tag{7.97}$$

$$I^f(T_1 - \tau_1, T_1, x, u) \leq I^f(T_1 - \tau_1, T_1, x_f, u_f) + \epsilon/4, \tag{7.98}$$

$$I^f(T_2, T_2 + \tau_2, x, u) \leq I^f(T_2, T_2 + \tau_2, x_f, u_f) + \epsilon/4. \tag{7.99}$$

It follows from (7.95), (7.96), (7.98), and (7.99) that

$$I^f(T_1 - \tau_1, T_2 + \tau_2, x, u)$$

$$= I^f(T_1 - \tau_1, T_1, x, u) + I^f(T_1, T_2, y, v) + I^f(T_2, T_2 + \tau_2, x, u)$$

$$\le I^f(T_1, T_2, x_f, u_f) - \epsilon + I^f(T_1 - \tau_1, T_1, x_f, u_f) + \epsilon/4$$

$$+ I^f(T_2, T_2 + \tau_2, x_f, u_f) + \epsilon/4$$

$$= I^f(T_1 - \tau_1, T_2 + \tau_2, x_f, u_f) - \epsilon/2.$$

Combined with (7.93) and (7.97), this contradicts (vi). The contradiction we have reached completes the proof of Lemma 7.25.

7.8 STP Implies (P1), (P2), and (P3)

Assume that STP and LSC property hold. Theorem 7.2 and STP imply (P2). We show that (P1) holds. Assume that $(x, u) \in X(-\infty, \infty)$ is (f)-good. Theorem 7.2 implies that there exists $M_0 > 0$ such that

$$\|x(t)\| \le M_0, \ t \in R^1, \tag{7.100}$$

for each $T > 0$,

$$I^f(-T, T, x, u) - I^f(-T, T, x_f, u_f) \le M_0. \tag{7.101}$$

Theorem 7.2 implies that there exists $M_1 > 0$ such that the following property holds:

(i) for each $T_1 < T_2$ and each $(y, v) \in X(T_1, T_2)$,

$$M_1 + I^f(T_1, T_2, y, v) \ge I^f(T_1, T_2, x_f, u_f).$$

Let $S_1 < S_2$. Choose a positive number T such that

$$-T < S_1 < S_2 < T. \tag{7.102}$$

By (7.101) and (7.102),

$$I^f(S_1, S_2, x, u) = I^f(-T, T, x, u) - I^f(-T, S_1, x, u) - I^f(S_2, T, x, u)$$

$$\le I^f(-T, T, x_f, u_f) + M_0 - I^f(T, S_1, x_f, u_f) + M_1 - I^f(S_2, T, x_f, u_f) + M_1$$

$$= I^f(S_1, S_2, x_f, u_f) + M_0 + 2M_1.$$

Thus for each pair of numbers $S_2 > S_1$,

$$I^f(S_1, S_2, x, u) \le I^f(S_1, S_2, x_f, u_f) + M_0 + 2M_1. \tag{7.103}$$

Let $\gamma > 0$. We show that there exists $T_\gamma > 0$ such that the following properties hold:

(ii) for each $S_1 < S_2 \le -T_\gamma$,

$$I^f(S_1, S_2, x, u) \le U^f(S_1, S_2, x(S_1), x(S_2)) + \gamma;$$

(iii) for each $S_2 > S_1 \ge T_\gamma$,

$$I^f(S_1, S_2, x, u) \le U^f(S_1, S_2, x(S_1), x(S_2)) + \gamma.$$

Assume the contrary. Then there exists a sequence of closed intervals $\{[a_i, b_i]\}_{i=1}^\infty$ such that for each integer $i \ge 1$ and each integer $j > i$,

$$[a_i, b_i] \cap [a_j, b_j] = \emptyset,$$

$$I^f(a_i, b_i, x, u) > U^f(a_i, b_i, x(a_i), x(b_i)) + \gamma.$$

By the relations above, there exists $(y, v) \in X(-\infty, \infty)$ such that

$$y(t) = x(t),\ v(t) = u(t),\ t \in R^1 \setminus \cup_{i=1}^\infty [a_i, b_i], \tag{7.104}$$

for every integer $i \ge 1$,

$$I^f(a_i, b_i, y, v) < I^f(a_i, b_i, x, u) - \gamma. \tag{7.105}$$

Let $q \ge 1$ be an integer and $T > 0$ satisfy

$$[a_i, b_i] \subset [-T, T],\ i = 1, \dots, q. \tag{7.106}$$

By (7.103)–(7.106), we have

$$I^f(-T, T, x, u) - I^f(-T, T, y, v)$$

$$\ge \sum_{i=1}^q (I^f(a_i, b_i, x, u) - I^f(a_i, b_i, y, v)) \ge q\gamma,$$

$$I^f(-T, T, y, v) \le I^f(-T, T, x, u) - q\gamma \le I^f(-T, T, x_f, u_f) + M_0 + 2M_1 - \gamma q.$$

Thus for every $T > 0$ satisfying (7.106),

$$I^f(-T, T, y, v) - I^f(-T, T, x_f, u_f) \le M_0 + 2M_1 - \gamma q.$$

Since q is any natural number, this implies that

$$I^f(-T, T, y, v) - I^f(-T, T, x_f, u_f) \to -\infty \text{ as } k \to \infty.$$

This contradicts Theorem 7.2. The contradiction we have reached proves that for each $\gamma > 0$, there exists $T_\gamma > 0$ such that properties (ii) and (iii) hold.

Let $\epsilon > 0$. By STP, there exist $\delta > 0$, $L > 0$ such that the following property holds:

(iv) for each $T_1 \in R^1$, each $T_2 \ge T_1 + 2L$, and each $(y, v) \in X(T_1, T_2)$ which satisfies

$$I^f(T_1, T_2, y, v) \le \min\{\sigma^f(T_1, T_2)+2M_0+2M_1,\ U^f(T_1, T_2, y(T_1), y(T_2))+\delta\},$$

we have

$$\|y(t) - x_f(t)\| \le \epsilon,\ t \in [T_1 + L, T_2 - L].$$

Let $T_\delta > 0$ be such that (ii), (iii) hold with $\gamma = \delta$, and let $k \ge 1$ be an integer. Properties (i)–(iii), (7.103), and the choice of T_δ imply that

$$I^f(T_\delta, T_\delta + 2L + k, x, u) \le U^f(T_\delta, T_\delta + 2L + k, x(T_\delta), x(T_\delta + 2L + k)) + \delta,$$

$$I^f(T_\delta, T_\delta + 2L + k, x, u) \le I^f(T_\delta, T_\delta + 2L + k, x_f, u_f) + M_0 + 2M_1$$

$$\le \sigma^f(T_\delta, T_\delta + 2L + k) + 2M_0 + 2M_1$$

and

$$I^f(-T_\delta - 2L - k, -T_\delta, x, u)$$

$$\le U^f(-T_\delta - 2L - k, -T_\delta, x(-T_\delta - 2L - k), x(-T_\delta)) + \delta,$$

$$I^f(-T_\delta - 2L - k, -T_\delta, x, u)$$

$$\le I^f(-T_\delta - 2L - k, -T_\delta, x_f, u_f) + M_0 + 2M_1$$

$$\le \sigma^f(-T_\delta - 2L - k, -T_\delta) + 2M_0 + 2M_1$$

Property (iv) and the relations above imply that

$$\|x(t) - x_f(t)\| \le \epsilon,\ t \in [T_\delta + L, T_\delta + L + k] \cup [-T_\delta - L - k, -T_\delta - L].$$

Since k is any natural number, we conclude that

$$\|x(t) - x_f(t)\| \le \epsilon,\ t \in [T_\delta + L, \infty) \cup (-\infty, -T_\delta - L].$$

Since ϵ is any positive number, (P1) holds.

We show that (P3) holds. In view of LSC property, for each integer $k \ge 1$, there exists $(x_k, u_k) \in X(-k, k)$ such that

$$x_k(-k) = x_f(-k),\ x_k(k) = x_f(k), \tag{7.107}$$

$$I^f(-k, k, x_k, u_k) = U^f(-k, k, x_f(-k), x_f(k)). \tag{7.108}$$

It follows from (A2) that for each integer $k \ge 1$, there exists $\delta_k \in (0, 2^{-k})$ such that the following property holds:

(v) for each $(T_i, z_i) \in \mathcal{A}$, $i = 1, 2$ satisfying

$$\|z_i - x_f(T_i)\| \le \delta_k,\ i = 1, 2$$

there exist $\tau_1, \tau_2 \in (0, b_f]$,

$$(x_1, u_1) \in X(T_1, T_1 + \tau_1),\ (x_2, u_2) \in X(T_2 - \tau_2, T_2)$$

such that

$$x_1(T_1) = z_1,\ x_1(T_1 + \tau_1) = x_f(T_1 + \tau_1),$$

$$I^f(T_1, T_1 + \tau_1, x_1, u_1) \le I^f(T_1, T_1 + \tau_1, x_f, u_f) + k^{-1},$$

$$x_2(T_2) = z_2,\ x_2(T_2 - \tau_2) = x_f(T_2 - \tau_2),$$

$$I^f(T_2 - \tau_2, T_2, x_2, u_2) \le I^f(T_2 - \tau_2, T_2, x_f, u_f) + k^{-1}.$$

Theorem 7.2 implies that there exists $M_0 > 0$ such that for each pair of numbers $S_2 > S_1$ and each $(y, v) \in X(S_1, S_2)$,

$$I^f(S_1, S_2, y, v) \ge I^f(S_1, S_2, x_f, u_f) - M_0. \tag{7.109}$$

By (7.108) and (7.109), for each pair of numbers $S_2 > S_1$ and each natural number $k > \max\{S_2, -S_1\}$,

$$\begin{aligned} I^f(S_1, S_2, x_k, u_k) &= I^f(-k, k, x_k, u_k) - I^f(-k, S_1, x_k, u_k) - I^f(S_2, k, x_k, u_k) \\ &\le I^f(-k, k, x_f, u_f) - I^f(-k, S_1, x_f, u_f) + M_0 - I^f(S_2, k, x_f, u_f) + M_0 \\ &= I^f(S_1, S_2, x_f, u_f) + 2M_0. \end{aligned} \tag{7.110}$$

It follows from STP, (7.108), and (7.109) that the following property holds:

(vi) for each $\epsilon > 0$, there exists $L > 0$ such that for each integer $k > L$,

$$\|x_k(t) - x_f(t)\| \le \epsilon,\ t \in [-k+L, k-L].$$

Fix an integer i. In view of (7.110), for each integer $k > |i| + 1$,

$$I^f(i, i+1, x_k, u_k) \le I^f(i, i+1, x_f, u_f) + 2M_0. \tag{7.111}$$

By LSC property, extracting subsequences and re-indexing, we obtain that there exists a subsequence $\{(x_{k_p}, u_{k_p})\}_{p=1}^{\infty}$ and that for each integer $i \ge 1$, there exists $(\tilde{x}_i, \tilde{u}_i) \in X(i, i+1)$ such that

$$I^f(i, i+1, \tilde{x}_i, \tilde{u}_i) \le \liminf_{p\to\infty} I^f(i, i+1, x_{k_p}, u_{k_p}), \tag{7.112}$$

$$x_{k_p}(t) \to \tilde{x}_i(t) \text{ as } p \to \infty \text{ for all } t \in [i, i+1]. \tag{7.113}$$

It follows from (7.113) and (vi) that for all integers i,

$$\tilde{x}_i(t) = x_f(t) \text{ for all } t \in [i, i+1]. \tag{7.114}$$

For all integers i and all $t \in [i, i+1]$ set

$$\tilde{u}_f(t) = \tilde{u}_i(t). \tag{7.115}$$

Clearly, $(x_f, \tilde{u}_f) \in X(-\infty, \infty)$. Let S be a natural number. By (7.110) and (7.112),

$$I^f(-S, S, x_f, \tilde{u}_f) \le \liminf_{p\to\infty} I^f(-S, S, x_{k_p}, u_{k_p}) \le I^f(-S, S, x_f, u_f) + 2M_0. \tag{7.116}$$

Theorem 7.2 and (7.116) imply that $(x_f, \tilde{u}_f)$ is (f)-good.

We show that $(x_f, \tilde{u}_f)$ is (f)-minimal. Assume the contrary. Then there exist $\gamma > 0$ and an integer $S_0 \ge 1$ such that

$$I^f(-S_0, S_0, x_f, \tilde{u}_f) > U^f(-S_0, S_0, x_f(-S_0), x_f(S_0)) + \gamma.$$

In view of the relation above, there exists $(y_*, v_*) \in X(-S_0, S_0)$ such that

$$I^f(-S_0, S_0, x_f, \tilde{u}_f) > I^f(-S_0, S_0, y_*, v_*) + \gamma, \tag{7.117}$$

$$y_*(-S_0) = x_f(-S_0),\ y_*(S_0) = x_f(S_0). \tag{7.118}$$

Lemma 7.25 implies that there exist $\delta_* > 0$ and $T_* > 0$ such that the following property holds:

(vii) for each $T_1 \geq T_*$, each $T_2 > T_1$, and each $(y, v) \in X(T_1, T_2)$ which satisfies

$$\|y(T_i) - x_f(T_i)\| \leq \delta_*, \ i = 1, 2,$$

the inequality

$$I^f(T_1, T_2, y, v) \geq I^f(T_1, T_2, x_f, u_f) - \gamma/16$$

holds and for each $T_1 < T_2 \leq -T_*$ and each $(y, v) \in X(T_1, T_2)$ which satisfies

$$\|y(T_i) - x_f(T_i)\| \leq \delta, \ i = 1, 2,$$

the inequality

$$I^f(T_1, T_2, y, v) \geq I^f(T_1, T_2, x_f, u_f) - \gamma/16$$

holds.

Choose

$$k_0 > 8\gamma^{-1} + S_0 + 8. \tag{7.119}$$

It follows from (7.112) to (7.114) that there exists a natural number q such that

$$k_q > k_0 + T_* + \lfloor b_f \rfloor + 4, \tag{7.120}$$

for each $t \in [-k_0 - T_* - \lfloor b_f \rfloor - 4, k_0 + T_* + \lfloor b_f \rfloor + 4]$,

$$\|x_{k_q}(t) - x_f(t)\| \leq \min\{\delta_{k_0}, \delta_*\}, \tag{7.121}$$

$$I^f(-k_0 - T_*, k_0 + T_*, x_f, \tilde{u}_f) \leq I^f(-k_0 - T_*, k_0 + T_*, x_{k,q}, u_{k,q}) + \gamma/16. \tag{7.122}$$

For all $t \in [-k_0 - T_*, k_0 + T_*] \setminus [-S_0, S_0]$ set

$$y_*(t) = x_f(t), \ v_*(t) = \tilde{u}_f(t). \tag{7.123}$$

In view of (7.118), (7.119), and (7.123), $(y_*, v_*) \in X(-k_0 - T_*, k_0 + T_*)$,

$$y_*(-k_0 - T_*) = x_f(-k_0 - T_*), \ y_*(k_0 + T_*) = x_f(k_0 + T_*), \tag{7.124}$$

$$I^f(-k_0 - T_*, k_0 + T_*, x_f, \tilde{u}_f) > I^f(-k_0 - T_*, k_0 + T_*, y_*, v_*) + \gamma. \tag{7.125}$$

Property (v) and (7.121) imply that there exist $\tau_1, \tau_2 \in (0, b_f]$,

$$(y_1, v_1) \in X(k_0 + T_* + b_f - \tau_1, k_0 + T_* + b_f),$$

$$(y_2, v_2) \in X(-k_0 - T_* - b_f, -k_0 - T_* - b_f + \tau_2)$$

such that

$$y_1(k_0+T_*+b_f-\tau_1) = x_f(k_0+T_*+b_f-\tau_1),\ y_1(k_0+T_*+b_f) = x_{k_q}(k_0+T_*+b_f), \tag{7.126}$$

$$I^f(k_0 + T_* + b_f - \tau_1, k_0 + T_* + b_f, y_1, v_1)$$

$$\leq I^f(k_0 + T_* + b_f - \tau_1, k_0 + T_* + b_f, x_f, u_f) + k_0^{-1}, \tag{7.127}$$

$$y_2(-k_0 - T_* - b_f) = x_{k_q}(-k_0 - T_* - b_f),$$

$$y_2(-k_0 - T_* - b_f + \tau_2) = x_f(-k_q - T_* - b_f + \tau_2), \tag{7.128}$$

$$I^f(-k_0 - T_* - b_f, -k_0 - T_* - b_f + \tau_2, y_2, v_2)$$

$$\leq I^f(-k_0 - T_* - b_f, -k_0 - T_* - b_f + \tau_2, x_f, u_f) + k_0^{-1}. \tag{7.129}$$

Define

$$\widehat{y}(t) = y_2(t),\ \widehat{v}(t) = v_2(t),\ t \in [-k_0 - T_* - b_f, -k_0 - T_* - b_f + \tau_2), \tag{7.130}$$

$$\widehat{y}(t) = x_f(t),\ \widehat{v}(t) = u_f(t),\ t \in [-k_0 - T_* - b_f + \tau_2, -k_0 - T_*), \tag{7.131}$$

$$\widehat{y}(t) = y_*(t),\ \widehat{v}(t) = v_*(t),\ t \in [-k_0 - T_*, k_0 + T_*], \tag{7.132}$$

$$\widehat{y}(t) = x_f(t),\ \widehat{v}(t) = u_f(t),\ t \in (k_0 + T_*, k_0 + T_* + b_f - \tau_1],$$

$$\widehat{y}(t) = y_1(t),\ \widehat{v}(t) = v_1(t),\ t \in (k_0 + T_* + b_f - \tau_1, k_0 + T_* + b_f]. \tag{7.133}$$

By (7.124), (7.126), (7.128), and (7.130)–(7.133),

$$(\widehat{y}, \widehat{v}) \in X(-k_0 - T_* - b_f, k_0 + T_* + b_f),$$

$$\widehat{y}(-k_0 - T_* - b_f) = x_{k_q}(-k_0 - T_* - b_f),\ \widehat{y}(k_0 + T_* + b_f) = x_{k_q}(k_0 + T_* + b_f). \tag{7.134}$$

It follows from (7.125), (7.127), and (7.129)–(7.133) that

$$I^f(-k_0-T_*-b_f, k_0+T_*+b_f, \widehat{y}, \widehat{v}) = I^f(-k_0-T_*-b_f, -k_0-T_*-b_f+\tau_2, \widehat{y}, \widehat{v})$$

$$+I^f(-k_0 - T_* - b_f + \tau_2, -k_0 - T_*, \widehat{y}, \widehat{v}) + I^f(-k_0 - T_*, k_0 + T_*, y_*, v_*)$$

$$+I^f(k_0+T_*, k_0+T_*+b_f-\tau_1, x_f, u_f)+I^f(k_0+T_*+b_f-\tau_1, k_0+T_*+b_f, y_1, v_1)$$

$$\le I^f(-k_0 - T_* - b_f, -k_0 - T_* - b_f + \tau_2, x_f, u_f) + k_0^{-1}$$

$$+I^f(-k_0 - T_* - b_f + \tau_2, -k_0 - T_*, x_f, u_f)$$

$$+I^f(-k_0 - T_*, k_0 + T_*, y_*, v_*) + I^f(k_0 + T_*, k_0 + T_* + b_f - \tau_1, x_f, u_f)$$

$$+I^f(k_0 + T_* + b_f - \tau_1, k_0 + T_* + b_f, x_f, u_f) + k_0^{-1}$$

$$\le I^f(-k_0 - T_* - b_f, -k_0 - T_*, x_f, u_f) + k_0^{-1} + I^f(-k_0 - T_*, k_0 + T_*, y_*, v_*)$$

$$+I^f(k_0 + T_*, k_0 + T_* + b_f, x_f, u_f) + k_0^{-1}$$

$$\le 2k_0^{-1}+I^f(-k_0-T_*-b_f, -k_0-T_*, x_f, u_f)+I^f(k_0+T_*, k_0+T_*+b_f, x_f, u_f)$$

$$+ I^f(-k_0 - T_*, k_0 + T_*, x_f, \tilde{u}_f) - \gamma. \tag{7.135}$$

By (7.122) and (7.135),

$$I^f(-k_0 - T_* - b_f, k_0 + T_* + b_f, \widehat{y}, \widehat{v})$$

$$\le 2k_0^{-1}+I^f(-k_0-T_*-b_f, -k_0-T_*, x_f, u_f)+I^f(k_0+T_*, k_0+T_*+b_f, x_f, u_f)$$

$$+I^f(-k_0 - T_*, k_0 + T_*, x_f, \tilde{u}_f) - \gamma$$

$$\le 2k_0^{-1}+I^f(-k_0-T_*-b_f, -k_0-T_*, x_f, u_f)+I^f(k_0+T_*, k_0+T_*+b_f, x_f, u_f)$$

$$- \gamma + I^f(-k_0 - T_*, k_0 + T_*, x_{k_q}, u_{k_q}) + \gamma/16. \tag{7.136}$$

Property (vii) and (7.121) imply that

$$I^f(-k_0-T_*-b_f, -k_0-T_*, x_{k_q}, u_{k_q}) \ge I^f(-k_0-T_*-b_f, -k_0-T_*, x_f, u_f)-\gamma/16, \tag{7.137}$$

$$I^f(k_0 + T_*, k_0 + T_* + b_f, x_{k_q}, u_{k_q}) \ge I^f(k_0 + T_*, k_0 + T_* + b_f, x_f, u_f) - \gamma/16. \tag{7.138}$$

In view of (7.119) and (7.136)–(7.138),

$$I^f(-k_0 - T_* - b_f, k_0 + T_* + b_f, \widehat{y}, \widehat{v}) - I^f(-k_0 - T_* - b_f, k_0 + T_* + b_f, x_{k_q}, u_{k_q})$$

$$\le 2k_0^{-1} - \gamma + \gamma/16 + \gamma/8 \le 2k_0^{-1} - \gamma/2 < -\gamma/4.$$

Together with (7.134), this contradicts (7.108). The contradiction we have reached proves that $(x_f, \tilde{u}_f)$ is (f)-minimal.

Assume that $(x, u) \in X(-\infty, \infty)$ is (f)-good and (f)-minimal. Then

$$\sup\{|I^f(-T, T, x, u) - I^f(-T, T, x_f, u_f)| : T \in (0, \infty)\} < \infty.$$

Property (P1) implies that

$$\lim_{t\to\infty} \|x(t) - x_f(t)\| = 0, \quad \lim_{t\to-\infty} \|x(t) - x_f(t)\| = 0.$$

Combined with STP this implies that $x(t) = x_f(t)$ for all $t \in R^1$. Thus (P3) holds.

7.9 An Auxiliary Result

Lemma 7.26. *Assume that (P1) holds and that $\epsilon > 0$. Then there exist $\delta, L > 0$ such that the following properties hold:*

(i) for each $T_1 \geq L$, each $T_2 \geq T_1 + 2b_f$, and each $(x, u) \in X(T_1, T_2)$ which satisfies

$$\|x(T_i) - x_f(T_i)\| \leq \delta, \ i = 1, 2, \tag{7.139}$$

$$I^f(T_1, T_2, x, u) \leq U^f(T_1, T_2, x(T_1), x(T_2)) + \delta, \tag{7.140}$$

the inequality

$$\|x(t) - x_f(t)\| \leq \epsilon, \ t \in [T_1, T_2] \tag{7.141}$$

holds;

(ii) for each $T_2 \leq -L$, each $T_1 \leq T_2 - 2b_f$, and each $(x, u) \in X(T_1, T_2)$ which satisfies (7.139) and (7.140), the inequality (7.141) holds.

Proof. By (A2), for each integer $k \geq 1$, there exists $\delta_k \in (0, 2^{-k})$ such that the following property holds:

(iii) for each $(T_i, z_i) \in \mathcal{A}$, $i = 1, 2$ satisfying $\|z_i - x_f(T_i)\| \leq \delta_k$, $i = 1, 2$ there exist $\tau_1, \tau_2 \in (0, b_f]$ and $(x_1, u_1) \in X(T_1, T_1 + \tau_1)$, $(x_2, u_2) \in X(T_2 - \tau_2, T_2)$ which satisfy

$$x_1(T_1) = z_1, \ x_1(T_1+\tau_1) = x_f(T_1+\tau_1), \ x_2(T_2) = z_2, \ x_2(T_2-\tau_2) = x_f(T_2-\tau_2),$$

$$I^f(T_1, T_1 + \tau_1, x_1, u_1) \leq I^f(T_1, T_1 + \tau_1, x_f, u_f) + 2^{-k},$$

$$I^f(T_2 - \tau_2, T_2, x_2, u_2) \leq I^f(T_2 - \tau_2, T_2, x_f, u_f) + 2^{-k}.$$

Assume that the lemma does not hold. Then for each integer $k \geq 1$ there exist

$$T_{k,1} \in R^1, \ T_{k,2} \geq T_{k,1} + 2b_f, \ (x_k, u_k) \in X(T_{k,1}, T_{k,2})$$

such that

$$\|x_k(T_{k,i}) - x_f(T_{k,i})\| \leq \delta_k, \ i = 1, 2, \tag{7.142}$$

$$I^f(T_{k,1}, T_{k,2}, x_k, u_k) \leq U^f(T_{k,1}, T_{k,2}, x_k(T_{k,1}), x_k(T_{k,2})) + \delta_k, \tag{7.143}$$

$$\sup\{\|x_k(t) - x_f(t)\| : t \in [T_{k,1}, T_{k,2}]\} > \epsilon, \tag{7.144}$$

either for all integers $k \geq 1$,

$$T_{k,1} \geq k + b_f \text{ or for all integers } k \geq 1, T_{k,2} \leq -k. \tag{7.145}$$

Extracting a subsequence and re-indexing, we may assume without loss of generality that for each integer $k \geq 1$,

$$\min\{|T_{k+1,2}|, |T_{k+1,1}|\} \geq \max\{|T_{k,1}|, |T_{k,2}|\} + 4b_f. \tag{7.146}$$

Let $k \geq 1$ be an integer. Property (iii) and (7.142) imply that there exist $\tau_{k,1}, \tau_{k,2} \in (0, b_f]$ and $(\tilde{x}_k, \tilde{u}_k) \in X(T_{k,1} - \tau_{k,1}, T_{k,2} + \tau_{k,2})$ such that

$$(\tilde{x}_k(t), \tilde{u}_k(t)) = (x_k(t), u_k(t)), \ t \in [T_{k,1}, T_{k,2}], \tag{7.147}$$

$$\tilde{x}_k(s) = x_f(s), \ s = T_{k,1} - \tau_{k,1}, \ T_{k,2} + \tau_{k,2}, \tag{7.148}$$

$$I^f(T_{k,1} - \tau_{k,1}, T_{k,1}, \tilde{x}_k, \tilde{u}_k) \leq I^f(T_{k,1} - \tau_{k,1}, T_{k,1}, x_f, u_f) + 2^{-k}, \tag{7.149}$$

$$I^f(T_{k,2}, T_{k,2} + \tau_{k,2}, \tilde{x}_k, \tilde{u}_k) \leq I^f(T_{k,2}, T_{k,2} + \tau_{k,2}, x_f, u_f) + 2^{-k}. \tag{7.150}$$

Property (iii) and (7.142) imply that there exist $\tau_{k,3}, \tau_{k,4} \in (0, b_f]$ and $(\widehat{x}_k, \widehat{u}_k) \in X(T_{k,1}, T_{k,2})$ such that

$$\widehat{x}_k(T_{k,1}) = x_k(T_{k,1}), \ \widehat{x}_k(T_{k,2}) = x_k(T_{k,2}), \tag{7.151}$$

$$(\widehat{x}_k(t), \widehat{u}_k(t)) = (x_f(t), u_f(t)), \ t \in [T_{k,1} + \tau_{k,3}, T_{k,2} - \tau_{k,4}], \tag{7.152}$$

$$I^f(T_{k,1}, T_{k,1} + \tau_{k,3}, \widehat{x}_k, \widehat{u}_k) \leq I^f(T_{k,1}, T_{k,1} + \tau_{k,3}, x_f, u_f) + 2^{-k}, \tag{7.153}$$

$$I^f(T_{k,2} - \tau_{k,4}, T_{k,2}, \widehat{x}_k, \widehat{u}_k) \leq I^f(T_{k,2} - \tau_{k,4}, T_{k,2}, x_f, u_f) + 2^{-k}. \tag{7.154}$$

In view of (7.151)–(7.153),

$$U^f(T_{k,1}, T_{k,2}, x_k(T_{k,1}), x_k(T_{k,2}))$$
$$\leq I^f(T_{k,1}, T_{k,2}, \widehat{x}_k, \widehat{u}_k) \leq I^f(T_{k,1}, T_{k,2}, x_f, u_f) + 2^{-k+1}. \tag{7.155}$$

By (7.143) and (7.155),

$$I^f(T_{k,1}, T_{k,2}, x_k, u_k) \leq I^f(T_{k,1}, T_{k,2}, x_f, u_f) + 2^{-k+2}. \tag{7.156}$$

It follows from (7.147), (7.149), (7.150), and (7.156) that

$$I^f(T_{k,1} - \tau_{k,1}, T_{k,2} + \tau_{k,2}, \tilde{x}_k, \tilde{u}_k)$$
$$\leq I^f(T_{k,1} - \tau_{k,1}, T_{k,1}, x_f, u_f) + I^f(T_{k,1}, T_{k,2}, x_k, u_k)$$
$$+ I^f(T_{k,2}, T_{k,2} + \tau_{k,2}, x_f, u_f) + 2^{-k+1}$$
$$\leq I^f(T_{k,1} - \tau_{k,1}, T_{k,2} + \tau_{k,2}, x_f, u_f) + 2^{-k+3}. \tag{7.157}$$

By (7.146) and (7.148), there exists $(x, u) \in X(-\infty, \infty)$ such that for every integer $k \geq 1$,

$$(x(t), u(t)) = (\tilde{x}_k(t), \tilde{u}_k(t)), \ t \in [T_{k,1} - \tau_{k,1}, T_{k,2} + \tau_{k,2}], \tag{7.158}$$

$$(x(t), u(t)) = (x_f(t), u_f(t)), \ t \in R^1 \setminus \cup_{k=1}^{\infty}[T_{k,1} - \tau_{k,1}, T_{k,2} + \tau_{k,2}]. \tag{7.159}$$

It follows from (7.145), (7.146), and (7.157)–(7.159) that for each integer $k \geq 1$, if $T_{i,1} \to \infty$ as $i \to \infty$, then

$$I^f(0, T_{k,2} + \tau_{k,2}, x, u) - I^f(0, T_{k,2} + \tau_{k,2}, x_f, u_f)$$

$$= \sum_{i=1}^{k}(I^f(T_{i,1} - \tau_{i,1}, T_{i,2} + \tau_{i,2}, \tilde{x}_i, \tilde{u}_i) - I^f(T_{i,1} - \tau_{i,1}, T_{i,2} + \tau_{i,2}, x_f, u_f))$$

$$\leq \sum_{i=1}^{k} 2^{-i+3} \leq 8$$

and if $T_{i,2} \to -\infty$ as $i \to \infty$, then

$$I^f(T_{k,1} - \tau_{k,1}, 0, x, u) - I^f(T_{k,1} - \tau_{k,1}), 0, x_f, u_f)$$

$$= \sum_{i=1}^{k} (I^f(T_{i,1} - \tau_{i,1}, T_{i,2} + \tau_{i,2}, \tilde{x}_i, \tilde{u}_i) - I^f(T_{i,1} - \tau_{i,1}, T_{i,2} + \tau_{i,2}, x_f, u_f))$$

$$\leq \sum_{i=1}^{k} 2^{-i+3} \leq 8.$$

Theorem 7.2, the relations above, (7.145), (7.158), and (7.159) imply that (x, u) is (f)-good. Together with (P1) this implies that $\lim_{t\to\infty} \|x(t) - x_f(t)\| = 0$ and $\lim_{t\to-\infty} \|x(t) - x_f(t)\| = 0$. On the other hand, in view of (7.144), (7.145), (7.147), and (7.158), at least one of the relation below holds:

$$\limsup_{t\to\infty} \|x(t) - x_f(t)\| > 0, \ \limsup_{t\to-\infty} |\|x(t) - x_f(t)\| > 0.$$

The contradiction we have reached completes the proof of Lemma 7.26.

7.10 The Main Lemma

Lemma 7.27. *Let LSC properties, (P1), (P2), and (P3), hold, $\tilde{u}_f = u$, $\epsilon \in (0, 1)$. Then there exists $\delta > 0$ such that for each pair of numbers $T_1 \in R^1$, $T_2 \geq T_1 + 3b_f$ and each $(x, u) \in X(T_1, T_2)$ satisfying*

$$\|x(T_i) - x_f(T_i)\| \leq \delta, \ i = 1, 2$$

$$I^f(T_1, T_2, x, u) \leq U^f(T_1, T_2, x(T_1), x(T_2)) + \delta,$$

the inequality $\|x(t) - x_f(t)\| \leq \epsilon$ for all $t \in [T_1, T_2]$.

Proof. Lemma 7.26 and (A2) imply that there exist $\delta_0 \in (0, \min\{1, \epsilon/4\})$ and $L_0 > 0$ such that the following properties hold:

(i) (A2) holds with $\epsilon = 1$ and $\delta = \delta_0$;
(ii) for each pair of numbers $T_1 \in R^1$, $T_2 \geq T_1 + 2b_f$ such that either $T_1 \geq L_0$ or $T_2 \leq -L_0$ and each $(x, u) \in X(T_1, T_2)$ satisfying

$$\|x(T_i) - x_f(T_i)\| \leq \delta_0, \ i = 1, 2$$

$$I^f(T_1, T_2, x, u) \leq U^f(T_1, T_2, x(T_1), x(T_2)) + \delta_0$$

we have

$$\|x(t) - x_f(t)\| \le \epsilon, \ t \in [T_1, T_2].$$

Theorem 7.2 implies that there exists $S_0 > 0$ such that for each pair of numbers $T_2 > T_1$ and each $(x, u) \in X(T_1, T_2)$,

$$I^f(T_1, T_2, x, u) + S_0 \ge I^f(T_1, T_2, x_f, u_f). \tag{7.160}$$

It follows from property (P2) that there exist $L_1 > 0$ and $\delta_1 \in (0, \delta_0)$ such that the following property holds:

(iii) for each $T \in R^1$ and each $(x, u) \in X(T, T + L_1)$ satisfying

$$I^f(T, T + L_1, x, u)$$

$$\le \min\{I^f(T, T+L_1, x_f, u_f)+2S_0+3, \ U^f(T, T+L_1, x(T), x(T+L_1))+\delta_1\}$$

we have

$$\min\{\|x(t) - x_f(t)\| : t \in [T, T + L_1]\} \le \delta_0.$$

(A2) implies that there exists a sequence $\{\delta_i\}_{i=1}^{\infty} \subset (0, 1)$ such that

$$\delta_i < 2^{-1}\delta_{i-1}, \ i = 2, 3, \ldots \tag{7.161}$$

and that the following property holds:

(iv) for each $(T, z) \in \mathcal{A}$ satisfying $\|z - x_f(T)\| \le \delta_i$, there exist $(y_1, v_1) \in X(T, T + b_f)$, $(y_2, v_2) \in X(T - b_f, T)$ which satisfy

$$y_1(T) = z, \ y_1(T + b_f) = x_f(T + b_f),$$

$$I^f(T, T + b_f, y_1, v_1) \le I^f(T, T + b_f, x_f, u_f) + i^{-1},$$

$$y_2(T - b_f) = x_f(T - b_f), \ y_2(T) = z,$$

$$I^f(T - b_f, T, y_2, v_2) \le I^f(T - b_f, T, x_f, u_f) + i^{-1}.$$

Assume that the lemma does not hold. Then for each integer $i \ge 1$, there exist

$$T_{i,1} \in R^1, \ T_{i,2} \ge T_{i,1} + 3b_f$$

and $(x_i, u_i) \in X(T_{i,1}, T_{i,2})$ such that

$$\|x_i(T_{i,1}) - x_f(T_{i,1})\| \le \delta_i, \ \ \|x_i(T_{i,2}) - x_f(T_{i,2})\| \le \delta_i, \tag{7.162}$$

$$I^f(T_{i,1}, T_{i,2}, x_i, u_i) \le U^f(T_{i,1}, T_{i,2}, x_i(T_{i,1}), x_i(T_{i,2})) + \delta_i, \tag{7.163}$$

and $t_i \in [T_{i,1}, T_{i,2}]$ for which

$$\|x_i(t_i) - x_f(t_i)\| > \epsilon. \tag{7.164}$$

Let i be a natural number. Property (ii) and (7.161)–(7.164) imply that

$$T_{i,1} < L_0, \ \ T_{i,2} > -L_0. \tag{7.165}$$

We show that

$$-2b_f - 1 - L_1 - L_0 \le t_i \le 2b_f + 1 + L_1 + L_0.$$

Property (i), (A2) with $\epsilon = 1$, $\delta = \delta_0$, (7.161), and (7.163) imply that there exists $(y_i, v_i) \in X(T_{i,1}, T_{i,2})$ such that

$$y_i(T_{i,1}) = x_i(T_{i,1}), \ \ y_i(T_{i,2}) = x_i(T_{i,2}), \tag{7.166}$$

$$y_i(t) = x_f(t), \ \ v_i(t) = u_f(t), \ \ t \in [T_{i,1} + b_f, T_{i,2} - b_f], \tag{7.167}$$

$$I^f(T_{i,1}, T_{i,1} + b_f, y_i, v_i) \le I^f(T_{i,1}, T_{i,1} + b_f, x_f, u_f) + 1, \tag{7.168}$$

$$I^f(T_{i,2} - b_f, T_{i,2}, y_i, v_i) \le I^f(T_{i,2} - b_f, T_{i,2}, x_f, u_f) + 1. \tag{7.169}$$

It follows from (7.161), (7.163), and (7.166)–(7.169) that

$$I^f(T_{i,1}, T_{i,2}, x_i, u_i) \le 1 + I^f(T_{i,1}, T_{i,2}, y_i, v_i) \le I^f(T_{i,1}, T_{i,2}, x_f, u_f) + 3. \tag{7.170}$$

In view of (7.160) and (7.170), for each pair of numbers $S_1, S_2 \in [T_{i,1}, T_{i,2}]$ satisfying $S_1 < S_2$,

$$I^f(S_1, S_2, x_i, u_i) = I^f(T_{i,1}, T_{i,2}, x_i, u_i) - I^f(T_{i,1}, S_1, x_i, u_i) - I^f(S_2, T_{i,2}, x_i, u_i)$$

$$\le I^f(T_{i,1}, T_{i,2}, x_f, u_f) + 3 - I^f(T_{i,1}, S_1, x_f, u_f) + S_0 - I^f(S_2, T_{i,2}, x_f, u_f) + S_0$$

$$= I^f(S_1, S_2, x_f, u_f) + 3 + 2S_0. \tag{7.171}$$

Assume that

$$t_i > 2b_f + 1 + L_1 + L_0. \tag{7.172}$$

Consider the restriction of (x_i, u_i) to the interval

$$[t_i - 2b_f - 1 - L_1, t_i - 1 - 2b_f] \subset [L_0, \infty). \tag{7.173}$$

Property (iii), (7.161), (7.163), (7.165), (7.171), and (7.173) imply that there exists

$$\tilde{t} \in [t_i - 2b_f - 1 - L_1, t_i - 1 - 2b_f] \tag{7.174}$$

such that

$$\|x_f(\tilde{t}) - x_i(\tilde{t})\| \le \delta_0. \tag{7.175}$$

Property (ii), (7.157), (7.162), (7.163), and (7.173)–(7.175) imply that

$$\|x_i(t) - x_f(t)\| \le \epsilon, \ t \in [\tilde{t}, T_{i,2}]$$

and in particular,

$$\|x_i(t_i) - x_f(t_i)\| \le \epsilon.$$

This contradicts (7.164). The contradiction we have reached proves that

$$t_i \le 2b_f + 1 + L_1 + L_0. \tag{7.176}$$

Assume that

$$t_i < -2b_f - 1 - L_1 - L_0. \tag{7.177}$$

Consider the restriction of (x_i, u_i) to the interval

$$[t_i + 2b_f + 1, t_i + 1 + 2b_f + L_1] \subset (-\infty, -L_0]. \tag{7.178}$$

Property (iii), (7.161), (7.163), (7.165), (7.171), and (7.178) imply that there exists

$$\tilde{t} \in [t_i + 1 + 2b_f, t_i + 1 + 2b_f + L_1] \tag{7.179}$$

such that

$$\|x_f(\tilde{t}) - x_i(\tilde{t})\| \le \delta_0. \tag{7.180}$$

Property (ii), (7.162), (7.163), and (7.178)–(7.180) imply that

$$\|x_i(t) - x_f(t)\| \le \epsilon, \ t \in [T_{i,1}, \tilde{t}]$$

and in particular,

$$\|x_i(t_i) - x_f(t_i)\| \le \epsilon.$$

This contradicts (7.164). The contradiction we have reached proves that

$$t_i \ge -2b_f - 1 - L_1 - L_0.$$

Together with (7.176) this implies that

$$-2b_f - 1 - L_1 - L_0 \le t_i \le 2b_f + 1 + L_1 + L_0. \tag{7.181}$$

In view of (7.8), (7.13), and (7.171), there exists $M_1 > 0$ such that the following property holds:

(v) for each integer $i \ge 1$ and each $S_2 > S_1$ satisfying $S_1, S_2 \in [T_{i,1}, T_{i,2}]$, $S_2 - S_1 \le 2b_f$,

$$I^f(S_1, S_2, x_i, u_i) \le M_1.$$

Lemma 7.20 and property (v) imply that there exists $M_2 > 0$ such that

$$\|x_i(t)\| \le M_2,\ t \in [T_{i,1}, T_{i,2}],\ i = 1, 2, \dots. \tag{7.182}$$

In view of (7.165) and (7.181), extracting a subsequence and re-indexing, we may assume without loss of generality that there exists

$$\tilde{T}_1 = \lim_{i\to\infty} T_{i,1} \in [-\infty, L_0],\ \tilde{T}_2 = \lim_{i\to\infty} T_{i,2} \in [-L_0, \infty] \tag{7.183}$$

$$\tilde{t} = \lim_{i\to\infty} t_i \in [-L_1 - L_0 - 2b_f - 1, L_0 + L_1 + 2b_f + 1]. \tag{7.184}$$

There exist a strictly decreasing sequence $\{T_1^{(k)}\}_{k=1}^{\infty}$ and a strictly increasing sequence $\{T_2^{(k)}\}_{k=1}^{\infty}$ such that

$$\tilde{T}_1 = \lim_{k\to\infty} T_1^{(k)},\ \tilde{T}_2 = \lim_{k\to\infty} T_2^{(k)}, \tag{7.185}$$

for all integers $k \ge 1$,

$$T_1^{(k)} < T_2^{(k)}. \tag{7.186}$$

If $\tilde{T}_2 = \infty$, then we may assume that

$$T_2^{(k)} = T_{k,2} \text{ for all integers } k \ge 1 \tag{7.187}$$

and if $\tilde{T}_1 = -\infty$, then we may assume that

$$T_1^{(k)} = T_{k,1} \text{ for all integers } k \geq 1. \tag{7.188}$$

By LSC property, property (v), and (7.171), extracting a subsequence and re-indexing, we may assume without loss of generality that there exist $\widehat{x} : (\tilde{T}_1, \tilde{T}_2) \to E, \widehat{u} : (\tilde{T}_1, \tilde{T}_2) \to F$ such that for each $t \in (\tilde{T}_1, \tilde{T}_2)$,

$$\widehat{x}(t) = \lim_{i\to\infty} x_i(t), \tag{7.189}$$

for each integer $k \geq 1$,

$$(\widehat{x}, \widehat{u}) \in X(T_1^{(k)}, T_2^{(k)}), \tag{7.190}$$

$$I^f(T_1^{(1)}, T_2^{(1)}, \widehat{x}, \widehat{u}) \leq \liminf_{i\to\infty} I^f(T_1^{(1)}, T_2^{(1)}, x_i, u_i), \tag{7.191}$$

for each integer $k \geq 1$,

$$I^f(T_1^{(k+1)}, T_1^{(k)}, \widehat{x}, \widehat{u}) \leq \liminf_{i\to\infty} I^f(T_1^{(k+1)}, T_1^{(k)}, x_i, u_i), \tag{7.192}$$

$$I^f(T_2^{(k)}, T_2^{(k+1)}, \widehat{x}, \widehat{u}) \leq \liminf_{i\to\infty} I^f(T_2^{(k)}, T_2^{(k+1)}, x_i, u_i). \tag{7.193}$$

By (7.171), (7.182), (7.189), and (7.191)–(7.193),

$$\|\widehat{x}(t)\| \leq M_2, \ t \in (\tilde{T}_1, \tilde{T}_2) \tag{7.194}$$

and for each pair of integers $k_1, k_2 \geq 1$,

$$I^f(T_1^{(k_1)}, T_2^{(k_2)}, \widehat{x}, \widehat{u}) \leq I^f(T_1^{(k_1)}, T_2^{(k_2)}, x_f, u_f) + 3 + 2S_0. \tag{7.195}$$

Lemma 7.21 and property (v) imply that the following property holds:

(vi) for every $\tilde{\epsilon} > 0$, there exists $\delta > 0$ such that for each integer $i \geq 1$ and each $t_1, t_2 \in [T_{i,1}, T_{i,2}]$ satisfying $|t_2 - t_1| \leq \delta$, we have

$$\|x_i(t_1) - x_i(t_2)\| \leq \tilde{\epsilon}$$

and that for each $t_1, t_2 \in R^1$ satisfying $|t_2 - t_1| \leq \delta$, we have

$$\|x_f(t_1) - x_f(t_2)\| \leq \tilde{\epsilon}.$$

Let $\tilde{\epsilon} > 0$ and $\delta > 0$ be as guaranteed by property (vi). It follows from property (vi), the choice of δ, and (7.189) that for each $t_1, t_2 \in (\tilde{T}_1, \tilde{T}_2)$ satisfying $|t_1 - t_2| \leq \delta$,

$$\|\widehat{x}(t_1) - \widehat{x}(t_2)\| \le \epsilon. \tag{7.196}$$

Property (vi), (7.161), (7.162), (7.189), and (7.196) imply that if $\tilde{T}_1 > -\infty$, then there exist

$$\lim_{t\to\tilde{T}_1^+} \widehat{x}(t) = \lim_{i\to\infty} x_i(T_{i,1}) = \lim_{i\to\infty} x_f(T_{i,1}) = x_f(\tilde{T}_1) \tag{7.197}$$

and we set

$$\widehat{x}(\tilde{T}_1) = \lim_{t\to\tilde{T}_1^+} \widehat{x}(t) \tag{7.198}$$

and if $\tilde{T}_2 < \infty$, then there exist

$$\lim_{t\to\tilde{T}_2^-} \widehat{x}(t) = \lim_{i\to\infty} x_i(T_{i,2}) = \lim_{i\to\infty} x_f(T_{i,2}) = x_f(\tilde{T}_2) \tag{7.199}$$

and we set

$$\widehat{x}(\tilde{T}_2) = \lim_{t\to\tilde{T}_2^-} \widehat{x}(t). \tag{7.200}$$

Property (vi), (7.162), and (7.164) imply that

$$\tilde{T}_1 < \tilde{t} < \tilde{T}_2. \tag{7.201}$$

Property (vi), (7.164), (7.184), (7.196), and (7.201) imply that

$$\|x_f(\tilde{t}) - \widehat{x}(\tilde{t})\| = \lim_{k\to\infty} \|x_f(t_k) - \widehat{x}(t_k)\| = \lim_{k\to\infty} \|x_f(t_k) - x_k(t_k)\| \ge \epsilon. \tag{7.202}$$

In view of (7.190), for every pair of numbers $\tau_1, \tau_2 \in (\tilde{T}_1, \tilde{T}_2)$ satisfying $\tau_1 < \tau_2$, $G(t, \widehat{x}(t), \widehat{u}(t))$, $t \in [\tau_1, \tau_2]$ is strongly measurable; if $\tilde{T}_1 > -\infty$, then for every $\tau \in (\tilde{T}_1, \tilde{T}_2)$, $G(\cdot, \widehat{x}(\cdot), \widehat{u}(\cdot))$ is strongly measurable on $[\tilde{T}_1, \tau)$; if $\tilde{T}_2 < \infty$, then for every $\tau \in (\tilde{T}_1, \tilde{T}_2)$, $G(\cdot, \widehat{x}(\cdot), \widehat{u}(\cdot))$ is strongly measurable on $[\tau, \tilde{T}_2]$.

Let $k_1 and k_2$ be natural numbers. By (7.7) and (7.8), there exists $\gamma_0 > 0$ such that

$$f(t, x, u) \ge 8(\|G(t, x, u)\| - a_0\|x\|)_+ \text{ for all } (t, x, u) \in \mathcal{M}$$

$$\text{satisfying } \|G(t, x, u)\| - a_0\|x\| \ge \gamma_0. \tag{7.203}$$

In view of (7.194),

$$\int_{T_1^{(k_1)}}^{T_2^{(k_2)}} \|G(t,\widehat{x}(t),\widehat{u}(t))\|dt$$

$$\leq \int_{T_1^{(k_1)}}^{T_2^{(k_2)}} a_0\|\widehat{x}(t)\|dt + \int_{T_1^{(k_1)}}^{T_2^{(k_2)}} (\|G(t,\widehat{x}(t),\widehat{u}(t))\| - a_0\|x(t)\|)_+dt$$

$$\leq a_0M_2(T_2^{(k_2)} - T_1^{(k_1)}) + \int_{T_1^{(k_1)}}^{T_2^{(k_2)}} (\|G(t,\widehat{x}(t),\widehat{u}(t))\| - a_0\|x(t)\|)_+dt. \quad (7.204)$$

Set

$$E_1 = \{t \in [T_1^{(k_1)}, T_2^{(k_2)}] : \|G(t,\widehat{x}(t),\widehat{u}(t))\| \geq a_0\|x(t)\| + \gamma_0\}, \quad (7.205)$$

$$E_2 = [T_1^{(k_1)}, T_2^{(k_2)}] \setminus E_1.$$

It follows from (7.8), (7.195), and (7.203)–(7.205) that

$$\int_{T_1^{(k_1)}}^{T_2^{(k_2)}} \|G(t,\widehat{x}(t),\widehat{u}(t))\|dt \leq a_0M_2(T_2^{(k_2)} - T_1^{(k_1)})$$

$$+ \int_{E_1} (\|G(t,\widehat{x}(t),\widehat{u}(t))\| - a_0\|\widehat{x}(t)\|)_+dt$$

$$+ \int_{E_2} (\|G(t,\widehat{x}(t),\widehat{u}(t))\| - a_0\|\widehat{x}(t)\|)_+dt$$

$$\leq a_0M_2(T_2^{(k_2)} - T_1^{(k_1)}) + \gamma_0(T_2^{(k_2)} - T_1^{(k_1)})$$

$$+ \int_{E_1} (\|G(t,\widehat{x}(t),\widehat{u}(t))\| - a_0\|\widehat{x}(t)\|)_+dt$$

$$\leq a_0M_2(T_2^{(k_2)} - T_1^{(k_1)}) + \gamma_0(T_2^{(k_2)} - T_1^{(k_1)}) + \int_{E_1} f(t,\widehat{x}(t),\widehat{u}(t))dt$$

$$\leq (a_0M_2 + \gamma_0)(T_2^{(k_2)} - T_1^{(k_1)}) + a_0(T_2^{(k_2)} - T_1^{(k_1)}) + I^f(T_1^{(k_1)}, T_2^{(k_2)}, \widehat{x}, \widehat{u})$$

$$\leq (T_2^{(k_2)} - T_1^{(k_1)})(a_0M_2 + \gamma_0 + a_0) + I^f(T_1^{(k_1)}, T_2^{(k_2)}, x_f, u_f) + 3 + 2S_0. \quad (7.206)$$

If $\tilde{T}_1 > -\infty$, $\tilde{T}_2 < \infty$, then in view of (7.206), there exists

$$\lim_{k\to\infty}\int_{T_1^{(k_1)}}^{T_2^{(k_2)}} \|G(t,\widehat{x}(t),\widehat{u}(t))\|dt$$

$$\le (\tilde{T}_2-\tilde{T}_1)(a_0M_2+\gamma_0+a_0)+I^f(\tilde{T}_1,\tilde{T}_2,x_f,u_f)+3+2S_0$$

and in view of Fatou's lemma

$$\int_{\tilde{T}_1}^{\tilde{T}_2}\|G(t,\widehat{x}(t),\widehat{u}(t))\|dt<\infty,$$

$G(\cdot,\widehat{x}(\cdot),\widehat{u}(\cdot))$ is Bochner integrable on $[\tilde{T}_1,\tilde{T}_2]$. If $\tilde{T}_1>-\infty,\ \tilde{T}_2=\infty$, in view of (7.206), then for every integer $i\ge 1$, there exists

$$\lim_{k\to\infty}\int_{T_1^{(k)}}^{T_2^{(i)}} \|G(t,\widehat{x}(t),\widehat{u}(t))\|dt$$

$$\le (T_2^{(i)}-\tilde{T}_1)(a_0M_2+\gamma_0+a_0)+I^f(\tilde{T}_1,T_2^{(i)},x_f,u_f)+3+2S_0$$

and in view of Fatou's lemma

$$\int_{\tilde{T}_1}^{T_2^{(i)}}\|G(t,\widehat{x}(t),\widehat{u}(t))\|dt<\infty,$$

$$G(\cdot,\widehat{x}(\cdot),\widehat{u}(\cdot)) \text{ is Bochner integrable on } [\tilde{T}_1,T_2^{(i)}]. \tag{7.207}$$

If $\tilde{T}_1=-\infty,\ \tilde{T}_2<\infty$, then for every integer $i\ge 1$, there exists

$$\lim_{k\to\infty}\int_{T_1^{(i)}}^{T_2^{(k)}} \|G(t,\widehat{x}(t),\widehat{u}(t))\|dt$$

$$\le (\tilde{T}_2-T_1^{(i)})(a_0M_2+\gamma_0+a_0)+I^f(T_1^{(i)},\tilde{T}_2,x_f,u_f)+3+2S_0$$

and in view of Fatou's lemma,

$$\int_{T_1^{(i)}}^{\tilde{T}_2}\|G(t,\widehat{x}(t),\widehat{u}(t))\|dt<\infty,$$

$$G(\cdot,\widehat{x}(\cdot),\widehat{u}(\cdot)) \text{ is Bochner integrable on } [T_1^{(i)},\tilde{T}_2]. \tag{7.208}$$

Let

$$\bar{\tau} \in (\tilde{T}_1, \tilde{T}_2).$$

Assume that $\tilde{T}_2 < \infty$. Then for each sufficiently large natural number k,

$$T_2^{(k)} > \bar{\tau} > T_1^{(k)},$$

in view of (7.190), $(\widehat{x}, \widehat{u}) \in X(\bar{\tau}, T_2^{(k)})$, for all $t \in [\bar{\tau}, T_2^{(k)}]$,

$$\widehat{x}(t) = \widehat{x}(\bar{\tau}) + \int_{\bar{\tau}}^{t} G(s, \widehat{x}(s), \widehat{u}(s))ds$$

and this implies that for all $t \in [\bar{\tau}, \tilde{T}_2)$,

$$\widehat{x}(t) = \widehat{x}(\bar{\tau}) + \int_{\bar{\tau}}^{t} G(s, \widehat{x}(s), \widehat{u}(s))ds \to \widehat{x}(\bar{\tau}) + \int_{\bar{\tau}}^{\tilde{T}_2} G(s, \widehat{x}(s), \widehat{u}(s))ds \text{ as } t \to \tilde{T}_2$$

and

$$(\widehat{x}, \widehat{u}) \in X(\bar{\tau}, \tilde{T}_2). \tag{7.209}$$

Assume that $\tilde{T}_1 > -\infty$. We show that $(\widehat{x}, \widehat{u}) \in X(\tilde{T}_1, \bar{\tau})$. It is sufficient to show that for all $t \in (\tilde{T}_1, \bar{\tau}]$,

$$\widehat{x}(t) = \int_{\tilde{T}_1}^{t} G(s, \widehat{x}(s), \widehat{u}(s))ds + \widehat{x}(\tilde{T}_1).$$

Let $t \in (\tilde{T}_1, \bar{\tau}]$ and $\epsilon > 0$. Then there exists an integer $k_0 \geq 1$ such that

$$T_1^{(k_0)} < t \leq \bar{\tau} < T_2^{(k_0)}, \tag{7.210}$$

$$\|\widehat{x}(T_1^{(k)}) - \widehat{x}(\tilde{T}_1)\| \leq \tilde{\epsilon}/4 \tag{7.211}$$

for all integers $k \geq k_0$,

$$\int_{\tilde{T}_1}^{T_1^{(k)}} \|G(s, x(s), u(s))\|ds < \tilde{\epsilon}/4 \text{ for all integers } k \geq k_0. \tag{7.212}$$

Let $k \geq k_0$ be an integer. Since $(\widehat{x}, \widehat{u}) \in X(T_1^{(k)}, T_2^{(k)})$, it follows from (7.190) and (7.210) that

$$\widehat{x}(t) = \widehat{x}(T_1^{(k)}) + \int_{T_1^{(k)}}^{t} G(s, \widehat{x}(s), \widehat{u}(s))ds. \tag{7.213}$$

By (7.211)–(7.213),

$$\|\widehat{x}(t) - \widehat{x}(\tilde{T}_1) - \int_{\tilde{T}_1}^{t} G(s, \widehat{x}(s), \widehat{u}(s))ds\|$$

$$\leq \|\widehat{x}(T_1^{(k)}) - \widehat{x}(\tilde{T}_1)\| + \int_{\tilde{T}_1}^{T_1^{(k)}} \|G(s, \widehat{x}(s), \widehat{u}(s))\|ds \leq \tilde{\epsilon}.$$

Since $\tilde{\epsilon}$ is any positive number, we conclude that

$$\widehat{x}(t) = \widehat{x}(\tilde{T}_1) + \int_{\tilde{T}_1}^{t} G(s, \widehat{x}(s), \widehat{u}(s))ds$$

and $(\widehat{x}, \widehat{u}) \in X(\tilde{T}_1, \bar{\tau})$. Together with (7.209), this implies that

$$\text{if } \tilde{T}_2 < \infty, \text{ then } (\widehat{x}, \widehat{u}) \in X(\bar{\tau}, \tilde{T}_2) \text{ and if } \tilde{T}_1 > -\infty \text{ then } (\widehat{x}, \widehat{u}) \in X(\tilde{T}_1, \bar{\tau}). \tag{7.214}$$

There are the following cases:

$$\tilde{T}_1 > -\infty,\ \tilde{T}_2 < -\infty; \tag{7.215}$$

$$\tilde{T}_1 > -\infty,\ \tilde{T}_2 = \infty; \tag{7.216}$$

$$\tilde{T}_1 = -\infty,\ \tilde{T}_2 < \infty; \tag{7.217}$$

$$\tilde{T}_1 = -\infty,\ \tilde{T}_2 = \infty. \tag{7.218}$$

Lemma 7.23 implies that the following property holds:

(vii) for each $\Delta \in (0, 1)$, there exists $\delta(\Delta) > 0$ such that for each $\tau_1 \in R^1$, each $\tau_2 \geq \tau_1 + 3b_f$, each $(y, v) \in X(\tau_1, \tau_2)$ satisfying

$$\|y(\tau_i) - x_f(\tau_i)\| \leq \delta(\Delta),\ i = 1, 2,$$

$$I^f(\tau_1, \tau_2, y, v) \leq U^f(\tau_1, \tau_2, y(\tau_1), y(\tau_2)) + \delta(\Delta),$$

each $t_1 \in [\tau_1, \tau_1 + \delta(\Delta)]$, each $t_2 \in [\tau_2 - \delta(\Delta), \tau_2]$,

$$I^f(t_1, t_2, y, v) \leq I^f(t_1, t_2, x_f, u_f) + \Delta/2.$$

Assume that (7.215) holds. It follows from (7.197) to (7.200), (7.202), (7.214), and (7.215) that

$$x_f(\tilde{T}_i) = \widehat{x}(\tilde{T}_i),\ i = 1, 2,\ \|\widehat{x}(\hat{t}) - x_f(\hat{t})\| \geq \epsilon. \tag{7.219}$$

Property (P3), the equality $\tilde{u}_f = u_f$, and (7.219) imply that

$$I^f(\tilde{T}_1, \tilde{T}_2, x_f, u_f) < I^f(\tilde{T}_1, \tilde{T}_2, \widehat{x}, \widehat{u}). \tag{7.220}$$

Let $\Delta \in (0, \min\{1, b_f/8\})$, $\delta(\Delta)$ be as guaranteed by property (vii) and let the following property hold:

(viii) (A2) holds with $\delta = \delta(\Delta)$, $\epsilon = \Delta/16$.

In view of (7.161), there exists an integer $k_0 \geq 1$ such that

$$\delta_k < \delta(\Delta) \text{ for all integers } k \geq k_0. \tag{7.221}$$

Let q be a natural number such that for all integers $i \geq q$,

$$|\tilde{T}_1 - T_1^{(i)}|,\ |\tilde{T}_2 - T_2^{(i)}| \leq \delta(\Delta)/4. \tag{7.222}$$

By (7.222), there exists an integer $k_1 \geq k_0$ such that for each integer $k \geq k_1$,

$$T_{k,1} < T_1^{(q)} \leq T_{k,1} + \delta(\Delta),\ T_{k,2} - \delta(\Delta) \leq T_2^{(q)} < T_{k,2}. \tag{7.223}$$

Property (vii), (7.162), (7.163), (7.221), and (7.223) imply that for each integer $k \geq k_1$,

$$I^f(T_1^{(q)}, T_2^{(q)}, x_k, u_k) \leq I^f(T_1^{(q)}, T_2^{(q)}, x_f, u_f) + \Delta/2. \tag{7.224}$$

By (7.191)–(7.193),

$$I^f(T_1^{(q)}, T_2^{(q)}, \widehat{x}, \widehat{u}) \leq \liminf_{k\to\infty} I^f(T_1^{(k)}, T_2^{(k)}, x_k, u_k)$$

$$\leq I^f(T_1^{(q)}, T_2^{(q)}, x_f, u_f) + \Delta/2.$$

Since the relation above holds for all integers $q \geq 1$ satisfying (7.222), we conclude that

$$I^f(\tilde{T}_1, \tilde{T}_2, \widehat{x}, \widehat{u}) \leq I^f(\tilde{T}_1, \tilde{T}_2, x_f, u_f) + \Delta/2.$$

Since Δ is any positive number, we conclude that

$$I^f(\tilde{T}_1, \tilde{T}_2, \widehat{x}, \widehat{u}) \leq I^f(\tilde{T}_1, \tilde{T}_2, x_f, u_f).$$

This contradicts (7.220). The contradiction we have reached proves that (7.215) does not hold.

Consider case (7.216) with $T_1 > -\infty$, $T_2 = \infty$. We may assume (see (7.187)) that

$$T_2^{(k)} = T_{k,2} \text{ for all integers } k \geq 1. \tag{7.225}$$

By (7.173), (7.191)–(7.193), and (7.225), for each integer $k \geq 1$,

$$I^f(T_1^{(k)}, T_{k,2}, \widehat{x}, \widehat{u}) = I^f(T_1^{(k)}, T_2^{(k)}, \widehat{x}, \widehat{u})$$

$$\leq \liminf_{i\to\infty} I^f(T_1^{(k)}, T_2^{(k)}, x_i, u_i) \leq I^f(T_1^{(k)}, T_{k,2}, x_f, u_f) + 3 + 2S_0. \tag{7.226}$$

In view of (7.214) and (7.226), for all sufficiently large natural numbers k,

$$I^f(\tilde{T}_1, T_{k,2}, \widehat{x}, \widehat{u}) \leq I^f(\tilde{T}_1, T_{k,2}, x_f, u_f) + 4 + 2S_0. \tag{7.227}$$

Set

$$\widehat{x}(t) = x_f(t),\ \widehat{u}(t) = u_f(t),\ t \in (-\infty, \tilde{T}_1). \tag{7.228}$$

Theorem 7.2, (7.197), (7.198), (7.214), (7.227), and (7.228) imply that $(\widehat{x}, \widehat{u}) \in X(-\infty, \infty)$ is (f)-good. By (P1),

$$\lim_{t\to\infty} \|\widehat{x}(t) - x_f(t)\| = 0. \tag{7.229}$$

Property (P3) and (7.202) imply that $(\widehat{x}, \widehat{u})$ is not (f)-minimal. Therefore there exist $S_* > 0$ and $\Delta \in (0, 1)$ such that

$$S_* > |\tilde{T}_1| + 1 + b_f, \tag{7.230}$$

$$I^f(-S_*, S_*, \widehat{x}, \widehat{u}) > U^f(-S_*, S_*, \widehat{x}(-S_*), \widehat{x}(S_*)) + 2\Delta. \tag{7.231}$$

Let $\delta(\Delta) \in (0, \min\{\Delta, b_f/8\})$ be such that properties (vii) and (viii) hold. In view of (7.161), there exists an integer $k(\Delta) \geq 1$ such that

$$\delta_k < \delta(\Delta) \text{ for all integers } k \geq k(\Delta). \tag{7.232}$$

Property (vii), (7.162), (7.163), and (7.232) imply that the following property holds:

(ix) for each integer $k \geq k(\Delta)$, each $\tau_1 \in [T_{k,1}, T_{k,1} + \delta(\Delta)]$, and each $\tau_2 \in [T_{k,2} - \delta(\Delta), T_{k,2}]$,

$$I^f(\tau_1, \tau_2, x_k, u_k) \leq I^f(\tau_1, \tau_2, x_f, u_f) + \Delta/2. \tag{7.233}$$

In view of (7.229), there exists $\tau_0 > 1$ such that

$$\|\widehat{x}(t) - x_f(t)\| \le \delta(\Delta)/4 \text{ for all } t \ge \tau_0. \tag{7.234}$$

Let q be a natural number such that

$$T_{q,2} > \tau_0 + S_*, \tag{7.235}$$

$$|I^f(\tilde{T}_1, T_1^{(q)}, \widehat{x}, \widehat{u})| < \Delta/8, \ |I^f(\tilde{T}_1, T_1^{(q)}, x_f, u_f)| < \Delta/8, \tag{7.236}$$

for all integers $i \ge q$,

$$|\tilde{T}_1 - T_1^{(i)}| \le \delta(\Delta)/2. \tag{7.237}$$

By (7.183), (7.189), and (7.234), there exists an integer $k_1 \ge k(\Delta) + q$ such that for each integer $k \ge k_1$,

$$T_{k,1} < T_1^{(q)} \le T_{k,1} + \delta(\Delta), \tag{7.238}$$

$$\|\widehat{x}(T_{q,2}) - x_k(T_{q,2})\| \le \delta(\Delta)/4. \tag{7.239}$$

Assume that an integer $k \ge k_1$. Then (7.238) and (7.239) hold. It follows from (7.162), (7.225), (7.232), (7.234), (7.235), (7.238), and (7.239) that

$$I^f(T_1^{(q)}, T_{q,2}, x_k, u_k) \le U^f(T_1^{(q)}, T_{q,2}, x_k(T_1^{(q)}), x_k(T_{q,2})) + \delta(\Delta), \tag{7.240}$$

$$\|x_k(T_{k,1}) - x_f(T_{k,1})\| \le \delta_k \le \delta(\Delta), \tag{7.241}$$

$$\begin{aligned} \|x_k(T_{q,2}) - x_f(T_{q,2})\| &\le \|x_k(T_{q,2}) - \widehat{x}(T_{q,2})\| + \|\widehat{x}(T_{q,2}) - x_f(T_{q,2})\| \\ &\le \delta(\Delta)/4 + \delta(\Delta)/4. \end{aligned} \tag{7.242}$$

Property (vii), (7.138), (7.163), (7.238), (7.241), and (7.242) imply that

$$I^f(T_1^{(q)}, T_{q,2}, x_k, u_k) \le I^f(T_1^{(q)}, T_{q,2}, x_f, u_f) + \Delta/2. \tag{7.243}$$

By (7.191)–(7.193) and (7.243),

$$\begin{aligned} I^f(T_1^{(q)}, T_{q,2}, \widehat{x}, \widehat{u}) &\le \liminf_{k\to\infty} I^f(T_1^{(q)}, T_{q,2}, x_k, u_k) \\ &\le I^f(T_1^{(q)}, T_{q,2}, x_f, u_f) + \Delta/2. \end{aligned} \tag{7.244}$$

In view of (7.236) and (7.244),

$$I^f(\tilde{T}_1, T_{q,2}, \widehat{x}, \widehat{u}) \le I^f(\tilde{T}_1, T_{q,2}, x_f, u_f) + 3\Delta/4 \tag{7.245}$$

for all integers $q \geq 1$ satisfying (7.235)–(7.237). Let a natural number q satisfy (7.235)–(7.237) and

$$q^{-1} < \Delta/4.$$

Property (viii), (7.231), and (7.234) imply that there exists $(\xi, \eta) \in X(-\infty, \infty)$ such that

$$\xi(t) = \widehat{x}(t),\ \eta(t) = \widehat{u}(t),\ t \in (-\infty, -S_*], \tag{7.246}$$

$$\xi(S_*) = \widehat{x}(S_*), \tag{7.247}$$

$$I^f(-S_*, S_*, \xi, \eta) < I^f(-S_*, S_*, \widehat{x}, \widehat{u}) - 2\Delta, \tag{7.248}$$

$$\xi(t) = \widehat{x}(t),\ \eta(t) = \widehat{u}(t),\ t \in (S_*, T_{q,2}], \tag{7.249}$$

$$\xi(T_{q,2} + b_f) = x_f(T_{q,2} + b_f), \tag{7.250}$$

$$I^f(T_{q,2}, T_{q,2} + b_f, \xi, \eta) \leq I^f(T_{q,2}, T_{q,2} + b_f, x_f, u_f) + \Delta/16. \tag{7.251}$$

In view of (7.228), (7.230), (7.246), and (7.250),

$$x_f(-S_*) = \widehat{x}(-S_*) = \xi(-S_*),\ \xi(T_{q,2} + b_f) = x_f(T_{q,2} + b_f). \tag{7.252}$$

Property (P3), equality $\tilde{u}_f = u_f$, (7.228), (7.235), (7.245), (7.248), (7.249), (7.251), and (7.252) imply that

$$0 \leq I^f(-S_*, T_{q,2} + b_f, \xi, \eta) - I^f(-S_*, T_{q,2} + b_f, x_f, u_f)$$

$$= I^f(-S_*, S_*, \xi, \eta) + I^f(S_*, T_{q,2}, \widehat{x}, \widehat{u})$$

$$+ I^f(T_{q,2}, T_{q,2} + b_f, \xi, \eta) - I^f(-S_*, T_{q,2} + b_f, x_f, u_f)$$

$$< I^f(-S_*, S_*, \widehat{x}, \widehat{u}) - 2\Delta + I^f(S_*, T_{q,2}, \widehat{x}, \widehat{u})$$

$$+ I^f(T_{q,2}, T_{q,2} + b_f, x_f, u_f) + \Delta/16 - I^f(-S_*, T_{q,2} + b_f, x_f, u_f)$$

$$\leq I^f(-S_*, T_{q,2}, \widehat{x}, \widehat{u}) - 2\Delta + \Delta/16 - I^f(-S_*, T_{q,2}, x_f, u_f)$$

$$= I^f(\tilde{T}_1, T_{q,2}, \widehat{x}, \widehat{u}) - 2\Delta + \Delta/16 - I^f(\tilde{T}_1, T_{q,2}, x_f, u_f)$$

$$\leq -2\Delta + \Delta/16 + 3\Delta/4 \leq -\Delta,$$

a contradiction. The contradiction we have reached proves that case (7.216) does not hold.

Consider case (7.217) with $T_1 = -\infty$, $T_2 < \infty$. We may assume that

$$T_1^{(k)} = T_{k,1} \text{ for all integers } k \geq 1. \tag{7.253}$$

By (7.171), (7.191)–(7.193), and (7.253), for each integer $k \geq 1$,

$$I^f(T_{k,1}, T_2^{(k)}, \widehat{x}, \widehat{u}) = I^f(T_1^{(k)}, T_2^{(k)}, \widehat{x}, \widehat{u})$$

$$\leq \liminf_{i\to\infty} I^f(T_1^{(k)}, T_2^{(k)}, x_i, u_i) \leq I^f(T_{k,1}, T_2^{(k)}, x_f, u_f) + 3 + 2S_0. \tag{7.254}$$

In view of (7.117), (7.214), and (7.254), for all sufficiently large natural numbers k,

$$I^f(T_{k,1}, \tilde{T}_2, , \widehat{x}, \widehat{u}) \leq I^f(T_{k,1}, \tilde{T}_2, x_f, u_f) + 4 + 2S_0. \tag{7.255}$$

Set

$$\widehat{x}(t) = x_f(t),\ \widehat{u}(t) = u_f(t),\ t \in (\tilde{T}_2, \infty). \tag{7.256}$$

Theorem 7.2, (7.199), (7.200), (7.214), (7.217), (7.255), and (7.256) imply that $(\widehat{x}, \widehat{u}) \in X(-\infty, \infty)$ is (f)-good. By (P1),

$$\lim_{t\to-\infty} \|\widehat{x}(t) - x_f(t)\| = 0. \tag{7.257}$$

Property (P3) and (7.202) imply that $(\widehat{x}, \widehat{u})$ is not (f)-minimal. Therefore there exist $S_* > 0$ and $\Delta > 0$ such that

$$S_* > |\tilde{T}_2| + 1 + b_f, \tag{7.258}$$

$$I^f(-S_*, S_*, \widehat{x}, \widehat{u}) > U^f(-S_*, S_*, \widehat{x}(-S_*), \widehat{x}(S_*)) + 2\Delta. \tag{7.259}$$

Let $\delta(\Delta) \in (0, \min\{\Delta, b_f/8\})$ be such that properties (vii) and (viii) hold. In view of (7.161), there exists an integer $k(\Delta) \geq 1$ such that

$$\delta_k < \delta(\Delta) \text{ for all integers } k \geq k(\Delta). \tag{7.260}$$

Property (vii), (7.162), (7.163), and (7.260) imply that the following property holds:

(x) for each integer $k \geq k(\Delta)$, each $\tau_1 \in [T_{k,1}, T_{k,1} + \delta(\Delta)]$, each $\tau_2 \in [T_{k,2} - \delta(\Delta), T_{k,2}]$,

$$I^f(\tau_1, \tau_2, x_k, u_k) \leq I^f(\tau_1, \tau_2, x_f, u_f) + \Delta/2.$$

In view of (7.257), there exists $\tau_0 > 1$ such that

$$\|\widehat{x}(t) - x_f(t)\| \le \delta(\Delta)/4 \text{ for all } t \le -\tau_0. \tag{7.261}$$

Let q be a natural number such that

$$T_{q,1} < -\tau_0 - S_*, \tag{7.262}$$

$$|I^f(T_2^{(q)}, \tilde{T}_2, \widehat{x}, \widehat{u})| < \Delta/8,\ |I^f(T_2^{(q)}, \tilde{T}_2, x_f, u_f)| < \Delta/8, \tag{7.263}$$

for all integers $i \ge q$,

$$|\tilde{T}_2 - T_2^{(i)}| \le \delta(\Delta)/2. \tag{7.264}$$

By (7.189) and (7.264), there exists an integer $k_1 \ge k(\Delta) + q$ such that for each integer $k \ge k_1$,

$$T_{k,2} - \delta(\Delta) \le T_2^{(q)} < T_{k,2}, \tag{7.265}$$

$$\|\widehat{x}(T_{q,i}) - x_k(T_{q,i})\| \le \delta(\Delta)/4,\ i = 1, 2. \tag{7.266}$$

Assume that an integer $k \ge k_1 + q$. Then (7.265) and (7.266) hold. It follows from (7.162), (7.163), (7.260)–(7.262), and (7.266) that

$$I^f(T_{q,1}, T_{k,2}, x_k, u_k) \le U^f(T_{q,1}, T_{k,2}, x_k(T_{q,1}), x_k(T_{q,2})) + \delta(\Delta), \tag{7.267}$$

$$\|x_k(T_{k,2}) - x_f(T_{k,2})\| \le \delta_k \le \delta(\Delta), \tag{7.268}$$

$$\begin{aligned}\|x_k(T_{q,1}) - x_f(T_{q,1})\| &\le \|x_k(T_{q,1}) - \widehat{x}(T_{q,1})\| + \|\widehat{x}(T_{q,1}) - x_f(T_{q,1})\| \\ &\le \delta(\Delta)/4 + \delta(\Delta)/4.\end{aligned} \tag{7.269}$$

Property (vii), (7265), and (7.267)–(7.269) imply that

$$I^f(T_{q,1}, T_2^{(q)}, x_k, u_k) \le I^f(T_{q,1}, T_2^{(q)}, x_f, u_f) + \Delta/2 \tag{7.270}$$

for all integers $k \ge k_1 + q$. By (7.191)–(7.193) and (7.270),

$$\begin{aligned}I^f(T_{q,1}, T_2^{(q)}, \widehat{x}, \widehat{u}) &\le \liminf_{k\to\infty} I^f(T_{q,1}, T_2^{(q)}, x_k, u_k) \\ &\le I^f(T_{q,1}, T_2^{(q)}, x_f, u_f) + \Delta/2.\end{aligned} \tag{7.271}$$

In view of (7.263) and (7.271),

$$I^f(T_{q,1}\tilde{T}_2, \widehat{x}, \widehat{u}) \le I^f(T_{q,1}, \tilde{T}_2, x_f, u_f) + 3\Delta/4 \tag{7.272}$$

for all integers $q \geq 1$ satisfying (7.262)–(7.264).

Let a natural number q satisfy (7.262)–(7.264). Property (viii), the choice of $\delta(\Delta)$, (7.259), and (7.261) imply that there exists $(\xi, \eta) \in X(-\infty, \infty)$ such that

$$\xi(t) = \widehat{x}(t),\ \eta(t) = \widehat{u}(t),\ t \in [S_*, \infty), \tag{7.273}$$

$$\xi(-S_*) = \widehat{x}(-S_*),\ I^f(-S_*, S_*, \xi, \eta) < I^f(-S_*, S_*, \widehat{x}, \widehat{u}) - 2\Delta, \tag{7.274}$$

$$\xi(t) = \widehat{x}(t),\ \eta(t) = \widehat{u}(t),\ t \in [T_{q,1}, -S_*), \tag{7.275}$$

$$\xi(T_{q,1} - b_f) = x_f(T_{q,1} - b_f), \tag{7.276}$$

$$I^f(T_{q,1} - b_f, T_{q,1}, \xi, \eta) \leq I^f(T_{q,1} - b_f, T_{q,1}, x_f, u_f) + \Delta/16. \tag{7.277}$$

$$\xi(t) = x_f(t),\ \eta(t) = u_f(t),\ t \in (-\infty, T_{q,1} - b_f). \tag{7.278}$$

In view of (7.256), (7.258), (7.273), and (7.276),

$$x_f(S_*) = \widehat{x}(S_*) = \xi(S_*),\ \xi(T_{q,1} - b_f) = x_f(T_{q,1} - b_f). \tag{7.279}$$

Property (P3), equality $\tilde{u}_f = u_f$, (7.258), (7.274), (7.275), and (7.277)–(7.279) imply that

$$\begin{aligned}
0 &\leq I^f(T_{q,1} - b_f, S_*, \xi, \eta) - I^f(T_{q,1} - b_f, S_*, x_f, u_f) \\
&= I^f(T_{q,1} - b_f, T_{q,1}, \xi, \eta) + I^f(T_{q,1}, -S_*, \xi, \eta) \\
&\quad + I^f(-S_*, S_*, \xi, \eta) - I^f(T_{q,1} - b_f, S_*, x_f, u_f) \\
&\leq I^f(T_{q,1} - b_f, T_{q,1}, x_f, u_f) + \Delta/16 + I^f(T_{q,1}, -S_*, \widehat{x}, \widehat{u}) \\
&\quad + I^f(-S_*, S_*, \widehat{x}, \widehat{u}) - 2\Delta - I^f(T_{q,1} - b_f, S_*, x_f, u_f) \\
&\leq I^f(T_{q,1}, S_*, \widehat{x}, \widehat{u}) - 2\Delta + \Delta/16 - I^f(T_{q,1}, S_*, x_f, u_f) \\
&= I^f(T_{q,1}, \tilde{T}_2, \widehat{x}, \widehat{u}) - 2\Delta + \Delta/16 - I^f(T_{q,1}, \tilde{T}_2, x_f, u_f) \\
&\leq -2\Delta + \Delta/16 + 3\Delta/4 \leq -\Delta,
\end{aligned}$$

a contradiction. The contradiction we have reached proves that case (7.217) does not hold.

Consider case (7.218) with $T_1 = -\infty$, $T_2 = \infty$. We may assume that

$$T_1^{(k)} = T_{k,1},\ T_2^{(k)} = T_{k,2} \text{ for all integers } k \geq 1. \tag{7.280}$$

By (7.171), (7.191)–(7.193), and (7.280), for each integer $k \geq 1$,

$$I^f(T_{k,1}, T_{k,2}, \widehat{x}, \widehat{u}) = I^f(T_1^{(k)}, T_2^{(k)}, \widehat{x}, \widehat{u})$$

$$\leq \liminf_{i \to \infty} I^f(T_1^{(k)}, T_2^{(k)}, x_i, u_i) \leq I^f(T_{k,1}, T_{k,2}, x_f, u_f) + 3 + 2S_0. \quad (7.281)$$

Theorem 7.2 and (7.281) imply that $(\widehat{x}, \widehat{u}) \in X(-\infty, \infty)$ is (f)-good. By (P1),

$$\lim_{t \to \infty} \|\widehat{x}(t) - x_f(t)\| = 0, \quad \lim_{t \to -\infty} \|\widehat{x}(t) - x_f(t)\| = 0. \quad (7.282)$$

Property (P3) and (7.202) imply that $(\widehat{x}, \widehat{u})$ is not (f)-minimal. Therefore there exist $S_* > 0$ and $\Delta > 0$ such that

$$S_* > 1 + b_f, \quad (7.283)$$

$$I^f(-S_*, S_*, \widehat{x}, \widehat{u}) > U^f(-S_*, S_*, \widehat{x}(-S_*), \widehat{x}(S_*)) + 2\Delta. \quad (7.284)$$

Let $\delta(\Delta) > 0$ be such that properties (vii) and (viii) hold. In view of (7.161), there exists an integer $k(\Delta) \geq 1$ such that

$$\delta_k < \delta(\Delta) \text{ for all integers } k \geq k(\Delta). \quad (7.285)$$

In view of (7.282), there exists $\tau_0 > b_f + 1$ such that

$$\|\widehat{x}(t) - x_f(t)\| \leq \delta(\Delta)/4 \text{ for all } t \in (-\infty, -\tau_0] \cup [\tau_0, \infty). \quad (7.286)$$

Let q be a natural number such that

$$T_{q,1} < -\tau_0 - S_*, \ T_{q,2} > \tau_0 + S_*. \quad (7.287)$$

By (7.189), there exists an integer $k_1 \geq k(\Delta) + q$ such that for each integer $k \geq k_1$,

$$\|\widehat{x}(T_{q,i}) - x_k(T_{q,i})\| \leq \delta(\Delta)/4, \ i = 1, 2. \quad (7.288)$$

Assume that an integer $k \geq k_1 + q$. Then (7.288) holds. It follows from (7.162) and (7.285)–(7.288) that

$$I^f(T_{q,1}, T_{q,2}, x_k, u_k) \leq U^f(T_{q,1}, T_{q,2}, x_k(T_{q,1}), x_k(T_{q,2})) + \delta(\Delta) \quad (7.289)$$

and for $i = 1, 2$,

$$\|x_k(T_{q,i}) - x_f(T_{q,i})\| \leq \|x_k(T_{q,i}) - \widehat{x}(T_{q,i})\| + \|\widehat{x}(T_{q,i}) - x_f(T_{q,i})\|$$

$$\leq \delta(\Delta)/4 + \delta(\Delta)/4. \quad (7.290)$$

Property (vii), the choice of $\delta(\Delta)$, (7.289), and (7.290) imply that

$$I^f(T_{q,1}, T_{q,2}, x_k, u_k) \le I^f(T_{q,1}, T_{q,2}, x_f, u_f) + \Delta/2. \tag{7.291}$$

In view of (7.191)–(7.193) and (7.291),

$$I^f(T_{q,1}, T_{q,2}, \widehat{x}, \widehat{u}) \le \liminf_{k\to\infty} I^f(T_{q,1}, T_{q,2}, x_k, u_k) \le I^f(T_{q,1}, T_{q,2}, x_f, u_f) + \Delta/2. \tag{7.292}$$

Property (viii), the choice of $\delta(\Delta)$, (7.283), (7.284), (7.286), and (7.287) imply that there exists $(\xi, \eta) \in X(-\infty, \infty)$ such that

$$\xi(-S_*) = \widehat{x}(-S_*),\ \xi(S_*) = \widehat{x}(S_*), \tag{7.293}$$

$$I^f(-S_*, S_*, \xi, \eta) < I^f(-S_*, S_*, \widehat{x}, \widehat{u}) - 2\Delta, \tag{7.294}$$

$$\xi(t) = \widehat{x}(t),\ \eta(t) = \widehat{u}(t),\ t \in [T_{q,1}, -S_*) \cup (S_*, T_{q,2}], \tag{7.295}$$

$$\xi(T_{q,1} - b_f) = x_f(T_{q,1} - b_f), \tag{7.296}$$

$$I^f(T_{q,1} - b_f, T_{q,1}, \xi, \eta) \le I^f(T_{q,1} - b_f, T_{q,1}, x_f, u_f) + \Delta/16, \tag{7.297}$$

$$\xi(T_{q,2} + b_f) = x_f(T_{q,2} + b_f), \tag{7.298}$$

$$I^f(T_{q,2}, b_f + T_{q,2}, \xi, \eta) \le I^f(T_{q,2}, b_f + T_{q,2}, x_f, u_f) + \Delta/16, \tag{7.299}$$

$$\xi(t) = x_f(t),\ \eta(t) = u_f(t),\ t \in (-\infty, T_{q,1} - b_f] \cup (T_{q,2} + b_f, \infty). \tag{7.300}$$

Property (P3), equality $\tilde{u}_f = u_f$, (7.287), (7.292), and (7.294)–(7.299) imply that

$$0 \le I^f(T_{q,1} - b_f, T_{q,2} + b_f, \xi, \eta) - I^f(T_{q,1} - b_f, T_{q,2} + b_f, x_f, u_f)$$

$$= I^f(T_{q,1} - b_f, T_{q,1}, \xi, \eta) + I^f(T_{q,1}, -S_*, \xi, \eta)$$

$$+ I^f(-S_*, S_*, \xi, \eta) + I^f(S_*, T_{q,2}, \xi, \eta)$$

$$+ I^f(T_{q,2}, T_{q,2} + b_f, \xi, \eta) - I^f(T_{q,1} - b_f, T_{q,2} + b_f, x_f, u_f)$$

$$\le I^f(T_{q,1} - b_f, T_{q,1}, x_f, u_f) + \Delta/16 + I^f(T_{q,1}, -S_*, \widehat{x}, \widehat{u})$$

$$+ I^f(-S_*, S_*, \widehat{x}, \widehat{u}) - 2\Delta + I^f(S_*, T_{q,2}, \widehat{x}, \widehat{u}) + I^f(T_{q,2}, T_{q,2} + b_f, x_f, u_f) + \Delta/16$$

$$- I^f(T_{q,1} - b_f, T_{q,2} + b_f, x_f, u_f)$$

$$= \Delta/16 - 2\Delta + I^f(T_{q,1}, T_{q,2}, \widehat{x}, \widehat{u}) - I^f(T_{q,1}, T_{q,2}, x_f, u_f)$$

$$\leq -2\Delta + \Delta/16 + \Delta/2 < -\Delta,$$

a contradiction. The contradiction we have reached proves that case (7.218) does not hold and completes the proof of Lemma 7.27.

7.11 Completion of the Proof of Theorem 7.8

Assume that properties (P1), (P2), and (P3) hold and $\tilde{u}_f = u_f$. Let $\epsilon, M > 0$. By Lemma 7.27, there exist $\delta_0 \in (0, \epsilon)$ such that the following property holds:

(i) for each $T_1 \in R^1$, each $T_2 \geq T_1 + 3b_f$ and each $(x, u) \in X(T_1, T_2)$ which satisfies

$$\|x(T_i) - x_f(T_i)\| \leq \delta_0, \; i = 1, 2,$$

$$I^f(T_1, T_2, x, u) \leq U^f(T_1, T_2, x(T_1), x(T_2)) + \delta_0$$

the inequality $\|x(t) - x_f(t)\| \leq \epsilon$ holds for all $t \in [T_1, T_2]$.

By Theorem 7.2, there exists $S_0 > 0$ such that for each $T_2 > T_1$ and each $(x, u) \in X(T_1, T_2)$, we have

$$I^f(T_1, T_2, x, u) + S_0 \geq I^f(T_1, T_2, x_f, u_f). \tag{7.301}$$

In view of (P2), there exist $\delta \in (0, \delta_0)$ and $L_0 > 0$ such that the following property holds:

(ii) for each $T \in R^1$ and each $(y, v) \in X(T, T + L_0)$ which satisfies

$$I^f(T, T + L_0, y, v)$$

$$\leq \min\{U^f(T, T+L_0, y(T), y(T + L_0))+\delta, \; I^f(T, T+L_0, x_f, u_f)+M+2S_0\}$$

there exists $S \in [T, T + L_0]$ such that $\|y(S) - x_f(S)\| \leq \delta_0$.

Set

$$L = 2L_0 + 2b_f + 2. \tag{7.302}$$

Assume that $T_1 \in R^1$, $T_2 \geq T_1 + 2L$ and that $(x, u) \in X(T_1, T_2)$ satisfies

$$I^f(T_1, T_2, x, u) \leq \sigma^f(T_1, T_2) + M, \tag{7.303}$$

$$I^f(T_1, T_2, x, u) \le U^f(T_1, T_2, x(T_1), x(T_2)) + \delta. \tag{7.304}$$

By (7.302) and (7.304),

$$I^f(T_1, T_1 + L_0, x, u) \le U^f(T_1, T_1 + L_0, x(T_1), x(T_1 + L_0)) + \delta, \tag{7.305}$$

$$I^f(T_2 - L_0, T_2, x, u) \le U^f(T_2 - L_0, T_2, x(T_2 - L_0), x(T_2)) + \delta. \tag{7.306}$$

In view of (7.303),

$$I^f(T_1, T_2, x, u) \le I^f(T_1, T_2, x_f, u_f) + M. \tag{7.307}$$

It follows from (7.301) and (7.306) that for each pair of numbers $S_1, S_2 \in [T_1, T_2]$ satisfying $S_1 < S_2$,

$$\begin{aligned} I^f(S_1, S_2, x, u) &= I^f(T_1, T_2, x, u) - I^f(T_1, S_1, x, u) - I^f(S_2, T_2, x, u) \\ &\le I^f(T_1, T_2, x_f, u_f) + M - I^f(T_1, S_1, x_f, u_f) + S_0 - I^f(S_2, T_2, x_f, u_f) + S_0 \\ &= I^f(S_1, S_2, x_f, u_f) + M + 2S_0. \end{aligned} \tag{7.308}$$

Property (ii), (7.305), (7.306), and (7.308) imply that there exist

$$\tau_1 \in [T_1, T_1 + L_0], \ \tau_2 \in [T_2 - L_0, T_2] \tag{7.309}$$

such that

$$\|x(\tau_i) - x_f(\tau_i)\| \le \delta_0, \ i = 1, 2. \tag{7.310}$$

If $\|x(T_2) - x_f(T_2)\| \le \delta$, then we may assume that $\tau_2 = T_2$, and if $\|x(T_1) - x_f(T_1)\| \le \delta$, then we may assume that $\tau_1 = T_1$. It follows from (7.302), (7.304), (7.310), and property (i) that

$$\|x(t) - x_f(t)\| \le \epsilon, \ t \in [\tau_1, \tau_2].$$

Thus STP holds and Theorem 7.8 is proved.

7.12 An Auxiliary Result for Theorem 7.9

Lemma 7.28. *Let $(x, u) \in X(-\infty, \infty)$ be (f)-good and $\epsilon > 0$. Then there exist $L > 0$ such that for each $T_2 > T_1$ satisfying either $T_1 \ge L$ or $T_2 \le -L$ the following inequality holds:*

$$I^f(T_1, T_2, x, u) \le U^f(T_1, T_2, x(T_1), x(T_2)) + \epsilon. \tag{7.311}$$

Proof. There exists $M_0 > 0$ such that for each $T > 0$,

$$|I^f(-T, T, x, u) - I^f(-T, T, x_f, u_f)| < M_0. \tag{7.312}$$

By Theorem 7.2, there exists $M_1 > 0$ such that for each $S_2 > S_1$ and each $(y, v) \in X(S_1, S_2)$, we have

$$I^f(S_1, S_2, y, v) + M_1 \ge I^f(S_1, S_2, x_f, u_f). \tag{7.313}$$

Assume that the lemma does not hold. Then there exists a sequence of closed intervals $\{[a_i, b_i]\}_{i=1}^{\infty}$ such that for each integer $i \ge 1$ and each integer $j > i$,

$$[a_i, b_i] \cap [a_j, b_j] = \emptyset, \tag{7.314}$$

$$I^f(a_i, b_i, x, u) > U^f(a_i, b_i, x(a_i), x(b_i)) + \epsilon. \tag{7.315}$$

By (7.314) and (7.315), there exists $(y, v) \in X(-\infty, \infty)$ such that

$$y(t) = x(t), \ v(t) = u(t), \ t \in R^1 \setminus \cup_{i=1}^{\infty}(a_i, b_i), \tag{7.316}$$

for every integer $i \ge 1$,

$$I^f(a_i, b_i, y, v) < I^f(a_i, b_i, x, u) - \epsilon. \tag{7.317}$$

Let $q \ge 1$ be an integer such that

$$q > (M_0 + M_1)\epsilon^{-1}. \tag{7.318}$$

Choose $T > 0$ such that

$$[a_i, b_i] \subset [-T, T], \ i = 1, \dots, q. \tag{7.319}$$

In view of (7.316) and (7.319),

$$I^f(-T, T, y, v) - I^f(-T, T, x, u)$$

$$\le \sum_{i=1}^{q}(I^f(a_i, b_i, y, v) - I^f(a_i, b_i, x, u)) \le -q\epsilon.$$

By (7.312), (7.313) and the relation above,

$$I^f(-T, T, x_f, u_f) - M_1 \le I^f(-T, T, y, v) \le I^f(-T, T, x, u) - q\epsilon$$

$$\le I^f(-T, T, x_f, u_f) + M_0 - q\epsilon,$$

$$q\epsilon \le M_0 + M_1.$$

This contradicts (7.318). The contradiction we have reached proves Lemma 7.28.

7.13 Proofs of Propositions 7.11 and 7.13

Proof of Proposition 7.11. By Theorem 7.2, there exists $S_0 > 0$ such that the following property holds:

(i) for each $T_2 > T_1$ and each $(y, v) \in X(T_1, T_2)$, we have

$$I^f(T_1, T_2, y, v) + S_0 \ge I^f(S_1, S_2, x_f, u_f).$$

Set

$$M = 2L_0 + 2a_0L_0 + S_0. \tag{7.320}$$

Assume that

$$(T_1, z_1) \in \mathcal{A}_{L_0},\ (T_2, z_2) \in \widehat{\mathcal{A}}_{L_0},\ T_2 \ge T_1 + 2L_0.$$

Clearly, there exist $\tau_1 \in (0, L_0]$, $\tau_2 \in (0, L_0]$ and $(x, u) \in X(T_1, T_2)$ such that

$$x(T_1) = z_1,\ x(T_1 + \tau_1) = x_f(T_1 + \tau_1),\ I^f(T_1, T_1 + \tau_1, x, u) \le L_0,$$

$$(x(t), u(t)) = (x_f(t), u_f(t)),\ t \in [T_1 + \tau_1, T_2 - \tau_2],$$

$$x(T_2) = z_2,\quad I^f(T_2 - \tau_2, T_2, x, u) \le L_0.$$

In view of the relations above, property (i) and (7.320),

$$U^f(T_1, T_2, z_1, z_2) \le I^f(T_1, T_2, x, u)$$

$$\le 2L_0 + I^f(T_1 + \tau_1, T_2 - \tau_2, x_f, u_f) \le 2L_0 + 2a_0L_0 + I^f(T_1, T_2, x_f, u_f)$$

$$\le 2L_0 + 2a_0L_0 + S_0 + \sigma^f(T_1, T_2) = M + \sigma^f(T_1, T_2).$$

Proposition 7.11 is proved.

Proof of Proposition 7.13. By Theorem 7.2, there exists $S_0 > 0$ such that property (i) of the proof of Proposition 7.11 holds.

Set

$$M = L_0 + a_0 L_0 + S_0. \tag{7.321}$$

Assume that

$$(T_1, z_1) \in \mathcal{A}_{L_0}, \ T_2 \geq T_1 + L_0.$$

Clearly, there exist $\tau_1 \in (0, L_0]$ and $(x, u) \in X(T_1, T_2)$ such that

$$x(T_1) = z_1, \ x(T_1 + \tau_1) = x_f(T_1 + \tau_1), \ I^f(T_1, T_1 + \tau_1, x, u) \leq L_0,$$

$$(x(t), u(t)) = (x_f(t), u_f(t)), \ t \in [T_1 + \tau_1, T_2].$$

In view of the relations above, property (i) and (7.321),

$$\sigma^f(T_1, T_2, z_1) \leq I^f(T_1, T_2, x, u)$$

$$\leq L_0 + I^f(T_1 + \tau_1, T_2, x_f, u_f) \leq L_0 + a_0 L_0 + I^f(T_1, T_2, x_f, u_f)$$

$$\leq L_0 + a_0 L_0 + S_0 + \sigma^f(T_1, T_2) = M + \sigma^f(T_1, T_2).$$

Proposition 7.13 is proved.

7.14 Proofs of Theorems 7.9 and 7.14

Proof of Theorem 7.9. Assume that property (P4) holds. Then (P1) follows from Theorem 7.2, Lemma 7.28, and (P4), while (P2) follows from (P4) and Theorem 7.2. Assume that (P1) and (P2) hold. Then (P4) follows from Theorem 7.2, Lemma 7.26, and property (P2).

Proof of Theorem 7.14. Assume that f has WTP. Then (P1) and (P2) hold.

Assume that (P1) and (P2) hold. We show that f has WTP. Let $\epsilon, M > 0$. By Theorem 7.2, there exists $S_0 > 0$ such that the following property holds:

(i) for each $\tau_2 > \tau_1$ and each $(y, v) \in X(\tau_1, \tau_2)$,

$$I^f(\tau_1, \tau_2, y, v) + S_0 \geq I^f(\tau_1, \tau_2, x_f, u_f).$$

Lemma 7.26 implies that there exist $\delta_0 \in (0, 1)$ and $L_0 > 2b_f$ such that the following property holds:

(ii) for each $T_1 \geq L_0$, each $T_2 \geq T_1 + 2b_f$ and each $(y, v) \in X(T_1, T_2)$ which satisfies

$$\|y(T_i) - x_f(T_i)\| \leq \delta_0,\ i = 1, 2, \tag{7.322}$$

$$I^f(T_1, T_2, y, v) \leq U^f(T_1, T_2, y(T_1), y(T_2)) + \delta_0 \tag{7.323}$$

we have

$$\|y(t) - x_f(t)\| \leq \epsilon,\ t \in [T_1, T_2] \tag{7.324}$$

and for each $T_2 \leq -L_0$, each $T_1 \leq T_2 - 2b_f$ and each $(y, v) \in X(T_1, T_2)$ which satisfies (7.322) and (7.323) inequality (7.324) hold.

Property (P2) implies that there exist $\delta \in (0, \delta_0)$ and $L_1 > 0$ such that the following property holds:

(iii) for each $T \in R^1$ and each $(y, v) \in X(T, T + L_1)$ which satisfies

$$I^f(T, T + L_1, y, v) \leq \min\{U^f(T, T + L_1, y(T), y(T + L_1)) + \delta,$$

$$I^f(T, T + L_1, x_f, u_f) + M + 2S_0\}$$

there exists $s \in [T, T + L_1]$ such that

$$\|y(s) - x_f(s)\| \leq \delta_0.$$

Set

$$l = 8L_0 + 8L_1 + 8b_f + 8. \tag{7.325}$$

Choose a natural number

$$Q \geq 6 + 6(M + S_0)\delta^{-1}. \tag{7.326}$$

Assume that $T_1 \in R^1$, $T_2 \geq T_1 + lQ$ and $(x, u) \in X(T_1, T_2)$ satisfies

$$I^f(T_1, T_2, x, u) \leq I^f(T_1, T_2, x_f, u_f) + M. \tag{7.327}$$

Property (i) and (7.327) imply that for each pair of numbers $\tau_1, \tau_2 \in [T_1, T_2]$ satisfying $\tau_1 < \tau_2$,

$$I^f(\tau_1, \tau_2, x, u) = I^f(T_1, T_2, x, u) - I^f(T_1, \tau_1, x, u) - I^f(\tau_2, T_2, x, u)$$

$$\leq M + I^f(T_1, T_2, x_f, u_f) - I^f(T_1, \tau_1, x_f, u_f) + S_0 - I^f(\tau_2, T_2, x_f, u_f) + S_0$$

$$= I^f(\tau_1, \tau_2, x_f, u_f) + M + 2S_0. \tag{7.328}$$

Set

$$t_0 = T_1. \tag{7.329}$$

If

$$I^f(T_1, T_2, x, u) \le U^f(T_1, T_2, x(T_1), x(T_2)) + \delta,$$

then we set

$$t_1 = T_2,\ s_1 = T_2. \tag{7.330}$$

Assume that

$$I^f(T_1, T_2, x, u) > U^f(T_1, T_2, x(T_1), x(T_2)) + \delta. \tag{7.331}$$

It is easy to see that for each $t \in (T_1, T_2)$ such that $t - T_1$ is sufficiently small, we have

$$I^f(T_1, t, x, u) - U^f(T_1, t, x(T_1), x(t)) < \delta.$$

Set

$$\tilde{t}_1 = \inf\{t \in (T_1, T_2] :\ I^f(T_1, t, x, u) - U^f(T_1, t, x(T_1), x(t)) > \delta\}. \tag{7.332}$$

Clearly, $\tilde{t}_1 \in (T_1, T_2]$ is well-defined, and there exist $s_1, t_1 \in [T_1, T_2]$ such that

$$t_0 < s_1 < \tilde{t}_1,\ s_1 \ge \tilde{t}_1 - 1/4, \tag{7.333}$$

$$I^f(t_0, s_1, x, u) - U^f(t_0, s_1, x(t_0), x(s_1)) \le \delta, \tag{7.334}$$

$$t_1 \le T_2,\ \tilde{t}_1 \le t_1 < \tilde{t}_1 + 1/4, \tag{7.335}$$

$$I^f(t_0, t_1, x, u) - U^f(t_0, t_1, x(t_0), x(t_1)) > \delta. \tag{7.336}$$

Assume that $k \ge 1$ is an integer, and we defined finite sequences of numbers $\{t_i\}_{i=0}^k \subset [T_1, T_2]$, $\{s_i\}_{i=1}^k \subset [T_1, T_2]$ such that

$$T_1 = t_0 < t_1 \cdots < t_k, \tag{7.337}$$

for each integer $i = 1, \ldots, k$,

$$t_{i-1} < s_i \le t_i,\ t_i - s_i \le 2^{-1}, \tag{7.338}$$

$$I^f(t_{i-1}, s_i, x, u) - U^f(t_{i-1}, s_i, x(t_{i-1}), x(s_i)) \le \delta, \tag{7.339}$$

and if $t_i < T_2$, then

$$I^f(t_{i-1}, t_i, x, u) - U^f(t_{i-1}, t_i, x(t_{i-1}), x(t_i)) > \delta. \tag{7.340}$$

(Note that in view of (7.329)–(7.336), our assumption holds for $k = 1$.)

By (7.337)–(7.340), there exists $(y, v) \in X(T_1, T_2)$ such that

$$y(t_i) = x(t_i),\ i = 1, \dots, k,$$

$$I^f(t_{i-1}, t_i, x, u) - I^f(t_{i-1}, t_i, y, v) > \delta \text{ for all integers } i \in [1, k] \setminus \{k\},$$

$$(y(t), v(t)) = (x(t), u(t)),\ t \in [t_{k-1}, T_2].$$

Property (i), the relations above, and (7.327) imply that

$$I^f(T_1, T_2, x_f, u_f) + M \ge I^f(T_1, T_2, x, u) \ge I^f(T_1, T_2, y, v) + \delta(k-1)$$

$$\ge I^f(T_1, T_2, x_f, u_f) - S_0 + \delta(k-1),$$

$$k \le 1 + \delta^{-1}(M + S_0). \tag{7.341}$$

If $t_k = T_2$, then the construction of the sequence is completed. Assume that $t_k < T_2$. If

$$I^f(t_k, T_2, x, u) \le U^f(t_k, T_2, x(t_k), x(T_2)) + \delta,$$

then we set $t_{k+1} = T_2$, $s_{k+1} = T_2$ and the construction of the sequence is completed. Assume that

$$I^f(t_k, T_2, x, u) > U^f(t_k, T_2, x(t_k), x(T_2)) + \delta. \tag{7.342}$$

It is easy to see that for each $t \in (t_k, T_2)$ such that $t - t_k$ is sufficiently small, we have

$$I^f(t_k, t, x, u) - U^f(t_k, t, x(t_k), x(t)) < \delta.$$

Set

$$\tilde{t} = \inf\{t \in (t_k, T_2] : I^f(t_k, t, x, u) - U^f(t_k, t, x(t_k), x(t)) > \delta\}.$$

Clearly, $\tilde{t}$ is well-defined, $\tilde{t} \in (t_k, T_2]$, and there exist $s_{k+1}, t_{k+1} \in [T_1, T_2]$ such that

$$t_k < s_{k+1} < \tilde{t},\ s_{k+1} > \tilde{t} - 1/4,\ \tilde{t} + 4^{-1} > t_{k+1} \geq \tilde{t},$$

$$I^f(t_k, s_{k+1}, x, u) - U^f(t_k, s_{k+1}, x(t_k), x(s_{k+1})) \leq \delta,$$

$$I^f(t_k, t_{k+1}, x, u) - U^f(t_k, t_{k+1}, x(t_k), x(t_{k+1})) > \delta.$$

It is not difficult to see that the assumption made for k also holds for $k+1$. In view of (7.341), by induction we constructed finite sequences $\{t_i\}_{i=0}^q \subset [T_1, T_2]$, $\{s_i\}_{i=1}^q \subset [T_1, T_2]$ such that

$$q \leq 2 + \delta^{-1}(M + S_0), \tag{7.343}$$

$$T_1 = t_0 < t_1 \cdots < t_q = T_2, \tag{7.344}$$

for each integer $i = 1, \ldots, q$,

$$t_{i-1} < s_i \leq t_i,\ t_i - s_i \leq 2^{-1}, \tag{7.345}$$

$$I^f(t_{i-1}, s_i, x, u) - U^f(t_{i-1}, s_i, x(t_{i-1}), x(s_i)) \leq \delta, \tag{7.346}$$

and if $t_i < T_2$, then

$$I^f(t_{i-1}, t_i, x, u) - U^f(t_{i-1}, t_i, x(t_{i-1}), x(t_i)) > \delta. \tag{7.347}$$

Assume that $i \in \{0, \ldots, q-1\}$,

$$t_{i+1} - t_i \geq 8L_0 + 8L_1 + 8b_f + 8 = l. \tag{7.348}$$

By (7.345) and (7.348),

$$s_{i+1} - t_i \geq 8L_0 + 8L_1 + 8b_f + 7. \tag{7.349}$$

Assume that

$$t_i \geq -2L_0 - 2L_1. \tag{7.350}$$

Property (iii), (7.328), (7.346), and (7.349) imply that there exist

$$t_{i,1} \in [t_i + 3L_0 + 2L_1, t_i + 3L_0 + 3L_1] \tag{7.351}$$

satisfying

$$\|x(t_{i,1}) - x_f(t_{i,1})\| \le \delta_0 \tag{7.352}$$

and

$$t_{i,2} \in [s_{i+1} - L_1, s_{i+1}] \tag{7.353}$$

satisfying

$$\|x(t_{i,2}) - x_f(t_{i,2})\| \le \delta_0. \tag{7.354}$$

It follows from (7.346), (7.349)–(7.354), and property (ii) that

$$\|x(t) - x_f(t)\| \le \epsilon, \ t \in [t_{i,1}, t_{i,2}]$$

and

$$\|x(t) - x_f(t)\| \le \epsilon, \ t \in [t_i + 3L_0 + 3L_1, t_{i+1} - L_1 - 1]. \tag{7.355}$$

Assume that

$$t_{i+1} \le 2L_0 + 2L_1 + 1. \tag{7.356}$$

Property (iii), (7.328), (7.346), and (7.349) imply that there exist

$$t_{i,1} \in [t_i, t_i + L_1] \tag{7.357}$$

satisfying

$$\|x(t_{i,1}) - x_f(t_{i,1})\| \le \delta_0 \tag{7.358}$$

and

$$t_{i,2} \in [s_{i+1} - 3L_1 - 3L_0 - 1, s_{i+1} - 3L_0 - 2L_1 - 1] \tag{7.359}$$

satisfying

$$\|x(t_{i,2}) - x_f(t_{i,2})\| \le \delta_0. \tag{7.360}$$

It follows from (7.346), (7.356)–(7.360), and property (ii) that

$$\|x(t) - x_f(t)\| \le \epsilon, \ t \in [t_{i,1}, t_{i,2}]$$

and

$$\|x(t) - x_f(t)\| \le \epsilon, \ t \in [t_i + L_1, t_{i+1} - 3L_1 - 3L_0 - 2].$$

In both cases, we have

$$\|x(t) - x_f(t)\| \le \epsilon, \ t \in [t_i + 3L_0 + 3L_1, t_{i+1} - 3L_1 - 3L_0 - 2]. \tag{7.361}$$

Assume that

$$t_i < -2L_0 - 2L_1, \ t_{i+1} > 2L_0 + 2L_1 + 1. \tag{7.362}$$

In view of (7.335) and (7.362),

$$s_{i+1} > 2L_0 + 2L_1. \tag{7.363}$$

Property (iii), (7.328), (7.346), (7.362), and (7.363) imply that there exist

$$t_{i,1} \in [L_0, L_0 + L_1] \tag{7.364}$$

satisfying

$$\|x(t_{i,1}) - x_f(t_{i,1})\| \le \delta_0 \tag{7.365}$$

and

$$t_{i,2} \in [s_{i+1} - L_1, s_{i+1}] \tag{7.366}$$

satisfying

$$\|x(t_{i,2}) - x_f(t_{i,2})\| \le \delta_0, \tag{7.367}$$

$$t_{i,3} \in [t_i, t_i + L_1] \tag{7.368}$$

satisfying

$$\|x(t_{i,3}) - x_f(t_{i,3})\| \le \delta_0 \tag{7.369}$$

and

$$t_{i,4} \in [-L_0 - L_1, -L_0] \tag{7.370}$$

satisfying

$$\|x(t_{i,4}) - x_f(t_{i,4})\| \le \delta_0. \tag{7.371}$$

It follows from (7.328), (7.346), (7.349), (7.362), (7.363), (7.364), (7.371), and property (ii) that

$$\|x(t) - x_f(t)\| \le \epsilon,\ t \in [t_{i,1}, t_{i,2}] \cup [t_{i,3}, t_{i,4}],$$

$$\|x(t)-x_f(t)\| \le \epsilon,\ t \in [L_0+L_1, t_{i+1}-1-L_1] \cup [t_i+L_1, -L_0-L_1] \quad (7.372)$$

and

$$\|x(t)-x_f(t)\| \le \epsilon,\ t \in [t_i+L_1, t_{i+1}-1-L_1] \setminus (-L_0-L_1, L_0+L_1). \quad (7.373)$$

By (7.361), (7.372), and (7.373),

$$\{t \in [T_1, T_2] :\ \|x(t) - x_f(t)\| \le \epsilon\}$$

$$\supset \cup\{[t_i + 3L_0 + 3L_1, t_{i+1} - 3L_0 - 3L_1 - 2] :\ i \in \{0, \dots, q-1\},$$

$$t_{i+1} - t_i \ge 8L_0 + 8L_1 + 8b_f + 8,\ t_i \ge -2L_0 - 2L_1 \text{ or } t_{i+1} \le 2L_0 + 2L_1 + 1\}$$

$$\cup\{[t_i + L_1, t_{i+1} - L_1 - 1] \setminus (-L_0 - L_1, L_0 + L_1) :\ i \in \{0, \dots, q-1\},$$

$$t_{i+1} - t_i \ge 8L_0 + 8L_1 + 8b_f + 8,\ t_i < -2L_0 - 2L_1,\ t_{i+1} > 2L_0 + 2L_1 + 1\}.$$

This implies that

$$\{t \in [T_1, T_2] :\ \|x(t) - x_f(t)\| > \epsilon\}$$

$$\subset \cup\{[t_i, t_{i+1}] :\ i \in \{0, \dots, q-1\},$$

$$t_{i+1} - t_i < 8L_0 + 8L_1 + 8b_f + 8\}$$

$$\cup\{[t_i, t_i + 3L_0 + 3L_1] \cup [t_{i+1} - 3L_0 - 3L_1 - 2, t_{i+1}] :\ i \in \{0, \dots, q-1\},$$

$$t_{i+1} - t_i \ge 8L_0 + 8L_1 + 8b_f + 8\}$$

$$\cup\{[t_i, t_i + L_1] \cup [t_{i+1} - L_1 - 1, t_{i+1}] \cup [-L_0 - L_1, L_0 + L_1] :$$

$$i \in \{0, \dots, q-1\},\ t_i < 2L_0 - 2L_1,\ t_{i+1} > 2L_0 + 2L_1 + 1]\}. \quad (7.374)$$

Clearly, the right-hand side of (7.374) is a finite union of intervals. In view of (7.326) and (7.343), their number does not exceed $3q < Q$. By (7.325), the maximal length of intervals in the right-hand side of the relation above does not exceed l. Theorem 7.14 is proved.

7.15 Auxiliary Results for Theorem 7.15

Recall that for each $z \in R^1$, $\lfloor z \rfloor = \max\{i \in R^1 : i \text{ is an integer}, i \le z\}$.

Proposition 7.29. *Assume that f has (P1), (P2) and LSC property and $(T_0, z_0) \in \cup\{\mathcal{A}_L : L > 0\}$. Then there exists an (f)-good and (f)-minimal pair $(x_*, u_*) \in X(T_0, \infty)$ such that $x_*(T_0) = z_0$.*

Proof. There exists $L_0 > 0$ such that

$$(T_0, z_0) \in \mathcal{A}_{L_0}. \tag{7.375}$$

It follows from Theorem 7.2 that there exists $S_0 > 0$ such that for each $T_2 > T_1$ and each $(x, u) \in X(T_1, T_2)$,

$$I^f(T_1, T_2, x, u) + S_0 \ge I^f(T_1, T_2, x_f, u_f). \tag{7.376}$$

Fix an integer $k_0 \ge L_0$. LSC property and (7.375) imply that for each integer $k \ge k_0$, there exists $(x_k, u_k) \in X(T_0, T_0 + k)$ satisfying

$$x_k(T_0) = z_0, \tag{7.377}$$

$$I^f(T_0, T_0 + k, x_k, u_k) = \sigma^f(T_0, T_0 + k, z_0). \tag{7.378}$$

In view of (7.8) and (7.375), for each integer $k \ge k_0$,

$$\sigma^f(T_0, T_0 + k, z_0) \le L_0 + I^f(T_0, T_0 + k, x_f, u_f) + a_0 L_0. \tag{7.379}$$

By (7.376), (7.378), and (7.379), for each integer $k \ge k_0$ and each pair of numbers s $T_1, T_2 \in [T_0, T_0 + k]$ satisfying $T_1 < T_2$,

$$\begin{aligned} I^f(T_1, T_2, x_k, u_k) &= I^f(T_0, T_0 + k, x_k, u_k) \\ &- I^f(T_0, T_1, x_k, u_k) - I^f(T_2, T_0 + k, x_k, u_k) \\ &\le I^f(T_0, T_0 + k, x_f, u_f) + L_0(1 + a_0) \\ &- I^f(T_0, T_1, x_f, u_f) + S_0 - I^f(T_2, T_0 + k, x_f, u_f) + S_0 \\ &= I^f(T_1, T_2, x_f, u_f) + 2S_0 + L_0(1 + a_0). \end{aligned} \tag{7.380}$$

By (7.370) and LSC property, extracting subsequences, using the diagonalization process, and re-indexing, we obtain that there exists a strictly increasing sequence of natural numbers $\{k_p\}_{p=1}^{\infty}$ such that $k_1 \ge k_0$ and for each integer $i \ge 0$, there exists

$\lim_{p\to\infty} I^f(T_0+i, T_0+i+1, x_{k_p}, u_{k_p})$ and there exists $(y_i, v_i) \in X(T_0+i, T_0+i+1)$ such that

$$x_{k_p}(t) \to y_i(t) \text{ as } p \to \infty \text{ for all } t \in [T_0+i, T_0+i+1], \tag{7.381}$$

$$I^f(T_0+i, T_0+i+1, y_i, v_i) \le \lim_{p\to\infty} I^f(T_0+i, T_0+i+1, x_{k_p}, u_{k_p}). \tag{7.382}$$

In view of (7.381), there exists $(x_*, u_*) \in X(T_0, \infty)$ such that for each integer $i \ge 0$,

$$(x_*(t), u_*(t)) = (y_i(t), v_i(t)),\ t \in [T_0+i, T_0+i+1]. \tag{7.383}$$

It follows from (7.380), (7.382), and (7.383) that for every integer $q \ge 1$,

$$I^f(T_0, T_0+q, x_*, u_*) \le \lim_{p\to\infty} I^f(T_0, T_0+q, x_{k_p}, u_{k_p})$$

$$\le I^f(T_0, T_0+q, x_f, u_f) + 2S_0 + L_0(1+a_0). \tag{7.384}$$

Theorem 7.6, (7.377), (7.381), and (7.384) imply that (x_*, u_*) is (f)-good and

$$x_*(T_0) = z_0. \tag{7.385}$$

In order to complete the proof of the proposition, it is sufficient to show that (x_*, u_*) is (f)-minimal. Assume the contrary. Then there exist $\Delta > 0$, an integer $\tau_0 \ge 1$, and $(y, v) \in X(T_0, T_0+\tau_0)$ such that

$$y(T_0) = x_*(T_0),\ y(T_0+\tau_0) = x_*(T_0+\tau_0), \tag{7.386}$$

$$I^f(T_0, T_0+\tau_0, x_*, u_*) > I^f(T_0, T_0+\tau_0, y, v) + 2\Delta. \tag{7.387}$$

By (A2) and Lemma 7.25, there exists $L_1, \delta > 0$ such that the following property holds:

(i) for each $(T, \xi) \in \mathcal{A}$ satisfying $\|\xi - x_f(T)\| \le \delta$, there exist $\tau_1 \in (0, b_f]$ and $(\tilde{x}_1, \tilde{u}_1) \in X(T, T+\tau_1)$ such that

$$\tilde{x}_1(T) = \xi\ , \tilde{x}_1(T+\tau_1) = x_f(T+\tau_1),$$

$$I^f(T, T+\tau_1, \tilde{x}_1, \tilde{u}_1) \le I^f(T, T+\tau_1, x_f, u_f) + \Delta/8$$

and there exist $\tau_2 \in (0, b_f]$ and $(\tilde{x}_2, \tilde{u}_2) \in X(T-\tau_2, T)$ such that

$$\tilde{x}_2(T-\tau_2) = x_f(T-\tau_2)\ , \tilde{x}_2(T) = \xi,$$

$$I^f(T-\tau_2, T, \tilde{x}_2, \tilde{u}_2) \le I^f(T-\tau_2, T, x_f, u_f) + \Delta/8;$$

if $T_2 > T_1 \ge L_1$, $(x,u) \in X(T_1,T_2)$, $\|x(T_i) - x_f(T_i)\| \le \delta$, $i = 1, 2$, we have

$$I^f(T_1, T_2, x, u) \ge I^f(T_1, T_2, x_f, u_f) - \Delta/8.$$

Theorems 7.2 and 7.9, (7.375), and (7.378) imply that there exists an integer $L_2 > L_0 + L_1 + 2|T_0|$ such that for each integer $k \ge k_0 + 2L_2$,

$$\|x_k(t) - x_f(t)\| \le \delta, \ t \in [|T_0| + L_2, T_0 + k - L_2]. \tag{7.388}$$

Set

$$q_0 = \lfloor |T_0| \rfloor + 1. \tag{7.389}$$

In view of (7.375), (7.381), (7.383), and (7.388),

$$\|x_*(t) - x_f(t)\| \le \delta \text{ for all } t \ge |T_0| + L_2. \tag{7.390}$$

By (7.382) and (7.383), there exists a natural number p_0 such that

$$k_{p_0} > k_0 + 2L_1 + 4L_2 + 2\tau_0 + 2 + 2b_f + q_0, \tag{7.391}$$

$$I^f(T_0, T_0+\tau_0+2L_2+q_0, x_*, u_*) \le I^f(T_0, T_0+\tau_0+2L_2+q_0, x_{k_{p_0}}, u_{k_{p_0}}) + \Delta/2. \tag{7.392}$$

Property (i), (7.385), (7.388), and (7.390) imply that there exists $(x,u) \in X(T_0, T_0 + k_{p_0})$ such that

$$(x(t), u(t)) = (y(t), v(t)), \ t \in [T_0, T_0 + \tau_0], \tag{7.393}$$

$$(x(t), u(t)) = (x_*(t), u_*(t)), \ t \in (T_0 + \tau_0, T_0 + \tau_0 + 2L_2 + q_0], \tag{7.394}$$

$$x(T_0 + \tau_0 + 2L_2 + b_f + q_0) = x_f(T_0 + \tau_0 + 2L_2 + b_f + q_0), \tag{7.395}$$

$$I^f(T_0 + \tau_0 + 2L_2 + q_0, T_0 + \tau_0 + 2L_2 + b_f + q_0, x, u)$$

$$\le I^f(T_0 + \tau_0 + 2L_2 + q_0, T_0 + \tau_0 + 2L_2 + b_f + q_0, x_f, u_f) + \Delta/8, \tag{7.396}$$

$$(x(t), u(t)) = (x_{k_{p_0}}(t), u_{k_{p_0}}(t)), \ t \in [T_0 + \tau_0 + 2L_2 + 2b_f + q_0, k_{p_0} + T_0],$$

$$I^f(T_0 + \tau_0 + 2L_2 + b_f + q_0, T_0 + \tau_0 + 2L_2 + 2b_f + q_0, x, u)$$

$$\le I^f(T_0+\tau_0+2L_2+b_f+q_0, T_0+\tau_0+2L_2+2b_f+q_0, x_f, u_f)+\Delta/8. \tag{7.397}$$

It follows from (7.377), (7.378), (7.381), (7.383), (7.385), and (7.393) that

$$I^f(T_0, T_0 + k_{p0}, x, u) \geq I^f(T_0, T_0 + k_{p0}, x_{k_{p0}}, u_{k_{p0}}). \tag{7.398}$$

By (7.327), (7.387), (7.392)–(7.396), (7.398), and property (i),

$$0 \leq I^f(T_0, T_0 + k_{p0}, x, u) - I^f(T_0, T_0 + k_{p0}, x_{k_{p0}}, u_{k_{p0}})$$

$$= I^f(T_0, T_0 + \tau_0, y, v) + I^f(T_0 + \tau_0, q_0 + T_0 + \tau_0 + 2L_2, x_*, u_*)$$

$$+I^f(T_0 + \tau_0 + 2L_2 + q_0, q_0 + T_0 + \tau_0 + 2L_2 + b_f, x_f, u_f) + \Delta/8$$

$$+I^f(T_0 + \tau_0 + 2L_2 + b_f + q_0, q_0 + T_0 + \tau_0 + 2L_2 + 2b_f, x_f, u_f) + \Delta/8$$

$$+I^f(T_0 + \tau_0 + 2L_2 + 2b_f + q_0, k_{p0} + T_0, x_{k_{p0}}, u_{k_{p0}})$$

$$-I^f(T_0, T_0 + \tau_0 + 2L_2 + q_0, x_{k_{p0}}, u_{k_{p0}})$$

$$-I^f(T_0 + \tau_0 + 2L_2 + q_0, T_0 + \tau_0 + 2L_2 + b_f + q_0, x_{k_{p0}}, u_{k_{p0}})$$

$$-I^f(T_0 + \tau_0 + 2L_2 + q_0 + b_f, T_0 + \tau_0 + 2L_2 + q_0 + 2b_f, x_{k_{p0}}, u_{k_{p0}})$$

$$-I^f(T_0 + \tau_0 + 2L_2 + 2b_f + q_0, T_0 + k_{p0}, x_{k_{p0}}, u_{k_{p0}})$$

$$\leq I^f(T_0, T_0 + \tau_0, y, v) + I^f(T_0 + \tau_0, T_0 + \tau_0 + 2L_2 + q_0, x_*, u_*)$$

$$+I^f(T_0 + \tau_0 + 2L_2 + q_0, T_0 + \tau_0 + 2L_2 + b_f + q_0, x_f, u_f)$$

$$+I^f(T_0 + \tau_0 + 2L_2 + q_0 + b_f, T_0 + \tau_0 + 2L_2 + 2b_f + q_0, x_f, u_f) + \Delta/4$$

$$-I^f(T_0, T_0 + \tau_0 + 2L_2 + q_0, x_{k_{p0}}, u_{k_{p0}})$$

$$-I^f(T_0 + \tau_0 + 2L_2 + q_0, T_0 + \tau_0 + 2L_2 + b_f + q_0, x_f, u_f) + \Delta/8$$

$$-I^f(T_0 + \tau_0 + 2L_2 + b_f + q_0, T_0 + \tau_0 + 2L_2 + 2b_f + q_0, x_f, u_f) + \Delta/8$$

$$\leq I^f(T_0, T_0 + \tau_0, y, v) + I^f(T_0 + \tau_0, T_0 + \tau_0 + 2L_2 + q_0, x_*, u_*)$$

$$-I^f(T_0, T_0 + \tau_0 + 2L_2 + q_0, x_{k_{p0}}, u_{k_{p0}}) + \Delta/2$$

$$< I^f(T_0, T_0 + \tau_0 + q_0 + 2L_2, x_*, u_*) - 2\Delta$$

$$-I^f(T_0, T_0+\tau_0+2L_2+q_0, x_{k_{p_0}}, u_{k_{p_0}}) + \Delta/4 < -\Delta,$$

a contradiction. The contradiction we have reached completes the proof of Proposition 7.29.

7.16 Proof of Theorems 7.15

Clearly, (i) implies (ii) and (ii) implies (iii). By Theorem 7.9, (iii) implies (iv). Evidently (iv) implies (v). We show that (v) implies (iii). Assume that $(x_*, u_*) \in X(S, \infty)$ is (f)-minimal and satisfies

$$\liminf_{t\to\infty} \|x_*(t) - x_f(t)\| = 0. \tag{7.399}$$

Since $(\tilde{x}, \tilde{u})$ is (f)-good, there exists $S_0 > 0$ such that for all numbers $T > S$,

$$|I^f(S, T, \tilde{x}, \tilde{u}) - I^f(S, T, x_f, u_f)| \le S_0. \tag{7.400}$$

By (A2), there exists $\delta > 0$ such that the following property holds:

(a) for each $(T, z) \in \mathcal{A}$ satisfying $\|z - x_f(T)\| \le \delta$, there exist $\tau_1 \in (0, b_f]$ and $(x_1, u_1) \in X(T, T+\tau_1)$ which satisfy

$$x_1(T) = z,\ x_1(T+\tau_1) = x_f(T+\tau_1),$$

$$I^f(T, T+\tau_1, x_1, u_1) \le I^f(T, T+\tau_1, x_f, u_f) + 1$$

and there exist $\tau_2 \in (0, b_f]$ and $(x_2, u_2) \in X(T-\tau_2, T)$ such that

$$x_2(T-\tau_2) = x_f(T-\tau_2),\ x_2(T) = z,$$

$$I^f(T-\tau_2, T, x_2, u_2) \le I^f(T-\tau_2, T, x_f, u_f) + 1.$$

In view of (7.399) and Theorem 7.9, there exists an increasing sequence $\{t_k\}_{k=1}^{\infty} \subset (S, \infty)$ such that

$$\lim_{k\to\infty} t_k = \infty,$$

$$\|x_*(t_k + 2b_f) - x_f(t_k + 2b_f)\| \le \delta \text{ for all integers } k \ge 1, \tag{7.401}$$

$$\|\tilde{x}(t) - x_f(t)\| \le \delta \text{ for all } t \ge t_0. \tag{7.402}$$

Let $k \geq 1$ be an integer. By property (a), (7.401) and (7.402), there exists $(y, v) \in X(S, t_k + 2b_f)$ such that

$$(y(t), v(t)) = (\tilde{x}(t), \tilde{u}(t)),\ t \in [S, t_k],$$

$$y(t_k + b_f) = x_f(t_k + b_f),$$

$$I^f(t_k, t_k + b_f, y, v) \leq I^f(t_k, t_k + b_f, x_f, u_f) + 1,$$

$$y(t_k + 2b_f) = x_*(t_k + 2b_f),\ I^f(t_k + b_f, t_k + 2b_f, y, v)$$

$$\leq I^f(t_k + b_f, t_k + 2b_f, x_f, u_f) + 1.$$

The relations above and (7.400) imply that

$$I^f(S, t_k + 2b_f, x_*, u_*) \leq I^f(S, t_k + 2b_f, y, v)$$

$$\leq I^f(S, t_k, \tilde{x}, \tilde{u}) + I^f(t_k + b_f, t_k + 2b_f, x_f, u_f) + 2$$

$$\leq I^f(S, t_k + 2b_f, x_f, u_f) + 2 + S_0.$$

This implies that (x_*, u_*) is (f)-good and (iii) holds.

We show that (iii) implies (i). Assume that (x_*, u_*) is (f)-minimal and (f)-good. Theorem 7.9 implies that

$$\lim_{t\to\infty} \|x_*(t) - x_f(t)\| = 0. \tag{7.403}$$

There exists $S_0 > 0$ such that

$$|I^f(S, T, x_*, u_*) - I^f(S, T, x_f, u_f)| \leq S_0 \text{ for all } T > S. \tag{7.404}$$

Let $(x, u) \in X(S, \infty)$ satisfy

$$x(S) = x_*(S). \tag{7.405}$$

We show that

$$\limsup_{T\to\infty}[I^f(S, T, x_*, u_*) - I^f(S, T, x, u)] \leq 0. \tag{7.406}$$

In view of Theorem 7.6, we may assume that (x, u) is (f)-good. Theorem 7.9 implies that

$$\lim_{t\to\infty} \|x(t) - x_f(t)\| = 0. \tag{7.407}$$

Let $\epsilon > 0$. By (A2) and Lemma 7.25, there exist $\delta \in (0, \epsilon)$ and $L_1 > 0$ such that the following property holds:

(b) for each $(T, z) \in \mathcal{A}$ satisfying $\|z - x_f(T)\| \leq \delta$, there exist $\tau_1 \in (0, b_f]$ and $(x_1, u_1) \in X(T, T + \tau_1)$ satisfying

$$x_1(T) = z\ , x_1(T + \tau_1) = x_f(T + \tau_1),$$

$$I^f(T, T + \tau_1, x_1, u_1) \leq I^f(T, T + \tau_1, x_f, u_f) + \epsilon/8$$

and then there exist $\tau_2 \in (0, b_f]$ and $(x_2, u_2) \in X(T - \tau_2, T)$ satisfying

$$x_2(T - \tau_2) = x_f(T - \tau_2)\ , x_2(T) = z,$$

$$I^f(T - \tau_2, T, x_2, u_2) \leq I^f(T - \tau_2, T, x_f, u_f) + \epsilon/8;$$

for each $T_1 \geq L_1$, each $T_2 > T_1$ and each $(y, v) \in X(T_1, T_2)$ which satisfies $\|y(T_i) - x_f(T_i)\| \leq \delta, i = 1, 2$, we have

$$I^f(T_1, T_2, y, v) \geq I^f(T_1, T_2, x_f, u_f) - \epsilon/8.$$

It follows from (7.403) and (7.407) that there exists $\tau_0 > |S_0|$ such that

$$\|x(t) - x_f(t)\| \leq \delta,\ \|x_*(t) - x_f(t)\| \leq \delta \text{ for all } t \geq \tau_0. \tag{7.408}$$

Let $T \geq \tau_0 + L_1$. Property (b) and (7.408) imply that there exists $(y, v) \in X(S, T + 2b_f)$ which satisfies

$$(y(t), v(t)) = (x(t), u(t)),\ t \in [S, T],\ y(T + b_f) = x_f(T + b_f), \tag{7.409}$$

$$I^f(T, T + b_f, y, v) \leq I^f(T, T + b_f, x_f, u_f) + \epsilon/8, \tag{7.410}$$

$$y(T + 2b_f) = x_*(T + 2b_f), \tag{7.411}$$

$$I^f(T + b_f, T + 2b_f, y, v) \leq I^f(T + b_f, T + 2b_f, x_f, u_f) + \epsilon/8. \tag{7.412}$$

By property (b) and (7.408),

$$I^f(T, T + 2b_f, x_*, u_*) \geq I^f(T, T + 2b_f, x_f, u_f) - \epsilon/8. \tag{7.413}$$

It follows from (7.405) and (7.409)–(7.413) that

$$I^f(S, T, x_*, u_*) + I^f(T, T + 2b_f, x_*, u_*) - \epsilon/8$$

$$\leq I^f(S, T + 2b_f, x_*, u_*) \leq I^f(S, T + 2b_f, y, v)$$

$$= I^f(S, T, x, u) + I^f(T, T + 2b_f, x_f, u_f) + \epsilon/4,$$

$$I^f(S, T, x_*, u_*) \leq I^f(S, T, x, u) + \epsilon \text{ for all } T \geq \tau_0 + L.$$

Since ϵ is any positive number, we conclude that (7.406) holds and (x_*, u_*) is (f)-overtaking optimal and (i) holds.

In order to complete the proof, we need only to apply Proposition 7.29.

7.17 Proof of Theorem 7.16

In view of Theorem 7.8, it is sufficient to show that f_r has (P1), (P2), and (P3) with $\tilde{u}_{f_r} = u_f$. Theorem 7.2 implies that there exists $M_0 > 0$ such that the following property holds:

(i) for each $S_2 > S_1$ and each $(x, u) \in X(S_1, S_2)$,

$$I^f(S_1, S_2, x, u) + M_0 \geq I^f(S_1, S_2, x_f, u_f).$$

We show that (P2) holds. Let $\epsilon, M > 0$. Property (i) of Section 7.4 implies there exists $\delta > 0$ such that the following property holds:

(ii)

$$\text{if } z \in E \text{ satisfies } \phi(z) \leq \delta, \text{ then } \|z\| \leq \epsilon.$$

Choose

$$L > \delta^{-1} r^{-1}(M_0 + M). \tag{7.414}$$

Assume that $T \in R^1$, $(x, u) \in X(T, T + L)$ satisfies

$$I^{f_r}(T, T + L, x, u) \leq I^{f_r}(T, T + L, x_f, u_f) + M. \tag{7.415}$$

By (7.415), property (i) and the definition of f_r,

$$I^f(T, T + L, x, u) + r \int_T^{T+L} \phi(x(t) - x_f(t))dt$$

$$= I^{f_r}(T, T+L, x, u) \leq I^{f_r}(T, T+L, x_f, u_f) + M \leq I^f(T, T+L, x, u) + M + M_0,$$

$$\int_T^{T+L} \phi(x(t) - x_f(t))dt \leq r^{-1}(M_0 + M).$$

Together with (7.414), this implies that

$$\inf\{\phi(x(t)-x_f(t)) : t \in [T, T+L]\} \le L^{-1}(M_0+M)r^{-1} < \delta.$$

Therefore there exists $\tau \in [T, T+L]$ such that $\phi(x(\tau)-x_f(\tau)) < \delta$. Together with property (ii) this implies that $\|x(\tau)-x_f(\tau)\| \le \epsilon$. Therefore (P2) holds.

Let us show that f_r has (P1). Let $(x, u) \in X(-\infty, \infty)$ be (f_r)-good. There exists $M_1 > 0$ such that

$$|I^{f_r}(-T, T, x, u) - I^{f_r}(-T, T, x_f, u_f)| < M_1 \text{ for all } T > 0. \tag{7.416}$$

We show that

$$\lim_{t\to\infty}(x(t)-x_f(t)) = 0, \quad \lim_{t\to-\infty}(x(t)-x_f(t)) = 0.$$

Assume the contrary. Then at least one of the following properties holds:

(iii) there exists a sequence $\{t_k\}_{k=1}^{\infty} \subset (0, \infty)$ such that for each integer $k \ge 1$, $t_{k+1} > t_k + 8$ and $\|x(t_k) - x_f(t_k)\| > \epsilon$;
(iv) there exists a sequence $\{t_k\}_{k=1}^{\infty} \subset (-\infty, 0)$ such that for each integer $k \ge 1$, $t_{k+1} < t_k - 8$ and $\|x(t_k) - x_f(t_k)\| > \epsilon$.

Property (i), the definition of f_r and (7.416) imply that for each $T > 0$,

$$M_1 > I^{f_r}(-T, T, x, u) - I^{f_r}(-T, T, x_f, u_f)$$

$$= r\int_{-T}^{T}\phi(x(t)-x_f(t))dt + I^f(-T, T, x, u) - I^f(-T, T, x_f, u_f)$$

$$\ge r\int_{-T}^{T}\phi(x(t)-x_f(t))d - M_0,$$

$$\Delta := \lim_{T\to\infty}\int_{-T}^{T}\phi(x(t)-x_f(t))d < (M_0+M_1)r^{-1}. \tag{7.417}$$

Property (i) and (7.416) imply that for each $T_2 > T_1$ and each $T > 0$ satisfying $[T_1, T_2] \subset [-T, T]$,

$$I^f(-T, T, x, u) < I^f(-T, T, x_f, u_f) + M_1,$$

$$I^f(T_1, T_2, x, u) = I^f(-T, T, x, u) - I^f(-T, T_1, x, u) - I^f(T_2, T, x, u)$$

$$\le I^f(-T, T, x_f, u_f) + M_1 - I^f(-T, T_1, x_f, u_f) + M_0 - I^f(T_2, T, x, u) + M_0$$

$$= I^f(T_1, T_2, x_f, u_f) + M_1 + 2M_0. \tag{7.418}$$

In view of (7.8), (7.13), and (7.418), for each $T \in R^1$,

$$I^f(T, T+1, x, u) \leq I^f(T, T+1, x_f, u_f) + 2M_0 + M_1 \leq 2M_0 + M_1 + 2\Delta_f + 2a_0. \tag{7.419}$$

Lemma 7.21 implies that there exists $\delta \in (0, 4^{-1})$ such that the following property holds:

(v) for each $T \in R^1$, each $(y, v) \in X(T, T+1)$ satisfying

$$I^f(T, T+1, y, v) \leq 2M_0 + M_1 + 2\Delta_f + 2a_0$$

and each $t_1, t_2 \in [T, T+1]$ satisfying $|t_1 - t_2| \leq \delta$ the inequality $\|y(t_1) - y(t_2)\| \leq \epsilon$ holds.

Property (i) of Section 7.4 implies that there exists $\gamma > 0$ such that the following property holds:

(vi)

$$\text{for each } \xi \in E \text{ satisfying } \|\xi\| \geq \epsilon/2, \text{ then } \phi(\xi) \geq \gamma.$$

Let $k \geq 1$ be an integer. It follows from (7.8), (7.13), and (7.419) that

$$I^f(t_k, t_{k+1}, x_f, u_f) \leq 2\Delta_f + 2a_0, \; I^f(t_k, t_{k+1}, x, u) \leq 2M_0 + M_1 + 2\Delta_f + 2a_0.$$

Properties (iii)–(vi) and the relations above imply that for all $t \in [t_k, t_k + \delta]$,

$$\|x(t) - x(t_k)\| \leq \epsilon/8, \; \|x_f(t) - x_f(t_k)\| \leq \epsilon/8,$$

$$\|x_f(t) - x(t)\| \geq \|x_f(t_k) - x(t_k)\| - \|x_f(t) - x_f(t_k)\| - \|x(t) - x(t_k)\| \geq \epsilon/2,$$

$$\phi(x(t) - x_f(t)) \geq \gamma.$$

Therefore

$$\int_{t_k}^{t_k+\delta} \phi(x(t) - x_f(t))dt \geq \gamma\delta.$$

This implies that

$$\lim_{T\to\infty} \int_{-T}^{T} \phi(x(t) - x_f(t))dt = \infty.$$

This contradicts (7.417). The contradiction we have reached proves that (P1) holds.

Let us show that (P3) holds. Clearly, (x_f, u_f) is (f_r)-minimal. Assume that $(x, u) \in X(-\infty, \infty)$ is (f_r)-good and (f_r)-minimal. Property (P1) implies that

$$\lim_{t\to\infty} \|x(t) - x_f(t)\| = 0, \quad \lim_{t\to-\infty} \|x(t) - x_f(t)\| = 0. \tag{7.420}$$

We show that $x(t) = x_f(t)$ for all $t \in R^1$. Assume the contrary. Then there exists $t_0 \in R^1$ such that

$$\gamma := \|x(t_0) - x_f(t_0)\| > 0. \tag{7.421}$$

By (7.421), there exists $\epsilon_0 \in (0, 1)$ such that

$$\|x(t) - x_f(t)\| > \gamma/2 \text{ for all } t \in [t_0, t_0 + \epsilon_0]. \tag{7.422}$$

Property (i) of Section 7.4 implies that there exists $\delta_0 > 0$ such that

$$\text{if } z \in E \text{ satisfies } \phi(z) \le \delta_0, \text{ then } \|z\| \le \gamma/4. \tag{7.423}$$

It follows from (7.422) and (7.423) that

$$\phi(x(t) - x_f(t))) > \delta_0 \text{ for all } \hat{t} \in [t_0, t_0 + \epsilon_0]. \tag{7.424}$$

By (A3) and Lemma 7.25, there exists $L_1 > 0$, $\delta_1 \in (0, \min\{\delta_0, \epsilon_0\})$ such that the following property holds:

(vii) for each $(S, z) \in \mathcal{A}$ satisfying $\|z - x_f(S)\| \le \delta_1$ there exist $\tau_1 \in (0, b_f]$ and $(y_1, v_1) \in X(S, S + \tau_1)$, $\tau_2 \in (0, b_f]$, $(y_2, v_2) \in X(S - \tau_2, S_2)$ such that

$$y_1(S) = z \text{ , } y_1(S + \tau_1) = x_f(S + \tau_1),$$

$$I^{f_r}(S, S + \tau_1, y_1, v_1) \le I^{f_r}(S, S + \tau_1, x_f, u_f) + 8^{-1}\epsilon_0\delta_0 r,$$

$$y_2(S - \tau_2) = x_f(S - \tau_2) \text{ , } y_2(S) = z,$$

$$I^{f_r}(S - \tau_2, S, y_2, v_2) \le I^{f_r}(S - \tau_2, S, x_f, u_f) + 8^{-1}\epsilon_0\delta_0 r,$$

$$\|y_1(t) - x_f(t)\| \le 8^{-1}\epsilon_0\delta_0, \ t \in [S, S + \tau_1],$$

$$\|y_2(t) - x_f(t)\| \le 8^{-1}\epsilon_0\delta_0, \ t \in [S - \tau_2, S_2];$$

for each $T_2 > T_1$ such that either $T_2 \le -L_1$ or $T_1 \ge L_1$ and each $(y, v) \in X(T_1, T_2)$ satisfying $\|y(T_i) - x_f(T_i)\| \le \delta_1$, $i = 1, 2$, we have

$$I^{f_r}(T_1, T_2, y, v) \ge I^{f_r}(T_1, T_2, x_f, u_f) - 8^{-1}\epsilon_0\delta_0 r.$$

In view of (7.422), there exists $T_0 > L_1$ such that

$$[-L_1 + L_0 - 2b_f - 2, L_0 + 1 + 2b_f + 2 + L_1] \subset [-T_0, T_0], \tag{7.425}$$

$$\|x(t) - x_f(t)\| \le \delta_1 \text{ for all } t \in (-\infty, -T_0] \cup [T_0, \infty). \tag{7.426}$$

Property (vii) and (7.426) imply that there exists $(y, v) \in X(-T_0 - b_f, T_0 + b_f)$ such that

$$y(-T_0 - b_f) = x(-T_0 - b_f), \tag{7.427}$$

$$(y(t), v(t)) = (x_f(t), u_f(t)),\ t \in [-T_0, T_0], \tag{7.428}$$

$$I^{f_r}(-T_0 - b_f, -T_0, y, v) \le I^{f_r}(-T_0 - b_f, -T_0, x_f, u_f) + 8^{-1}\epsilon_0\delta_0 r, \tag{7.429}$$

$$y(T_0 + b_f) = x(T_0 + b_f), \tag{7.430}$$

$$I^{f_r}(T_0, T_0 + b_f, y, v) \le I^{f_r}(T_0, b_f + T_0, x_f, u_f) + 8^{-1}\epsilon_0\delta_0 r. \tag{7.431}$$

By (7.427) and (7.430),

$$I^{f_r}(-T_0 - b_f, T_0 + b_f, y, v) \ge I^{f_r}(-T_0 - b_f, T_0 + b_f, x, u). \tag{7.432}$$

Property (vii), (7.426), (7.428), (7.429), (7.431), and (7.432) imply that

$$I^{f_r}(-T_0 - b_f, -T_0, x_f, u_f) - 8^{-1}\epsilon_0\delta_0 r$$

$$+ I^{f_r}(-T_0, T_0, x, u) + I^{f_r}(T_0, T_0 + b_f, x_f, u_f) - 8^{-1}\epsilon_0\delta_0 r$$

$$\le I^{f_r}(-T_0 - b_f, -T_0, x, u) + I^{f_r}(-T_0, T_0, x, u) + I^{f_r}(T_0, T_0 + b_f, x, u)$$

$$\le I^{f_r}(-T_0 - b_f, T_0 + b_f, y, v)$$

$$\le I^{f_r}(-T_0 - b_f, -T_0, y, v) + I^{f_r}(-T_0, T_0, y, v) + I^{f_r}(T_0, T_0 + b_f, y, v)$$

$$\le I^{f_r}(-T_0 - b_f, -T_0, x_f, u_f) + 8^{-1}\epsilon_0\delta_0 r + I^{f_r}(-T_0, T_0, x_f, u_f)$$

$$+ I^{f_r}(T_0, T_0 + b_f, x_f, u_f) + 8^{-1}\epsilon_0\delta_0 r,$$

$$I^{f_r}(-T_0, T_0, x, u) \le I^{f}(-T_0, T_0, x_f, u_f) + 2^{-1}\epsilon_0\delta_0 r. \tag{7.433}$$

Property (vii) and (7.426) imply that there exists $(\xi, \eta) \in X(-T_0 - b_f, T_0 + b_f)$ such that

$$\xi(-T_0 - b_f) = x_f(-T_0 - b_f), \tag{7.434}$$

$$(\xi(t), \eta(t)) = (x(t), u(t)), \ t \in [-T_0, T_0], \tag{7.435}$$

$$\xi(T_0 + b_f) = x_f(T_0 + b_f), \tag{7.436}$$

$$I^{f_r}(-T_0 - b_f, -T_0, \xi, \eta) \le I^{f_r}(-T_0 - b_f, -T_0, x_f, u_f) + 8^{-1}\epsilon_0\delta_0 r, \tag{7.437}$$

$$I^{f_r}(T_0, T_0 + b_f, \xi, \eta) \le I^{f_r}(T_0, b_f + T_0, x_f, u_f) + 8^{-1}\epsilon_0\delta_0 r. \tag{7.438}$$

By (7.434), (7.436), and (7.438),

$$I^f(-T_0 - b_f, T_0 + b_f, x_f, u_f) \le I^f(-T_0 - b_f, T_0 + b_f, \xi, \eta). \tag{7.439}$$

It follows from the definition of f_r, (7.435), (7.437), and (7.438) that

$$\begin{aligned} &I^f(-T_0 - b_f, T_0 + b_f, \xi, \eta) \\ &= I^f(-T_0 - b_f, -T_0, \xi, \eta) + I^f(-T_0, T_0, x, u) + I^f(T_0, T_0 + b_f, \xi, \eta) \\ &\le I^{f_r}(-T_0 - b_f, -T_0, \xi, \eta) + I^f(-T_0, T_0, x, u) + I^{f_r}(T_0, T_0 + b_f, \xi, \eta) \\ &\le I^f(-T_0 - b_f, -T_0, x_f, u_f) + 8^{-1}\epsilon_0\delta_0 r \\ &+ I^f(-T_0, T_0, x, u) + I^f(T_0, T_0 + b_f, x_f, u_f) + 8^{-1}\epsilon_0\delta_0 r. \end{aligned} \tag{7.440}$$

In view of (7.439) and (7.440),

$$I^f(-T_0, T_0, x_f, u_f) \le I^f(-T_0, T_0, x, u) + 4^{-1}\epsilon_0\delta_0 r. \tag{7.441}$$

By (7.424), (7.433), and (7.441),

$$I^f(-T_0, T_0, x_f, u_f) + 2^{-1}\epsilon_0\delta_0 r \ge I^{f_r}(-T_0, T_0, x, u)$$

$$= I^f(-T_0, T_0, x, u) + r\int_{-T_0}^{T_0} \phi(x(t) - x_f(t))dt = I^f(-T_0, T_0, x, u) + \epsilon_0\delta_0 r$$

$$\ge I^f(-T_0, T_0, x_f, u_f) + \epsilon_0\delta_0 r - 4^{-1}\epsilon_0\delta_0 r,$$

a contradiction. The contradiction we have reached proves $x(t) = x_f(t)$, $t \in R^1$. Thus (P3) holds and Theorem 7.16 is proved.

7.18 An Example

We consider a particular case of the problem considered in Section 7.1. Let $(E, \langle\cdot,\cdot\rangle)_E$ be a Hilbert space equipped with an inner product $\langle\cdot,\cdot\rangle_E$ which induces the norm $\|\cdot\|$, and let $(F, \langle\cdot,\cdot\rangle_F)$ be a Hilbert space equipped with an inner product $\langle\cdot,\cdot\rangle_F$ which induces the norm $\|\cdot\|_F$.

Let $A : E \to E$ is a bounded linear operator, and let $B : F \to E$ is a linear bounded operator, for all $(t, x, u) \in R^1 \times E \times F$,

$$G(t, x, u) = Ax + Bu.$$

Recall that for every $T > 0$,

$$\Phi_T \in \mathcal{L}(L^2(0, T; F), E)$$

is defined by

$$\Phi_T u = \int_0^T e^{A(T-s)} Bu(s)ds, \ u \in L^2(0, T; F).$$

Let $T_f > 0$, $\mathrm{Ran}(\Phi_{T_f}) = E$, $(x_f, u_f) \in X(-\infty, \infty)$,

$$\sup\{\|x_f(t)\| : t \in R^1\} < \infty.$$

Analogously to Theorem 6.50, we can prove the following result.

Theorem 7.30. *There exists a constant $c > 0$ such that for each $T \in R^1$, each $z^0, z^1 \in E$, there exist $u \in L^2(T, T + T_f; F)$ and $z \in C^0([T, T + T_f]); E)$ which is a solution of the initial value problem*

$$z'(t) = Az(t) + Bu(t), \ t \in [T, T + T_f] \ a.\ e.\ , z(T) = z^0$$

in E and satisfies $z(T + T_f) = z^1$ and

$$\|z(t) - x_f(t)\|, \|u(t) - u_f(t)\|$$

$$\le c(\|z^1 - x_f(T + T_f)\| + \|z^0 - x_f(T)\|), \ t \in [T, T + T_f].$$

We suppose that there exists $r_* > 0$ such that

$$\{(t, x, u) \in R^1 \times E \times F : \ \|x - x_f(t)\| \le r_*, \ \|u - u_f(t)\| \le r_*\} \subset \mathcal{M}.$$

Assume that $L : R^1 \times E \times F \to [0, \infty)$ is a Borelian function,

$$L(t, x_f(t), u_f(t)) = 0 \text{ for all } t \in R^1,$$

$\psi_0 : [0, \infty) \to [0, \infty)$ is an increasing function, $a_1 > 0$,

$$\lim_{t\to\infty} \psi_0(t) = \infty,$$

$$L(t, x, u) \geq -a_1 + \max\{\psi_0(\|x\|), \psi_0(\|Ax + Bu\|)\|Ax + Bu\|\}$$

for all $(t, x, u) \in \mathcal{M}$, $\mu \in R^1$, $\bar{p} \in E$. Let for all $(t, x, u) \in \mathcal{M}$,

$$f(t, x, u) = L(t, x, u) + \mu + \langle Ax + Bu, \bar{p}\rangle.$$

It is not difficult to see that f is Borelian; there exist a_0 and an increasing function $\psi : [0, \infty) \to [0, \infty)$ such that $\lim_{t\to\infty} \psi(t) = \infty$ and for all $(t, x, u) \in \mathcal{M}$,

$$f(t, x, u) \geq -a_0 + \max\{\psi(\|x)\|,$$

$$\psi(\|G(t, x, u)\| - a_0\|x\|)_+)(\|G(t, x, u)\| - a_0\|x\|)_+)\}.$$

We suppose that the following property holds:

(a) For each $\epsilon > 0$, there exists $\delta > 0$ such that for each $(t, x, u) \in \mathcal{M}$ satisfying $\|x - x_f(t)\|$, $\|u - u_f(t)\| \leq \delta$, we have

$$f(t, x, u) \leq f(t, x_f(t), u_f(t)) + \epsilon.$$

By property (a) and Theorem 7.30, (A3) holds. As in Section 6.19, we can show that (A1) holds too.

References

1. Ahmed, N.U.: Optimal control on infinite dimensional Banach spaces of neutral systems driven by vector measures. Commun. Appl. Nonlinear Anal. **16**, 1–14 (2009)
2. Ahmed, N.U.: Infinite dimensional uncertain dynamic systems on Banach spaces and their optimal output feedback control. Discuss. Math. Differ. Incl. Control Optim. **35**, 65–87 (2015)
3. Ahmed, N.U.: Optimal output feedback control law for a class of uncertain infinite dimensional dynamic systems. J. Abstr. Differ. Equ. Appl. **7**, 11–29 (2016)
4. Ahmed, N.U., Xiang, X.: Differential inclusions on Banach spaces and their optimal control. Nonlinear Funct. Anal. Appl. **8**, 461–488 (2003)
5. Anderson, B.D.O., Moore, J.B.: Linear Optimal Control. Prentice-Hall, Englewood Cliffs (1971)
6. Arkin, V.I., Evstigneev, I.V.: Stochastic Models of Control and Economic Dynamics. Academic Press, London (1987)
7. Aseev, S.M., Kryazhimskiy, A.V.: The Pontryagin maximum principle and transversality conditions for a class of optimal control problems with infinite time horizons. SIAM J. Control Optim. **43**, 1094–1119 (2004)
8. Aseev, S.M., Veliov, V.M.: Maximum principle for infinite-horizon optimal control problems with dominating discount. Dyn. Contin. Discrete Impuls. Syst. Ser. B **19**, 43–63 (2012)
9. Aseev, S.M., Krastanov, M.I., Veliov, V.M.: Optimality conditions for discrete-time optimal control on infinite horizon. Pure Appl. Funct. Anal. **2**, 395–409 (2017)
10. Atsumi, H.: Neoclassical growth and the efficient program of capital accumulation. Rev. Econ. Stud. **32**, 127–136 (1965)
11. Aubry, S.: Trajectories of the twist map with minimal action and connection with incommensurate structures. Common trends in particle and condensed matter physics (Les Houches, 1983). Phys. Rep. **103**(1–4), 127141 (1984)
12. Aubry, S., Le Daeron, P.Y.: The discrete Frenkel-Kontorova model and its extensions I. Physica D **8**, 381–422 (1983)
13. Bachir, M., Blot, J.: Infinite dimensional infinite-horizon Pontryagin principles for discrete-time problems. Set-Valued Var. Anal. **23**, 43–54 (2015)
14. Bachir, M., Blot, J.: Infinite dimensional multipliers and Pontryagin principles for discrete-time problems. Pure Appl. Funct. Anal. **2**, 411–426 (2017)
15. Ball, J.M.: Strong continuous semigroups, weak solutions and the variation of constants formula. Proc. Am. Math. Soc. **63**, 370-373 (1977)
16. Barbu, V.: Optimal Control of Variational Inequalities. Pitman Research Notes in Mathematics. Pitman, London (1984)

A. J. Zaslavski, *Turnpike Conditions in Infinite Dimensional Optimal Control*,
Springer Optimization and Its Applications 148,
https://doi.org/10.1007/978-3-030-20178-4

17. Barbu, V.: Analysis and Control of Nonlinear Infinite Dimensional Systems. Academic Press, Boston (1993)
18. Barbu, V., Precupanu, T.: Convexity and Optimization in Banach Spaces. Springer Monographs in Mathematics. Springer, Dordrecht (2012)
19. Bardi, M.: On differential games with long-time-average cost. In: Advances in Dynamic Games and Their Applications, pp. 3–18. Birkhauser, Basel (2009)
20. Baumeister, J., Leitao, A., Silva, G.N.: On the value function for nonautonomous optimal control problem with infinite horizon. Syst. Control Lett. **56**, 188–196 (2007)
21. Blot, J.: Infinite-horizon Pontryagin principles without invertibility. J. Nonlinear Convex Anal. 10, 177–189 (2009)
22. Blot, J., Cartigny, P.: Optimality in infinite-horizon variational problems under sign conditions. J. Optim. Theory Appl. **106**, 411–419 (2000)
23. Blot, J., Hayek, N.: Sufficient conditions for infinite-horizon calculus of variations problems. ESAIM Control Optim. Calc. Var. **5**, 279–292 (2000)
24. Blot, J., Hayek, N.: Infinite-Horizon Optimal Control in the Discrete-Time Framework. Springer Briefs in Optimization. Springer, New York (2014)
25. Bright, I.: A reduction of topological infinite-horizon optimization to periodic optimization in a class of compact 2-manifolds. J. Math. Anal. Appl. **394**, 84–101 (2012)
26. Brock, W.A.: On existence of weakly maximal programmes in a multi-sector economy. Rev. Econ. Stud. **37**, 275–280 (1970)
27. Carlson, D.A.: The existence of catching-up optimal solutions for a class of infinite horizon optimal control problems with time delay. SIAM J. Control Optim. **28**, 402–422 (1990)
28. Carlson, D.A., Jabrane, A., Haurie, A.: Existence of overtaking solutions to infinite dimensional control problems on unbounded time intervals. SIAM J. Control Optim. **25**, 1517–1541 (1987)
29. Carlson, D.A., Haurie, A., Leizarowitz, A.: Infinite Horizon Optimal Control. Springer, Berlin (1991)
30. Cartigny, P., Michel, P.: On a sufficient transversality condition for infinite horizon optimal control problems. Autom. J. IFAC **39**, 1007–1010 (2003)
31. Cellina, A., Colombo, G.: On a classical problem of the calculus of variations without convexity assumptions. Ann. Inst. H. Poincare Anal. Non Lineare **7**, 97–106 (1990)
32. Cesari, L.: Optimization - Theory and Applications. Springer, Berlin (1983)
33. Coleman, B.D., Marcus, M., Mizel, V.J.: On the thermodynamics of periodic phases. Arch. Rational Mech. Anal. **117**, 321–347 (1992)
34. Coron, J.M.: Control and Nonlinearity. AMS, Providence (2007)
35. Damm, T., Grune, L., Stieler, M., Worthmann, K.: An exponential turnpike theorem for dissipative discrete time optimal control problems. SIAM J. Control Optim. **52**, 1935–1957 (2014)
36. De Oliveira, V.A., Silva, G.N.: Optimality conditions for infinite horizon control problems with state constraints. Nonlinear Anal. **71**, 1788–1795 (2009)
37. Evstigneev, I.V., Flam, S.D.: Rapid growth paths in multivalued dynamical systems generated by homogeneous convex stochastic operators. Set-Valued Anal. **6**, 61–81 (1998)
38. Gaitsgory, V., Rossomakhine, S., Thatcher, N.: Approximate solution of the HJB inequality related to the infinite horizon optimal control problem with discounting. Dyn. Contin. Discrete Impuls. Syst. Ser. B **19**, 65–92 (2012)
39. Gaitsgory, V., Grune, L., Thatcher, N.: Stabilization with discounted optimal control. Syst. Control Lett. **82**, 91–98 (2015)
40. Gaitsgory, V., Mammadov, M., Manic, L.: On stability under perturbations of long-run average optimal control problems. Pure Appl. Funct. Anal. **2**, 461–476 (2017)
41. Gaitsgory, V., Parkinson, A., Shvartsman, I.: Linear programming formulations of deterministic infinite horizon optimal control problems in discrete time. Discrete Contin. Dyn. Syst. Ser. B **22**, 3821–3838 (2017)
42. Gale, D.: On optimal development in a multi-sector economy. Rev. Econ. Stud. **34**, 1–18 (1967)

43. Ghosh, M.K., Mallikarjuna Rao, K.S.: Differential games with ergodic payoff. SIAM J. Control Optim. **43**, 2020–2035 (2005)
44. Glizer, V.Y., Kelis, O.: Upper value of a singular infinite horizon zero-sum linear-quadratic differential game. Pure Appl. Funct. Anal. **2**, 511–534 (2017)
45. Grune, L., Guglielmi, R.: Turnpike properties and strict dissipativity for discrete time linear quadratic optimal control problems. SIAM J. Control Optim. **56**, 1282–1302 (2018)
46. Gugat, M., Trelat, E., Zuazua, E.: Optimal Neumann control for the 1D wave equation: finite horizon, infinite horizon, boundary tracking terms and the turnpike property. Syst. Control Lett. **90**, 61–70 (2016)
47. Guo, X., Hernandez-Lerma, O.: Zero-sum continuous-time Markov games with unbounded transition and discounted payoff rates. Bernoulli **11**, 1009–1029 (2005)
48. Hayek, N.: Infinite horizon multiobjective optimal control problems in the discrete time case. Optimization **60**, 509–529 (2011)
49. Hernandez-Lerma, O., Lasserre, J.B.: Zero-sum stochastic games in Borel spaces: average payoff criteria. SIAM J. Control Optim. **39**, 1520–1539 (2001)
50. Hille, E., Phillips, R.S.: Functional Analysis and Semigroups. AMS, Providence (1957)
51. Jasso-Fuentes, H., Hernandez-Lerma, O.: Characterizations of overtaking optimality for controlled diffusion processes. Appl. Math. Optim. **57**, 349–369 (2008)
52. Jiang, Y.-R., Huang, N.-J., Yao, J.-C.: Solvability and optimal control of semilinear nonlocal fractional evolution inclusion with Clarke subdifferential. Appl. Anal. **96**, 2349–2366 (2017)
53. Khan, M.A., Zaslavski, A.J.: On two classical turnpike results for the Robinson-Solow-Srinivisan (RSS) model. Adv. Math. Econ. **13**, 47–97 (2010)
54. Khlopin, D.V.: Necessity of vanishing shadow price in infinite horizon control problems. J. Dyn. Control. Syst. **19**, 519–552 (2013)
55. Khlopin, D.V.: On Lipschitz continuity of value functions for infinite horizon problem. Pure Appl. Funct. Anal. **2**, 535–552 (2017)
56. Kien, B.T., Thi, T.N., Wong, M.M., Yao, J.-C.: Lower semicontinuity of the solution mapping to a parametric optimal control problem. SIAM J. Control Optim. **50**, 2889–2906 (2012)
57. Kolokoltsov, V., Yang, W.: The turnpike theorems for Markov games. Dyn. Games Appl. **2**, 294–312 (2012)
58. Lasiecka, I., Triggiani, R.: Control Theory for Partial Differential Equations: Continuous and Approximation Theories; Vol 1: Abstract Parabolic Systems. Encyclopedia of Mathematics and Its Applications Series. Cambridge University Press, Cambridge (2000)
59. Lasiecka, I., Triggiani, R.: Control Theory for Partial Differential Equations: Continuous and Approximation Theories; Vol 2: Abstract Hyperbolic-Like Systems Over a Finite Time Horizon. Encyclopedia of Mathematics and Its Applications Series. Cambridge University Press, Cambridge (2000)
60. Leizarowitz, A.: Infinite horizon autonomous systems with unbounded cost. Appl. Math. Optim. **13**, 19–43 (1985)
61. Leizarowitz, A.: Tracking nonperiodic trajectories with the overtaking criterion. Appl. Math. Optim. **14**, 155–171 (1986)
62. Leizarowitz, A., Mizel, V.J.: One dimensional infinite horizon variational problems arising in continuum mechanics. Arch. Rational Mech. Anal. **106**, 161–194 (1989)
63. Leizarowitz, A., Zaslavski, A.J.: On a class of infinite horizon optimal control problems with periodic cost functions. J. Nonlinear Convex Anal. **6**, 71–91 (2005)
64. Li, X., Yong, J.: Optimal Control Theory for Infinite Dimensional Systems. Birkhauser, Boston (1995)
65. Lykina, V., Pickenhain, S., Wagner, M.: Different interpretations of the improper integral objective in an infinite horizon control problem. J. Math. Anal. Appl. **340**, 498–510 (2008)
66. Makarov, V.L., Rubinov, A.M.: Mathematical Theory of Economic Dynamics and Equilibria. Springer, New York (1977)
67. Malinowska, A.B., Martins, N., Torres, D.F.M.: Transversality conditions for infinite horizon variational problems on time scales. Optim. Lett. **5**, 41–53 (2011)

68. Mammadov, M.: Turnpike theorem for an infinite horizon optimal control problem with time delay. SIAM J. Control Optim. **52**, 420–438 (2014)
69. Marcus, M., Zaslavski, A.J.: On a class of second order variational problems with constraints. Isr. J. Math. **111**, 1–28 (1999)
70. Marcus, M., Zaslavski, A.J.: The structure of extremals of a class of second order variational problems. Ann. Inst. H. Poincaré Anal. Non Linéaire **16**, 593–629 (1999)
71. Marcus, M., Zaslavski, A.J.: The structure and limiting behavior of locally optimal minimizers. Ann. Inst. H. Poincaré Anal. Non Linéaire **19**, 343–370 (2002)
72. McKenzie, L.W.: Turnpike theory. Econometrica **44**, 841–866 (1976)
73. Mordukhovich, B.S.: Approximation Methods in Optimization and Control. Nauka, Moscow (1988)
74. Mordukhovich, B.S.: Minimax design for a class of distributed parameter systems. Autom. Remote Control **50**, 1333–1340 (1990)
75. Mordukhovich, B.S.: Existence Theorems in Nonconvex Optimal Control, Calculus of Variations and Optimal Control, pp. 175–197. CRC Press, Boca Raton (1999)
76. Mordukhovich, B.S.: Optimal control and feedback design of state-constrained parabolic systems in uncertainly conditions. Appl. Anal. **90**, 1075–1109 (2011)
77. Mordukhovich, B.S., Shvartsman, I.: Optimization and feedback control of constrained parabolic systems under uncertain perturbations. In: Optimal Control, Stabilization and Nonsmooth Analysis. Lecture Notes Control and Information Sciences, pp. 121–132. Springer, Berlin (2004)
78. Moser, J.: Minimal solutions of variational problems on a torus. Ann. Inst. H. Poincaré, Anal. Non Linéaire **3**, 229–272 (1986)
79. Ocana Anaya, E., Cartigny, P., Loisel, P.: Singular infinite horizon calculus of variations. Applications to fisheries management. J. Nonlinear Convex Anal. **10**, 157–176 (2009)
80. Pickenhain, S., Lykina, V., Wagner, M.: On the lower semicontinuity of functionals involving Lebesgue or improper Riemann integrals in infinite horizon optimal control problems. Control Cybernet. **37**, 451–468 (2008)
81. Porretta, A., Zuazua, E.: Long time versus steady state optimal control. SIAM J. Control Optim. **51**, 4242–4273 (2013)
82. Prieto-Rumeau, T., Hernandez-Lerma, O.: Bias and overtaking equilibria for zero-sum continuous-time Markov games. Math. Methods Oper. Res. **61**, 437–454 (2005)
83. Reich, S., Zaslavski, A.J.: Genericity in Nonlinear Analysis. Springer, New York (2014)
84. Rockafellar, R.T.: A growth property in concave-convex Hamiltonian systems. Hamiltonian dynamics in economics. J. Econ. Theory. **12**(1), 191196 (1976)
85. Rubinov, A.M.: Economic dynamics. J. Soviet Math. **26**, 1975–2012 (1984)
86. Samuelson, P.A.: A catenary turnpike theorem involving consumption and the golden rule. Am. Econ. Rev. **55**, 486–496 (1965)
87. Tonelli, L.: Fondamenti di Calcolo delle Variazioni. Zanicelli, Bolonia (1921)
88. Trelat, E., Zhang, C., Zuazua, E.: Optimal shape design for 2D heat equations in large time. Pure Appl. Funct. Anal. **3**, 255–269 (2018)
89. Trelat, E., Zhang, C., Zuazua, E.: Steady-state and periodic exponential turnpike property for optimal control problems in Hilbert spaces. SIAM J. Control Optim. **56**, 1222–1252 (2018)
90. Troltzsch, F.: Optimal Control of Partial Differential Equations. Theory, Methods and Applications. American Mathematical Society, Providence (2010)
91. Tucsnak, M., Weiss, G.: Observation and Control for Operator Semigroups. Birkhauser, Basel (2009)
92. von Weizsacker, C.C.: Existence of optimal programs of accumulation for an infinite horizon. Rev. Econ. Stud. **32**, 85–104 (1965)
93. Yosida, K.: Functional Analysis, 6th edn. Springer, Berlin (1980)
94. Zaslavski, A.J.: Ground states in Frenkel-Kontorova model. Math. USSR Izvestiya **29**, 323–354 (1987)
95. Zaslavski, A.J.: Optimal programs on infinite horizon 1. SIAM J. Control Optim. **33**, 1643–1660 (1995)

96. Zaslavski, A.J.: Optimal programs on infinite horizon 2. SIAM J. Control Optim. **33**, 1661–1686 (1995)
97. Zaslavski, A.J.: Dynamic properties of optimal solutions of variational problems. Nonlinear Anal. **27**, 895–931 (1996)
98. Zaslavski, A.J.: Turnpike theorem for convex infinite dimensional discrete-time control systems. J. Convex Anal. **5**, 237–248 (1998)
99. Zaslavski, A.J.: The turnpike property for extremals of nonautonomous variational problems with vector-valued functions. Nonlinear Anal. **42**, 1465–1498 (2000)
100. Zaslavski, A.J.: Existence and structure of optimal solutions of infinite dimensional control systems. Appl. Math. Optim. **42**, 291–313 (2000)
101. Zaslavski, A.J.: Turnpike theorem for nonautonomous infinite dimensional discrete-time control systems. Optimization **48**, 69–92 (2000)
102. Zaslavski, A.J.: The structure of approximate solutions of variational problems without convexity. J. Math. Anal. Appl. **296**, 578–593 (2004)
103. Zaslavski, A.J.: The turnpike property of discrete-time control problems arising in economic dynamics. Discrete Contin. Dyn. Syst. B **5**, 861–880 (2005)
104. Zaslavski, A.J.: Turnpike Properties in the Calculus of Variations and Optimal Control. Springer, New York (2006)
105. Zaslavski, A.J.: The turnpike result for approximate solutions of nonautonomous variational problems. J. Aust. Math. Soc. **80**, 105–130 (2006)
106. Zaslavski, A.J.: Turnpike results for a discrete-time optimal control systems arising in economic dynamics. Nonlinear Anal. **67**, 2024–2049 (2007)
107. Zaslavski, A.J.: Two turnpike results for a discrete-time optimal control systems. Nonlinear Anal. **71**, 902–909 (2009)
108. Zaslavski, A.J.: Stability of a turnpike phenomenon for a discrete-time optimal control systems. J. Optim. Theory Appl. **145**, 597–612 (2010)
109. Zaslavski, A.J.: Optimization on Metric and Normed Spaces. Springer, New York (2010)
110. Zaslavski, A.J.: Turnpike properties of approximate solutions for discrete-time control systems. Commun. Math. Anal. **11**, 36–45 (2011)
111. Zaslavski, A.J.: Structure of approximate solutions for a class of optimal control systems. J. Math. Appl. **34**, 1–14 (2011)
112. Zaslavski, A.J.: Stability of a turnpike phenomenon for a class of optimal control systems in metric spaces. Numer. Algebra Control Optim. **1**, 245–260 (2011)
113. Zaslavski, A.J.: The existence and structure of approximate solutions of dynamic discrete time zero-sum games. J. Nonlinear Convex Anal. **12**, 49–68 (2011)
114. Zaslavski, A.J.: A generic turnpike result for a class of discrete-time optimal control systems. Dyn. Contin. Discrete Impuls. Syst. Ser. B **19**, 225–265 (2012)
115. Zaslavski, A.J.: Existence and turnpike properties of solutions of dynamic discrete time zero-sum games. Commun. Appl. Anal. **16**, 261–276 (2012)
116. Zaslavski, A.J.: Nonconvex Optimal Control and Variational Problems. Springer Optimization and Its Applications. Springer, New York (2013)
117. Zaslavski, A.J.: Structure of Approximate Solutions of Optimal Control Problems. Springer Briefs in Optimization. Springer, New York (2013)
118. Zaslavski, A.J.: Necessary and sufficient conditions for turnpike properties of solutions of optimal control systems arising in economic dynamics. Dyn. Contin. Discrete Impuls. Syst. Ser. B Appl. Algorithms **20**, 391–420 (2013)
119. Zaslavski, A.J.: Turnpike properties of approximate solutions of nonconcave discrete-time optimal control problems. J. Convex Anal. **21**, 681–701 (2014)
120. Zaslavski, A.J.: Turnpike Phenomenon and Infinite Horizon Optimal Control. Springer Optimization and Its Applications. Springer, New York (2014)
121. Zaslavski, A.J.: Turnpike properties for nonconcave problems. Adv. Math. Econ. **18**, 101–134 (2014)
122. Zaslavski, A.J.: Structure of solutions of discrete time optimal control problems in the regions close to the endpoints. Set-Valued Var. Anal. **22**, 809–842 (2014)

123. Zaslavski, A.J.: Turnpike theory for dynamic zero-sum games. In: Proceedings of the Workshop "Variational and Optimal Control Problems On Unbounded Domains", Haifa. 2012. Contemporary Mathematics, vol. 619, pp. 225–247 (2014)
124. Zaslavski, A.J.: Turnpike properties of approximate solutions of dynamic discrete time zero-sum games. J. Dyn. Games **2014**, 299–330 (2014)
125. Zaslavski, A.J.: Stability of the Turnpike Phenomenon in Discrete-Time Optimal Control Problems. Springer Briefs in Optimization. Springer, New York (2014)
126. Zaslavski, A.J.: Convergence of solutions of optimal control problems with discounting on large intervals in the regions close to the endpoints. Set-Valued Var. Anal. **23**, 191–204 (2015)
127. Zaslavski, A.J.: Structure of approximate solutions of discrete time optimal control Bolza problems on large intervals. Nonlinear Anal. **123–124**, 23–55 (2015)
128. Zaslavski, A.J.: Discrete time optimal control problems on large intervals. Adv. Math. Econ. **19**, 91–135 (2015)
129. Zaslavski, A.J.: Turnpike Theory of Continuous-Time Linear Optimal Control Problems. Springer Optimization and Its Applications. Springer, Cham (2015)
130. Zaslavski, A.J.: Convergence of solutions of concave discrete optimal control problems in the regions close to the endpoints. Commun. Appl. Nonlinear Anal. **23**, 1–10 (2016)
131. Zaslavski, A.J.: Structure of solutions of optimal control problems on large intervals: a survey of recent results. Pure Appl. Funct. Anal. **1**, 123–158 (2016)
132. Zaslavski, A.J.: Linear control systems with nonconvex integrands on large intervals. Pure Appl. Funct. Anal. **1**, 441–474 (2016)
133. Zaslavski, A.J.: Bolza optimal control problems with linear equations and nonconvex integrands on large intervals. Pure Appl. Funct. Anal. **2**, 153–182 (2017)
134. Zaslavski, A.J.: Discrete-Time Optimal Control and Games on Large Intervals. Springer Optimization and Its Applications. Springer, Berlin (2017)

Index

A. J. Zaslavski, *Turnpike Conditions in Infinite Dimensional Optimal Control*, Springer Optimization and Its Applications 148,
https://doi.org/10.1007/978-3-030-20178-4

Printed in the United States
By Bookmasters